The Companion to Development Studies

Third edition

Edited by Vandana Desai and Robert B. Potter

Routledge
Taylor & Francis Group
LONDON AND NEW YORK

First edition published 2002 by Hodder Arnold
Second edition published 2008 by Hodder Education

Third edition published 2014
by Routledge
2 Park Square, Milton Park, Abingdon, Oxon OX14 4RN

and by Routledge
711 Third Avenue, New York, NY 10017

Routledge is an imprint of the Taylor & Francis Group, an informa business

British Library Cataloguing in Publication Data
A catalogue record for this book is available from the British Library

Library of Congress Cataloging in Publication Data
[CIP data]

ISBN: 978-0-415-82665-5 (hbk)
ISBN: 978-1-4441-6724-5 (pbk)
ISBN: 978-0-203-52898-3 (ebk)

Typeset in Bembo
by Sunrise Setting Ltd, Paignton, UK

Printed and bound by CPI Group (UK) Ltd, Croydon, CR0 4YY

Contents

List of Illustrations

Figures

Tables

Contributors

Giles Atkinson, London School of Economics and Political Science, London, UK.
Donald W. Attwood, Department of Anthropology, McGill University, Canada.
Mary Bachman DeSilva, Center for Global Health and Development, Boston University (BU) and Department of International Health, BU School of Public Health, Boston, MA, USA.
Matt Baillie Smith, Department of Social Sciences, Northumbria University, UK.
Hazel R. Barrett, Faculty of Business, Environment and Society, Coventry University, UK.
The late Baburao S. Baviskar, Institute of Social Sciences, New Delhi and Department of Sociology, University of Delhi, India.
Harriot Beazley, School of Social Sciences, University of the Sunshine Coast, Queensland, Australia.
Anthony Bebbington, Graduate School of Geography, Clark University, Worcester, MA, USA.
Subhes C. Bhattacharyya, Institute of Energy and Sustainable Development (IESD), De Montfort University, Leicester, UK.
Tony Binns, Department of Geography, University of Otago, New Zealand.
Matthew Louis Bishop, Institute of International Relations, University of the West Indies, St Augustine Campus, Trinidad and Tobago.
Richard Black, Sussex Centre for Migration Research, Department of Geography, University of Sussex, UK.
Saturnino M. Borras Jr, Agriculture, Food and Environmental Studies, International Institute of Social Studies (ISS) The Hague, Netherlands.
Emily Boyd, Department of Geography and Environmental Science, University of Reading, UK.
José Brambila-Macias, Trade and Agricultural Directorate, Organisation for Economic Co-operation and Development (OECD), Paris, France.
John Briggs, School of Geographical and Earth Sciences, University of Glasgow, UK.
Stephen Brown, School of Political Studies, University of Ottawa, Ontario, Canada.
Morten Bøås, Norwegian Institute of International Affairs.
Jessica Budds, School of International Development, University of East Anglia, UK.
Terry Cannon, Institute of Development Studies, University of Sussex, UK.
José E. Cassiolato, Economics Institute, Federal University of Rio de Janeiro, Brazil.
Sylvia Chant, Department of Geography and Environment, London School of Economics and Political Science, UK.
Ernestina Coast, Department of Social Policy, London School of Economics, UK.
Christopher Colclough, Centre for Education and International Development, University of Cambridge, UK.

Dennis Conway, Department of Geography, Indiana University, USA.
Sarah Cook, United Nations Research Institute for Social Development, Geneva, Switzerland.
Stuart Corbridge, London School of Economics and Political Science, UK.
Andrea Cornwall, School of Global Studies, University of Sussex, UK.
Ruth Craggs, School of Management and Social Sciences, St Mary's University College, UK.
Kavita Datta, Department of Geography, Queen Mary, University of London, UK.
Tim Daw, School of International Development, University of East Anglia, UK, and Stockholm Resilience Centre, Stockholm University, Sweden.
Vandana Desai, Department of Geography, Royal Holloway, University of London, UK.
Stephen Devereux, Institute of Development Studies, University of Sussex, UK.
Klaus Dodds, Department of Geography, Royal Holloway, University of London, UK.
Radha D'Souza, School of Law, University of Westminster, London, UK.
Jennifer A. Elliott, School of Environment and Technology, University of Brighton, UK.
Ruth Evans, Geography and Environmental Science, University of Reading, UK.
Denise Ferreira da Silva, School of Business and Management, Queen Mary University of London, UK.
Katherine E. Foo, Graduate School of Geography, Clark University, Worcester, MA, USA.
Des Gasper, International Institute of Social Studies, Erasmus University, Rotterdam, Netherlands.
Alan Gilbert, Department of Geography, UCL (University College London), UK.
Jonathan Glennie, Centre for Aid and Public Expenditure, Overseas Development Institute, London, UK.
Tom Goodfellow, Department of Town and Regional Planning, and Sheffield Institute of International Development, University of Sheffield, UK.
Mark Graham, Oxford Internet Institute and the School of Geography and the Environment, University of Oxford, UK.
David Greenaway, School of Economics, University of Nottingham, UK.
Ruth Hall, Institute for Poverty, Land and Agrarian Studies (PLAAS), School of Government, EMS Faculty, University of the Western Cape, South Africa.
Edward Heinemann, International Fund for Agricultural Development (IFAD), Rome, Italy.
Nikolas Heynen, Department of Geography, University of Georgia, USA.
Andrew Herod, Department of Geography, University of Georgia, USA.
Robin Hickman, Bartlett School of Planning, UCL (University College London), UK.
Jude Howell, Department of International Development, London School of Economics, UK.
Katja Hujo, Social Policy and Development Programme, United Nations Research Institute for Social Development, Geneva, Switzerland.
Susie A. Jacobs, Department of Sociology and Criminology, Manchester Metropolitan University, UK.
Gareth A. Jones, Department of Geography and Environment, London School of Economics and Political Science, UK.
Ray Kiely, School of Politics and International Relations, Queen Mary College, University of London, UK.
Madhu Purnima Kishwar, Centre for the Study of Developing Societies (CSDS), New Delhi, India.
Thomas Klak, Department of Environmental Studies, University of New England, US.

Dorothea Kleine, Department of Geography, Royal Holloway, University of London, UK.
Tiziana Leone, Department of Social Policy, London School of Economics, UK.
Philipp Lepenies, Institute for Advanced Sustainability Studies (IASS), Potsdam, Germany.
Sally Lloyd-Evans, Department of Geography and Environmental Science, University of Reading, UK.
Alex Loftus, Department of Geography, Kings College, University of London, UK.
Don D. Marshall, Sir Arthur Lewis Institute of Social and Economic Studies (SALISES), University of the West Indies, Cave Hill Campus.
Isabella Massa, International Economic Development Group, Overseas Development Institute, London, UK.
Emma Mawdsley, Department of Geography, University of Cambridge, UK.
Cheryl McEwan, Department of Geography, Durham University, UK.
Colin McFarlane, Department of Geography, Durham University, UK.
Cathy McIlwaine, Department of Geography, Queen Mary, University of London, UK.
Marion McNabb, Pathfinder International and Boston University School of Public Health (BUSPH), Boston, MA, USA.
Claire Mercer, Department of Geography and Environment, London School of Economics and Political Science, UK.
Paula Meth, Department of Town and Regional Planning, University of Sheffield, UK.
Chris Milner, School of Economics, University of Nottingham, UK.
Jayalaxshmi Mistry, Department of Geography, Royal Holloway, University of London, UK.
Giles Mohan, Development Policy and Practice, The Open University, UK.
Warwick E. Murray, Department of Geography, Victoria University of Wellington, New Zealand.
Patricia Northover, Sir Arthur Lewis Institute of Social and Economic Studies (SALISES), University of the West Indies, Mona Campus.
Ceri Oeppen, Department of Geography, University of Sussex, UK.
Phil O'Keefe, Environmental Management and Economic Development, Northumbria University, UK.
Chukwumerije Okereke, Department of Geography and Environmental Science, University of Reading, UK.
John Overton, School of Geography, Environment and Earth Science, Victoria, University of Wellington, New Zealand.
Ben Page, Department of Geography, UCL (University College London), UK.
Christof Parnreiter, Department of Geography, Hamburg University, Germany.
Jane Parpart, Department of Conflict Resolution, Human Security, and Global Governance, University of Massachusetts, Boston, USA.
Tulsi Patel, Department of Sociology, Delhi School of Economics, University of Delhi, India.
Robert B. Potter, Department of Geography and Environmental Science, School of Archaeology, Geography and Environmental Science, University of Reading, UK.
Marcus Power, Department of Geography, Durham University, UK.
Jules Pretty, School of Biological Sciences, University of Essex, UK.
Matin Qaim, Department of Agricultural Economics and Rural Development, Georg-August-University of Göttingen, Germany.
Shirin M. Rai, Department of Politics and International Studies, University of Warwick, UK.

Carole Rakodi, International Development Department, School of Government and Society, University of Birmingham, UK.
Michael Redclift, Department of Geography, King's College London, UK.
Jonathan Rigg, Department of Geography, National University of Singapore, Singapore.
Joanne Rose, ECT UK Ltd, North Shields, UK.
Susan Rose-Ackerman, Law School and Department of Political Science, Yale University, New Haven, CT, USA.
Barbara Rugendyke, School of Arts and Sciences, Australian Catholic University, Fitzroy, Victoria, Australia.
Lora Sabin, Center for Global Health and Development and Department of International Health, BU School of Public Health, Boston, MA, USA.
David Sapsford, School of Management, University of Liverpool, UK.
David Satterthwaite, International Institute for Environment and Development, London, UK.
Susanne Schech, Centre for Development Studies, Flinders University, Australia.
Frans J. Schuurman, Centre for International Development Issues Nijmegen (CIDIN), Radboud University, Netherlands.
Michel Seymour, Department of Philosophy, Université de Montréal, Canada.
Timothy M. Shaw, Department of Conflict Resolution, Human Security, and Global Governance, University of Massachusetts, Boston, USA.
Prakash Shetty, Institute of Human Nutrition, University of Southampton Medical School, Southampton, UK.
Deborah R. Sick, Department of Sociology and Anthropology, University of Ottawa, Canada.
James D. Sidaway, Department of Geography, National University of Singapore.
David Simon, Department of Geography, Royal Holloway, University of London, UK.
Kathleen Staudt, Department of Political Science, University of Texas at El Paso, USA.
Jonathan R. W. Temple, Department of Economics, University of Bristol, UK.
A. P. Thirlwall, School of Economics, University of Kent, UK.
Richard Tiffin, Centre for Food Security, School of Agriculture, Policy and Development, University of Reading, UK.
Emma Tomalin, Department of Theology and Religious Studies, University of Leeds, UK.
Jeemol Unni, Institute of Rural Management (IRMA), Anand, India.
Maya Unnithan-Kumar, Department of Anthropology, University of Sussex, UK.
Alison Van Rooy, Canadian public service, formerly with CIDA and the North-South Institute, Canada.
Ann Varley, Department of Geography, UCL (University College London), UK.
Eduardo Alcantara Vasconcellos, Brazilian Public Transport Association ANTP, São Paulo.
Rajesh Venugopal, Department of International Development, London School of Economics, UK.
Ben White, (Emeritus) International Institute of Social Studies (IIS), The Hague, The Netherlands.
Howard White, International Initiative for Impact Evaluation (3ie), Cairo, Egypt.
Katie D. Willis, Department of Geography, Royal Holloway, University of London, UK.
Annelies Zoomers, International Development Studies, Department of Geography and Planning, Faculty of Geosciences, Utrecht University, Netherlands.

Preface

We have been delighted to have the opportunity to thoroughly revise and update *The Companion to Development Studies*, in the form of this third edition. The first two editions were published in 2002 and 2008 respectively and have been very well received, both critically and in the marketplace. As we noted in the preface to the second edition, the major criticism we have encountered suggested that we had made it all too easy for students of development studies to read around their subject – a limitation that we can more than happily live with! Further, the volume has given rise to at least one other companion to development studies, and we take this as a sincere form of flattery.

Once again, our intention in the third edition has primarily been to bring the volume up to date. The existing structure has, therefore, largely been retained, and the chapters are divided into ten major parts, each prefaced by an editorial introduction. In addition to the ten editorial introductions, the third edition of *The Companion* consists of 109 chapters, around half of which are new contributions commissioned especially for this third edition. In a few cases, these new chapters represent an existing author being asked to prepare what amounts to an essentially new chapter under a modified remit or title.

The new chapters deal with pressing contemporary issues in development and include: development in global-historical context, the origins and nature of development studies, development as freedom, ethics and development, measuring development from Gross Domestic Product to the Human Development Index, BRICS and development, neoliberalism and the global financial crisis, an overview of globalization, globalization and localization, the knowledge-based economy and digital divide, corporate social responsibility, transnationality and migration, diaspora and development, rural poverty, food security, genetically modified crops, land reform, gender and land rights, global and world cities, studies in comparative urbanism, cities, crime and development, climate change and development, African climate change, ecosystems services and development, natural resource management, water and hydropolitics, transport and sustainability, migrant women, sexualities and development, disability and development, social protection, female participation in education, skill formation and training, international volunteering, fragile states, rights and social justice, the global war on terror, nationalism, ethnic conflict, foreign aid, aid conditionality and effectiveness.

Our goal as editors of *The Companion* has always been to bring together leading scholars from around the world in an effort to provide a truly international and interdisciplinary overview of the major issues that have a bearing on development theory and practice in the twenty-first century. From the outset, it was envisaged that the book would offer a one-stop reference guide for anyone with a practical, professional or academic interest in development studies. We hope that this revised edition will remain of relevance to those in the fields of development studies, sociology, government, politics and international relations, and economics, along with practitioners in NGOs, and those in donor agencies.

As the editors of *The Companion* we recognize the existence of numerous good general texts on development studies, including readers. However, this volume aims to perform a unique function in bringing together in an accessible format a wide range of concisely written overviews of the most important issues in the field. We hope that the book remains an invaluable course text, while with the exercise of critical judgment, it can be treated as a source of readings and discussion pieces in connection with higher-level options and training courses, for example, at the masters degree level. Thus, it remains our hope that students following certain courses may be able to make use of the volume over the duration of their studies and not just in connection with a single option or module. With this in mind, each chapter is brought to a close with suggestions for further reading, lists of references cited in the text, along with details of useful websites.

One of the principal strengths of the volume is that it has been written by well-known and respected authors from both the 'global South' and the 'global North'. For all editions, we have specifically targeted authors from around the world. As before, we were delighted – and just a little amazed – that our invitations to take part in the project were so overwhelmingly greeted with a positive response. We have felt sure each time that the excellent response from our invited contributors reflected the fact that there was a real gap in the development studies literature. We sincerely hope that the enthusiasm we encountered this third time round means that our invitees believe that *The Companion to Development Studies* has gone a long way to filling a genuine gap in the literature.

Over and above the contributors, a number of people have been vital in the production of the third edition. Lucy Winder inherited the project at Hodder Education and was enthusiastic to see a third edition commissioned. Also at Hodder, Beth Cleall covered a wide array of editorial tasks and made major logistical contributions from the very start of the project. When the book transferred from Hodder Education to Routledge in 2012, we found ourselves in the capable hands of Andrew Mold and Faye Leerink, who were delighted to add the title to the well-established and strong Routledge development studies list. Both have shown a very hands-on approach to working on the title, for which we are extremely grateful. Closer to completion, Lisa Salonen ably and calmly over saw the production process; and we were assisted by Alta Bridges who efficiently copyedited the manuscript. We hope that *The Companion* will go from strength to strength as part of the Routledge imprint.

Finally, the period since the delivery for publication of the second edition has been busy and demanding for both of us editors and the members of our respective families have made available sufficient time and space for us to be periodically obsessed with the commissioning and editing of this volume. *The Companion* is dedicated to them all once again with our sincere thanks.

Vandana Desai and Robert B. Potter
October 2013

Part 1

The nature of development and development studies

Editorial introduction

This first section of *The Companion* aims to provide an overview of the field of development studies in order to provide an introduction to the detailed chapters that make up the rest of the book. The chapters included in this first section explore and comment on two closely linked themes: first, the nature and progress of development studies as a distinct avenue of enquiry; and second, how development as a process and phenomenon can be conceptualized, defined and measured over time and across space. In addition, the section aims to introduce a number of important ongoing key issues in development studies such as the so-called New Institutional Economics (NIE), the Millennium Development Goals and BRICS (Brazil, Russia, India, China and South Africa).

At the outset, it is vital to stress the origins of development and underdevelopment in the global historical context of colonialism. The ideas and practices that underpinned post-war development had their origins earlier, in the late colonial period. Thus the campaign against poverty, so often thought to begin after World War II, in fact has its roots in earlier colonial development projects, and it is valuable to explore how these earlier incarnations of development were practiced and with what effects. As colonialism is often understood as a causal factor in contemporary poverty and inequality, then it is crucial to understand the connections between colonial and postcolonial development.

It is also salient to consider the origins of the so-called 'first', 'second' and 'third worlds' in the politics of the cold war, and the later transition to a North–South dichotomy following the work of the Brandt Commission. With the collapse of the Berlin Wall and the near total demise of the socialist Third World, some analysts have now strongly suggested that it is time to discard these descriptors and to talk instead of 'developing countries', and even more so perhaps, 'poor nations'.

Development studies as an academic subject examining such fundamental issues dates from the 1960s. Its origins lay in a number of British social scientists who were increasingly disillusioned by the insights being provided by existing approaches, including those provided by mainstream economics. The rise of development studies thereby reflected the perceived need for distinctly inter- and multidisciplinary approaches to the study of social, political and economic change. There seems little doubt that development studies as a discipline has achieved a good deal since its inception, notwithstanding the so-called 'impasse in development studies' in the 1980s. The impasse was generally attributed to the failure of development in Third World countries, the widespread imitation of Western policy together with the rise of the postmodern critique and the seemingly universal trends towards globalization.

Turning to early policy imperatives in development, these undoubtedly generally stressed catching up with, and generally imitating, the 'West'. Such formulations almost universally regarded development as the same thing as economic growth – being based on producing more goods and services and income. Since the 1970s/1980s, the issues on which a growing consensus appears to be emerging include the fact that economic growth is a necessary, but not sufficient condition for development. Without redistribution of income and wealth, inequalities are not going to be reduced, and there is much evidence that it is inequalities that hurt both people and societies in general. Thus, development must be regarded as synonymous with enhancing human rights and welfare, so that self-esteem, self-respect and improving entitlements become central concerns of the development agenda, not just growth. In this respect, in 1999 the Nobel Laureate economist Amartya Sen published his book under the title, *Development as Freedom*, arguing that development should be conceptualized as the expansion of the real freedoms that people desire and value. Such an approach would specifically work against longstanding racial- and gender-based inequalities in society, as a path to social change and justice.

There is, of course, a strong argument that matters of development, colonialism and race are inextricably linked. The early stages of what was seen as development assumed that Western ways of thinking and of doing things were the best and the most efficacious, thereby serving to 'other' (render as different) the remainder of the world along with traditional practices and ways of doing things. Since this period it is true to say that much of development theory and practice have needed to reconsider and re-evaluate this simple/deterministic set of assumptions on which the earliest development imperatives were based.

Thus, recent approaches to development have been somewhat more liberating in terms of the worldviews promulgated, and there are trends to link development studies with cultural studies, for example, in respect of the condition of postmodernity and the vital nature of issues of peace and security. Further, the trend of globalization, the reduction in the importance of the state, and the associated alienation of the state from civil society, all mean that development studies as a discipline faces a battery of issues, not least whether these trends are real and inescapable phenomena, or are constructs designed to legitimize the logic of the neoliberal market. If it is accepted that development is not just about economic growth, but the promotion of redistribution and the reduction of inequalities, then the value choices involved in such changes are vital and involve important and complex ethical issues.

In the period since the 1980s, a plethora of approaches to development have thereby been promulgated. For example, the 'New Institutional Economics' argues the existence of efficient institutions is a prerequisite for effective development. The approach basically attempts to incorporate a theory of institutions into development economics as a way forward. Whatever the theoretical or conceptual approach adopted, as development is something that is actively promoted by states and organizations, inevitably measuring development is important in monitoring and evaluating the effectiveness of development programmes and policies. The methods employed to measure development have reflected directly the principal conceptualizations of development as a process held at a given time or amongst those of a particular persuasion. Thus, early on, Gross Domestic/National Product (GDP/GNP) per capita was used as the invariant measures of income and thereby inferred economic development. From

1989 onward, the United Nations promoted the Human Development Index (HDI) as a wider measure of development, reflecting dimensions such as health, education as well as standard of living. The HDI has had wide currency since the late 1980s, and the approach was updated in 2010 by the United Nations.

If development is defined in terms of poor countries, an enduring issue remains the need to measure and understand poverty. Thus, poverty 'alleviation' and 'elimination' programmes, and the difficulties of Highly Indebted Poor Countries (HIPCs) are frequently stressed by the international development agencies. Poverty means that despite overall global trends, it still matters where you live, especially if you fall into the poorest third or so of people in the world, living in tropical Africa and Asia. Addressing poverty requires the political will, and many would argue that this remains the real obstacle to development. The Millennium Development Goals are major indicators that are being employed to assess the progress with specific development targets in the twenty-first century and show that although some progress has been made, much remains to be done in reducing gross levels of poverty and inequality.

In early 2000 the growth potential of Brazil, Russia, India and China, and later South Africa (BRICS) was increasingly recognized as a major development issue and these countries were collectively seen as representing the growth engines of the global economy, offering excellent investment opportunities with their large potential markets and rapidly growing middle-class populations. In 2002 the BRICS collectively accounted for some 25.9 per cent of world GDP. It is estimated that by 2040 the BRICS are likely to have a total GDP that is akin to that of the G6 nations, such is their growth trajectory.

1.1 Development in a global-historical context

Ruth Craggs

Introduction

Many texts locate the origins of development in the post-1945 era, alongside the emergence of the United States and the Soviet Union as global superpowers, anti-colonial movements, and decolonisation in much of the world. Although development as *global project* and *academic discipline* may have begun in this period, many scholars now argue that the ideas and practices that underpinned post-war development had their origins earlier, in the late colonial period. From this perspective,

> the post-war crusade to end world poverty represented not so much a novel proposal marking the dawn of a new age, as the zenith of decades, if not centuries, of debate over the control and use of the natural and human resources of colonized regions.
>
> *(Hodge, 2007: 3)*

If the campaign against poverty, so often thought to begin after World War II, in fact has its roots in earlier colonial development projects, then it is valuable to explore how these earlier incarnations of development were practiced and with what effects. As colonialism is often understood as a causal factor in contemporary poverty, inequality, and violence, it is crucial to understand the connections between colonial and postcolonial development.

This chapter provides the global-historical context for the development theories and practices explored in the rest of this volume. It explains the relevance of the colonial histories of places and people caught up in the nexus of development – as donors and recipients, ideologues and practitioners – to the ways that development was imagined, funded, practised, and received. It begins by exploring colonialism *as* development. It examines the ways in which colonial rule was presented as premised on the idea of developing and modernising colonies. This section highlights the ideologies of trusteeship and modernisation that underlay imperial rule and shaped colonial territories. The second section explores the material legacies of colonialism in the developing world and the continuing influence of colonial thought and practice in postcolonial development. For reasons of brevity, the focus falls on the British Empire (and its decolonisation), though some of the same trends can also be seen in other European contexts.

Colonialism as development

Enlightenment ideas of 'improvement' – or making a more efficient and orderly use of land – accompanied and legitimated colonial rule from at least the eighteenth century (Hodge, 2007).

Potential colonial land was understood as fair quarry for expanding European empires; cast as empty or ill exploited, it was seen as ripe for improvement by those with the expertise through which to make these transformations. Improvement entailed the development of infrastructure (where linked to European trade or settlement), the increase in economic output (benefitting metropolitan interests and markets) and the augmentation of the population with European settlers. In the nineteenth century, medical research aimed to bolster the colonial system by protecting the health of colonial servants, armies, and settler populations and prevent the 'degeneration' caused by tropical climates. Improvements were oriented towards metropolitan interests, rather than towards increasing the quality of life of local communities.

The late nineteenth and early twentieth century period saw a new model of imperialism in the French, British, and Dutch colonies, which placed more emphasis on development on humanitarian grounds for native colonial communities. This policy became known as 'trusteeship'. As Power (2003: 131) explains, 'Trusteeship in colonial administration was all about the mission to civilise others, to strengthen the weak, to give experience to the "child-like" colonial peoples who required supervision'. It therefore provided the mandate for European powers to help these territories develop through following a path towards Western modernity. In Britain, official policy was enshrined in the 1929 Colonial Development Act, providing British funding for economic development overseas for the first time, and the 1940 Colonial Development and Welfare Act, which ushered in state-led large-scale development for the purposes of improving welfare. Thus the 1930s onwards witnessed an increasing move towards interventionist development policies in colonial territories, resulting in a dramatic growth in the number, scale, and funding of colonial development projects. This trajectory became even more marked in the immediate post-World War II era, a period that has been called, as a result, the second colonial occupation (Low and Lonsdale, 1976).

Hodge (2007: 8) argues that this new push towards humanitarian development 'helped to reinvigorate and morally rearm the imperial mission in the late colonial epoch', providing continued legitimacy for empire in a rapidly changing geopolitical landscape between 1930 and 1950. This new style of imperialism was linked to an increasing concern over the poor conditions in the colonies, but also aimed to support the colonial system. Development hoped to stabilise colonial populations through the creation of an indigenous middle class invested in the colonial state, and to soothe growing local unrest during the period of depression. By creating products for European markets and markets for European goods, development also contributed to struggling European economies. As colonial planners began publically to discuss eventual decolonisation once colonies had 'progressed' enough, development policies became even more important, both as contributor to this colonial progress, and to ensure the creation of stable and amenable newly independent states.

Development projects in the late colonial period were closely allied with a belief in modernisation. This involved the linear progress of states towards a developed, modern (Western) society and economy. Official British colonial films of this era showcase this discourse of development as modernisation (see www.colonialfilm.org.uk). Projects were based on a growing faith in the role of science and technology to combat poverty and disease and focused on infrastructural improvement and the technical enhancement of agriculture, industry, and healthcare through new innovations. Higher yielding seeds, new crops, and intensive monoculture were encouraged (and indeed enforced), disease eradication programmes were rolled out, and new mining technology was introduced (Tilley, 2011). Late colonial housing and building projects drew on new materials such as concrete, scientific construction techniques and modernism in architectural design, imbricating notions of progress into the design of colonial landscapes (Crinson, 2003).

The high modernism of the Kariba Dam project (1955–1960) in what was then the Federation of Rhodesia and Nyasaland (present day Zambia and Zimbabwe) is illustrative of the discourse and practice of late colonial development (Tischler, 2012). A huge project to dam the Zambezi and provide hydroelectric power for the surrounding territories, its construction took 40 per cent of the colonial state's gross national product to complete and entailed the submergence of 57,000 local Gwembe Tonga homes (Tischler, 2012). Combining grand scale, new technology, a desire for industrialisation, a modern aesthetic and materials, and a notion of population crisis (to be solved through technological fix), this project exemplifies development as modernisation. As is clear in the words of Federal Prime Minister Godfrey Huggins, such modernisation often entailed the sacrifice of indigenous lands or ways of life towards the goal of national development:

> it is vital that we have this cheap power so that we can industrialise and employ our rapidly increasing African population. . . . The available land is limited but the African population is not. A permanent solution can only be found by industrialisation.
>
> *(Quoted in Tischler, 2012: 7)*

The Kariba Dam scheme aimed to stabilise the local indigenous population (at a time of increasing African nationalism) and to boost migration from Britain to Central Africa.

Figure 1.1.1 The Kariba Dam

Providing evidence of the Federation's modernisation, it was hoped the dam might cement the position of the white European settler community in the territory. As with earlier schemes of colonial improvement, the project laid claim to land through the notion of making it productive (Tischler, 2012).

Development became one of the fundamental tenets of colonial policy in the twentieth century. It aimed to both support colonialism and to improve the welfare of local populations, although it often failed on both counts, dispossessing people of land and contributing to anti-colonial critiques emerging in both colony and metropole. Although colonial development was imagined as a rational modernisation planned in Europe and put into practice in the colonies, this was never the case. Development policies were shaped by the specificities of colonial locations and reworked in connection to local knowledge and practices (Tilley, 2011). Development discourse and practice was constructed in negotiation in a colonial system.

We now turn to the legacies of colonialism in poorer nations that were once part of European empires, and to the legacies of colonialism present in contemporary development discourse and practice.

Legacies of colonialism in development

Issues that many contemporary development policies and programmes attempt to ameliorate have their roots, at least in part, in colonialism. Border disputes where colonial boundaries were pushed through previously united communities, ethnic tensions stoked by policies of 'divide and rule', and trauma from bloody wars of decolonisation are all elements of a legacy of colonialism. Many other issues can be traced back more specifically to colonial development policies, for example, unsustainable and environmentally damaging agricultural systems, polluting industrial sectors and inadequate workers rights, big infrastructure projects which disrupted communities and ecosystems, and arguments over land dispossession.

Less obvious are the colonial legacies that have shaped postcolonial development practice and ideology. The focus of state-led development in late colonialism fed into the postcolonial planning and policies of newly independent governments who often pursued vigorous large-scale state controlled development projects in areas of health, housing, industrial development, and power infrastructure. In addition, many postcolonial states relied, for the implementation of their development policies, on the input of colonial experts of various kinds – such as agriculturalists, other technical advisors, and colonial district officers (Hodge, 2007; Kothari, 2006a). These professionals were employed by the governments of newly independent states and made up a large proportion of the staff of international organisations formed in the wake of World War II such as the World Health Organisation, the United Nations Development Programme, the World Bank, the Commonwealth Development Corporation, and the government departments of former colonial powers, such as the Ministry of Overseas Development in the UK. They were also integral to the formation and staffing of the first development studies departments in UK universities (Hodge, 2007; Kothari, 2006a). They therefore also contributed to the shaping of the discipline of development as an academic subject in the second half of the twentieth century.

These continuities in policy and personnel underpinned – and were underpinned by – a continuation in broader development discourse. Crisis narratives about overpopulation and environmental degradation produced by late colonial experts fundamentally shaped post-war development as discipline and practice (Hodge, 2007). Problems continued to be depoliticised, and their solutions cast as scientific or technocratic. Ideologies which coded the West as developed and the rest as developing, Western as normal and non-Western as other, and which constructed a linear temporal path of development along which the West had travelled further continued (and continue) to hold sway after decolonisation. Ideologies of partnership and responsibility in development have replayed older colonial notions with relationships between donors and recipients, which continue to be less than equal (Noxolo, 2006). Power (2003: 131) has gone as far as to argue that 'Colonial humanitarianism has been reinvented after the formal end of colonial and imperial rule' as international development. Moreover, just as within colonialism, 'whiteness and the West provide symbols of authority, expertise and knowledge', in the postcolonial era, expertise continued (and continues) to be coded as Western and white, providing fundamental challenges for the theory and practice of development (Kothari, 2006b: 10).

Finally, although modernisation has now been discredited in much development discourse, it has continuing effects. As Ferguson (1999: 14) notes for the experience of Zambians living on the Copperbelt, 'the breakdown of certain teleological narratives of modernity . . . has occurred not only in the world of theory, but in the lived understandings of those who received such myths as a kind of promise'. Even if modernisation was always a myth, it was one that late colonial academics, policymakers, and ordinary people invested in, shaping experiences and imagined futures. Ferguson illustrates the devastating consequences when this idea of modernisation was 'turned upside down, shaken, and shattered' (1999: 13).

Conclusion

Though conceived and practiced differently within and between European empires, development was central to the colonial project, particularly from the late nineteenth and early twentieth centuries. Many of the ideas, policies, and priorities of postcolonial development can trace their genealogies to the colonial era, where they were shaped through metropolitan concerns to maintain and modernise colonies, and through contact with the local people, knowledge, and conditions. Colonialism therefore not only contributed to the material economic and social conditions in which development takes place today, but also fundamentally shaped the project of development itself, through continuities between the ideologies, people, and practices of colonial and postcolonial development.

References

Crinson, M. (2003) *Modern Architecture and the End of Empire*. Aldershot: Ashgate.

Ferguson, J. (1999) *Expectations of Modernity: Myths and Meanings of Urban Life on the Zambian Copperbelt*. London: University of California Press.

Hodge, J. M. (2007) *Triumph of the Expert: Agrarian Doctrines of Development and the Legacies of British Colonialism*. Athens: Ohio University Press.

Kothari, U. (2006a) 'From Colonialism to Development: Reflections of Former Colonial Officers' *Commonwealth and Comparative Politics* 44(1) 118–136.

Kothari, U. (2006b) 'An agenda for thinking about "race" in development' *Progress in Development Studies* 6(1) 9–23.

Low, D. A. and Lonsdale, J. M. (1976) 'Introduction: Towards the New Order 1945–1963' in D. A. Low and A. Smith (eds) *History of East Africa, Volume III*. Oxford: Clarendon Press.

Noxolo, P. (2006) 'Claims: a postcolonial geographical critique of "partnership" in Britain's development discourse' *Singapore Journal of Tropical Geography* 27(3) 254–269.

Power, M. (2003) *Rethinking Development Geographies*. London: Routledge.

Tilley, H. (2011) *Africa as Living Laboratory: Empire, Development, and the Problem of Scientific Knowledge, 1870–1950.* Chicago: Chicago University Press.
Tischler, J. (2012) 'Negotiating Development: The Kariba Dam Scheme in the Central African Federation' in P. Bloom, T. Manuh, and S. Miescher (eds) *Revisiting Modernization in Africa.* Bloomington: Indiana University Press.

Further reading

Ashton, S. R. and Stockwell, S. E. (eds) (1996) *Imperial Policy and Colonial Practice 1925–1945, Part II: Economic Policy, Social Policies and Colonial Research (British Documents on the End of Empire, Series A, Volume 1).* London: HMSO. A selection of primary sources drawn from British official archives which document the evolution and practice of British Colonial Development Policy between 1925 and 1945.

Websites

www.colonialfilm.org.uk/theme/empire-and-development. Films about the British colonies and development, with 150 available to view online. Many have accompanying critical essays analysing their contribution to understanding British colonialism.

1.2

The Third World, developing countries, the South, emerging markets and rising powers

Klaus Dodds

The power of words

In 2012, James Sidaway asked an important question in a short article in the journal *Professional Geographer.* He opined, 'Today, how useful is it to talk about the geography of development or of developing countries? What are and what remains of the geography of development and of the Third World?' Words, and in particular place-based labels, continue to matter but do so in a variety and at times bewildering manner. A generation of political and development geographers, often inspired by postcolonial and critical geopolitical theorising, continue to examine and interrogate the implications of terms such as Third World, middle- and low-income countries, majority world, southern periphery, two-thirds world, and/or rising powers with particular reference to countries such as Brazil, China and India (Slater 1993, Sidaway 2012).

In the midst of the cold war, the term 'Third World' was coined to signify a new geopolitical imagination based on a geography of global politics divided into three camps – the United States and its allies, the Soviet Union and the Communist world and a 'Third World' of postcolonial states in Africa, Asia and Latin America. Underpinned by an investment in technology and an ideological faith in free markets and enterprise, this tripartite division reflected American hegemony in a post-war financial and political international system US administrations helped to construct.

This chapter seeks to remind readers of this period characterised by cold war and superpower competition and consider how the fate of the Third World has changed over the last six decades. With the ending of the cold war, it was widely hoped that questions pertaining to development, poverty reduction and debt cancellation would enjoy a greater political profile. This has been achieved but not led to the profound changes hoped by many activists, campaigners and governments in the Third World and beyond. While the 2000s were dominated, in part, by the US-led War on Terror the most profound change has come with the emergence of China as a world economic power. Since reforms in 1978, China's ascent remains breathtaking in scale and scope. It is estimated that by 2020 China is likely to be the largest economy in the world, and thus anyone authoring an introductory chapter such as this will be concentrating, I suspect, increasingly on Chinese narratives about world politics and specifically development agendas (for a readable account of China's economic transformation see Fenby 2012).

The invention of the Third World

In the aftermath of the Korean conflict (1950–1953 and arguably ongoing), a new geopolitical imagination began to emerge as the conflict between the Soviet Union and the United States spread across the globe. Key geographical designations such as 'First World' and 'Third World' were deployed by Western social scientists in an attempt to highlight the profound differences between the United States and the Soviet Union. Newly decolonised countries in South Asia, Africa and Asia were seen as providing opportunities for both sides to project influence, extend trading opportunities and recruit for the purpose of defending particular parts of the earth's surface from the influence of the ideologies of either communism (in the case of the Soviets) or liberal democracy and capitalism (in the case of the United States). Billions of dollars and roubles were spent over the next fifty years in pursuit of that geopolitical objective. Both sides used the existence of this cold war to plan and implement development programmes, aid assistance, volunteer groups, trade stimulation, academic exchanges and/or arms sales.

During the cold war, the experiences of the Third World were never uniform as some countries and regions received greater attention than others (Westad 2005). In the case of Latin America, for instance, American administrations were adamant following the 1959 Cuban Revolution in particular that they would not tolerate any further socialist governments in the hemisphere. President Johnson ordered 20,000 US Marines to overthrow the government of the Dominican Republic in 1965 and President Nixon approved the overthrow of the socialist President of Chile, Salvador Allende, on 11 September 1973. During the 1960s, countless efforts were made either to assassinate or overthrow Cuba's Fidel Castro, especially in the aftermath of the Bay of Pigs fiasco, which witnessed US-backed anti-Castro forces being routed by the armed forces loyal to the socialist leader. When the United States was not attempting to promote revolution and/or turmoil, it was content to support violent anti-communist military regimes in Argentina, Brazil, Chile and Uruguay. In other parts of the world, regional allies such as Israel, Pakistan, South Korea and Taiwan received substantial financial and military forms of assistance because they were judged to be significant in the wider struggle to prevent the Soviet Union from extending its global influence.

While certainly not unique to the United States, the Soviets were also engaged in a programme of aid, development and intervention in an attempt to project a global communist revolution. Some Third World states welcomed Soviet largesse – India was one such beneficiary and many of its citizens were subsequently trained in Soviet universities and institutions. Soviet allies such as Cuba also assisted in this global mission – Cubans were based in Angola and played a vital role in buttressing the national security capability of the country in the 1970s when it faced South African forces who used South West Africa (Namibia) to launch covert raids against the country, which was consumed by civil war for much of its postcolonial existence. Elsewhere, the Soviet Union provided support for revolutionary movements in Central America, South East Asia and sub-Saharan Africa for much of the cold war period.

Not surprisingly, many members of the expanding Third World did not welcome the intensification of the cold war. In 1961, the Non-Aligned Movement (NAM) was created after an earlier Afro-Asian conference in Indonesia. NAM was important in so far as it signalled a resistance to the bipolar strictures of the cold war. Recently decolonised states resented and resisted pressures from the superpowers to align with one side or another. The purpose of NAM was to find a 'third way', one which avoided the plutocratic structures of the West and party-based authoritarianism of the East. It also sought to promote a different vision of development based on fairness, information and technological exchange and a reformed international political system. By the early 1970s, NAM was joined by the so-called 'Group of 77' in the United Nations and advocates of a New International Economic Order (NIEO). The NIEO was, alongside the term 'Third World', an expression of resistance against historic and contemporary forms of colonialism and oppression. Unfortunately for those campaigners, the early 1970s were turbulent years as rich countries such as the United States were rattled by oil shortages, the Vietnam debacle and the Watergate crisis.

It was evident by the latter stages of the cold war that the 'Third World' was a highly diverse group of countries that had enjoyed very different colonial-development trajectories. The political-economic condition of oil-producing states such as Nigeria, Saudi Arabia and Venezuela differed markedly from South East Asian states such as Singapore and Taiwan. sub-Saharan Africa and Central America probably contained some of the poorest states in the world, which were also immersed in damaging civil wars and/or damaged by corrupt and violent dictatorships and military regimes. The cold war provided an opportunity for these schisms and uncertainties to be exploited by the rich countries as they sought to extend ideological and economic influence. Tragically, millions perished as basic needs such as access to clean drinking water proved less politically attractive than expenditure on the latest tank, ship and/or missile. This meant, ultimately, that the potency associated with the NIEO/Third World movements dissipated and new terms such as 'the global South' found favour (Prashad 2007).

The global South

Amidst a general feeling of despair and pessimism amongst many development advocates and Third World academics, the United Nations-sponsored Brandt Commission reported on the state of the world in 1980 and 1983. Significantly, the Commission depicted a world divided between North and South and not a First, Second and Third World. In other words, whatever the ideological differences between the Soviet Union and the United States and their respective allies, the world was really divided between the rich North and poor South. The Commission called for the North in particular to recognise that the world was more interdependent than ever before (globalisation had not been coined as a term at that stage) and to promote a more equitable form of global political economy. The message of the two reports, although hard-hitting and significant, was a victim of geopolitical timing.

With the Soviet invasion of Afghanistan in 1979, the cold war appeared to have entered a new and more dangerous phase. A new US president, Ronald Reagan, was committed to confronting what he described as the 'evil empire'. American and Soviet military spending increased and American financial assistance was used to fund anti-Soviet forces in Afghanistan and anti-communist movements in places such as Nicaragua. The decision to fund anti-Soviet forces in Central Asia was particularly significant and arguably contributed to the emergence of the Al-Qaeda terror network in the 1980s and 1990s (Johnson 2000). What all this meant for the South was fairly straight forward – American and Soviet energies were directed towards this global struggle and any thought of reconstructing the world-economy and global development strategies was an irrelevance.

By the late 1980s, the Soviet Union was bankrupt and regimes in Eastern and Central Europe were crumbling. In 1989, Germans on both sides of the divide tore down the Berlin Wall, the most practical and symbolic illustration of a divided Europe. The so-called Velvet Revolution led to the undoing of all those former communist governments including the more brutal ones in Romania and East Germany. Elsewhere, military regimes in Latin America were crumbling and democratic governments emerged. The ending of the cold war was apparently completed when the Soviet Union folded in 1991. Some analysts such as Francis Fukuyama were swift to declare 'the end of history' in so far as it signalled the triumph of liberal democracy and market-based capitalism over state socialism and global communism (Fukuyama 1989). Whether or not that claim was justifiable, the ending of the cold war did begin to bring to an end a confrontation that had claimed millions of lives through proxy wars, government intervention and/or bombings. Democratic change in the 1990s was widespread and owed as much to the ending of the cold war as it did to a host of circumstances specific to particular places and regions.

Emerging markets and rising powers

Notwithstanding expressions of optimism regarding a global democratic revolution, the 1990s brought to the fore a stark realisation – the world remained highly divided despite all the attempts of governments and policy advocates to promote development. Endeavours to promote poverty reduction and/or inequalities were modest in scope and extent. Over the last twenty years, far more attention has been devoted towards inequalities and mal-development but the results have been mixed. On the one hand, international organisations such as the International Monetary Fund (IMF) have promoted structural adjustment and good governance in return for economic assistance and aid spending. The 1996 initiative designed to promote debt relief for a select number of Highly Indebted Poor Countries (HIPC) was criticised for demanding a range of measures such as privatisation in return for modest debt cancellation. This has led to accusations that these agencies are exercising a neo-colonial influence on countries in the global South. On the other hand, interest has grown in what was initially described as 'emerging markets' and later rising powers with particular emphasis on Brazil, Russia, India and China (the so-called BRICs).

'Emerging markets' was first used by the World Bank in the 1990s and later embraced by investment banks and fund managers. A decade later, the term once applied to countries such as Brazil and China transmogrified to another generic category 'rising powers' and then more specifically BRICs highlighting a small group of countries either possessing substantial resources such as oil and gas and/or demanding ever-greater resources as part of their national development trajectories. In the case of China, an export-led economic strategy also contributed to an ever-greater accumulation of foreign earnings and sponsor of foreign direct investment in Asia and Africa. China's investment in Africa has grown markedly – in 2010 trade was worth US$127 billion and Chinese investment valued at $40 billion (Fenby 2012: 256). In return for access to natural resources, China invests in infrastructural projects and lends money, unlike the World Bank, with minimal conditions and without having to take responsibility for the colonial and cold war legacies bequeathed by the United States and European countries such as Britain.

While it is possible to overstate China's economic power in the world, there is a sense, however, that the BRICs are beginning to make their presence felt not only in places like Africa but also in the very institutional architecture created by the United States in the late 1940s. It is now common to speak of G20 meetings rather than the G7 or G8. In July 2009, the first BRIC summit was held in Russia and in March 2012 the parties gathered again in India to discuss areas of common concern. One such area will be the dollar-centred monetary and financial international regime. If anything, the ongoing financial slowdown (2008 onwards) in North American and European economies stands in stark contrast to the economic performance of countries such as Brazil and China in particular.

What is noteworthy about the BRICs is their willingness to put forward alternative visions for the international economic order in the future. New words, new labels and new narratives are circulating with ever-greater force. In March 2009, for example, the governor of the People's Bank of China proposed that the US dollar should be replaced by a common currency created and managed by member countries. The intent was to help promote a more productive investment regime and stimulate economic growth more widely. While North American and European economies continue to grapple with financial slowdown and investment paralysis, the BRICs are becoming ever-more confident in their reformist agenda to the point that officials attached to the World Bank have warned that it will have to be taken very seriously indeed. The 'Third World', or at least part of it, has struck back. So-called 'emerging markets' such as Brazil and China are very much 'emerged' now.

Conclusions

There is a great deal more to be said on this topic and this short chapter has only touched on the cold war origins of development and the subsequent connections between North and South including contemporary interest in rising powers, BRICs and emerging markets (Power 2003). The ending of the cold war did not bring the profound changes that many campaigners hoped for in respect of substantial reforms in the terms and conditions of international trade and generous debt reduction. Anti-globalisation campaigners, alongside others such as Make Poverty History and the Jubilee Campaign, have helped to capture global media attention and pressurise Northern governments into making concessions on debt, trade and aid spending. However, these concessions are often modest and then often cancelled out by the terms and conditions attached to aid and debt cancellation. But this is only one part of the story regarding terms such as 'Third World' and more recent incarnations.

Coinciding with the financial crisis affecting North America and Europe (2008 onwards), there has been the most profound change involving countries such as Brazil, China and India. While terms such as 'emerging markets' and BRICs have been used in the last two decades to chart and classify their rising economic and political profile, their presence in the world is clearly being felt in terms of investment, production and consumption. These countries, with their own developmental strategies, are creating new patterns of engagement with the 'Third World' and with what was once thought to be the 'First World'. Moreover, it is striking to note how Chinese military and intelligence gathering Western defence analysts, in a manner reminiscent of the Soviet Union, are now tracking capabilities increasingly carefully in the 1970s and 1980s.

Let me conclude by introducing yet more labels, but these are my current favourites when it comes to making sense of a changing geopolitical landscape. The Bulgarian writer, Tzvetan Todorov, describes what he terms 'appetite' states such as China and 'anxiety' states such as the United States (Todorov 2010). This captures well, from a Western perspective at least, some of the ongoing concern about what these rapacious states might yet do in terms of consuming non-renewable resources such as oil and gas as well as changing the basic terms and conditions of the international financial and political order. All of this helps to remind us, especially those of us who have grown up in the shadow of American geopolitical power and the primacy of the English-speaking world, that there are other ways of knowing and making sense of the world around us.

References

Fenby, J. (2012) *Tiger Head, Snake Tails*. New York: Simon Schuster.
Fukuyama, F. (1989) 'The end of history'. *National Interest*. Available online at www.wesjones.com/eoh.htm
Johnson, C. (2000) *Blowback: The Costs and Consequences of American Empire*. New York: Henry Holt.
Power, M. (2003) *Rethinking Development Geographies*. London: Routledge.
Prashad, V. (2007) *The Darker Nations: A People's History of the Third World*. New York: New Press.
Sidaway, J. (2012) 'Geographies of development: New maps, new visions'. *Professional Geographer* 64: 49–62.
Slater, D. (1993) 'The geopolitical imagination and the enframing of development theory'. *Transactions of the Institute of British Geographers* 18: 419–437.
Todorov, T. (2010) *The Fear of the Barbarians*. Cambridge: Polity.
Westad, O. (2005) *The Global Cold War*. Cambridge: Cambridge University Press.

Further reading

Jacques, M. (2009) *When China Rules the World*. London: Verso. On the impact of China and its presence in the world.
Power, M. (2003) *Rethinking Development Geographies*. London: Routledge. For an insightful guide to development geographies and the contested geopolitics of neoliberalism.
Westad, O. (2005) *The Global Cold War*. Cambridge: Cambridge University Press. An excellent account of the impact of the cold war on the Third World.

Websites

G8: www.g8.co.uk
Make Poverty History: www.makepovertyhistory.org
World Trade Organisation: www.wto.org

1.3
The nature of development studies

Robert B. Potter

The development of development studies: An overview

The development of development studies as an academic subject that can be studied at university dates from the 1960s. In a review of the history of the field, Harriss (2005) puts its origins in a number of mainly British economists and other social scientists who were unhappy with the insights that were being provided at that time by existing social science subjects, notably traditional or classical economics. These traditional approaches basically emphasised the importance of quantitative paths to the study of societies and economies. These approaches were seen as reflecting a logical positivist orthodoxy that dominated the social sciences at that time, and which stressed the importance of hypothesis testing and statistical verification as the paramount sources for knowledge and advancement.

It was at this precise juncture in the 1960s that the 'new universities' were being established in the United Kingdom. These new tertiary educational institutions were premised on the idea of 'doing things differently'. In particular, the new universities were keen to promote multi- and inter-disciplinary studies that cut across the boundaries of existing, traditional disciplines. Changes in both thinking and practice at this point were also closely linked with the growth of radical Marxist approaches in the 1960s within society as a whole (Harriss, 2005; Kothari, 2005).

The Institute of Development Studies was established at the University of Sussex in 1966 and represented a founding institution. Seven years later in 1973, the first undergraduate teaching programme in the field of development studies opened at the University of East Anglia. In the meantime, development studies has spread as an academic discipline to other universities such as Oxford, Manchester, Bath, SOAS London, LSE and Birmingham among others, and has remained quite strongly British. In the words of Harriss (2005: 18): '[d]evelopment studies has been, institutionally, a distinctively British and to a lesser extent other European field of study.' Of course, over the years, scholars from developing countries have made fundamental contributions. Even to this day, the field means relatively little in the United States, for example, although similar issues are studied in cognate fields such as international relations, politics, economics and geography (see Potter *et al.* 2012).

It can be argued that, rather than being either inter- or multidisciplinary, development studies can make a strong claim to be cross-disciplinary in nature in that it serves to bring together a large number of fields in the study of poverty and inequality. This is represented in graphical summary form in Figure 1.3.1. The central concern of development studies may

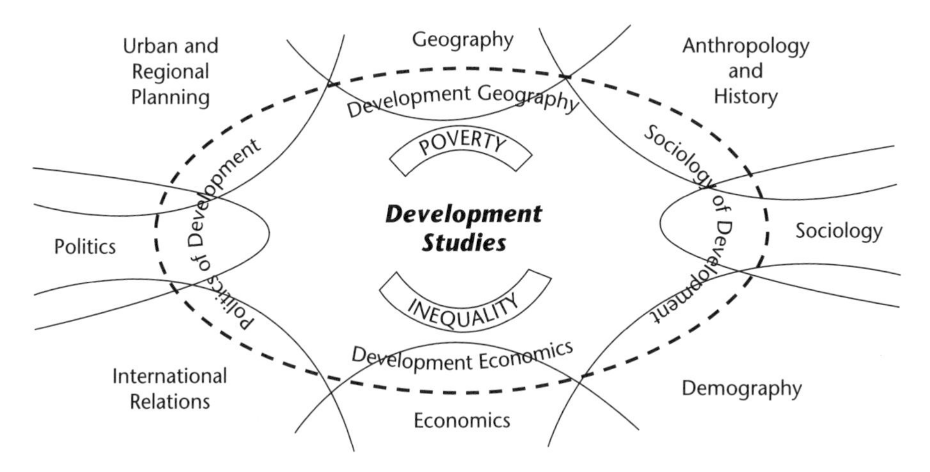

Figure 1.3.1 The various disciplines contributing to the cross-disciplinary field of development studies

be regarded as the existence and seemingly inexorable deepening of global poverty and inequality.

In its early stages, 'break-away' economists were strongly involved in the rise of development studies. Reflecting this in the core discipline, a distinct sub-discipline within economics can be recognised that is now conventionally referred to as 'development economics' (Figure 1.3.1). Geographers, with their strong tradition of regional and area studies represented another area of involvement and interest. In the same manner as in the case of economics, the rise of 'development geography' as a distinct field can be recognised.

Other mainstream social science disciplines, such as politics and sociology, also contributed to the rise of development studies and, equally, came to be characterised by home disciplinary patches, known respectively as the sociology of development and the politics of development (Figure 1.3.1). Attesting to the truly cross-disciplinary character of the field, subjects such as demography, international relations, anthropology and history, as well as urban and regional planning should also be identified as making a distinct contribution to development studies (Figure 1.3.1).

The evolution of thinking about development in relation to the development of the social sciences since the 1940s

Notwithstanding the mixed disciplinary genealogy of development studies, there is frequently a tendency to link its development to the evolution of human geography. For example, this was expressed by the alignment of the field of development studies as a sub-discipline of geography in successive research assessment exercises in the UK until 2008.

In a recent essay, Potter and Conway (2011) have attempted to summarise the evolution of thinking about development in relation to the advancement of geography and other social sciences since the 1940s. Presented in summary terms, such an account serves to stress the similarities and differences in focus that have characterised development practice, development geography and development studies over the past sixty years. In the account below, which draws on Potter and Conway (2011), we look at this decade by decade.

The 1940s

The modern roots of development practice can be traced to the immediate post-Second World War period and to the inaugural speech made by United States President Truman in 1947. In this Truman stated that it was the responsibility of rich nations to develop poorer countries in their own image. In this period, during which development was emerging, geography was characterised by forms of development-oriented enquiry that might be described as 'colonial', 'military', 'tropical' and 'regional' geographies. Interest in the 'great overseas' had been stimulated initially by consideration of the countries making up the British empire under colonialism. Then, between 1939 and 1945, a number of British geographers travelled to countries such as Singapore, Egypt, India and Ceylon (present-day Sri Lanka) as part of their wartime military service. The main statement on development came with the publication of the French geographer Pierre Gourou's (1947) text *Tropical Geography*. In addition, aspects of development had always formed part of the bread and butter of geography, in the guise of regional geography. But as already noted, there was no recognisable field of development studies *per se* at this juncture.

The 1950s

The decade of the 1950s was strongly involved with post-Second World War reconstruction and economic development. The ultimate goal was seen as the application of the historical-development experience of the rich nations in the development of the poorer nations. Development policy came to be strongly associated with classical economic theory during this period. This emphasised the importance of 'liberalising', or freeing-up trade at the global level. At the national level, the approach mainly advocated concentrating development around natural growth poles. At this time, underdevelopment was regarded as an initial state beyond which Western industrial nations had managed to progress. It was envisaged that the experience of the West could assist other countries in catching up by sharing capital and technology. Thus, the approach was very much a Western one and was top-down. In the field of geography, the period was associated with the spread of what is still referred to as the 'new geography'. This was based on the search for generalised explanations of the real world by means of the development of models and the use of quantitatively derived generalisations and laws. There was still no separate field of development studies as such.

The 1960s

The 1960s saw the emergence of radical political perspectives within both society as a whole as well as within mainstream social science subjects, albeit to a greater or lesser extent. A major development was the emergence of what came to be called 'dependency theory', which had its roots in Latin American and Caribbean development. Dependency theory essentially argued that the global pattern of Western-dominated development had served to keep the poor world poor, rather than serving to aid its accelerated development. Dependency theory thereby represented an almost complete rebuttal of classical and neo-classical economic approaches to the challenges of development, based on the 'spread' of the Western model. It argued that less developed countries would do better to de-link from the developed world and to follow an alternative development path. Despite this radical ferment in the wider social sciences in the 1960s, by contrast, it was the quantitative revolution that was coming to fruition in geography. But, by this time, the seeds were already being sown for the development of cross-disciplinary development studies, as already noted in terms of the establishment of the Institute for Development Studies at the University of Sussex in 1966.

The 1970s

It was the early 1970s that saw the emergence of more radical approaches in the field of geography. A landmark publication was David Harvey's *Social Justice and the City* in 1973. Harvey's and others' analyses gave rise to an increasing acceptance of political-economic, or structuralist approaches in human geography by the mid-1970s. Such an orientation was more encouraging to the emergence of a distinct development geography as a sub-discipline, although the term itself was little used at that stage. In a further series of developments, dissatisfaction with the quantitative revolution gave rise to avowedly humanistic approaches, which stressed the subjectivity of phenomena and knowledge. At the same time, alternative and more humanistically oriented approaches were coming to influence thinking about development practice, in what is often referred to as the emergence of 'another development'. This was anchored in a growing critique of urban-based, top-down, centre-out neo-classically inspired development policies. Moreover, it was at this very juncture that, as noted before, the first development studies undergraduate degree programme was introduced at the University of East Anglia in 1973.

The 1980s

In the 1980s, development practice and policy were broadly characterised by the rise of what can be referred to as the 'New Right' in Europe and 'Neoconservatism' in the United States. This is also referred to as the rise of the neoliberal agenda, the strong view that liberal free trade and unregulated free markets should be left to make economic decisions and that they will do so rationally and effectively. Thus the New Right sung the praises of the unrestrained power of the unregulated free market. In Britain, full-blown neoliberalism came in the form of Margaret Thatcher's 'popular capitalism', and in the United States it was witnessed by President Reagan's 'Reaganomics'. Both Thatcher and Reagan pushed for the extension of private market-inspired controls into the public sector. While development studies can be seen to have started consolidating as a field during the 1980s, and despite the emergence of common concerns between geography and development studies at this time, the focus of geography remained firmly in the fields of cultural and historical geography and the accent was mainly placed on Europe and North America. Surveys carried out at the time showed that what could be recognised as development geography was at best taught by one specialist member of staff in the majority of British geography departments (Potter and Unwin, 1988; Unwin and Potter, 1992).

The 1990s

Postmodernism emerged as an alternative paradigm for the social sciences at the start of the decade. The approach was associated with the rejection of meta-theories and meta-narratives – that is the big explanations that had come to be associated with modernising as the inevitable path to development. Instead, postmodernism suggested that emphasis should be placed on a wide range of possibly discordant and even contradictory views, voices and discourses. Thus, 'development' was one of the very meta-narratives that was to be questioned, giving rise to what have been referred to as 'anti-', 'post-' and 'beyond-' development stances. In particular, at this juncture, the question asked was whether the 'development mission' as posed in the 1940s, 1950s and beyond, could ever have been successful given its essentially Eurocentric stance and origins in Western experience and thought. In short, the move toward a distinctly postmodern turn might be interpreted as having given rise to doubts, uncertainties and reflections in both development studies and the social sciences in general. In that sense, the 1990s might be seen as having given rise to something of a greater degree of possible commonality between geography and development studies as academic fields. However, there is a critical, alternative interpretation of postmodernity, one that argues it is nothing more than the next logical stage in the progress of modernity. Such a view sees postmodernity as the latest manifestation of late capitalism, seeing individual and group choice being hailed and promoted in an essentially free-market setting.

The 2000s onward

Potter and Conway (2011) argue from the point of view of geography as an overall discipline, that since the turn of the new millennium, matters seem to have been changing for the better, and that there have emerged some grounds for optimism, both within the discipline and the development establishment more generally. Thus, there have been definite signs of a more positive view of development-related issues in geography as a discipline, albeit borne out of the pressing global developmental-geopolitical crises the world currently faces: '(t)he world is now so deeply unequal that the need for a truly global geography has never been greater' (Potter and Conway, 2011). There are signs that the geographical study of critical development-oriented issues is starting to become more valued and central to geography as a whole. This certainly needs to be the case in the twenty-first century as an era that faces the pressing realities of the global financial crisis, the other geographical realities of unregulated and unruly globalisation, transnationality, global conflict and environmental change and, in particular, climate change.

Bibliography and Further Reading

Gourou, P. (1947) *Les Pays Tropicaux: Principes d'une Geographie Humaine et Economique*. Paris: Presses Universitaires de France.

Harriss, J. (2005) 'Great promise, hubris and recovery: A participant's history of development studies'. In U. Kothari (ed.) *A Radical History of Development Studies: Individuals, Institutions and Ideologies*. London: Zed Books, pp. 17–46.

Harvey, D. (1973) *Social Justice and the City*. London: Arnold.

Kothari, U. (ed.) (2005) *A Radical History of Development Studies: Individuals, Institutions and Ideologies*. London: Zed Books.

Kothari, U. (2005) 'A radical history of development studies: Individuals, institutions and ideologies'. In *A Radical History of Development Studies: Individuals, Institutions and Ideologies*. London: Zed Books, pp. 1–14.

Potter, R. and Conway, D. (2011) 'Development'. In J. A. Agnew and D. N. Livingstone (eds) *The Sage Handbook of Geographical Knowledge*. London: Sage, pp. 595–609.

Potter, R. B. and Unwin, T. (1988) 'Developing areas research in British geography'. *Area* 20: 121–126.

Potter, R. B., Conway, D., Evans, R. and Lloyd-Evans, S. (2012) *Key Concepts in Development Geography*. Los Angeles: Sage.

Unwin, T. and Potter, R. B. (1992) 'Undergraduate and postgraduate teaching on the geography of the Third World'. *Area* 24: 56–62.

1.4

The impasse in development studies

Frans J. Schuurman

Introduction

Development studies is a relatively new branch of the social sciences. Coming into being in the late 1960s and early 1970s, it inherited many features of post-Second World War developments within the social sciences. Modernization theory contributed to its developmental orientation and its comparative methodology. From dependency theory it inherited its normative and progressive political character and its interdisciplinary conceptual frameworks.

In the 1970s, with dependency theory denouncing modernization theory as crypto-imperialist and modernization theorists hitting back by accusing dependency authors of being populist pseudo-scientists, development studies found fertile ground and grew into an increasingly accepted new discipline of the social sciences. Universities – often under pressure from leftist professors and students – created Third World Centres. Debates about the nature and impact of development assistance became popular, and the existence of many dictatorial regimes in the South led to numerous solidarity committees in the North. In the 1980s, things started to change for development studies. A number of occurrences in that decade, which will be dealt with in the following paragraphs, led to an increasingly uneasy feeling within the discipline that old certainties were fading away. It was felt that development theories in the sense of a related set of propositions of the 'if . . . then' kind, could ever less adequately explain experiences of development and underdevelopment. Whether it concerned modernization theories or neo-Marxist dependency theories, both sets of development theories were losing out in terms of their explanatory power. From the mid-1980s onwards, the so-called 'impasse in development studies' was talked about. The contours of this impasse were sketched for the first time in a seminal article by David Booth in 1985. In the years that followed, other authors continued the discussion, which took on new dimensions with the end of the cold war and the debate on globalization.

Reasons for the impasse

Three reasons can be held responsible for having changed the panorama for development studies to such an extent that it created this theoretical impasse. Chronologically they were: (a) the failure of development in the South and the growing diversity of (under)development experiences; (b) the postmodernist critique on the social sciences in general and on the normative characteristics of development studies in particular and, finally; (c) the rise of globalization in its discursive as well as in its ontological appearance. Each of these issues is considered in the account that follows.

The failure of development in the South

Although until the 1980s developing countries realized average improvements in life expectancy, child mortality and literacy rates, more recent statistics have shown, however, that these improvements were less valid for the poorest of these countries and, more specifically, for the lowest income groups. In fact, in the 1980s there was a reversal in some of the development indicators. It was realized that given the growth rates of that time, it would take another 150 years for Third World countries to achieve even half the per capita income of Western countries. Modernization theories failed to account for these figures and trends. Instead of a self-sustained growth (a much favoured concept of modernization), many developing countries were up to their ears in debt, which served to paralyse development initiatives.

Problems such as unemployment, poor housing, human rights offences, poverty and landlessness were increasing at alarming rates. UNICEF estimated a fall of 10–15 per cent in the income of the poor in the Third World between 1983 and 1987. In 1978, the Third World received 5.5 per cent of the world's income; in 1984 this had fallen to 4.5 per cent. The 'trickle-down' process (another favoured concept of modernization) had failed miserably. In 1960, the income ratio between the world's rich and poor countries was 20 : 1; in 1980 it increased to 46 : 1; and in 1989, the ratio was as high as 60 : 1.

Although dependency theory could certainly not be accused of an over-optimistic view concerning the developmental potentials of developing countries, it could not really account for the growing difference between Third World countries, nor were the developmental experiences of so-called socialist countries particularly enviable. In addition, Marxist and neo-Marxist development theories were dealt a heavy blow when the fall of the Berlin Wall meant the delegitimization of socialism as a political project of solving the problem of underdevelopment.

The postmodernist critique of the social sciences

The 1980s witnessed the advancement of postmodernism within the social sciences, bringing with it a tendency to undermine the 'great narratives' of capitalism, socialism, communism and so forth. The basic argument was that there is no common reality outside the individual. As such, political alternatives, which always exist by the grace of a minimum of common perception, were manoeuvred out of sight. Development theories based on meta-discourses or on the role of a collective emancipatory agency lacked, according to the postmodern logic, a sound basis. The Enlightenment ideal of the emancipation of humanity (shared by modernization and dependency theory alike) had not been achieved nor could it be achieved. In addition, in its quest for hidden metaphors, the postmodern method of deconstruction revealed that the notion of development contained a number of hidden and unwarranted evolutionist, universalist and reductionist dimensions which would definitely lead anyone working with this notion down the wrong path. As such, development studies became a direct target for a wide range of views furthering the notion of 'alternative development'. Under postmodernist, or perhaps better put, anti-modernist pressures, the central object of development studies – unequal access to power, to resources, to a humane existence – became increasingly substituted by something like 'socioeconomic diversity'. Apparently, the notion of diversity was considered to avoid the hidden universalist (read: Western or imperialist) and reductionist dimensions which inequality brought with it. At the same time, others considered this switch to a voluntarist and pluralist approach to the development problem not only as anathema, but also as inferior to a universalistic emancipation discourse.

Globalization

In the 1990s, the forces which had led to the impasse in development theories were joined by the discourse on globalization. Although the most recent factor, it probably represents the most important positive challenge to development studies. Whether globalization is a real phenomenon (cf. Hirst and Thompson, 1996), or nothing more than a discourse to legitimize neoliberal market logic, it is undeniable that it has had a major influence on development studies in the 1990s. To understand why this is so, it is important to realize the significance of the (nation-)state for social science theories in general, and for development studies in particular.

It is the declining, or at least changing, position and status of the (nation-)state which has been, and still is, at the core of the literature on globalization. As an interdisciplinary branch of the social sciences, theories within development studies try to connect economic, political and cultural aspects of inequality and development trajectories. The connection between these aspects is realized by using the (nation-)state as a linchpin. As such, theories of economic development became focused upon the workings of the national market and on economic relations between countries. In theories of political development, the role of the state and the process of nation-building were central objects of study. In more culturalistic development theories, the notion of a national identity was crucial in understanding the differences between development trajectories. This importance of the (nation-)state became visible in modernization theories, in neo-Marxist and Marxist development theories alike. Globalization changed all that. Many authors writing about globalization agree on the decreasing, or at least the changing, economic, political and cultural importance of (nation-) states. The central role of the state, it is said, is being hollowed out from above as well as from below. In a political sense, there is the increasing importance of international political organizations which interfere politically and also militarily in particular states. In this way, they relegate to the past the Westphalian principles about the sovereignty of (nation-)states and their monopoly on the use of institutionalized violence within their borders. The national state is hollowed out from below by the growing phenomenon of decentralization and local government.

Economically, the state is seen as disappearing as an economic actor through privatization supported by deregulation. Also, there is the growing importance of the global financial market where about $1,500 billion is shifted daily around the globe. Culturally, the idea of national identity as the central element in identity construction for individuals or groups is quickly eroding, in favour of cosmopolitanism on the one hand and/or the fortification of ethnic, regional and religious identities on the other.

It is not only that the globalization debate gives reason to suppose that the role of the (nation-) state has been, and still is, declining but also that, as a consequence, the former conjunctive dynamic (i.e. following the same spatial and time paths) of economy, polity and culture – on which the interdisciplinary character of many a development theory was based – has been replaced by a disjunctive dynamic (cf. Appadurai, 1990). Development studies has yet to redefine its object and its subject – as have the other social sciences – vis-à-vis globalization but this quest presents much more of a challenge than the former impasse ever did.

Conclusion

The impasse in development studies can in fact be traced back to a crisis of paradigms. The three reasons which were mentioned as being responsible for the impasse and its deepening – the lack of development and increasing diversity in the South, the postmodernist critique on 'grand narratives', and globalization – challenged, respectively, three post-Second World War developmental paradigms. These were:

1 The essentialization of the Third World and its inhabitants as homogeneous entities;
2 The unconditional belief in the enlightenment concepts of progress and the 'makeability' of society;
3 The importance of the (nation-)state as an analytical frame of reference and a political and scientific confidence in the state to realize progress.

Each of these paradigms came in for criticism, one after the other. Development theories related to these paradigms (such as modernization and dependency theories) became automatically tainted as well, initiating the so-called impasse in development studies.

However, in spite of this impasse, an important number of authors in the field of development studies have continued their work, some using more grounded theories, others trying to elaborate on new concepts like civil society, global governance and global social movements. Many feel that the growing inequality between, as well as within, North and South is enough of a reason to continue with development studies. To fit this effort in with the new reality shaped by globalization presents a new and exciting challenge, and one which relegates the impasse to a past period. Nevertheless, from well into the twenty-first century looking back upon the impasse, development studies has never been the same since. The aforementioned crisis in the three post-Second World War developmental paradigms served as academic warning flags, which, since the impasse, led to an almost continuous process of rethinking that needs to be explained, and the various explanations that have been used in development studies.

References

Appadurai, A. (1990) 'Disjuncture and difference in the global cultural economy', in M. Featherstone (ed.) *Global Culture, Nationalism, Globalization and Modernity*, London: Sage, pp. 295–311.

Booth, D. (1985) 'Marxism and development sociology: interpreting the impasse', *World Development* 13: 761–87.

Hirst, P. and Thompson, G. (1996) *Globalization in Question*, Cambridge: Polity Press.

Further reading

Corbridge, S. (1989) 'Marxism, post-Marxism and the geography of development', in R. Peet and N. Thrift (eds) *New Models in Geography*, Vol. 1, London: Unwin Hyman, pp. 224–54; identifies and elaborates upon three dimensions in Booth's critique of neo-Marxist development theories, that is, essentialism, economism and epistemology.

Edwards, M. (1989) 'The irrelevance of development studies', *Third World Quarterly* 11(1): 116–36; approaches development theories from the point of view of the practitioner.

Hearn, D. (ed.) (1999) *Critical Development Theory: Contributions to a New Paradigm*, London: Zed Books.

Leys, C. (1996) *The Rise and Fall of Development Theory*, London: James Currey; especially chapter 1 (pp. 3–45) provides a seminal overview of the post-Second World War development theories.

Munck, R. and O'Hearn, D. (eds) (1999) *Critical Development Theory: Contributions to a New Paradigm*, London: Zed Books.

Schuurman, Frans J. (ed.) (1993) *Beyond the Impasse: New Directions in Development Theory*, London: Zed Books; provides a general overview of the dimensions of the impasse and the attempts to develop new theories as well as the problems and possibilities of these attempts.

Simon, D. and Närman, A. (eds) (1999) *Development as Theory and Practice: Current Perspectives on Development and Development Co-operation*, Harlow: Longman.

Vandergeest, P. and Buttel, F. (1998) 'Marx, Weber, and development sociology: Beyond the impasse', *World Development* 16(6): 683–95; focuses on Booth's critique upon the underlying meta-theoretical assumptions of Marxism, pointing out the necessity of looking within the heterogeneity of developing countries for common denominators.
Willis, K. (2005) *Theories and Practices of Development*, London: Routledge.

1.5 Development and economic growth

A. P. Thirlwall

The economic and social development of the world's poorest countries is perhaps the greatest challenge facing society at the present time. Over one billion of the world's seven billion population live in absolute poverty; the same number suffer various degrees of malnutrition, and millions have no access to safe water, healthcare or education. This poverty is concentrated largely in countries described as 'developing', and coexists with the affluence enjoyed by the vast majority of people in countries described as 'developed'.

The standard of living of people is commonly measured by the total amount of goods and services produced per head of the population, or what is called gross domestic product (GDP) per capita (or gross national product (GNP) per capita if net income from abroad is added). This, in turn, is determined by the number of people who work, and their productivity. The basic proximate cause of the poverty of nations is the low productivity of labour associated with low levels of physical and human capital (education) accumulation, and low levels of technology.

Income per head in a country is naturally measured in units of its own currency, but if international comparisons of living standards are to be made, each country's per capita income has to be converted into a common unit of account at some rate of exchange. The convention is to take the US dollar as the unit of account and convert each country's per capita income into dollars at the official exchange rate. A country's official exchange rate, however, is not necessarily a good measure of the relative purchasing power of currencies, because it only reflects the relative prices of goods that enter into international trade. But many goods that people buy are not traded, and the relative price of these non-traded goods tends to be lower the poorer the country is, reflecting much lower relative labour costs. An exchange rate is required which reflects the purchasing power parity (PPP) of countries' currencies, and this is now provided by various international organizations, such as the World Bank, which uses US$1.25 per day measured at PPP to define the level of absolute poverty.

The economic growth of countries refers to the increase in output of goods and services that a country produces over an accounting period, normally 1 year. If a country is said to be growing at 5 per cent per annum, it means that the total volume of its domestic output (GDP) is increasing at this rate. If population is growing at 2 per cent per annum, this means that output per head (or the average standard of living) is growing at 3 per cent per annum.

Economic growth, however, is not the same as economic development. The process of economic (and social) development must imply a growth in living standards, but it is a much wider concept than the growth of per capita income alone. Growth, it might be said, is a necessary condition for the economic and social development of nations, but it is not a sufficient condition because an aggregate measure of growth or per capita income pays no attention to how that output is distributed amongst the population; it says nothing about the composition of output (whether the goods are consumption goods, investment goods or public goods such as education and health provision), and it gives no indication of the physical, social and economic environment in which the output is produced. In short, the growth rates of nations cannot be taken as measures of the increase in the welfare of societies because the well-being of people is a much more inclusive concept than the level of income alone.

If the process of economic and social development is defined in terms of an increase in society's welfare, a concept of development is required which embraces not only economic variables and objectives, but also social objectives and values for which societies strive. Many economists and other social scientists have attempted to address this issue, and here we mention the ideas of two prominent thinkers in the field: Denis Goulet and Amartya Sen (who in 1998 won the Nobel Prize for Economics for his work on the interface between welfare and development economics).

Goulet (1971) distinguishes three basic components or core values that he argues must be included in any true meaning of development which he calls life sustenance, self-esteem and freedom. Life sustenance is concerned with the provision of basic needs. No country can be regarded as fully developed if it cannot provide all its people with such basic needs as housing, clothing, food and minimal education. A country may have a relatively high average standard of living and an impressive growth performance over several years, but still have a poor provision of basic needs, leaving large sections of the population in an underdeveloped state. This issue is closely related to the distribution of income in societies measured by the share of total income going to the richest and poorest sections of society. The distribution of income is much more unequal in poorer developing countries than in richer developed countries, and it is perfectly possible for a poor country to be growing fast, yet its distribution of income to be worsening because the fruits of growth accrue to the rich. Such a country would have grown, but it would not have developed if the provision of basic needs for the poorest groups in the community had not improved.

Self-esteem is concerned with the feeling of self-respect and independence. A country cannot be regarded as fully developed if it is exploited by others, or cannot conduct economic relations on equal terms. In this sense, the colonization of large parts of Africa, Asia and South America kept the countries in these regions of the world in an underdeveloped state. Colonialism has now virtually ended, but some would argue that there are modern equivalents of colonialism, equally insidious and anti-developmental. For example, the International Monetary Fund (IMF) and World Bank dominate economic policy-making in many developing countries, and many of the policies that the countries are forced to pursue are detrimental to development. Also, multinational corporations that operate in many developing countries often introduce consumption patterns and techniques of production which are inappropriate to the stage of development of the countries concerned, and to that extent impair welfare. In international trade, poor and rich countries do not operate on a level playing field, and the strong may gain at the expense of the weak. The distribution of the gains from trade are not equitably distributed, not least because the terms of trade of primary producing developing countries (i.e. the price of their exports relative to the price of imports) tends to deteriorate through time (at an average rate of about 0.5 per cent per annum for at least the last century).

Freedom refers to the ability of people to determine their own destiny. No person is free if they are imprisoned on the margin of subsistence with no education and no skills. The great benefit of material development is that it expands the range of choice open to individuals and to societies at large. For the economic and social development of a country, however, all must participate and benefit from the process of growth, not just the richest few. If the majority are left untouched, their choices remain limited; and no person is free if they cannot choose.

Sen (1983, 1999) argues in a similar vein to Goulet that economic growth should not be viewed as an end in itself, but as the means to the achievement of a much wider set of objectives by which economic and social development should be measured. Development should focus on, and be judged by, the expansion of people's 'entitlements', and the 'capabilities' that these entitlements generate, and income is not always a good measure of entitlements. Sen defines entitlements as 'the set of alternative commodity bundles that a person can command in a society using the totality of rights and opportunities that he or she faces'. For most people, the crucial determinants of their entitlements depend on their ability to sell their labour and on the price of commodities. Employment opportunity, and the level of unemployment, must therefore be included in any meaningful definition of development. Entitlements also depend on such factors as what individuals can extract from the state (in the form of welfare provision); the spatial distribution of resources and opportunities, and power relations in society. Sen (1984) has analysed major world famines using the concept of entitlements and finds that several famines have not been associated with a shortage of food, but rather with a lack of entitlements because the food supply has been withdrawn from certain parts of the country or sections of society, or food prices have risen.

The thinking of Goulet, Sen and others has led to the construction of alternative measures of economic and social development to supplement statistics on growth rates and levels of per capita income of countries. The most notable of these measures are the Human Development Index (HDI) and the Human Poverty Index (HPI) compiled by the United Nations Development Programme (UNDP) and published in its annual *Human Development Report*. These alternative indices of the economic well-being of nations do not always correlate well with per capita income. The same growth rate and per capita income of countries can be associated with very different levels of achievement in other spheres such as life expectancy, death rates, literacy and education. As the UNDP says in its 1997 *Report*: 'although GNP growth is absolutely necessary to meet all essential human objectives, countries differ in the way that they translate growth into human development'.

The UNDP's Human Development Index is based on three variables: life expectancy at birth; educational attainment, measured as the geometric mean of the average and expected years of schooling; and the standard of living measured by real per capita income measured at PPP (see earlier). These variables are combined in a composite index that ranges from 0 to 1 (see Thirlwall, 2011, for details). Comparing the ranking of developing countries by their HDI and per capita income show some interesting divergences. Many oil-producing countries, for example, have much lower HDI rankings than their per capita income rank, while some poor countries rank relatively high by their HDI because they have deliberately devoted scarce resources to human development. Countries such as Cuba, Venezuela, Jamaica and some former states of the Soviet Union fall into this category.

The UNDP's multidimensional Human Poverty Index is based on indices of education; nutrition; child mortality; and access to safe water, sanitation and electricity. The ranking of countries by their HPI also shows some striking contrasts with their ranking by per capita income. The UNDP has calculated that the cost of eradicating poverty across the world is relatively small compared to global income – not more than 0.3 per cent of world GDP – and that political commitment, not financial resources, is the real obstacle to poverty eradication.

To conclude, economic growth is not the same as economic development. The annual growth rate of a country is a very precise measure of the growth of the total volume of goods and services produced in a country during a year but says nothing about its composition or distribution. Growth is a necessary condition for real income per head to rise, but it is not a sufficient condition for economic development to take place because development is a multi dimensional concept which embraces multifarious economic and social objectives concerned with the distribution of income, the provision of basic needs, and the real and psychological well-being of people. Many poor countries in the last 30 years have experienced quite a respectable rate of growth in living standards – averaging 2–3 per cent per annum – but the absolute number in poverty has continued to rise, and the distribution of income has become more unequal. Equally, at the global level, there is little evidence of the convergence of per capita incomes across nations. The poor countries have been growing, but the rich countries have been growing as fast, if not faster, in per capita terms. While the eradication of poverty, and the narrowing of the rich–poor country divide, remains one of the great challenges of the new millennium, economic growth in poor countries is not enough by itself for economic and social development to take place when viewed in a broader perspective.

Bibliography

Goulet, D. (1971) *The Cruel Choice: A New Concept on the Theory of Development*, New York: Atheneum.

Sen, A. (1983) 'Development: which way now?', *Economic Journal*, December.

Sen, A. (1984) *Poverty and Famines: An Essay in Entitlement and Deprivation*, Oxford: Clarendon Press.

Sen, A. (1999) *Development as Freedom: Human Capability and Global Need*, New York: Knopf.

Thirlwall, A. P. (2011) *Economics of Development: Theory and Evidence* (9th edn), London: Palgrave-Macmillan.

United Nations Development Programme (1997 and 1999) *Human Development Report*, New York: Oxford University Press.

1.6

Development and social welfare/human rights

Jennifer A. Elliott

Critical development studies

It is currently well-established that human well-being, including individual civil and political liberties, as well as meeting the physical and material needs of human society, are key concerns for development, both as outcomes and conditions for sustained progress. Issues of economic and social justice, democracy, empowerment, ethics and human dignity suffuse development theory and the activities of many institutions including the United Nations and the World Bank, donor organisations, NGOs and social movements alike. Understanding of poverty as a human rights issue, for example, has risen with recognition of the growing social and economic disparities at all scales and the multidimensional nature of poverty as encompassed within the Millennium Development Goals. Many different groups now use the discourse of human rights to shape their agendas and practices in development such as with women's rights and indigenous rights. For those adopting a rights-based approach to development, the overarching goal is the realisation of all human rights for all people. In short, there has been the 'insertion of a critical sensibility' (Radcliffe, 1999: 84) into development studies in recent decades through which 'many new problems *as well as* old ones' (Sen, 1999: xi, emphasis added) are being widely conceptualised in terms of human rights and freedoms. This chapter identifies some of the key changes through which the discourses and practice of development can be considered to have become more morally informed.

Rights and development as separate concerns

Although human rights and well-being were undoubtedly concerns in the 1940s and 1950s within international institutions, amongst governments of newly independent countries and in the emergent discipline of development studies, it has been argued that the predominant ideas and practices of development at that time were often devoid of ethical considerations and separate from those 'marked out for development' (Corbridge, 1999: 69). For example, ideas of progress during that period were generally synonymous with economic growth and the modernisation of traditional societies. In so far as welfare and rights issues were considered, it was assumed that these would follow as outcomes of processes of economic development.

The integration of human rights into development practice was also significantly compromised by the cold war. Whilst the 1948 Universal Declaration on Human Rights emphasised the universality of rights, the collective goal for humanity of realising those rights and the range of civil, political, economic, social and cultural rights for all, ideological divisions between Western liberal democracies and the Eastern bloc led to a separation of human rights activities within the UN itself and between countries. In 1966, for example, two separate international conventions (through which the states' accountability for human rights obligations was established in international law) were agreed; one referring to civil and political rights (CP) and another to economic, social and cultural rights (ESC). CP rights refer to the right to life, liberty and security, for example; the right to vote, to a free press and freedom of speech and on legal rights such as to due process of law and the presumption of innocence until proven guilty. ESC rights include the right to an adequate standard of living, the right to education, to work and equal pay, and the right of minorities to enjoy their own culture, religion and language. In short, Western countries emphasised CP rights and work towards the ratification and inscription of these into constitutional and legal frameworks. In contrast, socialist countries (and many developing nations) prioritised ESC rights, emphasising economic and social development and self-determination and 'criticizing the richest Western countries for their failure to secure these rights for all citizens' (United Nations Development Programme [UNDP], 2000: 3).

The basic needs approach

By the end of the 1960s, there was growing disillusionment with the practices and outcomes of development. Mounting evidence of increasing income inequality and rising poverty in many developing countries (despite overall economic growth in some) suggested that a more *direct* approach was required for the delivery of human welfare outcomes. What became known as the basic needs approach (BNA) drew together theorists and practitioners from a range of traditions, academic centres and institutions of development that searched for more human-centred and locally relevant processes and patterns of development. In short under the BNA, development was redefined as a broad-based, people-oriented and endogenous process, as a critique of modernisation and as a break with past development theory.

As a result of the influence of the BNA, the 1970s saw a wide variety of programmes focused on households and covering aspects of health, education, farming and reproduction practices designed to create a minimum level of welfare for the weakest groups in society. Development practice became characterised by district and regional planning (supported by major international donor institutions), by proliferating field bureaucracies and by development solutions through targeting (of social groups – particularly women and children – of sectors and of regions) to overcome the recognised inadequacies of the 'planning fantasies of the 1960s' (Chambers, 1993: 108).

Buying and selling welfare

However, whilst the BNA did much to put poverty, human needs and rights onto official development agendas in the 1970s, many assert that the decade of the 1980s was one of development 'reversals' rather than achievements with evidence, particularly within Africa, of falling school enrolments, literacy levels and life expectancies, for example. Similarly, development theory was proposed to have reached an 'impasse' (see Schuurman, 1993) as both modernisation thinking and neo-Marxist dependency theorists struggled to explain the growing diversity of development experiences on the ground. In the late 1980s, neoliberal development ideas became increasingly prominent and powerful (see Simon, 1999) with development policies promoting increased private sector and reduced direct state involvement. Progressively through the subsequent decade, basic human rights such as access to safe water and sanitation became 'commodities subject to the rigours of the market' (Bell, 1992: 85). Donors, for example, came under increasing pressure to find new methods of financing and providing welfare both 'at home' and abroad and governments of developing nations were required to cut state expenditures under conditions of structural adjustment programmes and for access to multilateral development finance. Whilst these pressures opened up spaces for new project types, processes and programmes in development, it has been suggested that the more radical aspects of the original BNA philosophies were often devalued in practice, 'reducing them from agendas for change and empowerment into little more than shopping lists that are hawked to donors for implementation, commonly more in line with donors' than recipients' priorities' (Simon, 1999: 27). For many governments within developing countries, their capacity to protect economic and social rights such as to ensure access to education became weakened.

Converging agendas through the 1990s

In 1986, the United Nations adopted the UN Declaration on the Right to Development, within which development itself was identified as an inalienable human right. The articles that supported the Declaration drew on wider debates about development at that time, in particular through engaging with the emerging critique of neoliberalism, with understandings of the uneven impacts and limits of globalisation and in the strengthening of notions of 'people centred' development and human empowerment (Manzo, 2003). Across many arenas of development, interest in working with and strengthening civil society rose very markedly through the 1990s (see Edwards and Gaventa, 2001) with civil society becoming for many 'the best hope for development and the improvement of human rights protection' (UNDP, 2000: 355).

An important contribution to understanding the meaning of and means for development through the decade came through the work of the UNDP and their reporting within their annual *Human Development Reports*. In short, development became strongly re-conceptualised in terms of components of 'human' development; the aspects of human well-being 'beyond income' and the processes through which people are empowered to shape their own development priorities and are able to exercise their human rights. The UNDP draws heavily on the work of Amartya Sen (1981, 1999). Sen considers human development in terms of individuals' capabilities to achieve, to flourish and live lives they have reason to value. In turn, poverty is considered as a set of interrelated 'unfreedoms' that constrain people's choices and opportunity to exercise their individual agency.

In 1990, the UNDP introduced the Human Development Index (HDI), a composite index designed to reflect achievements in the most basic human capabilities defined as leading a long life, being knowledgeable and enjoying a decent standard of living. In 1995, the Gender-related Human Development Index (GDI) and gender empowerment measure (GEM) were introduced encompassing the recognition that gender equality is a measure of and means for human and national development. Since 1997, the Human Poverty Index (HPI) has measured and monitored deprivation worldwide, in terms of the percentage of population not expected to live until the age of 40, illiteracy rates, the percentage of people lacking access to health services and safe water, and the percentage of children under five years who are moderately or severely underweight. In 2010, further measures were added 'adjusting' the HDI and GDI to expose the losses in human development occurring currently through inequality and disadvantage and confirming how issues of equity and empowerment are essential means for ensuring sustained human development for future generations.

It was within the 2000 *Human Development Report*, entitled *Human Rights and Human Development* that a significant step in understanding the relationship between the development and human rights agendas was made. It detailed how and why poverty was a human rights concern and how human rights are critical to achieving development and 'are not as has sometimes been argued a reward of development' (UNDP, 2000: iii). The turn of the millennium was also a point at which the UN was more broadly engaging in a process of internal reform. The Secretary General, Kofi Annan (2000), had committed to reassert the original mandate of the organisation as assisting in the realisation of human rights and fundamental freedoms and to work towards the mainstreaming of human rights as inalienable, interdependent and indivisible concerns throughout the work of the organisation. Such commitments were also central in the launch of the 'Millennium Development Goals' in the same year. Table 1.6.1 displays a number of quotations illustrating the converging agendas of human welfare and humans rights in international development into the twenty-first century.

Contemporary challenges

It has not been possible in this short chapter to do justice to the decades of work done in the fields of poverty, participation, gender and democracy, for example, which have all been extremely important in bringing about a more holistic and moral agenda within development.

Table 1.6.1 The multidimensional and interdependent nature of human rights and human development

- 'Political freedoms (in the form of free speech and elections) help to promote economic security. Social opportunities (in the form of education and health facilities) facilitate economic participation. Economic facilities (in the form of opportunities for participation in trade and production) can help to generate personal abundance as well as public resources for social facilities. Freedoms of different kinds can strengthen one another' (Sen, 1999: 11).
- 'Civil and social education will help people better understand their rights and increase their choices and income-earning capacity. At the same time, developing and implementing equal opportunity laws will empower people to gain more equitable access to productive resources' (UNDP, 1998: 10).
- 'Sustainable human development and human rights will be undone in a repressive environment where threat or disease prevails, and both are better able to promote human choices in a peaceful and pluralistic society' (UNDP, 1998: 6).
- 'A fundamental human freedom is freedom from want. Poverty is a human rights violation, and freedom from poverty is an integral and inalienable right' (UN Declaration on the Right to Development, 1986).
- 'Every step taken towards reducing poverty and achieving broad-based economic growth is a step towards conflict prevention' (Annan, 2000: 45).

However, this brief analysis has confirmed that such agendas require a shift in focus away from determining any particular means or 'specially chosen list of instruments' (Sen, 1999: 3) for development, towards more concern for the overarching ends of development and the processes through which these are defined and secured. Critically, these ends are necessarily plural and fluid as evidenced within continuing questions of the universality of human rights in the context of cultural diversity. But rather than debating the primacy of one human right, good, opportunity or resource over another, contemporary concerns are now more regularly focused on questions of appropriate entry points or sequencing in development interventions in recognition of the reinforcing and interdependent nature of these issues. As Sen has highlighted, the (interrelated) sources of people's 'unfreedoms' may be extremely varied. Development involves expanding these freedoms, as liberties to be valued in their own right and as the principal means (free agency, capability and choice) through which the overarching goals of development, for individuals to 'lead the kinds of lives they have reason to value' (Sen, 1999: 10), will be achieved.

References

Annan, K. (2000) *'We the Peoples': The Role of the United Nations in the Twenty-first Century*, Washington: United Nations.

Bell, M. (1992) 'The water decade valedictory, New Delhi 1990: Where pre- and post-modern met', *Area* 24(1): 82–89.

Chambers, R. (1993) *Challenging the Professions: Frontiers for Rural Development*, London: Intermediate Technology Publications.

Corbridge, S. (1999) 'Development, post-development and the global political-economy', in P. Cloke, P. Crang and M. Goodwin (eds) *Introducing Human Geographies*, London: Edward Arnold, pp. 67–75.

Edwards, M. and Gaventa, J. (2001) *Global Citizen Action*, London: Earthscan.

Manzo, K. (2003) 'Africa in the rise of rights-based development', *Geoforum* 34: 437–456. A paper that explores the rise of rights-based development (particularly within the critique of neoliberalism), the debate surrounding the concept and the contested ways in which it is endorsed by prominent organisations like DFID, the World Bank and the United Nations.

Radcliffe, S. A. (1999) 'Re-thinking development', in P. Cloke, P. Crang and M. Goodwin (eds) *Introducing Human Geographies*, London: Edward Arnold, pp. 84–92.
Schuurman, F. (ed.) (1993) *Beyond the Impasse: New Directions in Development Theory*, London: Zed Books.
Sen, A. (1981) *Poverty and Famine: An Essay on Entitlement and Deprivation*, Oxford: Clarendon.
Sen, A. (1999) *Development as Freedom*, Oxford: Oxford University Press.
Simon, D. (1999) 'Development revisited', in D. Simon and A. Narman (eds) *Development as Theory and Practice*, Harlow: Longman, pp. 17–54.
United Nations Development Programme (UNDP) (1998) *Integrating Human Rights with Sustainable Human Development*, New York: UNDP.
United Nations Development Programme (UNDP) (2000) *Human Development Report: Human Rights and Human Development*, New York.
United Nations General Assembly (1986) 'Declaration on the Right to Development'. Resolution 41/128.

Further reading

Burnell, P. and Randall, V. (eds) (2008) *Politics in the Developing World* (2nd edn), Oxford: Oxford University Press. Chapter 18 by Michael Freeman on Human Rights (pp. 353–372) provides a clear exposition of the history and detail of the international human rights regime and considers explicitly the problems of applying the concept of human rights to developing countries.
Mikkelsen, B. (2005) *Methods for Development Work and Research* (2nd edn), London: Sage. This book in chapter 6 provides a very good description of the origins and meaning of a rights-based approach to development, how this differs from a (more commonly found) rights-based *perspective* and how this approach is being put into practice within development.
The Office of the High Commissioner for Human Rights (OHCHR) is the department of the UN Secretariat that is mandated to promote and protect all rights established in the Charter of the United Nations and in international human rights laws and treaties. Their work can be followed via www.ohchr.org
Wolfe, M. (1996) *Elusive Development*, London: Zed Books. A readable book that reflects in some detail on the successes and failures of the Basic Needs Approach to development.

1.7

Development as freedom

Patricia Northover

Amartya Sen and the development imaginary

The discourse on 'development' – a central motif of narratives of modernity – addresses a highly contentious problem field and offers neither stable signifiers nor much common ground on the nature, tendencies, ethics and politics of *a process* of development in our contemporary world. Nevertheless, certain strands of thinking have been able to exercise a dominant influence on this discourse and have thus led the way in delineating the possibilities for imagining and even encountering development (Escobar 1995).

Development as Freedom (1999), the seminal text authored by the 1998 Nobel Laureate economist, Amartya Kumar Sen, expresses one such emergent dominant narrative that has found a global resonance in the twenty-first century. The revolutionary uprisings in the Middle East in 2011, for example, were predominantly seen as being driven by the aspirations for freedom, democracy and social justice; themes strongly embedded in the work of Sen. In *Development as Freedom* Sen argues for development as the expansion of the 'real freedoms' that people may enjoy and have reason to value. In the text, he rehearses his longstanding challenges to competing views on the meaning of social welfare and development and offers his work as a better analytical frame, than, for example, Utilitarian, Rawlsian, Marxist, or rights centric libertarian standpoints on development. Indeed, Sen posits that this framework of 'development as freedom' *should* act as a foundational and universal principle for all peoples in order to better facilitate the aspirations of the multifaceted forms of social contestation, protest and resistance against inequality – from race and gender struggles for equal freedoms, to de-colonial and popular uprisings for local and global social justice.

Practically and politically, Sen's conceptual and methodological approaches for assessing and valuing human well-being, and his message of 'development as freedom' have had an enormous impact on the development industry. His work, for example, has helped to underpin the rise of new development indicators, such as the Human Development Index, the Human Poverty Index and the Gender Empowerment Index, championed and developed through the UN Development Program (UNDP), and promoted through their *Human Development Reports*. Sen's work has also infused the theoretical study of gender and development, poverty, famine, social justice and democracy. The idea of *Development as Freedom*, moreover lends its weight to the transition to the Post-Washington Consensus from the so-called 'Washington Consensus' aid regime. The latter was a package of neoliberal aid reforms focused on 'getting the prices right' through a narrowly focused set of economic stabilization and structural adjustment policies, introduced under the context of economic crises to largely developing economies, and intended to return these countries to economic growth and neoliberal economic market equilibria. In the Post-Washington Consensus era – which focuses on a pro-poor and participatory approach to development, as well as on the institutions for 'good governance' – Sen's text, *Development as Freedom*, could easily become the new bible for guiding the current agenda for aid and development. Such is the power of this text.

'Development as capability expansion' – the road to *Development as Freedom*

Sen's influential work thus cuts across theoretical, philosophical and practical development domains, to espouse the powerful message of 'development as freedom'. This freedom is, in part, to be indexed by a capability approach to human development, where capabilities are more than skill sets, but rather speak to 'valuable beings and doings' that agents have reason to value and, most critically, are free to choose. This focus on capabilities or 'capability freedom' is represented as the key departure from utilitarian thought, and neo-classical strands of economic thought steeped in this tradition, where the focus is placed on either subjective desires and mental states of being, such as happiness, pleasure and desire fulfilment in the assessment of well-being. The capability approach to human development, is also distinguished from early growth and modernization theories for development and Rawls's theory of justice, since these other approaches to the development problematic tend to focus more on the commodities that can be marshalled, through a process of market driven growth, to support either consumption desires and needs or baskets of primary goods and resources.

However, as Sen has continuously emphasized, income is not necessarily correlated with well-being. In an earlier essay, where development was seen as capability expansion, Sen (1989) noted, for example, that income rich South Africa scored poorly in terms of quality of life for its citizens, as evident in lower life expectancy, while income poor China did much better on this development score. Although the later Sen (1999) seeks to set a higher development standard for China with his shift to the view of 'development as freedom', he has consistently sought to bring attention to the issue of *human well-being*, as the *quality of life* enjoyed through a specific set of *valued activities* for *individual agents*.

In particular, for Sen, *human* well-being must rely on both a more objective and a personal criteria of welfare, and to achieve this he draws on the Aristotelian notion of 'functionings'. Functionings describe various doings and beings that a person actually experiences or realizes. A specific functioning vector indicates which actual states of doing (e.g. reading or eating) and being (e.g. being well nourished, being literate, being part of a community and appearing without shame) have been realized. As Sen emphasizes, functionings can thus vary from elementary states to more complex ones. However, Sen has been reluctant, as Alkire (2002: 29) and Nussbaum (2003) have highlighted, to establish a list of human capability priorities or even to systematically defend a set of basic functionings, or capabilities, in line with basic needs (Alkire 2002: 157). This allows him to side-step the thorny issue of judging amongst valuable states of being and doing, but leaves his approach open to charges of policy impracticability, since no clear guidelines beyond an appeal for dialogue and participation are offered.

Sen's emphasis on Aristotelian functionings as the foundation for assessing well-being is furthermore deliberately (and problematically) set apart from a basic needs strategy for development (Alkire 2002: 166–174). Thus, rather than encouraging a focus on resources for needs to be met, attention is instead placed on a person's '*capability*' defined as 'a set of vectors of functioning, reflecting a person's freedom to live one type of life or another' (Sen 1992: 40). Capabilities reflect then 'effective possibilities': they describe what people *could do and what could be* achieved *even if* they are not actually chosen. Paying attention to human capabilities, or a person's capability set, helps one to distinguish between the unfreedom and deprivation of a starving child, and the liberty of the fasting monk, since despite the similar states of functioning experienced (i.e. a starving hunger), they reflect different states of capability and relatedly the presence/absence of choice.

A capability approach, however, also allows one to better and more equitably attune resources to support desired functioning vectors, as simply allocating the same basket of primary goods or resources to heterogeneous persons would not be equivalent to supporting equal well-being. This is because persons will differ in their ability to translate resources into functionings. The pregnant woman and disabled person would thus need different stocks of resources to attain their valued functionings, in comparison to the able-bodied man. Sen sharply illustrates the critical power of this capability approach to human development, when he demonstrates that there are around 60–100 million *missing women* in the world due to systematic gender discrimination (sex selective abortion, neglect for female health and nutrition especially during early childhood) resulting in severe capability deprivation for women (Sen 1999: 104–107).

Agency and power in *Development as Freedom*

In elaborating upon his ideas for promoting human and social well-being, which culminated in the text *Development as Freedom*, Sen is also keen to establish what he considers to be foundational analytical distinctions in how we may approach an understanding of well-being *through a valuing of individual agency*. He sets out his four elementary concepts as well-being achievement, well-being freedom, agency achievement and agency freedom, which are themselves derivative from the ideas of the 'well-being aspect' and the 'agency aspect' of a person (Sen 1992: 56–57; Alkire 2002: 9).

For well-being achievement, one looks *only* at the *actual states of valued functionings* that one realizes, while for well-being freedom one assesses the *ontological* or *real context of power* in which the achievement *regarding one's own well-being* is carried out. In other words, in assessing a state of well-being freedom, the relevant question to ask would be: was the agent operating in a *context of freedom*, so that choice reflects a real power that the agent can do, and could have done otherwise? If so, then we have well-being freedom, or a 'real opportunity to accomplish what we value' as wellness of being (Sen 1992: 31). Consider here again the distinction between the starving child and the fasting monk; the monk has exercised a well-being freedom whereas the child has exercised no such freedom, despite the similarity of well-being states. For grasping well-being freedom, one has to analyse then *well-being per se*, as well as *agency* – the ability of an agent to pursue and bring about certain goals – and finally the real *context of power* those actions are embedded in, and dependent on.

In contrast to well-being achievement, agency achievement is to be assessed on a wider set of objectives than *personal* well-being, and so can include other goals, such as seeking the independence of one's or another's country, that may conflict with *individual welfare*. In addition, like well-being freedom, agency freedom, which Sen defines as 'one's ability to promote goals that one has reason to promote' (1992: 60), is also to be assessed by referring to the *context of power* in which one's agency is exercised. In this case, however, the context of power, needed here for agency freedom to be recognized as something distinct from mere agency achievements, seems synonymous with the libertarian view of human freedom as a space of liberty. Agency freedom appears then to be calibrated by a liberal context that affords both negative liberty – absence from interference – and positive liberty – *one's own power* to achieve a desired goal or end state. This affinity between agency freedom and liberty reinforces Sen's persistent promotion of libertarian political ideals; ideals which he seeks to ground in something more than just a focus on rights, by including a concern for more objective moral considerations and consequences, such as individual advantage or well-being and the inequalities in the space of 'capability freedoms'. This is an important step as it allows Sen to avoid the complaint that libertarian priorities on rights ignore substantive inequalities in the command over resources and feasible well-being achievements. As Sen highlights in his discussion of famines, a concern with rights alone is insufficient to prevent human catastrophes (1999: 65–66).

In line with these conceptual and philosophical orientations, Sen thus emphasizes what he refers to as instrumental freedoms, five in all, that are vital to the project of promoting *Development as Freedom*. These instrumental freedoms are (a) political freedoms, (b) economic facilities, (c) social opportunities, (d) transparency guarantees and (e) protective security (Sen 1999: 10, 38–40). These are conceived of as 'distinct sets of rights and opportunities' that help to 'advance the general capability of a person' (Sen 1999: 10). All in all then, Sen advances his thesis of *Development as Freedom* by advocating that the expansion of freedom is not only (a) the primary end but also (b) the principal means of development (1999: 36). Development thus 'consists in the removal of various types of unfreedoms that leave people with little choice and little opportunity to exercise their reasoned agency' (1999: xii).

Critiques and limit points

Sen insists that the impulse to freedom and for freedom is not a peculiarly Western tradition, which is being copied by other peoples across time and space. And while Sen wishes to distance himself from the previous standard of 'development as modernization' that informed programs of development especially in the post-war period, he still seems rather wedded to the institutional scaffolding of Western liberal democracy and its plural economic freedoms entrenched in a capitalist market society. Yet, is Sen correct in making an empirical claim that development is invariably twinned with Liberal Freedoms in a capitalist market society? Sen seems to think so and argues the case based, in part, on his examination of the correlation he detects between famines and democratic freedom, and in particular the presence of a 'free press' (Sen 1999: 16, chapter 7). Others, such as Corbridge (2002), Selwyn (2011) and Northover (2012), arguing from rather different perspectives, suggest that it may indeed *not* be so well linked.

Why not? Well, first, as Corbridge (2002) highlights, because the evidence suggests that substantive well-being and structural transformations are often made under authoritarian systems, such as those in East Asia – China being the most striking case in point. Should such experiences not count as development? And have such outcomes not hinged on the presence of what is now commonly referred to as the 'developmental state'? That is, a state that has the capacity to exercise its authority to resolve and manage conflicts and direct the use of resources to promote broad development objectives.

Second, from the Marxian standpoint as embraced by Selwyn (2011), it may be argued that under capitalism one can only maintain a façade of liberty or insecure substantive freedoms in the context of commodity fetishism, labour exploitation and alienation, and the dialectical contradictions shaping the social relations of production. As such, capitalism itself is seen as a force that undermines the real freedoms possible in the present system due to the nature of capitalist relations of power. Indeed, rather than being linked to the absence of democratic institutions, famines are more associated with the historical conditions of structural inequality attendant with the real dispossessions needed for the creation and continued reproduction of a class of wage-labourers with the hegemonic rise of a capitalist world system. And finally, from a post-structuralist inflected critique of Sen, Northover (2012) suggests that Sen's thesis of development as freedom may be premised on a racial philosophy of place, that produces 'agency freedom' through a politics of abjection mirrored in the links between capitalism and slavery and with ongoing complex racialization processes. It is not surprising then that the message coming from some feminist critiques is that Sen has placed an undue emphasis on the abstract qualities of freedom, for example, qualities such as 'freedom of the people' and 'agency freedom' – Liberalism's liberty – which is hard to justify as Des Gasper and Van Staveren (2003) argue.

In the end, although Sen's message to the world and 'development community' is very attractive – especially in the present time of postcolonial and contemporary world crises given the events of 9/11 and the ongoing global economic crises sparked in 2008. In addition, despite Sen's intense and prolific efforts in popularizing this current narrative of development, his thesis of *Development as Freedom may not* hold the key to a brave new world, given the problematic history of freedom and its ongoing violent contexts of power.

References

Alkire, S. (2002) *Valuing Freedoms: Sen's Capability Approach and Poverty Reduction*, Oxford: Oxford University Press.
Corbridge, S. (2002) '*Development as Freedom*: The spaces of Amartya Sen', *Progress in Development Studies*, 2(3): 183–217.
Escobar, A. (1995) *Encountering Development: The Making and Unmaking of the Third World*, Princeton, NJ: Princeton University Press.
Gasper, D. and Van Staveren, I. (2003) '*Development as Freedom* and as what else?', *Feminist Economics*, 9(2–3): 137–161.
Northover, P. (2012) 'Abject blackness, hauntologies of development, and the demand for authenticity: A critique of Sen's *Development as Freedom*', *The global South*, 6(1): 66–86.
Nussbaum, M. (2003) 'Capabilities as fundamental entitlements: Sen and social justice', *Feminist Economics*, 9(2–3): 33–59.
Selwyn, B. (2011) 'Liberty limited? A sympathetic re-engagement with Amartya Sen's *Development as Freedom*', *Economic and Political Weekly*, 46(10): 68–76.
Sen, A. (1989) 'Development as capability expansion', *Journal of Development Planning*, 19: 41–58, reprinted in Sakiko Fukuda-Parr and A. K. Shiva Kumar (eds) (2003), *Readings in Human Development* (pp. 3–16). New York: Oxford University Press.
Sen, A. (1992) *Inequality Reexamined*, Cambridge, MA: Harvard University Press.
Sen, A. (1999) *Development as Freedom*, New York: Anchor Books.

Further reading

For a collection of Sen's foundational work on economic development issues, see Sen (1987) *Resources, Values and Development*, Cambridge, MA: Harvard University Press. This collection gathers his path breaking writings on institutions, social investment, ethics and well-being.

For a selection of critical perspectives on Sen and his relevance to feminist economics, see the (2003) special issue of *Feminist Economics*, edited by Bina Agarwal, Jane Humphries, Irene Robeyns. Irene Robeyns has also provided a critical theoretical overview of the capabilities approach in *Journal of Human Development* (2005), 6(1): 94–114.

For a useful collection of readings on the Human Development and Capability Approach, largely based on the Human Development Reports of the last decade, see S. Fukuda-Parr and A. Kumar (eds) (2005) *Readings in Human Development: Concepts, Measures and Policies for a Development Paradigm*, New York: Oxford University Press.

Websites

The Oxford Poverty and Human Development Initiative (OPHI), established in 2007, is an economic research centre within the Oxford Department of International Development, University of Oxford. The centre is led by Sabina Alkire. This site provides an excellent repository for many central works and research programmes on poverty and human development (available online at www.ophi.org.uk/).

The website of the Human Development and Capability Association, which promotes multidisciplinary research on poverty, justice and well-being (available online at www.capabilityapproach.com/index.php).

1.8
Race and development

Denise Ferreira da Silva

Forged in the late 1940s and early 1950s, the assemblage of the development apparatus coincided with the unfolding of the cold war, the decolonization in Asia and Africa, and the proliferation of military regimes in Latin America. Towards its goal, development mobilized personnel and institutions (financial, governmental, educational, etc.) to explain and solve the social, cultural, and economic troubles of the Third World – namely, overpopulation, the threat of famine, poverty, illiteracy, and so on (Escobar 1994). Both as a practical and an academic enterprise targeting the global space after WWII and decolonization (Power 2006), the development project included three aspects: (a) a *programme for intervention* involving a variety of institutions and international bodies, such as nation-states, banks, private corporations, the Bretton Woods's organizations (such as the IMF and the World Bank), funding agencies (such as the Ford and Rockefeller Foundations), and, at times, military forces; (b) the delineation of the *object of knowledge* relying on the concept of the cultural which, in the 1950s, would support sociology as a properly scientific field of inquiry, with a conceptual-methodological framework (structural functionalism) that linked the structural (institutional) and cultural (meanings) dimensions of the social; and finally (c) its *objective of the development project*, the elimination of the ills of underdevelopment.

Formulations of research questions, policy and strategies implicitly assumed racial and cultural difference, in the formulation of the meaning and direction for the development project, as its fulfilment required both structural and cultural change in the 'Third World' to make it more like the 'First'. Nevertheless, scholars in the field agree that silence dominates the theme of race and development but they do not provide a definite explanation as to why this is the case (White 2002; Kothari 2006b; Power 2006). Engaging this problematic, this chapter takes steps to delineate a possible reason for this silencing. First, it reviews critiques of development that expose the discontinuities and continuities between development and colonialism. Second, it traces the relationship between race, as a socially and historically constructed category (Omi and Winant 1994), and development back to the Enlightenment moral lexicon. What these steps show is how breaking the silence requires tracing the operations of racial power within the development project.

Beneath development as discourse

A productive move in critiques of development as discourse is to relate the operations of racial power to its relationship with colonialism. Focusing on the ideological level, such critiques highlight the intersection of the development and the colonial project in two moments. On the one hand, this highlights the *programmes for intervention* and *objectives*: as part of this scholar's focus on development as a practice and note how racial meanings reproduce a distinction that sustains the power differential necessary for the justification of both projects (White 2002;

Goudge 2003). On the other hand, regarding the *object of investigation* and *objectives of the development project*, critiques of development as discourse expose the workings of colonial constructs within development (Crewe and Fernando 2006; Escobar 1994; Kothari 2006b). Both moments register deployments of racial meanings to construct postcolonial Asia, Africa, and Latin America (the 'Third World') as an 'Other', one whose difference resides in its inability on its own to reach the same levels of scientific, technological, and economic development as white/Europeans and their descendants. Poverty, illiteracy, famine, and so on, become the identifiers of a 'Third World society', that is, one in need of intervention to foster cultural change and the economic growth necessary to resolve these social ills (Escobar 1994; Kothari 2006b).

These works reveal two truths in development discourse and show how, like its predecessor (the colonial project), the development project relies on an ethical and a political aspect. From the ethical point of view, establish: (a) moral deficiency – the target of the project (the 'Third World') emerges as inherently unable to achieve the *objective* to develop, to move forward, by themselves as they have a moral deficiency (registered in a lack of scientific knowledge and technological advances); and (b) an ethical imperative – both moments of the development project, intervention and investigation, assume that the movement is unidirectional; that is, that the difference between developed white/European and underdeveloped Third World (Asia/African/Latin American) 'others' is a difference of capacity, hence the former has the obligation to employ its superior capacity in order to help the latter (Escobar 1994). From the political point of view, these truths respond to how the development project reproduces global power differentials, that is, the socioeconomic inequalities that result from colonial, neo-colonial, and imperial relations between Europe and (later) the United States, and Third World countries: (a) the diagnostic that the 'Third World' is intellectually deficient registers the 'First World's' superior power, also expressed in scientific and technological achievements and economic prosperity and (b) the directive that it is the responsibility of the First World to help its neighbours to the South and to the East to catch up economically by providing them with a programme and the scientific, technological, personnel, educational, and financial means for economic development. Precisely these 'findings' raise a question about the workings of racial meanings: if the relationship is immediately traceable, why this silence? A possible answer: it results from the notion that development and the 'racial meanings' identified in these works are correlated.

For the most part, critiques of development as discourse do not elaborate on precisely how 'racial meanings' sustain the distinction between the West and the rest and justify power differentials in the postcolonial order. How racial difference operates in the development project – and hence within the development discourse/practice equation – becomes clear through a genealogy that refers both concepts to the post-Enlightenment moral lexicon. As another word for progress, development refers to the Enlightenment view of the human's both necessary (because of human nature) and morally desirable ability to use its unique intellectual abilities to transform the natural environment, thus improving conditions of life. For the philosopher Immanuel Kant, the 'natives' of the Americas, Asia, Africa, and the Pacific Islands are the 'natural men', lacking the basic moral traits that would allow them to transcend the forces at work in their immediate (natural) environment (Spivak 1998); they were 'affectable' minds (Silva 2007). Around the same time natural history's descriptive tools for measurement and classification articulated the 'others of Europe's' mental (intellectual and moral) deficiency, which it associated with the physical traits and their continents of origin. Somewhat later, another German philosopher, G. W. F. Hegel, postulated that progress is the direction of history, but only among European peoples and places. Inhabitants of the Americas and Africa, he stated, were outside the temporal movement of history (the trajectory of progress) and Asians were stuck in the early phases of historical development. None could move forward in time to realize, to materialize, humanity's superior attributes because they lacked moral traits, such as an experience and understanding of freedom and universality (Silva 2007).

Gathered in the arsenal of racial knowledge assembled in the mid-nineteenth century, the science of man, these speculative and descriptive truths became the basis for a scientific project for the knowledge of man and society, which included concepts ('racial type') and measures ('facial index') designed to show that they are indeed *natural* (scientific) truths, that is, expressions of how the laws of nature determine human physical and mental conditions. Early in the twentieth century the science of man came into disrepute, the concept of the cultural became the main referent of intellectual and moral difference, as it is expressed in language, religion, technologies, and social (legal, religious, economic) institutions. Nevertheless, bodily traits separate out those not born in Europe and/or of non-European descent, in regard to the 'others of Europe' and their descendants everywhere. Thus, racial difference would remain the first, immediate, signifier of cultural (moral and intellectual) traits. By the time the development apparatus was deployed, the ideas of racial and cultural difference, developed by anthropologists in the nineteenth and early twentieth century, were already part of political, sociological, and common sense views. Because they view physical and territorial characteristics as expressions of mental (intellectual and moral) traits. That is, racial and cultural difference defining the specificities of European 'civilization' were linked to bodily traits (whiteness), and to self-claimed intellectual superiority; thus explaining capacity (or lack of) for progress (to move forward, develop) as *natural* attributes of persons and places. This view also immediately prevented considerations of present and historical relationships, such as conquest and slavery. The wall of silence in race and development is a reflection of racial power/knowledge, as racial and cultural difference identifies the very subjects, objects, and sites for the development project.

How then to break this constitutive wall of silence? It should be evident now that the silence does not result from a lack of correspondence but from the opposite, namely, the excess of correspondence characteristic of a discursive situation in which one term (race) gives significance (meaning) to the other (development). For both race and development are referents of progress. First, race (as racial difference or cultural difference) allows the distinction between those who can and will and those who cannot and will not move forward (progress or develop). That is, racial (or cultural) difference explains why the Third World lacks the intellectual and moral capacity for developing. Second, development (progress) is precisely that which, according to philosophical and scientific statements, is the privilege of Europeans and their descendants elsewhere. Third, because development is the historical destiny of every human being, it is their (white/European) moral obligation to help others (who lack the capacity) to catch up with them. Every iteration of progress (as economic growth/cultural change) of the past 150 years or so repeats these racial truths: the idea of civilization (as in the civilizing mission), modernization (which named the knowledge aspect of development in the early years), and now globalization (which is characterized by many things, including the demand that the state stays away from economic affairs). To speak of race and development, then, is not to speak about two different concepts or two different discourses. For race operates within development. To speak of this relationship, we need to describe how racial discourse functions as a condition of possibility for every order of operation (project of intervention, object of investigation, objective) of the development project.

Conclusion

Framing of the development project, after WWII, became possible because racial difference and cultural difference functioned as discursive shortcuts, immediately communicating three necessary ethical and political truths: (a) the targets of the development project (illiteracy, poverty, famine) resulted from certain peoples' and places' *natural* incapacity to move forward on their own, (b) those who could develop, namely white/Europeans, had the moral obligation to help those (Asians, Africans, Latin Americans, and Pacific Islanders) who could not develop, and (c) the naturalization of the Third World's incapacity for development pre-empts attributions of the failures of the development project either to existing mechanisms of economic exploitation or to the effects of previous (colonial or imperial) unequal economic relations. Existing analyses of race and development show that 'racial meanings' operate in development discourse and practice. This chapter merely adds an account of how racial meanings (as referent to racial and cultural difference) work within the development project. These natural (biological or cultural) truths, which are signified in bodies and territories, serve to naturalize the effects of (past) colonial and (present) global capitalist architectures of economic exploitation.

Bibliography

Crewe, Emma and Fernando, Priyanthi (2006) 'The elephant in the room: Racism in representations, relationships, and rituals'. *Progress in Development Studies* 6(1): 40–54.
Escobar, Arturo (1994) *Encountering Development*. Princeton, NJ: Princeton University Press.
Goudge, Paulette (2003) *The Whiteness of Power: Racism in Third World Development and Aid*. London: Lawrence & Wishart.
Kothari, Uma (2006a) 'Critiquing "race" and racism in development discourse and practice'. *Progress in Development Studies* 6(1): 1–7.
Kothari, Uma (2006b) 'An agenda for thinking about race and development'. *Progress in Development Studies* 6(1): 9–23.
Omi, Michael and Winant, Howard (1994) *Racial Formation in the United States: From the 1960s to the 1990s*. New York: Routledge.
Power, Marcus (2006) 'Anti-racism, deconstruction and "overdevelopment"'. *Progress in Development Studies* 6(1): 24–39.
Silva, Denise Ferreira da (2007) *Toward a Global Idea of Race*. Minneapolis: University of Minnesota Press.
Spivak, Gayatri C. (1998) *Critique of Postcolonial Reason*. Cambridge, MA: Harvard University Press.
White, Sarah (2002) 'Thinking race, thinking development'. *Third World Quarterly* 23(3): 407–419.

1.9

Culture and development

Susanne Schech

Up until the 1980s, culture played an important role in underpinning modernization theory but was rarely scrutinized. Development was equated with modernization and assumed to be culture-neutral – anyone, anywhere could experience modernization (Watts, 2003: 434).

Then the 'cultural turn' finally arrived in development studies, having swept through other disciplines and created new fields of enquiry such as cultural studies and postcolonial studies along the way. Armed with new approaches and analytical tools, development scholars began to question the role of culture in development. It is now widely acknowledged that culture is intrinsic to economic, political, and social processes; indeed, we cannot understand development and change without taking 'the cultural factor' into account (Radcliffe and Laurie, 2006; Schech and Haggis, 2000). This applies to processes of change that are *immanent* to capitalist development and to *intentional* development, or the efforts to guide processes of change towards desirable outcomes through policies and projects. There are, however, different views about *how* culture matters, based on different ideas about the meaning of culture.

Modernization and culture

Some development scholars argue that cultural values, attitudes, orientations and opinions are a key variable in determining economic progress. According to this view, the developmental success of Western countries is based on the distinctive cultural institutions of Western civilization, and other countries should emulate these as much as possible. In the 1960s, Gunnar Myrdal (1968) painted a picture of Asia beset by abject poverty and corruption, which can only be rescued through international development assistance and the widespread adoption of the modernization ideals and attitudes. In his account, 'modern man' is defined by a set of attitudes including rationality, efficiency, orderliness, preparedness for change, energetic enterprise, integrity and self-reliance. Myrdal understood these 'modern' attitudes to be Western imports or impositions that would eventually displace the cultural traditions of Asia, albeit against popular resistance (Myrdal, 1968: 61–62). Like other modernization theorists, he perceived modern attitudes and patterns of social relations as a 'universal social solvent'.

From a modernization perspective, culture is what other societies possess, and is in most cases an obstacle to development (Watts, 2003). Traditional/modern is understood to be a hierarchical relationship whereby traditional cultural traits are destined to die out, or be 'bred out' of a people through more or less well-meaning policy interventions (Schech and Haggis, 2000: 18–19). Culture, in this view, is bounded and static, like a box handed down from one generation to the next that must be cast aside if it stands in the way of progress. Traditional societies exist outside of history, and any society that resists modernization and clings to tradition will remain underdeveloped. Only societies willing to give up their traditional values, institutions and cultural practices, or which happen to possess cultural traits that are favourable to modernization, will succeed in their quest for development (Harrison and Huntington, 2000).

Culture as a resource for development

Other scholars refute the proposition that culture is bounded and static. Amartya Sen (2004: 43) conceives culture as constantly changing, and any attempt to tie it down as futile. Cultural determinism, he argues, 'often takes the hopeless form of trying to fix the cultural anchor on a rapidly moving boat'. Cultures do not evolve in isolation as separate boxes but always through interaction with other cultures. Cultural interconnections go back far into history, produced by migration, conquest, trade, exploration, and pilgrimage. In the current era of globalization, the boundaries of people's lived experience are more permeable than ever before, informed by an awareness of other circumstances, experiences, images, and ways of living.

If the cultures of the global South are neither fixed nor internally homogeneous, it makes no sense to regard them as an obstacle to change. Indeed, many development scholars and practitioners see culture as a resource (Nederveen Pieterse, 2010). For some dependency theorists, selective traditions become the roots of a national culture, which will inspire an independent development path. From a neoliberal perspective, cultural practices generate stuff that can be marketed, such as Indian Bollywood films or African 'world music'. Culture can be used to brand a country in a competitive global market, or to shape human capital in ways that foster economic growth, through religious values, kinship networks, and reciprocal relationships.

Some post-development theorists see culture as a resource that empowers locally based opposition and alternatives to Western development interventions. For example, Indian scholar and activist Vandana Shiva argues that water cultures, or traditional indigenous knowledge about water as the source of all life, can help communities resist pressures to privatize water and defend fresh water sources as a public good (Opel and Shiva, 2008). Shiva and other post-development scholars have been criticized for romanticizing indigenous knowledge, portraying these knowledge systems as somehow outside of history, and ignoring the often very oppressive class and caste relations within which subsistence lifestyle and culture is placed (Schech and Haggis, 2000). Also working with the notion of culture as a resource, but one that is malleable, Appadurai (2004) argues that cultural practices can empower the poor. He gives the example of a pro-poor alliance of housing activists in India using housing exhibitions as a public space where poor people discuss their housing needs with politicians, donor agencies, local planners, architects, and professional builders. By employing what is essentially an upper-class cultural form and placing slum residents at its centre, the alliance enhances their visibility and recognition, as well as cleverly subverting the dominant class cultures in India.

The shift towards viewing culture as a resource for development has encouraged a variety of development actors to add culture to their development toolbox. In the social capital approach to development, culture is treated as a kind of glue that holds societies together and gives them a coherent structure that can be used for development interventions. The Danish development agency is one of many to subscribe to a culture and development approach, which affirms the right to cultural difference and the need to integrate culture into development programmes. The underpinning definition of culture 'as the total complex of spiritual, material, intellectual and emotional features that characterize a society or social group' (Danida, 2002: 5), however, does not sufficiently account for change and contestation.

Development as a cultural construct

Scholars who regard the resource approach as too limited turn to the new fields of cultural studies, postcolonial studies, and globalization for a much more expansive and dynamic definition of culture as:

> involved in all those practices . . . which carry meaning and value for us, which need to be meaningfully interpreted by others, or which depend on meaning for their effective operation. Culture, in this sense, permeates all of society.
>
> *(Hall, 1997: 3)*

Culture, from this perspective, is an active force in the production and reproduction of social life. Rather than seeing culture as one factor that shapes society alongside many others, such as class, gender, economic systems, and institutional arrangements (Sen, 2004), a constructivist approach considers all of these factors to be aspects of social life and thus culturally constructed. Development itself is a construct, a set of culturally embedded practices and meanings that are contested and changed over time.

This view of culture and development pays close attention to representation and power. From a cultural studies perspective, European colonization involved not only economic and political domination of the New World but also cultural domination: 'Europe brought its own cultural categories, languages, images, and ideas into the New World in order to describe and represent it' (Hall, 1992: 293–294). In his landmark contribution to postcolonial studies, Edward Said (1978) analyses Orientalism as not just a Western way of knowing the Orient but also 'a Western style of dominating, restructuring, and having authority over the Orient'. Colonial representations have long-lasting effects. They continue to frame how the world is seen through Western eyes and legitimize contemporary economic and geopolitical interventions (McEwan, 2009).

Arturo Escobar (1995) employs this approach in his analysis of development as a Western discourse that is globalized and constantly reproduced through powerful institutions. The World Bank, for example, presents itself as a knowledge bank capable of providing the (Western) knowledge and information required to dispel the darkness of poverty in the rest of the world. According to its statistics, countries are ranked from 'very high income' to 'low income' and 'least developed' and the latter are designated as needing development assistance. Their progress towards development goals is constantly monitored, like a patient in intensive care. In this construction of underdevelopment, citizens are represented as helpless victims living in ignorance and unable to do anything for themselves, and their governments are portrayed as incapable, fragile, corrupt, and requiring reform.

Culture and development now

Constructivist approaches to development are employed in the growing literature on governmentality. A good example is Tania Li's study of World Bank-funded community development programmes in Indonesia. Poor communities are treated as natural spaces for development interventions, and poor people as subjects who must be empowered to take responsibility for their own improvement. Poverty is rendered as a technical issue that can be solved through participatory planning and better governance. This leaves the unequal relations of production and appropriation outside the frame of view and intervention, and capitalist enterprise is seen 'only as a solution to poverty, not as a cause' (Li, 2007: 267).

Development scholars have started to engage with the shifting geographies and geopolitics of development. The rise of China, India and other countries of the South as global economic powers challenges the monopoly of the West on what it means to be modern and developed (Sidaway, 2012). It raises questions about the ways in which countries mobilize cultural power commensurate with their growing economic might on the global stage, and how cultural imagery is employed in rebranding themselves as prosperous nations. Other studies reveal how Northern economic ideas and policies travel and how they are interpreted, adapted, and resisted in the South (Ferguson, 2006). Bringing the economic together with culture and development challenges the assumption that economics is a science separate and immune from political and cultural enquiries (Pollard *et al.*, 2011). These studies of culture and development anchor culture within an analysis of the machinations of power and capitalism in development.

Other current work on culture and development involves detailed case studies that examine where, when, and how culture and development interact, and who is involved in these interactions. One fruitful place is the edges of development (Bhavnani *et al.*, 2009). The edges can be marginal positions from which any development interventions or programmes are experienced as a distant echo, or as acts of dispossession and oppression. They can also be liminal, or between, locations where new cultural meaning is created. They are inhabited by people who negotiate, contest, and blur the boundaries of modernity and tradition in their everyday lives. Such case studies show development as always cultural, specific, and culture as a terrain of struggle.

References

Appadurai, A. (2004) 'The capacity to aspire: Culture and the terms of recognition', in V. Rao and M. Walton (eds), *Culture and public action*. Stanford, CA: Stanford University Press, 59–84.

Bhavnani, K.-K., Foran, J., Kurian, P. and Munshi, D. (eds). (2009) *On the edges of development: Cultural interventions*. New York: Routledge.

Danida. (2002) *Culture and development: Strategies and guidelines*. Copenhagen: Danish Ministry of Foreign Affairs.

Escobar, A. (1995) *Encountering development: The making and unmaking of the Third World*. Princeton, NJ: Princeton University Press.

Ferguson, J. (2006) *Global shadows: Africa in the neoliberal world order*. Durham, NC: Duke University Press.

Hall, S. (1992) 'The West and the rest: Discourse and power', in S. Hall and B. Gieben (eds), *Formations of modernity*. Cambridge: Polity Press/Open University Press, 274–311.

Hall, S. (1997) 'The work of representation', in S. Hall (ed.), *Representation: Cultural representations and signifying practices*. London: Sage, 15–51.

Harrison, L. E. and Huntington, S. P. (eds). (2000) *Culture matters: How values shape human progress*. New York: Basic Books.

Li, T. M. (2007) *The will to improve: Governmentality, development, and the practice of politics*. Durham, NC: Duke University Press.

McEwan, C. (2009) *Postcolonialism and development*. London: Routledge.

Myrdal, G. (1968) *Asian drama: An inquiry into the poverty of nations* (3 vols). Harmondsworth, UK: Penguin.

Nederveen Pieterse, J. (2010) *Development theory*. Thousand Oaks, CA: Sage.

Opel, A. and Shiva, V. (2008) 'From water crisis to water culture'. *Cultural Studies*, 22(3–4): 498–509.

Pollard, J., McEwan, C. and Hughes, A. (eds). (2011) *Postcolonial economies*. London: Zed Books.

Radcliffe, S. A. and Laurie, N. (2006) 'Culture and development: Taking culture seriously in development for Andean indigenous people'. *Environment and Planning D: Society and Space*, 24(2): 231–248.

Said, E. W. (1978) *Orientalism: Western conceptions of the Orient*. New York: Pantheon Books.

Schech, S. and Haggis, J. (2000) *Culture and development: A critical introduction*. Oxford: Blackwell.

Sen, A. (2004) 'How does culture matter?', in V. Rao and M. Walton (eds), *Culture and public action*. Stanford, CA: Stanford University Press, 37–58.

Sidaway, J. D. (2012) 'Geographies of development: New maps, new visions?'. *The Professional Geographer*, 64(1): 49–62.

Watts, M. (2003) 'Alternative modern – development as cultural geography', in K. Anderson (ed.), *Handbook of cultural geography*. Thousand Oaks, CA: Sage, 433–453.

Further reading

Li, T. M. (2007) *The will to improve: Governmentality, development, and the practice of politics*. Durham, NC: Duke University Press. For an ethnographic study of development practices as forms of governmentality.

McEwan, C. (2009) *Postcolonialism and development*. London: Routledge. An accessible account of post colonial theory and its implications for development studies.

Schech, S. and Haggis, J. (2000) *Culture and development: A critical introduction*. Oxford: Blackwell. For an overall introduction to the topic.

1.10
Ethics and development

Des Gasper

The field of development ethics explores questions and debates concerning what is good development of societies and of the world, and good development for individual persons. Generations of experience suggest the inadequacy of the assumption that societal, world or personal development can be equated to economic growth and wealth. That assumption neglects issues of equity, security, personal relationships, natural environment, identity, culture and meaningfulness. In particular, equating national development to national economic growth neglects the welfare and rights of many groups of already disadvantaged people. Over ten million people a year, for example, are displaced from their home due to economic expansion, frequently with little or no compensation (Penz *et al.* 2011). An important alternative conception of development is 'human development', meaning achievement with respect to a wide range of well-reasoned values, not only those measured in money, and advancement of people's ability to achieve such well-reasoned values (Haq 1999; Nussbaum 2011). Development ethics tries to identify and systematically reflect on values and value-choices present in, or relevant to, cases and processes in the development of societies, persons, regions, and the globe.

Topics in development ethics include, amongst others: meanings and evolution of the idea of 'development', and the values that these meanings can contain about what is acceptable and desirable; concepts and evidence about human well-being and ill-being; assumptions and gaps in conventional economic evaluation; meanings and varieties of 'equity', and how equity can be neglected; and the significance of various types of human vulnerability and security (Goulet 1971), and their relationships to economic growth. Practically oriented development ethics looks at how and which values are or can be incorporated in systems of policy, laws, social routines, and public and individual actions.

Ethics of development: Why?

'Development' sounds self-evidently desirable. Why did a field of ethics of development arise? First, because of persistent undeserved removable poverty, sickness, insecurity and unhappiness despite economic growth. Rise of average incomes does not necessarily benefit ordinary and especially poor people. Despite enormous growth in human powers and economic turnover, hundreds of millions of people remain undernourished, leading to their physical and mental stunting, illness and premature death. As of 2004, UNICEF estimated that 30,000 babies and children under five years died every day from poverty-related causes; a third of the world's people lacked basic sanitation, almost a billion adults were illiterate, and 170 million children were engaged in hazardous work. Most of the people affected – including the babies and children – were not to blame for their own situation and had little unaided response-ability. As Nussbaum (e.g. 2011) asks: how far should the chance of being born in one nation rather than another determine the life chances of a baby?

Second, many people become harmed even within processes of economic development: they are made to bear the costs without sharing in the benefits, as for example in uncompensated displacement from their homes and livelihoods. The people displaced are nearly always poor, moved in order to clear the way for projects that very largely benefit people who are already better off. So, not only do some groups not share in benefits, they may be deliberately harmed, as were generations of slaves and many other workers. Even when not deliberately sacrificed, many suffer through increased exclusion and marginalisation. Issues of distribution and harm involve also future generations, notably in relation to damage to the natural environment, especially through human-induced climate change (UNDP 2007).

Development ethics looks at implications of the interconnections and in-built conflicts in socioeconomic development, within countries and internationally and over time. Increased carbon emissions, a core feature of modern development, indirectly eventually damage people in vulnerable environments around the world. Investment for future generations can be at the expense of present-day poor people; construction of infrastructure for some people's benefit results in displacement of others, typically with major economic, social and psychological disruption; and within markets, the increased wealth of some people competes away resources from poorer people, by forcing up the prices of goods such as land, housing and food. Famines and malnutrition have often been caused less by shortage of supply than by this mechanism, whereby wealth draws in resources from around the world (Davis 2001). Increased pressures of most sorts typically affect women especially, for they are the main caregivers and 'shock absorbers' in a society, that get noticed only when broken. The disproportionate concentration of costs of development upon some groups has been used as a mechanism of transformation – 'breaking eggs in order to make omelettes', in the words of Britain's Colonial Secretary Joseph Chamberlain and many others; industrialisation and the transformation of agriculture have typically partly occurred through processes by which many small agricultural, industrial and artisanal producers are forced out of business. In recent years around a quarter-million Indian farmers have committed suicide due to accumulation of unrepayable debts – here a case not of exclusion from modernisation processes, but of vulnerable people who are induced and/or choose to participate in types of economic modernisation and who then sometimes suffer severely when the risks and the 'small print' prove to be more than they can cope with.

Third, the gains in well-being through policy approaches that equate development to economic growth are sometimes very questionable. When and how far does acquisition of and preoccupation with material comforts and conveniences bring, or jeopardise, a fulfilling and meaningful life (Gasper 2007)? The French economist Louis-Joseph Lebret (1897–1966), who helped to found the subject of development ethics, spoke thus not only of 'development for all persons' but of 'development of all the person' and of 'putting the economy at the service of man'.

Fourth, besides outcomes, major questions arise about democratic participation in processes of decision-making (e.g. Chambers 1997; Ellerman 2005) and about responsibilities in relation to harm and undeserved suffering: who has responsibilities and to do what – to help, prevent, refrain, compensate – including in light of past injustices (such as slavery) and their consequences for present-day undeserved advantages and disadvantages?

Issues of development ethics become relevant because of arguments that better alternatives are possible compared to what has happened, and that real choices exist for the future too. Evidence of such alternatives may lie in the experiences of other countries (Drèze and Sen 1989) – for example, some East Asian countries that have combined a rapid climb out of poverty with a relatively high degree of population inclusion and sharing in benefits. Financial and economic calculations are also relevant; for example, that to attain the $2 per day per person income line for everyone would cost around 1.2 per cent of the gross national products of high-income countries, vastly less than they spend on military forces. A recent estimate of the extra costs required to achieve universal primary education equated to four days of the world's military spending.

Ethics in development: What and how?

The agenda of development ethics includes: to explore how the content of the idea of 'development' as societal improvement is value-relative; to highlight who bears the costs of various types of 'development', and to examine the value-choices that are implied and should be considered in development policy, programmes and projects; to present well-reasoned alternatives to mainstream habits regarding those choices, in particular to clarify the values behind evaluative and prescriptive arguments from economics, query the narrowness of using only values from the marketplace, introduce other relevant values and query therefore an automatic superior status for economics arguments, in relation, for example, to human rights arguments. The root concerns of development ethics – the insistence on not *equating* societal improvement to economic growth, and on identifying and comparing value and strategy alternatives; and the concern for not ignoring costs and their distribution – all apply not only to poor countries but with almost equal force in rich countries and for the globe.

In examining the value-choices in development, work in development ethics operates at the interface of ethics, development studies and development policy. It raises the question of which life-conditions and which effects are both unfair and avoidable? For those considered unfair and avoidable, what changes should be made? Who has what responsibilities – including to remedy the damage that they cause, and to respect and preserve local and global public goods? How far are national boundaries ethically relevant, in a world that is increasingly economically unified, such that people influence each other worldwide? When is international aid justified – for serving longer-term self-interest, or as praise-worthy but non-obligatory charity, or as obligation, including as an obligation of former colonial powers and beneficiaries from colonialism? What are the ethical requirements regarding its conduct?

Some development ethics work deals with basic issues of concepts and theory (e.g. Goulet 2006; Gasper 2004); some engages with specific sectors and policies (e.g. Drèze and Sen 2002). Some significant examples of the latter type are: (a) much work looks at health and the gross imbalances between health needs and health spending, including spending on research; diseases of the poor have been grossly neglected. Pogge amongst others has investigated both relevant ethical theory and health policy options; (b) the Jubilee 2000 debt-relief campaign studied the history of banking and found that all the countries that had long insisted on total repayment of debts by low-income countries, which had vastly escalated since the 1970s oil crises and the increased interest rates, had themselves had major earlier episodes of debt relief or repudiation; (c) Penz *et al.* (2011), building on the work of the World Commission on Dams, propose a detailed, principled approach to assessing, deciding on, and compensating for displacement, based on a synthesis of development ethics thinking, with attention to what is justifiable development and what are good procedures for resolving conflicts.

The field of development ethics is a meeting place of theory and practice, and of many disciplines and types of knowledge. It needs to use a broad vision, looking at the range of real experience of human joys and suffering (see e.g. Narayan *et al.* 2000), and at interconnections besides only those captured by markets and the categories of economics.

References

Chambers, R. 1997. *Whose Reality Counts? Putting the First Last.* London: Intermediate Technology Publications. About equity in participation in defining situations and providing information.

Davis, M. 2001. *Late Victorian Holocausts: El Niño Famines and the Making of the Third World.* London: Verso. A study of the implications and interactions of global power systems, global markets, and global climate systems in the late nineteenth century, especially the disastrous implications for vulnerable population groups.

Drèze, J. and A. Sen. 1989. *Hunger and Public Action.* Oxford: Clarendon. An overview of feasible public action to address basic human needs, under a variety of different political setups.

Drèze, J. and A. Sen. 2002. *India: Development and Participation.* Delhi: Oxford University Press. An illustration of a treatment of development policy evaluation and design that is systematically guided by an ethical standpoint – here, Sen's capability approach.

Ellerman, D. 2005. *Helping People to Help Themselves: From the World Bank to an Alternative Philosophy of Development Assistance.* Ann Arbor: University of Michigan Press.

Gasper, D. 2004. *The Ethics of Development.* Edinburgh: Edinburgh University Press.

Gasper, D. 2007. 'Uncounted or Illusory Blessings? Competing Responses to the Easterlin, Easterbrook and Schwartz Paradoxes of Well-Being'. *Journal of International Development*, 19(4): 473–492.

Goulet, D. 1971. *The Cruel Choice.* New York: Atheneum.

Goulet, D. 2006. *Development Ethics at Work: Explorations 1960–2002.* New York: Routledge. By a leading figure in formation of the field of development ethics.

Haq, M. 1999. *Reflections on Human Development*, 2nd edn. Oxford: Oxford University Press.

Narayan, D., R. Chambers, M. K. Shah and P. Petesch. 2000. *Voices of the Poor: Crying Out for Change.* New York: Oxford University Press. From over 10,000 interviews around the world.

Nussbaum, M. C. 2011. *Creating Capabilities: The Human Development Approach.* Cambridge, MA: Harvard University Press.

Penz, P., J. Drydyk and P. Bose. 2011. *Displacement by Development: Ethics, Rights and Responsibilities.* Cambridge: Cambridge University Press.

UNDP. 2007. *Human Development Report 2007/8 – Fighting Climate Change: Human Solidarity in a Divided World.* New York: United Nations Development Programme.

Further reading

Gasper, D. and A. L. St. Clair (eds). 2010. *Development Ethics.* Aldershot, UK: Ashgate. A collection of 28 papers from the last 35 years.

Pogge, T. 2008. *World Poverty and Human Rights*, 2nd edn. Cambridge: Polity Press.

Schwenke, C. 2008. *Reclaiming Value in International Development: The Moral Dimensions of Development Policy and Practice in Poor Countries.* Portsmouth, NH: Greenwood Press. A practically oriented textbook.

Websites

Human Development and Capability Association, HDCA: www.capabilityapproach.com
International Development Ethics Association, IDEA: http://developmentethics.org/

1.11 New institutional economics and development

Philipp Lepenies

New Institutional Economics (NIE) is an expansion of neo-classical economic theory. Its merits stem from the fact that it has identified efficient institutions to be a prerequisite for development. Yet, as neo-classical economics is increasingly rejected as a useful basis for social analysis, the prominent role of NIE in development will probably diminish over time. However, NIE still strongly influences development policy. Thus, it is useful to understand the theoretical context of NIE as well as its shortcomings.

What is NIE?

NIE attempted to incorporate a theory of institutions into economics. It was a deliberate attempt to make neo-classical economic theory more 'realistic'. Neo-classical economic theory assumed that information flows freely between the actors in competitive markets and that, as a result, institutions do not matter. In contrast, NIE postulated that information is distributed asymmetrically (*asymmetrical information*) and that market transactions come at a cost (i.e. the cost of gathering information or *transaction costs*). Consequently, institutions had to be formed to reduce these costs.

NIE retained the neo-classical assumptions that individuals seek to maximise their utility from scarce resources subject to budget constraints and that collective outcomes rest on the choices made by rational individuals (i.e. *methodological individualism*). However, it discards the concept of *instrumental rationality*, which implies that the choices made by each individual are completely foreseeable. With all information readily available to everyone (*perfect information*), there is no uncertainty in human actions. Institutions become unnecessary and efficient markets characterise economies.

The necessity for a modification of neo-classical theory arose from the fact that so-called *social dilemmas* could not be explained by it. Social dilemmas are situations in which the choices made by rational individuals yield outcomes that are socially irrational. This is obvious in the case of *market failures* that can be caused by *negative externalities* (i.e. a cost arising from an activity which does not accrue to the person or organisation carrying on the activity, e.g. pollution) and *public goods* (i.e. goods that are open to all, free of charge and thus not usually supplied by the market), but also applies to cases of *asymmetrical or imperfect information* (i.e. information is not fully available to everyone). Imperfect information might cause *moral hazards* (i.e. the danger that one of two parties of a contract knowingly alters her behaviour in order to maximise her utility at the other party's expense), *adverse selection* (e.g. imperfect health insurance contracts attract those who have high health risks) and/or *principal-agent dilemmas* (i.e. the problem of how a 'principal' can motivate an 'agent' to act for the principal's benefit rather than following his or her self-interest).

NIE's core argument is that institutions provide the mechanisms whereby rational individuals can transcend social dilemmas and economise on transaction costs (Bates 1995: 29). Institutions are thus

> the rules of the game of society . . . the humanly devised constraints that structure human interaction. They are composed of formal rules (statute law, common law, regulations), informal constraints (conventions, norms of behaviour and self-imposed rules of conduct), and the enforcement characteristics of both.
>
> *(North 1995: 23)*

The term 'New Institutional Economics' was coined in the 1970s by Oliver Williamson to distinguish it from an earlier attempt to incorporate institutions into economic theory at the beginning of the twentieth century, the so-called '(Old) Institutional Economics' whose main authors were Thorstein Veblen and John R. Commons.

NIE came into being in 1937, when Ronald Coase explained the existence of firms. Ironically, neo-classical theory could not explain why firms existed and why market transactions were not carried out solely by individuals as methodological individualism suggests. Coase departed from the Walrasian notion of market transactions being made costless on the spot by an invisible and omniscient auctioneer – that is, the idea of perfect information. Instead he assumed that 'the main reason why it is profitable to establish a firm would seem to be that there is a cost of using the price mechanism' (Coase 1937: 390). Transactions thus involve the cost of discovering what the relevant prices are.

With time, the idea of transaction costs, probably the single-most important concept of NIE, was developed further. Some authors distinguish different transaction costs in accordance with the three big areas of analysis within NIE, that is, the market, the firm and the State. Thus, there are:

- market transaction costs, which are those described by Coase;
- management transaction costs within a firm that come as a result of administrative procedures, strategic planning as well as supervision of the work force;
- political transaction costs, which are the costs of establishing, enforcing and utilising a political system.

Others identify transaction costs according to the process of transacting, that is, information and search costs, costs of negotiating contracts and the costs of enforcing them. Yet, all transaction costs 'have in common that they represent resources lost due to lack of information' (Dahlman 1979: 148).

It is important to point out that the NIE is not a homogeneous school of thought. Rather, it consists of a variety of theoretical writings by a large number of different authors. NIE includes research on transaction costs, political economy, contract theory, property rights, hierarchy and organisations, public choice and development.

NIE and development

Since the 1990s, NIE has had a tremendous impact on development policy and theory. This can be demonstrated by tracing out the obvious influence that NIE has had on the World Bank, by presenting Douglass C. North's NIE-inspired theory of development and by describing the relevance of NIE for development practitioners.

NIE and the World Bank

Since free markets alone cannot be relied upon to ensure development, NIE emphasises the necessity for development policy to design favourable growth-inducing institutional settings. This was reflected in the new role ascribed to the State in the World Bank's World Development Report (WDR) *The State in a Changing World* (1997). Therein, the State, after having been viewed as an obstacle to the functioning of competitive markets in the years before, was suddenly identified as an important facilitator of favourable institutional arrangements.

The appointment of Joseph Stiglitz, a major theorist of the NIE, as chief economist of the World Bank, also reflected the influence of the NIE on the Bank's policy. In 1986, Stiglitz stated that the assumptions of neo-classical economics were 'clearly irrelevant' for the analysis of developing countries. Instead, he showed that asymmetrical information prevailed in most markets (Stiglitz 1986: 257). During his spell as chief economist of the World Bank, the bank began to define itself as a 'knowledge bank' whose responsibility was to gather and disseminate information transparently on a global scale. NIE dominated the WDR 1998/1999 *Knowledge for Development*, which highlighted the general importance of overcoming asymmetrical information in development (World Bank 1998). The influence of NIE was also obvious in the WDR 1999/2000 *Entering the 21st Century*, which summarised the lessons learnt from the last 50 years of global development policy. One lesson plainly read: 'Institutions matter' (World Bank 1999: 1).

Explaining institutional change and underdevelopment

A major branch of NIE is concerned with the analysis of institutional change and underdevelopment. Its most prominent author is Douglass C. North who added a historical perspective to neo-classical economics. Historically, societies had to learn how to solve the problem of scarcity.

> The key . . . is the kind of learning that organisations acquired to survive. If the institutional framework made the highest pay-offs for organizations piracy, then organizational success and survival dictated that learning would take the form of being better pirates. If on the other hand productivity raising activities had the highest pay-off, then the economy would grow.
>
> *(North 1995: 21)*

Thus, developmental outcomes in the world differ according to how people learnt to cope with scarce resources.

For North, the Western capitalist system has been flexible enough to adapt itself to the institutional necessities induced by the higher division of labour, minute specialisation, impersonal exchange and worldwide interdependence. However, in a country with inefficient institutions, only a process of internal re-contracting can change the institutional setting. As long as those holding the bargaining power have an incentive to defend the status quo, and inefficiencies are perceived to be rewarding, the situation will not improve (*path dependence*). This is a major deviation from the neo-classical notion of long-term equilibrium – and a more pessimistic one where underdevelopment becomes plausible.

The still existing practical relevance of NIE for development

The practical relevance of NIE is twofold. First, concepts such as 'asymmetrical information', 'transaction costs', 'adverse selection', 'moral hazard' and 'principal-agent dilemmas' are currently widely used *tools for socioeconomic analysis*. Second, they also *serve as the basis for individual project design*. 'Institution-building' itself has become the raison d'être of many development projects in the last decade.

Parting from the definition of 'institutions', any attempt to establish 'rules of the game' and their enforcement characteristics (be it new laws, regulations or governance structures, etc.) can consequently be seen as an application of NIE.

A prominent case where concepts of NIE are used is that of the analysis of financial services for the poor. The banking sector usually does not offer financial services to the informal sector because information is asymmetrically distributed between the potential borrower and lender. The lender does not have sufficient information on the borrower whom he does not know personally, who usually does not keep written accounts or business plans and who cannot offer physical collateral. Thus, the lender cannot calculate the risk of default. As a result, credit to the informal sector is rationed since lenders are reluctant to give out credit. If financial services to the poor are to be provided, these problems have to be addressed with adequate institutional design.

In the absence of physical collateral, for instance, group-based lending could be an institutional design option to overcome the problems of asymmetrical information. By introducing peer-monitoring as a control mechanism of the borrowers and by linking future payments to group members to the repayment performance of the entire group during monitored weekly meetings, the risks posed by the lack of information described above are minimised.

However, just as NIE is a heterogeneous theory, there is also no such thing as a clear-cut NIE-approach to development. Few development practitioners or theorists who make use of NIE concepts would define themselves as being 'of the NIE'. Yet, the fact that elements of NIE are used so widely and that the importance of institution-building has been generally acknowledged is arguably the strongest sign of how much NIE has already become commonplace in development.

Critique and conclusion

NIE has 'challenged the dominant role ascribed to the market . . . [by highlighting that] neither State nor market is invariably the best way in which to organise the provision of goods and services' and that efficient institutions are the key to successful development (Harris, Hunter and Lewis 1995: 1). It is, without a doubt, the strongest merit of NIE to have put the issue of institutions on the development agenda.

However, NIE is not without limitations. As noted before, NIE is an attempt to change neo-classical economics 'from within'. This alone is praiseworthy. Yet, as it maintains the basic assumption that individuals rationally pursue the maximisation of their utility at all times, little or no room is given to any behaviour which might not be guided by the individual's rationally calculated quest for utility-maximisation. Hence, NIE is still not realistic enough as it maintains a simplistic and incomplete model of human behaviour.

The major flaw of the NIE, be it in development or elsewhere, is that many concepts of NIE are hard to measure, sometimes even hard to define as 'a clear cut definition of transaction costs does not exist' (Eggertsson 1990: 14). From this derives the difficulty in measuring exactly what transaction costs are. The same applies to the notion of asymmetrical information or the simple question, what 'information' means – especially when one takes into account that 'information' might mean different things to different people. Research which utilises concepts of NIE might thus bring forth insights for a special case. Nevertheless, it is often not comparable with other findings.

Laudably, NIE's historical analysis of development and institutional change rejects the simple idea of market-driven institutional progress. All the same, the attempt to explain persistent underdevelopment by analysing if and how institutions have used resources efficiently in the past is not as straightforward as it seems. A major problem arises out of the way in which history is interpreted. Different interpretations of the past might give rise to various interpretations of the present, especially of the reasons for underdevelopment. Therefore, a historical interpretation might not be shared by everyone. Any historical analysis is just one possible point of view – among many others.

In North's (1995) approach, the reasons for persistent underdevelopment are by assumption endogenous. As long as the bargaining power rests with those forces of society that have an interest in perpetuating inefficient institutions, no efficient institutions can emerge. Yet, the role that external factors (e.g. international political or economic power structures) can play in the explication of underdevelopment is not particularly highlighted, a severe omission given the global economic and political interdependencies. More problematic, still, is his notion of path dependence. With this idea, it seems that countries are trapped in their inefficiencies. This view is overly pessimistic and eclipses the possibility of active development or development cooperation.

Notwithstanding, NIE has rightly identified institution-building as a necessary developmental activity. Nevertheless, identifying a problem through NIE-inspired analysis does not automatically lead to infallibly designed institutions. As it is now generally accepted by the economics profession that humans do not behave as modelled in neo-classical economics, extreme caution is advised when attempting to create 'rules of the game' assuming an economically rational and utility-focused behaviour of the target group. NIE carries the risk that institutions are designed and planned by development experts from scratch – based on the inner logic of the NIE and less on an in-depth analysis of the complex social, political and cultural contexts and the subjective aspirations of a target population. If, however, one is conscious of the limitations of neo-classical economics and gives due weight to thorough social analysis, NIE can still provide important ideas.

References

Bates, R. H., 1995, 'Social Dilemmas and Rational Individuals: An Assessment of the New Institutionalism', in Harris, J., Hunter, J. and Lewis, C. M. (eds), *The New Institutional Economics and Third World Development*, London: Routledge, 27–48.

Coase, R., 1937, 'The Nature of the Firm', *Economica*, 4, 386–405.

Dahlman, C., 1979, 'The Problem of Externality', *Journal of Law and Economics*, 22, 141–162.

Eggertsson, T., 1990, *Ecomomic Behavior and Institutions*, Cambridge: Cambridge University Press.

Harris, J., Hunter, J. and Lewis, C. M., 1995, 'Introduction: Development and Significance of NIE', in Harris, J., Hunter, J. and Lewis, C. M. (eds), *The New Institutional Economics and Third World Development*, London: Routledge, 1–13.

North, D. C., 1995, 'The New Institutional Economics and Third World Development', in Harris, J., Hunter, J. and Lewis, C. M. (eds), *The New Institutional Economics and Third World Development*, London: Routledge, 17–26.

Stiglitz, J. E., 1986, 'The New Development Economics', *World Development*, 14, 257–265.

World Bank, 1997, *World Development Report 1997: The State in a Changing World*, Oxford: Oxford University Press.

World Bank, 1998, *World Development Report 1998/1999: Knowledge for Development*, Oxford: Oxford University Press.

World Bank, 1999, *World Development Report 1999/2000: Entering the 21st Century*, Oxford: Oxford University Press.

Further reading

Furubotn, E. G. and Richter, R., 1998, *Institutions and Economic Theory: The Contribution of the New Institutional Economics*, Ann Arbor: Michigan University Press. A very thorough and detailed analysis of practically all aspects of NIE.

1.12

Measuring development

From GDP to the HDI and wider approaches

Robert B. Potter

Both researchers and policymakers in the field of development studies have sought to find methods to measure levels and rates of change in development. Naturally, the approaches used to measure development have reflected directly the principal conceptualisations of development as a process that have been emphasised at various times.

During the 1950s through to the early 1980s, development was generally measured in terms of economic growth, and in particular, the growth of production and income. In the late 1980s, through to the 1990s, changes in the way development was being envisioned were directly recognised in the promotion of wider indices of human development and change. This trend towards recognising the multidimensional nature of development has been continued from the 1990s through to the start of the twenty-first century, whereby wider sets of factors, reflecting more subjective and qualitative dimensions, have increasingly been employed to define development. These have included wider measures of social welfare and human rights.

These three approaches to deriving measures of development provide the framework for this chapter, which looks specifically at: (i) measuring development in terms of economic growth, by means of GDP and GNP per capita; (ii) measuring development in terms of human development: the Human Development Index (HDI); and (iii) measuring development in terms of wider dimensions, including human rights and freedoms. Throughout, the account is strongly based on that recently provided in Potter *et al.* (2012).

Development as economic growth: GDP and GNP per capita as measures of development

Explained in simple terms, this approach uses 'income' per head of the population as a measure of development, suggesting that the higher the income of a country or territory, the greater its development. The approach sees development as being essentially the same thing as economic growth. During the era of unilinear development models and theories, the growth of GDP/GNP was taken as the surrogate measure of development. More accurately, in this approach, the standard of living of a country is used as a summary measure of development (Thirlwall, 2011). The GDP/GNP of a territory is directly affected both by the number of people working within a country and their overall level of productivity.

Gross Domestic Product (GDP) per capita – measures the value of all goods and services produced by a nation or a territory, whether by national or foreign companies. When calculated, the national total is divided by the total population, to give the value of goods and services produced per head of the population.

Gross National Product (GNP) per capita – this is Gross Domestic Product to which net income derived from overseas is added. In other words, income which is generated abroad is added, and payments made overseas are subtracted. This total is also then divided by the population. In recent years, international organisations like the World Bank have increasingly referred to this measure more directly as *Gross National Income (GNI) per capita*.

Through time from the 1950s, GDP and GNP/GNI have been used as measures of development. The measure has been popular as it makes possible the international comparison of living standards by using per capita incomes, customarily measured in United States dollars. Employing such an approach, the basic causes of poverty of any given nation are seen as the low productivity of labour that is associated with low levels of physical capital (natural resources) and human capital (for example, education) accumulation and low levels of technology. The economic growth of countries is measured by the increase in output of goods and services (GDP/GNP) that occurs over a given time period, normally a year (Thirlwall, 2011).

Development as human development: The Human Development Index (HDI) as a wider measure of development

But development is far wider than the growth of income alone. First, GDP/GNP takes no account of the distribution of national wealth and output between different groups of the population, or between different areas/regions. Further, such income-based measures do not take into acount the wider well-being of people, which includes more than goods, money and material well-being.

In the 1980s, it was increasingly recognised that non-economic factors are involved in the process of development. Reflecting this, in 1989, the United Nations Development Programme (UNDP) promoted the *Human Development Index* (HDI) as a wider measure of development. HDI data were published for the first time in 1990 in the inaugural *Human Development Report* (UNDP, 1990). In the original HDI the emphasis was placed on assessing human development as a more rounded phenomenon. There was still a measure of economic standing, but this was only one of three principal dimensions identified:

1. *A long and healthy life* (longevity) – originally measured by life expectancy at birth in years.
2. *Education and knowledge* – initially measured by the adult literacy rate and the gross enrolment ratio (the combined percentage of the population in primary, secondary and tertiary education).
3. *A decent standard of living* – originally measured by Gross Domestic Product per capita in US dollars, as outlined in the previous section.

In the *Human Development Report 2010* (UNDP, 2010), the formula was changed somewhat. The three dimensions of health, education and living standard are translated into four indicators: life expectancy, mean years of schooling, expected years of schooling and Gross National Income per capita. These are then summed to give a single Human Development Index. In the case of all the indicators, the measures are then transformed into an index ranging from 0 to 1, from the lowest to the highest levels of assessed human development, to allow equal weighting between each of the three dimensions.

Since 1990, the *Human Development Report* has been published by the UNDP every year. Within these reports, the HDI has been used to divide nations into what has come to be referred to as high-, middle- and low levels of human development. Recently, the classification has been extended to also include a very high-human development category.

It should be stressed that the HDI is a summary, and not a comprehensive measure of development. For example, over the years since its introduction, various methodological refinements and spin-offs have been made by the United Nations, including the Human Poverty Indexes 1 and 2, the Gender-related Development Index and the Gender Empowerment Measure. These are all variations on the basic Human Development Index. In each case, additional variables were brought in to reflect the revised index.

The use of HDI scores and their difference from GDP/GNP are shown if we look at some examples taken from the *Human Development Report 2009* (UNDP, 2009; see Potter *et al.*, 2012 for fuller coverage). The country with the highest HDI is Norway (0.971). Although its GDP was high, at US$53,433, this was not the highest by any means. Norway's first place standing on the HDI was a reflection of its high life expectancy at birth (80.5 years) and its high overall enrolment in education (96.6 per cent). In comparison, the USA was ranked as the 13th most developed nation (HDI = 0.956). Its GDP per capita was a little lower than that for Norway at US$45,592 per capita, but both life expectancy (79.1 years) and enrolment in education (92.4 per cent) were lower than for Norway. Both Norway and the USA fall into the very high development category.

Lower down in the overall HDI country listing, and falling into the high development category, some might be surprised to see Cuba in 51st place. Although its GDP at US$6,876 per capita is low for such a ranking, Cuba is pulled up by virtue of its relatively high life expectancy (78.5 years) and educational enrolment (100 per cent), levels that are very high for a nation with a relatively low income. In the low development category, Niger is to be found in 182nd place (HDI = 0.340). Not only is Niger's GDP per capita extremely low at US$627 per capita, its life expectancy stands at only 50.8 years at birth, and its combined educational enrolment is as low as 27.2 per cent.

Measuring development in terms of wider dimensions, including human rights and freedoms

A number of writers, have stressed the importance of self-esteem, basic freedoms and human rights as components of the development equation (e.g. Goulet, 1971; Sen, 1999). Such views represent specific recognition that wider aspects of development are vital, particularly those that relate to the quality of peoples' daily lives, their freedom from various inequalties, and the attainment of human rights and basic freedoms.

The Millennium Development Goals (MDG) are designed as instruments to steer the world to enhanced levels of development. For each MDG there are associated targets and detailed indicators. The indicators can be seen as dimensions that can be employed in order to assess the progress of nations and regions towards the goals and targets, and thereby represent measures of the wider dimensions of development, covering issues such as:

1 eradicating extreme poverty and hunger – measured by the percentage of the population living on less than $1 or $2 per day (now $1.25 and $2.50 per day);
2 achieving universal primary education;

3 promoting gender equality and empowering women;
4 reducing child mortality;
5 improving maternal welfare;
6 combatting diseases.

An impression of how such indicators can be used as measures of progress in development can be gained from the national reports covering progress with the MDGs that are available on the UNDP website. The 94 page report for India in 2009, for example, shows in considerable detail the mixed success achieved on the twelve targets that apply to it.

In terms of basic human rights, an interesting approach is to chart the extent to which countries have ratified the six major human rights conventions and covenants (for example, the *Rights of the Child*, *Against Torture*, etc.; see Potter *et al.*, 2008). In a similar manner, the *United Nations Human Development Report 2010* introduced the HDI-derived *Gender Inequality Index* (GII). The statistics input to the GII include the national female and male shares of parliamentary seats, and educational attainment. The GII also includes female participation in the labour market. Thereby, the GII represents a direct effort to measure the progress made by countries in advancing the standing of women in wider political and economic developmental terms.

Bibliography

Chant, S. and McIlwaine, C. (2009) *Geographies of Development in the 21st Century: An Introduction to the global South*, London: Edward Elgar.
Goulet, D. (1971) *The Cruel Choice: A New Concept on the Theory of Development*, London: Routledge.
Potter, R. B., Binns, T., Elliott, J. and Smith, D. (2008) *Geographies of Development*, 3rd edn, London: Pearson-Prentice Hall.
Potter, R. B., Conway, D., Evans, R. and Lloyd-Evans, S. (2012) *Key Concepts in Development Geography*, London: Sage.
Sen, A. (1999) *Development as Freedom: Human Capacity and Global Need*, Oxford: Oxford University Press.
Thirlwall, A. P. (2011) *Economics of Development: Theory and Evidence*, 9th edn, London: Palgrave Macmillan.
UNDP (1990) *Human Development Report 1990*, New York: Oxford University Press.
UNDP (2009) *Human Development Report 2009*, New York: Oxford University Press.
UNDP (2010) *Human Development Report 2010*, New York: Oxford University Press.
Willis, K. (2005) *Theories and Practices of Development*, London: Routledge.

Further reading and website

Potter, R., Conway, D., Evans, R. and Lloyd-Evans, S. (2012) *Key Concepts in Development Geography*, London: Sage. An extended treatment is provided including a table showing HDI and GDP data for a range of 17 countries among the various HDI categories and specifically showing the difference in rank for countries on the HDI and GDP (see table 1.2.1, p. 32)

UNDP website: http://hdr.undp.org/en/statistics/indices/gdi_gem/. This is a current source for the United Nations Development Programme's (UNDP) annual *Human Development Reports*. These explain in detail how the Human Development Index is calcuated, and also list the relevant data for the world's nations.

Further summary treatments are to be found in *Geographies of Development in the 21st Century* by Sylvia Chant and Cathy McIlwaine (2009) and *Theories and Practices of Development* by Katie Willis (2005), in both cases in the introductory chapter 1.

1.13 The measurement of poverty

Howard White

Introduction

The importance of the task of poverty reduction means that we must be clear as to what we mean by poverty, who the poor are and the best way to help them escape poverty. This chapter is concerned with the first of these points – the meaning and measurement of poverty. The next section outlines key concepts which underpin the various poverty measurements discussed in the subsequent section. Finally, some data on poverty trends are presented.

Poverty concepts

In everyday usage the term 'poverty' is synonymous with a shortage of income. But the development literature stresses the multidimensionality of poverty. In addition to material consumption, both physical and mental health, education, social life, environmental quality, spiritual and political freedom, and general well-being ('happiness') all matter. Deprivation with respect to any one of these can be called poverty.

Some dispute the use of multidimensionality, arguing that income-poverty (i.e. lack of material well-being) is what really matters. Arguments supporting this view include the high correlation between income and other measures of well-being such as health and education status and the view that governments can do something about income (i.e. support growth) but are less able to enhance spiritual well-being.

But there are good arguments in defence of multidimensionality. First, the correlation with income is not that strong for some indicators. Second, poor people themselves often rank other dimensions as being more important than income. Most famously, Jodha (1988) showed with Indian data that the welfare of the poor had risen by measures they considered important – such as wearing shoes and separate accommodation for people and livestock – whereas surveys showed their income to have fallen. Participatory approaches to poverty measurement seek to identify the things that matter to poor people. Different perceptions matter since the poverty concept adopted will influence policy. When poverty is defined solely in terms of income, then it is unsurprising that economic growth is found to be the most effective way to reduce poverty. But if basic needs such as health and education are valued then the development strategy is likely to put more emphasis on social policy. In the last few years governments of both developed and developing countries have sought to broaden summary measures of national welfare beyond income, the most extreme case being Bhutan in which the Planning Ministry has been renamed the Ministry of Happiness, a refocusing which also can prompt changes in social policy toward, for example, promoting social cohesion rather than merely income security.

Two further conceptual issues are: absolute versus relative poverty; and temporary versus permanent poverty. Absolute poverty is measured against some benchmark – such as the cost of getting enough food to eat or being able to write your own name for literacy. Relative poverty is measured against societal standards; in developing countries the basket of 'essentials' comprises food and a few items of clothing, whereas in developed countries it includes Christmas presents and going out once a month.

The distinction between the temporarily and the permanently poor is linked to the notion of vulnerability. The vulnerable are those at risk of falling into poverty. If there are poverty traps – such that once someone falls into poverty they cannot get out again – then there is a good case for anti-poverty interventions to prevent this happening.

Poverty measures

National-level measures

The most commonly reported development statistic is a country's GNP per capita. While a case may be made for using GNP as an overall development measure, it is not a good measure of poverty for two reasons. First, as an average, the statistic takes no account of distribution. Hence two countries can have the same level of GNP per capita, but in one of the two a far greater proportion of the population fall below the poverty line if income is less equally distributed. Second, GNP is of course an income measure which ignores other dimensions of poverty.

The most common income-poverty measure is the headcount, that is, the percentage of the population falling below the poverty line. However, this measure takes no account of how far people are below the poverty line – so that a rise in the income of the poor which leaves them in poverty appears to have no effect. Hence another measure, the poverty gap, is often used, which can be variously interpreted as the product of the headcount and the average distance of the poor below the poverty line (expressed as a percentage of the poverty line) and the benefit of perfect targeting. The poverty severity index is a similar measure which puts greater weight on those furthest below the poverty line. These three measures – the headcount, the poverty gap and the poverty severity index – are known collectively as the Foster-Greer-Thorbecke poverty measures and labelled P_0, P_1 and P_2 respectively.

Over the years a number of composite measures of development have been proposed – a composite being an average of a number of different measures. A previous measure, the Physical Quality of Life Index (PQLI), has been superseded in recent years by the UNDP's Human Development Index (HDI). The HDI is a composite of GDP per capita, life expectancy and a measure of educational attainment (which is an average of literacy and average enrolment rate for primary, secondary and tertiary education). However, just as income per capita takes no account of distribution neither does the HDI: schooling can increase by the already well educated extending their university education rather than expanding access amongst those with little or no education. However, UNDP has also proposed a Human Poverty Index (HPI), which focuses on deprivation. Specifically, the HPI is calculated as the average of the percentage of the population not expected to live to 40, the percentage who are illiterate and what is called the 'deprivation in living standard' (the average of those without access to water and healthcare, and the percentage of under-fives who are underweight).

Although the HDI is widely used there have been criticisms of its construction (which were summarized in a technical appendix to the 1996 *Human Development Report*), one of which concerns problems in using a composite. There are three main problems: which variables to put in the index; the necessarily arbitrary choice of weights in constructing the average; and that information is lost by combining three or four pieces of data into a single number. Thus it may be preferable to report a small range of social indicators, such as life expectancy, infant and child mortality and literacy, rather than attempt to combine these in to an overall poverty index.

The measurement of income poverty

The income poverty headcount is the percentage of the population whose income is below the poverty line. This calculation is fraught with difficulties.

First, poverty lines must be defined (it is common practice to use two lines), which is done either absolutely with reference to the cost of a basket of goods or relatively to mean income or a certain share of the population. In the former case the basket can be calculated either as the cost of acquiring a certain number of calories or of a basket of goods and services. In the first example, the resulting poverty line (food poverty line) is often used as the line for the extreme poor. It is then divided by the share of food in the budget of the poor (or the population as a whole, though strictly defined it should be that of a person on the poverty line) to get the upper poverty line.

In applying the poverty line, consumption (expenditure) is commonly used rather than income. First, because survey respondents will have a far clearer idea of their expenditure than their income. Second, when income is uneven households will smooth consumption (i.e. even it out over time) so that at any point in time current consumption is likely to be a more accurate measure of well-being than current income.

In practice, data are collected at the level of the household rather than the individual. Doing so ignores problems of intra-household allocation. There are no data on the number of women or children living in poverty (despite the tendency of some international organizations to report such figures), only data on the percentage of women and children living in households whose income is below the poverty line. The use of household-level data introduces problems of household composition and size. Household composition matters since the consumption requirements of different individuals varies – specifically children consume less than adults and, more controversially, women may need to consume less than men. This problem is catered for by the use of an adult equivalents scale, which expresses the consumption needs of women and children as a fraction of those of an adult male. Household size matters as there are economies of scale in household consumption – that is, two can live together more cheaply than they could apart as there are shared expenses (living space, utilities and many household items). Failure to take account of these economies will overstate poverty in large households.

Finally, prices vary across time and space. Allowance must be made for these price differences in order for the poverty line to be comparable. There are even greater difficulties in comparing between countries, partly since market exchange rates do not reflect differences in purchasing power. Rather, purchasing power parity (PPP) exchange rates should be used, which are not uniformly available.

Comparisons across time and space also require that consumption is measured in a comparable way. If survey designs differ greatly then 'aggregate consumption' may mean quite different things. It is commonly recognized that own-production should be measured as this is an important part of total consumption. But 'wild foods' (collected in nature) and festivals can also form an important source of food and are commonly overlooked. Similarly, sources of income from common property or the provision of free social services varies between countries and so introduces another source of incomparability.

It may seem from this discussion that measurement of income poverty is so difficult that it may be better to stick to some other measure. Certainly a small survey should stick to a proxy for income, at least in countries in which formal sector employment is low, such as housing

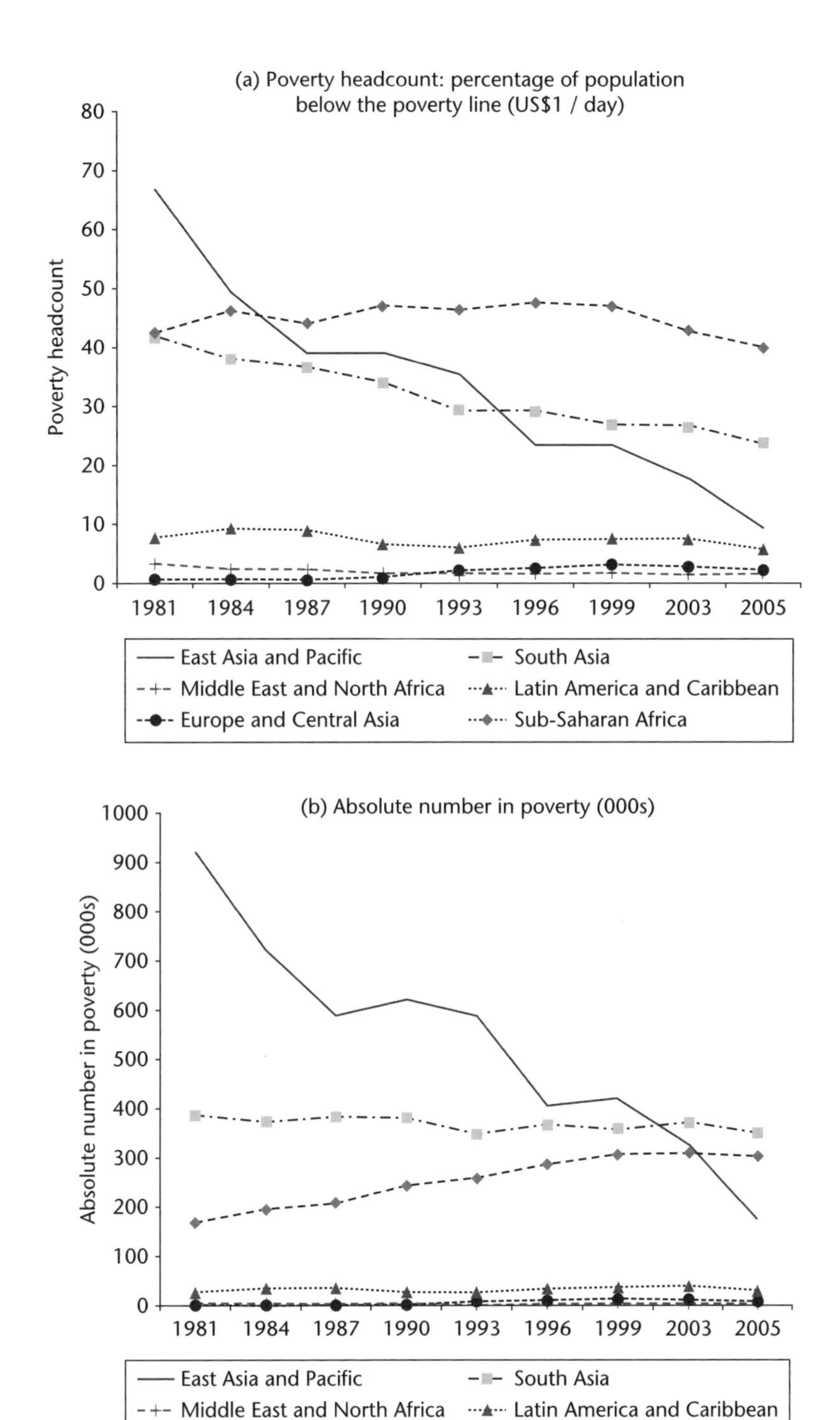

Figure 1.13.1 'Dollar a day' poverty by region, 1981–2005

Source: World Bank 2010 Poverty Estimates http://go.worldbank.org/4K0EJIDFA0 (accessed 17/6/12)

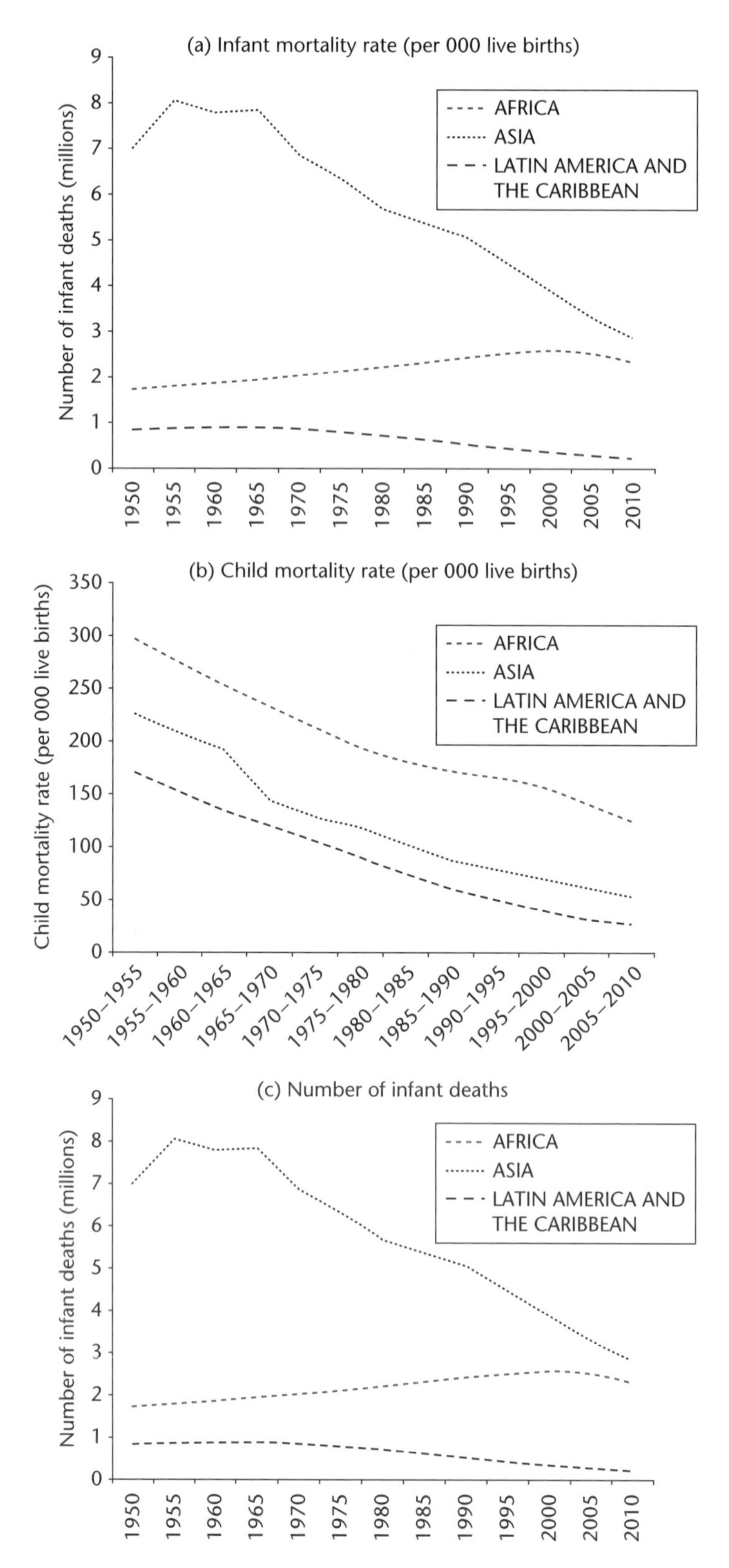

Figure 1.13.2 Trends in infant and child mortality

Source: Based on UN Population Projections 2010 revision (http://esa.un.org/wpp/Excel-Data/population.htm)

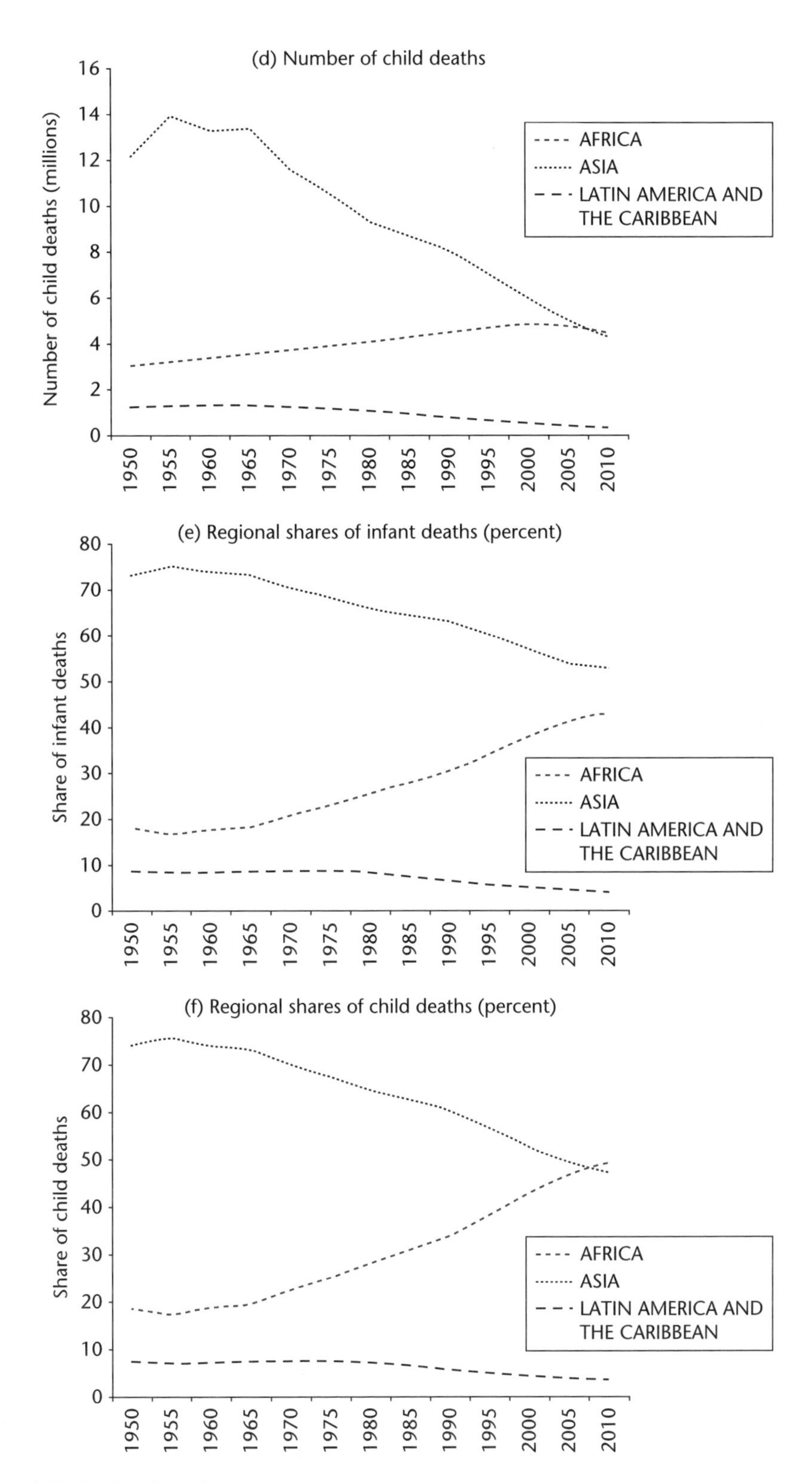

Figure 1.13.2 Continued

quality and ownership of a few household items. But other indicators are not without problems; indeed data quality is far worse for many social indicators than it is for income/expenditure.

Some data

Estimates of 'dollar a day' poverty are calculated only for the developing world, and are now available for poverty lines ranging from US$1 to US$2.50 per day. The lower threshold is used here. The proportion of absolutely poor in developed countries by this measure is nil or negligible. Figure 1.13.1 shows the evolution of 'dollar a day' poverty since the early 1980s. The most striking trend is the dramatic fall in poverty in East Asia, powered largely by reductions in the number of poor in the world's most populous country, China, but assisted by more recent declines in neighbouring Vietnam. There has been a slower, but still marked, decline in the poverty headcount in South Asia, including in the world's second largest country, India. Asia thus makes up the vast bulk of poverty reduction in the closing decades of the twentieth century. During the 1990s, global poverty fell from 29.8 per cent in 1990 to 20.3 per cent in 2003, but excluding China these figures were 24.4 and 20.7 per cent respectively. That is, over half the fall in poverty came from China alone. Indeed, sub-Saharan Africa, which has suffered economic hardship since the 1970s, saw a rise in income poverty in the 1990s, with close to half the people on the sub-continent living on less than a dollar a day in 1999, though the share had declined to just under 40 per cent by 2005.

From 1981 to 2005 the absolute number of poor fell dramatically in East Asia (922 to 176 million), and somewhat in South Asia (387 to 351 million people). But these global gains were partially offset by the rise in sub-Saharan Africa from 169 to 304 million people; and to a lesser extent by a rise in Europe and Central Asia from 3 to 10 million. Hence sub-Saharan Africa's share of those living on less than a dollar a day grew from just a tenth (11 per cent) in 1981 to over a third (35 per cent) by 2005.

These downward overall trends have been set back by the global recession which began with the 2008–2009 financial crisis. In the three years following the crisis over 50 million people are estimated to have fallen below the US$1.25 a day poverty line.

The Africanization of poverty is also evident when considering other poverty measures. Health is commonly measured by infant mortality (the number of children who die before their first birthday per 1,000 live births) and child mortality (deaths between first and fifth birthdays per 1,000 children); these two indicators are combined to make under-five mortality. The United Nations Population Division reports population data, including mortality, from 1950, with projections to 2050, though only those to 2010 are shown here. The positive news is that, in line with the long-run improvement in social indicators across the developing world, mortality rates have been falling across the world (Figure 1.13.2), though some African countries experienced a reversal in the 1990s as a result of HIV/AIDS and worsening health systems after three decades of economic decline. In Africa as a whole, the decline in mortality rates has been insufficient to keep up with population growth, so that the number of deaths continued to rise until very recently. But it is expected that the situation in Africa will improve less than that elsewhere, so the continent will account for close to two-thirds of the world's under-five deaths in the coming decades.

Bibliography

Baulch, B. (1996) 'The new poverty agenda: A disputed consensus', *IDS Bulletin* 27: 1–10.

Black, Richard and Howard White (2004) *Targeting Development: Critical Perspectives on the Millennium Development Goals*, London: Routledge.

Chambers, Robert (1995) 'Poverty and livelihoods: Whose reality counts?' *Discussion paper 347*, Brighton: Institute of Development Studies.
Chen, Shaohua and Martin Ravallion (2010) 'The developing world is poorer than we thought, but no less successful in the fight against poverty', *The Quarterly Journal of Economics*, November: 1577–1625.
Gordon, David and Paul Spicker (eds) (1999) *International Glossary on Poverty*, London: Zed Books.
Jodha, N. S. (1988) 'Poverty debate in India: A minority view', *Economic and Political Weekly* 22(45–47): 2421–2428.
Ravallion, M. (1992) 'Poverty comparisons: A guide to concepts and methods', *LSLS Working Paper 88*, Washington DC: World Bank.
White, Howard (1999) 'Global poverty reduction: Are we heading in the right direction?', *Journal of International Development* 11: 503–519.

Further reading

Most of the literature on poverty measurement concerns income poverty, the most comprehensive, though technical, treatment being Ravallion (1992). A critique of income measures is given by Jodha (1988) and a discussion of alternatives in Chambers (1995). More general analyses of both concepts and measurement is available in Baulch (1996) and White (1999). For a broader coverage of poverty issues consult the series of poverty briefings produced by ODI, including 'The meaning and measurement of poverty' by Simon Maxwell; these are available from www.odi.org/briefing. Finally, Gordon and Spicker (1999) is a useful resource.

The main data sources are the UNDP's *Human Development Report* and the World Bank's *World Development Indicators*, both of which are published annually and available on-line. Past issues of the *HDR* contain discussions of the various indices presented, including the HDI and HPI, whilst *WDI* contains useful information on data sources. The World Bank's poverty data are available from http://go.worldbank.org/4K0EJIDFA0; see also www.worldbank.org/poverty which includes many useful links to poverty-related material.

1.14 The millennium development goals

Jonathan Rigg

Deriving the goals

The Millennium Development Goals (MDGs) were adopted by the General Assembly of the United Nations on 18 September 2000. More than 190 countries have since signed up to the resolution. The eight goals, to be achieved by 2015, are linked to 18 targets and these, in turn, to 48 indicators (Table 1.14.1). The collective agreement by over 190 countries to strive to meet these goals was unprecedented. The MDGs arose from a wish that at the turn of the Millennium good intentions had to be matched by concrete actions. Under 'values and principles', the UN General Assembly agreed that:

> We believe that the central challenge we face today is to ensure that globalization becomes a positive force for all the world's people . . . only through broad and sustained efforts to create a shared future, based upon our common humanity in all its diversity, can globalization be made fully inclusive and equitable. . . . We are committed to making the right to development a reality for everyone and to freeing the entire human race from want.
>
> *(http://childrenandarmedconflict.un.org/keydocuments/english/unitednationsmil13.html)*

At the time of the Millennium Summit, United Nations Secretary-General Kofi Annan stated that: 'We will have time to reach the Millennium Development Goals – worldwide and in most, or even all, individual countries – but only if we break with business as usual'.

Progress towards the MDGs

More than a decade has passed since the MDGs were adopted, and the target date of 2015 is fast approaching. This provides us with an opportunity to assess whether and where the goals are likely to be achieved. Has the world community broken with 'business as usual'? The baseline year for measuring progress towards most of the targets is 1990 and the latest data come from 2005–2007 for some indicators and 2009–2011 for others. On this basis, the answer to the question of whether the goals will be achieved is that while there has been significant progress with respect to some of the goals, and in some parts of the world, there remain major gaps and shortfalls.

Table 1.14.2 summarises progress in 2011 towards goals 1–8 in nine regions of the world. Aggregating the data reveals that in only 40 per cent of cases are the targets likely to be met by 2015 at current rates of progress; in 47 per cent of cases, the target is not expected to be met given prevailing trends; while in 13 per cent there has been a deterioration in progress. The aggregate picture, however, is considerably more favourable than that painted in the first edition of this volume, based on 2005 data. At that time, the balance between the three progress categories was 16 : 47 : 37, while in 2011 it was 40 : 47 : 13. The number of regions showing a reversal of trends is sharply down, and those likely to meet the targets significantly up (Table 1.14.2).

What the aggregated data in Table 1.14.2 do not show is how this story of success, stagnation and failure unfolds in regional terms. Table 1.14.3 shows that there is a clear regional pattern to achievement and failure, which falls into three categories: those regions where progress towards the MDGs has been broad-based (North Africa and East Asia); those where progress has been substantial but with significant gaps (Southeast Asia, South Asia, Latin America, and the Caucasus and Central Asia); and those regions where the majority of targets and goals will not be met at prevailing rates of progress (sub-Saharan Africa, West Asia and Oceania). Importantly, what we cannot tell are *within* country variations, which are significant even in countries making good progress. As the 2011 MDG report puts it, 'despite real progress, we are failing to reach the most vulnerable . . . the poorest of the poor and those disadvantaged because of their sex, age, ethnicity, or disability' (UN 2011: 4). 'Disparities in progress between urban and rural areas', the report continues, 'remains daunting' (p. 4).

It is goal 8, relating to the derivation of a global partnership for development and its associated indicators and targets, which directs attention at the responsibility of the wider

Table 1.14.1 The MDGs, 1990–2015: 8 goals, 18 targets (and 48 indicators)

Goals (8)	*Targets (18)*
1. Eradicate extreme hunger and poverty	1. Halve the proportion of people living on less than a dollar a day 2. Reduce by half the proportion of people who suffer from hunger
2. Achieve universal primary education	3. Ensure that all boys and girls complete a full course of primary schooling
3. Promote gender equality and empower women	4. Eliminate gender disparity in primary and secondary education preferably by 2005, and at all levels by 2015
4. Reduce child mortality	5. Reduce by two thirds the mortality rate among children under five years of age
5. Improve maternal health	6. Reduce by three quarters the maternal mortality ratio
6. Combat HIV/AIDS, malaria and other diseases	7. Halt and begin to reverse the spread of HIV/AIDS 8. Halt and begin to reverse the incidence of malaria and other major diseases
7. Ensure environmental sustainability	9. Integrate the principles of sustainable development into country policies and programmes . . . 10. Reduce by half the proportion of people without sustainable access to safe drinking water 11. Achieve significant improvement in lives of at least 100 million slum dwellers, by 2020
8. Develop a global partnership for development	12. Develop further an open trading and financial system that is rule-based, predictable and non-discriminatory . . . 13. Address the least developed countries' special needs. . . 14. Address the special needs of landlocked and small island developing States 15. Deal comprehensively with developing countries' debt problems . . . 16. In cooperation with the developing countries, develop decent and productive work for youth 17. In cooperation with pharmaceutical companies, provide access to affordable essential drugs . . . 18. In cooperation with the private sector, make available the benefits of new technologies . . .

Source: Extracted and adapted from http://www.un.org/millenniumgoals/

international community and, particularly, richer countries. This is where the normativity of the MDGs is clearest. Again, the view from 2011 is more positive than it was in 2005, but with caveats. Aid has increased significantly in real terms (to 0.32 per cent of developed countries' GDP in 2010, or US$129 billion), although not by enough to meet the pledges made at the G8 summit in Gleneagles in 2005; there has been a modest reduction in tariffs in developed markets for goods produced in developing countries, but tariffs in emerging

Table 1.14.2 MDG progress chart (2011)

Goals and targets 1990–2015	*Number of regions in each category (out of 9 world regions)*			
	Target met or expected to be met by 2015	*Progress insufficient to reach the target if prevailing trends persist*	*No progress or deterioration in trends*	*Missing or insufficient data*
Goal 1: Halve, between 1990 and 2015, the proportion of people living on less than a dollar a day	3	4	1	1
Goal 1: Productive and decent employment	3	5	1	0
Goal 1: Reduce by half the proportion of people who suffer from hunger	4	2	2	1
Goal 2: Ensure that all boys and girls complete a full course of primary schooling	2	5	1	1
Goal 3: Eliminate gender disparity in primary education	8	0	1	0
Goal 3: Women's share of paid employment	3	6	0	0
Goal 3: Women's equal representation in national parliaments	0	8	1	0
Goal 4: Reduce by two-thirds the mortality rate among children under-fives	3	6	0	0
Goal 5: Reduce by three quarters the maternal mortality ratio	2	5	2	0
Goal 5: Access to reproductive health	1	7	0	1
Goal 6: Halt and begin to reverse the spread of HIV/AIDS	3	4	2	0
Goal 6: Halt and reverse the spread of tuberculosis	7	2	0	0
Goal 7: Reverse loss of forests	3	2	4	0
Goal 7: Reduce by half the proportion of people without sustainable access to safe drinking water	5	3	1	0
Goal 7: Reduce by half the proportion of people without sanitation	3	5	1	0
Goal 7: Achieve significant improvement in lives of slum dwellers	4	2	2	1
Goal 8: Internet users	5	4	0	0
Total (2011)	59	70	19	5
Total (2011) (%)	40	47	13	–
Total (2005) (%)	16	47	37	–

Note: The information is taken from the 2011 MDG progress report downloadable from: http://www.un.org/millenniumgoals/pdf/(2011E)_MDReport2011_ProgressChart.pdf.

Table 1.14.3 Progress chart by region (2011)

Goals and targets	*Number of regions in each category (out of 9 world regions)*			
	Target met or expected to be met by 2015	*Progress insufficient to reach the target if prevailing trends persist*	*No progress or deterioration in trends*	*Missing or insufficient data*
North Africa	11	5	1	0
Sub-Saharan Africa	2	14	1	0
East Asia	14	2	1	0
Southeast Asia	7	9	1	0
South Asia	6	9	2	0
West Asia	4	10	3	0
Oceania	2	4	7	4
Latin America and Caribbean	7	10	0	0
Caucasus and Central Asia	6	7	3	1
Total (2011)	59	70	19	5

Source: Extracted from http://www.un.org/millenniumgoals/pdf/(2011E)_MDReport2011_ProgressChart.pdf

markets remain and are often growing in significance; and debt service payments as a proportion of export revenues declined significantly across all regions between 2000 and 2009.

Criticising the self-evidently desirable

Few would contest that the objectives of the MDGs are creditable – they are self-evident 'goods', 'unimpeachably worthwhile' (Poku and Whitman 2011: 4). Thus, scholars have tended to shy away from their criticism. Critics have, however, questioned whether, first, the MDGs are 'fit for purpose'; second, whether we have the available data to measure the achievement of the targets identified; third, whether the targets adequately assess the goals to be achieved; fourth, whether the broad means set to achieve the goals are laudable; and, finally, whether there is a mechanism in place – beyond exhortation and moral persuasion – to support and propel the achievement of the MDGs, especially in relation to goal 8. It is significant the degree to which some world leaders appear to see the MDGs themselves *causing* progress. Secretary-General Ban Ki-Moon in the 2011 MDG report wrote that 'already, the MDGs have helped to lift millions of people out of poverty, save lives and ensure that children attend school' (UN 2011: 3).

James (2006) directs his criticism at a failure to distinguish between means and ends, or between *actual* achievements and *potential* achievements. Some of the MDG targets are ends manifested and measurable at the level of the individual. This applies to the targets associated with goals 4, 5 and 6 (all health related) and target 2 under goal 1 (referring to hunger) (see Table 1.14.1). But many of the other goals are means rather than ends. So, for example, James draws a distinction between completing primary school (target 3 under goal 2) and the acquisition of basic literacy and numeracy. The former (primary school education) may lead to the latter (literacy and numeracy), but if schooling is inadequate, as it so often is in the poorest countries, then this may not be achieved. In other words, the mere meeting of a target may not deliver the desired end of an adequate education.

In defence of this means-based approach, it has been suggested that cross-country data are simply not available to target ends and means-based measures are an acceptable proxy. But even here there are reasons to be cautious. Doubts have been expressed about whether we have sufficiently robust data to assess the achievement of the targets set, which is most acute in the poorest countries. Satterthwaite (2003: 184–185) writes of 'nonsense' statistics, such as the levels of urban poverty and urban service provision in Africa (linked to goal 7, targets 10 and 11).

A related criticism is whether the rather mechanical, target-based approach places a characteristically instrumentalist gloss on the achievement of the goals. The poverty target, for example (goal 1, target 1), is income-based and related to official data and basic needs ascertained by 'experts'. Other forms of deprivation (linked to social exclusion, political marginality and cultural rights) are ignored and inequalities in power overlooked (Satterthwaite 2003: 182). Moreover, it is not just *what* is done – reducing poverty, eradicating hunger, reducing maternal mortality – but *how* it is done. The general criticism that development has become a technocratic project informed by experts, driven by governments and multilateral agencies, and based on measures of success that pay little heed to local desires is equally apposite to the MDG initiative. Radical commentators would also draw attention to the market-based logic (free trade, private enterprise) that informs the MDG initiative.

Summary

The MDGs represent the first collective and integrated attempt to highlight the life conditions and life chances of the world's poor. With only a handful of years to run until 2015, it seems that most of the targets will not be met for most regions of the world. Some commentators choose to focus on the successes (the 'half-full' contingent) and others on the failures (the 'half-empties'). There is also a more fundamental debate about the goals themselves, the way that the targets have been framed and the reliability of the data on which progress is assessed. More fundamentally still, there is the question of whether the MDGs represent a set of appropriate measures of the achievement of development. These issues will be important when it comes to the question of whether there will be an MDG II.

References

James, Jeffrey (2006) 'Misguided investments in meeting Millennium Development Goals: A reconsideration using ends-based targets', *Third World Quarterly* 27(3): 443–458.

Poku, Nana and Whitman, Jim (2011) 'The Millennium Development Goals: Challenges, prospects and opportunities', *Third World Quarterly* 32(1): 3–8.

Satterthwaite, David (2003) 'The Millennium Development Goals and urban poverty reduction: Great expectations and nonsense statistics', *Environment and Urbanization* 15(2): 181–190.

UN (2011) *The Millennium Development Goals report 2011*, New York: United Nations. Available online at www.un.org/millenniumgoals/pdf/(2011_E)%20MDG%20Report%202011_Book%20LR.pdf.

Further reading

Jones, Phillip W. (2008) 'The impossible dream: Education and the MDGs', *Harvard International Review* 30(3): 34–38.

A great deal of valuable material on the MDGs can be gleaned from the UN, World Bank, UNDP and other websites. Two journal special issues have recently been published that focus on the MDGs: 'The Millennium Development Goals: challenges, prospects and opportunities' in *Third World Quarterly* (2011, 32, number 1) and 'Old Challenges and New Opportunities for the MDGs: Now and Beyond 2015' in the *Journal of the Asia Pacific Economy* (2011, 16, 4).

Websites

The main UN MDG website with background information on the initiative, the UN resolution, progress charts, the Secretary-General's interventions, supporting statistics, and annual reports can be found at www.un.org/millenniumgoals/.

The World Bank and UNDP also have web pages devoted to the MDGs with links to relevant studies, available online at www.worldbank.org/mdgs/ and www.undp.org/content/undp/en/home/mdgoverview.html respectively.

1.15

BRICS and development

José E. Cassiolato

Origins

On 29 March 2012 the leaders of Brazil, Russia, India, China and South Africa (BRICS), met in India, to lead the Fourth BRICS Summit. Only three days before, an article in the *International Herald Tribune* had dismissed BRICS as "an artificial bloc built on a catchphrase."

Analysts in BRICS countries, however, point out that the joint declaration of the summit "contains not only the most comprehensive criticism of the failures of the West that has been voiced by any group of countries since the end of the Cold War, but also the outlines of an alternative blueprint for managing our increasingly interdependent world" (Jha 2012).

A Goldman Sachs team published a report in 2003 which conceived the acronym BRICs (Brazil, Russia, China and India) and suggested that this group of countries would become the future "engines of growth and spending power" of the world economy, with a total GDP by 2040 similar to the combined one of the G6 countries. These countries presented excellent investment opportunities with their large potential domestic markets characterized by the prospective growth of their middle classes. Since then the acronym has attracted international attention with its explicit recognition of the bleak growth prospects of developed countries.

Also in 2003, the foreign ministers of India, Brazil and South Africa (IBSA) met to set up a forum to promote cooperation between themselves and other developing countries. Although they had conflicting interests in trade, they decided to negotiate jointly within the WTO. But IBSA aimed at much more, namely to act as "a coordinating mechanism among three emerging countries, three multiethnic and multicultural democracies, which are determined to contribute to the construction of a new international architecture" (Brasilia Declaration, item 6).

In 2006 a parallel initiative started with the four BRIC states mentioned in the Goldman Sachs report. It culminated in 2009 with the first formal meeting of the governments of these countries, now focusing on the reform of international financial institutions and improvement of the world economic situation, which was rapidly deteriorating with the onset of the financial crisis.

Both arrangements evolved in parallel until 2011 (the BRIC nations focusing on economic matters and the IBSA countries targeting social goals), when South Africa was formally included in the economic group. The new BRICS agenda advanced rapidly to include cooperation in innovation and technological development and, most important, proposals and actions and concrete economic initiatives. These included the development of a system of international payments among the members that bypassed the dollar and the creation of an alternative international bank, with the short run objective of shielding their economies from the currency instability of the West and the long-term political purpose of freeing themselves from the Western dominated international banking system.

The economic and social importance of BRICS

BRICS, collectively, were home to 42.2 per cent of the world population in 2010 and accounted for approximately 30 per cent of the earth's surface, holding significant reserves of natural resources such as energy and mineral resources, water and fertile lands. BRICS account for 24.3 per cent of world biodiversity: Brazil alone embraces 9.3 per cent of the total.

All BRICS are federal states, but Brazil is the only one with a common language. Three are democracies: India and South Africa are parliamentary democracies and Brazil is a presidential democracy. Russia's democracy is increasingly authoritarian and China is a one-party political system which rules a communist people's republic marked by state-controlled capitalism.

In 2002, BRICS accounted for 25.9 per cent of world GDP in PPP as compared to 14.5 per cent in 1990. The IMF estimates that by 2017 they will account for more than 30 per cent of world GDP. The economic performance of BRICS, however, has varied widely during the last few decades. China and India have been the fastest growing economies worldwide. Russia, after the severe 1990s crisis that resulted in a decline of 40 per cent in its real GDP, has recovered well. Brazil and South Africa have shown an irregular performance. However, they are all growing at higher rates than the developed countries in general.

Different performances were accompanied by significant changes in their productive structure, reflecting dissimilar development strategies. While in China the share of industry in GDP value added grew substantially to reach 48 per cent in 2009 (21.2% of world manufacturing, up from 3.2% in 1990) in India it remained around 26 per cent, while in the other nations it declined: in Brazil from 41.7 per cent in the early 1980 to 25.4 per cent in 2009, in Russia from 44.6 per cent to 32.9 per cent in 2009, in South Africa from 48.4 per cent to 31.4 per cent in 2009. In China, this was accompanied by a significant diversification of its industrial system with the share of technologically intensive sectors in total value added of the manufacturing industry reaching 42 per cent in 2009. In the other BRICS this share is around 15 per cent.

In Brazil, the high growth of services (from 50% to 68.5% in the same period) was accompanied by a renewed importance of agricultural goods and minerals, while although Russia's economy is heavily dependent on oil and gas, the defence-related industrial complex is significant together with non-electric machinery.

The Indian economy is essentially service-led, with a well-known capacity in ICT. Skills in the manufacturing sector are relatively modest and are concentrated in non-durable consumer goods and in the chemical-pharmaceutical and auto complexes.

Services (particularly finance and tourism) have an important role in the South African economy. Its economy is heavily based on natural resources and mining remains important with respect to employment and foreign trade.

The share of BRICS in total merchandise trade value more than doubled in the short period between 2000 and 2010, with exports rising from 7.5 to 16.4 per cent and imports from 6.2 to 14.9 per cent of the world total, in terms of value. China's exports mounted from 3.9 per cent to 10.4 per cent and imports increased from 3.4 per cent to 9.1 per cent of the world total in the same period. With the exception of South Africa, all other BRICS increased their share of world exports. Other than India and South Africa, the remaining BRICS countries managed to keep a surplus in their merchandise trade by 2010.

Although bilateral trade flows between BRICS countries have been relatively restricted in the past, since the first half of the 2000s there was a widespread increase of export and import flows between them. China is already the main trade partner of Brazil and South Africa and the second main trade partner of India and Russia. The opposite does not hold, however, as these four economies are neither the top import suppliers nor export destinations for China.

Investment rates differ significantly among BRICS. Pushed by government investment, a closed financial system and intertwined with high growth rates, China stands out by virtue of its investment rate constantly growing – from 37.4 per cent of GDP in 1992 to 48.3 per cent in 2011. India also experienced growing rates of investment in this period. The other BRICS, Brazil in particular, suffer from low investment rates.

The intensity and role of foreign direct investment (FDI) differs among BRICS. Brazil received the greatest share of FDI of all BRICS until the first half of the 1980s. Although China has surpassed Brazil since 1985, Brazil continued to be a major destination for FDI during the 1990s, most notably during the process of privatization. Since 2000, Russia and India have been strengthening their position as FDI inflow destinations. With the exception of South Africa, BRICS countries more than tripled their FDI outflows from 2005 to 2010, raising their participation in the world total from 3.6 per cent to 11.1 per cent.

The avowed FDI role and policies for the various members of BRICS also differ. In Brazil and South Africa, where the policy is more open towards FDI, its role is very significant, with transnational corporations (TNCs) dominating internal production in most industrial sectors (particularly in the auto and IT complex). In China, and India to a lesser extent, the role of FDI has always been subsidiary to development. In both countries the capital account was not liberalized and the financial sector remains controlled by state capital. As the ultimate goal of Chinese policy is to strengthen domestic firms (most with explicit or implicit tight links to the government), the state imposes stringent conditions that have to be met before subsidiaries are allowed to operate in the country. India still maintains significant restrictions in the operations of transnational capital. Brazil, Russia and South Africa (countries that liberalized their economies with few restrictions) got more portfolio investment, but most of the FDI received recently was used to buy up local companies.

The total foreign exchange reserves of BRICS are significant (about 40% of the world total in 2010) which is important given the independence it gives to monetary authorities to protect the local economy against the instability of global financial markets.

Such positive economic evolution was accompanied by increases in inequality levels in all BRICS with the exception of Brazil. Affirmative social policy in Brazil, however, did not prevent it from being, together with South Africa, still among the countries with the worst distribution of income. In addition, India and Russia are among those with the largest percentage of the population living below the poverty line.

Problems associated with high levels of poverty and the perverse distribution of income are common to the five countries, where a significant portion of the population lacks access to essential goods and services. This situation is reflected in the poor human development index scores recorded by the BRICS countries. The other undeniable challenges faced by BRICS are unemployment, poor quality employment and increasing informality of the economy.

Very large regional disparities and a big gap between the rural and urban population are also common problems. The wealthier regions are those that are more industrialized: the Brazilian southeast, the Chinese coastal provinces, the South African Gauteng province and Cape Town, the larger Russian cities such as Moscow and the southern regions of India.

The negative environmental impact of recent growth is another challenge to be faced by BRICS countries. In 2008, they were responsible for emitting 35.3 per cent of the total volume of CO_2, with China being the world's largest emitter.

The geo-politics of BRICS

There are marked differences between the insertions and interests of the BRICS nations within the worldwide system. The military and economic expansion of the 2000s already placed Russia among the great powers, with a growing intervention in conflicts in Central Asia and the Middle East and with the position of arms and military technology supplier to several countries.

China and India between them have 3,200 kilometers of common frontier. They have territorial disputes and are both atomic powers. Brazil and South Africa do not have territorial disputes, do not face internal or external threats to their security and are not relevant military powers. Since its democratization, South Africa has been involved in almost all peace negotiations in Africa, and has not ever presented an expansionary threat. Brazil has never been a state with expansionary objectives.

Russia has maintained its military arsenal and its seat on the UN Security Council, and was quickly incorporated into the G8. After 2000, Putin's government recentralized internal power, accelerated economic growth and is tenaciously chasing its restoration as a world power. After the 1990s China and India entered the world system as economic and military powers, and have clear hegemonic aspirations in their respective regions.

Despite these differences, the respective nations set up the BRICS forum stressing the need for changes in the rules of "management" of the world system and in its hierarchical and unequal distribution of power. They share a reforming agenda in relation to the UN System and to the composition of its Security Council and multilateral and trade liberalizing positions within WTO and the G20. So, their position as the epicentre for dynamizing the world economy that has been almost paralysed by the inadequacies of financial capitalism give them, collectively, more clout in the international arena. In other words, it is the disorganization of the world economy caused by finance-dominated globalization that has permitted the upsurge of BRICS as a collective geopolitical structure.

The BRICS nations have an important role to play in producing a reformist world agenda, as pointed out by India's President Singh at the 2011 BRICS summit in China: if we can cooperate, "our positions on some key areas such as sustainable development, balanced growth, energy and food security, reform of international financial institutions and balanced trade . . . will be to our advantage".

China will probably have a more prominent role in this respect as it is changing its "system of innovation" in order to address sustainability and social and environmental imbalances.

China's emerging "high-growth with low-carbon" strategy has been emphasized by recent policy decisions, in particular, its indigenous innovation strategy to make China less reliant on foreign technology through linking innovation to domestic needs and by giving increased priority to domestic consumption. For the other BRIC nations, Chinese success may lead to the identification of strategies towards strengthening domestic technological capabilities and fostering the technologies needed for the identification of a new techno-economic paradigm.

Bibliography

Armijo, L. (2007) The BRICS countries as analytical category: Mirage or insight? *Asian Perspective*, 31(4): 7–42.

Cassiolato, J. E. and Vitorino, V. (2010) *BRICS and Development Alternatives: Innovation Systems and Policies*. London: Anthem Press.

Jha, P. (2012) Delhi could be a turning point. *The Hindu*, 10 April.

O'Neill, J., Lawson, S. and Purushothaman, R. (2004) The BRICs and global markets: Crude, cars and capital. *CEO Confidential*, October.

Part 2
Theories and strategies of development

Editorial introduction

It is generally appreciated that ideas about how development can be put into practice have long been both controversial and highly contested. Development involves a range of actors, from international agencies, through the state, various organizations, down to the individual, all of whom have a vested interest in how change and development should proceed. Thus, all facets of development not only depend on political ideology, but on moral and ethical judgements and prescriptions too. Thus, ideas about development over time have tended to accumulate and accrue, rather than fade away. Right-wing, left-wing and liberal views remain whatever the political hegemony current at the time. These sorts of ideas are considered at the outset in this part of the book, before turning to some of the major theories and strategies of development that have been followed and popularized.

Right-wing stances on development can be regarded as having their origins in the Enlightenment and the era of modernity that followed. The eighteenth-century Enlightenment period saw an increasing emphasis placed on science, rationality and detailed empiricism. It also witnessed the establishment of the 'West' and 'Europe' as the ideal. It was during this period that the classical economists, Adam Smith and David Ricardo, writing in the 1700s, developed ideas surrounding the concept of comparative advantage, which stressed the economic efficacy of global free trade and, in many senses, gave rise to the earliest capitalist strategies of economic development. The approach envisaged little or no government restrictions on the operation of the economy.

Such approaches were followed by a plethora of dualistic and linear conceptualizations of the development process, including modernization theory, unbalanced and unequal growth, and top-down and hierarchical formulations. Together, such paths are generally referred to as 'neo-classical' or top-down development. Whatever one's critical view of modernization, the framework usefully pinpointed the salience of change as a necessary factor in the development equation.

These classically inspired approaches are, of course, still alive and kicking, in the form of the lineage of the so-called 'new right' orthodoxy of the 1980s, involving the 'magic of the market' and the neoliberal policies of structural adjustment, and more recently, poverty reduction strategies. The rise of so-called neoliberalism followed the global recession of 1978–1983, which witnessed a clear turn away from the ideas of Keynes, who in the Great Depression of the 1930s had argued that governments should spend their way out of recession. Policies of privatization, deregulation and public sector cutbacks, and the withdrawal of the state became the order of the day under the global influence of President Reagan and Prime Minister Thatcher in the 1980s. The IMF and World Bank enforced essentially the same generic types of policies on those developing nations seeking development assistance, in the form of structural adjustment programmes. While some analysts have inferred that the 2008 global economic recession has witnessed something of a break in the hegemony of the long wave of neoliberalism and right-wing development, other commentators are more circumspect in concluding that neoliberalism has survived and will continue to dominate the global economy.

The antithesis to classical and neo-classical views was provided by radical-dependency approaches in the 1960s. It is a reflection of the Eurocentricity of development theory that Andre Gunder Frank, a German-born and American-trained economist, has become the name most closely associated with dependency theory. This is despite the fact that the dependency approach essentially stemmed from the writings of structuralists working in Latin America and the Caribbean. In respect of process, dependency theory was couched in terms of inverted cascading global chains of surplus extraction, and it was again all too easy to reduce this to simple dichotomous terms, involving polar opposites such as 'core–periphery', 'rich–poor', 'developed–underdeveloped', and 'metropole–satellite'.

Although today many may reject pure dependency theory as an overblown reaction to right-wing free market orthodoxy, it is tempting to suggest that such ideas can inform decisions about development strategies in the 'more-realistic' setting of the mixed economy. For example, if tourism is followed as an explicit development strategy, how much dependence should be placed on foreign ownership and multinational capital? How much should local indigenous resources and capital be targeted in order to stimulate the indigenous economy? These are real development choices that need to be made even in a mixed economy context and one can argue that these choices are better understood in light of pure dependency theory.

It was left to world systems theory to stress that contemporary development has involved the emergence of a substantial semi-periphery in addition to the core and periphery identified by classic dependency theory. This semi-periphery largely consists of the newly industrializing countries (NICs) of East Asia and Latin America.

From the 1970s onward, what may be regarded as 'alternative views' on development were espoused in many quarters, including the need for participatory development and the need to listen to indigenous voices in development in the form of indigenous environmental knowledge. The era of postmodernity – stressing the rejection of high modernism – may not be regarded as fitting the realities of the developing world or poor countries in all respects, but the existence of these notions cannot be ignored in the analysis of the conditions faced by such nations. Early standpoints that took a less generic, less monumental and less linear view of the development process included what are referred to under the headings 'bottom-up' and 'agropolitan' approaches, which have come to include ideas of 'another' development. All these approaches stressed the importance of local indigenous knowledge and ways of doing things rather than running with the market.

More recently, the wider 'postist' stance afforded by postcolonialism has been added to the critique. This argues that the production of Western knowledge has been inseparable from the exercise of Western power, and critiques the outcomes of colonialism that have underpinned

Eurocentric worldviews on development. It is stressed that colonialism has been associated with insensitivity to the views, practices and conditions of other cultures. Post-development as an approach is similarly sceptical about the whole manner in which development has been framed in terms of grand narratives, suggesting that the whole question of development needs to be closely questioned, problematized and possibly rejected. Under such a perspective development is seen as contradictory at best. Finally, it is notable that evolving conceptualizations of the role of social capital underpin continuing debates concerning development theory and practice.

2.1

Theories, strategies and ideologies of development

An overview

Robert B. Potter

Introduction

A major characteristic of the multi-, inter- and cross-disciplinary field of development studies since its establishment in the 1940s has been a series of sea-level changes in thinking about the process of development itself. This search for new theoretical conceptualizations of development has been mirrored by changes in the practice of development in the field. Thus, there has been much debate and controversy about development, with many changing views as to its definition, and the strategies by means of which, however development is defined, it may be pursued. In short, the period since the 1950s has seen the promotion and application of many varied views of development. And the literature on development theory and practice has burgeoned (see, for example, Hettne, 1995; Preston, 1996; Cowen and Shenton, 1996; Potter *et al.*, 2008; Peet and Hartwick, 2009; Chant and McIlwaine, 2009; Nederveen Pieterse, 2010; Thirlwall, 2011; Potter *et al.*, 2012). A major theme is that ideas about development have long been controversial and highly contested.

It is also necessary to stress that development covers both theory and practice, that is, both ideas about how development should or might occur, and real-world efforts to put various aspects of development into practice. This is conveniently mapped into the nomenclature suggested by Hettne in his overview of *Development Theory and the Three Worlds* (1995). In reviewing the history of development thinking, he suggested that 'development' involves three things: *development theories*, *development strategies* and *development ideologies*.

Development theories

If a theory is defined as a set of logical propositions about how some aspect of the real world is structured, or the way in which it operates, *development theories* may be regarded as sets of ostensibly logical propositions, which aim to explain how development has occurred in the past, and/or how it should occur in the future (Potter *et al.*, 2008). Development theories can either be *normative*, that is they can generalize about what should happen or what should be the case in an ideal world; or *positive* in the sense of dealing with what has generally been the case in the past. This important distinction is broadly exemplified in the figure that accompanies this account (see Figure 2.1.1). Hettne (1995) remarks that 'development studies is explicitly normative', and that teachers, researchers and practitioners in the field 'want to change the world, not only analyse it' (Hettne, 1995: 12). The arena of development theory is primarily, although by no means exclusively, to be encountered in the academic literature, that is, in writing about development. It is, therefore, inherently controversial and contested.

Development strategies

On the other hand, *development strategies* can be defined as the practical paths to development which may be pursued by international agencies, states, non-government organizations and community-based organizations, or indeed individuals, in an effort to stimulate change within particular areas, regions, nations and continents. Thus, Hettne (1995) provides a definition of development strategies as efforts to change existing economic and social structures and institutions in order to find enduring solutions to the problems facing decision-makers. As such, Hettne argues that the term 'development strategy' implies an actor, normally the state. In order to sound less top-down, it is necessary to think in terms of a wider set of development-oriented actors, including all those listed above.

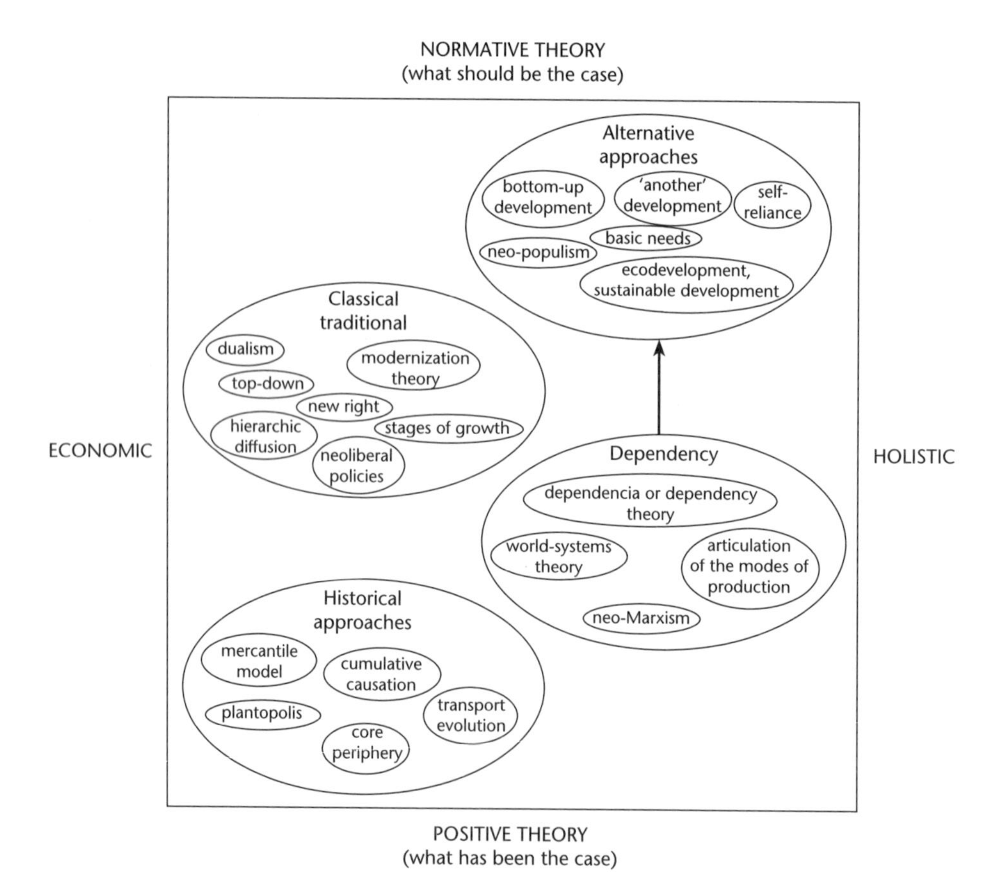

Figure 2.1.1 A framework for considering development theories

Source: Potter *et al.* 1999 Figure 3.2

Development ideologies

Different development agendas will reflect different goals and objectives. These goals will reflect social, economic, political, cultural, ethical, moral and even religious influences. Thus, what may be referred to as different *development ideologies* may be recognized. For example, both in theory and in practice, early perspectives on development were almost exclusively concerned with promoting economic growth. Subsequently, however, the predominant ideology within the academic literature changed to emphasize wider sets of political, social, ethnic, cultural, ecological and other dimensions of the wider human processes of development and change. Theories in development are distinctive by virtue of the fact that they involve the intention to change society in some defined manner. One of the classic examples is the age-old battle between economic policies which increase growth but widen income disparities, and those wider policy imperatives which seek primarily to reduce inequalities within society. All such efforts to effect change reflect some form of ideological base.

Development thinking

Perhaps the sensible approach is to follow Hettne (1995) and to employ the overarching concept of *development thinking* in our general deliberations. The expression 'development thinking' may be used as a catch-all phrase indicating the sum total of ideas about development, that is, including pertinent aspects of development theory, strategy and ideology. Such an all-encompassing definition is necessary due to the nature of thinking about development itself. As noted at the outset, development thinking has shown many sharp twists and turns during the twentieth century. The various theories that have been produced have not commanded attention in a strictly sequential-temporal manner. In other words, as a new set of ideas about development has come into favour, earlier theories and strategies have not been totally abandoned and replaced. Rather, theories and strategies have tended to stack up, one upon another, coexisting, sometimes in what can only be described as very convoluted and contradictory manners. Thus, in discussing development theory, Hettne (1995: 64) has drawn attention to the 'tendency of social science paradigms to accumulate rather than fade away'.

Development studies and disciplinary revolution/evolution

The characteristics of development studies as a distinct field of enquiry can be considered in a somewhat more sophisticated manner by referring to Thomas Kuhn's ideas on the *structure of scientific revolutions*. Kuhn (1962) argued that academic disciplines are dominated at particular points in time by communities of researchers and their associated methods, and they thereby define the subjects and the issues deemed to be of importance within them. He referred to these as 'invisible colleges', and he noted that these serve to define and perpetuate research which confirms the validity of the existing paradigm or 'supra-model', as he referred to it. Kuhn called this 'normal science'. Kuhn noted that only when the number of observations and questions confronting the status quo of normal science becomes too large to be dealt with by means of small changes to it, will there be a fundamental shift. However, if the proposed changes are major and a new paradigm is adopted, a scientific revolution can be said to have occurred, linked to a period of what Kuhn referred to as 'extraordinary research'.

In this model, therefore, scientific disciplines basically advance by means of revolutions in which the prevailing normal science is replaced by extraordinary science and, ultimately, a new form of normal science develops. In dealing with social scientific discourses, it is inevitable that the field of development theory is characterized by evolutionary, rather than revolutionary change. Evidence of the persistence of ideas in some quarters, years after they have been discarded elsewhere, will be encountered throughout the development literature. Given that development thinking is not just about the theoretical interpretation of facts, but rather about values, aspirations, social goals, and ultimately that which is moral, ethical and just, it is understandable that change in development studies leads to the parallel *evolution* of ideas, rather than *revolution*. Hence, conflict, debate, contention, positionality and even moral outrage are all inherent in the discussion of development strategies, and associated plural and diverse theories of development.

Approaches to development thinking

There are many ways to categorize development thinking through time. Broadly speaking, it is suggested here that four major approaches to the examination of development thinking can be recognized, and these are shown in Figure 2.1.1 (see Potter *et al.*, 2008). The framework first maps in the distinction previously made between *normative development theories* (those focusing on what *should be* the case), and *positive theories* (those which ponder on what has *actually been* the case). Another axis of difference between theories is seen as relating to whether they are *holistic* or *partial*, and most partial theories emphasize the economic dimension. This is also intimated in the figure.

These two axes can be superimposed on one another to yield a simple matrix or framework for the consideration of development thinking, strategies or theories, as shown. Following Potter *et al.* (2008: ch. 3), as noted, four distinct groupings of development theory can be recognized by virtue of their characteristics with regard to the dimensions of holistic–economic and normative–positive. The approaches are referred to here as: (i) the classical–traditional approach; (ii) the historical–empirical approach; (iii) the radical political economy–dependency approach; and, finally, (iv) bottom-up and alternative approaches. Following the argument presented in the last section, each of these approaches may be regarded as expressing a particular ideological standpoint, and can also be identified by virtue of having occupied the centre stage of the development debate at particular points in time. Classical–traditional theory, embracing dualism, modernization theory, top-down conceptualizations, the new right and neoliberal imperatives, is seen as stressing the economic and, collectively, existing midway between the normative and positive poles. In direct contrast, according to this framework, radical–dependency approaches, embracing neo-Marxism, and the articulation of the modes of production, are seen as being more holistic. At the positive end of the spectrum exist those theories which are basically historical in their formulation, and which purport to build upon what has happened in the past. These include core–periphery frameworks, cumulative causation and models of transport evolution, especially the mercantile model. In contrast, once again, are theories which stress the ideal, or what should be the case. These are referred to as 'alternative approaches', and basic needs, neo-populism, 'another development', ecodevelopment and sustainable development may be included in this category.

Conclusion

Many diverse and varied approaches to development remain in currency today, and in many different quarters. Hence, in development theory and academic writing, left-of-centre socialist views may well be more popular than classical and neo-classical formulations, but in the area of practical development strategies and policies, the 1980s and beyond have seen the implementation of neoliberal interpretations of classical theory, stressing the liberalization of trade, along with public-sector cut-backs. Such plurality and contestation are an everyday part of the field of development studies. In the words of Hettne (1995: 15), 'theorizing about development is therefore a never-ending task'.

References

Chant, S. and McIlwaine, C. (2009) *Development Geographies in the 21st Century*, London: Edward Elgar.

Cowen, M. P. and Shenton, R. (1996) *Doctrines of Development*, London: Routledge.

Hettne, B. (1995) *Development Theory and the Three Worlds: Towards an International Political Economy of Development* (2nd edn), Harlow, UK: Longman.

Kuhn, T. (1962) *The Structure of Scientific Revolutions*, Chicago: University of Chicago Press.

Nederveen Pieterse, J. (2010) *Development Theory* (2nd edn), London: Sage.

Peet, R. and Hartwick, E. (2009) *Theories of Development: Contentions, Arguments, Alternatives* (2nd edn), London: Guilford Press.

Potter, R. B., Binns, T., Elliott, J. and Smith, D. (2008) *Geographies of Development: An Introduction to Development Studies* (3rd edn), New York: Pearson-Prentice Hall.

Potter, R. B., Conway, D., Evans, R. and Lloyd-Evans, S. (2012) *Key Concepts in Development Geography*, London: Sage.

Preston, P. W. (1996) *Development Theory: An Introduction*, Oxford: Blackwell.

Thirlwall, A. P. (2011) *Economics of Development: Theory and Evidence* (9th edn), London: Palgrave Macmillan.

Further reading

Hettne, B. (1995) *Development Theory and the Three Worlds: Towards an International Political Economy of Development* (2nd edn), Harlow, UK: Longman. Briefly introduces the concepts of development theories, strategies and ideologies (pp. 15–16), before presenting an overview of Eurocentric development thinking, the voice of the 'other', globalization and development theory, and 'another development'.

Peet, R. and Hartwick, E. (2009) *Theories of Development: Contentions, Arguments, Alternatives* (2nd edn), London: Guilford Press. The two substantive sections, making up chapters 2–7, review what are referred to as conventional (Keynesian economics to neoliberalism, modernization) and non-conventional/critical (Marxism, socialism, post-structural, postcolonialism, feminist) theories of development.

Potter, R. B., Binns, T., Elliott, J. and Smith, D. (2008) *Geographies of Development* (3rd edn), Harlow, UK: Prentice Hall. Seeks to stress the plural and contested nature of development theory and practice. Chapter 3 overviews theories and strategies of development, stressing their diversity and value-laden character. The structure of the account is based on the figure employed in the present chapter.

Preston, P. W. (1996) *Development Theory: An Introduction*, Oxford: Blackwell. The first part of the book treats social theory in general terms. Thereafter, contemporary theories of development are summarized followed by what are referred to as new analyses of complex change.

2.2

Smith, Ricardo and the world marketplace, 1776 to 2012

Back to the future and beyond

David Sapsford

Introduction

Why do countries trade with one another? What determines the terms on which trade between countries is conducted in the world marketplace? These two questions are perhaps the most fundamental to be considered in any analysis of international trade, be it trade *between* developed and developing countries or trade *amongst* countries in either the developing or the developed world. These questions are of special importance in the context of economic development, since if there are 'gains from trade' to be had, the distribution of such gains between trading partners carries important implications for living standards and economic welfare within the participating countries.

The classical economists, most notably Adam Smith (1723–90) and David Ricardo (1772–1823) considered these two questions, and their analyses are outlined in the following section. Subsequent sections consider the available evidence regarding the changes that have occurred over the long run in the terms on which trade between developed and developing nations has been conducted, and explore the implications of this for economic development in the Third World.

Absolute and comparative advantage

The foundations of the economic theory of international trade were laid by Adam Smith in *The Wealth of Nations* (1776). Smith's analysis of division of labour is well known and to a large extent he saw the phenomenon of international trade as a logical extension of this process, with particular regions or countries (rather than particular individuals) specializing in the production of particular commodities. Smith's view is clearly demonstrated by the following quotation:

> It is the maxim of every prudent master of a family, never to attempt to make at home what it will cost him more to make than buy . . . What is prudence in the conduct of every private family, can scarce be folly in that of a great kingdom. If a foreign country can supply us with a commodity cheaper than we ourselves can make it, better buy of

> them with some part of the produce of our own industry, employed in a way in which we have some advantage.
>
> *(Smith, 1776: 424)*

Thus, according to Smith, countries engage in trade with one another in order to acquire goods more cheaply than they could produce them domestically, paying for them with some proportion of the output that they produce domestically by specializing according to their own 'advantage'. Central to this view is the notion that relative prices determine trade patterns, with countries buying abroad when foreign prices are below domestic ones. In addition, Smith argued that by expanding the size of the market, international trade permits greater specialization and division of labour than would otherwise have been possible. This is perhaps one of the earliest arguments in favour of globalization as a process by which the size of the world marketplace is increased.

Economics textbooks abound with simple two-country/two-good examples that illustrate Smith's argument. Suppose that the world consists of only two countries (say, the UK and the USA) and only two goods (say, food and clothing). Within this (over)simplified framework let us assume that the USA is more efficient than the UK at producing food (in the sense that fewer resources are needed to produce a unit of food in the USA than in the UK) and (in the same least resource-cost sense) that the UK is more efficient than the USA at producing clothing. In economists' terminology, this example represents the case where the UK possesses *absolute advantage* in the production of clothing, while the USA possesses *absolute advantage* in the production of food. To further simplify, let us assume that labour is the only factor of production and that within each country it is mobile between the two industries. Assume also that wages are the same in both countries and that transport costs are zero. Based on this battery of assumptions, the USA will be the cheaper source of food and the UK of clothing. It is a matter of simple arithmetic to show that if both countries are initially producing some of each good, it is always possible to increase output of both goods if each country specializes in the production of that good for which it possesses absolute advantage. It also follows that by trading, each country can consume the bundle of clothing and food that it consumed in the absence of trade (that is, under *autarky*) while still leaving some of each product over! Each country thus has the potential to increase its consumption of both goods and, assuming that more of each good is preferable to less, trade can, in principle, allow both trading partners to increase their *economic welfare*. As already noted, the distribution of this surplus (that is, the distribution of the *gains from trade*) between the two countries is an important matter, especially in the context of economic development. We return to this issue in the following section.

The case analysed by Adam Smith considered, quite naturally at the time he was writing, the situation where one country possesses absolute advantage in the production of one good, while the other country possesses it in the production of the other good. Writing four decades later, David Ricardo considered the rather more tricky analytical case in which one of the two countries (say the UK) is more efficient at producing *both* goods. According to Adam Smith's absolute advantage argument, both goods should be produced by the UK. However, this situation can clearly not represent a feasible state of affairs in the long run since although the USA will seek to purchase both goods from the UK, the UK will not wish to buy anything from the USA in return. Ricardo (1817) was the first economist to provide a formal analysis of this case and by so doing he derived his famous *Law of Comparative Advantage*.

Table 2.2.1 Labour requirements matrix

	Labour per unit of output	
	UK	*USA*
Food	5	6
Clothing	2	12

According to Ricardo's Law of Comparative Advantage, which encompasses Adam Smith's analysis of absolute advantage as a special case, world output and therefore (on the basis of the assumption discussed above) world economic welfare will be increased if each country specializes in the production of that good for which it possesses *comparative* advantage. The concept of comparative advantage is basically concerned with comparative efficiency and Ricardo's law follows from recognizing the fact that differences in the relative prices of the two goods between the two countries opens up the possibility of mutually beneficial trade. To take a concrete example, suppose that the labour required to produce 1 unit of each good in each country is as set out in Table 2.2.1. Notice that the UK requires less labour than the USA in both industries.

On the basis of these figures (and assuming that labour productivity in each industry does not alter with the level of output) we can see that in the absence of trade each unit of food within the UK trades for 2.5 units of clothing since each is equivalent to the output of five people. Likewise in the USA 1 unit of food trades for 0.5 unit of clothing, each being the output of six people. It is the difference between these two relative prices (or internal terms of trade) that opens up the possibility for mutually beneficial trade. For example, if US prices prevail in the world outside the UK, a British person in possession of 1 unit of food can exchange this within the UK for 2.5 units of clothing, which could then be sold in the USA for 5 units of food; thereby providing a gain equal to 4 units of food. Likewise, if British relative prices prevail, an American producer employing 12 people to make 1 unit of clothing could switch to the food industry and thereby produce 2 units of food, which could then be sold in the UK for 5 units of clothing; thus realizing a gain of 4 units of clothing. At intermediate relative prices (or terms of trade) both countries can gain from trade, although not to the extent shown in the respective examples given above.

In a nutshell, according to Ricardo's analysis each country shifts its production mix towards the good for which it possesses comparative advantage. In our example, the UK has comparative advantage in the production of clothing, whereas the USA's comparative advantage is in food, where it is *less inefficient*. Reading across the rows in Table 2.2.1 we see that this follows because the UK requires five-sixths of US unit inputs in food, but only one-sixth in clothing.

Who gains from trade?

While the elegance of Ricardo's analysis and its logical correctness within the confines of its own assumptions can not be faulted, it does beg a question that is vitally important in the context of trade that takes place between countries of the developed/industrialized world and countries of the Third World. While the analysis demonstrates quite clearly the *potential* benefits to trading partners from engaging in international trade in the world marketplace, it has nothing whatsoever to say about the division of these potential gains between them. As we saw in the preceding example, if relative prices in the world marketplace were equal to

US relative prices then the UK would effectively appropriate all of the gains from trade for herself whereas, at the opposite end of the spectrum, the USA would scoop all of the gains if British relative prices prevailed.

In order to focus ideas let us consider trade between the countries of the developed/industrialized world and those of the developing world and, for simplicity, assume that the former produce manufactured goods while the latter produce primary commodities. The fact that Ricardo's analysis did not shed any light on the issue of how the potential gains from trade are shared out in practice did not seem to constitute a problem in the minds of classical economists since in a related context Ricardo, like Smith before him, had argued that as an inevitable consequence of the twin forces of diminishing returns in the production of primary commodities from a fixed stock of land (including mineral resources) as population increased, and the downward pressures on production costs in manufacturing generated by the moderating influences of surplus population and urbanization upon wages, the price of primary products would rise over the long run in relation to the price of manufactured goods, thereby giving rise to an upward drift in the net barter terms of trade between primary commodities and manufactured goods. On the above assumptions, this movement will translate into an improvement in the terms of trade of developing countries vis-à-vis the developed countries. On the basis of this argument there was little, if any, reason to be concerned about the plight of developing countries in the context of their trading relations with the industrialized world since it predicted that over the long run, the terms of trade would shift steadily in their favour, with the result that they would enjoy an increasing proportion of the potential gains from trade.

However, in the early 1950s the classical prediction of a secular improvement was challenged by both Prebisch (1950) and Singer (1950). Both argued forcefully that in direct contravention of the then still prevailing classical prediction, the terms of trade had actually, as a matter of statistical fact, been historically subject to (and could be expected to continue to be subject to) a declining trend. Both analyses therefore implied that contrary to the classical view, developing countries were actually obtaining a falling proportion of the potential gains from their trade with the countries of the developed world. Recent statistical evidence (Sapsford, Bloch and Pfaffenzeller, 2010) indicates that the declining trend observed by both Prebisch and Singer persisted into the first decade of the twenty-first century.

A number of theoretical explanations have been put forward in the literature to account for the observed downward trend in the terms of trade of developing countries, relative to developed countries, and these can be conveniently summarized under the following four headings:

1 *differing elasticities of demand for primary commodities and manufactured goods* (with the inelastic nature of the former resulting in a tendency for increases in the conditions of commodity supply to be felt more strongly in price decreases than in quantity increases);
2 *differing rates of growth in the demands for primary commodities and manufactured goods* (with the demand for primaries expanding less rapidly than the demand for manufactures due to their lower income elasticity of demand – especially so in the case of agricultural commodities due to the operation of Engel's Law – plus the development of synthetic substitutes and the occurrence in manufacturing of technical progress of the raw materials-saving sort);
3 *technological superiority* (the argument being that the prices of manufactured goods rise relative to those of primaries because they embody both a so-called Schumpeterian rent

element for innovation, plus an element of monopolistic profit arising from the monopoly power of multinational producers);

4 *asymmetries in market structure* (the argument here is that differences in market structure – with primary commodities typically being produced and sold under competitive conditions, while manufacturing in industrialized countries is often characterized by a high degree of monopoly by organized labour and monopoly producers – mean that while technical progress in the production of primary commodities results in lower prices, technical progress in manufacturing leads to increased factor incomes as opposed to lower prices).

Policy implications

Although space constraints do not allow the discussion in any detail of the policy implications of the observed worsening trend in the terms on which trade is conducted in the world marketplace between primary commodity-producing countries and manufacturing countries, it is important to note that the Prebisch–Singer hypothesis is sometimes advanced as one argument in favour of development policies of the import-substituting industrialization as opposed to export promotion variety (Sapsford and Balasubramanyam, 1994). However, the policy issues here are not clear-cut and the fact that all four of the above explanations relate as much, if not more, to the characteristics of different types of countries as to the characteristics of different types of traded goods highlights the need to devise and implement policies that address differences and imbalances of the former as opposed to the latter sort.

It is now the case that at least some of the international agencies involved in the world trading system have come to accept that primary commodity producers in developing countries do face real and significant uncertainties and risks regarding the prices that they will actually receive for their products when they come to the world market. See Morgan (2001) and Toye (2010) for useful discussions of the array of, largely unimplemented or failed, schemes that have been put forward over the last seven decades in order to address the price risks and volatilities faced by primary commodity producers in developing countries.

1776 to 2012: Back to the future?

Some 236 years have elapsed since Adam Smith laid the initial foundations of trade theory as we know it today. It is testament to the logical correctness of his analysis, especially as extended by Ricardo, that this theoretical framework is still pivotal to twenty-first century thinking in both trade theory and policy formulation. As we have seen, the central prediction of this approach is that provided the world terms of trade lie within the limits imposed by domestic opportunity cost ratios, international specialization and exchange via trade provides an opportunity for both trading partners to benefit from increased output (and therefore economic welfare) with given resource/factor endowments. However, we have also seen that there is a school of thought surrounding the Prebisch–Singer hypothesis suggesting that in practice, the actual terms of trade have drifted, within the range delineated by the Ricardian analysis, in favour of the industrialized (manufacturing) nations to such an extent that these nations have appropriated for themselves the lion's share of the gains from trade, leaving only small pickings for the (primary commodity dependent) poorer countries of the developing world.

What does current experience tell us in relation to the fundamental question of 'who has gained what from participating in international trade'? The basic structure of international institutions that currently oversee/govern the day-to-day conduct of international trade and commerce were laid down in 1944 at the famous Bretton-Woods Conference. Prominent amongst these institutions is what is now known as the World Trade Organisation[1] (WTO) whose mission, in a nutshell, is to provide an arena and set of processes and rules designed to achieve multilateral reductions in trade barriers. The underlying philosophy here, squarely in the spirits of both Smith and Ricardo, is to maximize the potential global gains from trade by minimizing (if not completely eliminating) impediments to free trade, such as import tariffs, quotas and so on.

We have now accumulated almost seven decades of experience of the operation of this process of tariff reduction via multilateral trade negotiations under the auspices of the WTO and its predecessors. Although advocates of free trade see the WTO as having achieved considerable success in its mission to reduce average tariff levels, experience over the last decade or so might be interpreted as suggesting an altogether less rosy picture when one comes to ask the important question as to who has actually harvested the global gains generated by this move closer to free trade in the sense understood by Smith and Ricardo. Although a detailed discussion of the operation of the WTO is beyond the scope of the current chapter, it should be noted that it seeks to achieve multilateral reductions in tariff (and other non-tariff) barriers via a series of negotiating rounds. In 1994, the trade deal that came out of the so-called Uruguay round of negotiations was signed, although the negotiations did appear to be on the verge of collapse as late as 1990. One major factor that surfaced during the Uruguay round was the view of poor, primary commodity dependent countries that the proposed package would bestow substantial benefits upon the industrialized countries, while offering them very little. In 1999, pressure to offer a better deal to poor countries lead to a summit meeting in Seattle, which ended in failure (accompanied by public disorder). In 2001, in an attempt to reinvigorate the process of multilateral tariff reductions, WTO members agreed to launch fresh talks, known as the Doha Development Round. However, despite this initiative, the 2003 ministerial summit in Cancun, Mexico, collapsed in acrimony over the developed countries' intransigency over the issue of removal of subsidies paid to their farmers. Despite several subsequent attempts to inject new life into the Doha round the process has, effectively, ground to a halt.

What is to be made of this tale? At the time of writing (March 2012) the picture is clearly one where the very continued existence of the process of tariff reductions via multilateral negotiations is hanging by little more than a thread. The current stumbling block from the perspective of the poor countries is the persistent refusal of the major developed countries (including both the EU and US) to remove the trade barrier imposed by the still substantial subsidies paid to their farmers. However, there is a wider view, according to which this particular issue is but a symptom of a more fundamental problem: namely that after participating in the process of multilateral tariff reduction for at least half a century, the poor countries of the world have continually seen the gains from the trade being appropriated by their richer trading partners. Indeed, some commentators are predicting that such is their degree of dissatisfaction with a process which has delivered so little to them relative to their richer trading partners, that group(s) of poor countries are on the verge of withdrawing altogether from the process in favour of going it alone.

Whether the thread eventually breaks altogether remains to be seen!

Beyond

In primary commodity markets, as in life, things can – and sometimes do – change very rapidly. Today's world economy is a *very* different place to that which existed at the turn of the twenty-first century. Three factors have been at work: first, the spectacular growth rates achieved by China and India (plus Brazil) since 2005; second, the sub-prime loans/credit crunch crisis of 2008; and third, the post-2010 eurozone debt crisis. Each of these factors carries major implications for developing countries dependent on primary commodity exports and it remains to be seen how they will interact and work out in the future years and decades. However, a number of implications are already clear (Sapsford, Bloch and Pfaffenzeller, 2010). Not surprisingly, the spectacular growth rates of China and India led to marked upward movements in the real price of primary commodities in both 2006 and 2007, especially strongly in the cases of raw materials and metals. What is, however, surprising is the speed at which these gains in terms of trade of commodity dependent developing countries evaporated in the wake of the 2008 credit crunch. International Monetary Fund data indicate that while commodity prices peaked in March 2008 they fell with alarming rapidity thereafter. Between March and November 2008, a period of only eight months, the IMF's index of non-fuel primary commodity prices fell by 32 per cent, with corresponding falls in the prices of industrial inputs and food and beverages equalling 35 and 28 per cent respectively! This is a truly breathtaking rate of collapse – clearly good times can disappear as quickly as they arrive. By mid-2011, these falls had been largely reversed but as the eurozone crisis unfolded prices seem to be moving downwards again. It remains to be seen where the eurozone crisis will ultimately end but if, as does not seem unlikely, Greece exits the Euro then it is not inconceivable that the ensuing downward effects on industrial output in both the eurozone and elsewhere may well generate price falls of a similar magnitude to those observed in 2008.

It remains to be seen whether this order of increased volatility in the primary commodity terms of trade, with all of the associated difficulties it raises for primary commodity dependent LDCs, will become a feature of twenty-first century economic life.

I wonder what Adam Smith and David Ricardo would make of this!

Note

1 Previously known as the General Agreement on Tariffs and Trade (GATT), although originally named (by Keynes as the principal architect of the Bretton-Woods system) the International Trade Organisation (ITO). See Chen and Sapsford (2005).

References

Chen, J. and Sapsford, D. (eds) (2005) *Global Development and Poverty Reduction: The Challenge For International Institutions*, Cheltenham, UK: Edward Elgar.

Morgan, W. (2001) 'Commodity futures markets in LDCs: A review and prospects', *Progress in Development Studies* 1(2): 139–50.

Prebisch, R. (1950) 'The economic development of Latin America and its principal problem', UN ECLA; also published in *Economic Bulletin for Latin America* 7(1) (1962): 1–22.

Ricardo, D. (1817) *On the Principles of Political Economy and Taxation*, London: Penguin (reprinted 1971).

Sapsford, D. and Balasubramanyam, V. N. (1994) 'The long-run behavior of the relative price of primary commodities: Statistical evidence and policy implications', *World Development* 22(11): 1737–45.

Sapsford, D., Bloch, H. and Pfaffenzeller, S. (2010) 'Commodities still in crisis?' In M. Nissanke and G. Mavrotas (eds), *Commodities, Governance and Economic Development under Globalization*, London: Palgrave Macmillan, pp. 99–115.

Singer, H. (1950) 'The distribution of gains between investing and borrowing countries', *American Economic Review, Papers and Proceedings* 40: 473–85.

Smith, A. (1776) *The Wealth of Nations*, London: Penguin (reprinted 1961).

Toye, J. (2010) 'Commodities, cooperation and world economic development: The mission of Alfred Maizels, 1996–2006'. In M. Nissanke and G. Mavrotas (eds), *Commodities, Governance and Economic Development under Globalization*, London: Palgrave Macmillan, pp. 3–17.

Further reading

Detailed discussion of both the theoretical arguments and statistical evidence underlying the declining trend in terms of trade hypothesis can be found in the following:
Bloch, H. and Sapsford, D. (2011) 'Terms of trade movements and the global economic crisis', *International Review of Applied Economics* 25(5): 503–17.
Sapsford, D. (2008) 'Terms of trade and economic development'. In A. Dutt and J. Ros (eds), *International Handbook of Development Economics*, Cheltenham, UK: Edward Elgar, pp. 16–29.
Sapsford, D., Sarkar, P. and Singer, H. (1992) 'The Prebisch–Singer terms of trade controversy revisited', *Journal of International Development* 4(3): 315–32.
Singer, H. (1987) 'Terms of trade and economic development'. In J. Eatwell, M. Milgate and P. Newman (eds), *The New Palgrave: A Dictionary of Economics*, London: Macmillan, pp. 626–8.
Spraos, J. (1983) *Inequalizing Trade?* Oxford: Oxford University Press.
A comprehensive discussion of a wide range of issues relating to the relationship between economic development and international trade may be found in the following:
Greenaway, D. (ed.) (1988) *Economic Development and International Trade*, London: Macmillan.
Highly accessible discussion of the main issues involved in the globalization debate, seen from either side of the fence, can be found in the following two references:
Bhagwati, J. (2004) *In Defense of Globalization*, London: Oxford University Press.
Stiglitz, J. (2003) *Globalization and its Discontents*, London: Penguin Books.

2.3

Enlightenment and the era of modernity

Marcus Power

Introduction: The 'rough and tumble' of early industrialism

Just as light cuts through darkness, the philosophy of the Enlightenment was seen as something that would open the eyes of the world's poor and free them from unjust rule. The 'age of Enlightenment' is most often traced to the eighteenth century and represented a catalyst for the development of particular styles of social thought in the form of a movement or a programme in which reason was used in order to achieve freedom and progress, and during which hostility to religion was omnipresent. In its simplest sense, the Enlightenment was the creation of a new framework of ideas and secure 'truths' about the relationships between humanity, society and nature which sought to challenge traditional worldviews dominated by Christianity. Science, and the scientific approach, became the tool to investigate the world, instead of theological dogmas. According to Gay (1973: 3), at this time educated Europeans experienced 'an expansive sense of power over nature and themselves: the pitiless cycles of epidemics, famines, risky life and early death, devastating war and uneasy peace – the treadmill of human existence – seemed to be yielding at last to the application of critical intelligence'. Fear of change began to give way to fear of stagnation. It was a century of commitment to enquiry and criticism, of a decline in mysticism, of growing hope and trust in effort and innovation (Hampson, 1968). One of the primary interests was social reform, and the progression and development of societies built around an increasing secularism and a growing willingness to take risks (Gay, 1973).

There is no monolithic 'spirit of the age' that can be discerned, however, and the Enlightenment does not represent a set of ideas which can be clearly demarcated, extracted and presented as a list of essential definitions. There were, however, many common threads to this patchwork of Enlightenment thinking: the primacy of reason/rationalism, a belief in empiricism, the concept of universal science and reason, the idea of progress, the championing of new freedoms, the ethic of secularism and the notion of all human beings as essentially the same (Hall and Gieben, 1992: 21–2). Thinkers such as Kant, Voltaire, Montesquieu, Diderot, Hume, Smith, Ferguson, Rousseau and Condorcet found a receptive audience for their 'new style of life' (Hampson, 1968) producing a large collection of novels, plays, books, pamphlets and essays for the consumption of nobles, professionals (especially lawyers), academics and the clergy. New cultural innovations in writing, painting, printing, music, sculpture and architecture, and new technological innovations in warfare, agriculture and manufacture had a major impact on the *philosophes*, the free-thinking intellectuals or 'men of letters' that had brokered this enlightened awakening in France. The *philosophes* sought to redefine what was considered as socially important knowledge, to bring it outside the sphere of religion and to provide it with a new meaning and relevance. For Hall and Gieben (1992: 36) four main areas distinguish the thought of the *philosophes* from earlier intellectual approaches:

- anti-clericalism;
- a belief in the pre-eminence of empirical, materialist knowledge;
- an enthusiasm for technological and medical progress;
- a desire for legal and constitutional reform.

There is thus clearly a risk of applying the term 'the Enlightenment' too loosely or too widely, as if it had touched every intellectual society and every intellectual elite of this period equally. The Enlightenment is thus best considered as an amorphous, dynamic and variegated entity (Porter, 1990). More than simply a predominantly French movement centred around a small group of *philosophes*, scholars have recently begun to consider the complex spatiality of 'the Enlightenment' as a cosmopolitan process, to view it in its international context (where its key ideas and views were transmitted across borders) and thus to identify a number of different 'Enlightenments'. Reaching its climax in the mid-eighteenth century in Paris and Scotland, but with foundations in many countries (including several outside of Europe such as the USA), 'the Enlightenment' was thus a sort of intellectual fashion or 'a tendency towards critical inquiry and the application of reason' (Black, 1990: 208) rather than a singular coherent intellectual movement or institutional project. The *philosophes* of the eighteenth-century Enlightenment in France, for example, did not act in concert and neither should they be seen as a unified family for their views were too disparate (Porter, 1990).

It is also important to remember that the new 'style of life' championed by Enlightenment intellectuals was in the main reserved for the fortunate and the articulate – the rural and urban masses had little share. It was not until the eve of the French Revolution in the 1780s that a new social group emerged concerned with popularizing Enlightenment ideas. Similarly, though many women played a major part in the development and diffusion of Enlightenment ideas, applying such ideas to their social conditions meant negotiating a number of contradictory positions within patriarchal societies. The emancipatory potential of this knowledge thus turned out to be limited in that it was conceived of as abstract and utilitarian, as a mastery over nature which thus becomes characterized by power. As Doherty (1993: 6) has argued:

> Knowledge is reduced to technology, a technology which enables the *illusion* of power and of domination over nature. It is important to stress that this is an illusion. This kind of knowledge does not give actual power over nature. . . . What it does give in the way of power is, of course, a power over the consciousness of others who may be less fluent in the language of reason. . . . Knowledge thus becomes caught up in a dialectic of mastery and slavery.
>
> *(Emphasis in original)*

The Enlightenment was also closely linked to the rise of modernity and provided an important crucible for the invention of the modern idea of 'development' which began to emerge 'amidst the throes of early industrial capitalism in Europe' (Cowen and Shenton, 1996: 5). The metaphor of the 'light of reason' shining brightly into all the dark recesses of ignorance and superstition in 'traditional' societies was a powerful and influential one at this time. In Europe, the light that the process of 'development' brought was intended to 'construct order out of the social disorders of rapid urban migration, poverty and unemployment' (Cowen and Shenton, 1996: 5). Many Enlightenment thinkers also viewed the remedy for the disorder brought on by industrialization as related to the 'capacity' to use land, labour and capital in the interests of society as a whole. Only certain kinds of individuals could be 'entrusted' with such a role (Cowen and Shenton, 1996). Property, for example, needed to be placed in the hands of 'trustees' who would decide where and how society's resources could be most effectively utilized. In eighteenth-century France, the prevailing social orders were represented as three 'estates' – clergy, nobility and the 'third estate', which comprised everyone else, from wealthiest bourgeois to poorest peasant (Hall, 1992). This 'dialectic of mastery and slavery' and this gap between the *philosophes* (who were often members of the second estate) and the peasantries of European eighteenth-century societies, are both important parts of the historical context of Enlightenment thinking. Although they appeared to represent a threat to the established order, these ideas and writings sought evolutionary rather than revolutionary change, arguing that progress and development could come about within the existing social order through the dissemination of ideas among 'men of influence' (Hall and Gieben, 1992).

'Modernity' and the rise of the social sciences

The influential economist John Maynard Keynes (1936: 570), once wrote that 'practical men, who believe themselves to be quite exempt from any intellectual influences, are usually the slaves of some defunct economist'. So it is with much development thinking today. A variety of twentieth-century movements including neo-classicism (of which Keynes was an important part) and liberalism can trace their origins back to the Enlightenment. The foundations of many modern disciplines (including development studies) were intimately bound up with the Enlightenment's concept of progress and the idea that development could be created through the application of reasoned and empirically based knowledge. The Enlightenment had forged the intellectual conditions in which the application of reason to practical issues could flourish through such 'modern' institutions as the academy, the learned journal and the conference. In turn, a 'modern' audience was constituted for the dissemination of social and political ideas alongside a class of intellectuals that could live from writing about them (Hall, 1992). Through the Enlightenment, state bureaucracies began to use social statistics to provide the evidence necessary for 'rational' choices in the allocation of resources. This process of labelling people was part of a wider intellectual paradigm that considered categorization, quantification and measurement as integral to rational and objective decision-making. These 'official' labels were – and still are – generally portrayed and accepted as objective facts, though many are rooted in intensely political processes. For example, many conventional racial and group classifications were created in the imperial and colonial periods, when authorities counted, categorized, taxed and deployed slave, servile and forced labour, often over vast geographical areas (IDS, 2006: 1).

The emergence of an idea of 'the West' was also important to the Enlightenment in that it was a very European affair, which put Europe and European intellectuals at the very pinnacle of human achievement. This view sees 'the West' as the result of forces largely internal to Europe's history and formation (Hall, 1992) rather than as a 'global story' involving other cultural worlds. In the making of nineteenth-century European 'modernity', Europeans had a sense of difference from other worlds (e.g. 'Africa'), which shaped the ways in which they were viewed as distant, uncivilized and immature stages in the progress of humanity. The establishment of modern modes of scientific enquiry, of modern institutions and the modern 'development' of societies in nineteenth-century Europe thus partly incorporated a contrast with the 'savage' and 'uncivilized' spaces of the non-Western world. The emergence of area studies disciplines in the twentieth century can also be traced back to Enlightenment efforts to support theories of human progress by comparing Europe to other regions of the world and in elaborating the contrast between Europe and other areas (Ludden, 2003). This tradition of universal comparison and ranking has also arguably continued to be a feature of 'development thinking' in the twenty-first century.

Modernist reason was not as inherently good as the 'enlightened' thinkers believed and has been used for a wide variety of purposes. Reason can be imperialist and racist (as in the making of the idea of 'the West'), taking a specific form of consciousness for a universal, a standard that all must aspire to reach. Reason was also a potent weapon in the production of social normativity during 'the Enlightenment', driving people towards conformity with a dominant and centred 'norm' of behaviour (Doherty, 1993). Modernist reason was therefore dependent on the 'othering' of non-conformists, of cultures and societies that were not informed by this reason and social norms and were thus banished to the lower echelons of humanity, defined as 'backward', 'undeveloped' or 'uncivilized'. The emergence of new ideas about social, political and economic development was therefore bound up with these pressures to conform to particular notions of knowledge, reason and progress, and with the making of a 'Third Estate' or 'Third World' of non-conformity as the alter ego of a developed 'West'.

Conclusions: Completing the Enlightenment beyond Europe

Much contemporary development thinking has its roots in the Enlightenment as the 'age of reason', which shaped concepts of progress, growth and social change. Modernist thought also envisaged a process of enlightenment, of becoming more modern and less traditional and also saw a group of enlightened Western intellectuals and scientists 'guiding' the paths to progress of distant others. Arturo Escobar (1995: 2–4) has even argued that the post-1945 development project is 'the last and failed attempt to complete the Enlightenment in Asia, Africa and Latin America' (Escobar, 1995: 221). After 1945, modernization theorists in the United States also 'saw their project as the Enlightenment writ large' (Gilman, 2003: 8) and even the vision of the modern developed under Soviet Communism (albeit with a very different collectivist, anti-religious and anti-capitalist belief system) was similarly a product of the Enlightenment. 'Development' thus has complex roots in the emergence of 'the Enlightenment', in the dawn of industrial capitalism in Europe and America and in the rise and formation(s) of modernity. It is also important to remember that the self-identification of European and Western countries as 'developed' has partly been produced through a contrasting of modernity with the tradition and backwardness of the 'Third World' as Other.

The work of Enlightenment thinkers like Adam Smith (with his free market economics) remains very relevant to 'international development' today for some observers. Examples of this can be found in some of the key global development institutions like the World Bank, that see their (neo-classical) knowledges as potentially enlightening. Consider the following quotation from a speech given by the World Bank President James Wolfensohn in 1996: 'Knowledge is like light. Weightless and intangible, it can easily travel the world, enlightening the lives of people everywhere. Yet billions of people still live in the darkness of poverty – unnecessarily' (quoted in Patel, 2001: 2).

Thus the knowledge and expertise of contemporary development practitioners is seen as something almost universal that easily traverses borders extinguishing the darkness of poverty wherever it shines. For some theorists and practitioners of development today, people and places can become 'developed' simply though acquiring scientific and technical knowledge about the 'normal' or correct series of developmental stages. If only it were that simple.

References

Black, J. (1990) *Eighteenth-century Europe 1700–1789*, London: Macmillan.

Cowen, M. P. and Shenton, R. W. (1996) *Doctrines of Development*, London: Routledge.

Doherty, T. (1993) 'Postmodernism: An introduction', in T. Doherty (ed.), *Modernism/Postmodernism*, Hemel Hempstead, UK: Harvester Wheatsheaf, pp. 1–31.

Escobar, A. (1995) *Encountering Development: The Making and Unmaking of the Third World*, Princeton, NJ: Princeton University Press.

Gay, P. (1973) *The Enlightenment: A Comprehensive Anthology*, New York: Simon and Schuster.

Gilman, N. (2003) *Mandarins of the Future: Modernization Theory in Cold War America*, Baltimore, MD: Johns Hopkins University Press.

Hall, S. (1992) 'The West and the rest: Discourse and power', in S. Hall and B. Gieben (eds), *Formations of Modernity*, Cambridge: Open University/Polity, pp. 275–331.

Hall, S. and Gieben, B. (1992) *Formations of Modernity*, Cambridge: Open University/Polity.

Hampson, N. (1968) *The Enlightenment*, London: Penguin.

Institute of Development Studies (IDS) (2006) 'The power of labelling in development practice', *IDS Policy Briefing*, 28 (April).

Keynes, J. M. (1936) *The General Theory of Employment, Interest and Money*, London: Macmillan.

Ludden, D. (2003) 'Why area studies?' in A. Mirsepassi, A. Basu and F. Weaver (eds), *Localizing Knowledge in a Globalizing World: Recasting the Area Studies Debate*, New York: Syracuse University Press, pp. 131–7.

Patel, R. (2001) 'Knowledge, power, banking', *Znet Magazine*, 20 July.

Porter, R. (1990) *The Enlightenment*, London: Macmillan.

Further reading

Cowen, M. P. and Shenton, R. W. (1996) *Doctrines of Development*, London: Routledge. Provides an accessible discussion of enlightenment ideas, exploring their bearing on the construction of particular development approaches and doctrines.
Doherty, T. (1993) 'Postmodernism: An introduction', in T. Doherty (ed.), *Modernism/Postmodernism*, Hemel Hempstead, UK: Harvester Wheatsheaf, pp. 1–31. Offers clear and accessible definitions of modernism.
Hall, S. and Gieben, B. (1992) *Formations of Modernity*, Cambridge: Open University/Polity. Focuses on the making of modernity in the non-Western world.
Rist, T. (1997) *History of Development*, London: Routledge. For an excellent introduction and overview to early development discourses and ideas.

2.4

Dualistic and unilinear concepts of development

Tony Binns

The development imperative

After the Second World War, Europe embarked on a massive programme of reconstruction, instrumental to which was the Marshall Plan, launched by the US government on 5 June 1947. While the Marshall Plan was heralded as US financial help to the devastated economies and infrastructures of Western Europe, this 'goodwill gesture' was also designed to stimulate markets for America's burgeoning manufacturing sector. The Marshall Plan, which injected US$17 billion mainly into the UK, France, West Germany and Italy between 1948 and 1952, generated much confidence in the role of overseas economic aid (Hunt, 1989; Rapley, 1996). Another landmark in the recognition of the need for richer countries to play an active role in the development of poorer countries came less than two years later, on 20 January 1949, when US President Truman in 'Point Four' of his Inaugural Address proclaimed:

> we must embark on a bold new program for making the benefits of our scientific advances and industrial progress available for the improvement and growth of underdeveloped areas. More than half the people of the world are living in conditions approaching misery. Their food is inadequate. They are victims of disease. Their economic life is primitive and stagnant. Their poverty is a handicap and a threat both to them and to more prosperous areas. For the first time in history, humanity possesses the knowledge and skill to relieve the suffering of these people . . . I believe that we should make available to peace-loving peoples the benefits of our store of technical knowledge in order to help them realize their aspirations for a better life.
>
> *(Public Papers of the Presidents of the United States, 1964: 114–15)*

'Point Four' probably inaugurated the 'development age' and 'represents a minor masterpiece . . . in that it puts forward a new way of conceiving international relations' (Rist, 1997: 71–2).

The neo-classical paradigm

The so-called 'neo-classical paradigm' dominated much thinking about development in the two or three decades after the Second World War. Adam Smith, the founding father of the classical school, writing in his *Wealth of Nations* (1776) in the early years of the Industrial Revolution, saw manufacturing as capable of achieving greater increases in productivity than agriculture. He emphasized the expansion of markets as an inducement for greater productivity which would, he believed, lead to greater labour specialization and productivity. A century later in 1890, Alfred Marshall, in his influential book, *Principles of Economics*, spelt out the 'neo-classical perspective', emphasizing the desirability of maximizing aggregate economic welfare, whilst recognizing that this was dependent on maximizing the value of production and raising labour productivity (Marshall, 1890). Technological change was recognized as being vital to raising productivity and meeting the demands for food and raw materials from a growing population. There was also a strong belief that free trade and the unimpeded operation of the market were necessary for maximizing efficiency and economic welfare (Hunt, 1989).

Dualism

Another theme that emerged in the post-war period was that underdeveloped economies were characterized by a 'dichotomous' or 'dualistic' nature, where advanced and modern sectors of the economy coexisted alongside traditional and backward sectors. A strong proponent of the dualistic structure of underdeveloped economies was the West Indian economist Arthur Lewis, whose seminal paper 'Economic development with unlimited supplies of labour' was published in 1954. Like others who followed him, Lewis did not differentiate between economic growth and development. The paper, which significantly opens with the statement, 'This essay is written in the classical tradition', envisages a division of the economic system into two distinct sectors, the capitalist and the subsistence. The subsistence sector, according to Lewis, consists predominantly of small-scale family agriculture and has a much lower per capita output than the capitalist sector, where manufacturing industry and estate agriculture, either private or state-owned, are important elements. The process of development, Lewis suggested, involves an increase in the capitalists' share of the national income due to growth of the capitalist sector at the expense of the subsistence sector, with the ultimate goal of absorption of the latter by the former. Since most labour for the capitalist sector would come from underemployed labour in subsistence agriculture, changes within the latter sector were seen as essential for the process of overall economic development.

The Lewis model had a significant influence on development thinking in the 1950s and 1960s, but it has been criticized for failing to appreciate the positive role of small-scale agriculture in the development process. With such agronomic successes as the Green Revolution, it was realized that raising the productivity of the rural subsistence sector could actually be an important objective rather than a constraint in development policy.

The concept of dualism is also apparent in some early spatial development models, focusing on the different qualities and potential of contrasting regions, rather than economic sectors as in the Lewis model. While some would argue that the development of certain areas at the expense of others is likely to inhibit the growth of the economy as a whole, others regarded initial regional inequality as a prerequisite for eventual overall development. Both Gunnar Myrdal and Albert Hirschman, for example, advocated strategies of 'unbalanced growth'. Myrdal's 'cumulative causation' principle (1957) suggested that once particular regions have by virtue of some initial advantage moved ahead of others, new increments of activity and growth will tend to be concentrated in already expanding regions because of their derived advantages, rather than in other areas of the country. Thus, labour, capital and commodities move to growing regions, setting up so-called 'backwash effects' in the remaining regions which may lose their skilled and enterprising workers and much of their locally generated capital. However, Myrdal recognized that such less dynamic areas may benefit from centrifugal 'spread effects', in that by stimulating demand in other, particularly neighbouring regions, expansion in the growing areas may initiate economic growth elsewhere.

Hirschman (1958), working independently of Myrdal, followed similar thinking, proposing a strategy of 'unbalanced growth', and suggesting that the development of one or more regional centres of economic strength is essential for an economy to lift itself to higher income levels. He envisaged spatial interaction between growing 'Northern' and lagging 'Southern' regions in the shape of 'trickle-down' and 'polarization' effects, similar to Myrdal's spread and backwash effects. Keeble (1967) argued that Hirschman's model,

> far from assuming a cumulative causation mechanism, implies that if an imbalance between regions resulting from the dominance of polarization effects develops during earlier stages of growth, counter-balancing forces will in time come into operation to restore the situation to an equilibrium position. Such forces, chief of which is government economic policy, are not to be thought of as intensified trickling-down effects, but as a new element in the model, arising only at a late stage in development. Their inclusion, together with the exclusion of any cumulative mechanism, represents the model's chief structural differences from that of Myrdal.
>
> *(Keeble, 1967: 260)*

A significant policy implication of Hirschman's unbalanced growth model is that governments should not necessarily intervene to reduce inequalities, since the inevitable search for greater profits will lead to 'a spontaneous spin-off of growth-inducing industries to backward regions' (Potter *et al.*, 2008: 84).

The spatial models of Myrdal and Hirschman have strong parallels with the work of François Perroux and other French economists in the 1940s and 1950s, who pointed out that growth did not appear everywhere simultaneously, but instead is frequently located in a 'growth centre or pole' (*pôle de croissance*). In essence, the growth centre model depicts the transmission of economic prosperity from a centre, most commonly an urban-industrial area, as a result of the interplay of spread and backwash effects. The model singles out crucial variables in the development of spatial variation in economic prosperity within a region and specifies how they operate. A particular 'growth industry', such as motor manufacturing, is likely to attract other linked industries, such as those which supply it with inputs and/or derive their inputs from it. Other agglomeration economies may encourage further growth, whilst technological change is encouraged through close proximity and interaction between the various industrial enterprises.

Unilinear models

Much post-war development thinking was strongly 'Eurocentric' in that, often inappropriately, 'theories and models [were] rooted in Western economic history and consequently structured by that unique, although historically important, experience' (Hettne, 1995: 21). Walt Rostow's 'uni-linear' model (1960; see Figure 2.4.1) is probably the best-known attempt to show how a country's economy and society progress through a series of stages, and is firmly based on the Euro-American experience. It was undoubtedly the most influential modernization theory to emerge in the early 1960s. It is interesting to note that Rostow entitled his book *The Stages of Economic Growth: A Non-communist Manifesto* and, '[his] perception of the purpose of the United States' promotion of economic development in the Third World was governed by a strongly anti-communist stance' (Hunt, 1989: 96). Indeed, early in his book Rostow asserts that he is aiming to provide 'an alternative to Karl Marx's theory of modern history' (Rostow, 1960: 2). The key element in Rostow's thinking was the process of capital formation, represented by five stages through which all countries pass in the process of economic growth.

- *Stage 1, Traditional society*: Characterized by primitive technology, hierarchical social structures, production and trade based on custom and barter, as in pre-seventeenth-century Britain.
- *Stage 2, Preconditions for take-off*: With improved technology and transport, increased trade and investment, economically based elites and more centralized national states gradually emerged. Economic progress was assisted by education, entrepreneurship and institutions capable of mobilizing capital. Often traditional society persisted side by side with modern economic activities, as in seventeenth- and eighteenth-century Britain, when the so-called 'agricultural revolution' and world exploration (leading to increased trade) were gaining momentum. While the preconditions for take-off were actually endogenous in Britain, elsewhere they were probably the result of 'external intrusion by more advanced societies' (Rostow, 1960: 6).
- *Stage 3, Take-off*: The most important stage, covering a few decades, when the last obstacles to economic growth are removed. 'Take-off' is characterized by rapid economic growth, more sophisticated technology and considerable investment, particularly in man-

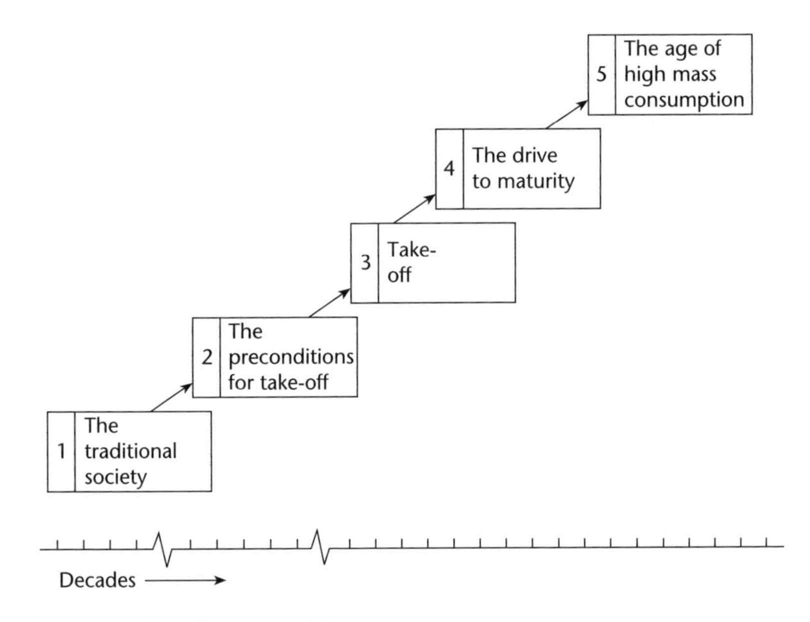

Figure 2.4.1 Rostow's unilinear model

ufacturing industry. The share of net investment and saving in national income rise from 5 per cent to 10 per cent or more, resulting in a process of industrialization, as in early nineteenth-century Britain. Agriculture becomes increasingly commercialized and more productive with increasing demand from growing urban centres.

- *Stage 4, Drive to maturity*: A period of self-sustaining growth, with increasing investment of between 10 and 20 per cent of national income. Technology becomes more sophisticated, there is greater diversification in the industrial and agricultural sectors and falling imports, as in late nineteenth- and early twentieth-century Britain.
- *Stage 5, Age of high mass consumption*: The final stage, characterized by the increasing importance of consumer goods and services, and the rise of the welfare state. In Britain and Western Europe, this stage was not reached until after the Second World War (post-1945), but in the USA mass production and consumption of consumer goods, such as cars, fridges and washing machines, came earlier, during the 1920s and 1930s.

Despite its considerable influence on development planning at the time, Rostow's model has been strongly criticized for a number of reasons. First, it is a 'unilinear' model, implying that 'things get better' over time, which is by no means always true as, for example, the experience of many sub-Saharan African countries indicates. Increases in per capita income have scarcely kept pace with world trends and the HIV/AIDS pandemic has had a devastating effect on mortality and life expectancy rates. Most sub-Saharan African countries are relatively worse off in the early twenty-first century than in the 1960s when many gained their independence. Second, it is a 'Eurocentric' model, suggesting that all countries will imitate the experience of Europe and America. It is quite inappropriate to apply such a model to countries which have been subjected to colonial rule and whose economies and societies have been manipulated to serve the demand for agricultural and mineral resources from the growing manufacturing sectors in the metropolitan countries. Third, the model suggests that all countries progress through these stages in the same sequence as happened in Europe and North America. But in some developing countries the sequence of events has not been so straightforward, with rapid change, for example, in the agricultural, industrial and service sectors happening at the same time, rather than sequentially. Whilst modern consumer goods, schools and hospitals, may be present in towns and cities, in remote rural areas these facilities are frequently absent, and poor farmers still use simple technology to produce food for their families. Finally, it is often wrongly seen as a 'development' model, whereas it is actually an 'economic growth' model. Rostow was concerned more with economic progress and increasing industrial investment, rather than human welfare and other non-economic indicators of development. Some countries have experienced periods of rapid economic growth, yet much of the population has felt little benefit from this – what might be called 'growth without development' (Binns, 1994; Binns *et al.*, 2012). The real significance of the Rostow model was that it seemed to offer every country an equal chance to develop.

From dualism to basic needs

The lack of distinction and explanation drawn by Rostow and others between the processes of 'growth' and 'development' led some writers to try to clarify the situation. There was also growing concern that economic growth, which had been the main preoccupation of Lewis, Hirschman, Myrdal and Rostow, did not necessarily eliminate poverty, and that the so-called 'trickle-down' effects of growth generally failed to benefit the poor in both spatial and social terms. Dudley Seers provided much-needed clarification on the meaning of development, suggesting that poverty, unemployment and inequality should be key foci in the development debate and that there should be greater concern for the fulfilment of basic needs (notably food, health and education) through the development process (Seers, 1969, 1972). The basic needs approach gained momentum in the mid-1970s. The International Labour Organization's 1976 conference on World Employment adopted the 'Declaration of Principles and Programme of Action for a Basic Needs Strategy of Development', highlighting poverty alleviation as a key objective for all countries in the period up to the year 2000. Possibly the main weakness of the basic needs strategy was its 'top-down' approach, 'which made it vulnerable to changing fashions in the international aid bureaucracy' (Hettne, 1995: 180). In spite of such limitations, the debates surrounding the meaning and process of development and the question of basic needs did much to move development thinking and policy away from earlier dualistic, unilinear, and essentially Eurocentric, approaches of the 1950s and 1960s.

Bibliography

Binns, T. (1994) *Tropical Africa*, London: Routledge.
Binns, T., Dixon, A. and Nel. E. L. (2012) *Africa: Diversity and Development*, London: Routledge.
Hettne, B. (1995) *Development Theory and the Three Worlds: Towards an International Political Economy of Development*, London: Longman.
Hirschman, A. O. (1958) *The Strategy of Economic Development*, New Haven, CT: Yale University Press.
Hunt, D. (1989) *Economic Theories of Development*, London: Harvester Wheatsheaf.
Keeble, D. E. (1967) 'Models of economic development', in R. J. Chorley and P. Haggett (eds) *Socio-economic Models in Geography*, London: Methuen, pp. 243–305.
Lewis, W. A. (1954) 'Economic development with unlimited supplies of labour', *The Manchester School of Economic and Social Studies* 22(2), May; reprinted in A. Agarwala and S. Singh (eds) (1958) *The Economics of Underdevelopment*, Oxford: Oxford University Press, pp. 400–49.
Marshall, A. (1890) *Principles of Economics* (8th edn), London: Macmillan (reprinted 1920).
Myrdal, G. (1957) *Economic Theory and Underdeveloped Regions*, London: Duckworth.
Potter, R. B., Binns, T., Elliott, J. A. and Smith, D. (2008) *Geographies of Development*, London: Pearson.
Public Papers of the Presidents of the United States (1964) *Harry S. Truman, Year 1949*, 5, Washington DC: United States Government Printing Office.
Rapley, J. (1996) *Understanding Development: Theory and Practice in the Third World*, London: UCL Press.
Rist, G. (1997) *The History of Development*, London: Zed Books.
Rostow, W. (1960) *The Stages of Economic Growth: A Non-communist Manifesto*, Cambridge: Cambridge University Press.
Seers, D. (1969) 'The meaning of development', *International Development Review* 11(4): 2–6.
Seers, D. (1972) 'What are we trying to measure?' *Journal of Development Studies* 8(3): 21–36.
Smith, A. (1776) *The Wealth of Nations* (2 vols), London: Methuen (reprinted 1961).
Willis, K. (2011) *Theories and Practices of Development*, London: Routledge.

Further reading

For detailed consideration of development theory, see Hettne (1995) and Hunt (1989). Keeble's chapter (1967) in Chorley and Haggett's *Socio-economic Models in Geography*, though written over 30 years ago, is still helpful. A more recently written overview is provided in Chapter 3 of Potter *et al.*'s *Geographies of Development* (2008). Hirschman (1958), Lewis (1954), Rostow (1960) and Smith (1961) are justifiably regarded as 'classic' texts, whilst Alfred Marshall's *Principles of Economics* was a key undergraduate textbook for over 50 years. Willis (2011) provides a useful introduction to theory and practice in development. Binns *et al.* (2012) consider various aspects of development theory and practice in the context of Africa.

2.5

Neoliberalism

Globalization's neoconservative enforcer of austerity

Dennis Conway

Following the global recession of 1978–1983, "a concerted, long-term and highly effective ideological effort on the part of identifiable actors" (George 1999: 1) which we now call *neoliberalism*, brought about a dramatic turn away from Keynesian economic thinking and political-economic practices just about everywhere in the globalizing world (Conway and Heynen 2006). Harvey (2005: 3) succinctly depicts the pervasiveness of its ascendency:

> Deregulation, privatization, and the withdrawal of the state from many areas of social provision have been all too common. Almost all states, from those newly minted after the collapse of the Soviet Union to old-style social democracies and welfare states such as New Zealand and Sweden, have embraced, sometimes voluntarily and in other instances in response to coercive pressures, some version of neoliberal theory and adjusted at least some policies and practices accordingly . . . Neoliberalism has, in short, become hegemonic as a mode of discourse.

More severe in his condemnation, Bourdieu (1998) ridicules neoliberalism as a free market system built upon the *structural violence* of unemployment, of the insecurity of job tenure and the menace of the lay-off. A new global class of privileged elites – characterized in contemporary American political discourse as the "1%," have been the project's beneficiaries these past thirty years or so. Furthermore, this global political-economic "right-wing," or "conservative" project has not only perpetuated previous inequalities, but exacerbated the global divide. To Conway and Heynen (2006: 20), the poor and "new poor" of the peripheral global South are being made to suffer through another round of the same bitter medicine they suffered under colonialism and postcolonialism; namely a neoliberal modernization version of the *dependistas* "development of underdevelopment." They assess neoliberalism's destructive, disciplinary assault this way:

> The common collective interest and the public good has been negotiated away by ideological, political, social and economic power-plays, which privilege individual accumulation and self-interest among internal elites over communal obligation and societal responsibility for ones' fellow human beings; neighbors, citizens, guests, alike.

The global majority (labor and dependents together), at the same time, are being duped, co-opted and coerced by the power and persuasion of neoliberalism's theological message that "the Market is God" (Cox 1999). And George (1999: 2–3) insightfully depicted neoliberalism's ascendency:

> [T]he neoliberals and their funders have created a huge international network of foundations, institutes, research centers, publications, scholars, writers and public relations hacks to develop, package and push their ideas and doctrine relentlessly . . .
>
> So, from a small, unpopular sect with virtually no influence, neoliberalism has become the major world religion with its dogmatic doctrine, its priesthood, its law-giving institutions and perhaps most important of all, its hell for heathen and sinners who dare to contest the revealed truth.

Summarizing this economic "instigator and partner" to contemporary globalization, neoliberalism is essentially about making trade between nations easier for the most powerful. It is all about the free movement of goods, commodities, resources, and commercial enterprises so that cheaper resources can be accessed to maximize profits and efficiency. To help accomplish this, neoliberalism requires the removal of various controls deemed as barriers to free trade, such as tariffs, regulations, certain standards, laws, legislation, and regulatory measures. Most importantly, it requires the removal of restrictions on capital flows and investment across national boundaries, so that global South markets and resource stocks can be respectively penetrated and exploited by global North capitalist corporations, or transnational corporations either independently of their client-nations, or in partnership with them. And, the major "theological faith" of neoliberalism advocates the following:

- The rule of the market – freedom for capital, goods, and services, where the market is self-regulating allowing the "trickle down" notion of wealth distribution. It also includes the de-unionizing of labor forces and removals of any impediments to capital mobility, such as regulations. The freedom is from the state, or government.
- Reducing public expenditure for social services, such as health and education, by the government.
- Deregulation, to allow market forces to act as a self-regulating mechanism.
- Privatization of public goods, resources and services (ranging from water, power, transportation to information dissemination and exchange, internet-use, communication).
- Changing perceptions of public and community good to individualism and individual responsibility.

Neoliberalism's ascendency

English economist, Adam Smith's 1776 *Wealth of Nations* text was the exemplary benchmark of the first "liberal" economic model, which promoted a free market ideology with no government restrictions on manufacturing production and no tariff barriers to trade and commerce. Government intervention in economic matters should be supportive of such commercial entrepreneurialism, not regulatory. And, imperial Great Britain certainly practiced this economic liberalism to great effect, as it expanded its global reach beyond its colonies into Latin America, East Asia, and beyond.

The 1930s' Great Depression, however, exposed this ideological model's shortcomings, so that in accordance with the structural prerogatives such a crisis in capitalism brings, a new national economic orthodoxy would come to the fore, labeled "Keynesianism" after its economic author, John Maynard Keynes. Accordingly, during the post-World War II long wave of advanced capitalism – from the 1950s to the 1970s – Keynesianism, in contrast to liberalism, was the dominant economic orthodoxy. It mandated a much more central role for government (and central bank) intervention and involvement, and furthermore argued that full-employment was necessary for capitalism to grow and for people to prosper. The belief that the government should intervene where the private market was loathed to go, subsidize capital, provide public welfare services and support a social safety net for the citizenry at large, was a dramatic pendulum swing in economic thinking and practice. Both in advanced capitalist nations of the global North – Europe, Japan, and North America and in these core countries' global South peripheries – colonies and postcolonial dependencies – these state-interventionist and regulatory ideas greatly influenced political and economic agendas during the post-World War II period of advanced capitalism from the 1950s to the end of the 1970s.

At the same time, and predictably, this social-democratic model of advanced capitalism experienced its own economic contradictions, structural limitations and resultant financial crises in which "galloping inflation" was targeted as a main reason for the recessional crisis of the late 1970s and thereby precipitated its fall from grace. International events and international affairs starting in the early 1970s brought this long wave's crisis to a head and helped precipitate its "recessional" conclusion. There was the unraveling of the 1948 Bretton-Woods currency agreement of fixed exchange rates when, in response to the burgeoning trading of Euro-dollars, President Nixon took the US dollar off the gold standard in 1971 and major currencies became speculative commodities. In the major core countries of the global North, inflationary pressures, government overspending, high taxation rates, continued high military budgets, and general downturns in consumer confidence were some of the main features of this long wave's stagnation. Keynesianism – especially its mandate for widespread state intervention in economic matters – was discredited. Finally, two OPEC-driven oil price hikes in 1974–1975 and again in 1978–1979 effectively raised the price of a barrel of oil eight-fold, dramatically raising energy costs, and contributing to widespread indebtedness, particularly in developing countries of the global South.

Neoliberal institutions' answers to indebtedness – austerity

As agents of neoliberalism from their inception, the IMF and the World Bank have been the stalwart "enforcers" of neoliberal policies of austerity and fiscal servitude; notably their Structural Adjustment Programs (SAPs) and more recently their Poverty Reduction Programs. Immediate implementation of "stabilization measures" to reduce public spending, remove subsidies on basic foods and other local commodities, reduce the government's wage bill, plus devaluations of local currencies were IMF-imposed austerity measures. In addition, there should be a phased-in implementation of economic adjustment measures to open markets, remove tariffs and barriers, even bring about tax reductions for the local elites to encourage their entrepreneurialism. The World Bank and its "development look-alikes" – the regional development banks, Asian Development Bank, Inter-American Development Bank – added their institutional might through their "conditionalities" for development loans, so that these financial giants did their part in this "pact with the neoliberal devil," and continued to promote the opening up of countries to more free trade and corporate penetration.

Accordingly, from 1979/1980 onwards, "supply side economics" solutions to economic recovery – Reaganomics and Thatcherism, for example – were offered as alternatives to combat inflation, reduce government overspending, reduce public sector workforces, and roll back wages. Neoliberalism was to be the way forward. In Britain, Prime Minister Margaret Thatcher would jubilantly trumpet her convincing acronym, TINA, or "*T*here *Is* *N*o *A*lternative," in defense of her government's conservative economic policies and privatization plans. In the United States, Ronald Reagan's succession to the Presidency in 1981 also signaled the ascension of his brand of "pragmatic conservatism" and Washington's ideological right-turn away from Keynesianism and its state-intervention practices.

Neoliberalism as an ideological, right-wing discourse and narrative and as an unchallenged model of economic efficiency and capitalist enterprise prevailed through the 1980s and into the 1990s, as the era of neoliberal globalization appeared to have no end. Blind faith in the market was preached with a religious fervor that resonated well in the United States (Cox 1999). In this now sole-remaining Super Power (with the geopolitical demise of the USSR and its break-up), the Clinton administration's embracing of neoliberalism appeared to be a resounding political-economic success. The 1990s, for many in the US, were relatively promising. Credit and mortgages were too easy to acquire from a now de-regulated banking sector, and economic expansion in many productive sectors received the benefits of technological innovations and logistics development as the IT era brought computer technology into the reach of everyone but the impoverished. Mislead by superficial appearances of consumer-oriented largesse and easy credit, a "culture of contentment" appeared to embrace American society (Galbraith 1992). Forgotten in the mix, was the widening income inequality between the top 5 percent and the bottom half of the working population, including the youthful middle classes' growing disillusionment with ever achieving the "American Dream" in the US. Between 1992 and 2007 (during Presidents Clinton and Bush's administrations) the real income of the bottom 90 percent of US families rose by 13 percent, while for the top 400 families it rose by 399 percent during the same period. As evidence, the total *income* of the top 400 families – the "investor class" grew close to $140 billion in 2007 – the crest of this latest neoliberal capitalist long wave.

The "Great Recession" of 2008–2012: Neoliberal capitalism's "bust"

The largely unforeseen end of this neoliberal capitalist long wave would then prove the overly optimistic pundits wrong, as the housing bubble burst in America leading the rest of the globalizing world into a major, precipitous economic downturn, not seen (or experienced) since the 1930s' "Great Depression" (Stiglitz 2010). Though the collapse of Lehman Brothers in September 2008 turned out to be the bellwether event that heralded the crisis, warning signs had been around since the summer of 2007. There were large current deficits in the US, UK, and many European economies – PIIGS (Portugal, Italy Ireland, Greece, and Spain) – being financed by the excess savings of emerging economies in the global South and oil producers (demonstrating an unsustainable global current-account imbalance). Monetary policy had been loosened, most notably in the US in the wake of that nation's mild downturn in 2001. And, by the time the crisis occurred in late 2007, there was an overall dearth of financial regulation accountability and oversight of financial institutions' speculative practices, fund managements, and derivative transactions. Careful, financial risk-taking had given ground to searches for yields and high-risk ventures, so that when US housing prices dropped nationally, this rapidly deteriorating situation in such a pivotal sector of the economy exposed liquidity, generated sub-prime loan defaults, caused credit markets to freeze and uncovered the extent of the global dispersal of derivative loans, so that the global financial system caved in soon thereafter (Verick 2010).

Concluding cautions – neoliberalism's tenacity

Driven aggressively by US geopolitical expediency and leadership as an "American project" to foster globalization (Agnew 2005), there has been an extensive refashioning of the US and global regulatory regimes such that the international mobility of finance capital is still by and large unfettered. The exponential expansion of the resultant dysfunctional global "financial-credit" economy has created debt peonage within advanced capitalist countries and mounting debt burdens internationally. International debt combined with absence of capital controls accentuated boom and bust cycles of debtor countries, such that recurrent international debt crises have occurred often, and are still with us today (Verick 2010).

Viewing the 2008–2012 global crisis and its aftermath from a progressive perspective, Peck *et al.* (2012) find neoliberalism has survived "this near-death experience, and some of its new strains seem even more aggressive." These progressive critics of neoliberalism's reactionary record of recurrent crises and structural re-organization prior to 2007, then pose two cautionary questions (Peck *et al.* 2012: 267):

> But what if the enduring contradictions of neoliberalism have, rather perversely, become *drivers* of this rolling program, which increasingly takes the form of an evolving pattern of (crisis-driven and crisis-exploiting) reregulation – at the same time reactively opportunistic and proactively experimental? What if the vulnerabilities and limits of neoliberalism ultimately account for its long-term tenacity as a regulatory (dis)order?

It appears that progressive alternatives to neoliberalism will be a hard sell in today's geopolitical discourses that have been ideologically created and distorted by several decades of cumulatively entrenched neoliberalization. Still trumpeting neoliberalism's suitability as TINA, a largely uncontested bundle of pro-market and pro-corporate rationalities of "soft capitalism" has become deeply intermixed with resilient re-formulations of social, corporate-financial, and state power (Conway 2012; Peck *et al.* 2012), that "brook no opposition." Harvey's (2005: 3) view of neoliberalism's devastating structural power as "creative destruction" of everything from "prior institutional powers, divisions of labor, welfare provisions, ways of life and thought, reproductive activities, attachments to land and habits of the heart" has turned out to be prophetic, unfortunately. Austerity for the impoverished majority at the bottom of the global divide appears to be the price to pay to ensure that the financial interests of the top 1 percent – the investor classes, neoconservatives, neoliberal corporate elites and their acolytes – remain favored in geo-economic and geopolitical negotiations about global futures.

Bibliography

Agnew, J. (2005) *Hegemony: The New Shape of Global Power*, Philadelphia, PA: Temple University Press.

Bourdieu, P. (1998) "The essence of neoliberalism," *Le Monde Diplomatique*, December.

Conway, D. (2012) "Neoliberalism and globalization." In R. Potter, D. Conway, R. Evans, and S. Lloyd-Evans (eds) *Key Concepts in Development Geography*, London: Sage, pp. 82–91.

Conway, D. and Heynen, N. (2006) *Globalization's Contradictions: Geographies of Discipline, Destruction and Transformation*, New York: Routledge, pp. 17–34.

Cox, H. (1999) "The market as god: Living in the new dispensation," *Atlantic Monthly*, March: 18–23.

Galbraith, J. K. (1992) *The Culture of Contentment*, Boston: Houghton Mifflin.

George, S. (1999) *A Short History of Neoliberalism*. Paper presented at the Conference on Economic Sovereignty in a Globalising World, March 24–26: Global Policy Forum; available online at www.globalexchange.org/resources/econ101/neoliberalismhist.

Hardt, M. (2004) *Multitude: War and Democracy in the Age of Empire*, New York: Penguin Press.
Harvey, D. (2005) *A Brief History of Neoliberalism*, Oxford: Oxford University Press.
Mahmud, T. (2011) "Is it Greek or déjà vu all over again? Neoliberalism and winners and losers of international debt crises," *Loyola University Chicago Law Journal*, 42: 630–712.
New American Century Report (2000) *Rebuilding America's Defenses: Strategy, Forces and Resources for a New Century*, Washington, DC: Project for the New American Century.
Peck, J., Theodore, N., and Brenner, N. (2012) "Neoliberalism resurgent? Market rule after the Great Recession," *The South Atlantic Quarterly*, 111(2): 265–288.
Stiglitz, J. E. (2010) *Free Fall: America, Free Markets and the Sinking of the World Economy*, New York: W. W. Norton.
Verick, S. (2010) *The Great Recession of 2008–2009: Causes, Consequences and Policy Responses*, Bonn, Germany: Institute for the Study of Labor IZA; Discussion Paper, No. 4934, May.

2.6

Dependency theories

From ECLA to Andre Gunder Frank and beyond

Dennis Conway and Nikolas Heynen

Dependency Theory, more than a theoretical construct, is a way of understanding historically embedded, political-economic relations of peripheral capitalist countries, especially Latin American countries, within the broader context of the global economy. It is, essentially, a *critique* of the development paths, policies, and strategies followed in Latin America and elsewhere in the peripheral global South. *Dependency Theory* emerged as a critical lens through which the history of Latin American development, marginalized as it was by Western hegemony, could be better understood; the "development of underdevelopment," no less. The initial theorization was a *structuralist* perspective by economists who were associated with the United Nations Economic Commission for Latin America (ECLA). This was soon transformed, and informed, by more critical *dependency* notions and the spread of Marxist and neo-Marxist critiques of imperialism (Chilcote 1984).

Perhaps, one of *Dependency Theory's* most important characteristics is that it was a product of Latin American scholarship (much of it written in Spanish) rather than Western or North American/European scholars. These authorities theorized on the Latin American condition as "insiders," as erstwhile, often passionate native sons. This gave rise to a more informed, and more involved, appreciation of the reasons for Latin American underdevelopment as *Dependistas* dealt with the context of various countries' specific national circumstances, and theorized about Latin America's structures of social organization and localized behaviors.

More widely, it was the publication of the writings of Andre Gunder Frank (and the collection and translation of other Latin American original contributions by North American Latin Americanists) that brought the *Dependency School's* ideas to the notice of North American and European development studies.

Prior to World War II, Latin American countries' economic strategies primarily revolved around a development path based on the export of natural resources and primary commodities to core countries. Many, including Argentinian Raúl Prebisch, Brazilians Paul Singer and Celso Furtado, and Chilean Osvaldo Sunkel, felt that Latin America's historical marginalization and resultant underdevelopment were perpetuated by such unequal commercial arrangements. While free-market notions of "comparative advantage" might suggest Latin America should benefit from providing their primary goods to the industrialized countries, Prebisch (1950) posited there were short-term fluctuations in the terms of trade in Latin American countries, deteriorations in the long-term and improved terms of trade in the advanced countries. Such *structuralist* assessments had core countries, particularly Britain and the United States, benefitting at Latin America's expense.

Consequently, Prebisch and other ECLA *structuralists* felt that major structural changes in development policy were needed to improve Latin America's economic situation. They proposed structural changes which favored switching to more domestic production under tariff protection as a means of replacing industrial imports. In line with this strategy, capital goods, intermediate products, and energy would be purchased with national income revenues from primary exports, and technology transfer would be negotiated with transnational corporations. This development strategy – often referred to as import-substitution industrialization (ISI) – became widely practiced throughout Latin America, the Caribbean, and the Third World/global South in general.

Although the ECLA *structuralist* analyses recognized some of the problems underlying Latin American underdevelopment, the proposed import-substitution industrialization (ISI) remedies brought other, more problematic, forms of dependency. Multinational and transnational corporate power and authority over technology transfer and capital investment emerged as a new form of neo-colonial dependency. Fernando Henrique Cardoso (1973) pointed this out in his assessments of power and authority in Brazil, and he preferred to characterize the situation in such peripheral economies as *associated dependent development*. Indeed, Cardoso felt that the dependent capitalist process of "industry-by-invitation" occurred mostly under authoritarian regimes, and further, that state policies would favor multinational capital at the expense of labor.

Prebisch's identification of core–periphery relations as the global historical heritage behind unequal development meant Latin America continued to face a formidable structural reality. Imperialism and colonialism were to be challenged more rigorously. Capitalism, or more specifically peripheral capitalism, was not the answer for Latin American development. Accordingly, alternative critical commentary, more deeply rooted within Marxist and neo-Marxist ideologies, emerged to better explain Latin America's subordinate place within the global economy and to better understand the processes that led to such exploitive *and dependent* relations. ECLA *structuralism* was recast in *dependencia* terms.

Baran's influential (1957) *Political Economy of Growth* described the reasons for Latin America's underdevelopment within a Marxist framework as being a consequence of advanced nations forming special partnerships with powerful elite classes in less developed or pre-capitalist countries of the global South. Such alliances were of course detrimental to the capitalist development of such "backward" economies since they benefitted the minority class of Latin American elites rather than advancing economic development at large. Such "partnerships," according to Baran, perpetuated the ability of core countries to maintain traditional systems of surplus extraction, thereby making domestic resources continuously available to them, and making the economic development of Latin American countries unlikely, since any surplus generated was appropriated by the elites. Thus, the imperial core countries would keep Latin America subordinate, and maintain their monopoly power, to ensure a steady outflow of cheap primary resources.

Andre Gunder Frank further developed Baran's ideas, by focusing upon the dependent character of peripheral Latin American economies. In Frank's (1966) prognosis, the "development of underdevelopment" was the concept which best characterized the capitalist dynamics that both developed the core countries and at the same time caused greater levels of underdevelopment and dependency within Latin American countries. Frank used this conceptual framework to explain the dualistic capitalist relations that had occurred, and which he felt would continue to occur between Latin American and core counties, as a result of the continued domination of these core countries in Latin America.

Although there was a popular perception that Third World countries regained some sense of self-determination following decolonization, Frank argued this was a fallacy. Exploitation of many Third World/global South countries by colonial and neo-colonial core countries intensified following their achievement of political "independence," further contributing to greater unequal relations. Thus, given the class-based stratification of Latin American society, which Baran blamed for the development of ties between Latin American elites and capitalist and political leaders from core countries, revolutionary action to remove such elites from power would be needed to forge a reformulation of international capitalist relations. Frank (1979) suggested this was only possible through revolutionary action which strove to install socialist ideals within the political systems of the dependent countries.

Besides arguing that the dependent core–periphery relationship was best articulated at the national scale, Frank also posited that a similar metropolis-satellite relationship occurred at smaller (regional) scales. In particular, he described similar dependent circumstances occurring between cities in Latin American countries and colonies and their non-urban peripheries. He illustrates this relationship within the context of the privilege that has always existed for colonial Latin American cities. As the place of administration for colonial powers, the city has always been the power-base from which the expansion of capitalism has spread. Within this more localized scenario, the city and its peripheral hinterland becomes increasingly polarized as a result of the capitalist relations between them, namely the metropolis exploiting its satellites. Given the localized nature of this relationship, dense networks of metropolis-satellite combinations form what Frank referred to as "constellations across national space."

As an explanation for Latin America's peripheral position in terms of modern versus traditional structures, Frank contended that this dualist perspective failed to truly comprehend the historical significance and transformative impact of capitalism's penetration of the continent's economic, political, and social structures. However, the dependent relationship Frank posited as a counter explanation to such dualist notions drew sharp criticism from many. Laclau's (1971) analysis is perhaps the most notable.

Laclau asserted that Frank's analytical method has significant shortcomings because it was based on an erroneous characterization of Marx's notion of modes of production. Instead of basing the construction of a mode of production on social or class relations, as Marx did, Laclau claims that Frank's reliance on market relations as the defining quality of the processes under which production occurs is inherently flawed. As a consequence, Laclau faults Frank for constructing a circular concept of capitalism which is inherently imbalanced. Laclau concludes that as a result of the flawed interpretation of the mode of production, Frank's analysis offers little more than an account of a history that is well reported; in effect, he contributes nothing to theoretical explanation in terms of determining conditions.

The resultant tensions within Frank's analytical framework as a result of arguably incorrect, or less than accurate, usage of Marxist ideology, led the way to other neo-Marxist investigations of the linkages and possible reconciliation between *Dependency Theory* and Marxism. Seeking to "resolve the debate," Chilcote (1984) effectively situated the various capitalist and socialist approaches to the "development of underdevelopment" – *structuralism, dependencia, internal colonialism, neo-Marxism,* even *Trotskyism* – as a full set of alternative theories and perspectives on development and underdevelopment. He also found a place for Wallerstein's more worldly focus in this collection of alternatives.

Indeed, Wallerstein (1974, 1980) adapted dependency notions to comment on the commercial relations between the core countries and Latin America, and examined world historiography in terms of the dominant and subordinate relations that successive emerging cores, their peripheries, and semi-peripheries experienced. This account started with the "long sixteenth century," passing through successive eras of capitalism to the present neoliberal era of globalization (the post-1980s). Wallerstein's "World Systems Theory" complements and expands upon Frank's ideas, providing a more comprehensive global stage appropriate for understanding the wider reach and more diverse spatial realignments of commercial capitalist relations in contemporary times. More recent "world systems" explanations of geopolitical eras detail the transformations of the world's hegemonic relationships of core–periphery relationships to the present global era that continues into the second decade of the twenty-first century (Conway and Heynen 2006). Amin (2003) also offers a much more critical view of contemporary geopolitical times than Wallerstein.

Ghosh (2001) further provided a contemporary critical appraisal and overview of contemporary thoughts on the full set of alternative dependency theories, pointing out the significant "inter-temporal paradigm shifts" in the theory's wider application in our rapidly globalizing world. As Ghosh (2001: 133) reminds us:

> There are indeed many issues and areas of development where dependency plays a major role. Some of these are; aid dependency, technological dependency, dependency for foreign capital investment, trade dependency, dependency for better human capital formation and so forth.

There are obvious connections between the divergent trajectories of capitalism's expansion in the global North as opposed to the global South. Equally obvious, "unequal competition" remains an extremely powerful, dependency relationship in globalization's transformative, disciplinary, and destructive influences (Conway and Heynen 2006). Just as the imperialism of old imposed colonialism fostered dependency and underdevelopment, modern globalization of the post-1980s has several salient features that are de facto, neoliberal successors to these imperial mechanisms. They represent: (a) a programme of binding individuals, institutions, and nations into a common set of market relationships; (b) a calculated economic strategy of the capitalist economies, corporations, and international financial institutional systems to encourage and stimulate capitalist growth for "winners" – core and emerging markets – not the "losers" with no comparative advantages, weak or failed states, or the corruption-weakened; and (c) a means of extracting surplus through the exploitation of cheap labor, high quality manpower, and resources of the global South (Ghosh 2001: 158).

"Dependency thinking" has come a long way since its initial Latin American interpretations, but even in today's "globalizing world" the geopolitical and geo-economic struggles underway in Latin America are anything but predictable, and can no longer be so easily framed in the centuries-old structural terms of core US hegemony and Latin American dependency. The evolving world system of core–periphery relationships has entered a new advanced phase of "modernity" in which there are new dependency relationships, ecological uncertainty, rapid technological change, and a multiplicity of cross-cutting flows of information, cultural messages, and knowledge exchange. They occur at multiple scales and scopes of influential power and authority – ranging from the global to the local, from the exceptional to the ordinary, and from the elites to the bourgeoisie and working classes.

Furthermore, and as a concluding recommendation, "dependency thinking" today requires us to confront the power hierarchies of the recent past (and present) using much more informed critical perspectives on the geo-economic power of transnational corporations, the emergence of BRICs in the global South, and the comparative declines of traditional cores' hegemonic authority in the global forum – the G20, the UN, and such (Amin 2003). Marxist theory may no longer be sufficient in and of itself to explain contemporary processes of global North and global South interdependencies, but the derivative critical perspectives drawn from such structural, neo-Marxist analysis by the likes of Amin, Wallerstein, and Frank can still help in the formulation of progressive alternatives for developing societies.

References

Amin, S. (2003) *Obsolescent Capitalism: Contemporary Politics and Global Disorder.* New York: Zed Books.

Baran, P. (1957) *The Political Economy of Growth.* New York: Monthly Review Press.

Cardoso, F. H. (1973) "On the characterization of authoritative regimes in Latin America." In A. Stepan (ed.) *The New Authoritarianism in Latin America.* Princeton, NJ: Princeton University Press, pp. 34–57.

Chilcote, R. H. (1984) *Theories of Development and Underdevelopment.* London: Westview Press.

Conway, D. and Heynen, N. (2006) *Globalization's Contradictions: Geographies of Discipline, Destruction and Transformation.* New York: Routledge.

Frank, A. G. (1966) "The Development of Underdevelopment." *Monthly Review* 18(4): 17–31.

Frank, A. G. (1979) *Dependent Accumulation and Underdevelopment.* New York: Monthly Review Press.

Ghosh, B. N. (2001) *Dependency Theory Revisited.* Burlington, VT: Ashgate.

Laclau, E. (1971) *Feudalism and Capitalism in Latin America.* London: New Left Review.

Prebisch, R. (1950) *The Economic Development of Latin America and its Principal Problems.* New York: United Nations Dept. of Economic Affairs.

Wallerstein, I. (1974) *The Modern World System. Volume 1: Capitalist Agriculture and the Origins of the European World-Economy in the Sixteenth Century.* New York: Academic Press.

Wallerstein, I. (1980) *The Modern World System. Volume 2: Mercantilism and Consolidation of the European World-Economy, 1600–1750.* New York: Academic Press.

Further reading

Blomstrom, M. and Hettne, B. (1984) *Development Theory in Transition: The Dependency Debate and Beyond – Third World Responses.* London: Zed Books.

Cardoso, F. H. (1979) "On the Characterization of Authoritarian Regimes in Latin America." In D. Collier (ed.) *The New Authoritarianism in Latin America.* Princeton, NJ: Princeton University Press, pp. 34–57.

Hettne, B. (1990) *Development Theory and the Three Worlds.* New York: John Wiley & Sons and Longman Scientific & Technical.

Kay, C. (1989) *Latin American Theories of Development and Underdevelopment.* New York: Routledge.

Palma, G. (1978) "Dependency: A Formal Theory of Underdevelopment or a Methodology for the Analysis of Concrete Situations of Underdevelopment?" *World Development* 14(3): 881–924.

2.7 The New World Group of dependency scholars

Reflections of a Caribbean avant-garde movement

Don D. Marshall

This chapter neither aspires to a chronology nor historical sequencing of events. Instead it retrospectively examines the rise and demise of an intellectual movement in the Anglophone Caribbean under the animating force of decolonisation. Allowance is made for a foray into the reasons behind the thwarted impulses of that age and the present decline of radical critique in the modern neoliberal period.

Introduction: Post-New World intellectual currents

Since the emergence of the New World movement in the early 1960s, it might be reasonable to expect that gathering forces in the international system – shaped by the imperatives of globalisation – would once more present the spectre of the emergence of vital new political forces. Then, as now, the region was thrown back into contemplation on the relevance of its development strategy. With the benefit of the backward glance, 'New World' was first founded in Georgetown towards the end of 1962 against the backdrop of a long general strike and growing racial conflict between African-Guyanese and Indian-Guyanese. The early founders aspired to invent an indigenous view of the region, convinced that the modernisation ideologies very much in vogue neither inhered a strategy for real, independent development nor an understanding of the political economy legacy of the Caribbean, of which more later.

Currently, Caribbean intellectuals in the main, particularly its social scientists, take on the colour of their historical environs: if neoliberal capitalism cannot be successfully challenged, then to all intents and purposes it does not exist; all that remains is the challenge of massaging a link between market liberalisation and populist statism. To be sure, this concern among Caribbean scholars and commentators does not preclude expression of despair in some quarters over the sustainability of the island-national project of the Caribbean. This forecast is based on an understanding of the export impetus girding contemporary capitalism and the difficulties associated with making the transition in political economies dominated by merchant capital.

Decolonisation and the rise of the New World

The New World movement in the Anglophone Caribbean was marked by an optimism of will and intellect. Newly independent governments were seen to be in pursuit of development guideposts to chart a self-reliant future. At the popular level, claims for social equality through redistribution became intensely salient as an expression of justice. And knowledge producers both within the academic and literary community, no longer under the heel of colonial power, focused energies either on transformative or ameliorative development agendas. Social dialogue and action seemed governed by an impulse towards West Indian self-definition manifested in discussions on race, class, culture and the question of ownership and control of the region's resources. The general decolonisation horizon within which such mood and thought moved was also marked by raging debates occurring in the academic world between modernisation theorists and neo-Marxist scholars. The New World Group made up of largely historians and social scientists would come to draw from, and intervene in, these debates, combining serious inquiry into the development possibilities under capitalism, with integrative, normative and programmatic thinking on nation building.

Considered by their pragmatic counterparts in government, media and academy as 'radicals', this cluster of writers and commentators across the Caribbean came to be known as the New World Group (NWG). Their thoughts and ideas on socialism, national self-determination and the delimiting horizons of capitalism reached a West Indian mass audience through public lecture series, various national fora, and newspapers and newsletters of their creation. The *New World*, a Jamaica-based magazine, first appeared in 1963 and was published fortnightly under the editorship of Lloyd Best with assistance from a host of University of the West Indies (UWI, Mona Campus) scholars, George Beckford, Owen Jefferson, Roy Augier, Derek Gordon, Don Robotham and Trevor Munroe, to list a few. From 1965, *New World* was published as a quarterly. Bearing the imprint of the UWI, the 'New World' would serve as a loose association attaching its name to anti-imperialist consciousness-raising activity across the region. Indeed NWGs were said to be formed in St. Vincent, St. Lucia, Washington DC, Montreal, St. Kitts, Trinidad, Barbados, Anguilla, Jamaica and Guyana. Other publications that appeared either as complements to or refinements of *New World*'s mission included *Moko* and *Tapia*, Trinidad based weekly newspapers appearing in 1969, *Abeng*, a Jamaican newsletter launched in the same year and the 1970 St. Lucia-based *Forum*.

The first issue of *New World Quarterly* (NWQ) focused on Guyana's development dilemma. The analysis therein moved beyond conventional state-centric explanations about the country's savings gap, low technologies, unskilled, undifferentiated labour markets and inadequate infrastructure. Guyana's and indeed the Caribbean's limited development, it was argued, was a function of the region's structural dependent linkages with Europe in terms of its value system and its economic relations. This point of view resonated with the dependency perspective first advanced by Paul Baran and subsequently extended by others who specialised in Latin American area studies. It was certainly a more assimilable 'angle' for Norman Girvan and Owen Jefferson to deploy in their doctoral theses explaining Jamaican underdevelopment (circa 1972) than the market-deficiency arguments of neoclassical proponents. As Girvan and Jefferson saw it, the move towards self-government and independence could not arrest the process of underdevelopment so long as the domestic economies remained dependent on foreign capital and terms of trade set under colonial rule.

Principally, the path of resistance for New World associates was forged out of opposition to Arthur Lewis' (1955) import-substitution industrialisation (ISI) model, favoured by Caribbean governments in the 1960s and 1970s. Briefly, the ISI programme required state provision of incentives to transnational enterprises in order to attract offshore industrial operations. The various budgetary and fiscal preparatory statements placed emphasis on the prospects for increased employment, technology transfer and stimulated markets for local inputs.

Beckford (1972), and Best and Levitt (1968) levelled a critique of Lewis' model that was representative of the dominant positions New World associates adopted on the question of Caribbean capitalist development. With epistemic insights drawn from orthodox Marxists and Latin American structuralists, their research fitted the growing canon of work seeking to establish dependency as the source of persistent underdevelopment. Beckford and others in the NWG would enrich this stock argument by anchoring the dependency concept within the plantation experience of Caribbean societies.

Dependency theory and plantation economy

Beckford's (1972) *Persistent Poverty* defined the historic plantation slave economy as a quintessential dependent economy, the units of which included Caribbean land, African unfree labour and European capital. This is Best and Levitt's (1968) 'pure plantation economy' as no other economic activity occurred outside the sugar plantation. Beckford's work was as much a repudiation of Caribbean development strategies, as it was a paradigmatic challenge to the liberal fallacy of 'progress'. For him, the mode of accumulation in the region remained a modified plantation economy variant, as dependent investment and aid ties with London and other metropolitan cities persisted. After lamenting the disarticulation between branch-plant production and the rest of the host economy, and the general mono-product character of local economies, Beckford and, later, Best and Levitt outlined other structural features of plantation economy which generated underdevelopment:

1. Land requirements of plantation production tended to restrict domestic food production.
2. Terms of trade often deteriorated as rising food and other imports presented balance of payment difficulties.
3. And stagnant educational levels tended to foreclose on product diversification options and improvements.

Havelock Brewster (1973), seized by the plantation economy argument, argued that foreign capital could not possibly champion industrialisation in accordance with common needs and the utilisation of the internal market. This was so, he surmised, because the gridlocked nature of a plantation economy with its lack of an internal dynamic, its reliance on outdated technologies and hierarchical management practices guaranteed for the region a subordinate role in its relationships with core firms and countries.

We may gather from this that unlike their dependency counterparts in Latin America, most New World associates relied less on external-determinist explanations to explain Caribbean underdevelopment. They focused on the *internal* workings of Caribbean economies to account for the region's structural dependency, even as they were careful to note that the characteristics of these economies extended back to colonial relations between Britain and the West Indies. *Dependentistas* and structuralists, on the other hand, placed the centre–periphery relations they depict within the context of macro-historical forces intent on locking peripheral societies into an unyielding spiral of exploitation and poverty.

Interestingly enough, Walter Rodney, a Guyanese historian, and Trevor Munroe, a Jamaican political scientist, could be said to have framed Caribbean development in such deterministic terms except that they singled out the social legacy of the plantation experience as especially debilitating for non-white races. Both were inspired by Marx's historical materialist method but Rodney was inclined to argue that nation building in the region had to be about renewing spirits, constructing grounds for black liberation and pursuing self-reliance.

Trevor Munroe's perspective was expressed in more classical but nuanced terms as he was mindful of the plantation slavery experience. As he would frame it, underdevelopment in the region was the predictable outcome of undeveloped class formation – itself partly perpetuated by those mix of domestic policies which threw the territories back on traditional activity and on traditional metropolitan dependence. The extent of the lag in technological, market, infrastructural and resource development will pose a challenge to aspirant Caribbean societies committed to constructing a capitalist economy.

Of the NWG, however, Best's dependency perspective evinced a deep-seated ambivalence towards Western discourses on development. Perhaps he was self-conscious of the post-colonial scholar's place in such literary transactions, of the dangers of succumbing to the neoclassical association between open economies and automatic economic growth. In the context of plantation economies, such assumptions muddled an already complex situation, Best argued. His dependency perspective was consistently embedded in extended and detailed analyses of ruling circles. Apart from providing address to the aforementioned features of neo-colonial dependency in the region, he singled out the shared outlook of Caribbean elites and Western development planners as a major brake on effecting meaningful socioeconomic transformation. Not surprisingly, his appeal was for a shift in the register of social consciousness on the part of the ruling elite. The colonial hangover apart, Best failed to draw sufficient attention to the degree of class conflict inherent in decolonisation as new class forces move to reorient the social system and the values that define that system.

The demise of the New World

As the 1970s dawned, the New World movement shuffled to a halt as division arose over strategies, tactics and modes of resistance to neo-colonialism. By this time, Best was especially critical of the group, decrying what he saw as New World's fatal attraction for governments, and a tendency to substitute policy-oriented research for contemplative scholarship. Increasingly, such knowledge products, he argued, amounted to exercises in self-justification, and as such were quite explicit disclosures of governmental discourse in action. He was also resistant to the idea that New World could move towards the formation of a political party or organisation contending for power. In a polemic entitled, 'Whither New World', Best (1968) spoke of the tensions of the group offering the following observation: 'There is among us, much unwitting intolerance, little cool formulation, hardly any attentive listening and even less effective communication.' Munroe would come to lament their facetious pursuit of class unity and vowed to distance himself from what he termed the 'bourgeois idealism' of New World.

The disintegration of the NWG was in part a result of the attention given by many to the immediate realms of the policy process. Mona-based economists, in particular, played key advisory roles in the Michael Manley Administration of the 1970s, while others across the region responded to appeals from governments for technical and project management assistance. But there are some scholars that instead place emphasis on the internal arguments between Best and others on the question of New World's relevance and its activist orientation. Their analysis, in my view, falls short precisely because they insufficiently recognise that New World, as any avant-garde movement, became compromised not so much by bourgeois acceptance as by *absorption into the intelligentsia.* Attendance to career, administration, and public service would spawn a culture marked by keynote address, cocktail attendance and doctoral authority. Consequently, the new radicals were to be found on the outskirts of black power movements, drawn less to its ideology as to the struggle for worker freedom and justice.

On a wider intellectual plane, New World could be said to suffer the slump it did largely because the dependency concept itself lacked lasting explanatory power. Overall, there was a circularity in the dependency argument: dependent countries are those that lack the capacity for autonomous growth and they lack this because their structures are dependent ones. Other scholars have also made the point about development in the world economy being in fact *dependent* development, pointing to foreign investment relationships between core states and firms. By the late 1970s, the emphasis among neo-Marxists shifted away from an independent weight placed on 'dependency' as undesirable, towards either a normative condemnation of state capitalism or an appeal to Third World states to *negotiate* the scope of their dependency.

Summary: Back to the future

If we posit that openings for dissent are as necessary to democracy as securing of consent, then Caribbean civil and uncivil society can continue to offer sites for objection and challenge. But there has been no New World equivalent emerging out of the tensions of the present neoliberal period. True the rise and influence of non-governmental organisations (NGOs), particularly women's organisations, trade unions and the galvanising work of the Caribbean Policy Development Centre along with that of critical scholars have served to exert pressure on increasing public transparency and inclusion. To be sure it is not at all clear that NGOs constitute an intrinsically virtuous force for the collective good. These can run a similar course to that of the New World. Beyond a certain point NGOs may lose the critical element that caused them into existence as they render services to governance agencies, take funds from them or 'cross over' to work for government institutions and organisations that they previously challenged. Currently, market mentalities predominate in government bureaucracies, business firms and in academia. From various nostrums, academicians from the UWI, particularly social scientists, are exhorted by media, business and government commentators to give advice and attention to the technicality of social control or constitutional and other reforms. In most issue spaces, ruling discourses of technocratic expertise seem to arbitrarily suppress alternative perspectives. The UWI's role in this is not entirely surprising as the university's struggle for relevance and its sensitivity to budget efficiency do make for a climate where conformity to the prevailing common sense seems the best course for research programming. Hegemony-affirming research thus continues to triumph. Political and intellectual challenges are foreclosed in the prevailing environment where priority of survival continues to be asserted both as an operating principle and as a rationale for the absence of radical critique. This is the 'bourgeois villainy' Best would speak of when the case was hardly self-evident among intellectuals of New World. The associates then at least managed a discussion of Caribbean dependency that was enriched by site characteristics of plantation production relations. This added colour to parallel debates in Latin America. For New World associates, the dependency concept had operative power; it encouraged an interesting entry point for challenging the colonial mode of accumulation. It also fashioned an intellectual *cachet* of dissent in the region, illuminating history and social fact as economic paradigms came under challenge.

Bibliography

Beckford, G. (1972) *Persistent Poverty: Underdevelopment in Plantation Economies of the Third World*, London: Maroon Publishing House and Zed Press.

Best, L. (1968) 'Forum: Whither New World', in *New World*, IV(1): 1–6.

Best, L. and K. Levitt (1968) 'Outlines of a Model of Pure Plantation Economy', in *Social and Economic Studies*, 17(3), September: 283–326.
Blomstrom, M. and B. Hettne (1984) *Development Theory in Transition*, London: Zed Books Ltd.
Brewster, H. (1973) 'Economic Dependence: A Quantitative Interpretation', in *Social and Economic Studies*, 22(1): 90–95.
Dookeran, W. (ed.) (1996) *Choices and Change: Reflections on the Caribbean*, Washington, DC: Inter-American Bank.
Hendriks, C. M. (2006) 'Integrated Deliberation: Reconciling Civil Society's Dual Role in Deliberative Democracy', in *Political Studies*, 54: 486–508.
Lewis, R. C. (1998) *Walter Rodney's Intellectual and Political Thought*, Detroit: The Press University of the West Indies and Wayne State University Press.
Lewis, W. A. (1955) *The Theory of Economic Growth*, London: Allen and Unwin.
Marshall, D. D. (2000) 'Academic Travails and a Crisis-of-Mission of UWI Social Sciences: From History and Critique to Anti-Politics', in G. Howe (ed.) *Higher Education in the Anglophone Caribbean: Past, Present, and Future Directions*, Mona: University of the West Indies Press, pp. 59–84.
Munroe, T. (1990) *Jamaican Politics: A Marxist Perspective in Transition*, Kingston and Boulder, Colorado: Heinemann Publishers (Caribbean Limited) and Lynne Rienner Publishers.
Rodney, W. (1972) *How Europe Underdeveloped Africa*, London: Penguin Books.

2.8

World-systems theory

Core, semi-peripheral, and peripheral regions

Thomas Klak

Definition

World-systems theory (WST) argues that any country's development conditions and prospects are primarily shaped by economic processes, commodities chains, divisions of labour, and geopolitical relationships operating at the global scale. World-systems theorists posit the existence of a single global economic system since at least the outset of European industrialization around 1780–90. According to WST doyen Immanuel Wallerstein and others, the global system dates back even further, to at least 1450, when international trade began to grow, and when Europe embarked on its 'age of discovery' and colonization (Frank and Gills, 1993). Contrary to much social science thinking, WST stresses the futility of a 'statist orientation' – that is, the attempt to analyse or generate development by focusing at the level of individual countries, each of which is profoundly shaped by world-system opportunities and constraints (Bair, 2005).

WST has identified a number of regularly occurring historical cycles associated with the level and quality of global business activity. These cycles account for economic booms and busts of various durations. The main economic periods for WST are *Kondratieff cycles*, named after the Russian economist who discovered them in the 1920s. Each cycle or *long wave* lasts about 50–60 years and represents a qualitatively different phase of global capitalism, not just a modification of the previous cycle. Kondratieff cycles are themselves divided into a period of expansion and stagnation. There is first an A-phase of upswing, economic expansion, and quasi-monopolistic profitability, fuelled by technological innovations and organized by new asymmetrical institutional rules. Price inflation increases during the A-phase. This then leads into a B-phase of increased competition, profit decline, economic slowdown, and price deflation. The profit squeeze towards the end of the B-phase motivates capitalists and policymakers to create new and innovative ways to accumulate capital. They work to shift investment out of established economic sectors, regulated environments, and production locations, and thereby create the conditions for a new Kondratieff cycle (Knox *et al.*, 2003).

The previous Kondratieff cycle began in the 1940s, expanded until 1967–73 (A-phase), and then contracted through the 1980s (B-phase). Each cycle's organizing institutions and rules are both economic and political. For that cold war cycle, key economic rules and structures included the US dollar as the global currency, and supranational bodies such as the World Bank, the IMF, and the G7. Political structures included the UN and the geopolitical divisions brokered at the Yalta conference. It divided Europe into US- and Russian-dominated zones, pitted global capitalism against Russian-led state socialism (communism), and presented the Third World as ideologically contested turf. The early twenty-first century found the world in a cycle shaped by the WTO, neoliberal free trade, and global financial liberalization aimed at ensuring quasi-monopolistic profitability and global power for core countries. As in the cold war cycle, the United States remains the pre-eminent core (and thus global) power, but its hegemony is now contested by other strengthening core countries and semi-peripheral countries, notably China. The global financial crisis since 2008 may signal the dawn of new long wave shaped by such factors as information technologies, resource scarcity, and climate change (Moody and Nogrady, 2010).

Scholars and disciplines influencing, and influenced by, WST

WST is deeply linked to its principal architect, Immanuel Wallerstein (born 1930). Indeed, few influential theoretical perspectives are so intertwined with one contemporary scholar. WST's conceptual roots are largely in Marxism. Wallerstein (1979) says that WST follows 'the spirit of Marx if not the letter'. Evidence of Marx's spirit includes WST's emphasis on class, the state, imperialism, and control over the means of production and labour power. WST's objections to classical Marxism include concern over a theoretical component known as *developmentalism*. This is the idea that societies move sequentially through feudalism, capitalism, and socialism to communism, and that they can be analysed and transformed individually and separately from the world system. WST's alternative view – that there has been for centuries but one world economy driven by capital accumulation – employs a concept of mode of production closer to that of Karl Polanyi than to Marx.

WST has much interdisciplinary relevance, and has therefore attracted both supporters and detractors from across the social sciences. WST complements political-economic analysis rooted in the traditions of *dependency theory* (Cardoso and Faletto, 1979), *uneven development* (Smith, 1984), and *dependent development* (Evans, 1979). A conceptually overlapping but perhaps less economistic and highly influential alternative to WST is *the regulation school*. Usually applied at a more local level than WST (i.e. to national or subnational systems), regulation theory seeks to identify phases of capitalism of variable length based on relations between a particular prevailing method of accumulating capital, and an associated 'mode of regulation', that is, a set of state regulations and behavioural norms (O'Hara, 2003).

The geography of WST: Three groups of nation-states

WST's temporal cycles of systemic integration, order, turbulence, transition and reconstitution of the global economy play out variably across geographical space. The world system is very unequal. Despite (or, world-system theorists argue, *because of*) several centuries of worldwide economic integration and trade, and more than sixty-five years of World Bank-led international development, global inequalities continue to rise, and at an increasing pace. The difference in per capita income separating the richest and poorest countries was 3 : 1 in 1820, 35 : 1 in 1950, 72 : 1 in 1992, 108 : 1 in 2004, and 384 : 1 in 2011 (UNDP, 1999, 2006, 2011). Within this highly unequal world order are place-specific dynamics. At times, regions can rise and fall in terms of power, development, and economic potential. WST describes this globally differentiated space with reference to nation-states, regional groupings thereof, and regions within nation-states. These fall into three categories (see Figure 2.8.1).

Scholars disagree over which variables best define a country's positions in the world system (Mahutga *et al.*, 2011). With this caveat in mind, general geographical features can be described. Countries of the *core* or *centre* are the sites of global economic (and especially industrial) control and wealth, and the associated political and military strength and influence. Core countries feature higher-skill, capital-intensive production. Politically, they collectively establish and enforce the rules of the global order and, through these advantages, appropriate surplus from non-core countries. The *semi-periphery* is positioned between the core's strengths and periphery's weaknesses. It mixes characteristics of the core (e.g. industry, export power, prosperity) and the periphery (e.g. poverty, primary product reliance, vulnerability to core decision-making). The semi-periphery is the most turbulent category, in that its members most frequently rise or fall in the global hierarchy. In semi-peripheral countries, there is much hope for development and joining the core countries, and narrow windows of opportunity to do so. But there are also intense interactions with core countries bent on fostering their own capital accumulation by maintaining the hierarchical status quo. The *periphery* is the backwater of the world system. It provides low-skill production and raw materials for industries elsewhere. It has poor living conditions and bleak development prospects. The semi-periphery versus periphery distinction for non-core regions is important. It avoids grouping such a heterogeneous set of countries with respect to development, industrialization, trade, resource control, and geopolitics. Still, putting the world's 200 countries into just three groups inevitably glosses over much intra-group heterogeneity. Note the regional clustering of countries in the three categories in the figure. At present the core is mainly North America, Western Europe, and Japan. The semi-periphery is essentially East Asia, Latin America's larger countries, and most of the former Soviet realm. The periphery is everything else, particularly Africa (Wallerstein's empirical focus).

A nation-state's position in the world system is historically *path dependent*, but not deterministically so. Nation-states can move between categories over time, depending on their accumulation regimes, development strategies, and international aid and alliances. Indeed, WST is quite useful for analysing the upward and downward movement of countries over time. There is no agreement over each country's categorization, depending on the defining characteristics and their interpretation. In addition, relative positions *within* each of the three categories can also shift over time.

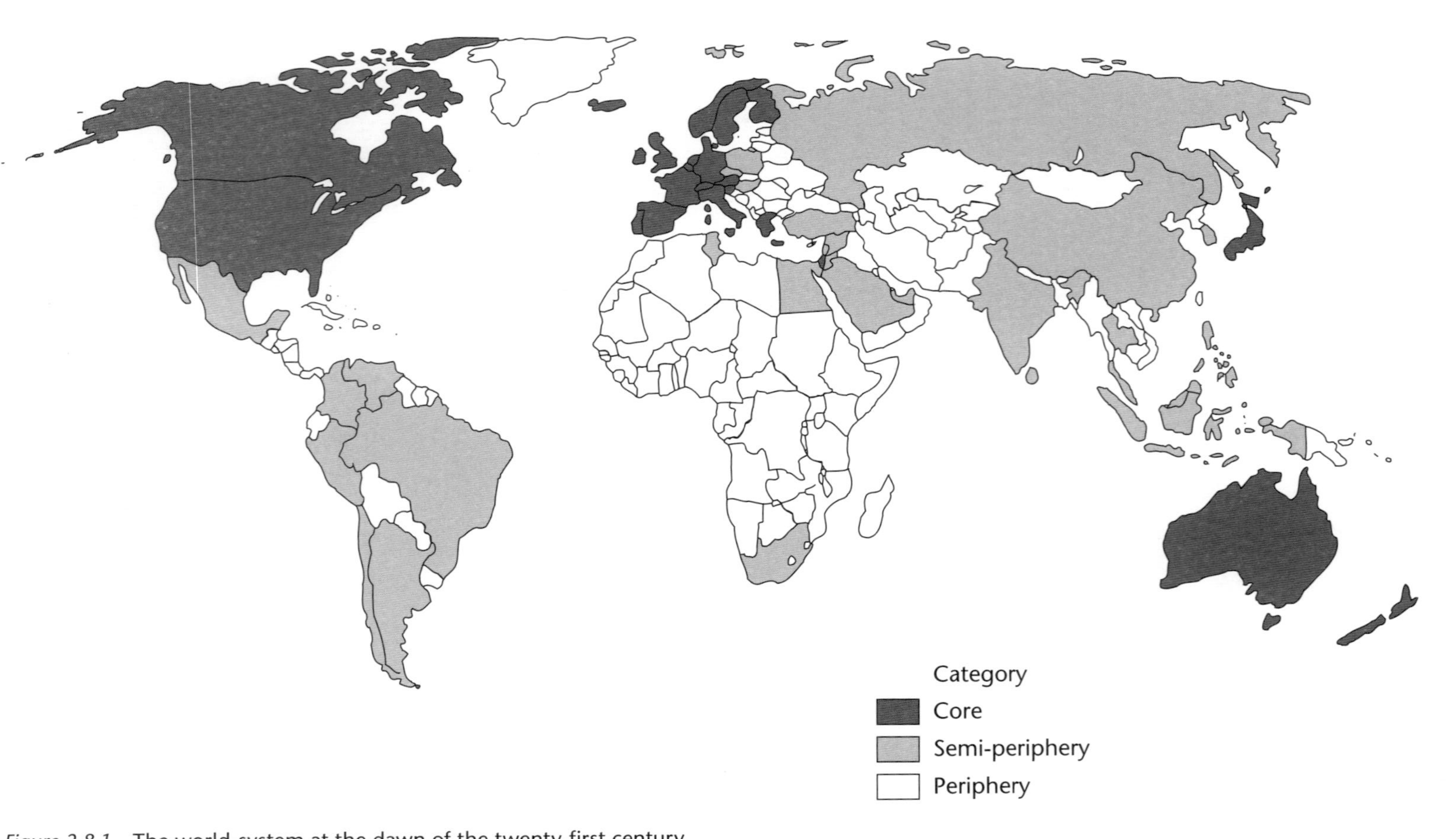

Figure 2.8.1 The world-system at the dawn of the twenty-first century

Notes: For an explanation of the country-level classification system shown in this figure, please see Gwynne *et al.* 2003.

East Asia illustrates the semi-periphery's potential and turbulence. Following massive US aid and industrial export growth in recent decades, South Korea has recently been knocking on the core's gate, although it was set back considerably by the 1997 Asian financial crisis. Indonesia has traditionally been peripheral, but in recent decades it has arguably joined the semi-periphery. Its increased clout derives from economic growth based on industrial exports for Nike and others, large resource endowments including oil exports, and its status as the world's fourth most populous country (see Figure 2.8.1). China's industrial export boom and associated capital accumulation since the 1980s drove it into the semi-periphery. Now many Japanese and US leaders fear China's global resource hunger and ambitions as a soon-to-be core country (Zweig and Jianhai, 2005).

Criticisms of WST

One capitalist world economy, divided by Kondratieff cycles, since at least 1450?

Need we subscribe to WST's totalizing global history to employ it effectively to understand recent development? Compared to Wallerstein, few writers employing a WST framework are as deeply historical, and few treat economic activities during previous centuries in such a globally holistic way. Much work, for example, has been done to identify the evolving features of capitalism associated with five Kondratieff cycles extending back only to 1789. Many other WST-influenced scholars focus on the dynamics of contemporary capitalism. WST purists may reject these approaches as insufficiently historical.

While Kondratieff cycles have considerable historical and empirical support (Mandel, 1980), they remain controversial. Others have assembled evidence to cast doubt on the existence and significance of long waves, and to suggest instead that capitalism moves through phases of differing lengths, problems, and features (e.g. Maddison, 1991). As mentioned earlier, the regulation school is one alternative conceptualization of contemporary capitalist dynamics.

Metatheory?

Beyond the considerable empirical analysis of Kondratieff cycles and their associated production and technological features, many WST claims remain untested and are perhaps untestable. Most WST-influenced scholarship focuses on the contemporary global political economy and the lack of time series data limits testing. Further, how could the simple three-category spatial division of the world system be tested? WST-inspired writing tends to read like an open-ended analysis of unfolding world events. Critics can claim that this method allows one to find and fit the data anecdotally to the theory. Better to think of a world-system *approach*, *analysis*, or *perspective* than a world-system *theory*.

Neglect of the local?

Operating mainly at the global level and concerned with economic cycles over decades if not centuries, WST is too holistic to account for local dynamics. Indeed, WST underplays the generative role of local activities, initiatives, social movements, and people.

Conclusion

World-systems theory, with its keen sense of historical, cyclical, technological, and geographical patterns, has undoubtedly deepened our understanding of the global political economy.

It is a satisfying antidote to the reductionism, ahistoricism, and superficiality in most popular interpretations of economic change. WST's historical and holistic perspective and level-headedness serve to counter the recent hyperbole about the uniqueness of globalization and inevitability of neoliberalism.

In practice, many scholars employing a WST perspective downplay the details and measurement of the cycles of upswing and downswing in the global economy. They focus instead primarily on contemporary trends, and adopt a qualitative approach to understanding business cycles, global systemic change, and the associated realignments of geopolitical and economic power, constraints, and potential. Many economists and some WST purists would judge a more qualitative version of WST to be insufficiently rigorous and therefore theoretically deficient. WST defenders would counter that a more qualitative approach is suitable, given their aim to see the 'big picture' and to decipher and rectify contemporary economic and political institutions and options.

References

Bair, J. (2005) 'Global capitalism and commodity chains: Looking back, going forward'. *Competition and Change*, 9: 153–80.

Cardoso, F. and Faletto, E. (1979) *Dependency and Development in Latin America*, Berkeley: University of California Press.

Evans, P. (1979) *Dependent Development: The Alliance of Multinational, State, and Local Capital in Brazil*, Princeton, NJ: Princeton University Press.

Frank, A. G. and Gills, B. (eds) (1993) *The World System: Five Hundred Years or Five Thousand?* London: Routledge.

Gwynne, R., Klak, T. and Shaw, D. (2003) *Alternative Capitalisms: Geographies of 'Emerging Regions'*, London: Hodder Arnold.

Knox, P., Agnew, J. and McCarthy, L. (2003) *The Geography of the World Economy* (4th edn), London: Hodder Arnold.

Maddison, A. (1991) *Dynamic Forces in Capitalist Development*, Oxford: Oxford University Press.

Mahutga, M., Kown, R. and Grainger, G. (2011) 'Within-country inequality and the modern-world system: A theoretical reprise and empirical first step'. *Journal of World-Systems Research*, XVII: 279–307.

Mandel, E. (1980) *Long Waves of Capitalist Development*, Cambridge: Cambridge University Press.

Moody, J. and Nogrady, B. (2010) *Welcome to the Sixth Wave*, North Sydney: Random House Australia.

O'Hara, P. (2003) 'Deep recession and financial instability or a new long wave of economic growth for US capitalism? A regulation school approach'. *Review of Radical Political Economics*, 35: 18–43.

Smith, N. (1984) *Uneven Development: Nature, Capital, and the Production of Space*, New York: Blackwell.

UNDP (1999, 2006, 2011) *Human Development Report*, New York: Oxford University Press.

Wallerstein, I. (1979) *The Capitalist World-Economy*, Cambridge: Cambridge University Press.

Zweig, D. and Jianhai, B. (2005) 'China's global hunt for energy'. *Foreign Affairs*, September/October.

Further reading

Knox, P., Agnew, J. and McCarthy, L. (2008) *The Geography of the World Economy* (5th edn), London: Hodder Education. Couples WST with economic geography to explore the workings of the contemporary global economy.

Shannon, T. R. (1996) *An Introduction to the World-System Perspective* (2nd edn), Boulder: Westview Press. Useful overview of WST, endorsed by Wallerstein.

Wallerstein, I. (ed.) (2004) *The Modern World System in the Longue Durée*, Boulder, CO: Paradigm Publishers. Wide-ranging chapters by leading world-system scholars providing a useful overview of WST.

Wallerstein, I. (2004) *World-System Analysis: An Introduction*. Durham, NC: Duke University Press. Wallerstein's own introduction to the field.

Websites

Fernand Braudel Center for the Study of Economies, Historical Systems, and Civilizations, Binghamton University. Available online at www.binghamton.edu/fbc/. The Center aims to 'explain systematically and coherently what is fundamentally a single occurrence, the development of the modern world system'.

Journal of World-Systems Research. Available online at http://jwsr.ucr.edu/index.php. Free online journal devoted to WST.

Wallerstein, Immanuel. His official website, available online at www.iwallerstein.com/.

2.9

Indigenous knowledge and development

John Briggs

Interest in indigenous knowledge systems particularly developed during the 1980s, primarily in response to dissatisfaction with modernisation as a means of improving living standards for the majority of the population of the global South. Modernisation, through the diffusion of formal scientific and technical knowledges from the North to the global South, has been seen to be an effective way of eradicating poverty. Consequently, development has frequently been conceptualised as a fundamentally technical issue, driven by the dominant science discourses from Europe and North America. By the 1980s, however, it had become clear that this transfer had not been wholly successful in transforming the lives of many, and especially so in Africa.

Alternatives were sought, and in promoting local-level, even anti-development, approaches, Escobar (1995, 98) perhaps captures the spirit best when he writes: 'the remaking of development must start by examining local constructions, to the extent that they are the life and history of the people, that is, the conditions of and for change'. This highlights the importance of local-level histories, geographies and sociocultural constructs in understanding community level development, as well as the need for a more explicit acknowledgement of indigenous knowledge as a valid body of knowledge. Despite this, much current development thinking still reflects the dominance of formal science; development remains a technical challenge and the voices of the poor and dispossessed are still little heard. However, the challenge for a new vision remains, and there is an increasing sympathy for the view that 'there is now an explicit understanding among many promoters and practitioners that farmer participatory research has clear advantages for the development of appropriate, environmentally friendly and sustainable production systems' (Okali *et al*., 1994, 6).

The first major discussions of indigenous knowledge in development can be traced to a collection of papers in the IDS Bulletin in 1979 (see, for example, Howes, 1979). This was followed by an important landmark work edited by Brokensha *et al.* (1980). Richards (1985) took the debate forward with a study that showed how African farmers used their own knowledge systems as the basis for successful agricultural production. Interestingly, Richards's study raises the issue as to whether these local knowledge systems are complementary to formal science, or whether they are rather more radical alternatives, which better reflect the needs, aspirations and priorities of local people. Based on much of this pioneering work, indigenous knowledge has become increasingly important in discussions on sustainable development because of the ways in which such knowledge has apparently allowed people to live in harmony with nature while still being able to make a living. If indigenous knowledge has indeed become an accepted part of development, then this can be seen as 'a shift from the preoccupation with the centralised, technically oriented solutions of the past decades that failed to alter life prospects for a majority of the peasants and small farmers of the world' (Agrawal, 1995, 414).

A number of development agencies have been keen to try to deploy indigenous knowledge in their development strategies. The World Bank (1998: i), for example, argues that there is a need 'not only to help bring global knowledge to the developing countries, but also to learn about indigenous knowledge (IK) from these countries, paying particular attention to the knowledge base of the poor'. Although the broad thrust is very welcome, indigenous knowledge is still seen as little more than a list of easily identifiable, mostly technical and discrete knowledges. There is little sense of dealing with embedded knowledges as part of a wider economy and society. Indigenous knowledge in the World Bank's conceptualisation is not allowed to offer a fundamental challenge to development, but simply to offer the opportunity for some technical, place-specific solutions where indigenous knowledge can hopefully be integrated into World Bank-supported programmes. This can only come about once the validity of indigenous knowledge has been confirmed through the lens of formal science. Only then can indigenous knowledge be judged to be worthy of serious investigation and dissemination (Briggs and Sharp, 2004).

Herein, therefore, lies a fundamental problem in deploying indigenous knowledge in development, that of the tensions between formal Western science and indigenous knowledge, a people's science, a tension which may be referred to as the binary divide. Western science is rational, controlled, rigorous and universal; indigenous knowledge is irrational, imbued with folklore and too place-specific to offer any meaningful solution to underdevelopment. The danger with this position is that if modern Western science is located at one end of the development spectrum, indigenous knowledge is located at the other. It is, however, increasingly apparent that such polar extremes are in reality untenable, and there is greater sympathy for the view that indigenous knowledge represents a complementary, not competing, knowledge, and that it represents a sense of additionality (Reij *et al.*, 1996).

A problem, however, is that if an overdependence on modernisation approaches has failed to deliver significantly improved living standards for the bulk of the world's population over the last fifty years or so, then an overdependence on indigenous knowledge as an alternative, at the other extreme, may also fail to deliver meaningful development results. The tensions between the two knowledge systems have been exacerbated by the resistance of modernisation theorists and practitioners to using indigenous knowledge systems in development. For them, the problem of poverty is to be treated by technology transfer, by capital investment, and by the release of productive forces. The development agenda is defined in the corridors of power in the 'North', and, in this, the voice of the 'South' is largely unheard. For example, for many in the 'North', dryland areas have to be managed in a rational, technocratic manner, as befits fragile and vulnerable environments, using knowledges rooted firmly in Western science and technology. No worthwhile contribution can be made by the inhabitants of such areas themselves, as they have little meaningful to offer; indeed, left to their own ways, their management, if that is what it can be called, will only result in further degradation. The voice of the South is to be ignored as it has no worthwhile contribution to make. Proponents of indigenous knowledge argue that the indigenous knowledge of those people resident in particular places can be of equal, or even greater, value than more formal Western scientific knowledge. However, if this argument is followed through, then Western science loses its universal hegemonic position, a position of power, and becomes one of a range of competing and contested knowledge systems. Pretty (1994, 38) has observed that 'the trouble with normal science is that it gives credibility to opinion only when it is defined in scientific language, which may be inadequate for describing the complex and changing experiences of farmers and other actors in rural development'. Consequently, knowledge that is not rooted in Western science is still seen by many in the development community as flawed, other than in instances where straightforward and uncontroversial indigenous technical solutions can be incorporated into development practice.

An additional problem with indigenous knowledge in the context of development is its empirical nature and the extent to which it is place-specific, and hence not easily transferable over geographic space. Methods of indigenous soil and water management in particular have attracted considerable interest, as has research on medicinal plant use. There is now a better understanding of how local people make sense of soil properties and characteristics, using attributes such as colour and feel, rather than chemical factors that Western pedologists might employ. Ostberg (1995), in a fascinating study in Tanzania, for example, discusses how farmers talk about 'cool' land, which is good for cultivation, and then land which becomes 'tired', and then 'hot', and which should no longer be cultivated. Interesting as these micro-narratives are, there is still a sense of frustration among development practitioners as to how useful they are in the bigger scheme of things.

In a similar vein, the fact that indigenous knowledge is differentiated within communities makes it difficult to use in development. Although it would be attractive to generate the concept of a community knowledge, shared by all its members, in reality this rarely, if ever, exists, because such knowledge is fragmentary and is cut across by factors such as wealth, production priorities, household circumstances and so on. All these factors impact on an individual's access to knowledge and that individual's ability to deploy such knowledge. Various studies have also shown that there can be clear gender differences in indigenous knowledge acquisition and how such knowledge is deployed (Briggs *et al.*, 2003). With this kind of fragmentation and differentiation, it becomes a challenge for indigenous knowledge to be successfully and effectively deployed across a range of rural settings.

The power relations associated with indigenous knowledge are no less problematic, particularly at the local scale. Bluntly put, whose knowledge counts? There is a tendency among some to take the view that indigenous knowledge is an inherently 'good thing', but, of course, this need not be the case at all. An example of local meetings in Tanzania showed that it tended to be a small group of the same male voices that were constantly being heard (Cleaver, 1999). Under these circumstances, indigenous knowledge can become the product (and property) of only a small group of powerful individuals.

There has sometimes been a tendency to romanticise indigenous knowledge as a static, unchanging, pristine and untainted knowledge system. Hence, the trick becomes one of how to tease out these knowledges, which will then provide the key to a sustainable development. The danger with this approach is that it privileges indigenous knowledge in the same way that modernisation privileges Western science. However, it is clear that rural people in the global South are very open to a whole range of ideas, regardless of their origins, as long as they make economic sense and are culturally acceptable. The notion that indigenous knowledge is static, unchanging, pristine and untainted is very difficult to sustain; instead, it is fluid, dynamic and provisional. People will adopt and experiment with new ideas if they will improve their livelihoods, and so it may be that instead of the term indigenous knowledge, a better term might be local mediated knowledges, deliberately in the plural (Briggs, 2005).

There is also the issue of the contextualisation of indigenous knowledge. It does not exist separate from the society in which it is found, but is very much embedded in everyday practice, reflecting the economic, social, cultural and political characteristics of that village, society, and so on. This is particularly awkward for development because it makes the deployment of indigenous knowledge difficult over different geographic spaces, highlighting one of the key differences between indigenous knowledge and Western science. Whereas the former is deeply embedded within its context, the latter is separated, almost disembodied from its context, and is, therefore, presumably much more universally applicable. This line of reasoning leads inexorably to the conclusion that indigenous knowledge cannot be successfully developed into a development tool, because it has little relevance or applicability outside its immediate area.

The seeming inability of indigenous knowledge to be 'scaled up', and used beyond its immediate locality, has led to a sense of frustration, perhaps best summarised by Sillitoe (2010, 12), when he writes: '[After] two decades or so, the indigenous knowledge (IK) in development initiative has not, frankly, had the success some of us expected'. There is little doubt that there has been a plethora of locally based studies over the last two decades that demonstrate the strength and value of indigenous knowledge at the local level, and these have been important in their own right to validate indigenous knowledge in relation to scientific knowledge. However, there is now a view developing that the focus of indigenous knowledge research needs to turn from content more towards process, to focus on ways of knowing at the local level and to recognise such knowledges not as a tool, but more as a perspective on development at this local level (Berkes, 2009). This implies more nuanced understandings being developed of indigenous ways of knowing, of complex power relations associated with knowledge at the local level, and of how empowerment can be enhanced without the problems associated with participatory methods (Cooke and Kothari, 2001). This is without doubt a challenge, but one that needs to be addressed before indigenous knowledge becomes an important element of the development armoury.

References

Agrawal, A. (1995) Dismantling the divide between indigenous and scientific knowledge. *Development and Change*, 26, 413–439.

Berkes, F. (2009) Indigenous ways of knowing and the study of environmental change. *Journal of the Royal Society of New Zealand*, 39, 151–156.

Briggs, J. (2005) The use of indigenous knowledge in development: Problems and challenges. *Progress in Development Studies*, 5, 99–114.

Briggs, J. and Sharp, J. (2004) Indigenous knowledges and development: A postcolonial caution. *Third World Quarterly*, 25, 661–676.

Briggs, J., Sharp, J., Hamed, N. and Yacoub, H. (2003) Changing women's roles, changing environmental knowledges: Evidence from Upper Egypt. *Geographical Journal*, 169, 313–325.

Brokensha, D., Warren, D. and Werner, O. (1980) (eds) *Indigenous knowledge systems and development.* New York: University Press of America.

Cleaver, F. (1999) Paradoxes of participation: Questioning participatory approaches to development. *Journal of International Development*, 11, 597–612.

Cooke, B. and Kothari, U. (2001) (eds) *Participation: The new tyranny?* London: Zed Books.
Escobar, A. (1995) *Encountering development: The making and unmaking of the Third World*. Princeton, NJ: Princeton University Press.
Howes, M. (1979) The use of indigenous technical knowledge in development. *IDS Bulletin*, 10, 12–23.
Okali, C., Sumberg, J. and Farrington, J. (1994) *Farmer participatory research: Rhetoric and reality*. London: Intermediate Technology Publications/Overseas Development Institute.
Ostberg, W. (1995) *Land is coming up: The Burunge of central Tanzania and their environments*. Stockholm: Stockholm Studies in Anthropology.
Pretty, J. N. (1994) Alternative systems of enquiry for a sustainable agriculture. *IDS Bulletin*, 25, 37–48.
Reij, C., Scoones, I. and Toulmin, C. (1996) (eds) *Sustaining the soil: Indigenous soil and water conservation in Africa*. London: Earthscan.
Richards, P. (1985) *Indigenous agricultural revolution: Ecology and food production in West Africa*. London: Hutchinson.
Sillitoe, P. (2010) Trust in development: Some implications of knowing in indigenous knowledge. *Journal of the Royal Anthropological Institute*, 16, 12–30.
World Bank (1998) *Indigenous knowledge for development: A framework for action*. Knowledge and Learning Centre, Africa Region, World Bank. Available online at www.worldbank.org/afr/ik/ikrept.pdf (accessed 30 May 2003).

Further reading

Ellen, R., Parkes, P. and Bicker, A. (2000) (eds) *Indigenous environmental knowledge and its transformations: Critical anthropological perspectives*. Amsterdam: Harwood Academic Publishers.
Leach, M. and Mearns, R. (1996) (eds) *The lie of the land: Challenging received wisdom on the African environment*. London: International African Institute.
Pottier, J., Bicker, A. and Sillitoe, P. (2003) (eds) *Negotiating local knowledge: Power and identity in development*. London: Pluto Press.

2.10

Participatory development

Giles Mohan

Introduction

Over the past 30 years a wide range of organizations have started involving local people in their own development, so much so that it has become a new 'orthodoxy' (Cornwall 2002a). This chapter begins by looking at different definitions of participatory development and examines through what sorts of organization it is achieved. As there are many possible approaches, I have included case studies which demonstrate different facets of participation. This brings us on to a critique and an overview of where things seem to be heading, particularly linking participation to citizenship.

Participatory development in theory

The emergence of participatory development (PD) is tied into critiques of both theory and practice.

The emergence of participation

According to the strongest advocates of PD, 'normal' development is characterized by Eurocentrism, positivism, and top-downism which constitute 'epistemological disenfranchisement' (Connell, 2007: 107). The tendency is to equate development with the modernity achieved by 'Western' societies with the flipside that 'non-expert' local people were sidelined.

As it became apparent that programmes had yielded limited benefits, the volume of criticism grew. In the 1970s, radicals such as Paulo Freire (1970) advocated participatory action research which created new learning environments for people to express their needs and achieve development. Green (2010) shows how in the case of Tanzania, participation in development became a deep-rooted political culture in the post-independence period allied to experiments in African socialism. Even mainstream organizations like the World Bank pushed for basic needs which targeted marginalized groups. Added to this were academics, most notably Robert Chambers, who argued that 'putting the last first' was necessary for rural development.

Contested definitions

Participation is generally deemed 'a good thing', but it has multiple meanings, which makes it amenable to different interpretations and uses while appearing to be speaking about the same thing. Thus, Green (2010) terms it a 'boundary object', which creates 'the possibilities for groups with divergent perspectives and interests to enter into temporary collaborations around shared objects of management' (p. 1242). Therefore, defining these divergent meanings is important for assessing its possible (ab)uses and impacts. In terms of development, a key question is if people participate, what are they aiming to gain by participating? One view is about *efficiency and effectiveness* of 'formal' development programmes (Cornwall, 2002a). The goals of development are valid although the institutions are malfunctioning, but can be improved by involving the beneficiaries. Another view concerns *mutual learning*, in which participation entails understanding where others are coming from and, ideally, learning from one another to achieve a better outcome (Chambers, 1997). Others take this further in seeing participation as more *transformative* (Hickey and Mohan, 2005). That is, 'development' is flawed and only by valorizing other voices can meaningful social change occur. It is in this sense that the recent emphasis on participation as citizenship, which I discuss later, is aimed.

Despite these differences, there has been a growing acceptance regarding the importance of local involvement from both neoliberals and radicals, what Dagnino (2008) terms a 'perverse confluence'. Underlying this 'consensus' is the belief in not relying solely on the state. So, it is not accidental that PD gained popularity around the same time as the neoliberal counter-revolution of the 1980s with its discourse of self-help and individualism and has remained popular under the slightly more state-friendly 'inclusive liberalism' of the new millennium (Golooba-Mutebi and Hickey, 2010).

Powerful processes

It needs emphasizing that whichever approach to participation we adopt, PD is fundamentally about power (Nelson and Wright, 1995). Cornwall (2002b) usefully distinguishes between 'invited' and 'claimed' spaces of participation. Invited spaces are the more formal events where development agents create forums for stakeholders to contribute and, ideally, reach a consensus. By contrast, claimed spaces are more organic and involve the poor taking control of political processes, without necessarily being invited in. In practice, political struggle usually has elements of both invitation and claiming with subaltern agents resisting and subverting these political processes in creative ways (Shakya and Rankin, 2008).

Participatory development in practice

In this section I discuss the institutional arrangements involved in PD and the processes through which it attempts to change power relations.

Grassroots civil society

In rejecting the statism and top-downism of 'normal' development, the focus for PD has become the grassroots level which permits a plurality of developmental goals to be realized as well as giving communities the self-determination they need. Hence, PD has become associated with civil society. If state structures are bureaucratic and unaccountable, then civil society organizations are believed to be more accountable and hands-on. Although civil society has multiple meanings, it has largely been interpreted as the realm of non-governmental organizations (NGOs), with many southern-based ones relying on funding and institutional support from northern partners and increasingly states use NGOs as vehicles for certain forms of social development (Dagnino, 2008).

New knowledges

The first step in reversing the biases marginalizing the poor concerns rethinking knowledge generation. The expert systems of modernity relied upon scientific approaches so that the recipients of development were treated as passive. PD reverses this. The research methods for accessing local knowledges were inspired by Paulo Freire and have grown into a veritable industry (Chambers, 1997), but all centre upon trying to see the world from the point of view of those directly affected by the developmental intervention.

The most widely used methodology is participatory rural appraisal (PRA). As Chambers (1997: 103) explains:

> The essence of PRA is change and reversals – of role, behaviour, relationship and learning. Outsiders do not dominate and lecture; they facilitate, sit down, listen and learn. Outsiders do not transfer technology; they share methods which local people can use for their own appraisal, analysis, planning, action, monitoring and evaluation. Outsiders do not impose their reality; they encourage and enable local people to express their own.

PRA relies on many visual and oral techniques for generating knowledge because it is felt that the medium of written language is prejudicial to free expression. So, PD seeks out diversity rather than treating everybody as uniform objects of development.

Participation in action

So far I have outlined the theory of PD, but what happens when it is practised in the 'real' world? These brief case studies demonstrate different facets of PD. The Aga Khan Rural

Support Programme (India) has used participation to enhance the effectiveness of pre-determined projects. The participatory approach aimed at 'consensus-building' and 'to find a meeting ground to negotiate terms of collaboration' (Shah, 1997: 75). In a dam scheme the farmers were not given an option regarding water payments, but the participatory exercise helped reach mutually agreeable solutions. As Shah (1997: 77) concludes, this was 'certainly not true empowerment where villagers decide and prioritize development proposals with minimal external support and facilitation'. Shah suggests that while transformatory participation might be desirable it is rarely viable where external agents are time-bound and accountable to funders. But that is not to say they are dictatorial and that the lack of true empowerment detracts from real benefits. As Corbidge (2007: 201) argues, also based on Indian examples, 'good practical (indeed political) arguments can be made in favour of particular development policies that might seem reformist or hopelessly pragmatic'. Corbridge makes a plea for not basing our analysis and prescriptions on a normative 'high ground' of politics divorced from the messy realities of actually existing struggle; a theme I return to later.

A similar issue is raised where participatory approaches have been 'scaled up'. In the mid-1990s, the major donors initiated poverty reduction strategies (PRSs), which responded to the criticism of structural adjustment programmes for being imposed on countries. Instead, formulation of PRSs is to be 'owned' by the countries concerned, which means scaling up the invited spaces in which citizens and their representative organizations have a voice (Lazarus, 2008). However, these mass participation exercises are often piecemeal, late in the policy process, and involve only 'safe' civil society organizations who will not question the neoliberal logic of PRSs.

By contrast, Esteva and Prakash (1998) see the Mexican Zapatistas as a political force pushing for a different understanding of development through novel forms of participation. However, more low key and less combative approaches focus on civic engagement in urban service delivery through such things as school boards whereby participation in one institution has knock on effects that transform the process of local governance (Fung, 2004). Taken together, it becomes clear that these different uses of participation are not exclusive and means that in any given situation we need to be realistic and specific about the nature of participation that is either envisaged and/or possible.

The problems of participatory development

Having looked at these case studies it is worth drawing together some of the interrelated problems that have emerged with PD.

The first is tokenism. As PD has become popular, some agencies use the rhetoric of participation with limited empowerment. In many cases PRA has become so routinized that many agencies treat it as a 'rubber stamp' to prove their participatory credentials. In the PRS process, which champions participation and ownership, most representatives of civil society are in fact midde-class activists who are hand-picked to ensure agreement (Golooba-Mutebi and Hickey, 2010). Allied to this is that reliance on a toolkit approach to knowledge creation has tended to produce endless local studies, which should reflect local contexts, but end up looking very similar (Lange, 2010).

Second, much PD has treated communities as socially homogeneous (Robins *et al.*, 2008). While community empowerment might be an improvement on unresponsive bureaucracies, there have been cases where support for 'the community' has meant that resources have passed to elites.

Third, the emphasis on civil society can create competition and overlap between local organizations. With aid being channelled through such organizations, it is the better organized or more acceptable which capture resources, often those run by and for middle classes. The result is that weaker organizations or those more genuinely championing the poor are further undermined. Allied to this is that many 'partnerships' between northern and southern NGOs are heavily loaded in favour of the former. Such problems are repeated when the relationships are not NGO-to-NGO but state-to-NGO, for as Lange (2010) shows, participatory schemes in Tanzania established parallel political structures which by-passed local government actors and institutions who are, despite all the weaknesses of electoral democracy, accountable to local people.

Fourth is whether participation is an end in itself or also a means to an end. From a democratic perspective, simply being able to participate is a major achievement, but for the poor their lack of resources means that any participatory process must yield tangible benefits. Furthermore, as Brett (2003) warns, simply participating is meaningless unless there is some institutionalized accountability. He argues that we should focus on 'the nature of the institutional constraints that determine how much leverage users can exercise over agencies, whether these operate in the state, market or voluntary sector' (Brett, 2003: 18).

The final problem is broader and relates to the causes of underdevelopment. PD seeks to give local people control, but many processes affecting their (or our) lives are often not readily tackled at the local level. For example, it is very hard for a small cooperative in Africa to change the rules governing international trade when the World Trade Organization is dominated by the developed economies. The emphasis on grassroots society can leave important structures untouched and do nothing to strengthen states and make them more accountable to their citizens (Green, 2010).

Taken together these operational critiques of participation add up to a 'depoliticization' of the idea. It promises some quite radical approaches to development but ends up disconnecting genuine needs and political struggles from circumscribed interventions of development agents. It is, as Murray Li argues (cited in Green, 2010: 1251) about 'rendering technical' the conflictual realities of poverty and its amelioration.

Citizenship and the future of participatory development

It becomes clear that while PD has brought benefits to some communities it has been abused and does little to address extra-local processes. This recognition that development will involve broader questions of citizenship and sovereignty has been part of the 'inclusive liberalism', which sees agencies building the capacity of the state rather than bypassing it and empowering civil society (Golooba-Mutebi and Hickey, 2010). This involves bolstering citizenship.

This reframing of participation as citizenship gained ground from the turn of the millennium and situates PD in a broader range of sociopolitical practices, or expressions of agency (Gaventa, 2002), through which people extend their status and rights as members of particular political communities, thereby increasing their control over socioeconomic resources. This unites a 'liberal' theory of citizenship, stressing formal rights and political channels, with 'civic republican' approaches that emphasize the collective engagement of citizens in the determination of their community affairs. The focus here is on substantive rather than procedural forms of citizenship, a participatory notion that offers the prospect that citizenship can be claimed 'from below' through the efforts of the marginalized.

While the citizenship turn promised to break from the voluntarism of PD as discussed above, it too suffers from depoliticization and neoliberal co-option. In many cases citizenship action was reduced to being a consumer within a market and so undermined communal notions of rights and moralities (Dagnino, 2008). Like PD it tended to imagine a 'pure' world of citizenship where the poor could relatively straightforwardly secure access to state services (Robins *et al.*, 2008).

The reality for the poor in the global South is one of 'indeterminancy' and struggle which means that singular and theoretically pure forms of political practice are impossible. As Robins *et al.* (2008: 1079) argue 'In the scramble for livelihoods and security, poor people tend to adopt plural strategies; they occupy multiple spaces and draw on multiple political identities, discourses and social relationships, often simultaneously'. It is in these multiple practices that the poor can leverage gains from more formal participatory schemes as Corbidge (2007) and Golooba-Mutebi and Hickey (2010) note. In turn, this means we should not necessarily ditch PD based on a universal condemnation of its depoliticizing effects, but rather, as Lazarus (2008) notes, start with the political realities of poor people and not a normative ideal, however well-meaning.

References

Brett, E. A. (2003) 'Participation and accountability in development management', *Journal of Development Studies*, 40(2): 1–29.

Chambers, R. (1997) *Whose Reality Counts? Putting the First Last*, London: Intermediate Technology Publications.

Connell, R. (2007) *Southern Theory: The Global Dynamics of Knowledge in Social Science*, London: Polity Press.

Corbridge, S. (2007) 'The (im)possibility of development studies', *Economy and Society*, 36(2): 179–211.

Cornwall, A. (2002a) *Beneficiary, Consumer, Citizen: Perspectives on Participation for Poverty Reduction*, SIDA Studies no. 2, Stockholm: Swedish International Development Cooperation Agency.

Cornwall, A. (2002b) 'Making spaces, changing places: Situating participation in development', *IDS Working Paper 170*. Brighton: Institute of Development Studies.

Dagnino, E. (2008) 'Challenges to participation, citizenship and democracy: Perverse confluence and displacement of meanings', in A. J. Bebbington, S. Hickey, and D. Mitlin (eds) *Can NGOs Make a Difference? The Challenge of Development Alternative*, London: Zed Books, 55–70.

Esteva, G. and Prakash, M. (1998) *Grassroots Post-Modernism: Remaking the Soil of Cultures*, London: Zed Books.

Freire, P. (1970) *The Pedagogy of the Oppressed*, New York: The Seabury Press.

Fung, A. (2004) *Empowered Participation: Reinventing Urban Democracy*, Princeton, NJ: Princeton University Press.

Gaventa, J. (2002) 'Exploring citizenship, participation and accountability', *IDS Bulletin*, 33(2): 1–11.

Golooba-Mutebi, F. and Hickey, S. (2010) 'Governing chronic poverty under inclusive liberalism: The case of the Northern Uganda Social Action Fund', *Journal of Development Studies*, 46(7): 1216–1239.

Green, M. (2010) 'Making development agents: Participation as boundary object in international development', *Journal of Development Studies*, 46(7): 1240–1263.

Hickey, S. and Mohan, G. (2005) 'Relocating participation within a radical politics of development', *Development and Change*, 36(2): 237–262.

Lange, S. (2010) 'The depoliticisation of development and the democratisation of politics in Tanzania: Parallel structures as obstacles to delivering services to the poor', *Journal of Development Studies*, 44(8): 1122–1144.

Lazarus, J. (2008) 'Participation in poverty reduction strategy papers: Reviewing the past, assessing the present and predicting the future', *Third World Quarterly*, 29(6): 1205–1221.

Nelson, N. and Wright, S. (1995) 'Participation and power', in N. Nelson and S. Wright (eds) *Power and Participatory Development: Theory and Practice*, London: Intermediate Technology Publications, 1–18.

Robins, S., Cornwall, A. and von Lieres, B. (2008) 'Rethinking "citizenship" in the postcolony', *Third World Quarterly*, 29(6): 1069–1086.

Shah, A. (1997) 'Developing participation', *PLA Notes*, 30: 75–8.

Shakya, Y. and Rankin, K. (2008) 'The politics of subversion in development practice: An exploration of microfinance in Nepal and Vietnam', *Journal of Development Studies*, 44(8): 1214–1235.

2.11 Postcolonialism

Cheryl McEwan

What is postcolonialism?

Postcolonialism is a difficult and contested term not least because it is far from clear that colonialism has been relegated to the past. Its meaning is not limited to 'after-colonialism' or 'after-independence', but refers to ways of criticizing the material and discursive legacies of colonialism (Radcliffe, 1999: 84). Broadly speaking, therefore, postcolonial perspectives can be said to be *anti-colonial*. They have become increasingly important across a range of disciplines over the last 20 years.

A number of core issues underpin postcolonial approaches. First, they stress the need to destabilize the dominant discourses of imperial Europe (e.g. history, philosophy, linguistics and 'development'), which are unconsciously ethnocentric, rooted in European cultures and reflective of a dominant Western worldview. Postcolonial studies problematize the very ways in which the world is known, challenging the unacknowledged and unexamined assumptions at the heart of European and American disciplines that are profoundly insensitive to the meanings, values and practices of other cultures.

Second, postcolonial critiques challenge the experiences of speaking and writing by which dominant discourses come into being. For example, a term such as 'the Third World' homogenizes peoples and countries and carries other associations – economic backwardness, the failure to develop economic and political order, and connotations of a binary contest between 'us' and 'them', 'self' and 'other' – which are often inscribed in development writings. These practices of naming are not innocent. Rather they are part of the process of 'worlding' (Spivak, 1990), or setting apart certain parts of the world from others. Said (1978) has shown how knowledge is a form of power, and by implication violence; it gives authority to the possessor of knowledge. Knowledge has been, and to a large extent still is, controlled and produced in 'the West'. Global economic power might be starting to shift, but the power to name, represent and theorize is still located in 'the West', a fact which postcolonialism seeks to disrupt.

Third, postcolonialism invokes an explicit critique of the spatial metaphors and temporality employed in Western discourses. Whereas previous designations of the Third World signalled both spatial and temporal distance – 'out there' and 'back there' – a postcolonial perspective insists that the 'other' world is 'in here'. The Third World is integral to what 'the West' refers to as 'modernity' and 'progress'. It contributes directly to the economic wealth of Western countries through its labour and economic exploitation. In addition, the modalities and aesthetics of the Third World have partially constituted Western languages and cultures. Postcolonialism, therefore, attempts to rewrite the hegemonic accounting of time (history) and the spatial distribution of knowledge (power) that constructs the Third World.

Finally, postcolonialism attempts to recover the lost historical and contemporary voices of the marginalized, the oppressed and the dominated through a radical reconstruction of history and knowledge production. Postcolonial theory has developed this radical edge through the works of scholars such as Spivak and Said who, in various ways, have sought to recover the agency and resistance of peoples subjugated by both colonialism and neo-colonialism.

These core issues form the fabric of the complex field of inquiry of postcolonial studies, based in the 'historical fact' of European colonialism and the diverse material effects to which this phenomenon has given rise.

Postcolonialism and development

The possibility of producing a truly decolonized, postcolonial knowledge in development studies became a subject of considerable debate during the 1990s, culminating in new dialogue between the two approaches that continues today. In theoretical terms, postcolonialism has been greatly influenced by Marxism and post-structuralism, drawing on the political-economy approaches of the former and the cultural and linguistic analyses of the latter. The politics of postcolonialism diverge sharply from other discourses and, although it shares similarities with dependency theories, its radicalism rejects established agendas and accustomed ways of seeing. This means that postcolonialism is a powerful critique of 'development' and an increasingly important challenge to dominant ways of apprehending North–South relations.

Critiquing discourses of development

Postcolonialism challenges the very meaning of development as rooted in colonial discourse depicting the North as advanced and progressive and the South as backward, degenerate and primitive. Early postcolonial writers, such as Van der Post, challenged this assumption by referring to hunter-gatherers as the first affluent peoples. Postcolonialism has prompted questions about whether such indigenous systems of equity, reciprocity and communalism are more advantageous to peoples of the South than the pursuit of capitalism, with its emphasis on individual wealth and incorporation into the global economy. The superiority of modern industrialization and technological progress is increasingly questioned, creating alternative knowledges to reshape perceptions of non-Western societies and their environments.

Critics argue that to subject development to postcolonial critique is a form of intellectual faddism; as long as there are pressing material issues such as poverty in the world, concerns with the *language* of development are esoteric. However, language is fundamental to the way we order, understand, intervene and justify those interventions (Escobar, 1995a). As Crush argues, postcolonialism offers new ways of understanding what development is and does, and why it is so difficult to think beyond it. The texts of development are written in a representational language – metaphors, images, allusion, fantasy and rhetoric – the imagined worlds bearing little resemblance to the real world. Development writing often produces and reproduces misrepresentation. Postcolonialism seeks to remove negative stereotypes about people and places from such discourses. It challenges us to rethink categories such as 'Third World' and 'Third World women', and to understand how location, economic role, social dimensions of identity and the global political economy differentiate between groups and their opportunities for development.

As Crush suggests, the texts of development are 'avowedly strategic and tactical', promoting and justifying certain interventions and delegitimizing and excluding others. Power relations are clearly implied in this process; certain forms of knowledge are dominant and others are excluded. The texts of development contain silences. It is important to ask who is silenced, and why? Ideas about development are not produced in a social, institutional or literary vacuum. A postcolonial approach to development literature, therefore, can say a great deal about the apparatuses of power and domination within which those texts are produced, circulated and consumed. Development discourse promotes and justifies real interventions with material consequences. It is, therefore, imperative to explore the links between the words, practices and institutional expressions of development, and between the relations of power that order the world and the words and images that represent the world. By doing so, postcolonial approaches have possibilities for effecting change.

Agency in development

Postcolonialism challenges the notion of a single path to development and demands acknowledgement of a diversity of perspectives and priorities. The politics of defining and satisfying needs is a crucial dimension of current development thought, to which the concept of agency is central. Postcolonial approaches question who voices the development concern, what power relations are played out, how participants' identities and structural roles in local and global societies shape their priorities, and which voices are excluded as a result? They attempt to overcome inequality by opening up spaces for the enactment of agency by non-Western peoples. However, poverty and a lack of technology make this increasingly difficult; non-Western academics, for example, rarely have the same access to books and technologies of communication as their Western counterparts.

Despite this, postcolonial critique has led to a questioning of authorization and authority. By what right and on whose authority does one claim to speak on behalf of others? On whose terms is space created in which they are allowed to speak? Are we merely trying to incorporate and subsume non-Western voices into our own canons? It is no longer feasible to represent the peoples of the Third World as passive, helpless victims. Their voices are now being heard, and their ideas are increasingly being incorporated into grassroots development policies. Postcolonial critics have also had impact on development studies, particularly within gender and development. They have forced a move away from totalizing discourses and a singular feminism (based upon the vantage point of white, middle-class Western feminists, which failed to acknowledge the differences between women) towards the creation of spaces where the voices of black women and women from the South can be heard (see, for example, Mohanty, 1988; McEwan, 2001). Postcolonial feminisms allow for competing and disparate voices among women, rather than reproducing colonialist power relations where knowledge is produced and received in the West, and white, middle-class women have the power to speak for their 'silenced sisters' in the South.

New dialogues and approaches in development

One of the major criticisms of postcolonialism has been that it is too theoretical and not rooted enough in material concerns; emphasis on discourse detracts from an assessment of material ways in which colonial power relations persist; consequently, postcolonialism is ignorant of the real problems characterizing everyday life in the global South. However, recent work at the interface of postcolonialism and development actively refutes these charges. Postcolonial critiques of economic development, challenge the amnesia about (neo) colonialism within development and question its blind loyalty to scientific progress and universal economic prescriptions (Kapoor, 2008; McEwan, 2009). Fundamental questions, rooted in both postcolonial and political-economic theory, are being asked about how capitalism reproduces inequality in the name of development and how it is that the deepening of capitalist social relations comes to be taken as development (Wainwright, 2008). Clearly, postcolonialism does not concede the space of materiality – the provisioning of livelihoods, tangible constraints on life, relations of production and distribution – to economics (Pollard *et al.*, 2011). Rather, it suggests radically different ways of understanding and responding to these issues. Economists, economic models and dominant/orthodox notions of development erase the richness of human agency and experience in response to economic and other crises through their drive to produce 'development aggregates', which then often fail to produce adequate responses to these crises. In contrast, postcolonial approaches emphasize the need to understand development through the eyes of local people who are making daily livelihood decisions in situations of conflict, despair, uncertainty, ambivalence, hope and resistance (Sylvester, 2011). This different approach is producing innovation in the sources and methods used within development studies. This includes the reading of postcolonial stories (e.g. novels in the case of Sylvester (2011), or poetry in the case of Madge and Eshun (2013)) as part of development theory, training and practice in the field as a means of understanding the thoughts and actions of those subject to development interventions.

Postcolonialism is a significant advancement in development studies. It demonstrates how the production of Western knowledge forms is inseparable from the exercise of Western power. It also attempts to loosen the power of Western knowledge and reassert the value of alternative experiences and ways of knowing (Thiong'o, 1986; Bhabha, 1994). It articulates some difficult questions about writing the history of 'development', about imperialist representations and discourses surrounding 'the Third World', and about the institutional practices of development itself. It has the potential to turn critique of conventional development into productive 're-learning to see and reassess the reality' of the global South (Escobar, 2001, 153). It has been an important stimulus to alternative formulations such as 'indigenous' and 'alternative modernities' and rights-based approaches to development (Simon, 2006). And, precisely because of their divergent traditions, increasing dialogue between postcolonialism and development studies offers new ways of conceptualizing and doing development (Sylvester, 2006).

Postcolonialism has an expansive understanding of the potentialities of agency. It shares a social optimism with other discourses, such as gender and sexuality in Western countries, and rethinking here has helped generate substantial changes in political practice. Emerging dialogues between postcolonialism and development studies have the potential to engage postcolonial theory in considering questions of inequality of power and control of resources, human rights, global exploitation of labour, child prostitution and genocide, helping to translate the theoretical insights of postcolonialism into action on the ground and a means of tackling the power imbalances between North and South. They might also inspire critical reshaping of postcolonial futures and counter new forms of orientalism that continue to disadvantage the developing world. The challenge now, as Simon (2006) contends, is to link postcolonialist concerns with local identities, practices and agendas to broader campaigns and projects for progressive and radical change that are substantively postcolonial and critically developmental. This is beginning to emerge with new North–South alliances, alternative and progressive trading structures such as fair and ethical trade, and critical analysis of the role of agencies and institutions. Therefore, despite the seeming impossibility of transforming North–South relations by the politics of difference and agency alone, postcolonialism is a much-needed corrective to the Eurocentrism and conservatism of much writing on development. It is playing an important role in re-imagining critical development studies and generating new dialogue and action. Through its focus on the politics of knowledge production and problematizing power relations between different actors engaged in the development nexus, it is also revivifying within development studies a longstanding concern with the moral imperatives underpinning development research, the ethics of research, and an ethos of solidarity with others (McEwan, 2009).

Bibliography

Bhabha, H. (1994) *The Location of Culture*, London: Routledge.
Escobar, A. (1995b) 'Imagining a post-development era', in J. Crush, (ed.) *Power of Development*, London: Routledge, pp. 211–27.
Escobar, A. (2001) 'Culture sits in places: Reflections on globalism and subaltern strategies of localization'. *Political Geography* 20: 139–74.
Kapoor, I. (2008) *The Postcolonial Politics of Development*, London: Routledge.
Madge, C. and Eshun, G. (2013) ' "Now let me share this with you": Exploring poetry for postcolonial geography research'. *Antipode* 44(4): 1395–1428.
McEwan, C. (2001) 'Postcolonialism, feminism and development: Intersections and dilemmas', *Progress in Developing Studies* 1(2): 93–111.
Mohanty, C. (1988) 'Under Western eyes: Feminist scholarship and colonial discourses', *Feminist Review* 30: 61–88.
Pollard, J., McEwan, C. and Hughes, A. (eds) (2011) *Postcolonial Economies*, London: Zed.
Radcliffe, S. (1999) 'Re-thinking development', in P. Cloke, P. Crang and M. Goodwin (eds) *Introducing Human Geographies*, London: Arnold, pp. 84–91.
Said, E. (1978) *Orientalism*, London: Routledge and Kegan Paul.
Simon, D. (2006) 'Separated by common ground? Bringing (post)development and (post)colonialism together', *The Geographical Journal* 172(1): 10–21.
Spivak, G. (1990) *The Postcolonial Critic: Interviews, Strategies, Dialogue*, London: Routledge.
Spivak, G. (1999) *A Critique of Postcolonial Reason*, Cambridge, MA: Harvard University Press.
Sylvester, C. (2006) 'Bare life as a development/postcolonial problematic', *The Geographical Journal* 172(1): 66–77.
Sylvester, C. (2011) 'Development and postcolonial takes on biopolitics and economy', in J. Pollard, C. McEwan and A. Hughes (eds) *Postcolonial Economies*, London: Zed, pp. 185–204.
Thiong'o, Ngugi wa (1986) *Decolonising the Mind*, London: James Curry.
Wainwright, J. (2008) *Decolonizing Development*, Oxford: Blackwell.

Guide to Further Reading

Crush, J. (ed.) (1995) *Power of Development*, London: Routledge. A collection of essays exploring the language of development, its rhetoric and meaning within different political and institutional contexts.
Escobar, A. (1995a) *Encountering Development: The Making and Unmaking of the Third World*, Princeton, NJ: Princeton University Press. A provocative analysis of development discourse and practice.
The Geographical Journal (2006) 'Postcolonialism and development: New dialogues', special issue, 172(1): 6–77.
McEwan, C. (2009) *Postcolonialism and Development*, London: Routledge. A comprehensive account of the significance of postcolonial theory within development theory and practice.
Schwarz, H. and Ray, S. (eds) (2005) *A Companion to Postcolonial Studies*, Oxford: Blackwell. A wide-ranging volume of essays by leading postcolonial scholars that cuts across themes, regions, theories and practices of postcolonial study.

Websites

www.fairtrade.org.uk/. UK website of the Fairtrade Foundation, which aims to offer independent guarantees that disadvantaged producers in the developing world are getting a better deal.
www.fsm2013.org/en. Website of the World Social Forum, where social movements, networks, NGOs and other civil society organizations opposed to neoliberalism and all forms of imperialism come together to share and debate ideas and to network for effective action.
http://us.oneworld.net/. US website that encourages people to discover their power to speak, connect, and make a difference by providing access to information and enabling connections between thousands of organizations and millions of people around the world.
http://web.worldbank.org/wbsite/external/countries/africaext/extindknowledge/0,,menuPK:825562~pagePK:64168427~piPK:64168435~theSitePK:825547,00.html. The World Bank's website on indigenous knowledge and its role in the development process.

2.12 Postmodernism and development

David Simon

Postmodernism: Panacea, placebo or perversity?

Postmodernism became a major social scientific theoretical paradigm during the 1980s and 1990s, although its popularity has now waned. In development studies it gained prominence as one of the routes for transcending the so-called theoretical 'impasse' that emerged in the mid- to late 1980s. However, the concept assumed diverse meanings, a factor contributing substantially to the often heated but unenlightening debates over its usefulness in the context of development.

The raft of new development textbooks appearing since the late 1990s, when such debate was at its zenith, has devoted surprisingly little attention to postmodernism (Simon, 1999: 38–43). Some make no mention of it or of other 'post-' or 'anti-developmentalist' approaches at all, while others include only a few pages or a single chapter, almost as an afterthought. Very few give fuller coverage, with the result that most current students continue to have little exposure to these debates.

Postmodernism first emerged in art, architecture and literature in the mid-1970s. The concepts of ideals, absolutes, order and harmonization, which had given rise to increasing alienation of the individual, were challenged, and the objective became a celebration of diverse forms and sharp contrasts, in order to rupture conventional expectations. This is generally achieved through the juxtaposition of radically different styles in street façades.

In Latin America, writers like Gabriel García Márquez and Carlos Fuentes pioneered a literary style that broke with the established tradition of a single, chronological flow to novels, and replaced it with multiple, cross-cutting strands, flashbacks, forward leaps and previews in structurally much more complex forms. It won the authors prizes but has not proven a durable literary form.

In the social sciences, postmodernism gained a foothold as part of the ferment in discourse that included post-structuralist rejection of modernist meta-theories and grand narratives of a single mode of explanation or 'truth'. Ahluwalia (2010) demonstrates that post-structuralism derives from a complex blend of colonial, anti-colonial and postcolonial roots, strongly linked to the Algerian liberation struggle. The work of Michel Foucault, Henri Lefebvre, Jacques Derrida, Jean Baudrillard and Frederic Jameson looms large in the foundations of postmodernism. Among the most widely reputed social scientific accounts of postmodernism are those of Jean-François Lyotard (1984), David Harvey (1989) and Ed Soja (1991). They situate it explicitly within the (cultural) logic of late capitalism, as part of the search for profit accumulation in a context of globalized production and consumption. As such, they are critical and see it as having limited social explanatory value, serving mainly to justify conspicuous self-expression, rather than representing a profound new paradigm.

Significantly, several leading advocates of postmodernism and postcolonialism in cultural studies, sociology and allied disciplines, including Homi Bhabha, Trin Minha, Gyatri Spivak and the late Edward Said, hail from the global South, even though generally now working in Northern universities. Among geographers, sociologists and development specialists, and especially those working in Latin America, some of the most trenchant critics of conventional development espoused postmodernism as the way forward during the 1990s – in particular, Santiago Colás, Arturo Escobar, Gustavo Esteva and David Slater. However, the book that established Escobar's (1995) anti-development reputation is principally a critique of 'the development project', offering little insight into a revisioned future beyond an invocation of new social movements. By contrast, Colás (1994) interprets postmodern developments in different spheres of Argentinian society, while Esteva and Prakash (1998) provide one of the very few detailed expositions of regional and local-level postmodernism in practice as social action. Slater (1992, 1997) has taken forward geopolitical and development debates across the North–South divide. Other authors have been cautious about the relevance of postmodernism relative to postcolonialism. Escobar's more recent work has moved substantially beyond anti-development, forming part of the increasing consensus around the need for alternative approaches, or post-development (see Simon, 2007).

Nevertheless, most social scientists working in, or concerned with, poor regions of the world have tended to ignore postmodernism or to dismiss it as an irrelevance on the grounds that:

- Postmodernism literally means 'after the modern'; however, in the global South, the majority of people are still poor and struggling to meet basic needs and to enjoy the fruits of modernization so powerfully held up to them as the outcome of development. In such situations, modernity has yet to be widely achieved, so that which follows on from the modern can have little relevance.
- Postmodernism was merely a temporarily fashionable Northern paradigm, which found expression mainly in aesthetic/architectural terms and as playful, leisured heterodoxes and new forms of consumption centred on individualism, which can best be described as self-indulgence by the well-off. Such preoccupations are seen as irrelevant to the global South, if not actually harmful in terms of distracting attention from the urgent priorities of the poor majority, namely survival, the ability to meet basic needs and related 'development' agendas, as well as the broader structural forces and processes which impact upon them.

A conceptual schema

The following typology distinguishes the different connotations of postmodernism/postmodernity in order to facilitate understanding. Postmodernity describes the 'condition' or manifestation, while postmodernism is the ideology or intellectual practice. At least three broad interpretations of the postmodern can be distinguished in the vast and multidisciplinary literatures.

The chronological approach

This is the most literal interpretation, in terms of which the postmodern necessarily follows the modern. In practice, however, no clean break between eras can be distinguished: there was no dramatic event to act as signifier, and there has been no agreement on the basis of transition. At best, one might be able to conceptualize a transitional phase of some years' duration.

In terms of globalization and mass consumption, for instance, the traditional mass-market air package holiday would be modern, whereas the more differentiated and personalized small-group luxury tour, complete with ecotourist credentials and/or sanitized versions of conflictual local histories in distant countries for the benefit of international tourists, might be conceived of as postmodern.

The aesthetic approach

The second basic understanding of postmodernism is as a form of expression in the creative and aesthetic disciplines like art, architecture and literature. This perspective reflects the considerations – and is exemplified by the authors – cited in the introduction. Inevitably, perhaps, most such attention has been centred on elite and middle-class consumption, especially in terms of leisure activities but also increasingly in the working environment and public spaces. Terms most frequently associated with this movement include pastiche, mélange, playfulness, commodity-signs, imaginaries, and spectacles. Theme parks, pleasure domes and other purpose-built leisure complexes that offer decontextualized time–space representations of various places and experiences (Featherstone, 1995; Watson and Gibson, 1995), often in sanitized form, are characteristic of this approach in much the same way as great exhibitions of global exploration, scientific discoveries and industrial achievements were hallmarks of Victorian modernity.

Postmodernism as intellectual practice

The approach of postmodernism as *problematique* or intellectual practice is the most relevant from the perspective of development studies. Here, the postmodern supposedly represents fundamentally different ways of seeing, knowing and representing the world. The modern approach, rooted in Enlightenment thinking about rationality, is concerned with a search for universal truths, linked to positivist scientific methodology and neo-classical economics. Such universalizing, globalizing approaches are referred to as meta-narratives.

Postmodern practice rejects such singular explanations in favour of multiple, divergent and overlapping interpretations and views. Simplicity should give way to complexity and pluralism, in terms of which these different accounts are all accorded legitimacy. The privileging of official and formal discourses should be replaced by approaches lending credence to both the official and unofficial, formal and informal, dominant and subordinate, central and marginal groups, and to their discourses and agendas. Top-down development, so closely associated with official national and international agendas of modernization, has been discredited over a long period (but nevertheless still proves remarkably persistent); instead, bottom-up approaches or some hybrid of the two should be encouraged.

Hence, postmodernism represents a potentially fruitful approach for addressing the conflictual and divergent agendas of social groups, be it in relation to access to productive resources and/or the bases for accumulating social power, mediating the impacts of large development schemes, evolving complementary medical services that harness the most appropriate elements from both Western and indigenous systems, or addressing longstanding conflicts between statutory and customary legal systems (Esteva and Prakash, 1998; Simon, 1998, 1999). Empowerment of the poor and powerless should be the objective.

Such discourses have much in common with earlier, liberal pluralism, basic needs and grassroots development paradigms, although the emphasis on coexistence and multiple modes of explanation is different. Equally, there are considerable areas of overlap with some strands of postcolonialism, which is centrally concerned with the cultural politics and identities of previously subordinated groups.

Extreme postmodernism can become almost indistinguishable from anarchism, in that all forms of social or collective action prove impossible due to the inability to agree – or even to conceive of agreeing – on any shared rationality or basic rules of what is, and is not, acceptable behaviour. Extreme relativism means that everyone's views are equally valid; without some decision-making rules, any social action not gaining unanimity or consensus becomes impossible.

Accordingly and because of the emphasis by some authors on playful, leisured self-fulfilment, postmodernism is sometimes criticized as a conservative ideology embedded within late capitalism – and hence rejected. Somewhat prematurely, Ley (2003) even wrote its epitaph because of its perceived fad status and because of problems operationalizing it in meaningful distinction from the modern, meaning recent or present-day. Postmodernism may indeed have lost much of its social scientific prominence since the early 2000s in favour of the closely related umbrella term of postcolonialism, but that does not deny that it retains much of value, especially in terms of understanding contemporary social dynamics and complexities. Gabardi's (2001) articulation of 'critical postmodernism' 'not only as a product of the modern-postmodern debate, but also as a theoretical and ideological response to our current late modern/postmodern transition and a practical tool for negotiating this transition' (p. xxi) remains one of the most detailed and spirited articulations of that perspective.

Conclusion

Postmodern discourses arose within changing intellectual and geopolitical circumstances. Its multiple uses and meanings have contributed to confusion and misinterpretation. It found favour among some leading artistic and intellectual voices, who contributed greatly to its refinement and prominence. However, it is also true that many Northern writers linking postmodernism to globalization and other politico-economic changes have, either implicitly or explicitly, simply assumed their Northern research and arguments to have global relevance.

'Moderate' forms of postmodern intellectual practice do indeed have global relevance in the cause of problem analysis, development promotion and empowerment. They lend legitimacy to different social groups and their voices rather than merely seeking a compatible mouthpiece to support external interventions in the name of 'development'. This may help to transcend the shortcomings of discredited official rapid 'modernization as development' and to facilitate local communities to develop according to their own conceptions, governed by acceptable rules of conduct. Such processes are not free from conflict, nor can nostalgia for long-dead traditions and heritages substitute for tackling present-day problems imaginatively in relation to today's dynamic realities (Simon, 2007).

Extreme postmodernism is unduly relativistic and permissive; it may preclude any social contract or action and should be rejected.

References

Ahluwalia, P. (2010) *Out of Africa: Post-structuralism's Colonial Roots*, London: Routledge.
Colás, S. (1994) *Postmodernity in Latin America: The Argentine Paradigm*, Durham, NC: Duke University Press.
Escobar, A. (1995) *Encountering Development: The Making and Unmaking of the Third World*, Princeton: Princeton University Press.
Esteva, G. and Prakash, M. S. (1998) *Grassroots Postmodernism: Remaking the Soil of Cultures*, London: Zed Books.
Featherstone, M. (1995) *Undoing Culture: Globalization, Postmodernism and Identity*, London: Sage.
Gabardi, W. (2001) *Negotiating Postmodernism*, Minneapolis: Minnesota University Press.
Harvey, D. (1989) *The Condition of Postmodernity*, Oxford: Blackwell.
Ley, D. (2003) 'Forgetting postmodernism? Recuperating a social history of local knowledge', *Progress in Human Geography* 27(5): 537–560.
Lyotard, J.-F. (1984) *The Postmodern Condition: A Report on Knowledge*, Minneapolis: Minnesota University Press.
Simon, D. (1998) 'Rethinking (post)modernism, postcolonialism and posttraditionalism: South–North perspectives', *Environment and Planning D: Society and Space* 16(2): 219–245.
Simon, D. (1999) 'Development revisited: Thinking about, practising and teaching development after the cold war', in D. Simon and A. Närman (eds) *Development as Theory and Practice: Current Perspectives on Development and Development Co-operation*, Harlow: Longman.
Simon, D. (2007) 'Beyond anti-development: Discourses, convergences, practices', *Singapore Journal of Tropical Geography* 28(2): 205–218.
Slater, D. (1992) 'Theories of development and politics of the post-modern – exploring a border zone', *Development and Change* 23(3): 283–319.
Slater, D. (1997) 'Spatialities of power and postmodern ethics – rethinking geopolitical encounters', *Environment and Planning D: Society and Space* 15(1): 55–72.
Soja, E. (1991) *Postmodern Geographies: The Reassertion of Space in Critical Social Theory*, London: Verso.
Watson, S. and Gibson, K. (eds) (1995) *Postmodern Cities and Spaces*, Oxford: Blackwell.

Further reading

Esteva, G. and Prakash, M. S. (1998) *Grassroots Postmodernism: Remaking the Soil of Cultures*, London: Zed Books. An extended treatment of postmodern practice, linking new social movements, grassroots organizations and regionally based rebellions against inequitable and oppressive governments in Latin America and Asia.

Harvey, D. (1989) *The Condition of Postmodernity*, Oxford: Blackwell. This remains an important reference guide to postmodernity as a function of late capitalism, although focused on the North.

Simon, D. (1998) 'Rethinking (post)modernism, postcolonialism and posttraditionalism: South–North perspectives', *Environment and Planning D: Society and Space* 16(2): 219–245. A detailed exposition of the themes outlined here.

Slater, D. (1997) 'Spatialities of power and postmodern ethics – rethinking geopolitical encounters', *Environment and Planning D: Society and Space* 15(1): 55–72. Explores post-cold war geopolitical change and its implications across the North–South divide, including postmodern concerns for distant strangers.

Watson, S. and Gibson, K. (eds) (1995) *Postmodern Cities and Spaces*, Oxford: Blackwell. An important collection of essays, addressing different approaches to postmodernism in Northern and Southern urban contexts.

2.13
Post-development

James D. Sidaway

> Instead of the kingdom of abundance promised by theorists and politicians in the 1950s, the discourse and strategy of development produced its opposite: massive underdevelopment and impoverishment, untold exploitation and repression. The debt crisis, the Sahelian famine, increasing poverty, malnutrition, and violence are only the most pathetic signs of the failure of forty years of development.
>
> *(Escobar, 1995: 4)*

> Development occupies the centre of an incredibly powerful semantic constellation . . . at the same time, very few words are as feeble, as fragile and as incapable of giving substance and meaning to thought and behavior.
>
> *(Esteva, 1992: 8)*

> Along with 'anti-development' and 'beyond development', post-development is a radical reaction to the dilemmas of development. Perplexity and extreme dissatisfaction with business-as-usual and standard development rhetoric and practice, and disillusionment with alternative development are keynotes of this perspective. Development is rejected because it is the 'new religion of the West . . . it is the imposition of science as power . . . it does not work . . . it means cultural Westernisation and homogenisation . . . and it brings environmental destruction. It is rejected not merely on account of its results but because of its intentions, its world-view and mindset. The economic mindset implies a reductionist view of existence. Thus, according to Sachs, 'it is not the failure of development which has to be feared, but its success' (1992: 3).
>
> *(Nederveen Pieterse, 2000: 175)*

Jan Nederveen Pieterse goes on to explain how, from these critical perspectives, 'development' has often required the *loss* of 'indigenous' culture, or the destruction of environmentally and psychologically rich and rewarding modes of life. Development is also criticized as a particular vision that is neither benign nor innocent. It reworks, but is never entirely beyond prior colonial discourses (see Kothari, 2005). Development comprises a set of knowledges, interventions and worldviews (in short a 'discourse') and hence powers – to intervene, to transform and to rule. It embodies a geopolitics (see Slater, 1993), in that its origins are bound up with Western power and strategy for the Third World, enacted and implemented through local elites. Therefore, American power and the cold war containment of communism structured the meaning of development in the second half of the twentieth century, and were associated with ideas of modernization (see Engerman *et al.*, 2003). Such modernization strategies have deeper roots in American prototypes from earlier in the century (see Ekbladh, 2010; Sewell, 2010). Western agencies, charities and consultants long dominated development's agendas (see Jackson, 2005; Stirrat, 2000, 2008). Development came to the fore during the cold war as a powerful combination of policy, action and understanding. Related to concepts of anti-development and postcolonial criticisms, post-development arose in the 1990s as a critique of the standard assumptions about progress, who possessed the keys to it and how it might be implemented. Such critique also proved suggestive for those studying the politics of local and regional development in Western Europe (see Donaldson, 2006).

Of course, as a number of people have pointed out, many of the critiques associated with post-development reformulate scepticisms and calls for alternatives that have long been evident. According to Marshall Berman (1983), an example is the myth of Faust, which crops up repeatedly in European cultures. Faust is a man who would develop the world and himself, but must also destroy all that lies in his path to this goal and all who would resist him. The myth of Faust, who sells his soul for the earthly power to develop, bears witness to a very long history of critics of progress and modernity. Throughout the twentieth century, populist ideas of self-reliance and fulfilling 'basic needs' have also been sceptical of many of the claims of development, particularly when the latter takes the forms of industrialization and urbanization (see Kitching, 1989). Subsequently, the history of ideas of dependency has been, in part, a rejection of Western claims of development as a universal panacea to be implemented in a grateful Third World. From Latin American roots (see Kay, 1989), dependency ideas were widely disseminated and sometimes took the form of a rejection of Western modernization/development as corrupting and destructive (see Blomstrom and Hettne, 1984; Leys, 1996; Rist, 1997) or as a continuation of colonial forms of domination (Rodney, 1972). In particular, writers from predominately Islamic countries (most notably Iran) saw the obsession with development as part of a misplaced 'intoxification' with the West (see Dabashi, 1993). Either way, the Third World was not simply a passive recipient of development, but became a project and site for liberation and struggle (Prashad, 2007; Sidaway, 2007). Likewise, more conventional Marxist accounts have long pointed to the 'combined and uneven' character of development and its highly contradictory consequences (see Lowy, 1980). Feminist writings have also criticized the ways in which the so-called 'Third World woman' is represented as needing 'development' and Western-style 'liberation' (Mohanty, 1988), and have opened up alternative ways of conceptualizing the economic and social change of 'development'.

Some critics have therefore complained that 'post-development' was never really beyond, outside or subsequent to development discourse. In this view, 'post-development' was merely the latest version of a set of criticisms that have long been evident *within* writing and thinking about development (Curry, 2003; Kiely, 1999). Development has always been about choices, with losers, winners, dilemmas and destruction as well as creative possibility. Gavin Kitching (1989: 195), who is concerned to put post-Second World War debates about development into a longer historical perspective (stressing how they also reproduce even older narratives from the nineteenth century), argues:

> It is my view that the hardest and clearest thinking about development always reveals that there are no easy answers, no panaceas whether these be 'de-linking', 'industrialization', 'rural development', 'appropriate technology', 'popular participation', 'basic needs', 'socialism' or whatever. As I have had occasion to say repeatedly in speaking on and about this book, development is an awful process. It varies only, and importantly, in its awfulness. And that is perhaps why my most indulgent judgements are reserved for those, whether they be Marxist-Leninists, Korean generals, or IMF officials, who,

> whatever else they may do, recognize this and are prepared to accept its moral implications. My most critical reflections are reserved for those, whether they be Western liberal-radicals or African bureaucratic elites, who do not, and therefore avoid or evade such implications and with them their own responsibilities.

In this sense, perhaps post-development's scepticism towards grand narratives about development is less original than the theoretical frames (the analysis of discourse) that it brings to bear in problematizing these. Yet according to some post-development writers, not only are there 'no easy answers', but the whole question of 'development' should be problematized and/or rejected.

There are a number of more fundamental objections to post-development. The first is that it overstates the case. Such arguments usually accept that development is contradictory (that it has winners and losers), but refuse to reject all that goes under its name. For to reject all development is arguably a rejection of the possibility for progressive transformation; or it is to ignore the tangible improvements in life chances, health, wealth and material well-being evident in some places, notably the 'developmental states' of East or Southeast Asia (Rigg, 2003). The changing global map of production, consumption and finance in recent decades is also redrawing the map of more and less developed spaces and of global wealth and power (see Sidaway, 2012; Bräutigam and Xiayang, 2012). Moreover, development itself has long been so varied and carried so many meanings (see Williams, 1976) that critiques need to be specific about what they mean when they claim to be anti- or 'post-development'.

In this context, Escobar's (1995) work, in particular, was often criticized. One objection has been that he understates the potential for change within development discourse (see Brown, 1996). Escobar's work reflects his experiences as an anthropologist in Colombia. As an account based on experiences of twentieth century Colombia, Escobar's critique of development could seem suggestive. Colombia has experienced periods of brutal civil war and foreign intervention. It became a major source of cocaine, connected to often violent smuggling networks extending northwards into the United States and Europe. Yet there is a risk that Escobar's text obscures the diversity of experiences of development, not all of which are as troubled as the Colombian experience.

The second objection involves rejecting post-development as yet another intellectual fad of limited (or no) relevance to the poor in the Third World. Sometimes this objection draws attention to the fact that many of those who write about or disseminate post-development ideas live precisely the cosmopolitan, middle-class, relatively affluent lives that development promises to deliver. Such questions parallel the critique of *postcolonialism* as an intellectual fashion most useful to the careers of Western-based intellectuals.

However, a few counter-points are in order here. First, a whole set of writings and ideas are grouped together under the rubric of post-development. Michael Watts (2000: 170) explained that:

> There is of course a polyphony of voices within this post-development community – Vandana Shiva, Wolfgang Sachs, Arturo Escobar, Gustavo Esteva and Ashish Nandy, for example, occupy quite different intellectual and political locations. But it is striking how intellectuals, activists, practitioners and academics within this diverse community participated in a global debate.

Moreover, it is important to point out that for Escobar (1995) and others exploring the (geo) politics of development, to criticize development is not necessarily to reject change and possibility. Rather, it is to make us aware of the consequences of framing this as 'development'. It stresses that development is (for good and bad) always about power. Hence, as Clive Gaby (2012: 1249) notes, projects such as the Millennium Development Goals involve 'a logic of ambitious social, cultural and spatial engineering'. Moreover, alternative visions considering, for example, democracy, popular culture, resourcefulness and environmental impacts would transform the imagined map of more or less developed countries. Recognition that development is but one way of seeing the world (and one that carries certain consequences and assumptions) can open up other perspectives. What happens, for example, to the perception of Africa when it is seen as *rich* in cultures and lives whose diversity, wealth and worth are not adequately captured by being imagined as more or less *developed*? Alternatively, why are poverty and deprivation (or for that matter, excessive consumption amongst the affluent) in countries like the United States, New Zealand or the United Kingdom not issues of 'development' (see Jones, 2000; Kurian and Munshi, 2012)? What is taken for granted when the term 'development' is used? For it often seems that, in Escobar's (1995: 39) words, development has 'created a space in which only certain things could be said or even imagined'. Post-development literatures teach us not to take this 'space' and its contours for granted.

References

Berman, M. (1983) *All That Is Solid Melts Into Air: The Experience of Modernity*, London: Verso.

Blomstrom, H. and Hettne, B. (1984) *Development Theory in Transition: The Dependency Debate and Beyond: Third World Responses*, London: Zed Books.

Bräutigam, D. and Xiayang, T. (2012) 'Economic statecraft in China's new overseas special economic zones: Soft power, business or resource security?' *International Affairs* 88(4): 799–816.

Brown, E. (1996) 'Deconstructing development: Alternative perspectives on the history of an idea', *Journal of Historical Geography* 22(3): 333–339.

Curry, G. N. (2003) 'Moving beyond postdevelopment: Facilitating indigenous alternatives for "development"', *Economic Geography* 79(4): 405–423.

Dabashi, H. (1993) *Theology of Discontent: The Ideological Foundation of the Islamic Revolution in Iran*, New York and London: New York University Press.

Donaldson, A. (2006) 'Performing regions: Territorial development and cultural politics in a Europe of the regions', *Environment and Planning A* 38(11), 2075–2092.

Ekbladh, D. (2010) *The Great American Mission: Modernization and the Construction of an American World Order*, Princeton, NJ: Princeton University Press.

Engerman, D. C., Gilman, N., Haefele, M. H. and Latham, M. E. (eds) (2003) *Staging Growth: Modernization, Development and the Global Cold War*, Amherst and Boston: University of Massachusetts Press.

Escobar, A. (1995) *Encountering Development: The Making and Unmaking of the Third World*, Princeton, NJ: Princeton University Press.

Esteva, G. (1992) 'Development', in W. Sachs (ed.) *The Development Dictionary: A Guide to Knowledge as Power*, London: Zed Books, pp. 6–25.

Gaby, C. (2012) 'The Millennium Development Goals and ambitious developmental engineering', *Third World Quarterly* 33(7): 1249–1265.

Jackson, J. T. (2005) *The Globalizers: Development Workers in Action*, Baltimore, MD: Johns Hopkins University Press.

Jones, P. S. (2000) 'Why is it alright to do development "over there" but not "here"? Changing vocabularies and common strategies of inclusion across the "First" and "Third" Worlds', *Area* 32(2): 237–241.

Kay, C. (1989) *Latin American Theories of Development and Underdevelopment*, London and New York: Routledge.

Kiely, R. (1999) 'The last refuge of the noble savage? A critical account of post-development', *European Journal of Development Research* 11(1): 30–55.

Kitching, G. (1989) *Development and Underdevelopment in Historical Perspective: Populism, Nationalism and Industrialization* (revised edition), London and New York: Routledge.
Kothari, U. (2005) 'From colonial administration to development studies: A post-colonial critique of the history of development studies', in U. Kothari (ed.) *A Radical History of Development Studies: Individuals, Institutions and Ideologies*, London: Zed, pp. 47–66.
Kurian, P. A. and Munshi, D (2012) 'Denial and distancing in discourses of development: Shadow of the "Third World" in New Zealand', *Third World Quarterly* 33(6): 981–999.
Leys, C. (1996) *The Rise and Fall of Development Theory*, London: James Currey.
Lowy, M. (1980) *The Politics of Combined and Uneven Development: The Theory of Permanent Revolution*, London: New Left Books.
Mohanty, C. P. (1988) 'Under Western eyes: Feminist scholarship and colonial discourses', *Feminist Review* 30: 61–88.
Nederveen Pieterse, J. (2000) 'After post-development', *Third World Quarterly* 21(2): 175–191.
Prahshad, V. (2007) *The Darker Nations: A People's History of the Third World*, New York: The New Press.
Rigg, J. (2003) *Southeast Asia: The Human Landscape of Modernization and Development*, second edition, London and New York: Routledge.
Rist, G. (1997) *The History of Development: From Western Origins to Global Faith*, London: Zed Books.
Rodney, W. (1972) *How Europe Underdeveloped Africa*, London: Bogle L'Ouveture.
Sewell, B. (2010) 'Early modernisation theory? The Eisenhower administration and the foreign policy of development in Brazil', *English Historical Review* 125: 1449–1480.
Sidaway, J. D. (2007) 'Spaces of Postdevelopment', *Progress in Human Geography* 31(3): 345–361.
Sidaway, J. D. (2012) 'Geographies of development: New maps, New visions?', *The Professional Geographer* 64(1): 49–62.
Slater, D. (1993) 'The geopolitical imagination and the enframing of development theory', *Transactions of the Institute of British Geographers NS* 18: 419–37.
Stirrat, R. L. (2000) 'Cultures of consultancy', *Critique of Anthropology* 20(1): 31–46.
Stirrat, R. L. (2008) 'Mercenaries, missonaries and misfits. Representations of development personnel', *Critique of Anthropology* 28(4): 406–425.
Watts, M. (2000) 'Development', in R. J. Johnson, D. Gregory, G. Pratt and M. Watts (eds) *The Dictionary of Human Geography* (4th edn), Oxford: Blackwell, pp. 167–171.
Williams, R. (1976) *Keywords*, London: Fontana.

Further reading

Crush, J. (ed.) (1995) *Power of Development*, London and New York: Routledge. An introduction and collection of 14 essays that examines the 'power' of development discourses. The essays show how claims of development to being a solution to problems of national and global poverty, disorder and environmental degradation, are sometimes illusions.

Escobar, A. (1995) *Encountering Development: The Making and Unmaking of the Third World*, Princeton, NJ: Princeton University Press. Written by a Colombian anthropologist and drawing on the trajectory of that country (whilst making more general claims), this critique uses the ideas of Michel Foucault to understand 'development' as a discourse and therefore as a particular (Western) regime of truth, power and knowledge. It remains influential.

Ferguson, J. (1990) *The Anti-Politics Machine: 'Development', Depoliticization and Bureaucratic Power in Lesotho*, Cambridge: Cambridge University Press. Like Escobar, another often cited book-length critique of the discourse of development written by an anthropologist. Less sweeping in its claims than Escobar, but no less persuasive and influential. Subsequent texts by this author; (2006) *Global Shadows: Africa in the Neoliberal World Order*, and (1999) *Expectations of Modernity: Myths and Meanings of Urban Life on the Zambian Copperbelt*, are also rewarding.

Rahnema, M. and Bawtree, V. (eds) (1997) *The Post-Development Reader*, London: Zed Books. An introduction plus 440 pages comprising of 37 short extracts (and an afterword) from thinkers, politicians and activists who problematized development. Each reading has a short introduction that helps to contextualize it (written by the editors). This remains amongst the best places to start a course of further reading and/or to get a flavour of 'post-development'. Another edited collection of material on post-development appeared a decade on. It also offers a valuable route into the literatures and debates: Ziai, A. (ed.) (2007) *Exploring Post-Development: Theory, Practice, Problems and Perspectives*, London and New York: Routledge.

Sauders, K. (ed) (2002) *Feminist Post-Development Thought: Rethinking Modernity, Post-colonialism and Representation*, London: Zed Books. Seventeen essays examining intersections between feminism, post-development and post-colonialism.

Sidaway, J. D. (2007) 'Spaces of postdevelopment', *Progress in Human Geography* 31(3), 345–361. This article is a wider review of the literatures on post-development and a reconsideration of development as socio-spatial transformation that grew out of an earlier version of this chapter.

2.14 Social capital and development

Anthony Bebbington and Katherine E. Foo

Introduction

The concept of social capital relates social norms, rules, and reciprocal obligations to patterns of social and economic action (Woolcock, 1998). For James Coleman, "Social capital is defined by its function. It is not a single entity but a variety of different entities, with two elements in common: they all consist of some aspect of social structures, and they facilitate certain actions of actors" (Coleman, 1988: S98). Meanwhile, Pierre Bourdieu defined social capital as the "aggregate of the actual or potential resources which are linked to possession of a durable network of more or less institutionalized relationships of mutual acquaintance and recognition . . . which provides each of its members with the backing of the collectively owned capital" (Bourdieu, 1986: 21). Economists have used the concept as a way of describing the "social something" (Hammer and Pritchett, 2006) that their econometric tools could otherwise not handle: the social relationships through which information is exchanged, risk managed, cooperation made possible, and so on (Hammer and Pritchett, 2006; Durlauf and Fafchamps, 2005; Fafchamps, 2006). Other social scientists have used social capital to explore how social relationships affect governance, democracy, livelihood and collective action (Woolcock, 2010).

Engaging these differences, Uphoff (1999) distinguishes cognitive and structural definitions of social capital. Cognitive social capital pertains to the domain of values, trust and perceptions. This conceptualization is apparent, for instance, in attitudinal survey research which gives quantitative measures to levels of trust in society and relates this trust to other indicators, in particular ones of economic performance. A structural conception of social capital leads researchers to focus on social relations, networks, loose associations and formal organizations. Within development studies this structural conception has gained most attention, with social capital referring to the resources – information, reputations, credit – that flow through and are made available by social networks. Some writers view social capital as the interpersonal relationships that individuals mobilize to enhance their wealth and status (Bourdieu, 1977) while others understand it as properties of social organization that facilitate coordinated, collective action (Putnam, 1993; Woolcock, 2010). This latter approach also considers how social networks can be shaped so that they are conducive to building more democratic, supportive, and inclusive communities.

Why did social capital become prominent in development?

While there are continuities between the concept of social capital and themes in nineteenth century classical sociology (Woolcock, 1998), and although the concept had been deployed in urban planning (Jacobs, 1992[1961]), sociology (Bourdieu, 1986) and economics (Loury, 1977), it was Robert Putnam's work in political science that popularized the concept as an independent variable in economic and political development while at the same time giving it a quite particular meaning (Putnam, 1993). In his study of regional government performance in Italy, Putnam argued that, *ceteris paribus*, Italy's local governments were more effective and responsive to their citizens, and its sub-national economies more dynamic in those regions exhibiting higher rates of participation in civic associations. Through involvement in these associations people learnt citizenship and developed networks of civic engagement (*social capital*) that, in their aggregate, fostered greater levels of accountability and responsibility in society and more efficiency in the economy. Putnam thus tied social capital to coordination, cooperation and aggregate development performance, a quite distinct conceptualization from that of prior approaches. This conception proved to be much more intuitively accessible to a range of audiences than was the case for earlier renditions.

The visibility of Putnam's work in academic and popular outlets drove collective debate of his argument in sociology and political science, especially in the USA. It also caught the attention of senior figures in the World Bank, where both the economic research and the social development communities began exploring the relevance of social capital for their own understandings of development (Bebbington *et al.*, 2004, 2006). This link to the World Bank is important because, while the concept was set to be widely debated within academic social science and North American community development, its passage into development studies was accelerated and amplified by its usage within the World Bank.

Within the World Bank the concept proved especially helpful to those communities who already questioned the value of formal economic approaches to development. They saw in social capital a means of bringing social organizations, relationships and empowerment into the institution's narrative on development in a way that would still allow conversations with the Bank's economists. Development became understood as a function of different "forms of capital" at scales that ranged from the nation to the household. Early statements on sustainability and the "wealth of nations" (Serageldin and Steer, 1994) argued that the sustainability of development could be understood as a function of the mixes and trade-offs among produced capital, natural capital, human capital and social capital. A "weak" concept of sustainability would consider development as sustainable as long as the overall capital stock increased; an "absurdly strong" notion of sustainability would not allow draw-down in any of these forms of capital; and "sensible" sustainability would hold total capital stock intact and avoid depletion of any capital beyond critical levels.

If national development was a function of capital mixes and substitutions, then it was only a few short steps to using similar approaches to the study of poverty, welfare and livelihoods at the household and individual levels. Work at the World Bank analyzed household poverty as a function of household access to human, social, natural and financial capital, and social capital was identified as an especially critical determinant (Grootaert, 1999; Narayan and Pritchett, 1999). Other development agencies' approaches to livelihoods followed a similar tack (Carney, 1998). These approaches argued that social capital – understood, broadly, as the networks, organizations and relations to which the person or household had access – facilitated access to other assets, or to the institutions providing those assets, and in that way reduced poverty and vulnerability. This argument has been used in micro-financial services literature and practice, in which social capital (in the form of group membership) is taken as a guarantee that loans will be repaid. Another strand in this writing (and also at the World Bank) has seen social capital as an important safety net, a means of reducing vulnerability. Here social relationships (formal or informal) are valued for the role that they can play in helping people recover from or cope with crisis, violence or other sources of risk and perturbation (Moser, 1998).

Two points merit comment here. The underlying influence on the use of the concept came from neo-classical economic approaches to production functions and, to a lesser extent, ideas in ecological economics about stocks of natural capital. Notably absent was Bourdieu's (1977) notion that the distributions of forms of capital (economic, cultural, symbolic, social) have to be understood as interrelated and in large measure mutually reinforcing. In his conception, for example, social capital serves to consolidate control of economic capital and relationships of power. There was no necessary reason why such conceptions could not have influenced development thinking (Bebbington, 2007). Second, even if the broader model at work here was underpinned by frameworks from economics rather than sociology, social development professionals latched onto the idea quickly. This type of asset-based framework allowed a development narrative that saw participatory processes and strong organizational fabrics as assets of equal importance to education, finance or infrastructure. Capital based approaches to sustainable development offered the prospect of incorporating what had typically been local, idiographic and operational concerns into wider theories of development in which the social was as important as the economic.

Criticisms and elaborations

A good case can be made that development studies research has tended to overstate the potential that social capital holds as a resource for poor people (Cleaver, 2005), and understate the extent to which local, national and international political economy structures their ability to accumulate more assets and to get ahead. A social theoretical lens would conceptualize social capital as embedded in multiple historical and geographical scales, as both constituted by and constitutive of wider relations of political and cultural economy. Indeed, social capital has been subject to penetrating critique in both social and political science as well as in development studies (e.g. Fine, 2001). These criticisms have been many and varied. Critics note that conceptualizations of social capital: can refer to so many dimensions of social life as to become relatively meaningless (Portes, 1998); do not allow for clear identification of causality; perpetuate romanticized notions of community (Cleaver, 2005; Portes, 1998); facilitate the further colonization of social science by neoliberal economics (Fine, 2001); turn social relations into objects of financial calculation; and ignore questions of political economy, power, and politics. In considerable measure such criticism reflects the extent to which early adoption of the concept was underlain by the production function approaches just noted as well as its association with the World Bank.

Even when some claim that such criticisms have been repetitive (Woolcock, 2010), there are indications that social capital research has recognized and responded to some of the points made. For example, Jamal (2009) highlights the tangled and negative dimensions of associational life in an authoritarian context, and the ways in which the forms of social capital they involve are prone towards clientelism and patronage. In other studies, methodological progress has been made in developing multi-level approaches to social capital in order to better address the relationships between social structure, well-being and health (Kawachi, 2008). Meanwhile studies of social capital in local organizations have sought to combine the insights of experimental economics (to understand the emergence and effects of trust and reciprocity) with those of critical social science (to address the effects of power asymmetries on cooperation) (Serra, 2011).

Of course, not all new research has been so self-reflective, and more generally publications on (and citations of) social capital continue to boom (Woolcock, 2010; Serra, 2011; Svendsen and Svendsen, 2009). At the same time, the term has found its way into everyday discussions of development (and not only in the English language). Indeed, Woolcock (2010) argues that one of the great strengths of social capital is that the term facilitates many different conversations – both outside and within academia – among groups who otherwise would be unlikely to talk to each other about the relationships between social organization, development and democracy. In these different senses "social capital" may have some affinity with that other slippery development concept, "sustainability." Each manages to bundle into a single term something that is at once conceptual, normative and intuitive. Perhaps for that very same reason, each appeals across a wide disciplinary and political spectrum and has traction in scholarly, policy *and* popular debate while at the same time being difficult to pin down with great precision. These qualities are simultaneously sources of great strength and great weakness. They may also prove to assure that both concepts will have a long shelf life in development studies even when many who use them feel some discomfort in doing so.

References

Bebbington, A. 2007. Social capital and development studies II: Can Bourdieu travel to policy? *Progress in Development Studies* 7(2): 155–162.

Bebbington, A., Guggenheim, S., Olson, E. and Woolcock, M. 2004. Exploring social capital debates at the World Bank. *Journal of Development Studies* 40(5): 33–64.

Bebbington, A., Woolcock, M., Guggenheim, S. and Olson, E. (eds) 2006. *The search for empowerment: Social capital as idea and practice at the World Bank.* West Hartford, CT: Kumarian.

Bourdieu, P. 1977. *Outline of a theory of practice.* Cambridge: Cambridge University Press.

Bourdieu, P. 1986. The forms of capital. In J. Richardson (ed.) *Handbook of theory and research for the sociology of education.* Westport, CT: Greenwood, pp. 241–258.

Carney, D. (ed.) 1998. *Sustainable rural livelihoods: What contribution can we make?* London: Department for International Development.

Cleaver, F. 2005. The inequality of social capital and the reproduction of chronic poverty. *World Development* 33(6): 893–906.

Coleman, J. 1988. Social capital in the creation of human capital. Supplement: Organizations and institutions: Sociological and economic approaches to the analysis of social structure. *The American Journal of Sociology* 94: S95–S120.

Durlauf, S. and Fafchamps, M. 2005. Social capital. In P. Aghion and S. Durlauf (eds) *Handbook of economic growth* (Vol. 1B). New York: Elsevier, pp. 1639–1699.

Fafchamps, M. 2006. *Development and social capital* (Global Poverty Research Group, Working Paper Series 007). Manchester/Oxford: Global Poverty Research Group.

Fine, B. 2001. *Social capital versus social theory: Political economy and social science at the turn of the millennium.* London: Routledge.

Grootaert, C. 1999. *Social capital, household welfare and poverty in Indonesia.* (Local Level Institutions, Working Paper, No. 6). Washington, DC: World Bank, Social Development Department.

Hammer, J. and Pritchett, L. 2006. Scenes from a marriage: World Bank economists and social capital. In A. Bebbington, M. Woolcock, S. Guggenheim and E. Olson (eds) *The search for empowerment: Social capital as idea and practice at the World Bank*, pp. 63–90.
Jacobs, J. 1992[1961]. *The death and life of great American cities*. New York: Vintage Books.
Jamal, A. A. 2009. *Barriers to democracy: The other side of social capital in Palestine and the Arab world*. Princeton, NJ: Princeton University Press.
Kawachi, I. 2008. Social capital and health. In C. Bird, P. Conrad, A. Fremont and S. Timmermans (eds) *Handbook of medical sociology* (6th edn). Nashville, TN: Vanderbilt University Press.
Loury, G. 1977. A dynamic theory of racial income differences. In P. A. Wallace and A. LaMond (eds) *Women, minorities and employment discrimination*. Lexington, MA: Lexington Books, pp. 153–186.
Moser, C. 1998. The asset vulnerability framework: Reassessing urban poverty reduction strategies. *World Development* 26(1): 1–19.
Narayan, D. and Pritchett, L. 1999. Cents and sociability: Household income and social capital in rural Tanzania. *Economic Development and Cultural Change* 47(4): 871–897.
Portes, A. 1998. Social capital: Its origins and applications in modern sociology. *Annual Review of Sociology* 24: 1–24.
Putnam, R. 1993. *Making democracy work: Civic traditions in modern Italy*. Princeton, NJ: Princeton University Press.
Serageldin, I. 1996. *Sustainability and the wealth of nations*. Washington, DC: World Bank.
Serageldin, I. and Steer, A. (eds) 1994. *Making development sustainable: From concepts to action* (Environmentally Sustainable Development, Occasional Paper Series No. 2). Washington, DC: World Bank.
Serra, R. 2011. The promises of a new social capital agenda. *Journal of Development Studies* 47(8): 1109–1127.
Svendsen, G. and Haase Svendsen, G. (eds) 2009. *Handbook of social capital: The troika of sociology, political science and economics*. Northampton, MA: Edward Elgar.
Uphoff, N. 1999. Understanding social capital: Learning from the analysis and experience of participation. In P. Dasgupta and I. Serageldin (eds) *Social capital: A multifaceted perspective*. Washington, DC: World Bank.
Woolcock, M. 1998. Social capital and economic development: Towards a theoretical synthesis and policy framework. *Theory and Society* 27(2): 151–208.
Woolcock, M. 2010. The rise and routinization of social capital, 1988–2008. *Annual Review of Political Science* 13: 469–487.

Part 3
Globalisation, employment and development

Editorial introduction

We are living through an era that many commentators maintain is characterised by globalisation. This increasingly global remit seems to apply in the fields of industrialisation and employment in particular. The sets of interrelated changes involved have often been referred to under the umbrella title 'global shifts'. On the one hand, there has been a shift whereby some parts of the so-called 'Developing World' have become newly industrialising countries (NICs), although it is vital to stress that this is true of a very limited number of nations. On the other hand, there has been another shift that has witnessed the increasing globalisation of production via the activities of transnational corporations (TNCs), economic units that are to be found operating across boundaries in more than one country.

The so-called new international division of labour (NIDL) has to be seen as a vital aspect of globalisation, pinpointing shifts in production by world region, and affecting both manufacturing and producer services. At least three NIDLs can be recognised: at the time of European colonisation, the industrial development of certain semi-developed areas at the end of the nineteenth century, and the present era, in which foreign direct investment (FDI) has expanded greatly.

But ideas concerning globalisation have to be qualified. In the sphere of production, for example, the shifts that have occurred have only witnessed the incorporation of a limited number of new locations. Thus, commentators have referred to a process of 'divergence', which is leading to increasing differentiation between the places that make up the global economic system in terms of the things that they produce. Thus, the thesis of hyper-mobility can be overstretched, especially in respect of productive capital. According to this argument, globalisation is giving rise to new forms of localisation – and what is sometimes rather inelegantly referred to as new forms of 'glocalisation', a combination of globalisation and new forms of localisation.

Realities such as these have given rise to the growing appreciation, in certain quarters at least, that rather than becoming more uniform, the world is becoming more differentiated and unequal. Thus, while many governments continue to state their invariant faith in globalisation as a macroeconomic global policy, the protests of the anti-globalisation movement have been increasingly heard along with the call for policies that may be less detrimental to the poor and to poor regions. In contrast, key aspects of consumption and consumer tastes show signs of becoming increasingly uniform at the global scale, and this process of relative homogenisation may be seen as giving rise to global 'convergence'.

Such changes need to be seen in a context where perspectives on trade and industrial policy in developing countries have altered greatly over the last twenty-five years. Recent trends have seen the wholesale promotion of market deregulation and liberalisation, after an early platform which emphasised protectionism – hence, the clarion call for fair trade policies as opposed to free trade policies. In the neoliberal approach, export-processing zones and free trade zones are important parts of the so-called new international division of labour, and represent what are seen as relatively easy paths to industrialisation. By the end of the twentieth century, over ninety countries had established export-processing zones as part of their economic strategies.

Knowledge and information are increasingly central commodities in the operation of the contemporary global economy. Issues such as who produces and reproduces, who has access to, and who is responsible for information in the global economy are thus vital. And once again, it is concluded by many analysts that rather than homogenising and democratising platforms of knowledge, the Internet seems to be reinforcing the power of the global North in the global digital division of labour.

Since the 1970s, corporate social responsibility (CSR) has been a term used to describe socially responsible behaviours by businesses, for example, multinational corporations working in poorer nations with sources of non-unionised cheap labour. This is also very important in circumstances where local environmental regulations are either weak or entirely absent. As such, CSR is now a vital component of global development.

In the context of all of these market-oriented changes, the informal sector – where people essentially provide employment for themselves – has generally responded by providing more jobs. In effect, the informal sector has compensated for public-sector cutbacks, recession and neoliberal programmes of economic restructuring. Questions of regulation loom large in this regard, and the high incidence of child labour in Africa, Asia and South America has been a notable point of debate over recent years. Increased awareness has seen child labour become a priority issue for global institutions concerned with human rights, with the International Labour Organization declaring child labour an urgent challenge to human rights.

A major debating point has been whether globalisation serves only the interests of the global capitalist system – in particular large corporations, the rich and elites – rather than the poor and relatively disadvantaged around the world. This has given rise to the policy-related argument that globalisation should be advanced to serve the interests of ordinary citizens, as strongly promulgated in the 2000 United Kingdom Government's White Paper under the title *Eliminating World Poverty: Making Globalisation Work for the Poor.* In this, the UK Government pledged to manage globalisation in a manner that would reduce poverty and promote directly the international development targets. It was also stressed that globalisation should promote economic growth that is both equitable and environmentally sustainable. Many would argue that nothing like this has happened thus far in the course of globalisation.

Between 1970 and 2010, the number of international migrants more then doubled to 213.9 million. Although the majority of these moves occurred within Europe, international migrants made up the largest proportion of the population in both Oceania and North America. Such migration has become a vital component in the development process: for example, in connection with the brain drain of young able nationals, remittances from those based overseas, and possible brain gains via return migration and the diasporic overseas community. The diaspora is vital – the individuals and groups living overseas who maintain emotional and economic ties with the homeland and who can be seen as a source of money, ideas and skills for future development. India, China and Mexico are frequently cited in terms of diasporas that are regarded as contributing significantly to development.

3.1 Globalisation

An overview

Andrew Herod

In the past two decades the word 'globalisation' has become ubiquitous. However, as a descriptor of changes taking place in the planet's contemporary political economy 'globalisation' is a highly contested term. Moreover, the connection between 'globalisation' as a discursive construct and how we understand the material processes which are linking places across the globe together ever more tightly is central in much of the debate about what globalisation may or may not be and about how it is supposedly playing out historically and geographically. Hence, are growing flows of cross-border trade and information evidence of 'globalisation' or merely the latest phase in processes of economic, political, and cultural 'internationalisation' that have been unfolding over millennia? Such a question is important, for how we define the growing connections across space – as evidence of 'globalisation' or instead of 'internationalisation' – dramatically shapes what we think is going on.

In this context, here I want to address four things:

1. distinctions between 'globalisation' and 'internationalisation';
2. whether globalisation is a fairly novel process or something that has been unfolding for centuries;
3. whether or not globalisation is an inevitable process; and
4. the relationship between 'the global' scale and other geographical scales of social organisation.

Finally, I briefly discuss why these questions are important for thinking about development.

Globalisation or internationalisation – one or the other (or both)?

For some, there is no difference between the concept of globalisation and that of internationalisation – they see these two terms as equivalents. For others, however, there are significant distinctions between them. Thus, whereas by definition *inter*nationalism takes as its scalar referent the nation-state and looks at the relationship between various nation-states, globalisation instead takes as its scalar referent the globe. Arguably, the division between these two ways of viewing what is seen to be happening in the early twenty-first century – internationalisation or globalisation – has been most clearly articulated by, on the one hand, Japanese management guru Kenichi Ohmae and, on the other, the British academics Paul Hirst and Grahame Thompson.

For his part, Ohmae argued in his provocatively titled 1995 book, *The End of the Nation State*, that the nation-state as a politico-economic institution is becoming defunct. Thus he claimed that growing cross-border economic linkages are leading to its evisceration as a market regulator and its replacement with regional economic assemblages that pay little attention to national boundaries (in this regard, see also his 1990 book, *The Borderless World*). By way of contrast, Hirst and Thompson (1996) argue that nation-states are still important and that growing planetary economic integration has not been 'global' but, rather, has been quite geographically uneven – the most significant economic relationships, they suggest, are still between North America, Western Europe, and East Asia. For them, 'hyperglobalists' like Ohmae have overplayed their hand in describing what is going on and nation-states remain very powerful economic regulators. Indeed, nation-states are actually necessary to facilitate the kinds of economic integration that Ohmae sees as being evidence of globalisation – only nation-states have the legal authority to sign free trade agreements into law, for instance.

In light of these differences of opinion, it is important to note that debates about the emergence of globalisation frequently discursively counterpose the hyperglobalists and the globalisation sceptics to generate an either/or proposition – *either* the nation-state is becoming increasingly irrelevant as an economic actor, *or* it still retains significant structural capacities. Such a representation means that for some the era of globalisation follows that of internationalisation – there is a distinct historical break and globalisation emerges out of the ashes of the international system of strong nation-states which has dominated the planet's political economy for about two centuries. For others, though, contemporary developments represent little more than a continuation of longstanding processes and the argument that we are now in a 'globalised' world is exaggerated. There has been, in other words, no stark historical break with the past. However, casting the debate in terms of what is happening to *the* (singular) nation-state in the early twenty-first century ignores the complexity of the situation, for it fails to recognise that different nation-states have different structural capacities. Rather than thinking of what is happening to *the* nation-state, then, it is perhaps more productive to think about what is happening to nation-state*s*. Doing this means that we can see some nation-states as still strong or even getting stronger and so quite capable of shaping how the global economy unfolds whereas others are relatively weak and/or are becoming weaker and are largely condemned to ride on the waves generated by the planet's powerful economic actors. This latter perspective allows for a more nuanced position, for it recognises that processes and practices of both globalisation and internationalisation may be playing out simultaneously, if in historically and geographically uneven ways for different nation-states. Consequently, instead of an either/or proposition, we can have a both/and understanding of the relationship between globalisation and internationalisation.

Finally, much analysis has centred upon the question of what processes of globalisation are doing *to* the nation-state – are they weakening it or not? However, it is important also to recognise that nation-states can be significant drivers of globalisation. It is national governments, for instance, that legislate the deregulation of financial markets. Recognising this fact means that rather than viewing the relationship between 'globalisation' and the nation-state as unidirectional – globalisation impacts the nation-state – it is important to see it as a two-way relationship, one in which globalisation impacts nation-states' structural capacities but also one in which nation-states concomitantly shape how globalisation plays out.

Is globalisation new?

One of the key points of debate between hyperglobalists and globalisation sceptics is whether globalisation is a relatively new phenomenon or one which has been unfolding for hundreds, if not thousands, of years. For example, Ohmae (2005: 28) has portrayed the last two decades of the twentieth century as marking an 'event horizon' of sorts, with the end of the cold war and several other developments coalescing to sow the seeds 'of a variety of plant not previously grown [globalisation], belonging to a totally novel and unknown genus and species'. Indeed, he has even argued (p. 245) that if companies are to succeed in the new global economy they must be 'genetically different' from the companies of the pre-global age – they must have 'a different set of chromosomes.' For hyperglobalists, then, the emergence of 'globalisation' is not simply the continuation of processes that have been ongoing for centuries but, rather, it represents a fundamental break with the past.

Critics of such a position, though, have argued that the processes that we today see as marking the ongoing march of globalisation have ancient origins. Andre Gunder Frank and Barry Gills (1992), for instance, have maintained that the world system, in which different parts of the planet have been increasingly connected together through trade, investment, and migration, is at least 5,000 years old. Political scientist David Wilkinson (2003) has likewise argued that the origins of globalisation are quite old, suggesting that there are different time-frames for the assorted elements that have contributed to the contemporary state in which humanity finds itself – the 'discovery' of North America by Europeans 500 years ago, the colliding 3500 years ago of two localised civilisations, one which had arisen in the Nile Valley and the other in Mesopotamia, to become a single expansionary civilisation that continued to engulf others with which it came into contact, or even the movement out of East Africa by *Homo erectus*.

By way of contrast, Immanuel Wallerstein (1974; 1980) has suggested that contemporary capitalist globalisation really has its origins in the early sixteenth century emergence of what he calls a world system, in which the activities of entities like the British East India Company increasingly connected Europe to Asia and the Americas. If this spread of merchant capital in the sixteenth and seventeenth centuries represented the first tranche in what we would today call globalisation, then the growth of cross-border financial speculation in the nineteenth century, facilitated by the extension of a planetary telegraph cable network by which capital could easily be transferred from place to place, marked a second (Herod, 2009). For many such writers, then, it is only in the twentieth century that what commentators like Ohmae have called 'globalisation' – the cross-border integration of manufacturing processes – has really emerged.

Is globalisation inevitable?

For neoliberal writers, the end of the cold war and the supposed victory of free-market capitalism that it augured have resulted in a world in which the future will always be more globalised than was the past. Hence, Lowell Bryan and Dianna Farrell (1996: 10) have maintained that we are now at a stage of globalisation from which '[w]e cannot go backward', whereas Ohmae (2005: 18) has contended that '[t]he global economy . . . is going to grow stronger rather than weaker . . . It is irresistible'. Such market triumphalism has been represented by the neoconservative intellectual Francis Fukuyama (1992: xii) as marking the 'end of history', wherein the triumph of Western liberal democracy and capitalism over Sovietism means that free-market capitalism will now literally have the entire planet across which to unfold its inevitable global end game. Significantly, this TINA (there is no alternative) argument that neoliberal globalisation is inevitable has often been presented in the language of biology (à la Ohmae above), an attempt, perhaps, to naturalise capitalist economic and political processes.

In contradistinction to those who see neoliberal economic globalisation as an unstoppable juggernaut, critics of the TINA advocates have argued that globalisation is not, in fact, inevitable. For instance, anti-globalisation activist and political scientist Susan George (2004) has argued for the proposition TATA – 'there are thousands of alternatives' – whilst the alter-globalisation World Social Forum group has declared that 'another world is possible'. Still others have suggested that 'there are many alternatives' (TAMA). For these activists and writers, efforts to make globalisation seem inevitable are part of a strategy of undermining opposition to it, of seeking to create a self-fulfilling prophesy – if people believe something is inevitable they are unlikely to expend much energy challenging it. The hyperglobalists' critics, then, point to all sorts of organising that seeks either to challenge globalisation per se (e.g. various nationalist groups) or which seeks to implement non-neoliberal versions of globalisation à la Karl Marx's 'workers of all lands, unite!' model.

Conceiving the global as a scale of social organisation

In much of the rhetoric around globalisation the global has been taken as the *primus inter pares* of geographical scales, the spatial resolution from which there is no escape. There are (at least) two important things to consider in this representation. First, whereas some have assumed that the global exists 'out there' as a scale of social organisation, just waiting to be discovered and used, others – typically of a more materialist persuasion – have argued that the global scale is not a scale inherent to human social organisation but is one that must be actively created by myriad social actors who link the different parts of the planet together through their labours. The global scale upon which globalisation turns, they insist, is made and not simply revealed. Recognising this is important because it undermines any sense of humans inexorably reaching for some preformed global scale in the manner suggested by those who see globalisation as predestined.

Second, though, globalisation has often been presented in terms of bringing about the 'delocalisation' and/or 'denationalisation' of economic and political life. This raises important questions about how the global is imagined to relate to other scales of social existence like the national and the regional. In particular, such scales have tended to be viewed either as being discrete entities or in networked terms. With regard to the first perspective, the global is sometimes imagined in verticalist terms (with the global scale sitting atop all other scales as if it were the highest rung on a ladder) or in horizontalist terms (with scales viewed as interlocked concentric circles, wherein the global is the outermost ring containing other scales that become progressively smaller as one moves towards the centre of the set of circles). In both cases, though, the global is portrayed as a discrete resolution of social life, one that is either 'above' all other scales (the rung metaphor) or which 'contains' them (the circle metaphor) (for a diagrammatic representation, see Herod, 2010: 15). However, French social theorist Bruno Latour (1996: 370) has suggested that, rather than seeing scales in discrete areal terms, such that the boundary between, say, the national and the global scales is easily discernible, it is more productive to view scales as 'fibrous, thread-like, wiry, stringy, ropy, [and] capillary' – that is to say, as rhizomic, with the global, the national, or the urban viewed not as separate spatial arenas but as locations along various parts of networks, as a terminology for distinguishing shorter and less-connected networks from longer and more-connected ones. Such an approach results in quite different metaphors for describing scalar relations. Hence, rather than scales being conceived of as larger or smaller circles or higher or lower rungs on a ladder, they may be conceived of in terms of locations on tree roots or perhaps a spider's web, where the end of one part of the network and the beginning of another is more difficult to differentiate (Herod, 2010: 49–52).

What does this mean for thinking about development?

The issues discussed above have great import for how we think about development. For example, supposing that nation-states are relatively weak and that the world has moved from an era of internationalisation to one of globalisation means that statist policies to encourage economic development might be presumed to be less effective than they once were, thereby eliminating consideration of the nation-state as a potentially useful agent of development. For its part, imagining globalisation to be inevitable suggests that economic decision-makers have little flexibility to challenge the onward march of neoliberal capitalism and should just get out of the market's way – any limits placed on the market, in other words, are doomed to fail, such that market-led development becomes the only option considered viable. Likewise, supposing that globalisation is new serves to cut off present patterns of development from what went before and so may cause us to misunderstand how these patterns originated – it may lead decision-makers to imagine that they can implement in a relatively short space of time things that actually took decades or even centuries in other parts of the world to come about. Moreover, it tends to minimise the impacts of imperialism which have shaped the global economy and so may end up 'blaming the victim' – the underdevelopment of places like the Central African Republic or Bangladesh become interpreted as the result of conditions inherent in those countries (e.g. high birth rates) rather than as the result of their historical relationships with Europe. Foreshortening the historical narrative, then, can lead policymakers to implement quite different plans to stimulate economic development than if they had a longer historical view – urging population control strategies rather than policies aimed at compensating for the colonial legacy through favoured access to former colonial powers' domestic markets or perhaps making transfer payments to build infrastructure. By the same token, how we conceive of scales and the manner in which they are related profoundly shapes how we understand economic and political possibilities. Thus, do we view firms like Apple, General Motors, and McDonald's as 'global' or as 'multi-locational'? This question is important because viewing a corporation as 'global' may imbue it with more imagined power, which could lead economic actors to believe that it is less regulatable, than might viewing it as simply 'multi-locational'. Such beliefs have significant implications for how development strategies are advanced and implemented.

Bibliography

Bryan, L. and Farrell, D. (1996) *Market Unbound: Unleashing Global Capitalism*. New York: Wiley.

Frank, A. G. and Gills, B. K. (1992) 'The five thousand year world system: An interdisciplinary introduction'. *Humboldt Journal of Social Relations* 18(1): 1–79.

Fukuyama, F. (1992) *The End of History and the Last Man*. New York: Free Press.

George, S. (2004) *Another World Is Possible If . . .* London: Verso.

Herod, A. (2009) *Geographies of Globalization: A Critical Introduction*. Chichester, UK: Wiley-Blackwell.

Herod, A. (2010) *Scale*. London: Routledge.

Hirst, P. and Thompson, G. (1996) *Globalization in Question: The International Economy and the Possibilities of Governance*. Cambridge: Polity.

Latour, B. (1996) 'On actor-network theory: A few clarifications'. *Soziale Welt* 47: 369–381.

Ohmae, K. (1990) *The Borderless World: Power and Strategy in the Interlinked Economy*. New York: HarperBusiness.

Ohmae, K. (1995) *The End of the Nation State: The Rise of Regional Economies*. New York: McKinsey and Company.

Ohmae, K. (2005) *The Next Global Stage: Challenges and Opportunities in Our Borderless World*. Upper Saddle River, NJ: Wharton School Publishing.

Wallerstein, I. (1974) *The Modern World-System. Volume I: Capitalist Agriculture and the Origins of the European World-Economy in the Sixteenth Century.* New York: Academic Press.
Wallerstein, I. (1980) *The Modern World-System. Volume II: Mercantilism and the Consolidation of the European World-Economy, 1600–1750.* New York: Academic Press.
Wilkinson, D. (2003) 'Globalizations: The first ten, hundred, five thousand and million years'. *Comparative Civilizations Review* 49 (Fall): 132–145.

3.2

The new international division of labour

Alan Gilbert

For at least two decades, both the academic literature and the popular media have been obsessed with globalization. In the process, a New International Division of Labour (NIDL) has been created. But while the World Bank (1995: 1) announces that: 'these are revolutionary times in the global economy' and the ILO (1995: 68–69) declares that: 'globalization has triumphed', it is less obvious what precisely has changed. Certainly, the world is still not flat (Friedman, 2005) and while many parts of the world participate actively in the NIDL, others still play a rather peripheral role. What few really understand, although many claim to, is what effect the NIDL is having on our lives. Perhaps the only certain answer is that it depends on who you are and where you live; some people are doing very well in the NIDL whereas others are most certainly not.

What is the NIDL?

According to Held and McGrew (2002: 1)

> Globalization, simply put, denotes the expanding scale, growing magnitude, speeding up and deepening impact of transcontinental flows and patterns of social interaction. It refers to a shift or transformation in the scale of human organization that links distant communities and expands the reach of power relations across the world's regions and continents.

This has produced an NIDL that is difficult to define precisely but incorporates the following ingredients:

First, most areas of the world now constitute part of the global market and in 2008 exports of goods and services made up one-third of world GDP (UNCTAD, 2009: 18). Increasingly we all consume similar products and are bombarded with the same kinds of advertising. It is doubtful whether many people around the world would fail to recognize the names Coca-Cola, Nike, Ford and Sony. The products and images of these companies dominate our television screens and our streets.

Second, manufacturing production is no longer confined to a relative handful of industrialized countries. The production of clothes, shoes, bicycles and televisions has become global. Transnational companies increasingly produce their goods in countries where labour is cheap and political conditions are stable, for example, China, Indonesia, Korea, Thailand, Mexico and the Dominican Republic. As a result the manufactured exports from poorer countries have increased greatly.

Third, the investment and portfolio capital flowing across the globe has grown immensely (Munck, 2005). Between 1982 and 2007 the total stock of foreign direct investment increased from US$790 billion to $14,909 billion (UNCTAD, 2009: 18). International financial transactions rose even more rapidly. According to Dicken (2010: 369), daily foreign exchange transactions in 1973 were roughly twice that of world trade; by 2007 they were a hundred times larger. Many of those financial transactions are speculative in nature, hence Strange's (1986) creation of the term 'casino capitalism'.

Fourth, large companies have become more important players in the world economy; 'Multinational corporations now account, according to some estimates, for at least 25 per cent of world production and 70 per cent of world trade, while their sales are equivalent to almost 50 per cent of world GDP' (Held and McGrew, 2002: 53). Today, most large companies operate globally and few retain a close allegiance to a single country. Most Volkswagen cars are no longer made in Germany; the company has plants in Brazil, China, Mexico, Slovakia and many other places. Major firms have also emerged in the world of banking, law and accountancy (Dicken, 2010). Transnational corporations are the new global brokers, responsible for most of the investment flows flushing around the world system.

Finally, NIDL has increased mobility and some 214 million people lived abroad in 2010 (World Bank, 2011). Skilled and unskilled workers have been crossing borders in search of work, and sometimes for protection. Although these flows are still proportionately smaller than those that occurred in the nineteenth century, in 2006 approximately 19 per cent of all Mexicans, 16 per cent of all Salvadorans, and 11 per cent of all Cubans and Dominicans were living in the United States. As a result, most countries have become ethnically more diverse (Bidwai, 2006). Another consequence is that most migrants send money home, a sum estimated at US$440 billion in 2010 (World Bank, 2011). In 2009, remittances from emigrants provided more than a quarter of GDP in Tajikistan and Tonga and more than one-fifth in Lesotho, Moldova, Nepal, Lebanon and Samoa (World Bank, 2011).

Nevertheless, the amount of globalization should not be exaggerated. Despite claims that the world is now flat, globalization has not affected every country or national region equally (Friedman, 2005). As Bidwai (2006: 31) puts it:

> Today's globalised world is deeply contradictory. On the one hand, there is growing interdependence, exchange and interaction between many different parts of the globe. On the other hand, there are huge swathes of land that are virtually excluded from any meaningful interaction with the rest of the world. They have experienced stagnation or decline, want and insecurity, mounting social chaos, and even outright economic and political devastation through war and famine. About two-fifths of the world's people live in such societies.

Northern Mexico is clearly part of NIDL, Myanmar and North Korea are definitely not. One of the great concerns about the NIDL is how it has supposedly marginalized substantial parts of the world. While many Africans occasionally buy global products, very few global products are made there.

And even in countries that have clearly been strongly globalized, many elements of life retain their national characteristics. According to Ghemawat (2011) only 2 per cent of students study at universities outside their home countries and only 3 per cent of people live outside their country of birth. In the United States only 7 per cent of directors of S&P 500 companies are foreign. In addition, even if we seemingly consume the same things, food tastes remain national and regional. The British may now eat curries and sweet-and-sour chicken but they still eat fish and chips.

Is NIDL new?

The technological innovations that have allowed the development of rapid transport links and instant electronic communication are definitely new. But many of the changes have a longer history. They may be occurring on a larger scale and at a faster pace than ever before but the massive movement of capital, agricultural products, manufactures, people and ideas has actually been under way for centuries (Gilbert, 1990). In this sense NIDL is definitely not new. It is merely the latest in a series of major restructurings of the world economy. Walton (1985) has argued that at least two NIDLs have preceded the present one. NIDL mark one was brought about by Europe dividing the world into colonies and reorganizing production and markets in the new colonies. NIDL mark two occurred when previously semi-developed areas of the world began to industrialize from the end of the nineteenth century. The process of import substituting industrialization created major industrial concentrations in countries such as Argentina, Brazil, China, India, Mexico and South Africa. Despite their continuing poverty, these countries contained some of the world's largest industrial economies in 1960.

The current situation, which I will call NIDL mark three, is a highly significant development on those earlier shifts. NIDL mark one led to the decimation of aboriginal peoples in Latin America and Australasia, the slave trade across the Atlantic, the incorporation of new food products into the European diet and their production in the colonies, certainly constituting as significant a change to the world as the events of the past thirty years or so. Think of the diet of the average Briton before the potato, the banana, tobacco, sugar cane, tea and coffee reached these shores. Around 10 per cent of the world's population left one country for another between 1870 and 1914 (World Bank, 2002: 3). NIDL mark two led to the growth of industry and major cities in the periphery of Europe, North Africa, South Africa, India, China and much of Latin America. Again, this represented a major shift in the organization of world production.

What effect has NIDL mark three had on the world at large?

Authors like Bhagwati (2004) and Wolf (2004) view NIDL's impact to be positive whereas UNRISD (1995) and Milanovic (2003) give greater emphasis to its downsides. The difference in opinion is partly due to the variability of NIDL3's effects: whether you gain or lose depends on who you are and where you live. But it is also down to perspective.

To its critics globalization has increased inequality. For, if the world is becoming a richer place, dire poverty remains well entrenched and in many places people are actually becoming poorer. The unleashing of fierce competition between nations has led to Western Europe,

North America and parts of the Far East increasing their wealth while most of Africa and parts of Asia and Latin America have been losing out (Held and McGrew, 2002: 1). But this interpretation is not supported by estimates of global inequality which show that over the last thirty or so years the world's Gini coefficient has diminished (World Bank, 2005). Although Wade (2004) has reservations about the calculations that underpin this conclusion there can be little doubt that the rise of BRIC nations (Brazil, Russia, India and China) has changed the world map and brought greater equality between rich nations and at least some former poor ones.

The rise of India and China is also producing more equality in another way, through their seemingly insatiable demand for raw materials. While this has generally caused problems for rich countries, it has generated an export boom for many producers in Africa and Latin America. At the same time as Chinese exports have brought problems for many manufacturers in Argentina, Brazil, Colombia and Peru, those economies have prospered as a result of the commodity boom.

But if a case can be made that the gap between many rich and poor countries is declining, globalization, in association with neoliberal economic practice, has been changing the distribution of income within countries. If poverty is not actually increasing in most places, there is no doubt that virtually every country in the world is becoming more unequal (Gilbert, 2007; World Bank, 2005). After decades of more equitable growth, since 1980 most countries in the world have seen the incomes of the rich leap ahead. In the United States,

> between 1970 and 2008 the Gini coefficient . . . grew from 0.39 to 0.47. In mid-2008 the typical family's income was lower than it had been in 2000. The richest 10% earned nearly half of all income, surpassing even their share in 1928, the year before the Great Crash.
>
> *(The Economist, 2010)*

And, despite its claims to be a nation that provides easy routes to upward social mobility, the paths appear to be narrowing (Hutton, 2002: 152). See also Figure 3.2.1.

Figure 3.2.1 Income share of the rich in the US, excluding capital gains (%)

Source: Adapted from Piketty and Saez (2003)

The effect on the state

To some, a worrying feature of NIDL3 is the change it has brought in the role of the state. Power has shifted from the nation-state to the transnational corporation.

> The world's 37,000 parent trans-national corporations and their 200,000 affiliates now control 75 per cent of all world trade in commodities, manufactured goods and services. One-third of this trade is intrafirm – making it very difficult for governments and international trade organizations to exert any control.
>
> *(UNRISD, 1995: 27)*

> The decisions of private investors to move private capital across borders can threaten welfare budgets, taxation levels and other government policies. In effect, the autonomy of states is compromised as governments find it increasingly difficult to pursue their domestic agendas without cooperating with other agencies, political and economic, and above and beyond the state.
>
> *(Held and McGrew, 2002: 23)*

Worse still is that many governments have simply lost control to new and often sinister groups.

> Everything from gangs and criminal cartels, narco-trafficking networks, mini-mafias and favela bosses, through community, grassroots and non-governmental organizations, to secular cults and religious sects proliferate. These are the alternative social forms that fill the void left behind as state powers, political parties, and other institutional forms are actively dismantled or simply wither away as centres of collective endeavour and of social bonding.
>
> *(Harvey, 2005: 171)*

However, in many countries the role of the state has changed rather than diminished. The experience of the East Asian 'tigers' shows that global markets are entirely compatible with strong states (Held and McGrew, 2002: 47). State power is still intact because transnational corporations cannot run the world alone and rely on national governments to perform a series of important local tasks (Dicken, 2010). The state is now concerned less with protecting its national citizens than with creating the conditions which will attract foreign investment. As Mittelman (1994: 431) puts it: 'the state no longer serves primarily as a buffer or shield against the world economy. Rather, the state . . . increasingly facilitates globalization, acting as an agent in the process'.

Some argue that the power of transnational corporations has forced states to lower the level of taxation and corporate 'tax termites' have been eating away at national budgets through the use of off-shore financial centres and dubious forms of 'transfer pricing'. Yet, Bhagwati (2004: 101) points out that 'the total tax burden of the members of the OECD has in fact increased over the last thirty years, from 26 per cent of GDP in 1965 to 37 percent of GDP in 1997, despite Reagan, Thatcher, and globalization.' Figure 3.2.2 strongly supports that point.

Is NIDL3 stable?

The World Bank (1995: 55) claims that: 'Global markets are not only larger than any single domestic market but generally more stable as well'. But, as Barber (1992) long ago claimed: 'All

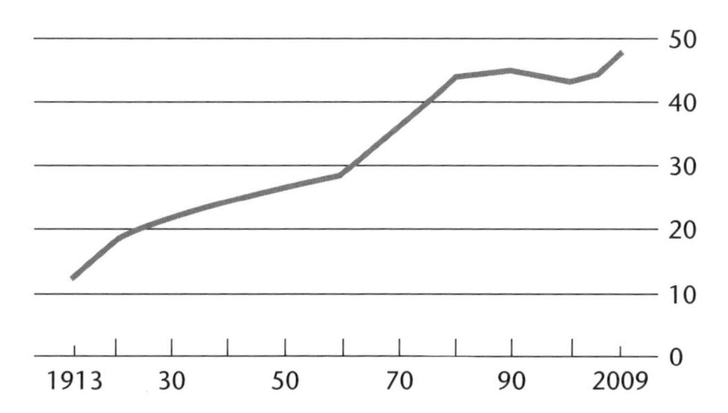

Figure 3.2.2 Average of government spending in 13 rich-world countries as % of GDP

Source: Adapted from Micklethwait (2011)

national economies are now vulnerable to the inroads of larger, transnational markets within which trade is free, currencies are convertible, access to banking is open, and contracts are enforceable under law'. The fortunes of individual countries are likely to ebb and flow in a highly competitive world market. Chinese exports are undermining industrial activity almost everywhere and Colombia's position as the world's second largest exporter of cut flowers could easily disappear in the face of competition from Ecuador and Kenya.

Financial flows are a still more problematic form of globalization, worrying even the defenders of NIDL3 (Stiglitz, 2005; Bhagwati, 2004). As Fernández-Arias and Hausmann (2000: 3) put it:

> Financial liberalization and integration have not worked out as advertised . . . Instead, emerging markets have been rattled by financial turmoil, especially during the past two to three years. Depending on one's viewpoint as optimist or pessimist, financial integration and globalization have either generated excessive volatility or run amok.

According to Harvey (2005: 94): 'unregulated financialization plainly posed a serious danger of contagious crises'. And, as Dicken (2010: 373) notes, the history of the twentieth century provides clear evidence that 'the inherent instability of financial markets creates periodic large-scale upheavals'.

Certainly, the economic crises that hit Mexico in 1994, the Far East in 1997–1998 and Brazil in 1999 demonstrate that in an era of global financial markets, economic disaster always lurks just around the corner. And, of course, the events of 2008–2009 have demonstrated just how unstable financial systems and capital flows are in a globalized world. While it can be justifiably claimed that the problems of Europe and the United States were caused by uncontrolled private and public spending at home, globalization helped that process. Without the trade surpluses of China, Germany and Japan being reinvested in dollars and government bonds the growing deficits of the United States and southern Europe would not have been possible (Lanchester, 2009; Milanovic, 2011).

Some would also claim that political discontent will spread under NIDL3. Since globalization has encouraged the spread of democracy, governments are arguably less stable because voters regularly throw them out. This was clearly seen in Mexico in 2000 with the fall from grace of the PRI, a party that had held power constantly since 1928. This transition occurred peacefully but elsewhere populations have been more radical and hostile. Is this what underlay the protests in the Arab Spring of 2011? Increasing knowledge of how other countries are governed and growing awareness of the corruption of their leaders has led to local populations rebelling against authoritarian rule in Egypt, Libya, Syria and Tunisia.

Or is political conflict also motivated by the effects of globalization unleashing atavistic feelings that encourage people to return to their supposed roots; 'subterranean cultural pluralism' as Mittelman (1996: 8) describes it? Arguably, religion, language, ethnicity and culture are becoming increasingly influential; dividing countries, turning native populations against immigrants and unleashing terrorist attacks across the world. The result may be a

> Lebanonization of national states in which culture is pitted against culture, people against people, tribe against tribe – a Jihad in the name of a hundred narrowly conceived faiths against every kind of interdependence, every kind of artificial social cooperation and civic mutuality.
>
> *(Barber, 1992)*

If globalization is causing resentment against new compromises and commitments, economic downturns may also stimulate discontent especially when the causes are thought to lie externally. The 2011–2012 protests in Greece, for example, with its undercurrent of feeling against German economic bullying, are not dissimilar to the austerity riots that greeted many governments' acceptance of IMF structural adjustment agreements in the 1980s (Walton, 1998).

The need for multinational governance

During the twentieth century most people in developed countries were gradually protected by the emergence of a welfare state; and Keynesian economic management enabled national governments to control the worst excesses of economic instability. But now that the world economy is global, 'the role of "moderator" can no longer belong to the nation-state, but to international (global) actors. It is where the international financial institutions, such as the World Bank, enter' (Milanovic, 2003: 679).

Unfortunately, some argue that these global actors do not act fairly. Stiglitz (2005) claims that globalization has favoured the special interests in the North rather than those of the poor 'At the IMF, it is only finance ministers and central bankers whose voices are heard; in trade negotiations, it is the trade ministers, often with close links of commercial and financial interests, who set the agenda' (Stiglitz, 2005: 228–229).

If the poor and indeed the majority are to be protected from the unfavourable side of globalization, there is increasing agreement that a new form of world governance is needed (ODI, 1999; ILO, 1995). To counter the power of the transnational corporations, new forms of multinational government must be created. The emergence of major political alliances and trade blocs like the European Union and NAFTA were meant to be a step in this direction. But others would argue that we need world institutions that can control financial flows and tax capital movements. Only in this way will the undesirable face of NIDL be controlled. Without the creation of such a level of government, tax havens, drug flows, international crime, environmental devastation and labour exploitation will get wholly out of control. In this sense, the role of the state with respect to the market has not changed. It is just that NIDL3 had created a situation where a global form of government is required to help the disabled national state look after those who are less able to compete in the new competitive world.

Bibliography

Barber, B. R. (1992) 'Jihad vs. McWorld', *Atlantic Magazine*, March.

Bhagwati, J. (2004) *In Defense of Globalization*, New York: Oxford University Press.

Bidwai, P. (2006) 'From what now to what next: Reflections on three decades of international politics and development', *Development Dialogue* 47, 29–64.

Dicken, P. (2010) *Global Shift: Transforming the World Economy* (6th edn), London: Sage Paul Chapman.

The Economist (2004) 'Economics focus: Monetary lifeline', *The Economist*, 31 July 2004.

The Economist (2010) 'Upper bound', *The Economist*, 15 April.

Fernández-Arias, E. and Hausmann, R. (2000) *What's wrong with international financial markets?* (Inter-American Development Bank, Research Department Working Paper No. 429). New York: Inter-American Development Bank.

Friedman, T. L. (2005) *The World Is Flat: A Brief History of the Twenty-first Century*, New York: Farrar, Straus & Giroux.

Ghemawat, P. (2011) *World 3.0: Global Prosperity and How to Achieve it*, Boston, MA: McGraw-Hill.

Gilbert, A. G. (1990) 'Urbanisation at the periphery: Reflections on the changing dynamics of housing and employment in Latin American cities', in D. Drakakis-Smith (ed.) *Economic Growth and Urbanisation in Developing Areas*, London: Routledge, pp. 73–124.

Gilbert, A. G. (2007) 'Inequality and why it matters', *Geography Compass* 1(3), 422–447.

Harvey, D. (2005) *A Brief History of Neoliberalism*, Oxford: Oxford University Press.

Held, D. and McGrew, A. (2002) *Globalization/Anti-globalization*, London: Polity.

Hutton, W. (2002) *The World We're in*, London: Little Brown.

ILO (International Labour Office) (1995) *World Employment: An ILO Report: 1995*, Geneva: ILO.

Lanchester, J. (2009) *Whoops: Why Everyone Owes Everyone and No One can Pay*, New York: Allen Lane.

Mickelthwait, J. (2011) 'Taming Leviathan', *The Economist*, 19 March, 3–6.

Milanovic, B. (2003) 'The two faces of globalization: Against globalization as we know it', *World Development* 31(4), 667–683.

Milanovic, B. (2011) *The Haves and the Have nots*, New York: Basic Books.

Mittelman, J. H. (1994) 'The globalisation challenge: Surviving at the margins', *Third World Quarterly* 15, 427–443.

Mittelman, J. H. (ed.) (1996) *Globalization: Critical Reflections*, Boulder, CO: Lynne Rienner Publishers.

Munck, R. (2005) *Globalization and Social Exclusion: A Transformative Perspective*, Bloomfield, CT: Kumarian Press.

ODI (1999) 'Global governance: An agenda for the renewal of the United Nations?' *ODI Briefing Paper* (July).

Piketty, T. and Saez, E. (2003) 'Income inequality in the United States, 1913–1998', *Quarterly Journal of Economics* 118(1), 1–39.

Stiglitz, J. (2005) 'The overselling of globalization', in M. M. Weinstein (ed.) *Globalization: What's New?* New York: Columbia University Press, pp. 228–261.

Strange, S. (1986) *Casino Capitalism*, Oxford: Blackwell.

UNCTAD (2009) *World Investment Report 2009: Transnational Corporations, Agricultural Production and Development*, New York: United Nations.

UNRISD (1995) *States of Disarray*, Geneva: UNRISD.

Wade, R. H. (2004) 'Is globalization reducing poverty and inequality?' *World Development* 32, 567–589.

Walton, J. (ed.) (1985) *Capital and Labour in an Industrializing World* (ch. 1), London: Sage.

Walton, J. (1998) 'Urban conflict and social movements in poor countries: Theory and evidence of collective action', *International Journal of Urban and Regional Research* 22, 460–481.

Wolf, M. (2004) *Why Globalization Works*, London and New Haven: Yale University Press.

World Bank (1995) *World Development Report 1995*, Oxford: Oxford University Press.

World Bank (2002) *Globalization, Growth, and Poverty: Building an Inclusive World Economy*, New York: World Bank and Oxford University Press.

World Bank (2005) *World Development Report 2006: Equity and Development*, New York: Oxford University Press.

World Bank (2011) *Migration and Remittances Factbook 2011*, Washington, DC: World Bank.

3.3

Global shift

Industrialization and development

Ray Kiely

The last 30 years have seen a global shift in the international division of labour, in which some parts of the former 'Third World' have become newly industrializing countries (NICs). This is most visible in the case of East Asia, and particularly China in recent years, but can also be seen in other parts of the developing world. The old, colonial-based division of labour in which it was said that the 'advanced' capitalist countries produced the industrial goods and the 'Third World' produced the primary goods was always an oversimplification; now it is simply inaccurate. Thus, according to United Nations Development Programme (UNDP) figures, by the late 1990s almost 50 per cent of manufacturing jobs were located in the developing world and over 60 per cent of developing country exports to the so-called 'First World' were of manufactured goods, a 1,200 per cent increase since 1960 (UNDP 1998: 17). The share of the developing world in global manufacturing exports has increased substantially, from 4.4 per cent in 1965, to 30.1 per cent in 2003 (Glyn 2006: 91).

One explanation for these changes can be found in the growing globalization of production. Transnational companies (TNCs), which operate in at least one country beyond that of origin, are major agents in this globalization process. They may invest beyond their own country to take advantage of market access, cheap labour, lack of regulation (such as rules concerning the environment) or access to raw materials. Apologists argue that TNCs are therefore developmental, providing host countries with income, employment, technology and so on. Critics argue that TNCs are agents of exploitation, and that they distort the development of nation-states. For instance, the use of cheap labour amounts to super-exploitation, and intra-firm trade and capital mobility allows these companies to evade tax payments.

This debate over the character of TNCs is also reflected in disputes over the nature of the changing international division of labour. Apologists such as neoliberals argue that the rise of some newly industrializing countries is evidence that the global economy is a level playing field on which any nation may develop as long as they follow the correct policies. In this way, more open investment, trade and financial policies allow developing countries to draw on the opportunities presented by globalization, thereby reducing global poverty (World Bank 2002).

While it is undoubtedly true that there have been enormous changes in the world economy over the last 30 years, many advocates who claim a substantial global shift of industrial production overstate their case. In particular, one sometimes has the impression that *productive* capital is as hyper-mobile as *financial* capital (which itself is selective in terms of where it locates), and that its movement from one part of the globe to another is a relatively unproblematic task. However, capital continues to concentrate in certain areas. This is because capital faces a number of 'sunk costs' which constitute significant barriers to exit. These may include start-up costs, access to local suppliers, and the acquisition of local trust and acceptance. Once established, growth tends to be cumulative, as earlier developers tend to monopolize technology and skills, established markets and access to nearby suppliers.

Of course, these advantages are never absolute, and later developers may leapfrog earlier outmoded production techniques. So, for example, Korean steel and shipbuilding industries developed and ultimately became more competitive than those of Britain in these industrial sectors. This was not because of a hyper-mobile productive capital that relocated from high-cost Britain to low-cost Korea, but was instead a product of a successful alliance between the Korean state and local capital in developing these industries.

Similarly, the partial move away from Fordist mass production to smaller-batch post-Fordist niche production may give some potential to late developers. However, the key point is that these changes do not entail the end of capital's tendency to agglomerate in certain parts of the world and thereby marginalize others, thus maintaining uneven development. Indeed, global, 'post-Fordist' flexible accumulation may intensify this tendency as suppliers locate even more closely to final producers as their stock is delivered on a regular just-in-time basis, as opposed to the old irregular just-in-case system.

In some sectors, the barriers to exit are less significant and so industrial capital is more mobile. This is especially the case in labour-intensive industries such as clothing, textiles and semi-conductors. In these and other sectors, fixed costs are lower as technology is not so advanced. This provides former Third World countries with potential competitive advantages over the 'advanced' capitalist countries, and is the source of growing concerns in the US, particularly in relation to its growing trade deficit with China. But even this advantage is a mixed blessing as employment (often of young women) may involve work for low wages in poor conditions. Employers in these factories are just as likely to be local capitalists as TNCs. Sometimes these employers may be suppliers to Western retailers, who focus their activities on design and marketing. It could be argued, as neoliberals contend, that this focus on low cost, labour-intensive production is a necessity for poorer countries, and that upgrading to higher value production will occur over time, just as it did for the now developed countries. The recent rapid growth of China is therefore good news, and it clearly shows the benefits of globalization in contrast to the closed policies of the Maoist era. One need not apologise for the Maoist period to question these upbeat claims. The key point is precisely that while low barriers to entry constitute a competitive advantage for developing countries over their more established competitors, precisely because the barriers are low the chances are that many enter a highly competitive environment. The risk then is of a race to the bottom, whereby many developing countries compete on the basis of cost-cutting, to the detriment of the potentially developmental dynamic experienced by those countries who upgraded in an earlier era (and through protectionist policies – see Chang 2002). This problem is exacerbated by a global reserve army of labour which can keep wages low, and substantial global over-capacity in many sectors which further drives down prices and leads competitors into yet another round of competitive cost-cutting (Kaplinsky 2005; Kiely 2007). Indeed, while there is much talk about US decline in the context of the global economic downturn from 2008 onwards, serious questions need to be asked about China's dependence on exports to the US market and the willingness of the Chinese state to continue to finance US deficits, which may be a sign of Chinese dependence at least as much as Chinese strength (Hung 2011).

Industrial production can therefore be said to be increasingly globalized, but the networks of production processes which lead to a finished commodity remain hierarchically structured. Gereffi (1994: 219) distinguishes two kinds of production process or commodity chain, which, although perhaps over-simplified, remains useful for understanding the nature of continued global hierarchies. First, producer commodity chains exist where the site of production is relatively immobile and so production agglomerates in favoured locations. Second, buyer commodity chains exist where there is greater mobility and labour intensity of production. This may give so-called peripheral areas certain advantages in terms of low labour costs, but in these industries barriers to entry exist at the level of brand name merchandising and retail levels. In the first case, marginalization occurs through absence of industrial investment; in the case of the latter, the value added by industrial production tends to be low, at least compared to the marketing and design stages.

The continued concentration of higher-value industrial production in selected areas is reflected in the figures on foreign investment. Most direct foreign investment is located in the 'developed' world – the global FDI share of developing countries for the period from 2003–5, was approximately 35 per cent, with Asia and Oceania's share standing at 21 per cent and South, South-East and East Asia's share standing at 18.4 per cent, compared to Latin America and the Caribbean's 11.5 per cent and Africa's share of just 3 per cent (UNCTAD 2006: 6–7). Moreover, investment in the developing world is itself highly concentrated. While the global FDI share of the top five developing countries has increased substantially, from around 11 per cent in 2000, to around 18 per cent in 2005 (UNCTAD 2006: 4), the share for most other developing countries has declined. Recent data suggest an increasing share of FDI for developing countries (UNCTAD 2010), but this is in the context of falling global amounts of FDI. Moreover, history suggests that shares tend to diverge once new foreign investment booms occur, with developed countries receiving the lion's share of FDI.

Foreign investment figures alone do not tell the whole story, as TNCs may raise investment capital from a variety of sources (international money markets, equities and so on), and production may involve cross-border production networks between formally independent firms – the buyer commodity chains discussed above, for example. However, the evidence suggests that in these areas there is also a high rate of concentration, not least in the trade in manufactured goods. But, perhaps most telling, *the type* of manufacturing that is generally occurring in the developing world is not necessarily overcoming underdevelopment. Since the early 1980s, while the developed countries' share of manufacturing exports fell (from 82.3 per cent in 1980 to 70.9 per cent by 1997), its share of manufacturing value added actually *increased* over the same period, from 64.5 per cent to 73.3 per cent. Over the same period, Latin America's share of world manufacturing exports increased from 1.5 per cent to 3.5 per cent, but its share of manufacturing value added fell from 7.1 per cent to 6.7 per cent (Kozul Wright and Rayment 2004: 14). For developing countries as a whole, manufacturing output's contribution to GDP has barely changed since 1960: it stood at 21.5 per cent in 1960, and increased to just 22.7 per cent in 2000.

The above outline contrasts sharply with neoliberal and (some) dependency perspectives outlined earlier. Both neoliberal and dependency theories assume that productive capital is hyper-mobile but go on to draw very different conclusions. For neoliberalism, this mobility means that capital will move from areas of abundance to scarcity in order to take advantage of lower costs in the latter. In the long run, so long as there is a global free market unhindered by the operations of interventionist nation-states, this will lead to a system of perfect competition between free and equal producers each exercising their respective comparative advantages. For some versions of dependency theory, this mobility means that capital can move to areas of lower costs in order to increase the rate of exploitation without promoting national development. The exposition above suggests that such a scenario does exist in certain sectors, but it cannot be generalised across the board. The result is the continuation of a core–periphery division of the world, as peripheral industrializers suffer from new forms of dependence.

Similarly, while neoliberals regard TNCs as modernizing agents, dependency-oriented writers regard them as agents of underdevelopment.

As already made clear, *both* positions exaggerate the degree of mobility of capital and this weakness leads to other problems. Clearly, given the tendency for capital to concentrate in certain regions and (relatively) marginalize others, neoliberal optimism concerning a level playing field in the global economy is seriously misplaced. On the other hand, the tendency of some dependency theorists to regard the newly industrializing countries as being simply peripheral industrializers is also inadequate. To conceptualize the world on the basis of a timeless core–periphery divide is ahistorical, and one is left with the feeling that whatever happens in the 'Third World' (for instance, industrialization or lack of industrialization) occurs because of the will of an all-powerful core, simply pulling the strings of a passive periphery.

Similarly, the debate on the developmental effects of TNCs is too black and white, with one side (neoliberalism) assuming that the effects are unproblematically favourable while the other (some dependency approaches) assumes that they are all bad. The impact of TNCs will depend on a number of specific factors, such as the particular sector in which the TNC operates, the role of the state in regulating TNC behaviour, and local resistance to the potentially bad effects of a particular transnational company. In a capitalist-dominated world, the question then moves away from a simple one of can or should particular countries open up or do without TNC investment, and instead becomes one of finding the best strategies for dealing with companies which may have different interests from those of the local population. However, given the global concentration of capital, the desperation of countries to attract foreign investment, and neoliberal hegemony in the international order, it is fair to say that the capacity (or perhaps even willingness) of states in the developing world to regulate (foreign and local) companies to behave in 'developmental' ways is seriously compromised.

The global economy therefore continues to be characterized by polarization, with some people and regions at the cutting edge of globalization while others are marginalized. Transnational companies tend to be highly selective in their choice of investment location, concentrating in parts of the former First World or selected parts of the former periphery, and being highly selective in the kinds of economic activity within particular locations. There is no longer a clear division of the world between core and periphery (though in fairness this division may never have been as clear as some underdevelopment theorists implied), but at the same time, this does not mean the end of uneven and unequal development. The world is divided into many cores and peripheries, many of which can be located *within* nation-states. At the same time, the rise of manufacturing in the developing world has not narrowed the gap between the developed and most developing countries.

References

Chang, H. J. (2002) *Kicking Away the Ladder*, London: Anthem. Important analysis of how developed countries themselves industrialized through protectionist policies, and an examination of the implications for contemporary developing countries.

Gereffi, G. (1994) 'Capitalism, development and global commodity chains', in L. Sklair (ed.) *Capitalism and Development*, London: Routledge, pp. 211–31. A useful summary of the theory of global commodity chains.

Glyn A. (2006) *Capitalism Unleashed*, Oxford: Oxford University Press.

Hung, Ho-Fung (2011) 'Sinomania: Global crisis, China's crisis?', in L. Panitch, G. Albo and V. Chibber (eds) *The Socialist Register 2011*, London: Merlin, pp. 217–34. Important article that links the globalization of production and China's rise to the current global economic crisis, and in doing so suggests good reasons to point to the limits of China's rise.

Kaplinsky, R. (2005) *Globalization, Poverty and Inequality*, Cambridge: Polity. Very useful examination of the links between global commodity chains, uneven development and poverty and inequality.
Kiely, R. (2007) *The New Political Economy of Development*, Basingstoke: Palgrave Macmillan. Examination of the current state of the international political economy (and geopolitics) of development, challenging upbeat claims made for the links between globalization and development.
Kozul Wright, R. and Rayment, P. (2004) 'Globalization reloaded: An UNCTAD perspective', UNCTAD Discussion Papers no. 167, pp. 1–50.
UNCTAD (2006) *World Investment Report 2006*, Geneva: United Nations. Annual report on investment flows in the global economy.
UNCTAD (2010) *World Investment Report 2010*, Geneva: United Nations. Annual report on investment flows in the global economy. Its generally upbeat nature becomes blurred after a close reading of the figures, and even at times the implicit argument, made more explicit in other UNCTAD reports and publications. See also other reports such as UNCTAD (2006) above.
UNDP (1998) *Globalization and Liberalization*, New York: United Nations Development Programme. Report on the social effects of development in the global economy.
World Bank (2002) *Globalization, Growth and Poverty*, Oxford: Oxford University Press. Well-known, if thinly argued, upbeat case for the claim that poverty reduction is a reality, which has been caused by pro-globalization policies.

Further reading

The following items afford an introduction to global shifts in industrialization.

Dicken, P. (2011) *Global Shift*, London: Sage, 6th edition.
Gereffi, G. and Korzeniewicz, M. (eds) (1994) *Commodity Chains and Global Capitalism*, Westport, CT: Greenwood Press.
Glyn A. (2006) *Capitalism Unleashed*, Oxford: Oxford University Press.
Held, D., McGrew, A., Goldblatt, D. and Perraton, J. (1999) *Global Transformations*, Cambridge: Polity.
Hoogvelt, A. (2001) *Globalisation and the Postcolonial World*, London: Macmillan.
Kiely, R. (1998) *Industrialization and Development: A Comparative Analysis*, London: UCL Press.
Kozul Wright, R. and P. Rayment (2004) 'Globalization reloaded: An UNCTAD perspective', UNCTAD Discussion Papers no. 167, pp. 1–50.

3.4

Globalisation/localisation and development

Warwick E. Murray and John Overton

Introduction

The concepts of globalisation and development at the local scale are intimately linked and yet there has been little attempt in the literature to explore these links theoretically or empirically. There are a number of controversies around which studies of this topic can be organised (see Murray, 2006, for an entry into these discussions). Underlying this is the central question: does globalisation make development more or less even? It is our contention here that the agendas and processes that comprise globalisation perpetuate differences between localities. This means that understanding the relationship between globalisation and local development is very complex and, therefore, regulating and reforming it is a difficult task.

Definitions

Most researchers agree that globalisation involves the stretching of social relations across space in ways that create local to local articulations which increasingly transcend national borders. This de-territorialisation through time-space distantiation gives rise to new networks of inclusion and pockets of exclusion. Driving the expansion of globalisation is the unfolding of capitalism and its inherent need to reduce the turnover time of capital resulting in agendas, technologies and resultant processes that create time-space compression (Harvey, 2003). Underlying the expansion of globalisation are cultural agendas and imperatives associated with Westernisation. We define globalisation as:

> a collection of dialectical human agency driven processes which create local–local and person–person networks of inclusion/compression that increasingly transcend territorial/national borders and stretch to become global in proportion. The processes are dialectical in that the relative social distance of those not on the net from those on the net widens as the intensity and extensity of the process increase. Thus globalization simultaneously creates spaces of exclusion/marginalization leading to an increase in social, economic, political and cultural unevenness across space . . . the rise of such processes is intimately tied to the rise and expansion of capitalism and thus goes back to the first 'global' empires . . . From a political viewpoint globalization has often been recast as an agenda or a discourse where the self-interest of certain groups leads them to accord normative or moral status to the processes.
>
> *(adapted from Murray and Overton, 2014)*

There are three central themes that can be drawn from the above. The first is that globalisation has proceeded through various historical waves that have had differentiated implications for local patterns of development. Second, globalisation does not homogenise local landscapes of development; rather it creates new networks of privilege and patterns of exclusion. Finally, globalisation can be conceived of as a wittingly constructed agenda comprised of various processes that have local impacts, and the exact nature of this combination will evolve according to the dominant development discourse of the time.

Waves of globalisation

Processes involving the outward imperial expansion of cultures, polities and economies over large areas of the globe have occurred for centuries, as in, for example, the spread of Chinese trading networks over much of Asia. Yet it was with the rise and expansion of post-Enlightenment European capitalism after about AD 1500 that such reach became more recognisably global.

Table 3.4.1 Waves of globalization – a framework

Wave	*Period (approximate dates)*	*Restructuring crisis*
Wave 1	**Colonial Globalization (c.1500–c.1945)**	
	Mercantilist phase (c.1500–c.1800)	Industrial revolution
	Industrialist phase (c.1800–c.1945)	Great Depression and WWII
Wave 2	**Postcolonial globalization (c.1945–)**	
	Modernization phase (c.1945–c.1980)	
	Neoliberal phase (c.1980–)	Oil crises

Source: Murray 2006, p. 88

Table 3.4.1 summarises the two major waves of globalisation, each associated with two distinct phases. The first wave saw the expansion of European power to much of the New World. The first phase of this, from *c*.1500 to *c*.1800 was associated with mercantilism, the penetration of new trading networks across the globe by European companies. The industrial revolution from the late eighteenth century began to replace mercantile capitalism with a new form – industrial capitalism. This led to a new phase of globalisation, that led by industrialism and associated with the acquisition of territories overseas by European powers and the formation of colonies of settlement, exploitation of local physical and human resources, and increasing rivalry between colonial powers.

The First World War represented the apex of this phase when colonial powers clashed, yet it was a crisis of capitalism in the Great Depression of the 1930s and the Second World War that caused the demise of this first wave. After 1945 a new world order was put in place that replaced the old colonial systems with one constructed against the backdrop of superpower rivalry between the capitalist West on one hand and the communist bloc of the Soviet Union and China on the other. The first phase of this second wave was associated with modernisation – decolonisation and the establishment of state-centred developmentalism, whereby newly independent states constructed infrastructure and welfare systems but, whilst some state involvement in economic activity was in evidence, the main strategy for development involved the deepening of capitalism through TNCs and the wider diffusion of the culture of capitalism.

The oil crises of the 1970s and the subsequent debt crisis, led to the fourth phase, that of neoliberalism and the one most commonly associated with contemporary globalisation. Neoliberalism saw an attack on the state and its replacement primarily by global capitalism through the promotion of trade liberalisation. In this it was accompanied by a growing role for both global institutions, such as the International Monetary Fund and World Trade Organisation, and civil society, both filling the vacuum left by a retreating state sector.

Yet whilst this neoliberal phase is continuing, there are signs that it may itself be coming to an end. The financial crisis of 2007 saw capitalism under threat, yet it seems as if the essential elements of capitalism and its global reach with respect to conditioning local development patterns have not been diminished.

Networks of privilege and local enclaves

Globalisation can be conceptualised as 'a pattern of multifibred networks, characterised by multidirectional flows, nestled within a larger system (capitalism) which creates at its nodes a mosaicking' (Murray, 2006, p. 52) and enclaving of space. Thus despite rhetoric, localisation processes and resultant geographies remain salient. In this sense, it is wrong to assume that globalisation is homogenising global society; difference is perpetuated and society fragmented – rather like Cardoso and Faletto's (1979) notion that the expansion of capitalism integrates globally but leads to disintegration locally. Global processes encounter infinitely differentiated localities in political, economic and cultural senses resulting in what geographers refer to as 'glocalisation'. In theory the local is blended with the global creating ever more heterogeneous space. In reality, in the case of peripheral regions, however, it is often global processes which attain ascendency given that poorer societies and localities are less well equipped to regulate and resist the worst implications of globalisation.

Globalisation in development theory and practice

The temporal and spatial dimensions of globalisation noted in the above two sections have had a profound effect on the way we conceive and practice local development. Over the past 70 years, since the rise and pursuit of the modern discourse of development (Rist, 1997), we have seen the ebb and flow of both development theory and development strategies. Also, at various stages, particularly at points of crisis, there have been forms of resistance to globalisation which have produced both local and global reactions and reformations. The links between theories and concepts of globalisation are presented in Table 3.4.2 and are discussed in what follows.

The modernisation phase of globalisation after 1945 was closely linked to development theories which promoted state-centred development, yet development that was still strongly pro-capitalism and pro-trade (McMichael, 2008). Thus Rostow's notions of the take-off foresaw a conventional capitalist path to 'progress' and the following of Western models of industrialisation, infrastructure development and education. Yet the apparent failure of this Western model, whether through the political reactions to the Vietnam War or Watergate or the economic fall-out of the oil crises, helped spurn the rise of new more radical Latin American structuralist-inspired development theories, particularly dependency theory, in the 1970s. These argued that development and underdevelopment were inseparable and that capitalism was the cause of poverty for many as well as the means of enrichment for a few. It

Table 3.4.2 Theses of globalization and development theories – a schema

Development theory	*Perspective on globalization*	*Definition of development*	*Explanation for lack of development*	*Outcome of development*	*Main strategy and policy*
Neoliberal	Pro-globalization hyperglobalist	Market-based economic growth modernization	State intervention corruption isolation	Convergence in incomes. Liberal democracy	Liberalization, deregulation, marketization
Structuralist/ neostructuralist	Alter-globalization transformationalist	Holistic income growth that is sustainable	Nature of insertion into global system	Depends on how practised and regulated	Selective intervention for equity and sustainability
Dependency/ post-development	Anti-globaliztion sceptical	Discourse to perpetuate capitalism	Exploitation by imperial and neo-imperial etites	Perpetuation of underdevelopment and marginality	Withdraw from capitalism. Alternative lifestyles

Source: Murray 2006, p. 266

marked a major point of resistance to the notion that globalisation was beneficial, desirable and attainable and it called for much more self-sufficient and inward-looking state-centred development strategies.

Yet out of these challenges to global capitalism, particularly the debt crisis, emerged a new development ideology, one that both harked back to modernisation and suggested a new more fundamental role for capitalism and globalisation. This 'global project' (McMichael, 2004) mounted an attack on the state as a key economic agent, it overturned barriers to trade and global financial flows and it aggressively sought to restructure economies, polities, societies and cultures. In the 1980s and 1990s neoliberalism promoted a much more assertive form of globalisation, opening the way for even deeper penetration of global trade, investment and migration. Furthermore, state regulation was largely replaced by a new regulatory regime based around global institutions and treaties.

Neoliberalism itself has been challenged and resisted in the past 20 years at the local scale. Realisation that early neoliberal reforms had severe social and economic consequences and were associated with heightened poverty and inequality led not only to active political reaction – as seen in the anti-globalisation movements or the end poverty campaigns – but also to the repackaging of structuralist development theories and policies. Neostructuralism, seen particularly in Latin American approaches to development, has sought to temper the excesses of neoliberalism with a supposed concern for poverty and inequality through a less passive state role. It has also been seen in new aid regimes, which have ostensibly been used to alleviate poverty, but also through reconstructing state institutions so that they provide basic services and permit the continued advance of capital investment and global trade (Murray and Overton, 2011). Thus we are witnessing the continual interplay of political and economic forces, manifested in both theory and practice, which are constructing and reconstructing new forms of globalisation and development involving changing relationships amongst global capital, state institutions and civil society.

Conclusions

Globalisation has forged and perpetuated inequities in local human development across time and space. During the first wave of globalisation this differentiation was manifested in the colonial division of labour and was more explicitly core/periphery in its spatiality. During the more recent postcolonial wave, however, the geometry of privilege and marginalisation has become ever more complex. Through the operation of globalisation as currently practised, that being neoliberalism, new networks and flows have evolved that create patterns of development and underdevelopment that cannot be neatly mapped within national borders. In this sense one of the central consequences of globalisation is that old First and Third World and core/periphery concepts are less relevant and local patterns become more relevant. There are thus local pockets of deprivation in wealthier societies, and pockets of privilege in poorer societies. The pockets of privilege are sewn together through conduits of economic, political and cultural flows creating globalised enclaves where they meet.

Globalisation is created through human activity and can be altered, it is therefore possible to regulate it and alter it in order to optimise its benefits for poorer societies. However, this has rarely happened and thus inequalities within and between localities have been perpetuated. Reform will involve action on the part of enlightened transnational bodies and nation states that engage with grassroots organisations and an informed local civil society in order to convey the needs and desires of communities. There has been a flourishing in such movements of resistance over the recent past evolving from the anti-neoliberal movements of Latin America in the 1980s to the Occupy movement of the 2010s. These movements challenge the dominant discourses of neoliberal globalisation and seek to utilise the potential spaces of globalisation for progressive local development. However, reform of global institutions and their fuller democratisation is sorely required. Despite the rhetoric of the level playing field, it is clear that we exist in a world of enormous inequalities within and between localities – uneven local development *is* the capitalist system. The task for researchers is to get inside the new networks of power driving this unevenness in order to inform progressive policy.

Bibliography

Cardoso, F. H. and Faletto, E. (1979) *Dependency and Development in Latin America*, Berkeley: University of California Press.

Harvey, D. (2003) *The New Imperialism*, Oxford: Oxford University Press.

McMichael, P. (2004) *Development and Social Change*, Newhaven, CT: Pine Press.

McMichael, P. (2008) *Development and Social Change: A Global Perspective*, Los Angeles: Pine Forge Press.

Murray, W. E. (2006) *Geographies of Globalization*, London: Routledge.

Murray, W. E. and Overton, J. (2011) 'Neoliberalism is dead, long live neoliberalism? Neostructuralism and the international aid regime of the 2000s', *Progress in Development Studies* 11(4): 307–319.

Murray, W. E. and Overton, (2014) *Geographies of Globalization*, London/New York: Routledge.

Rist, G. (1997) *The History of Development: From Western Origins to Global Faith*, London: Zed Books.

Sidaway, J. D. (2012) 'Geographies of development: New maps, new visions', *The Professional Geographer* 64(1): 49–62.

3.5

Trade and industrial policy in developing countries

David Greenaway and Chris Milner

Introduction

The last quarter of the twentieth century witnessed a substantial change of attitude in academic and policy circles about the appropriate form of trade and industrial policy for economic development. In this chapter we consider the nature, extent and consequences of the resulting liberalization of trade policies in developing countries that this has induced.

Recent liberalization in developing countries

Defining liberalization

In a stylized two-sector world defining liberalization is straightforward: removal of a tariff, or indeed any other intervention, which restores the free trade set of relative prices is unambiguously trade liberalization. However, in practice, things are more complicated and at least two other concepts are used: changes in policy which reduce anti-export bias and move the relative prices of tradables towards neutrality; and the substitution of more efficient for less efficient forms of intervention. These are overlapping, but they do not map on to one another on a one-to-one basis. It is possible to engineer a more neutral set of relative prices by introducing an export subsidy with a pre-existing import tariff, or by lowering the tariff, but the resource allocation effects of the two may differ. Although trade theory points to some striking non-equivalences between tariffs and quotas, a theorist might not regard the replacement of the latter with the former as liberalization. Policy analysts do, and this particular reform is a standard ingredient of World Bank liberalization packages.

Rationale for liberalization

There are very powerful economic arguments in favour of free trade (Dornbusch, 1992; Krueger, 1997). It is not too difficult to show that, in the absence of imperfections, free trade is optimal for a small open economy, as most developing countries are. Of course, we do not live in a world free of imperfections and there are a great many arguments for second best intervention. However, trade policy is rarely the most efficient form of intervention and even where it is, as the recent analysis of strategic trade policy has shown, the results are not easily generalizable. So the theoretical case for liberal trade policies appears to be a robust one. Why, however, in the early 1980s did it suddenly become so much more persuasive and more acceptable to developing countries to adopt reform measures?

The accumulation of empirical evidence relating to the costs of protection/benefits of liberalization was a factor. Evidence on the former was certainly comprehensive and also fairly convincing. Several influential cross-country studies (Krueger, 1981; Balassa, 1982), together with a multitude of country-specific studies, emphasized the consequences of long-term reliance on import substitution regimes in the form of high and complex patterns of protection, high resource costs, pervasive rent-seeking behaviour, poor macroeconomic performance and stagnating growth (see Greenaway and Milner, 1993).

More controversial was the evidence which appeared to suggest that liberal trade policies were also growth-enhancing; the key piece of data here being the export performance and growth of the so-called 'gang of four': Hong Kong, Taiwan, Korea and Singapore. Although placing great emphasis on the experience of these countries was a little disingenuous, given that only one (Hong Kong) followed a free trade policy, it was nevertheless influential since the others did pursue explicit export promotion policies. Moreover, their growth performance may have had more to do with their ability to react to key macroeconomic shocks in the 1970s, rather than their trade policies.

Role of the Bretton Woods agencies and other sources of liberalization

Policy conditionality is routinely applied by the IMF in connection with stabilization loans (SLs). World Bank policy conditionality dates from the launch of its structural adjustment programme (SAP) in 1980. This involved the disbursement of staged support in the form of structural adjustment loans (SALs) or sector adjustment loans (SECALs), typically on concessional terms, which were conditional upon reforms, often involving trade policy.

More recently, there have been additional sources of or expressions of commitment to trade liberalization in developing countries. Many countries have sought to anchor their early reforms through membership of the World Trade Organisation (WTO) and to participate more actively in regional trading arrangements. The further evidence of successful opening up of domestic markets and export-led growth (e.g. by China and India) has increased the desire and need for developing countries to integrate further, both regionally and globally.

Ingredients of trade reform programmes

All episodes of trade policy reform typically included measures to reduce anti-export bias, be they import liberalization or export support measures. Tariff reductions, quota elimination, relaxation of import licensing and so on all figure prominently. Note also that measures designed to rationalize and improve the transparency of the protective structure have been common: conflating shadow tariffs into actual tariffs; reducing tariff exemptions; and substituting tariffs for quotas.

Although the menu of reforms has been fairly standard, the manner of implementation has varied. This is partly due to the fact that the agreement of a package is the outcome of a bargaining process, and relative bargaining power differs from one case to another; and also partly due to a growing recognition that initial conditions and infrastructural support (Milner and Zgovu, 2006) are vital to the prospects for sustainability, and these vary from case to case (see Dean, 1995).

Evaluating experience with liberalization

It is possible that in the short run liberalization has some undesirable, but inevitable, side effects. Specifically, during the transition, unemployment may rise and/or trade tax revenue may fall. Both are invariably fears on the part of liberalizing governments. A minority of analysts, most notably Michaely *et al.* (1991), claim that such fears are unfounded. In practice, however, they do occur in many cases: it all depends on initial circumstances and the sequencing of reform.

The evidence

There are some rather important complications associated with conducting an evaluation of a liberalization. First, what is the counterfactual? Should one just assume a continuation of pre-existing policies and performance? Second, how does one disentangle the effects of trade reforms from other effects? Third, supply responses will differ from economy to economy: how long should one wait before conducting an assessment? For a review of these issues and of the evidence on outward orientation and performance, see Edwards (1998) and Greenaway (1998).

The evidence suggests that reform programmes tend to be associated with an improvement in the current account of the balance of payments and with an improvement in the growth rate of real exports. Some countries that have undergone adjustment show a subsequent improvement in investment but some experienced a slump. Finally, on balance, the impact on growth may be positive, in the sense that there are more cases of a positive growth impact than a negative growth impact, although growth does sometimes deteriorate.

What can one say overall? There have certainly been notable adjustment successes and failures. Some adjustment programmes where trade liberalization has figured prominently have resulted in rapid adjustment, a rapid supply side response and sustainable growth. In many others, especially in sub-Saharan Africa, stabilization has turned out to be a false dawn as a significant supply response failed to materialize. Are there general lessons?

Design of trade policy reform

Timing and sequencing issues

Initial timing

A number of arguments have been put forward for conducting any required macroeconomic stabilization in advance of structural adjustment policies. The stabilization programme, for example, may reduce the burdens on the export sector of an overvalued exchange rate. Against these arguments, evidence suggests that the aggregate adjustment costs of trade reform are relatively small alongside those associated with stabilization. One might be able to reap efficiency gains from trade reform quickly and before political resistance builds up.

Sequencing

The general view is that liberalization of the capital account should be held back until well into the process of trade reform and that the initial stages of trade reform should see import quota reform before tariff liberalization. The costs of rent-seeking and monopoly associated with quotas, and greater transparency and increased tariff revenue are often cited in the ranking of tariffs over quotas. On export incentives, the general consensus appears to be in favour of giving exporters access to inputs at world prices, and against export subsidization.

Speed of liberalization

There are a number of general arguments for rapid reform. First, it gives strong signals to economic agents, demonstrates government commitment and thereby increases the effectiveness and credibility of reforms. Second, it restricts the time and opportunities for resistance from affected lobby groups. There are, however, arguments for gradualism. First, government revenue may decline too rapidly if trade taxes are eliminated in advance of non-trade tax reforms. Second, adjustment costs may justify gradualism on political economy grounds, especially if gradualism slows down the pace of income redistribution. Third, although rapid/radical reform may be viewed as a means of signalling commitment, overambitious reforms may also lack credibility if the government already lacks a 'reputation' for good governance or sustaining policies. Finally, given limited foreign exchange reserves and the external credit-worthiness of many developing countries, it is important that liberalizations are compatible with other policy changes. Abrupt liberalization may require abrupt exchange rate depreciation. If this is not politically feasible, then credibility may require gradual trade liberalization (see Falvey and Kim, 1992, for a discussion).

Sustainability and credibility issues

The private sector is likely to be sceptical about sustainability and credibility where governments are pressured into trade reform. Commitment is uncertain and external circumstances may change, or internal reaction to reform may undermine resolve. A lack of credibility both blunts the incentives to adjust, for example, deterring reallocation of factors to the export sector or deferring investment, and sets in motion forces that undermine sustainability. If consumers expect reforms to be reversed they have an incentive to consume or speculatively accumulate more now of the temporarily cheaper imports. This increases the current account deficit and the probability of policy reversal.

Mitigating strategies can be designed. The need for macroeconomic stability and consistency is obvious. It may also be inadvisable to remove capital controls until trade reforms are fully consolidated. Where lack of credibility is associated with fear of reversion to previous policies once the private sector has reacted to reform – governments need to design strategies to build reputation and demonstrate commitment (see Rodrik, 1989).

Trade policy and macro stability

Direct links between trade policy and macroeconomic stability are limited. Trade policy determines the functional openness of the economy (e.g. trade-to-GDP ratio), but the trade balance is determined by the balance between national income and expenditure. It is exchange rate overvaluation (and fiscal deficit) that is the important link with macroeconomic balances and stability. Although trade reform (if sufficiently radical) can signal government commitment to inflation control, it can also interfere with the prevention of real exchange rate appreciation. Countries liberalizing trade policies often devalue their currency to compensate for the liberalization impact on the balance of payments. The potential inflationary effects of depreciation are likely to constrain the use of nominal exchange rate policy, hence sustained trade liberalization is likely to involve some deterioration in the external balance until there is an export response.

Trade reform and stabilization are linked through trade taxes. Given the high dependence of many developing economies on trade taxes and the slowness of any non-trade tax reforms, fiscal effects must be borne in mind. Replacement of quotas by tariffs, greater simplicity and uniformity in tariff structures, which reduce tax evasion through smuggling and under-invoicing are likely to be fiscal-enhancing.

Conclusions

Economic perspectives on trade and industrial policy in developing countries have changed profoundly over recent decades. The current consensus is that deregulation and liberalization can help in growth promotion, but are not in themselves a panacea. Trade policy reform may be necessary, but not sufficient, to reap the growth benefits of greater openness and outward orientation (see Milner, 2006; Chang *et al.*, 2009): the macroeconomic environment, the broader infrastructural and social context are all equally important to fashioning the outcome of reform.

References

Balassa, B. (1982) *Development Strategies in Semi-Industrialised Economies*, Baltimore, MD: Johns Hopkins University Press.

Chang, R., Kaltani, I. and Loayza, N. J. (2009) 'Openness can be good for growth: The role of policy complementarities', *Journal of Development Economics* 90: 33–49.

Dean, J. M. (1995) 'The trade policy revolution in developing countries', *The World Economy*, Global Trade Policy, Oxford: Blackwell.

Dornbusch, R. (1992) 'The case for trade liberalisation in developing countries', *Journal of Economic Perspectives* 6: 69–85.
Edwards, S. (1998) 'Openness, productivity and growth: what do we really know', *Economic Journal* 108: 383–98.
Falvey, R. and Kim, C. D. (1992) 'Timing and sequencing issues in trade liberalisation', *Economic Journal* 102: 908–24.
Greenaway, D. (1998) 'Does trade liberalisation promote economic development?' *Scottish Journal of Political Economy* 45: 491–511.
Greenaway, D. and Milner, C. R. (1993) *Trade and Industrial Policies in Developing Countries*, London: Macmillan.
Krueger, A. O. (ed.) (1981) *Trade and Employment in Developing Countries*, Chicago, IL: University of Chicago Press.
Krueger, A. O. (1997) 'Trade policy and economic development: How we learned', *American Economic Review* 87: 1–22.
Michaely, M., Papageorgiou, D. and Choksi, A. (eds) (1991) *Liberalising Foreign Trade*, Oxford: Blackwell.
Milner, C. R. (2006) 'Making NAMA work: Supporting adjustment and development', *The World Economy* 29: 1409–22.
Milner, C. R. and Zgovu, E. (2006) 'A natural experiment for identifying the impact of "natural" trade barriers on exports', *Journal of Development Economics* 80: 251–68.
Rodrik, D. (1989) 'Credibility of trade reform – a policy maker's guide', *The World Economy* 12: 1–16.

Further reading

Greenaway, D. (1993) 'Liberalising foreign trade through rose tinted "glasses"', *Economic Journal* 103: 208–22.
Greenaway, D. and Milner, C. R. (1993) *Trade and Industrial Policies in Developing Countries*, London: Macmillan. This book explains the analytical toolkit with which the economist has to measure and evaluate trade policies and their impacts, with evidence relating to trade policies and structural adjustment lending (SAL) reforms in the 1980s.
Michaely, M., Papageorgiou, D. and Choksi, A. (eds) (1991) *Liberalising Foreign Trade*, Oxford: Blackwell. This large multi-country study, commissioned by the World Bank, evaluates the impact and success of trade liberalizations during structural adjustment programmes. Though arguably not wholly impartial and using a methodology open to challenge (see Greenaway, 1993), it remains a valuable source of information on pre- and post-reform trade regimes in many developing countries.
Milner, C. R. (2006) 'Making NAMA work: Supporting adjustment and development', *The World Economy* 29: 1409–22. Useful for further information.
Morrissey, W. O. and McGillivray, M. (eds) (2000) *Evaluating Economic Liberalisation*, London: Macmillan. This collection of essays on economic liberalization issues from across the developing world includes essays on trade liberalization and on the linkages between wider policy reforms sponsored by the World Bank.
Rodrik, D. (1998) 'The new global economy and developing countries: globalisation, social conflict and economic growth', *The World Economy* 21: 143–58. Also useful.
Rodrik, D. (1999) *The New Global Economy and Developing Countries: Making Openness Work*, Washington, DC: Overseas Development Council. This essay considers the set of domestic policies and institutions required to take advantage of the opportunities created by globalization and to reduce the costs associated with rapid economic and social change.
Santos-Paulino, A. (2005) 'Trade liberalisation and economic performance: Theory and evidence for developing countries', *The World Economy* 28: 783–821. This is a comprehensive survey article on the theoretical and empirical literature on trade liberalization and economic performance in developing countries.

3.6

The knowledge-based economy and digital divisions of labour

Mark Graham

> The new international economy creates a variable geometry of production and consumption, labor and capital, management and information.
>
> *(Castells (1989: 348) in Downey 2008)*

Information is the raw material for much of the work that goes on in the contemporary global economy, and there are few people and places that remain entirely disconnected from international and global economic processes (Castells 1996). Information, and ultimately knowledge, is the carrier for the myriad signals needed for such markets to be constantly enacted, performed and understood.

As such, it is important to understand who produces and reproduces, who has access, and who and where are represented by information in our contemporary knowledge economy. This chapter discusses inequalities in traditional knowledge and information geographies, before moving to examine the Internet-era potentials for new and more inclusionary patterns. It concludes that rather than democratizing platforms of knowledge sharing, the Internet seems to be enabling a digital division of labour in which the visibility, voice and power of the North is reinforced rather than diminished.

Information geographies

> Information is not knowledge, Knowledge is not wisdom, Wisdom is not truth.
>
> *(Frank Zappa 1979)*

As Frank Zappa points out it is important to distinguish between information, knowledge, and other signals, representations and understandings. While this chapter is not the right venue for a detailed discussion and problematization of the differences between such terms, it is important to clarify what is meant by *information* and *knowledge*. *Information* is generally used to refer to codified descriptions that can answer questions such as 'who', 'what', 'where', and 'why'. *Knowledge*, in contrast, usually refers to the structuring, process, organizing, or internalization of information.

Traditionally, information and knowledge about the world have been highly geographically constrained. The transmission of information required either the movement of people or media capable of communicating that knowledge. Historical maps offer perhaps the best illustration of the geographic limitations to knowledge transmission. The thirteenth-century *Carta Pisana* (the world's oldest navigational chart), for instance, which was produced somewhere on the Italian peninsula,[1] depicts relatively accurate information about the Mediterranean, less accurate information about the fringes of Europe and no information about any parts of the world that are farther afield.

The example of the *Carta Pisana* starkly illustrates the constraints placed on knowledge by distance. Thirteenth-century transportation and communication technologies (e.g. ships and books) allowed some of the constraints of distance to be overcome by the map's Italian cartographers. But, in the thirteenth century those technologies were not effective enough to allow detailed knowledge about the Americas, East Asia, and much of the world to be represented on the map.

These highly uneven geographies of information matter. They shape what is known and what can be known, which in turn influences the myriad ways in which knowledge is produced, reproduced, enacted, and re-enacted. Importantly, it is not just artefacts from the Middle Ages that display such uneven patterns. Almost all mediums of information (e.g. book publishing, newspaper publications and patents) in the early twenty-first century are still characterized by huge geographic inequalities: with the global North producing, consuming and controlling much of the world's codified knowledge, and the global South largely left out of these processes.

Figure 3.6.1 starkly illustrates some of these patterns by visualizing the locations in which academic journals are published. The cartogram uses data from all 9500 journals included in the Web of Knowledge Journal Citation Reports (JCR) database and visualizes each country with a box that is sized according to the number of journals published from within it. The shading of each country indicates the average impact factor (a measure of how often articles within a journal are cited) of all journals within that country. The JCR database is an especially crucial metric not only because its owners claim[2] that it offers a 'systematic, objective means to critically evaluate the world's leading journals', but also because it forms an important part of the ways that academics, departments, and universities are evaluated (i.e. non-JCR publications are generally considered to be less valuable than those in the JCR database).

The map reveals a staggering amount of inequality in the geography of the production of academic knowledge. The United States and the United Kingdom publish more indexed journals than the rest of the world combined. Western Europe, in particular, Germany and the Netherlands, also scores relatively well. Most of the rest of the world then scarcely shows up in these rankings. One of the starkest contrasts is that Switzerland is represented at more than three times the size of the entire continent of Africa. The global South is not only under-represented in these rankings, but also ranks poorly on average citation score measures. Despite the large number and diversity of journals in the United States and United Kingdom, those countries manage to maintain higher average impact scores than almost all other countries.

These geographies of information reveal how knowledge and economic power are closely intertwined, and undoubtedly both reflect and reproduce positionalities of centrality and marginality in the global knowledge economy. Despite the entrenchment of much of the world's codified knowledge in the global North, many people are pointing to the potential for significant changes in such patterns. The Internet and other information and communications technologies (ICTs) provide and enable possibilities for fundamentally different communications media, methods, platforms and practices. In other words, while movements and control of information were previously constrained by the

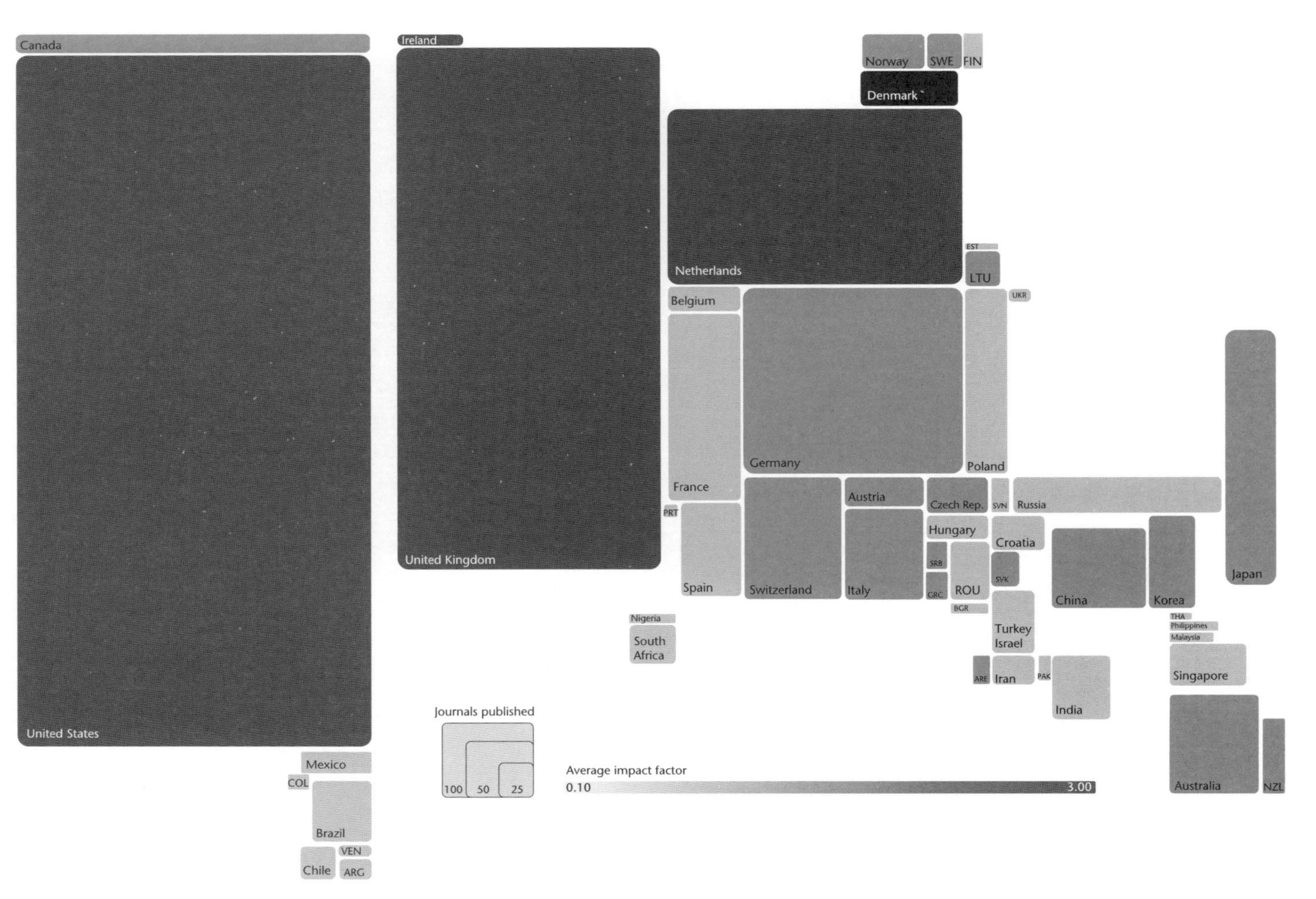

Figure 3.6.1 A cartogram of all journals in Thompson Reuters' Web of Knowledge

Source: Graham *et al.* (2011)

significant limitations of communication and transportation technologies, such constraints rarely apply in the Internet age. Movements of information are almost instantaneous and can be transmitted across the world for minimal costs. As such, there are very real potentials for the geographic and temporal frictions that traditionally constrained and limited the movements of information to be overcome.

The potentials of ICTs and reconfigured information economies

Access to ICTs is no longer confined to an elite few. In 2013, there were over six billion mobile phones in use. This means that most people on our planet now have some form of access to telecommunications services, and indeed, most mobile devices are now in use in the global South. There are also approximately two billion Internet users around the world. In other words, almost one in three human beings has some form of online access.

Concomitant with this broadening of access to communication technologies has been a fairly widespread belief that now, for the first time in human history, many of the geographic frictions that traditionally contributed to concentrations of information can be overcome. For example, at the 2003 World Summit on the Information Society, Harvard Law Professor Lawrence Lessig asserted[3] that '[f]or the first time in a millennium, we have a technology to equalize the opportunity that people have to access and participate in the construction of knowledge and culture, regardless of their geographic placing.'

The central idea here is that the Internet is able to bring into being an ethereal alternate dimension with two key characteristics. First, a 'space' that is infinite and everywhere (because everyone with an Internet connection can enter); and, second, one that is simultaneously fixed in a distinct (albeit non-physical) location that allows all willing participants to arrive into, and interact in, the same virtual space (Graham 2011). It is thus important to examine closely the difference that the Internet has made in bringing about potentially new information geographies.

Unfortunately, what most contemporary mappings of information demonstrate is that the Internet has failed to enable a more distributed geography of codified information creation and use. Figure 3.6.2, for instance, maps contributions to Wikipedia, which is one of the world's largest online platforms of user-generated content. Despite the fact that the platform is potentially available and open to most[4] of the two billion people on earth with an Internet connection, that hundreds of thousands of people have contributed, and that hundreds of thousands of places around the world have been described, we still see an incredibly concentrated geography of codified knowledge. For example, there is more than twice as much content created about France than the entire continent of Africa. It is not just Wikipedia that displays such skewed patterns of online information geographies. Many other platforms, repositories of content and online databases exhibit similar spatial cores and peripheries of knowledge (Graham and Zook 2011).

While earlier information sources (like the *Carta Pisana*) had more apparent lacunae, absences and local origins, online platforms can be more duplicitous in their appeals to be neutral, objective, and comprehensive. Despite the many ways of understanding Internet geographies (Zook 2007), there remains a widespread assumption that the Internet is a neutral space facilitating many-to-many relationships and allowing access to what Wikipedia's founder refers to as 'the sum of all human knowledge' and Google's founders describe[5] as their 'unbiased and objective' results. However, the Internet has only enabled amplifications of earlier unequal patterns of information geographies.

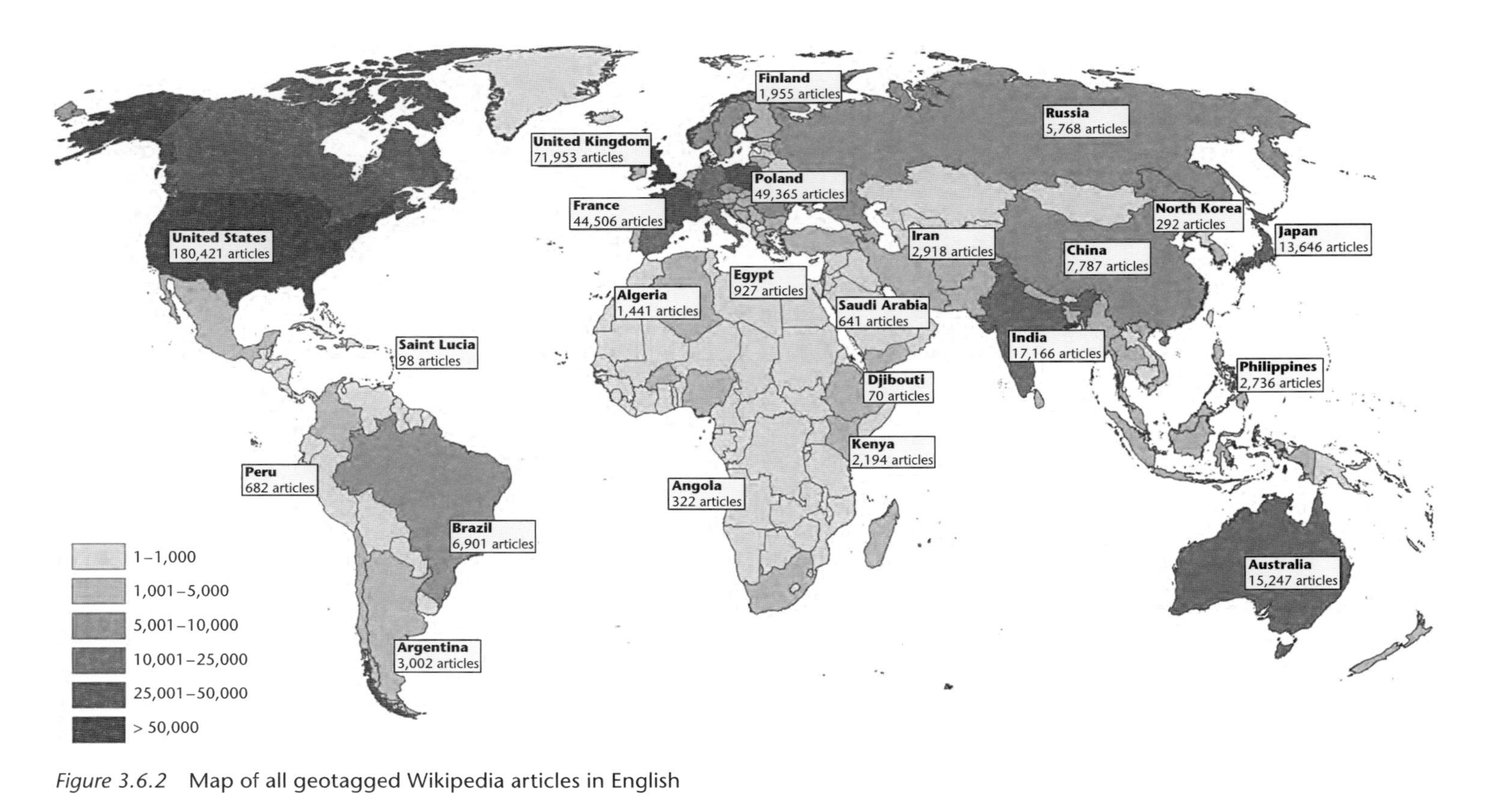

Figure 3.6.2 Map of all geotagged Wikipedia articles in English

Source: Mark Graham (www.zerogeography.net)

Digital informational divides

The dense clouds of information, or 'repositories of experience' (Grabher 2002), about some parts of the world are likely self-reinforcing because of the ways that exchanges of both codified and tacit knowledge are significantly facilitated by spatial proximity (Benner 2003). These initial uneven geographies of information were brought into being by the spatial fixes of physical telecommunication networks, rates of literacy, patterns of access to existing knowledge, capital and other resources necessary to produce and publish, and a range of other social, economic and political patterns, practices, and processes. Moreover, despite the evolving ways in which space is produced through spatial fixes and changing geographies of literacy, knowledge and access, dense clusters of information persist in many places because of the self-perpetuating nature of knowledge transfer discussed above.

The stickiness of information cores and peripheries, even in an age of supposed friction-free communications, is concerning because of Harley's (1989) observation that spatial configurations of information both have power and reproduce power. Because of its uneven geographies, the power/knowledge nexus is thus inherently inclusionary and empowering for some people and places and inherently exclusionary and disempowering for others. Knowledge clusters that are reinforced by repeated rounds of spatial fixes thus result in, and reinforce, a landscape of uneven geographic development (Downey 2008). While the earlier 'New International Division of Labour' heralded a movement of production from the global North to the global South (Dicken 2010), we now seem to be witnessing a new digital division of labour in which much of the world's knowledge work is produced in the global cores.

Ultimately, despite a rapid growth in education and Internet access for much of the world, most people on our planet are still entirely disconnected from global platforms of knowledge sharing. Even amongst those two and a half billion that are now online, a significant proportion of those that are connected are still left out of global networks, debates and conversations. Digital divisions cannot be simply bridged through connections and open online platforms, and much more work needs to be done to overcome inequalities in visibility, voice and power in an increasingly networked world. In other words, while connectivity is clearly a prerequisite for participation in twenty-first-century platforms of knowledge sharing and participation, connectivity and access are by no means a determinant of knowledge access, creation and sharing.

Notes

1 The precise origins of the map are unclear.
2 http://thomsonreuters.com/products_services/science/science_products/a-z/journal_citation_reports/
3 www.itu.int/wsis/docs/pc2/visionaries/lessig.pdf
4 Wikipedia is sporadically censored in some countries, most notably China.
5 www.sec.gov/Archives/edgar/data/1288776/000119312504142742/ds1a.htm#toc59330_1

Bibliography

Benner, C. (2003) Learning communities in a learning region: The soft infrastructure of cross-firm learning networks in Silicon Valley. *Environment and Planning A*, 35: 1809–1830.
Castells, M. (1996) *The Rise of the Network Society*. Oxford: Basil Blackwell.
Dicken, P. (2010) *Global Shift*. London: Sage.
Downey, G. (2008) Human geography and information studies. *Annual Review of Information Science and Technology*, 41(1): 683–727.

Fröbel, F., Heinrichs, J. and Kreye, O. (1978) The new international division of labour. *Social Science Information*, 17: 123–142.
Gibbs, J., Kraemer, K. L. and Dedrick, J. (2003) Environment and policy factors shaping global e-commerce diffusion: A cross-country comparison. *The Information Society*, 19(1): 5–18.
Grabher, G. (2002) Fragile sector, robust practice: Project ecologies in new media. *Environment and Planning A*, 34: 1911–1926.
Graham, M. (2011). Time machines and virtual portals. *Progress in Development Studies*, 11(3): 211–227.
Graham, M. and Zook, M. (2011) Visualizing global cyberscapes: Mapping user-generated placemarks. *Journal of Urban Technology*, 18(1): 115–132.
Graham, M., Hale, S. A. and Stephens, M. (2011) *Geographies of the World's Knowledge*. London: Convoco Edition.
Harley, J. B. (1989) Deconstructing the map. *Cartographica*, 26–1–20.
Zook, M. (2007). The geographies of the Internet. *Annual Review of Information Science and Technology*, 40(1): 53–78.

Further reading

Castells, M. (1989) *The Informational City: Information Technology, Economic Restructuring, and the Urban-Regional Process*. New York: Blackwell.
Dicken, P. (2010) *Global Shift*. London: Sage.
Graham, M. (2011) Cloud-collaboration: Peer-production and the engineering of the Internet. In S. Brunn (ed.) *Engineering Earth: The Impact of Megaengineering*. New York: Springer, pp. 67–83.
Graham, M. and Haarstad, H. (2011) Transparency and development: Ethical consumption and economic development through Web 2.0 and the Internet of things. *Information Technologies & International Development*, 7(1): 1–18.
Zook, M. (2006) The geographies of the Internet. *Annual Review of Information Science and Technology*, 40(1): 53–78.

3.7

Corporate social responsibility and development

Dorothea Kleine

Definitions of corporate social responsibility (CSR)

The last decades have seen an increase in global flows of goods, capital, people, services and ideas, thus an enhanced pace of globalisation. Powerful actors in this global economic order are multinational corporations (MNCs), which orchestrate their business to take advantage of a global division of labour. In the resulting global production networks, income-poor countries of the global South are frequently used as a source of cheap labour and, in addition, local social and environmental regulation tends to be weak or weakly enforced. Nation-states may be unable or unwilling to monitor MNC business practices in their countries (Dicken 2010). Meanwhile, the rise of campaign non-governmental organisations (NGOs) paired with news media and significant interest by Northern consumers, has led to exposure of "sweatshop" practices and environmental damage caused by MNCs. High-profile examples include criticism of Nike in the 1990s and recently of Apple supplier Foxconn in China, Shell's Brent Spar affair and activities in Nigeria, protests over labour relations at Coca-Cola suppliers in Colombia and the boycott of Nestlé following the companies' efforts to sell baby milk in sub-Saharan Africa.

While debates around the role and responsibilities of businesses in society are much older, CSR, since the 1970s, has been a positive term to describe socially responsible behaviour by businesses. Keith Davis defined it in 1973 as "the firm's consideration of, and response to, issues beyond the narrow economic, technical and legal requirements of the firm" (Davis 1973: 312). Originally focused primarily on social aspects, it now increasingly includes environmental issues, resulting in its occasional reformulation as corporate responsibility. While several definitions exist, it is vital to recognise that CSR is a conceptual space, rather than a single concept. This conceptual space has become increasingly prominent and negotiated between different stakeholders, including companies, states, NGOs, citizens and consumers (Crane *et al.* 2008). As Matten and Moon (2004: 335) explain, it is a "cluster concept which overlaps with such concepts as business ethics, corporate philanthropy, corporate citizenship, sustainability, and environmental responsibility. It is a dynamic and contestable concept that is embedded in each social, political, economic and institutional context".

Operating in a contested conceptual space

A primary distinction needs to be made between different models of CSR. Models like corporate philanthropy see social responsibility as a sort of add-on. In this model, companies operate "normally" and in pursuit of profit, and once these profits have been generated, a percentage of these gains is then reinvested in charitable causes, for example, in the communities the Northern consumers live in or indeed in poverty reduction projects in the global South.

Another model has to do with systemic responsibility – consideration of the so-called externalities of economic practices is integrated in all decision-making processes and the core business is meant to be operated in a socially and environmentally sustainable way. Investment in communities occurs as part of an overall systemic approach to ethical business practices.

CSR thus covers different levels of responsible business behaviour which go beyond the legal minimum. Milton Friedman, one of the most prominent opponents of the CSR movement, argued that: "There is one and only one social responsibility of business – to use its resources and engage in activities designed to increase its profit as long as it stays within the rules of the game, which is to say, engages in open and free competition, without deception and fraud" (Friedman 1970). It is worth noting that even Friedman's expectation of businesses' behaviour would have been severely tested in the recent financial crisis of 2007/8 and its aftermath. As a supporter of the idea, Carroll (1991) offered a pyramid of CSR: the base is formed by the economic responsibility to be profitable, leading to, at the next level, the legal responsibilities of obeying the law and playing by the rules of the game, above which rest the ethical responsibilities of doing what is right, just and fair and avoiding harm, while at the top are philanthropic responsibilities to be a good corporate citizen, such as contributing resources to the local community and improving quality of life.

To sum up several such multilevel models, at the base of a ladder of CSR is thus responsible management and honest reporting. For the global South this would mean that MNCs would pay the taxes and license fees they owe the respective host countries, and do not pay bribes.

On the next level, there is corporate philanthropy and charitable investment. Examples of this abound, for example, the $145 million given by Google (Google 2011) and $102 million given by Coca-Cola (Coca-Cola 2011) to charitable causes in 2010.

Level three is the decent treatment of staff and adherence to and improving on local state labour standards. Here many MNCs or their suppliers have been accused of undercutting state regulation, for example, by not allowing union representation in the workplace even where there is a right to unionise. Level four has companies considering the impact of their operations on the local environment and community, for example, by reducing or preventing pollution.

A further level then concerns the treatment of suppliers and the behaviour of MNCs in value chains, where they often are the dominant actor with the greatest negotiating power. Fair dealings with suppliers can be self-certified, but, together with labour standards and minimum prices, also forms a key element of third-party certification systems such as the Fairtrade label. A much further, and much rarer step is the emergence of alternative business models, ranging from giving employees a modest share of the profits to actual employee ownership (practised by, for example, John Lewis, Waitrose and producer cooperatives in the global South such as Kuapa Kokoo) or consumer-member ownership (such as the UK Co-operative Group). A final level of CSR could be described as using corporate funds to try to influence public opinion and lobby government towards tighter measures for corporate governance and environmental sustainability. For example, the UK Co-operative Group spent £200,000 to support The Wave demonstration on 5 December 2009 in London, which called for a "safe and fair deal for developing countries" ahead of the Copenhagen Climate Change summit.

According to DFID (2003), business can help reduce poverty in income-poor countries by investing, producing, and paying wages and taxes, thus contributing to economic growth. MNCs would have to respect labour standards, create jobs and could provide skills and training. Further, business could support local communities through local sourcing.

Drivers of CSR

There are several key drivers behind the trend towards corporate social responsibility. MNCs who have much of their value locked into consumer-facing brands, such as Nike and Apple, are particularly alert to threats to their brand reputation when activists refer to environmental or social wrong-doing. For instance, the campaign NGO Greenpeace used this brand sensitivity when running an explicit "Green my Apple" campaign against toxins in hardware production. Workers in the global South are affected because the majority of e-waste recycling takes place in countries with low-wage labour and limited health and safety regulation and enforcement.

The flipside to negative reputational risk up to and including boycotts is a growing trend towards positive ethical consumption in major countries with a high or medium human development index. For example, according to the yearly Ethical Consumerism Report, the UK market in products with ethical claims grew from £13.5 billion (1999) to £46.8 billion (2011) in 12 years (The Cooperative Group 2011). Ethical consumption movements have also been building momentum in some countries in the global South such as Brazil and Chile (Agloni and Ariztía 2011; Bartholo *et al.* 2011).

There are existing international agreements such as the OECD guidelines for multinational enterprises that cover employment, industrial relations, human rights, health and safety, bribery and the environment. The core conventions of the International Labour Organization (ILO) form a set of texts upon which many corporate codes of conduct draw. The UN Global Compact is a multi-stakeholder network of organisations that have signed a set of universal principles based on human rights, labour standards and the environment (www.unglobalcompact.org).

Companies frequently look to fixed standards and business-compatible procedures for switching to sustainable sourcing. Certification schemes exist, such as FLO Fairtrade or Forest Stewardship Council (FSC) and Marine Stewardship Council (MSC) for sustainably sourced wood and fish respectively, as well as several animal welfare and organic certification systems. In the case of outsourcing in global value chains, large auditing companies such as PWC and KPMG now also offer social auditing services, for example, in the garments industry – this entails the auditor, paid by the MNC, checking on paper and in situ whether the supplier is conforming to labour standards set out by law and by the MNC corporate policy. Corporate codes have been shown to result in better outcome standards for workers but not necessarily improved process rights (Barrientos and Smith 2007).

Another significant driver is sustainable public procurement – the way the state buys goods and services. While some public actors such as the European Parliament have a long-standing commitment to sourcing their consumables as Fairtrade certified, there is increasingly a wider discussion about introducing social and environmental criteria in public procurement. The trend towards ethical investment is another driver. Institutional investors such as major public pension funds increasingly use policies of negative and positive screening when choosing which companies to invest in. Once these investors own shares, they can also raise social and environmental issues with the company board.

As Matten and Moon (2004) point out, CSR is embedded in each social, political, economic and institutional context, and this includes key public discourses. Public awareness, linked to media coverage, is a major driver for changes in business practice. Apart from conventional media, blogs and social media such as Facebook, YouTube and Twitter are becoming increasingly significant. There are pilot projects which use smartphones to give consumers in the shop direct access to third-party information about the company behind the product and the social and environmental impacts of its production (www.fairtracing.org, Kleine 2008). ICTs can allow producers and consumers to communicate much more directly.

Criticisms of CSR

Eighty-one per cent of larger companies in the EU and 40 per cent in the USA now publish CSR reports (PWC 2010) as part of their reporting cycle, but this mainstreaming of CSR and its use in advertising for the company has led to a growing degree of scepticism and cynicism among consumers. In particular, where there is self-reporting without third-party audit, accusations of "whitewash" (painting an overly positive picture about social conditions) and "greenwash" (about environmental issues) can arise.

Critics also point out that CSR should not replace state regulation. Where multinational companies operate in countries in the global South with weak regulation on labour standards and environmental protection, company CSR policy might become a more potent driver of management decisions than local legislation. Critics argue that individual companies' policies are too weak a basis to rely on and propose strengthening local legislation and its enforcement instead.

While state and third-party regulation are seen as more rigorous and credible than company self-regulation, with them come costs for registration and external audits. Regulation emanating from the global North can affect production standards, and thus the lives of producers in the global South. Such external, top-down standards might not fit with local conditions, for example, where the paperwork needed to prove adherence to a standard can overwhelm small producers with more informal record-keeping, especially when they also have to pay for certification. DFID highlight a further risk related to child labour. Where concerned MNCs withdraw contracts immediately when there is evidence of child labour, children and their families can be left in a worse condition than before. Instead, companies should engage and help raise wage levels so that families do not depend on also using their children's labour for survival. DFID argues for such principled CSR engagement rather than disinvestment.

To conclude, CSR is a contested conceptual ground on which citizens and campaign NGOs can challenge businesses to have a positive development impact. It is a topic which transcends dichotomies of global North and South, and instead demonstrates very clearly how in an interconnected world, not just MNC's economic activities, but also debates and alliances around ethical business practice reach beyond national borders.

References

Agloni, N. and Ariztía, T. (2011). *Consumo ético en Chile: una revisión de la investigación existente.* First Report: Leveraging Buying Power for Development, Santiago de Chile, Universidad Diego Portales. Available online at www.sustainablechoices.info (accessed 22 December 2013).

Barrientos, S. and Smith, S. (2007). "Do workers benefit from ethical trade? Assessing codes of labour practice in global production systems." *Third World Quartely* 28(4): 713–729.

Bartholo, R., Afonso, R. and Pereira, I. (2011). *Consumo Ético no Brasil.* First Report: Leveraging Buying Power for Development, Rio de Janeiro, COPPE – Universidad Federal do Rio de Janeiro.

Carroll, A. B. (1991). "The pyramid of corporate social responsibility: Towards the moral management of organizational stakeholders." *Business Horizons* 34(4): 39–48.

Co-operative (2011). *The Ethical Consumerism Report.* Available online at www.co-operative.coop/PageFiles/416561607/Ethical-Consumerism-Report-2011.pdf.

Crane, A., Matten, D. and Spence, L. J. (eds) (2008). *Corporate Social Responsibility – Readings and Cases in a Global Context.* Oxford: Routledge.

Davis, K. (1973). "The case for and against business assumption of social responsibilities." *The Academy of Management Journal* 16(2): 312–322.

DFID (2003): *DFID and Corporate Social Responsibility. An Issues Paper.* London. Available online at http://webarchive.gov.uk/+/http://dfid.gov.uk/pubs/files/corporate-social-resp.pdf (accessed 22 December 2013).

Dicken, P. (2010). *Global Shift: Transforming the World Economy* (6th edn), London: Sage, Paul Chapman.

Friedman, M. (1970). "The social responsibility of business is to increase profits." *The New York Times Magazine*, 13 September.

Kleine, D. (2008). "Negotiating partnerships, understanding power: Doing action research on Chilean Fairtrade wine value chains." *The Geographical Journal* 174(2): 109–123.

Matten, D. and Moon, J. (2004). *"Implicit" and "Explicit" CSR: A Conceptual Framework for Understanding CSR in Europe.* I.R.P.S. 29–2004. Nottingham: University of Nottingham.

PWC (2010). *CSR Trends 2010: Stacking up the Results.* New York: PriceWaterhouseCoopers. Available online at www.pwc.com/ca/en/sustainability/publications/csr-trends-2010-09.pdf (accessed 6 June 2012).

3.8

The informal economy in cities of the South

Sylvia Chant

What is the urban informal economy?

The urban 'informal economy' has traditionally been equated with a heterogeneous range of precarious, low-productivity, poorly remunerated income-generating activities in cities of the South. Informal employment usually prevails in commerce and services, but also occurs in manufacturing production. Although many people in the informal economy work on their own account in street-vending, the running of 'front room' eateries, stalls or shops, the operation of domestic-based industrial units, and the transport of passengers and goods (see Figures 3.8.1 and 3.8.2), other informal workers are subcontracted by large firms, especially in labour-intensive industries such as toys, footwear and clothing.

The term 'informal economy' has long been in existence but was – and often still is – more commonly referred to as the 'informal sector'. The latter made its major debut into the academic and policy literature in the early 1970s and is normally attributed to the economic anthropologist Keith Hart on the basis of his fieldwork in Ghana. Hart's denomination of the 'informal sector' described economic activities which fell outside the boundaries of state regulation (so-called 'formal' generally large-scale and/or corporate concerns such as factories, public services and commercial chains) and further subdivided these into 'legitimate' and 'illegitimate' varieties. The former comprised ventures that made a contribution to economic growth, albeit in small ways, such as petty commerce, personal services and home-based production. 'Illegitimate' informal activities, alternatively, if not necessarily 'criminal' in nature, were of questionable worth to national development, such as prostitution, begging, pickpocketing and scavenging (Hart, 1973).

Hart's terminology was enthusiastically embraced by the International Labour Organization (ILO) in its 1972 'Kenya mission', whose criteria for distinguishing formal and informal activities comprised relative ease of entry, size, nature of enterprise ownership, type of production and levels of skill, capital and technology. The single most important factor that has persisted in definitions of 'informality' over time, however, is regulation (Hart, 2010).

Regulation primarily implies legality, recognizing that legality comprises different dimensions, and that three are particularly pertinent to the demarcation between formal and informal activities, especially at the enterprise level (Tokman, 1991: 143):

1 legal recognition as a business activity (which involves registration, and possible subjection to health and security inspections);
2 legality concerning payment of taxes;
3 legality vis-à-vis labour matters such as compliance with official guidelines on working hours, social security contributions and fringe benefits.

Figure 3.8.1 Informal breakfast business: Fajara, The Gambia

Photo: Sylvia Chant

Figure 3.8.2 Pedal power: Informal transport in Mexico City

Photo: Sylvia Chant

Social security tends to be the most costly aspect of legality, so while micro-enterprises may well register themselves as businesses with the relevant authorities, they may simultaneously avoid paying social security contributions for themselves and their workers (Tokman, 1991: 143). Yet in the context of a mounting 'informalization' of urban economies over time, the spotlight has been placed on the fact that a broad spectrum of small and larger-scale firms employ labour on an informal basis, with the ILO accordingly changing its definition of informal employment between 1993 and 2003 to encompass people employed on an informal basis in the formal economy (Jütting *et al.*, 2008; see also Chen, 2010). In turn, the term 'informal economy' has been deemed more appropriate than 'informal sector' since it shifts the focus away from firms per se to labour arrangements.

Recent trends in informal employment

Data on informal employment need to be treated with caution, not only on account of the intrinsically irregular and/or clandestine nature of informal work, but because of shifting classificatory schema by different governments and regional and global organizations (Thomas, 1995). Indeed, in 2009, the ILO introduced two new terms: 'working poor' and 'vulnerable employment'. The former refers to people working for less than $1.25 per day (extreme working poverty), while the latter is the sum of own-account workers and contributing family workers.

Acknowledging that terminological and classificatory changes make temporal and geographical comparisons difficult, there is considerable evidence to suggest that informal employment has increased in recent decades. In Latin America, for example, an estimated 7 out of 10 new jobs created during the 1990s were informal, and between 1990 and 2002, the share of the non-agricultural labour force employed informally rose from 43 per cent to 51 per cent. In Asia, the share of the workforce informally employed in 2002 was even higher, at 71 per cent, and in sub-Saharan Africa, 72 per cent, with countries such as Benin and Chad reporting levels of 90 per cent (see Chen, 2010; Heintz, 2006).

From the 1980s onwards, the major driving factors in informal employment growth seem to have been recession and neoliberal economic restructuring. People have been pushed into informal work through cutbacks in public employment, the closure of private firms and the mounting tendency for formal employers to resort to subcontracting arrangements and/or to casualize in-house work. Another significant process has been for smaller firms to move wholesale into informality as a result of declining ability to pay registration, tax and labour overheads. As Thomas (1996: 99) summarizes, the 'top-down' informalization promoted by governments and employers has been matched by a 'bottom-up' informalization stemming from the need for retrenched formal sector workers and newcomers to the labour market to create their own sources of earnings and/or to avoid the punitive costs attached to legal status.

Informality in relation to recession, restructuring and gender

In light of the above, it is hardly surprising that the informal economy has become increasingly competitive in recent times. Indeed, although the informal economy continued to expand during some of the hardest years of crisis and restructuring in the 1980s and early 1990s, it was not able to absorb all the job losses in formal enterprises. This has led to the notion that growth of the informal economy may not be as much *counter-cyclical* – expanding in periods of slump – as *pro-cyclical*, and thereby contingent upon health in the formal economy. As summarized by UNRISD (2010: 112), 'contrary to the conventional wisdom that

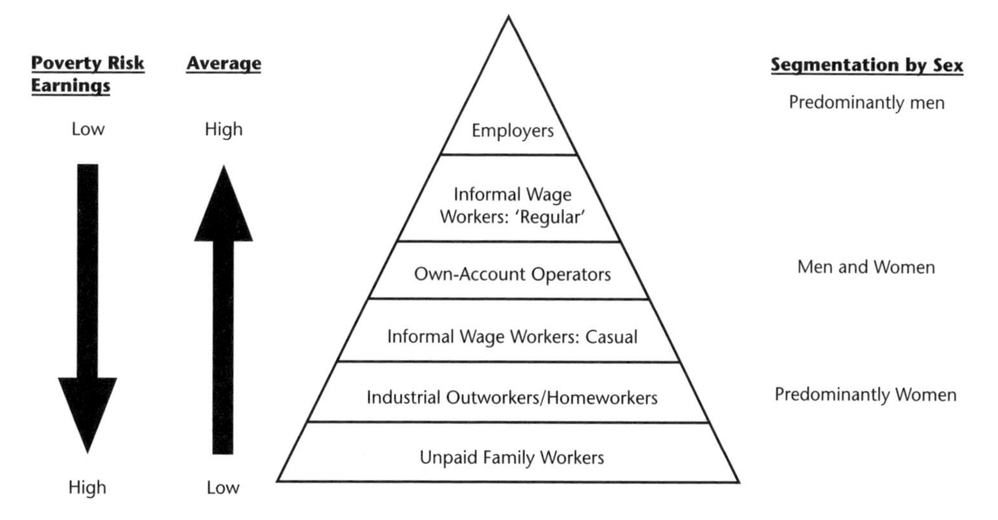

Figure 3.8.3 Segmentation by sex within the informal economy
Source: Chen (2010, p. 468, fig. 71.1)

the informal economy serves as a cushion for formal workers who lose their jobs, economic downturns affect the informal economy in similar ways as they do the formal economy'. The heterogeneity of informal activities is also pertinent here, with Jütting *et al.* (2008) noting that the behaviour of informal activities may well depend on whether these are in the competitive 'upper tier' where people 'choose' to be informally employed, or 'lower tier' where people have no choice other than to be informally employed.

Although there are a number of benefits as well as drawbacks of informal employment, for example, some informal workers earn more than salaried workers, self-employment can be a source of pride or prestige, informality permits flexibility and prompt adaptation to changing demand and family circumstances, and people often acquire skills in the formal economy which can subsequently be used to advantage in their own businesses (see Jütting *et al.*, 2008), for the most part it appears that informality is a strategy of 'last resort', being a fragile means of basic subsistence in situations where social welfare provision for those outside the formal labour force is minimal or non-existent (Thomas, 1996). Indeed, that women workers are disproportionately represented in informal activities compared with men (Chant and Pedwell, 2008; Heintz, 2006), and within the informal economy to be confined to the lowest and least paid tiers of informal activity (see Figure 3.8.3 on sex segmentation in the informal economy) means that they often bear the biggest brunt of economic deterioration.

Links between formal and informal activities

The interconnectedness of the formal and informal economies has been made ever more visible during the last thirty years, and in particular, the dependence of the latter on the former for contracts, supplies and economic viability. This has rendered redundant previous notions of labour market dualism.

An early attempt to resist the construction of the formal and informal economies as discrete and autonomous entities was Caroline Moser's seminal neo-Marxian exposition on 'petty commodity production'. Unlike the 'dualist model', this theorized urban labour markets as a continuum of productive activities in which large formal firms benefited from the existence of micro-entrepreneurs (Moser, 1978). This provided fodder for a thesis of 'stucturalist articulation', which views urban labour markets as 'unified systems encompassing a dense network of relationships between formal and informal enterprises'. Although it is recognized that links between formal and informal activities are often exploitative, it is also acknowledged that some opportunities may be opened up for informal enterprises by globalization and neoliberal strategies of export promotion (Portés and Itzigsohn, 1997: 240–1). In many respects this has encouraged recommendations for more active and sympathetic policy stances towards informality.

The informal economy and policy

The fact that there was no explicit policy towards the informal economy in most developing countries until recently was partly due to anticipation that labour surpluses would eventually be absorbed by formal industry and services, and partly because of the reluctance of economists and civil servants to acknowledge informal activities as anything other than a 'parasitic', 'unproductive' form of 'disguised unemployment' (Bromley, 1997: 124). Even where informal entrepreneurs may have wanted to 'become legal', prohibitive costs and convoluted bureaucratic procedures have usually dissuaded them from so doing. In addition to indirect discrimination resulting from state subsidies to the large-scale capital-intensive sector, informal entrepreneurs have often been subjected to harassment or victimization (Thomas, 1996: 56–7).

Although the latter scenario persists in a number of places, officialdom has gradually come round to the notion that the informal economy is more of a seedbed of entrepreneurial potential than a 'poverty trap'. Theoretical weight has been added to this shift by the Peruvian economist, Hernando de Soto, whose controversial book *The Other Path*, first published in English in 1989, argues that the informal economy is a product of excessive and unjust regulation created by governments in the interests of the society's powerful and dominant groups (Bromley, 1997: 127). Emphasizing the ways in which informality relieves unemployment, provides a gainful alternative to crime, and harnesses the creativity and business acumen of the disaffected masses, De Soto asserts that 'illegality' is a perfectly justifiable response on the part of the urban poor. Governments should consider tolerating 'non-conformity' more widely, and give informal entrepreneurs greater encouragement, protection and freedom (Bromley, 1997: 127).

These ideas have been hotly debated in the literature. One problem with De Soto's eulogization of informality is that it gives a misleading impression of a segment of the urban economy, which is perhaps better understood as 'a picture of survival rather than a sector full of entrepreneurial talent to be celebrated for its potential to create an economic miracle' (Thomas, 1995: 130). Another set of problems arises from the prospectively perverse implications of proactive informal policies. One outcome of advocating decontrol of enterprises is that a precedent is set for greater deregulation in formal firms, which has often been harmful to low-income groups in developing regions.

Nonetheless, accepting that the informal economy is likely to persist for the foreseeable future, measures are arguably needed to help informal activities to operate more efficiently and with better conditions for their workers. This is especially pertinent in light of the 'decent work' agenda advanced by the ILO since 2001, and its incorporation into the first Millennium Development Goal (MDG) in 2008.

In terms of how 'decent work deficits' in the informal economy might be addressed, Martha Chen (2009: 215) argues that

> informality may not be the problem, and formalization may not be the answer . . . . Imposing the existing narrow set of formal economic institutions on the large diverse set of informal economic activities may not be desirable nor feasible. What is needed are twenty-first century institutional arrangements – a creative mix of formal and informal – for twenty-first century economic realities.

With reference to the Caribbean, Portés and Itzigsohn (1997: 241–3) suggest that much could be done to diminish the constraints faced by the informal economy such as lack of working capital through limited access to mainstream financial institutions, concentration in highly competitive low-income markets, the social atomization of informal entrepreneurs due to the irregular and/or chaotic nature of supplies, and the existence of a 'craftsman ethic' which prevents some informal entrepreneurs, particularly in artisanal production, from changing their traditional methods of production.

While specific policy initiatives in different developing countries are discussed in detail elsewhere (see for example, Chen *et al.*, 2004), a much-favoured intervention on the institutional/macroeconomic side of the labour market is the repeal of regulations and policies which inhibit entrepreneurship without serving any legitimate public regulatory purpose. There has also been advocacy for governments to consider simpler and diminished requirements and/or allow for progressive implementation (Chant and Pedwell, 2008; Tokman, 1991: 155).

On the supply side of the labour market, there has been interest in, and/or support for, policies geared to education and training to promote the diversification of the informal sector, to enhance access to credit, to provide assistance in management, marketing and packaging, and to introduce measures to promote greater health and safety. There has also been advocacy for orienting policies away from individual firms or workers as a means of utilizing the social networks and social capital (reciprocity, trust, social obligations among kin, friends, neighbours and so on) which so frequently fuel the operation of the informal economy (Portés and Itzigsohn, 1997: 244–5).

Given that a disproportionate number of female workers are engaged in informal activities, it is also important that to advance initiatives that help women specifically, as in Chen's (2010) '3V' ('Voice, Visibility and Validity') framework which exhorts institutions to make visible the participation of women workers, to recognize and validate them, and to ensure their better representation in labour market organizations.

Prospects for the informal economy

Regardless of policies which may be implemented by governments and agencies, it is likely that the informal economy will continue to be a significant feature of urban labour markets in the South. One important reason for this is demographic pressure, not only on account of new young entrants to the labour force, especially in sub-Saharan Africa, but also due to the ageing of populations which, coupled with exiguous state welfare and declining household incomes, may prevent older people from exiting employment. The potential 'crowding-out' of the informal economy is likely to be exacerbated by ongoing increases in the number of women in employment.

On the demand side of the labour market, the current climate of deregulation is likely to provoke further contraction in public employment and to foster increasingly 'flexible' labour contracts in the formal economy as firms face ever-tougher global competition.

Recognizing that policies to bolster the informal economy will have to address a wide range of concerns simultaneously, one key area is that of extending and enhancing systems of public education and training which encompass commercial and managerial skills, alongside instruction in cutting-edge developments such as information and communications technology (ICT). Policies geared to supporting people's efforts to sustain their livelihoods should also take due steps to consult the groups concerned. The fact that the informal economy has survived so well through three decades of periodic crisis and ongoing restructuring in developing regions testifies to the fact that there are valuable lessons to be learned 'from below'.

References

Bromley, R. (1997) 'Working in the streets of Cali, Colombia: Survival strategy, necessity or unavoidable evil?' in J. Gugler (ed.), *Cities in the Developing World: Issues, Theory and Policy*, Oxford: Oxford University Press, pp. 124–38.

Chant, S. and Pedwell, C. (2008) *Women, Gender and the Informal Economy: An Assessment of ILO Research, and Suggested Ways Forward*, Geneva: International Labour Organization (http://www.ilo.org/wcmsp5/groups/public/-dgreports/-dcomm/documents/publication/wcms_091228.pdf).

Chen, M. (2009) 'Informalisation of labour markets: Is formalisation the answer?' in S. Razavi (ed.) *The Gendered Impacts of Liberalisation: Towards 'Embedded Liberalism'?* Geneva: UNRISD, pp. 191–218.

Chen, M. A. (2010) 'Informality, poverty, and gender: Evidence from the global South', in S. Chant (ed.) *The International Handbook of Gender and Poverty: Concepts, Research, Policy*, Cheltenham, UK: Edward Elgar, pp. 463–71.

Chen, M. A., Carr, M. and Vanek, J. (2004) *Mainstreaming Informal Employment and Gender in Poverty Reduction*, London: Commonwealth Secretariat.

Hart, K. (1973) 'Informal income opportunites and urban employment in Ghana', in R. Jolly, E. de Kadt, H. Singer and F. Wilson (eds) *Third World Employment*, Harmondsworth, UK: Penguin, pp. 66–70.

Hart, K. (2010) 'The informal economy', in K. Hart, J.-L. Laville and A. D. Cattani (eds) *The Human Economy: A Citizen's Guide*, Cambridge: Polity Press, pp. 142–53.

Heintz, J. (2006) *Globalisation, Economic Policy and Employment: Poverty and Gender Implications*. Geneva: ILO, Employment Strategy Paper 2006/3. Employment Policy Unit, Employment Strategy Department (www.ilo.org).

Jütting, J., Parlevliet, J. and Xongiani, T. (2008) *Informal Employment Re-loaded*, OECD Development Centre Working Paper No.266, Paris: OECD (http://www.oecd.org/dataoecd/4/7/39900874.pdf).

Moser, C. (1978) 'Informal sector or petty commodity production? Dualism or dependence in urban development', *World Development* 6: 135–78.

Portés, A. and Itzigsohn, J. (1997) 'Coping with change: The politics and economics of urban poverty', in A. Portés, C. Dore-Cabral and P. Landoff (eds) *The Urban Caribbean: Transition to a New Global Economy*, Baltimore, MD: Johns Hopkins University Press, pp. 227–48.

Thomas, J. J. (1995) *Surviving in the City: The Urban Informal Sector in Latin America*, London: Pluto.

Thomas, J. J. (1996) 'The new economic model and labour markets in Latin America', in V. Bulmer-Thomas (ed.) *The New Economic Model in Latin America and its Impact on Income Distribution and Poverty*, Basingstoke, UK: Macmillan, in association with the Institute of Latin American Studies, University of London, pp. 79–102.

Tokman, V. (1991) 'The informal sector in Latin America: From underground to legality', in G. Standing and V. Tokman (eds) *Towards Social Adjustment: Labour Market Issues in Structural Adjustment*, Geneva: International Labour Office (ILO), pp. 141–57.

United Nations Research Institute for Social Development (UNRISD) (2010) *Combating Poverty and Inequality: Structural Change, Social Policy & Politics*, Geneva: UNRISD (http://www.unrisd.org).

Further reading

Chant, S. and Pedwell, C. (2008) *Women, Gender and the Informal Economy: An Assessment of ILO Research, and Suggested Ways Forward*, Geneva: International Labour Organization (http://www.ilo.org/wcmsp5/groups/public/-dgreports/-dcomm/documents/publication/wcms_091228.pdf). This report provides a review and analysis of the ILO's research on women, gender and the informal economy.
Chen, M. A., Carr, M. and Vanek, J. (2004) *Mainstreaming Informal Employment and Gender in Poverty Reduction*, London: Commonwealth Secretariat. A comprehensive review of the role of informal employment in reducing poverty, including mainstream debates on the urban informal economy as well as dedicated discussion on gender therein.
Jütting, J., Parlevliet, J. and Xongiani, T. (2008) *Informal Employment Re-loaded*, OECD Development Centre Working Paper No. 266, Paris: OECD (http://www.oecd.org/dataoecd/4/7/39900874.pdf). A wide-ranging publication which draws particular attention to the hetereogeneity of informal employment and to its benefits and drawbacks.
Portés, A., Dore-Cabral, C. and Landoff, P. (eds) (1997) *The Urban Caribbean: Transition to a New Global Economy*, Baltimore, MD: Johns Hopkins University Press. A book which combines discussion of debates on the informal sector, urbanization and globalization, with case studies from the capital cities of Costa Rica, Haiti, Guatemala, Jamaica and the Dominican Republic.
Thomas, J. J. (1995) *Surviving in the City: The Urban Informal Sector in Latin America*, London: Pluto. A very detailed account of the nature of informal employment in Latin American cities.

Websites

International Labour Organization: http://www.ilo.org
UN Economic Commission for Latin America and the Caribbean – an excellent source of up-to-the-minute data on this region: http://www.cepal.org
Women in Informal Employment: Globalizing and Organizing: http://www.wiego.org

3.9
Child labour

Sally Lloyd-Evans

Child labour as a global issue

One of the most hotly debated issues in the development agenda over the last 25 years has been the high incidence of child labour in Asia, Africa, Latin America and the Caribbean as work has become a key feature of many childhoods (Bass, 2004; Ansell, 2005). In the 1990s, heightened concern over the future welfare of millions of the world's poorer children largely developed from media coverage of child-related issues, such as the murder of Brazilian street children by police, and increased documentation on child work by non-governmental organizations (NGOs) and international institutions such as the United Nations Children's

Fund (UNICEF), the World Bank and the International Labour Organization (ILO). Since 2000, a more child-centred development agenda was linked to the Millennium Development Goals' (MDGs) focus on the provision of 'decent work' for youth, universal primary education and the global anti-child labour programme (International Programme on the Elimination of Child Labour – IPEC), while NGOs had brought child trafficking and forced labour to the attention of the public. Increased awareness has escalated child labour as a priority issue for global institutions concerned with human rights, equity and civil society, with the ILO declaring child labour as an urgent human rights challenge. Recent concerns over the effectiveness of international child labour legislation largely stems from three inter-related global factors: the economic crisis, the dismal employment prospects of youth and adolescents, and the rise in exploitative and hazardous work as a result of the deregulation of formal labour markets.

In 2012, the ILO estimated that 215 million children across the world 'are trapped' in child labour, with over 50 per cent exposed to the 'worst forms of child labour' in mining or quarrying, slavery and forced labour, illicit activities such as prostitution, or armed conflict (ILO, 2011). Hazardous work is defined as work that jeopardises the safety, health or morals of children, and its abolition has been the main focus of international legislation since the implementation of Convention 182 on the elimination of the 'worst forms of child labour' by the ILO in 1999. While recent figures present a significant decrease in child labour since the mid 1990s, they hide marked disparities between age, gender and geography. First, while gains in reducing hazardous work have been made for the under 14s, there has been a 20 per cent increase in the total number of youth labourers aged 15 to 17 in hazardous jobs, from 52 to 62 million between 2004 and 2008 (ILO, 2011). As the likelihood of children engaging in hazardous work increases with age, global concern over the livelihoods of 1.2 billion adolescents in a world of high youth unemployment, exploitation and social exclusion has increased. Universal statistics also mask significant gendered social forces that channel girls into invisible spaces of domestic labour, unpaid household and care work, while the proportion of adolescent boys engaged in hazardous work is rapidly increasing. Although the total number of children employed in hazardous jobs remains highest in Asia and the Pacific, the greatest incidence of hazardous work relative to the total number of working children is found in sub-Saharan Africa.

Child labour is rooted in poverty, history, culture and global inequality. Although the fundamental reason why children work is poverty, there are other important drivers such as the role of sociocultural norms attached to the importance of child work in family socialisation and the prevalence of inadequate education systems that push children into work at an early age. While global institutions argue that the incidence of child labour will decline as a country's per capita GDP rises, child labour is also seen to be a serious consequence of neoliberalism and unequal trade. Kielland and Tovo (2006: 6) argue that in Africa modern child labour is a result of the rapid and rather violent encounter between a rural farming society and a modern urban culture where extended family survival strategies are disintegrating.

As a result, the debate over working children is marked by moral indignation and sympathy, which detracts attention from an analysis of the processes that draw children into work (Aitken *et al.*, 2008; Potter *et al.*, 2012). Furthermore, prevailing perceptions of children as helpless 'victims' often undervalue the essential contribution they make to household incomes, and denies them the 'right' or agency to help their families in the struggle for a more equitable distribution of resources. Hence, the question over whether child labour should be abolished or accepted within a broader rights based approach continues to be contested. As this chapter will highlight, child labour is an extremely complex and multifaceted subject, as 'work' can simultaneously be seen as both harmful to a child's development and yet essential for providing for their basic needs and socialization into adulthood.

Child labour and child work: Conceptual issues

What is child labour?

Child labour takes many forms, from paid work in factories to street selling in the informal economy, which are particularly characteristic in cities, to unwaged labour in the household or on the land. International campaigns have recently highlighted the plight of the world's 'invisible' children, many of whom have been trafficked into slavery or sold into debt bondage. Major geographical differences in the incidence and nature of child work can become blurred in the uniform category of 'child labour'. Of the 215 million child labourers, the ILO (2011) estimates that 113.6 million are working in the Asia-Pacific region, 65.1 million in sub-Saharan Africa and 14.1 million in Latin America and the Caribbean. Geographical disparities appear to have increased over the last ten years, with Latin America and the Caribbean substantially reducing the proportion of children in work to 1 in 10 in contrast to an increase in sub-Saharan Africa to 1 in 4. Here, rising poverty and a diminishing adult labour force, due to HIV/AIDS, have driven more children into both paid labour and unpaid care roles (Robson, 2004).

Child labour still evades a universally agreed definition, not least because the common description of an adult worker as 'someone who sells their labour power' does not apply to children working alongside their parents within the household (Wells, 2009: 99). Most definitions are centred on whether work has a 'detrimental impact' on a child's physical, mental or moral development and current debates focus on whether there is a clear distinction between 'child labour' (usually waged) and 'child work' (unwaged), which is socialization undertaken in the course of everyday life. The ILO defines child labour as work performed by children who are under the minimum age that legislation specified for that job or work which is deemed detrimental for children and is therefore prohibited (ILO, 2011). Although there is growing acceptance that excessive work is bad for the under 12s, as it harms their social development and prevents them from attending school, the decision to exclude domestic chores from some child labour classifications has been widely criticized due to the excessive working hours and conditions experienced by many children. With an estimated 60 per cent of working children engaged in agriculture, commentators have asked why unpaid household labour is considered to be morally 'neutral', compared to waged work in industry, when both can be equally detrimental to the development of children (Nieuwenhuys, 1994). Here, gender is salient, as girls rather than boys are usually expected to undertake a greater proportion of invisible household activities. As unpaid children's work is largely hidden within invisible economies of care and reproduction (Wells, 2009; Nieuwenhuys, 2007), child labour needs to be deconstructed in its appropriate social and cultural context.

Child labour and development: Contemporary thinking

'Western' perceptions of childhood

It has been suggested that the global preoccupation with Third World child workers is further evidence of the enforcement of Western codes of conduct in developing nations, as much literature on child labour stems from a Western and predominantly middle-class construction of 'childhood' (Aitken, 2001). The idea of 'working children' challenges traditional meanings of childhood, which are predominantly based on Western norms of behaviour. These have considerable geographical, gendered and cultural implications in relation to whether it is acceptable for children to make decisions about their lives (White, 1994). New geographies of global childhood have been important in highlighting the heterogeneity of childhood experiences in different parts of the world, and research on 'other childhoods' has sought to critique constructions of childhood which dominate understanding of child labour in the global North (Bourdillon, 2006; Skelton, 2009). Not only has recent work highlighted the important contributions made by children to their own livelihoods, it has offered new understandings of how children's roles in household (re)production provide a route to understanding new forms of socio-spatial inequalities brought about by globalization.

Socio-spatial dimensions

There are socio-spatial dimensions to child labour which may explain why some categories of 'child work' are deemed to be more undesirable than others. In particular, many definitions of hazardous 'child labour' are defined by spatial parameters. As Jones (1997) argues, societal views on child labour and related issues will depend on the meanings people attach to public 'spaces' and what they see as appropriate places for children. In Latin America, millions of street children are perceived as 'criminal' and a 'threat' to family and social order. Across the world, city streets and industrial factories are seen as 'unnatural spaces' for child workers (Van Blerk, 2005; Bromley and Mackie, 2009), while the private household space is deemed to be safe. The only public space deemed acceptable for children is school, regardless of the quality of education available to the child. Such conceptual dilemmas regarding children's spatial identities impact upon the development and implementation of global policies which address child labour.

Child labour from a rights perspective

It is not surprising that issues surrounding child labour have taken precedence in development campaigns. Closely aligned to contemporary campaigns over human rights, social protection and fairer globalization, multilateral institutions, such as the World Bank and the ILO, want to 'make child labour history' and they have largely adopted a 'paternalistic' approach to child workers which regards them as passive victims of an unfair global system in need of protection (Jones, 1997). Although the existence of child labour is a visible indication of uneven development and poverty, most global institutions believe that hazardous child labour can be eliminated independently of poverty reduction. Schooling is universally regarded as the best way to eliminate child labour but this one-dimensional approach masks the interdependence between work and education that shapes the livelihood trajectories of many young people (Chant and Jones, 2005).

Despite the ILO's optimistic belief that the end of child labour is within reach, many commentators question whether the total abolition of child labour serves the best interests of the child. A commonly held assumption is that the most successful way of protecting children from harmful work is to exclude them from all employment, but critics argues that children should have a right to benefit from work that is appropriate to their age as it can be important for self-esteem, socialization and household maintenance. Although progress has been made in protecting children's rights following the international adoption of the United Nations

Convention on the Rights of the Child in 1989, campaigners argue that children's rights to work remain contested.

While there is reticence by institutions and governments to abolish all forms of child work for the reasons outlined earlier, the ILO's goal to eliminate the worst forms of child labour by 2016 has been accepted by 95 per cent of member states who have ratified Convention 182. From 2012, the ILO states that anti-child labour policies will be embedded within a global workplace rights agenda. Although positive, international solutions to abolish child labour take decision-making away from child workers and they are regarded by many as unfeasible (Robson, 2004). By contrast, many NGOs and grassroots organizations have attempted to implement small-scale programmes which recognize children as rational individuals who can be empowered to take control of their own lives. Grassroots initiatives, such as street drop-in centres endeavour to give children the opportunity to work in safe environments while also providing time for schooling and recreation. Kielland and Tovo (2006) have highlighted the use of 'conditional transfers' (food or fuel) to poor families in exchange for removing their children from hazardous work in Africa.

Over the last decade, child labour has been increasingly represented as an urgent human rights challenge, and yet there is continued scepticism over whether policy initiatives will enhance children's quality of life unless there are changes to the unequal distribution of global wealth and trade. While the international community have embraced a framework of international regulations that seek to prevent children from working, there is now an understanding of the need to adopt a more flexible approach to child work that embraces the positive aspect of employment while protecting their rights.

References

Aitken, S. (2001) 'Global crises of childhood: Rights, justice and the unchildlike child', *Area* 33, 119–127.

Aitken, S., Lund, R. and Kjørholt, A. T. (eds) (2008) *Global Childhoods: Globalisation, Development and Young People*, London: Routledge.

Ansell, N. (2005) *Children, Youth and Development*, London: Routledge.

Bass, L. (2004) *Child Labour in sub-Saharan Africa*, Boulder, CO: Lynne Rienner.

Bourdillon, M. (2006) 'Children and work: A review of current literature and debates', *Development and Change* 37 (6), 1201–1226.

Bromley, R. D. F. and Mackie, P. K. (2009) 'Child experiences as street traders in Peru: Contributing to a reappraisal for working children', *Children's Geographies* 7 (2), 141–158.

Chant, S. and Jones, G. (2005) 'Youth, gender and livelihoods in West Africa: Perspectives from Ghana and the Gambia', *Children's Geographies* 3 (2), 185–199.

ILO (2011) *Children in Hazardous Work: What We Know, What We Need to Do*, Geneva: IPEC /ILO.

Jones, G. A. (1997) 'Junto con los ninos: street children in Mexico', *Development in Practice* 1 (1), 39–49.

Kielland, A. and Tovo, M. (2006) *Children at Work: Child Labour Practices in Africa*, London: Lynne Rienner.

Nieuwenhuys, O. (1994) *Children's Lifeworlds: Gender, Welfare and Labour in the Developing World*, London: Routledge.

Nieuwenhuys, O. (2007) 'Embedding the global womb: Global child labour and the new policy agenda', *Children's Geographies* 5 (1–2), 149–163.

Potter, R., Conway, D., Evans, R. and Lloyd-Evans, S. (2012) *Key Concepts in Development Geography*, London: Sage.

Robson, E. (2004) 'Hidden child workers: young carers in Zimbabwe', *Antipode* 36 (2), 227–248.

Skelton, T. (2009) 'Children's geographies/geographies of children: Play, work, mobilities and migration', *Geography Compass* 3 (4), 1436–1448.

Van Blerk, L. (2005) 'Negotiating spatial identities: Mobile perspectives on street life in Uganda', *Children's Geographies* 3 (1), 5–21.

Wells, K. (2009) *Childhood in a Global Perspective*, Cambridge: Polity Press.
White, B. (1994) 'Children, work and "child labour": Changing responses to the employment of children', *Development and Change* 25 (4), 849–878.

Further reading

Bourdillon, M. (2006) 'Children and work: A review of current literature and debates', *Development and Change* 37 (6), 1201–1226. An informed review of conceptual debates.
Bromley, R. D. F. and Mackie, P. K. (2009) 'Child experiences as street traders in Peru: Contributing to a reappraisal for working children', *Children's Geographies* 7 (2), 141–158. A reappraisal of child work that highlights the benefits of street trading to children's empowerment.
Nieuwenhuys, O. (2007) 'Embedding the global womb: Global child labour and the new policy agenda', *Children's Geographies* 5 (1–2), 149–163. A contemporary critique of child labour, globalization and social production.
Wells, K. (2009) *Childhood in a Global Perspective*, Cambridge: Polity Press. Provides detailed discussions of global constructions of childhood.

3.10

Migration and transnationalism

Katie D. Willis

Introduction

International migration has become one of the key characteristics of the increasingly interconnected world of the early twenty-first century (Castles & Miller, 2009). While many of these migrants are fleeing persecution or natural disasters, economic migration is highly significant. Between 1970 and 2010, the number of international migrants more than doubled to 213.9 million. While Europe had the largest share of the world's international migrants in 2010, largely as a result of movement within the European Union by EU members, international migrants made up the largest percentage of the population in Northern America and Oceania (Table 3.10.1).

Because of improved communications technology and transport, links across national borders are easier than they were in the past. This means that many migrants now live what have been termed 'transnational lives'. According to Basch *et al.* (1994: 6) 'transnationalism' in relation to migration is 'the process by which transmigrants, through their daily activities, forge and sustain multi-stranded social, economic, and political relations that link together their societies of origin and settlement, and through which they create transnational social fields that cross national borders'. This ability to live both 'here' and 'there' has implications

Table 3.10.1 The world's international migrants, 1970–2010

	No. of international migrants (millions)		*International migrants as % of population*	*% of international migrants by region*
	1970	*2010*	*2010*	*2010*
More developed regions	38.4	127.7	10.3	59.7
Less developed regions	43.0	86.2	1.5	40.3
Africa	9.9	19.3	1.9	9.0
Asia	27.8	61.3	1.5	28.7
Latin America and Caribbean	5.7	7.5	1.3	3.5
Northern America	13.0	50.0	14.2	23.4
Oceania	3.0	6.0	16.8	2.8
Europe	21.9	69.8	9.5	32.6
WORLD	81.4	213.9	3.1	100.0

Source: Compiled from data from UNDESA (2009)

for the construction of migrant identities, but the focus of this chapter will be on the role of transnational migration in development practices in the global South.

Brain drain

While emigration may be a positive strategy for individuals and their families, for communities and countries in the global South such migration has often been represented as a loss of resources; a so-called 'brain drain'. This is particularly because more educated, skilled and dynamic individuals are over-represented within economic migrant flows. Within agricultural communities, this outmigration (to both national and international destinations) can leave insufficient labour to maintain agricultural production.

The medical sector has received particular attention in relation to the brain drain and international migration (Connell, 2010). Training doctors, nurses and other medical personnel is very expensive for governments, but it is viewed as a wise investment as these workers will contribute to a country's social and human development once they have been trained. However, as migration becomes more affordable and logistically possible, medical workers may choose to migrate overseas for better salaries and working conditions. Jobs are available due to shortages in hospitals in the global North, or the unmet demand for care workers for the elderly, infirm and children. The impacts of the migration of medical personnel are particularly acute in sub-Saharan Africa, where formal healthcare resources are already limited due to insufficient funding for staff, infrastructure and medicines.

Remittances

Remittances are money and goods sent back 'home' by migrants. This process has become much easier due to developments in the banking system and international money transfer services such as Western Union. Remittances may also be returned through less formal means, such as being sent with friends or relatives on a trip 'home'. The transnational connections, both virtual and real, help facilitate the flows of remittances.

Remittances can clearly be used by family members for personal consumption, such as home improvements and property, health or education expenditure. Such expenditure is

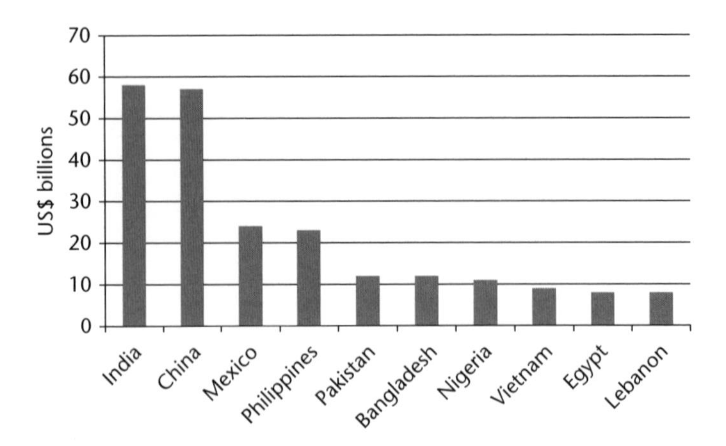

Figure 3.10.1 Top ten recipients of migrant remittances, 2011 (US$ billions)

Source: Adapted from Mohapatra *et al.* (2011, box fig. 1, p.3)

often characterised as being 'economically non-productive', particularly if money is spent on imported goods. However, in some cases, money received from abroad may be invested in small businesses, so contributing to local level economic development. Improvements in housing and expenditure on health and education may also have wider development benefits. While this remittance expenditure benefits the recipients, it may lead to increasing inequalities within communities (De Haas, 2006).

As the level of international migration and flows of money have increased, remittances have attracted greater attention from governments and development policymakers. While it is impossible to provide an exact figure for the level of remittances due to informal and illegal transfers, the World Bank estimates that global remittances reached US$483 billion in 2011, with US$351 billion of this going to developing countries (Mohapatra *et al.*, 2011: 15). To put this into perspective, in 2010 overseas development assistance (ODA) from the countries of the Development Assistance Committee (DAC) of the Organization for Economic Cooperation and Development (OECD) was US$128.5 billion (OECD, 2012: table 13). This means that remittances far outweighed the amount of bilateral aid from the world's richest countries. India and China are the recipients of the largest amount of remittances (see Figure 3.10.1), but as a proportion of gross domestic product (GDP), smaller countries are reliant on remittances. For example, in 2010, remittances represented 31 per cent of Tajikstan's GDP and 29 per cent of Lesotho's (Mohapatra *et al.*, 2011: 3). Civil unrest (as in North Africa in 2011–2012) and a global economic crisis (2008–2010) can have significant effects on migrant job opportunities, and therefore on remittances (IOM, 2011; Wright & Black, 2011).

Given the size of these money flows, it is not surprising that governments in certain key labour exporting countries have sought to encourage migrants to remit. This may be in the form of individuals remitting, or through community organisations investing in projects. International migrants from the same village or urban district often end up migrating to the same location overseas. Hometown associations (HTAs) or equivalent may be set up to provide support for migrants overseas, but they may also raise money for projects 'at home' (Mercer *et al.*, 2008). These can be crucial for local infrastructure developments such as schools or health clinics, but it is important not to over-romanticise these organisations and the capacity for exploitation or enhanced inequality which may be involved (Page and Mercer, 2012).

Flows 'back home' through transnational networks may also be in the form of intangible 'social remittances' (Levitt, 2001) incorporating ideas and practices regarding politics, environment and gender relations, for example. These social remittances are transferred not only through individuals, but also collectively through HTAs, as Levitt and Lamba-Nieves (2011) demonstrate through their work with Dominican migrants in the USA.

Return migration

International migration can contribute to development processes of the 'home' country through more than just remittances. While the brain drain involves a loss of talent and human resources, skills and experiences gained during migration can provide very positive inputs if migrants return. The influx of social, cultural and human capital, as well as financial capital, can be an important trigger to economic development. This process has been represented as a 'brain gain'.

Of course, not all return migrants contribute in this way. For some, return may represent the end of their working lives, while in other cases, the reality of coming 'home' fails to live up to their expectations and further mobility follows as they cannot settle. Return migration may also be associated with the transference of potentially destructive practices, such as the development of US-linked criminal gangs in Mexico and Central America. However, here are many cases of positive outcomes for both individuals and communities as return migrants invest in new businesses, provide an impetus in local politics or add to the pool of skilled labour (Conway and Potter, 2006). In some cases, governments have been so keen to encourage certain kinds of return migration, that they have offered tax incentives (UNDP, 2009).

As with many forms of migration, it is impossible to identify return migration definitively, as a migrant may move away again in the future. In certain parts of the world, transnational connections have developed around businesses, thus creating migration circuits. These businesses often started by providing goods, such as food, for immigrant communities who were missing their home comforts (Alvarez, 2005).

The state and transnational migration

Increasing international mobility and the intensification of transnational exchanges does not mean the end of the nation-state; in fact without the continued presence of the nation-state the concept of 'transnationalism' would be redundant. However, processes of transnationalism and globalisation have required national governments to change their development strategies and the relationships between states and citizens. As outlined above, international migration provides opportunities for development capital and expertise. This has meant governments paying increased attention to citizens living elsewhere and encouraging continued links. This is a break from past practices where governments focused almost completely on citizens living within the national boundaries.

Given the financial possibilities provided by remittances, governments are increasingly facilitating temporary labour outmigration through formal government-approved schemes. Such activities are not new, for example, thousands of Mexicans migrated to the USA as part of the Bracero Program 1942–1964, but the current scale of government involvement in such schemes is unprecedented in countries of the global South. The government of the Philippines, for example, regulates the activities and fees of agencies recruiting workers through the Philippine Overseas Employment Administration (POEA). There is also a strong political rhetoric of the overseas Philippine workers being heroes and heroines of the nation, so encouraging these migrants to feel a continued attachment to the country.

Such attachment is also encouraged through the increased relaxation of dual nationality regulations. While some countries, such as India, do not allow dual citizenship, many others are allowing their nationals to take on passports of another country. There is also increased flexibility in relation to voting in elections and standing for political office. All of these legal changes reflect national government attempts to seize the opportunities presented by transnational migration (UNDP, 2009).

Conclusions

Migration is often viewed as a response to development failure, as individuals seek to find a better life for themselves and their families elsewhere. Increasing international migration flows and greater possibilities for transnational economic, social and political practices have meant that emigration does not mean the severing of ties with home that it often did in the past. Rather, international migration is increasingly being seen as a positive contributor to development in both the 'sending' and 'receiving' countries. This perspective has meant national governments, international development organisations and donor agencies have begun to develop policies to maximise the benefits accruing from international migration. Of course, international mobility can involve great hardships for the people involved and those left behind, and it can exacerbate existing inequalities at a local, national and global scale, but there is also considerable potential for migration to contribute to improvements in the standard of living and quality of life.

References

Alvarez, Jr, R. R. (2005) *Mangos, Chiles and Truckers: The Business of Transnationalism*, Minneapolis: Minnesota University Press.

Basch, L. G., Glick Schiller, N. and Blanc-Szanton, C. (1994) *Nations Unbound: Transnational Projects, Post-Colonial Predicaments, and Deterritorialized Nation-States*, Longhorne, PA: Gordon and Breach.

Castles, S. and Miller, M. (2009) *The Age of Migration: International Population Movements in the Modern World*, Fourth Edition, Basingstoke: Palgrave Macmillan.

Connell, J. (2010) *Migration and the Globalisation of Health Care*, Cheltenham: Edward Elgar.

Conway, D. and Potter, R. B. (2006) 'Caribbean transnational return migrants as agents of change', *Geography Compass*, 1, 25–50.

De Haas, H. (2006) 'Migration, remittances and regional development in Southern Morocco', *Geoforum*, 37, 565–580.

International Organization for Migration (IOM) (2011) *World Migration 2011: Communicating Effectively About Migration*, Geneva: IOM. (Available online at www.iom.int)

Levitt, P. (2001) *The Transnational Villagers*, Berkeley: University of California Press.

Levitt, P. and Lamba-Nieves, D. (2011) 'Social remittances revisited', *Journal of Ethnic and Migration Studies*, 37 (1), 1–22.

Mercer, C., Page, B. and Evans, M. (2008) *Development and the African Diaspora: Place and the Politics of Home*, London: Zed.

Mohapatra, J., Ratha, D. and Silwal, A. (2011) 'Outlook for remittance flows, 2012–14', *Migration and Development Brief*, 17. World Bank Migration and Remittances Unit. (Available online at www.worldbank.org/migration)

Organization for Economic Cooperation and Development (OECD) (2012) 'OECD statistics'. (Available online at www.oecd.org, accessed 18 March 2012)

Page, B. and Mercer, C. (2012) 'Why do people do stuff? Reconceptualising remittance behaviour in diaspora-development research and policy', *Progress in Development Studies*, 12 (1), 1–18.

United Nations, Department of Economic and Social Affairs (UNDESA), Population Division (2009) *World Migrant Stock: The 2008 Revision Population Database*. (Available online at http://esa.un.org/migration)

United Nations Development Programme (UNDP) (2009) *Human Development Report 2009: Overcoming Barriers: Human Mobility and Development*, Basingstoke: Palgrave Macmillan. (Available online at www.undp.org)
Wright, K. and Black, R. (2011) 'Poverty, migration and human well-being: Towards a post-crisis research and policy agenda', *Journal of International Development*, 23 (4), 555–564.

Further reading

De Haas, H. (2010) 'Migration and development: A theoretical perspective', *International Migration Review*, 44 (1), 1–38. A clear overview of the changing debates within migration and development.
International Development Planning Review (2010) Special issue on 'Mobility, migration and development', *International Development Planning Review*, 32 (3–4). A useful collection of articles dealing with development implications of internal and international migration.
United Nations Development Programme (UNDP) (2009) *Human Development Report 2009: Overcoming Barriers: Human Mobility and Development*, Basingstoke: Palgrave Macmillan. (Available online at www.undp.org) Overview of key policy debates in relation to human development and mobility, including excellent statistical data.

Websites

http://esa.un.org/migration United Nations Department of Economic and Social Affairs, World Migrant Stock database.
www.iom.int International Organization for Migration. An inter-governmental organisation focusing on migration policies and the impacts of migration. *World Migration Reports* provide excellent overviews of policies and statistics.
http://migration.ucdavis.edu/mn/ Migration News website run by University of California, Davis.
www.worldbank.org/migration World Bank migration and remittances site.
www.migrationinformation.org Migration Information Source run by the Migration Policy Institute, a Washington DC based think-tank.
http://migratingoutofpoverty.dfid.gov.uk/ Migrating Out of Poverty research programme consortium funded by the UK Department for International Development (DFID).

3.11

Diaspora and development

Claire Mercer and Ben Page

Diasporas have become an increasingly important feature of the development lexicon over the last decade, which is a consequence of the way the policy debate on the 'migration-development nexus' has evolved (Mohan 2002). Individuals and groups who retain an emotional tie to a distant homeland in the global South are now seen as a potential source of money, ideas and skills for international development. India, China and Mexico are often cited as places where diasporas have made a significant contribution to recent development.

Development economists have been analysing remittances for many decades. But it was only at the very end of the twentieth century when policymakers realised what a large volume of money was being transferred to the global South by workers in the global North that remittances became a high priority in development debates. The first wave of policy responses from international development agencies and national governments sought to (a) increase the volume, accuracy and coverage of *data* on remittances and (b) map and enumerate *diaspora associations*. The second wave of policies aimed to find ways to encourage migrants to send more remittances and to send them through *formal channels* (often by engaging their associations). The third (and current) wave of policies seeks to capitalise on the wealth (rather than the income) of diasporas – particularly through the use of *new financial products*. The significance given to these policies in recent years underpins the claim that diasporas are amongst the 'new agents of development'.

In this chapter we will consider some of the ways in which diasporas can contribute to the development of public goods and services, focussing particularly on collective activities and new financial instruments. We will finish by raising some critical issues in relation to this development strategy.

Defining diasporas

The word 'diaspora' originates from the Greek *speiro*, meaning 'to sow' or 'to disperse'. A diaspora is a group of people who are dispersed across multiple sites from an 'originary' homeplace. This is a better term than 'original' because it suggests the constructed nature of the origins of a people rather than an essentialised narrative with a prehistoric starting point. Diasporas are more than just a group of migrants because members consciously maintain their shared commitment to each other and to a homeplace. For example, the Armenian diaspora has major communities in Argentina, Azerbaijan, France, Georgia, Iran, Russia, Syria, Ukraine and the USA, and many people in the diaspora maintain a collective identity on the basis of their shared relationship to Armenia despite their global distribution. There has been much debate about the definition and meaning of diaspora (Cohen 2008) in relation to: the nature of dispersal (was it forced or voluntary?), the length of time since leaving the homeland (are they first generation migrants or had their family left the homeplace long ago?), the nature of the commitment to home (is it emotional or material?), and the nature of the home itself (is it an existing territory or an imagined homeland?). Diaspora is not a simple concept and various definitions emerge from different disciplinary traditions.

In the context of development studies, 'diaspora' usually indicates a dispersed group of people whose identities are shaped by an emotional connection to a shared homeplace, regardless of how, when or why they left home. So researchers working in this field describe first generation economic migrants (and even migrants who plan to return to their home when they retire) as 'diasporas' (Conway *et al.* 2012). From this perspective, an emotional connection to home is demonstrated by the actions taken by members of the diaspora to secure the well-being of the homeplace and the well-being of those who reside there. The ideas that permeate academic research do not emerge independently of the world of policy and this definition of diaspora is a by-product of current policies, which aim to encourage precisely these sentiments in order to generate particular development outcomes. In other words, development policymakers are often trying to *create* diasporas that conform to their own vision of what a diaspora should be. In this development studies literature there is sometimes a tendency to assume that all diasporas (and all members of any given diaspora) have an interest in development, but this is not the case. There is no automatic relationship between diasporas and development.

The private and public distinction in diaspora/development debates

The vast majority of remittances are sent by individuals to individuals, usually their families or households. This money is commonly spent on school fees, healthcare, clothing, food, land and housing. It is a *private* matter. For development policymakers, such flows have a series of benefits both for recipients (poverty reduction, sustaining livelihoods, social protection and investing in the productivity of future generations) and also for national economies (improving debt-to-export ratios, generating tax revenues, and increasing foreign currency receipts). However, the concern is that such capital flows do not necessarily translate into investment into the *public* goods and services that are the basis of many conceptions of development (roads and transport infrastructure, power supply, water supply, education and health infrastructure). In the next two sections we consider two ways in which that anxiety is addressed: first through the conscious 'development work' undertaken by diaspora associations and second through the emerging possibilities for governments in the South to finance development by leveraging diaspora wealth.

Engaging diaspora associations

Diasporas were involved with the development of their homeplace long before they came to the attention of the international development community. In our own research, for example, we have traced such processes back to at least the 1930s in Cameroon (Mercer *et al.* 2008). One of the almost universal features of diasporas is that they form associations in order to socialise and support each other. These associations might be based on religious, professional, sporting, ideological or alumni affiliation, however, one of the most common forms is the hometown association (HTA). HTAs are clubs for migrants who come from a specific hometown, village, region or sometimes a country. They emerged in response to the challenge of looking after one's peers in new environments away from home (Moya 2005). In addition to their social function, HTAs sometimes pool resources for earmarked development projects at home. Variations on the HTA have been documented for many migrant groups worldwide (most of whom were labour migrants) since at least the end of the sixteenth century. For example, in the early twentieth century, burial societies were established across sub-Saharan Africa in the context of colonial labour migration in order to ensure the social welfare of members in the diaspora, and to repatriate bodies to the homeplace in the event of death. More recently, HTAs have raised money in the diaspora for development projects at home. These projects include income-generating schemes; the construction or rehabilitation of public services such as schools, health facilities and water supplies; and the provision of cultural and leisure facilities for traditional ceremonies and sports (Levitt and Lamba-Nieves 2011).

International development policy has struggled to engage HTAs. This arises partly from the fear that support for HTAs might be divisive, and partly from practical difficulties. The fears about divisiveness are based on the assumption that HTA membership is ethnically homogeneous so that funding them might unwittingly fuel sectarian or identity politics in home countries. This may sometimes be the case, since a shared sense of connection to place is often the basis for HTA activity. However, research on African HTAs has demonstrated that membership is not always homogeneous, and that, anyway, there is no automatic relationship between the shared identity that HTAs capitalise on, and a divisive territorial politics at home

(Mercer *et al.* 2008). From the perspective of the development industry, the practical problems of engaging the diaspora relate to the small size, the fractious nature, the limited capacity, the lack of transparency and the amateur character of HTAs. From the perspective of HTAs, the problem is whether or not they want to be engaged.

Despite these challenges there have been some successful strategies that have enabled policymakers to support a diaspora's development initiatives. The '3 × 1 Program for Migrants' (launched in 2004) provides matching funding from the Mexican government for approved HTA projects. It is called 3 × 1 because for every dollar donated by the diaspora, another three are donated – one by the municipal government, one by the state government and one by the federal government – in Mexico. The programme has supported over 6000 projects proposed by 1000 HTAs in 27 different states in Mexico, receiving US$42 million from the government in 2008 (Aparacio and Meseguer 2011). The provision of public goods has dominated these projects (potable water, sewage systems, electrification, road paving, community centres, improvements to health and education), although income-generating projects, investment and scholarships have also been supported.

New financial instruments and diaspora wealth

More recently policymakers have recognised that while remittances tap migrants' *incomes*, diaspora *wealth* offers far greater potential as a source of financial resources for the governments of developing countries. Current policy debates are thus concerned with diasporas' activities in entrepreneurship, capital markets, the businesses of 'nostalgia trade' and 'heritage tourism' in addition to the continuing interest in philanthropy, volunteering, and advocacy (Newland 2010). Such debates focus on the developmental possibilities that could accrue from those highly skilled diasporas whose business activities transfer knowledge, skills, technology, business acumen and investment back to their homeplaces. For example, the Chinese, Ghanaian, Indian and Malaysian governments have developed policies to incentivise their highly skilled workers overseas not only to share their skills, but to invest in businesses at home.

There is also great interest in leveraging diaspora wealth (including savings, real estate, retirement funds, stocks, bonds and trust funds) for development financing (Terrazas 2010). Foreign or domestic currency savings accounts in banks at home, the securitisation of remittances, diaspora bonds and diaspora mutual funds can raise financing either directly for infrastructural spending or indirectly by facilitating governments' access to more favourable lending terms on global financial markets (Ketkar and Ratha 2009). Of these new instruments, diaspora bonds – long-dated sovereign debt agreements that provide the issuer with access to fixed-term funding at lower interest rates (Terrazas 2010) – have proved very popular, with Rwanda, Kenya and Nigeria the latest governments to issue diaspora bonds, following Bangladesh, Ethiopia, Ghana, India, and Sri Lanka.

Evidently there is scope to experiment with ways to link diasporas to the provision of public goods and services either by engaging diaspora associations or by accessing diaspora wealth using new financial instruments. Yet, both of these strategies are more talked about than acted on. The specific examples of successes are much cited because they are few in number and much of this field remains speculative.

Some critical analysis of diasporas as agents of development

This brings us to some of the problems associated with the current enthusiasm for diaspora-led development. There has always been some concern over the political impact of diasporas in their home countries either because of their role in the perpetuation of local conflicts from the safety of a distant diaspora or because of the reactionary role played by Machiavellian individuals in the diaspora (for example, in fomenting dissent and supplying military hardware), but the list of concerns about this development approach is also growing. We only have space to flag a few points here and for a more extensive critique readers should see some of the 'further reading' listed below.

First, diasporas' developmental activities might perpetuate existing spatially uneven patterns of development at different scales. For example, concerns have been raised about the geographical unevenness of HTA support under the 3 × 1 programme, since there is evidence that the *Partido de Accion Nacional* (PAN), which won presidential elections in 2000 and 2006, used it to reward municipalities and states that had voted for them (Aparacio and Meseguer 2011). Further, the towns and villages that are already relatively wealthy tend to have well-organised HTAs (mostly in the USA) and so were more likely to generate successful project proposals than poorer places. Second, policymakers assume that diasporas will engage in the development of their homeplace in particular ways, but not all diasporas are willing to go along with policymakers' enthusiasms. Third, the engagement of diasporas as agents of development raises important questions about responsibility. Why should diasporas – the majority of whom are not highly paid – be expected to contribute to the development of their village, town or country? Fourth, the increasingly important role played by diasporas can further undermine the social compact in which national governments are expected to provide goods and services in exchange for taxes from residents. In this way it continues a broader process in which development projects are financed and delivered from external sources and generate dependency on exogenous institutions. Fifth, the work done to publicise the benefits of diaspora-led development provides an ideological shield behind which the exploitative practices of the brain-drain can be hidden. Finally, the whole edifice rests on the assumption that remittances from diasporas will continue to flow.

References

Aparacio, F. J. and C. Meseguer (2011) 'Collective remittances and the state: The 3 × 1 program in Mexican municipalities', *World Development*, 40, 1, 206–222.

Cohen, R. (2008) *Global Diasporas: An Introduction*, second edition, Routledge, Oxford.

Conway, D., R. Potter and G. St Bernard (2012) 'Diaspora return of transnational migrants to Trinidad and Tobago: The additional contributions of social remittances', *International Development Planning Review*, 34, 2, 189–209.

Ketkar, S. and D. Ratha (eds) (2009) *Innovative Financing for Development*, World Bank, Washington DC.

Levitt, P. and D. Lamba-Nieves (2011) 'Social remittances revisited', *Journal of Ethnic and Migration Studies*, 37, 1, 1–22.

Mercer, C., B. Page and M. Evans (2008) *Development and the African Diaspora: Place and the Politics of home*, Zed, London.

Mohan, G. (2002) 'Diaspora and development', in J. Robinson (ed.) *Development and Displacement*, Oxford University Press, Oxford, 77–140.

Moya, J. C. (2005) 'Immigrants and associations: A global and historical perspective', *Journal of Ethnic and Migration Studies*, 31, 5, 833–864.

Newland, N. (2010) *Diasporas: New Partners in Global Development Policy*, USAID and Migration Policy Institute, Washington DC.

Terrazas, A. (2010) 'Diaspora investment in development and emerging country capital markets: Patterns and prospects', in Newland (ed.) *Diasporas: New Partners in Global Development Policy*, USAID and Migration Policy Institute, Washington DC, 60–93.

Further reading

Bakewell, O. (2009) 'Migration, diasporas and development: Some critical perspectives', *Jahrbücher f. Nationalökonomie u. Statistik (Journal of Economics and Statistics)*, 229, 6, 787–802.

De Haas, H. (2010) 'Migration and development: A theoretical perspective', *International Migration Review*, 44, 1, 1–38.

Mohan, G. (2006) 'Embedded cosmopolitanism and the politics of obligation: The Ghanaian diaspora and development', *Environment and Planning A*, 38, 5, 867–883.

Østergaard-Nielsen, E. (2011) 'Codevelopment and citizenship: The nexus between policies on local migrant incorporation and migrant transnational practices in Spain', *Ethnic and Racial Studies*, 34, 1, 20–39.

Page B. and C. Mercer (2012) 'Why do people do stuff? Reconceptualising remittance behaviour in diaspora-development research and policy', *Progress in Development Studies*, 12, 1, 1–18.

Raghuram, P. (2009) 'Which migration, which development? Unsettling the edifice of migration and development', *Population, Space and Place*, 15, 2, 103–117.

Websites

African Foundation for Development – the most well established, active and influential Pan-African diaspora/development organisation active in the UK: www.afford-uk.org/

Common Ground Initiative – a DFID-funded scheme run by the NGO Comic Relief that aims to support the development initiatives of African diaspora organisations: www.comicrelief.com/apply for-a-grant/programmes/common-ground-initiative

Dilip Ratha (lead economist at World Bank on migration and remittances) 'People Move' blog, which is excellent for up-to-date stories. You can sign up for email alerts or RSS feed: http://blogs.worldbank.org/peoplemove/blog/36

Ghana Transnet – a great source of case study material from an extended study on relations between Ghanaians in the Netherlands and Ghana: http://ghanatransnet.org/

Migration Policy Institute: www.migrationpolicy.org/research/migration_development.php

Remittances and Development Dialogue of the Inter-American Dialogue – lots of good recent reports in different country contexts: www.thedialogue.org/page.cfm?pageID=80

The World Bank's site on Migration and Remittances: http://go.worldbank.org/SSW3DDNLQ0

Part 4
Rural development

Editorial introduction

Rural poverty persists and remains a concern of many developing countries despite impressive advances in agricultural technology such as GM crops, which led to an increase in agricultural productivity and successful agricultural systems. Different strategies and policies have been adopted in the last fifty years. The UN has taken initiatives in outlining the Millennium Development Goals in order to make reduction of poverty the most important goal in the twenty-first century. Agriculture is still the main source of income and employment in rural areas. Rural livelihoods are increasingly under stress and rural poverty has intensified in many regions. Rapid urbanisation and improvement of communications and transport technology have resulted in a significant increase in mobility. Family life is increasingly individualised and many villages are experiencing erosion of community life. On the other hand women in rural areas are dependent on agriculture for a living and play an active role and are responsible for agricultural/food production and processes in many communities especially with small land holdings, yet in many countries women lack legal rights to own land. Access to land is crucial to poor people's capacity to construct viable livelihoods and overcome rural poverty and hence land policies and particularly land reform have gained a lot of prominence in recent rural social and political movements and agrarian conflict.

In order to survive, many farmers are exploiting the land beyond its carrying capacity. For these reasons critics argue that yield benefits cannot be extended or even sustained. An alternative system advocates integrated management. Input use can be cut substantially if farmers substitute knowledge, labour and management skills. It is important to explore how different means have affected farmer's livelihoods and whether or not they are likely to deliver food security for hungry people.

Intensive agricultural methods such as agrochemicals and pesticides degrade agricultural resources. This leads to accelerated deforestation, soil degradation, damage to biodiversity, pollution and vulnerability to pest attacks and contributes to environmental stresses and extreme weather conditions, leading to deteriorating food security. Food security in the developing countries must not come to be dependent on surpluses from the industrialised countries or, worse, food aid. Priority must be given to the future nutritional needs of their people and to ways and means of meeting those needs locally.

To understand environmental degradation, we need to consider people's livelihoods, for these establish the relationship between economic activity and local environment. Characteristically, forms of environmental degradation are generated by livelihoods based on primary commodity production, such as those of wage labourers in agriculture, or petty commodity producers, sometimes degraded to subsistence producers. The vulnerability of rural people, created by shifting seasonal constraints, short-term economic shocks, longer-term trends of change (such as trade liberalisation and globalisation), the spread of AIDS, ethnic rivalry and conflicts influence institutional structures and processes which encourage them to pursue diverse livelihood strategies to combat rural poverty and vulnerability. All these reasons have led to modern famines, which are more complex and challenged the role of governments, donors and humanitarian organisations in this respect.

Sustainable development and poverty alleviation in rural areas depend on effective common resources management and local governance which hinges on adaptations to local agroecological and social conditions. Varied types of cooperatives have helped people cope with various economic, social and environmental problems. A successful example of NGOs promoting cooperatives is, for example, the Fairtrade movement which guarantees coffee producers a minimum price. There is a need to sustain natural systems and achieve greater equity (e.g. providing more secure and affordable access to land for the poor and credit) in economic, social and political dimensions, to form part of a broader strategy of pro-poor growth.

4.1
Rural poverty

Edward Heinemann

Rural and urban populations

Defining 'rural' populations is less straightforward than it seems. International statistics rely on national definitions of the terms 'rural' and 'urban', and these vary from country to country. As a result, in many situations, areas classified as urban may have rural characteristics, particularly in terms of their reliance on agriculture, and this can lead to a significant undercounting of the rural population. In addition, in recent years the economic linkages between rural and urban areas have grown ever more extensive, and the locations of many millions of people and households in developing countries span the two worlds rather than being exclusively rural or urban.

However defined, the rural population of the developing world is declining as a proportion of its total population. According to the UN Department of Economic Affairs (2009), in 2010 around 55 per cent of the total population was classified as rural, and by 2025 the figure will be below 50 per cent. The rural population itself continues to increase, however, from 2.4 billion in 1980 to around 3.1 billion in 2010; but it is doing so ever more slowly – currently 0.4 per cent per year, and shortly after 2020 it will peak at around 3.2 billion and then start to fall.

There is substantial variation across regions. In Latin America and the Caribbean the rural population has been in decline since the late 1980s, and only 20 per cent of the population was still rural in 2010. At the other extreme, in sub-Saharan Africa and South-central Asia between 60 and 70 per cent of the population was still rural in 2010, and in both regions a majority of the population will be rural for another 20 years or more.

Levels and locations of rural poverty

Migration to urban areas is the key driver of the rural population's relative decline. Poverty is in turn one key driver of migration. And poverty rates are considerably higher in the rural areas than in the urban: of the United Nations (2011) estimate of 1.4 billion people living on less than US$1.25/day in 2005 (since updated to 1.3 billion in 2010 by the World Bank (Chen and Ravallion, 2012), approximately one billion – around 70 per cent – lived in rural areas (IFAD, 2011); and a majority of the world's poor will be rural for many decades to come. Again, there is much variation by region: in Latin America and the Caribbean, and the Middle East and North Africa, a majority of the poor now live in urban areas, while in South Asia, South East Asia and sub-Saharan Africa, over three-quarters of the poor still live in rural areas. Looking at the location of poverty through a different lens, Sumner (2011) makes two important points: first, as many developing countries have progressed from low-income country status to become middle-income countries (MICs), fully three-quarters of the world's poor now live in MICs; and, second, less than a quarter of the world's poor now live in fragile and conflict-affected countries.

Around 35 per cent of the total rural population of developing countries live on less than US$1.25 a day, down from around 54 per cent in 1988 (IFAD, 2011). This decline is mainly due to massive reductions in rural poverty rates in China and South East Asia. The two regions where today rural poverty remains most widespread are South Asia, which has by far the largest *number* of poor rural people – over 500 million; and sub-Saharan Africa, which has the highest *rate* of rural poverty – over 60 per cent of the rural population, and an absolute number of rural poor, around 300 million people, that is still increasing.

Within countries too, rural poverty levels vary considerably. The highest rates of rural poverty are often found in remote areas, characterised by an unfavourable natural resource base, poor infrastructure, weak state and market institutions and political isolation. Yet high rates of rural poverty and large numbers of poor rural people do not always coincide. In Vietnam, for example, the highest rates of poverty are found in relatively remote hill areas in the north-west and central highlands, yet greater *numbers* of the poor live in the more densely populated, better-off delta lowlands (Benson *et al.*, 2010). Many other countries show similar patterns. In Latin America, the geography of rural poverty reflects a long history of the poor, many of them indigenous peoples, being pushed into areas of low agricultural potential, which have subsequently received only limited public investments.

Characteristics of poor rural households

Valdés *et al.* (2008) analyse the determinants of rural poverty and the characteristics of poor rural households across 15 countries in Africa, Asia, Eastern Europe and Central Asia, and Latin America. Across the board they find that, compared to the non-poor, poor rural households are significantly larger and have a higher share of (non-working age) dependants. For lack of alternatives, agriculture is critical to their livelihoods, yet they typically own, or have access to, significantly smaller plots of land and many hold or rent land under tenure arrangements which offer little security. In many developing countries the poorest are the landless. They also own less livestock and have less durable assets to help them ride out shocks. On average, poor households have significantly less education, and substantially less access to running water and electricity. They also have less access to markets as well as to agricultural and financial services.

Rural poverty is also usually rooted in historical factors and economic, social and political relations within societies. It may be reflected in various forms of exclusion, discrimination and disempowerment, and unequal access to, and control over, assets. These limit people's opportunities to improve their livelihoods and undermine their efforts to do so, and increase the risks they face. In some cases these may be the main feature of poverty: indeed, in parts of Latin America and Asia rural poverty can be defined primarily in these terms. These affect some groups of people more than others in each society, but across rural societies the distribution of power often works particularly against women, against youth and against indigenous peoples; and all three of these groups make up a disproportionate share of the rural poor.

Rural livelihoods

Depending on local opportunities and constraints, and their own profiles and characteristics, rural households can derive their livelihoods from a range of sources: their own on-farm production (crops and livestock); common property resources to which they have access (forests and fisheries); employment (agricultural and non-agricultural); self-employment; and transfers, including remittances from migrant relatives and social transfers.

Diversified livelihoods are common among rural households, reflecting their strategies to reduce and manage risks of failure in any single income source. In most of the countries studied by Valdés *et al.* (2008), between 30 and 60 per cent of rural households depended on at least two sources of income to generate 75 per cent of their total income. However, there are variations across regions and countries. In sub-Saharan Africa, a majority of rural households probably still earn at least 75 per cent of their income from on-farm sources, while in many Asian and Latin American countries, non-farm income sources now make up more than half of total rural incomes.

Across all developing regions the share of non-farm income in total rural household income is increasing (IFAD, 2011). Yet agriculture continues to play a key role in the economic portfolios of rural households: in most of the countries studied by Valdés *et al.* (2008) about 80 per cent of rural households engage in farm activities. Typically, it is the poorer rural households that derive the highest proportion of their incomes from farming and agricultural labour, while the better-off households derive the most from non-farm activities. In all cases, income gains are associated with a shift out of agriculture, towards more non-agricultural wages and self-employment.

Rural-to-urban and international migration are also important livelihood strategies for many rural households. Migration can provide opportunities for more secure incomes and for better access to education; and remittances have become a significant element of household incomes in much of the developing world. Although wealthier households generally gain more in absolute terms, poor households count remittances as a vital component of their income and a key element of their risk mitigation and coping strategies.

Moving in and out of poverty

By no means are all poor rural people stuck in poverty as a permanent state. Some become poor, some formerly poor move out of poverty, and some may move in and out of poverty several times in their lives. Dercon and Shapiro (2007) show that across countries and regions there are more people who are sometimes poor than always poor; while IFAD (2011), looking at nine countries in Asia, sub-Saharan Africa and Latin America (Figure 4.1.1), find that it is common for 10 to 20 per cent of the rural population to fall into, or move out of, poverty (defined by national poverty lines) within a five to ten year period. Narayan *et al.* (2009) confirm that rural people do not resign themselves to poverty: they repeatedly take initiatives to improve their lives; and while they may face enormous problems of access to opportunity, they do not generally see themselves as trapped in poverty.

Households often fall rapidly into poverty as a result of a shock, such as exposure to illness, market volatility, failed harvests, natural disasters or conflict – and those most likely to fall are large households with high dependency ratios. Moving out of poverty is a slower process, based on successful enterprise or employment; and shaped by ownership of land, livestock or other productive assets; education; participation in non-farm wage labour and self-employment. Good health is a precondition for moving out of poverty; and Narayan *et al.* (2009) find that feeling confident and empowered is both a factor behind, and a consequence of, moving out of poverty. Local context also matters: upward mobility is harder in communities where there are limited economic opportunities, where social divisions limit access to the opportunities that exist, or where local solidarity networks are weak.

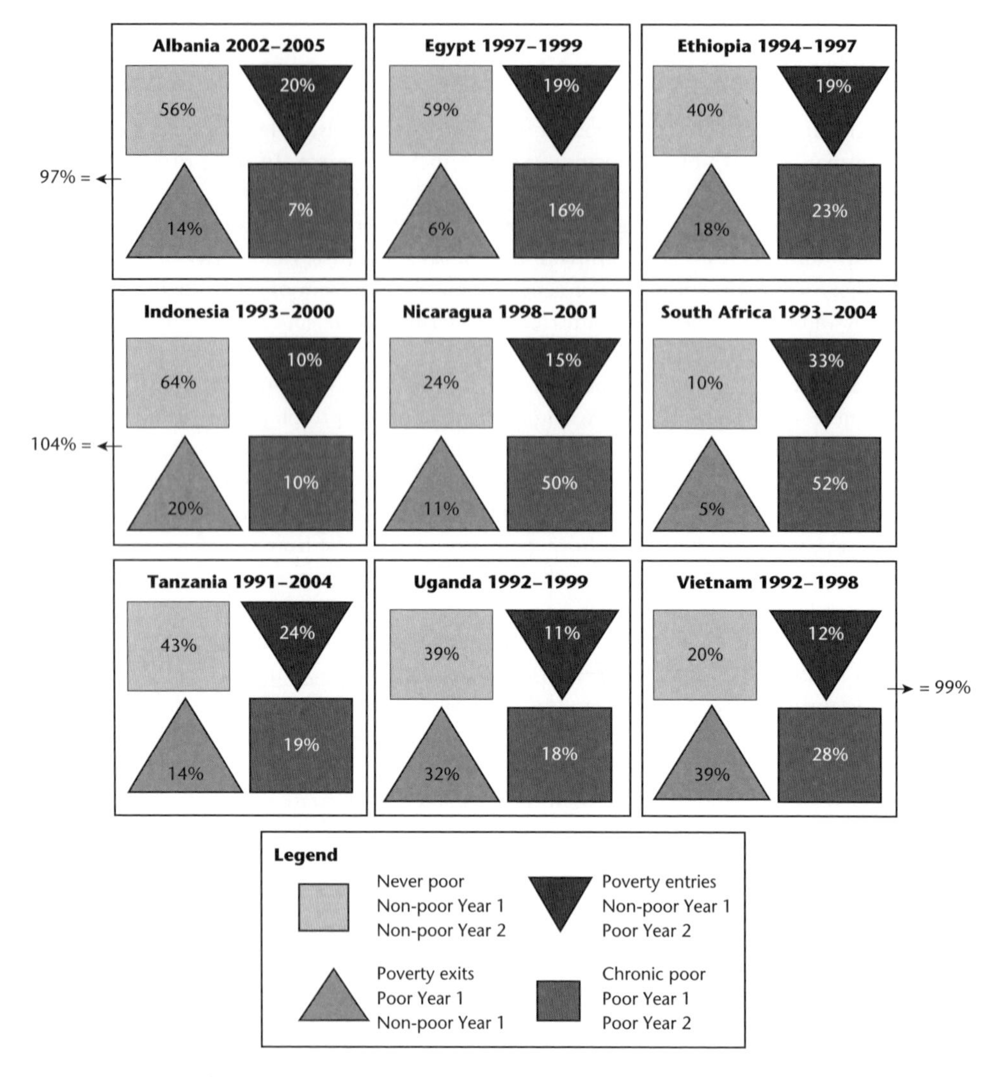

Figure 4.1.1 Rural poverty dynamics

Bibliography

Benson, T., M. Epprecht, and N. Minot. 2010. Mapping where the poor live. In *The poorest and hungry: Assessments, analyses, and actions.* J. von Braun, R. Vargas Hill, and R. Pandya-Lorch (eds). Washington, DC: International Food Policy Research Institute.

Chen, S. and M. Ravallion. 2012. 'An update to the World Bank's estimates of consumption poverty in the developing world'. Development Research Group, Briefing Note, World Bank, 3 January 2012. Available online at http://siteresources.worldbank.org/INTPOVCALNET/Resources/Global_Poverty_Update_2012_02-29-12.pdf

Dercon, Stefan and J. S. Shapiro. 2007. *Moving on, staying behind, getting lost: Lessons on poverty mobility from longitudinal data.* Global Poverty Research Group (GPRG) Working Paper No. 075. Available online at www.gprg.org/pubs/workingpapers/pdfs/gprg-wps-075.pdf

International Fund for Agricultural Development (IFAD). 2011. *Rural poverty report 2011. New realities, new challenges: New opportunities for tomorrow's generation.* Rome: IFAD. Available online at www.ifad.org/rpr2011/report/e/print_rpr2011.pdf

Losch, B., S. Fréguin-Gresh and E. White. 2011. *Rural transformation and late developing countries in a globalizing world: A comparative analysis of rural change*. Washington, DC: World Bank. Available online at http://siteresources.worldbank.org/AFRICAEXT/Resources/258643-1323805221801/RuralStruc_final_report_v2_hd.pdf

Narayan, D., L. Pritchett and S. Kapoor. 2009. *Moving out of poverty. Vol. 2: Success from the bottom up*. Washington, DC: World Bank/Basingstoke, UK: Palgrave Macmillan.

Sumner, A. 2011. *Where do the poor live? An update*. Available online at www.ids.ac.uk/files/dmfile/WheredothepoorliveupdateApril20112.pdf

United Nations. 2011. *The Millennium Development Goals report*. New York: United Nations. Available online at http://www.un.org/millenniumgoals/pdf/(2011_E) MDG Report 2011_Book LR.pdf

United Nations Department of Economic and Social Affairs. 2009. *World urbanization prospects: The 2009 revision population database*. Available online at http://esa.un.org/wup2009/unup/index.asp?panel=1

Valdés, A., W. Foster, G. Anríquez, C. Azzarri, K. Covarrubias, B. Davis, S. DiGiuseppe, T. Essam, T. Hertz, A. P. de la O, E. Quiñones, K. Stamoulis, P. Winters and A. Zezza. 2008. *A profile of the rural poor*. Background paper for the IFAD *Rural poverty report 2011*. Available online at http://www.ifad.org/rural/rpr2008/background.htm

World Bank (2007). *World development report 2008: Agriculture for development*. Washington, DC: World Bank.

Further reading

Ahmed, A. U., R. Vargas Hill, L. C. Smith, D. M. Wiesmann, T. Frankenberger, K. Gulati, W. Quabili and Y. Yohannes. 2007. *The world's most deprived: Characteristics and causes of extreme poverty and hunger*. Washington, DC: International Food Policy Research Institute. Analyses the characteristics of the half billion or so people living on less than US$0.75 a day. Available online at www.ifpri.org/sites/default/files/publications/vp43.pdf

IFAD. 2010. *Introducing the Multidimensional Poverty Assessment Tool (MPAT): A new framework for measuring rural poverty*. Available online at www.ifad.org/mpat/resources/flyer.pdf

Websites

An update to the World Bank's estimates of consumption poverty in the developing world: http://siteresources.worldbank.org/INTPOVCALNET/Resources/Global_Poverty_Update_2012_02-29-12.pdf

Statistics and key facts about indigenous peoples: www.ruralpovertyportal.org/web/guest/topic/statistics/tags/indigenous_peoples

IFAD gender-related publications: www.ifad.org/gender/pub/index.htm

4.2 Rural livelihoods in a context of new scarcities

Annelies Zoomers

Introduction

Rural development – along with other issues such as the appearance of the food, energy and climate crises as well as the booming amount of large-scale land investments since 2008 – is back on the policy agenda of international organizations and governments in Africa, Asia and Latin America. After a long period of non-intervention, there is increasing concern on how to deal with rural land and how to guarantee food security on a global level while stimulating the production and use of renewable energy and/or taking care of climate-sound agriculture as well as rural poverty.

Along with the booming interest in large-scale land investments, rural livelihoods are increasingly under stress, especially in areas with a growing competition for using 'empty' land for global food production, fuel, dams, urbanization and so on (Zoomers 2010). According to the IFAD's rural poverty report of 2011 'global poverty remains a massive and predominantly rural phenomenon – with 70 per cent of the developing world's 1.4 billion extremely poor people living in rural areas, key areas of concern being sub-Saharan Africa and South Asia' (IFAD 2011); increasing numbers of rural people inhabit ecologically vulnerable zones, and the incidence of extreme poverty is more prominent than before. What do we know about 'rural livelihoods' – how should policymakers deal with rural poverty in this context of land 'grabbing' and new scarcities?

The livelihood approach

Efforts have been made since the 1990s to gain a better understanding of rural livelihoods and bring rural development strategies more in line with the aspirations and priorities of the rural people. This *livelihood approach* was a response to the disappointing results of former approaches in devising effective policies to encourage development and/or to alleviate poverty; the central objective was 'to search for more effective methods to support people and communities in ways that are more meaningful to their daily lives and needs, as opposed to readymade, interventionist instruments' (Appendini 2001).

There are many different approaches to sustainable livelihood, but the most common definition is the one given by Chambers and Conway (1991):

> A livelihood comprises the capabilities, assets (including both material and social resources) and activities required for a means of living. A livelihood is sustainable when it can cope with and recover from stresses and shocks and maintain or enhance its capabilities and assets both now and in the future, while not undermining the natural resource base.

Livelihood research, analysing in detail how rural people build their livelihoods on the basis of specific combinations of capitals (e.g. natural, financial, social, human, etc.) in a particular context, helped to uncover a number of dimensions of rural livelihood that up to then had not been very clear. It helped to get a more holistic understanding of livelihood, showing that livelihood does not concern material well-being, but rather that it also includes non-material well-being. According to Bebbington (1999: 2022)

> a person's assets, such as land, are not merely means with which he or she makes a living: they also give meaning to that person's world. Assets are not simply resources that people use in building livelihoods; they are assets that give them the capability to be and act. Assets should not be understood only as things that allow survival, adaptation and poverty alleviation, they are also the basis of agent's power to act and to reproduce, challenge or change the rules that govern the control, use and transformation of resources.
>
> *(in De Haan and Zoomers 2005)*

The livelihood approach represents a multidisciplinary view of poverty, acknowledging that poverty is not an economic problem, but that it involves political, cultural, social and ecological aspects as well (Kaag 2004: 52).

In addition, most livelihood studies have in common that – in contrast to the earlier tendency to conceive poor people as passive victims – they highlight the active, and even proactive, role played by the (rural) poor. The emphasis is on seeing people as agents actively shaping their own future, focusing not on what poor people lack, but rather on what they have (their capitals) and on their capability (Sen 1981; Chambers and Conway 1991).

The livelihood approach is grounded in the idea that people's livelihood largely depends on the opportunity to access 'capitals' (which form the basis for their livelihood strategies). These capitals are 'human capitals (skills, education), social capital (networks), financial capital (money), natural capital (land, water, minerals) and physical capital (houses, livestock, machinery). Sometimes a cultural capital is added; or the physical or financial capitals are replaced by produced capital (Bebbington 1999). Moser, in her vulnerability framework (1998) makes a distinction between labour and household relations – and housing – as additional assets. According to Bebbington, social capital is the most critical asset to rural people because it enables access, and access is needed to diversify and expand their assets portfolio. Social capital enables people to widen access to resources and actors, to make living meaningful and to modify power structures and rules (Bebbington 1999). Social capital plays an increasingly important role and is also a direct consequence of (international) migration (and the 'new economics of migration' theory).

In the livelihood approach, the emphasis is on the flexible combinations and trade-off between capitals (all capitals are linked to each other), for example, if a person does not have land to cultivate (natural capital), he/she will try to acquire a plot through purchasing a parcel (financial capital), or entering into sharecropping relations through their network of social relations (social capital). Capital combinations will evolve in time. Somebody who is forced to migrate because of an emergency (e.g. acute need for cash in case of disease of death) might initially be able to compensate the loss of human capital by mobilizing 'social capital' (i.e. neighbours and relatives working their land). Over time, however, due to their prolonged absence, people rely less on this free help (financial capital replacing social capital). They will be forced to spend more money on hiring labour for working their fields. 'A positive increase in access to one capital can mean a decrease in access to another capital, this is because the input of the one capital can mean a decrease in access to another capital'

Table 4.2.1 An overview of dominant concepts related to rural development policies since the 1950s

Period	*Dominant concepts*	*Dominant actor*	*Examples of interventions*
1950s–	Green revolution	National governments and private sector	Investments in agricultural extension, mechanization, irrigation, etc.; stressing the need for introducing Western technology allowing for 'modernization'.
1960s–1970s	Land reforms and/or Agricultural colonization	National governments and international donors	Redistribution of the land/expansion of agricultural area; stressing the need to redistribute the land and/or providing the rural poor with more and better land ('growth with distribution').
1970s–1980s	Integrated rural development	International donors, national governments and NGOs	Area-based investment in infrastructure, agriculture, small-scale industry, irrigation schemes, etc.; stressing the need for a bottom up and integrated approach with sufficient participation ('basic needs, bottom up development, appropriate development').
1980s–today	Sustainable development	International donors, NGOs and local governments	'*A development that meets the needs of the present generation without compromising the ability of future generations to meet their needs*' stressing the need to sustain the natural system and achieve greater equity in economic, social and political dimensions. More emphasis than before was put on the long-term environmental and distributional aspects of development.
1990s–today	Pro-poor growth/achievement of MDGs	Market forces/international donors, local governments, private sector and NGOs	Stressing the need to create an enabling environment and facilitating market forces, e.g. • helping smallholders to become involved in non-traditional agro-export • the creation of an active, free and transparent land market • micro-credit and saving-systems • sustainable (rural) tourism
2007/2008–today	Land governance, land titling, codes of conduct and responsible investment (in the context of land grabbing)	Market forces/international donors, central governments, private sector and NGOs	Stressing the need for new investments (dealing with the food, energy and climate crises).

(Bebbington 1999). To the extent that rural livelihoods are described in terms of 'vulnerability', a distinction can be made between sensitivity (the magnitude of a system's response to an external event) and resilience (the ease and rapidity of a system's recovery from stress) (Moser 1998: 3). Increasing attention has been given recently to vulnerability and, in particular, resilience in the context of studies on climate change.

A weakness of the livelihood approach is that poverty is sometimes 'romanticized', focusing on the ability, or flexibility, of people to cope with crisis and paying less attention to the fact that poverty is 'bad'. The emphasis is placed on whether or not the poor are able to keep their position rather than on the possibility of social upward mobility (and finding a way out of poverty).

Another important drawback of the livelihood approach is the relative neglect of structural limitations. The livelihood framework is centred on people's assets and capabilities, and by speaking about 'capitals' it seems that people have access to assets; and, indeed, people might have access to land, but the majority have only access to small plots of eroded land. Calling this their 'natural capital' is rather artificial – it will not help people to find a way out of poverty and many people are not capable of exchanging this for other capitals. In the livelihood approach much attention is given to 'agency' but the success of the rural poor will in practice be determined by structural – often locational – limitations (and not so much by strategic actions).

An evaluation

The livelihood approach has helped to gain a better understanding of the realities of rural life and the dynamics of rural livelihoods. It made clear that rural livelihoods cannot be understood without opening the 'black box' of the household. Within the family, members are selectively involved in decision making, and are occupied in different activities, each of them having their own negotiating power. Along with the growing complexity of labour division, the interests of a household's individual members will not always be consistent with the family goal (if there is any); variations in personal capacities and motivations affect the interrelationships among the various activities as well as the degree of internal cohesion. Conflict and competition may arise between activities and among members of the household. In many rural areas, family life is increasingly individualized, and many villages are experiencing erosion of community life. Young people, in particular, aspire to having an urban life; villages and community organizations are increasingly used as a type of safety valve.

Moreover, livelihood studies have ascertained that during the last decade increasing numbers of people in rural areas have opted for a development path characterized by multitasking and income diversification (Bryceson *et al.* 2000; Reardon *et al.* 2001). There is a tendency towards livelihood diversification, namely 'a process by which . . . households construct an increasingly diverse portfolio of activities and assets in order to survive and to improve their standard of living' (Ellis 2000: 15). In many cases, the bulk of income of the rural poor no longer originates from agriculture; people have multiple income sources. Distinctions between rural and urban livelihoods are increasingly difficult to make: on the one hand, rural people (who are formally registered as 'villagers') live and work in the city for most of the year (staying with family members and working in construction, housekeeping, etc.) and the better-off in the rural areas buy parcels in the cities in order to be able to give their children a better education (exit-strategies). On the other hand, people in the urban sphere start growing food crops in the cities (new types of urban agriculture), whereas urban elites are increasingly expanding into the rural sphere by becoming the owners of rural lands. Affluent urbanites obtain the land by foreclosing on loans. Increasing land values have led them to look upon land as an attractive commodity for investment purposes.

In addition to multitasking and the blurring of the rural-urban interface, there is a trend in which rural people increasingly develop multi-local livelihoods. Rapid urbanization and the improvement of communications and transport technology have resulted in a significant increase in mobility. Increasing numbers of rural poor now engage in urban and rural life, commuting from the countryside to urban centres on a daily basis. Poor people also supplement their income by travelling large distances to earn additional money as temporary migrants; and, in turn, international migration (South–South and South–North) is rapidly increasing. Considerable numbers of rural poor are no longer rooted in one place; although they maintain relations with their home communities, they are also attached to other places and function in larger networks.

Conclusion

Livelihood research has made it clear that policymakers and the rural populations do not always operate according to the same logic, which may be a major reason why many policy interventions have disappointing results. First, with respect to policymakers' persistent focus on the *via campesina*, the majority of the rural poor are no longer investing in agriculture, but have a broad range of different income sources, whereas their land is mainly used for subsistence farming and not for generating income. Policymakers often base their intervention on solving problems in situ (i.e. the area where poverty concentrates – poverty mapping) whereas many of the poor search for outside solutions (such as migration). Rural life is in many cases no longer agricultural or village based. The *via campesina* of development is only a realistic option for a happy few in areas with good market access, especially those with sufficient land and capital (and knowledge of international markets). The majority of the rural poor are now trapped in a multi-local livelihood and the possibilities for a better life are difficult to find: low prices, environmental degradation, erosion and land fragmentation restrict the possibilities for locally based livelihood improvements, and there is no clear way out of this situation.

Finally, how should policymakers deal with rural poverty in this context of land 'grabbing' and new scarcities? Given the diversity and dynamics of livelihood 'strategies' problems cannot be resolved by a fixed selection of programmes or projects. Priorities and needs will differ by socioeconomic class as well as by gender and age; moreover, needs change over time. Instead of assuming that it is possible to reach different types of target groups with one and the same policy, it would be useful to adopt a more flexible approach with an array of choices. Current strategies to protect local people and instil codes of conduct might help to prevent exclusion and/or displacement but are too limited for making rural livelihoods 'sustainable' and solving rural poverty.

Bibliography

Appendini, K. (2001) Land and livelihood: What do we know and what are the issues. In A. Zoomers (ed.) *Land and sustainable livelihood in Latin America*. Amsterdam: KIT Publications, pp. 23–38.

Bebbington, A. (1999) Capital and capabilities: A framework for analysing peasant viability, rural livelihoods and poverty. *World Development* 27(12): 2021–2044.

Bryceson, D. F., Kay, C. and Mooij, J. (2000) *Disappearing peasantries? Rural labour in Africa, Asia and Latin America*. London: Intermediate Technology Publications.

Chambers, R. and Conway, G. (1991) *Sustainable rural livelihoods: Practical concepts for the 21st century*. Brighton, UK: IDS.

De Haan, L. and Zoomers, A. (2005) Exploring the frontier of livelihood research. *Development and Change* 36(1): 27–47.
Ellis, F. (2000) *Rural livelihoods and diversity in developing countries*. Oxford: Oxford University Press.
International Fund for Agricultural Development (IFAD) (2011) *Rural poverty report 2011 – new realities, new challenges: New opportunities for tomorrow's generation*. Available online at www.ifad.org/rpr2011/
Kaag, M. (2004) Ways forward in livelihood research. In D. Kalb, W. Pansters and H. Siebers (eds) *Globalization & development: Themes and concepts in current research*. Dordrecht, Netherlands: Kluwer Publishers, pp. 49–74.
Moser, C. (1998) The asset vulnerability framework: Reassessing urban poverty reduction strategies. *World Development* 26(1): 1–19.
Reardon, T., Berdegué, J. and Escobar, G. (2001) Rural non-farm employment and incomes in Latin America: Overview and policy implications. *World Development* 29(3): 395–409.
Scoones, I. (1998) *Sustainable rural livelihoods. A framework for analysis* (IDS working paper 72). Available online at www.ids.ac.uk/ids/bookshop/wp.
Sen, A. (1981) *Poverty and famines: An essay on entitlements and deprivation*. Oxford: Oxford University Press.
Zoomers, A. (2010) Globalization and the foreignisation of space: Seven processes driving the current global land grab. *The Journal of Peasant Studies* 37(2): 429–447.

Websites

www.ifad.org
www.livelihoods.org
www.odi.org.uk

4.3 Food security

Richard Tiffin

Introduction

Modern definitions of food security emphasise that simply producing enough food for everyone is only part of the challenge. The definition that has been most widely accepted since it was proposed at the World Food Summit in 1996 is: 'when all people at all times have access to sufficient, healthy, safe, nutritious food to maintain a healthy and active life', which leads to the identification of the three dimensions of food security: food availability, food access and food use. Food availability requires that sufficient quantities of food are available on a consistent basis, food access is delivered when individuals have sufficient resources to obtain appropriate foods for a nutritious diet and food use recognises that ensuring that the energy and nutrient intakes are adequate depends in part on factors such as feeding and cooking practices as well as the way in which food is allocated between individuals in a household.

In this chapter we review some of the reasons why food security is of increasing concern and some of the routes to guarantee continued supply.

Drivers of change

The heightened interest in food security is the result of a recognition that a number of important factors come together to threaten our ability to adequately feed the planet's population. First there is the population itself. Figure 4.3.1 shows the United Nation's predictions for urban and rural population in developed, less developed and least developed countries. The rapid increase in global populations, as shown in this figure, have given rise to a form of neo-Malthusianism in some quarters with resultant calls for population curbs as a contribution to ensuring global food security. Tiffin (2012) argues against this on the grounds that the rapid population growth of the last century is unlikely to continue and that where growth is expected to be rapid it is an integral part of the process of economic development which will itself contribute to increases in food production. The fact that the world's population is predicted to exceed nine billion in 2050 is well known, but Figure 4.3.1 also emphasises that this growth is not uniform across regions and between the urban and rural sectors. The most significant demographic change over the next forty years is predicted to be a major shift from rural to urban populations in the less-developed countries. As populations become more urbanised the composition of the diet is expected to change. Evidence of this is already becoming apparent and is shown in Figure 4.3.2, which plots the FAO food availability data for Africa.[1] The data indicate that cereal consumption has been more-or-less static since the early 1980s, but at the same time meat – more specifically poultry meat – consumption has increased. This change of diet is indicative of a phenomenon which has been termed the

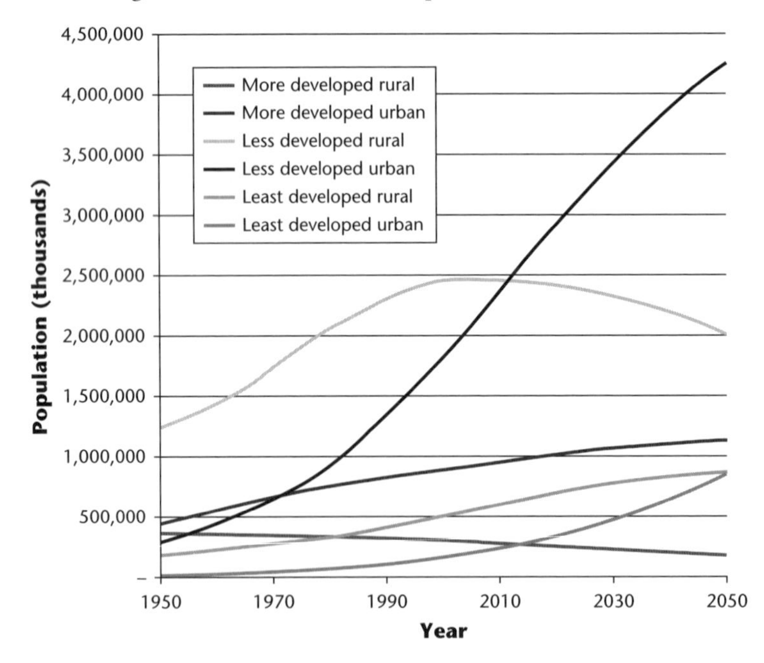

Figure 4.3.1 Urban and rural population growth in developed, less developed and least developed countries

Source: United Nations (2011)

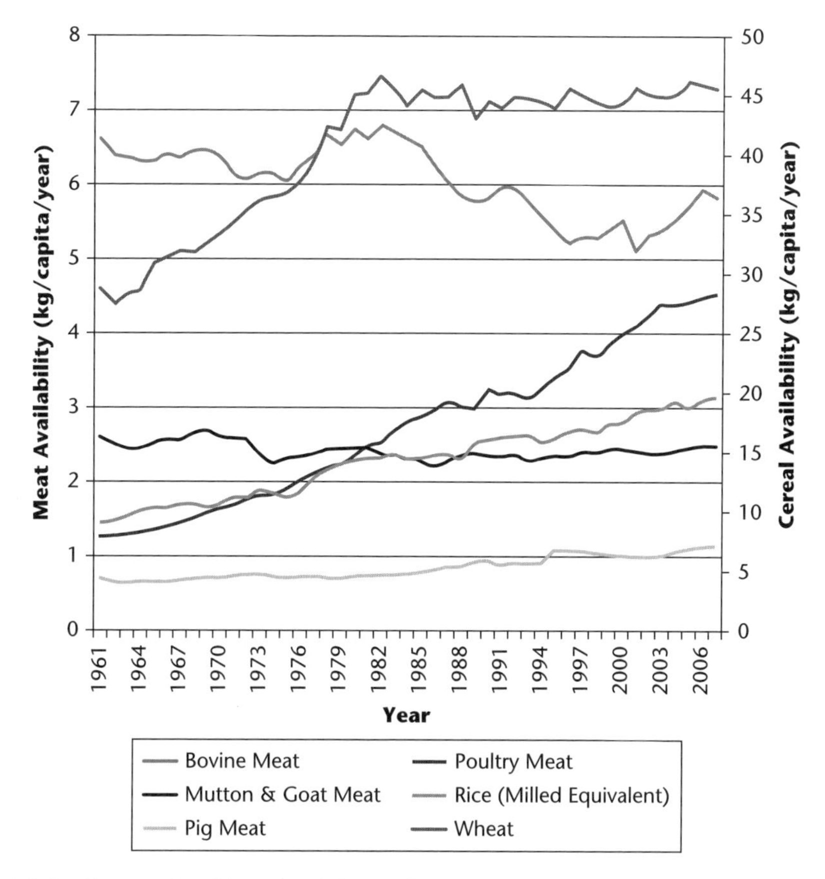

Figure 4.3.2 Changes in African food demand

Source: FAOStat (2012)

nutrition transition (Popkin and Gordon-Larsen 2004), which highlights the fact that the problems of guaranteeing nutritious food and healthy life are not confined to avoiding malnutrition but in some cases will increasingly be concerned with what might be termed over-nutrition. When it is also recognised that many of the health-related problems in the developing world are already associated with the quality of the diet, for example, Figure 4.3.3 shows the prevalence of vitamin A deficiency, the complexity of ensuring that food security extends way beyond just providing sufficient food is apparent.

The second major driver of change that leads to global food security concern is climate change. The linkages between agricultural productivity and the climate are complex. Because of the extreme images that it produces, perhaps the most easily identified of the impacts of climate change on food availability is through drought. For example, Gornall *et al.* (2010) argue that the areas sown with the major crops have all seen an increase in the percentage area affected by drought – from 5–10 per cent to 15–25 per cent since the 1960s. Ambient temperature has a direct impact on crop growth, for example, Ellis *et al.* (1988) show that the reciprocal of the time taken from sowing to awn emergence in barley is linearly related to mean diurnal temperature. Additionally, extreme temperature events at critical periods of the crop's development will have an impact on crop productivity. Wheeler *et al.* (1996) show

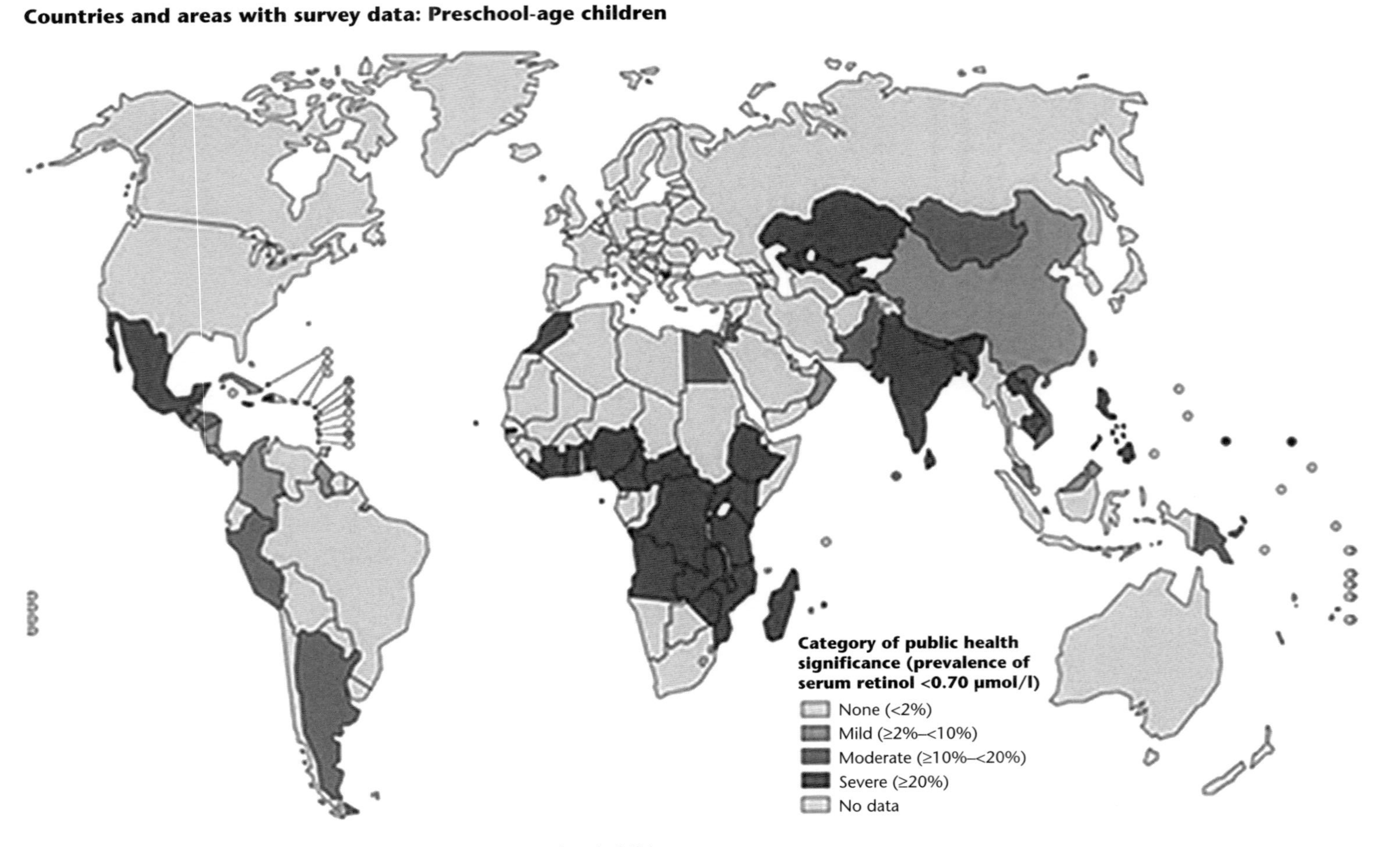

Figure 4.3.3 Prevalence of vitamin A deficiency in preschool children

Source: WHO (2009)

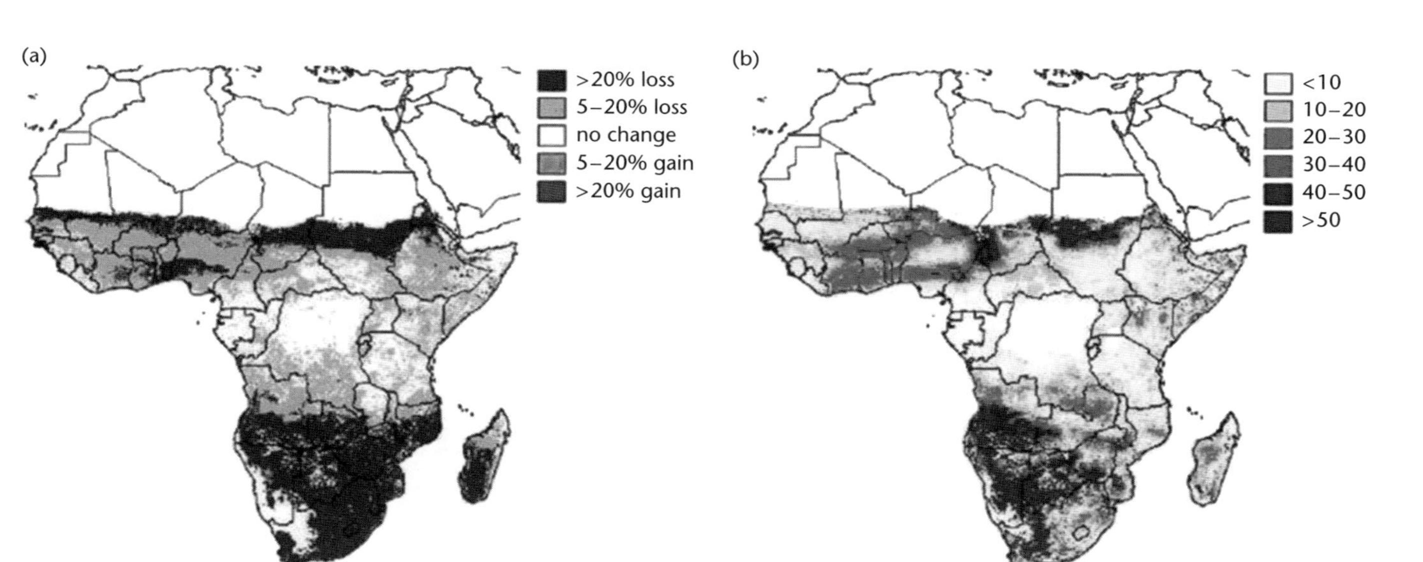

Figure 4.3.4 Length of growing period in the 2090s compared with the present. (a) Mean percentage change for an ensemble of 14 global climate models (GCMs). (b) Coefficient of variation (%) of the change in length of growing period for an ensemble of 14 GCMs

Source: Thornton et al. (2011)

that grain numbers in wheat are reduced by temperatures over 31°C immediately before flowering. One way of summarising the complex interactive effects of climate change on agricultural productivity is to measure the length of growing period (LGP). The LGP is a measure of the suitability of an area for agriculture and takes into account climate and water availability. Thornton *et al.* (2011) discuss the challenges that will face agriculture and the food system in sub-Saharan agriculture in a world where average temperatures rise by 4°C using the LGP. Figure 4.3.4 reproduces one of their figures and shows the predicted changes in the LGP in sub-Saharan Africa. The figure shows how most of the region will experience a reduction in the LGP in many cases by as much as 20 per cent. The only areas that will see an increase are small and located in East Africa. Even this apparently good news is tempered by a further result, reported by Thornton *et al.* (2011), that in these areas maize yields are expected to fall by 19 per cent in a 'four degree' world. The relationship between food production and climate change does not just concern the impacts of climate on production. Agricultural production itself contributes to the emissions of greenhouse gases (GHG). For example, Defra (2011) estimate that agriculture accounts for 9 per cent of the total UK GHG emissions. These emissions include nitrous oxide (55%), which stems largely from the application of synthetic fertiliser, methane (36%) from enteric fermentation and manure and carbon dioxide (9%) from energy use. A further important source of GHG emissions at the global scale arises from the clearance of land for agricultural production. Tilman *et al.* (2011) estimate this loss to be 1.195 tonnes of carbon per hectare for crop years 1–20 following clearance.

The final major driver of change is the increasing scarcity of resources including energy and key plant nutrients such as phosphates. Arguably, however, the most important issues arise because of the increased scarcity of land and water. Agriculture is globally the major user of extracted water resources with the percentage of withdrawn water used in agriculture varying between 32 per cent in Europe and 86 per cent in Africa (see Figure 4.3.5). The FAO report from which this figure is taken goes on to note that demographic growth and economic development are placing unprecedented pressure on water resources with an estimatation that two-thirds of the global population will live in conditions of drought stress by 2025.

Figure 4.3.6 shows the distribution of agricultural land use across the globe. It can be seen that in all of the areas where agriculture is a realistic proposition the rate of use is at least 70 per cent. This fact together with the increasing pressure on alternative land uses and the implications for carbon cycling noted above mean that there is little or no prospect of increasing the amount of land that is available for food production. This in turn means that we will only be able to feed the growing population by using the existing agricultural land more intensively. It is recognised, however, that a lot of the intensification in food production that has occurred in the past has come at the expense of the wider ecosystem. As a result the concept of sustainable intensification is increasingly emphasised as the route to ensuring food security (Royal Society 2009). Central to the concept of sustainable intensification is the recognition that food production is just one of an array of interlinked ecosystem services that is delivered by the planet. The Millennium Ecosystem Assessment (MEA 2005) described ecosystem services in four categories: supporting, provisioning, regulating and cultural. Food production is categorised under provisioning but the ecosystem services-approach emphasises the fact that an increase in this service may be achieved at a cost to services in the other categories. Sustainable intensification therefore demands a careful appraisal of the available options in order to identify those where the benefits of increased food production are not achieved at a greater cost in terms of the delivery of the other services.

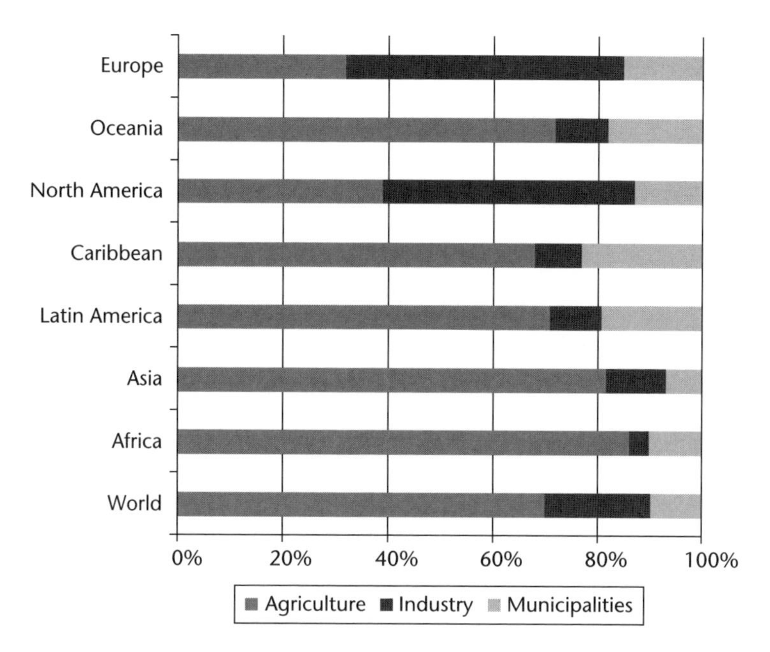

Figure 4.3.5 Distribution of water withdrawal between sectors

Source: FAO (2007)

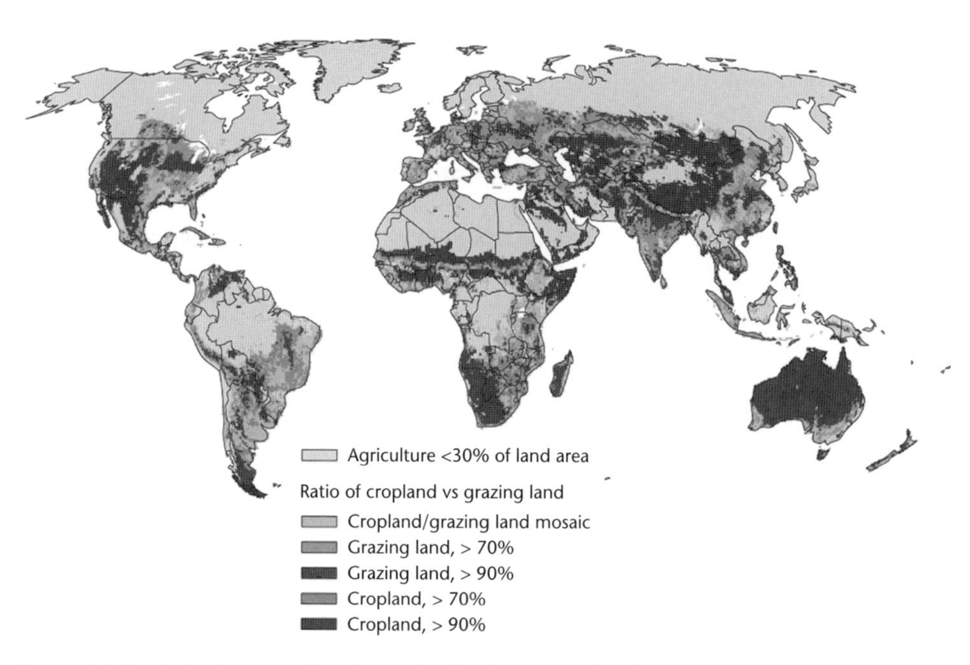

Figure 4.3.6 Agriculture and land use distribution – croplands and pasture lands

Source: Ahlenius (2011)

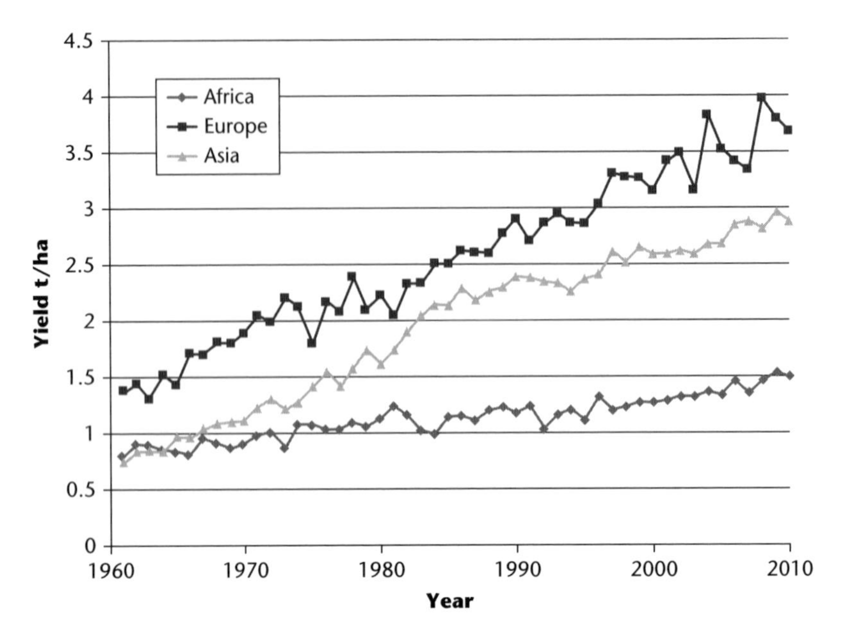

Figure 4.3.7 Wheat yields in Africa, 1960–2010

Source: FAOSTAT (2012)

Addressing the food security challenge

The preceding section highlights the reasons for the heightened interest in food security. Looking to the future, the highly influential UK government's Foresight report on the Future of Food and Farming (2011) identifies five challenges which need to be met in providing solutions:

1 Balancing future demand and supply sustainably (to ensure that food supplies are affordable);
2 Ensuring that there is adequate stability in food supplies (and protecting the most vulnerable from the volatility that does occur);
3 Achieving global access to food and ending hunger (this recognises that producing enough food in the world so that everyone can potentially be fed is not the same thing as ensuring food security for all);
4 Managing the contribution of the food system to the mitigation of climate change;
5 Maintaining biodiversity and ecosystem services while feeding the world.

The emphasis is on taking steps which recognise that food production must be increased but that it must be done in a sustainable manner. It is emphasised that increased isolationism is unlikely to be helpful and noted that the imposition of export bans in the food price spike of 2011 may have worsened the situation. While the development of new technology, to, for example, increase yields, is undoubtedly part of the solution, much can also be achieved with existing technology. Figure 4.3.7 compares the growth in wheat yields that has occurred in Africa, Asia and Europe and shows how the rate of progress has been significantly faster in Europe and Asia. Evenson and Gollin (2003) argue that the considerable gains in agricultural productivity, which the green revolution has brought to many parts of the world, have not been experienced in Africa. As a result there is considerable potential for increasing food production in Africa by translating existing technology. It is also argued that reductions in food waste offer the potential to partially fill the gap in food production. For example, Gustavsson *et al.* (2011) estimate that roughly one-third of total food production globally is lost or wasted. Important differences exist between regions where these losses occur. In Europe and North America, around a third of the losses occur in the household while in Africa and Asia household losses are less than 10 per cent. Improving the food supply chain in Africa and Asia is clearly important and is therefore, as far as ensuring food security, an issue which is likely to become increasingly challenging as these countries become more urbanised.

Conclusion

The challenge of meeting the food demands of our growing population is likely to be at the top of the international agenda for at least the first half of the twenty first century. This chapter has sought to draw out some of the complexity which lies behind this challenge and in the solutions to it. Feeding a growing population is a small part of what confronts us. We need to feed an increasingly wealthy and more urbanised population the right type of food to ensure healthy lives. This does not just mean producing enough food, it also means producing the right type of food, producing it sustainably and ensuring that people are able to access it. Finally, we cannot just think of the challenge as being one of matching demand and supply with policies acting on both sides to achieve equilibrium. To do so runs the risk of curbing the processes of economic development which have and will continue to underpin the ability of developed societies, with their expanded populations, to feed themselves.

Note

1 Food availability statistics measure the amount of food that is available for human consumption. It serves as an approximation to actual food consumption.

References

Ahlenius, H. (2011). Agriculture land use distribution – croplands. Available online at www.grida.no/graphicslib/detail/agriculture-land-use-distribution-croplands-and-pasture-land_8afa

Defra (2011). *Greenhouse gas emission projections for UK agriculture to 2030*, London: Department of the Environment Food and Rural Affairs.

Ellis, R. H., Roberts, E. H., Summerfield, R. J. and Cooper, J. P. (1988). Environmental control of flowering in barley (*Hordeum vulgare* L.). II, Rate of development as a function of temperature and photoperiod and its modification by low-temperature vernalization, *Annals of Botany* 62(2): 145–158.

Evenson, R. E. and Gollin, D. (2003). Assessing the impact of the green revolution, 1960 to 2000, *Science* 300(5620): 758–762.

FAO (2007). Agriculture and water scarcity: A programmatic approach to water use efficiency and agricultural productivity, Food and Agriculture Organisation of the United Nations, Rome.

FAOSTAT (2012). Available online at http://faostat3.fao.org/home/index.html

Foresight. The future of food and farming (2011). Final project report. London: The Government Office for Science.

Gornall, J., Betts, R., Burke, E., Clark, R., Camp, J., Willett, K. and Wiltshire, A. (2010). Implications of climate change for agricultural productivity in the early twenty-first century, *Philosophical Transactions of the Royal Society B: Biological Sciences* 365(1554): 2973–2989.

Gustavsson, J., Cederberg, C., Sonesson, U., Van Otterdijk, R. and Meybeck, A. (2011). *Global food losses and food waste*, Rome: Food and Agriculture Organisation of the United Nations.

MEA (2005). Ecosystems and human well-being, Washington, DC: Island Press.
Popkin, B. and Gordon-Larsen, P. (2004). The nutrition transition: Worldwide obesity dynamics and their determinants, *International Journal of Obesity and Related Metabolic Disorders* 28(Suppl 3): S2–S09.
Royal Society (2009). Reaping the benefits: Science and the sustainable intensification of global agriculture, London: The Royal Society.
Thornton, P. K., Jones, P. G., Ericksen, P. J. and Challinor, A. J. (2011). Agriculture and food systems in sub-Saharan Africa in a 4°C+ world, *Philosophical Transactions of the Royal Society A* 369: 117–136.
Tiffin, R. (2012). Economists are not dismal, the world is not a petri dish and other reasons for optimism, *Food and Energy Security* 1(1): 3–8. Available online at http://dx.doi.org/10.1002/fes3.7
Tilman, D., Balzer, C., Hill, J. and Befort, B. L. (2011). Global food demand and the sustainable intensification of agriculture, *Proceedings of the National Academy of Sciences* 108(50): 20260–20264.
United Nations (2011). *World population prospects: The 2010 revision, highlights and advance tables* (working paper), New York: United Nations, Department of Economic and Social Affairs, Population Division.
Wheeler, T., Batts, G., Ellis, R., Hadley, P. and Morison, J. (1996). Growth and yield of winter wheat (*Triticum aestivum*) crops in response to CO_2 and temperature, *The Journal of Agricultural Science* 127(1): 37–48.
WHO (2009). *Global prevalence of vitamin A deficiency in populations at risk, 1995–2005*, Geneva: World Health Organisation.

4.4
Famine

Stephen Devereux

Introduction

'Famine' can be defined as a widespread and protracted disruption in access to food, which will result in acute malnutrition and mass mortality unless alternative sources of food are available. An estimated 70 million people died in famines during the twentieth century (Devereux, 2000). On the other hand, by the 1970s, famine was eradicated from historically famine-prone Europe (Russia) and most of Asia (China, India, Bangladesh). Nonetheless, since the 'Band Aid' famine in Ethiopia in 1984, a number of famines have occurred in Africa (Malawi, Niger, Somalia, Sudan) and Asia (North Korea). The persistence of famines into the twenty-first century is paradoxical, given recent impressive advances in agricultural technology, communications, early warning systems and international humanitarianism.

Historical famine trajectories

Pre-twentieth century famines were often triggered by natural disasters that destroyed the subsistence basis of agrarian communities, whose vulnerability was exacerbated by underdevelopment – weak markets, undiversified livelihoods, no food aid system. Since the late nineteenth century, improvements in transport and communications integrated isolated communities into the wider economy, allowing governments and traders to respond promptly to food crises. Another factor that reduced famine vulnerability was the emergence of nation-states, but this also exposed previously autarchic communities to potent global forces. During the colonial period in India, catastrophic famines occurred that cost millions of lives, even after the British introduced the Famine Codes in the 1880s to prevent further famines and legitimise their rule (Davis, 2001). In parts of Africa starvation was used as a way of crushing initial resistance to colonisation. The penetration of colonial capitalism into subsistence-oriented economies – the commodification of food crops and expansion of 'cash crops' – was blamed for causing the 1970s famine in the West African Sahel (Meillassoux, 1974).

After independence, historically famine-prone countries took one of two routes. Some, like India, made progress in reducing vulnerability factors, through strengthening political accountability for famine prevention, and through improvements in food production associated with 'Green Revolution' biotechnology. There has been no major famine in south Asia since Bangladesh in 1974. By contrast, independence in Africa was associated with increased political instability and the emergence of 'war famines', the first occurring in Biafra, Nigeria, in the 1960s. Subsequently, many African countries suffered food crises related to conflict and militarisation, including Angola, Mozambique, Liberia, Sierra Leone and Uganda. In the Horn of Africa, the lethal combination of wars and droughts contributed to the persistence of famine to the present (Von Braun *et al.*, 1998).

Theories of famine

Theoretical explanations for famine tend to reflect the disciplinary specialisation of their authors. This section reviews the main theories by discipline: demography (Malthusianism), economics ('entitlements', market failure), politics (democracy, international relations) and conflict (war).

Malthusianism

The most famous theory of famine was conceived in the 1790s by an English priest, Thomas Malthus, who argued in his *Essay on the Principle of Population* (Malthus, 1798) that population could not grow indefinitely in a world of limited natural resources. Eventually, the number of people needing food would exceed global production capacity, when famine would intervene to regulate population growth and balance the demand and supply of food. Although Malthusianism remains extremely influential, as a theory of famine it has many limitations (Dyson and Ó Gráda, 2002). Malthus developed his theory before the industrial revolution moved vulnerable people out of agriculture and into urban centres, before advances in transport and communications allowed food surpluses to be shipped around the world, and before scientific advances in agricultural research dramatically increased crop yields.

Moreover, in no country has population growth ever been 'regulated' by famine. When 30 million Chinese died in the 'Great Leap Forward' famine around 1960, the population of China was 650 million; it now exceeds one billion. In poor countries, high fertility rates and 'dependency ratios' persist as a vulnerability factor, but it is economic growth that solves this problem, not food crises. In wealthy countries, a 'demographic transition' to zero population growth (when births equal deaths) has already occurred. Global population is projected to stabilise during the twenty-first century, at a total well within global food production capacity.

Economics and Sen's 'entitlement approach'

Amartya Sen's *Poverty and Famines* was published in 1981 (Sen, 1981), and was immediately recognised as the most influential contribution to famine thinking since Malthus. Sen argued that a person's 'entitlement to food' derives from four sources – production, trade, labour, and gifts – and that famine is not primarily determined by national food availability, but by failures of access to food at the level of groups or individuals. Even when a drought causes crop failure, only some groups suffer 'entitlement failure' and face starvation – wealthy families and urban residents are rarely affected. Paradoxically, farmers who produce food are often most vulnerable to famine; this is because their 'entitlement' derives mainly from a single unreliable source – rain-fed agriculture.

The entitlement approach complements economic explanations for famines that are based on poverty and market failures. When food production or market supplies of food fall, prices rise and, because food is essential for survival, people who cannot afford to buy the food they need face starvation. In Bangladesh in 1974, rapid food price rises caused by expectations of harvest failure made rice unaffordable for landless labourers, because wealthy people and traders hoarded rice and created an artificial scarcity that caused 1.5 million deaths (Ravallion, 1987).

Politics: Famine as 'act of man'

Famines have always been 'political', but the role of political factors in either creating or failing to prevent famine is increasingly recognised. First, famines affect people who are politically and economically marginalised. Most famines occur in countries that have little global economic or geopolitical significance – Ethiopia, Sudan, Malawi, North Korea. Within countries that suffer famine, the entire population is never at risk – invariably, minority ethnic groups and poor people living in remote rural districts with negligible political influence are most vulnerable.

Second, famines are related to lack of democracy. Apart from his 'entitlement approach', Sen also made an invaluable contribution to the political analysis of famines. Sen argued that two features of democracies protect people against famine: a vigilant free press, so that emerging food crises are neither ignored nor covered up; and free and fair elections to ensure state accountability, because a failing government can be dismissed by the electorate (Drèze and Sen, 1989). To illustrate this argument, Sen contrasted the experiences of India and China. Since achieving independence in 1947, India has effectively prevented famine, while China suffered the world's worst famine in history in 1958–1962. Sen attributes India's success largely to its democratic institutions of campaigning journalism and active opposition politics, both of which were absent in communist China, where lack of information meant the famine was not predicted, while lack of accountability allowed the state to escape unpunished.

Another famine theorist, Alex de Waal, adapted Rousseau's concept of the 'social contract' to argue that post-independence governments in India have upheld an 'anti-famine contract' with their population (De Waal, 1997). Conversely, the persistence of famine in other countries might be explained by the absence of such a 'contract'. If the state faces no pressure to prioritise the basic needs of its citizens, human rights abuses carry no political cost, and this explains why famines are more likely to occur in authoritarian regimes (Stalin's Soviet Union, Mao's China, Mengistu's Ethiopia, Kim Jong Il's North Korea) or during wars, rather than in stable democracies. The famines in the Soviet Union were attributable to punitive government policies, such as forced collectivisation and grain seizures. The

1984 famine in Ethiopia was concealed by the military government, which was fighting a civil war against drought-affected Tigray at the time, until it was exposed by the foreign media. The 1990s famine in North Korea occurred under a repressive regime that does not respect basic human rights and is not responsive to international pressure.

A third political factor in contemporary famines is the role of aid donors. The emergence since World War II of an international humanitarian community that assumes some responsibility for protecting lives across the world has played a significant role in reducing famine deaths, and has taken the pressure off governments for ensuring the food security of their citizens. When food aid is not delivered in time to prevent a food crisis, the national government blames the international donors, and the donors blame the government. Ultimately, no one is held accountable. Political tensions between aid donors and national governments have also played a role – food aid has even been used as a political weapon. During the 1974 famine in Bangladesh, the United States withheld food aid because Bangladesh was trading with Cuba in violation of US trade sanctions. In 1984, the United States delayed sending food aid to Ethiopia in an attempt to undermine the Marxist Dergue regime, and in 2000 emergency food aid to the Somali region was delayed because the donors disapproved of Ethiopia's border war with Eritrea and feared that food would be diverted to feed Ethiopian soldiers. Failures of accountability by national governments and international donors have been identified as a fundamental cause of the persistence of famine in twenty-first-century Africa (Devereux, 2009).

War and 'complex political emergencies'

Conflict has become a feature of many African famines, and in the 1990s the phrase 'complex political emergencies' was coined to characterise these famines. But the relationship between war and famine is not new. Famine has been used as a weapon of war at least since the Middle Ages, when cities were besieged until their inhabitants either surrendered or starved. During World War II, parts of the Netherlands were blockaded by the German army, food stores were seized and all imports were prohibited into cities like The Hague and Leiden, where an estimated 10,000 people died. More recently, the town of Juba was subjected to similar starvation tactics during a civil war in Sudan.

Famine conditions can also be created as an unintended consequence of conflict, because of its devastating multiple impacts on food production, marketing systems and relief interventions. First, conflict disrupts agricultural production – farmers are displaced, conscripted, disabled or killed; food crops and granaries are destroyed; livestock are raided or slaughtered. Second, war undermines food marketing – trade routes are disrupted, markets are bombed, traders stop operating because they fear for their safety. Third, relief interventions are undermined by logistical constraints and security risks – governments often ban humanitarian agencies from operating in conflict zones, food convoys are attacked and aid agencies withdraw their staff. People in conflict areas become cut off from any source of food, and starvation follows (De Waal, 1997; Duffield, 2001).

The 2011 famine in Somalia is the most recent example of a 'complex emergency' famine, where food production was first disrupted by drought, then trade and aid responses were undermined by conflict and insecurity. The combination of a weak national state, weak international donors that withdrew under pressure from the militant group Al-Shabaab and United States counter-terrorism laws that effectively criminalised aid flows into southern Somalia, and the obstructive practices of Al-Shabaab, all contributed to a lethal failure of humanitarian response, despite excellent early warning information. Unless accountability for allowing this famine to occur is enforced, it seems likely that similar avoidable tragedies will occur in future, in Somalia or elsewhere.

Future famines

Modern famines are less widespread and less severe than historical famines: fewer countries are vulnerable, and fewer people die. On the other hand, recent famines have been exacerbated by a number of new factors – including flawed processes of economic liberalisation and political democratisation, rising prevalence of HIV/AIDS, and problematic relationships between national governments and international donors. Modern famines are also more complex, being caused by multiple failures – of the weather, 'coping strategies', markets, local politics, national governments and the international community. The Malawi famine of 2002 is typical of these 'new famines' (Devereux, 2007): though triggered by erratic rains, food subsidies and parastatal marketing outlets had been abolished under structural adjustment reforms, food prices spiralled as traders failed to respond, high HIV-prevalence had undermined community coping capacity, and donors intervened too late, trying to force the government to admit that it had corruptly sold the national strategic grain reserve.

All contemporary famines are fundamentally political. Decisions taken by governments, donors and humanitarian organisations either contribute to creating famine conditions, or are responsible for failures to prevent famine. It follows that political will to prevent famine is essential, at every level. The 'right to food' is enshrined in the Universal Declaration of Human Rights of 1948, but too little progress has been made in enforcing this right to date. Some writers have argued that famine should be criminalised as a crime against humanity in international law (Edkins, 2000). Until an 'anti-famine contract' is established at the global level, whether voluntarily or enforced under international law, future famines will continue to be tolerated, rather than eradicated.

References

Davis, M. (2001), *Late Victorian Holocausts: El Niño Famines and the Making of the Third World*, London: Verso.

Devereux, S. (2000), 'Famine in the Twentieth Century', *IDS Working Paper*, 105, Brighton: Institute of Development Studies.

Devereux, S. (ed.) (2007), *The New Famines: Why Famines Persist in an Era of Globalisation*, London: Routledge.

Devereux, S. (2009), 'Why does Famine Persist in Africa?', *Food Security*, 1(1): 25–35.

De Waal, A. (1997), *Famine Crimes: Politics and the Disaster Relief Industry in Africa*, Oxford: James Currey.

Drèze, J. and Sen, A. (1989), *Hunger and Public Action*, Oxford: Oxford University Press.

Duffield, M. (2001), *Global Governance and the New Wars*, London: Zed Books.

Dyson, T. and Ó Gráda, C. (eds) (2002), *Famine Demography: Perspectives from the Past and Present*, Oxford: Oxford University Press.

Edkins, J. (2000), *Whose Hunger? Concepts of Famine, Practices of Aid*, Minneapolis: University of Minnesota Press.

Malthus, T. (1798, 1976 edition), *An Essay on the Principle of Population*, New York: W. W. Norton.

Meillassoux, C. (1974), 'Development or Exploitation: Is the Sahel Famine Good Business?', *Review of African Political Economy*, 1: 27–33.

Ravallion, M. (1987), *Markets and Famines*, Oxford: Oxford University Press.

Sen, A. (1981), *Poverty and Famines: An Essay on Entitlement and Deprivation*, Oxford: Clarendon Press.

Von Braun, J., Teklu, T. and Webb, P. (1998), *Famine in Africa: Causes, Responses, and Prevention*, Baltimore: Johns Hopkins University Press.

Further reading

Devereux, S. (1993), *Theories of Famine*, Hemel Hempstead: Harvester Wheatsheaf. This book summarises the main theoretical explanations for famine causation, including climate shocks, Malthusianism, 'entitlement', market failure, government policy, and war.

Devereux, S. (ed.) (2007), *The New Famines: Why Famines Persist in an Era of Globalisation*, London: Routledge. This edited collection critically reviews recent developments in famine theory and includes studies of recent famines in Ethiopia, Madagascar, Malawi, Sudan, Iraq and North Korea.

De Waal, A. (1997), *Famine Crimes: Politics and the Disaster Relief Industry in Africa*, Oxford: James Currey. Alex de Waal takes an explicitly political approach to explaining contemporary African famines, blaming African governments and failures of the international community.

Sen, A. (1981), *Poverty and Famines: An Essay on Entitlement and Deprivation*, Oxford: Clarendon Press. The most influential book on famine since Malthus's *Essay on the Principle of Population*, in which Sen introduces his 'entitlement approach' to the analysis of poverty and famines.

4.5

Genetically modified crops and development

Matin Qaim

Introduction

A genetically modified (GM) crop is a plant into which one or several genes coding for desirable traits have been inserted through genetic engineering. The basic techniques of plant genetic engineering were developed in the early 1980s, with the first GM crops becoming commercially available in the mid-1990s. Since then, GM crop adoption has increased rapidly. In 2011, GM crops were already grown on more than 10 per cent of the global arable land. Half of this area is located in developing countries (Figure 4.5.1).

The crop traits targeted through genetic engineering are not completely different from those pursued by conventional breeding. However, since genetic engineering allows the direct transfer of genes across species boundaries, some traits that were previously difficult or impossible to breed, can now be developed more effectively. Examples include resistance to pests, tolerance to drought, or higher levels of micronutrients in food crops.

The potentials to contribute positively to development are manifold. Against the background of a dwindling natural resource base and climate change, productivity increases in agriculture are important to ensure sufficient availability of food and other raw materials for a growing population. Furthermore, new seed technologies can cause income growth in the small farm sector, on which many of the poor depend for their livelihoods. In spite of these

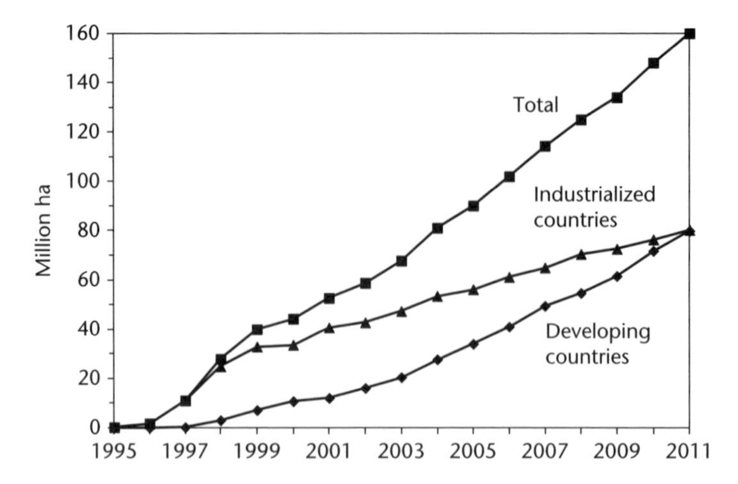

Figure 4.5.1 Global area grown with GM crops (1995–2011)

Source: James (2011, cover page)

potentials, GM crops face significant public opposition. Concerns are related to possible environmental risks, such as undesirable gene flow, which could potentially harm biodiversity, or negative impacts on non-target organisms. Some also have health worries in terms of toxic or allergenic effects of the genes and traits newly introduced to the crop plants. Finally, there are fears about adverse social implications. In particular, given the dominance of a few multinational companies, it is believed that GM crops may contribute to the exploitation of smallholder farmers.

This chapter summarizes the empirical evidence with a focus on socioeconomic effects in developing countries. In terms of environmental and health risks, comprehensive assessments suggest that the GM crops available so far are safe. Specific risks may occur for future GM crops. However, such risks do not seem to be connected to the technique of genetic engineering but would be present for any conventionally produced crops with the same heritable traits.

Impacts of commercialized GM crops

GM crops commercialized so far mostly involve herbicide tolerance or insect resistance traits.

Herbicide-tolerant crops

Herbicide-tolerant (HT) crops are tolerant to certain broad-spectrum herbicides, which are more effective, less toxic, and usually cheaper than selective herbicides. HT is so far mostly used in soybean, maize, cotton, and canola in North and South America. HT-adopting farmers benefit through lower herbicide expenditures. Total herbicide quantities applied were reduced in some situations, but not in others. In Argentina, herbicide quantities were even increased significantly. This is largely due to the fact that herbicide sprays were substituted for tillage. Reduced tillage has environmental advantages through lower soil erosion, lower fuel use, and lower greenhouse gas emissions. In terms of crop yield, there is no significant difference between HT and conventional crops in most cases.

HT technology reduces the cost of production, but the innovating companies charge a technology fee on seeds, which varies between crops and countries. Several studies for HT crops in North America showed that the net profit effects for farmers were small. In South American countries, the profit effects are larger, because – due to weaker intellectual property rights (IPRs) – the technology fee charged on seeds is lower. In Argentina, the mean profit gain through HT soybean adoption is around US$25 per hectare. The technology is so attractive for farmers that HT is now used on almost 100 per cent of the Argentine soybean area. In Brazil and other South American countries, adoption rates have also increased rapidly (Qaim 2009).

Most soybeans in North and South America are grown on large and fully mechanized farms. So far, HT crops have not been widely adopted in the small farm sector. Smallholders often weed manually, so that HT crops are less suitable, unless labor shortages justify conversion to chemical practices.

Insect-resistant crops

Insect-resistant GM crops involve genes from *Bacillus thuringiensis* (Bt) that make the plants resistant to certain pest species. The most widely used examples are Bt maize and Bt cotton, which are grown in a number of different countries. If insect pests are effectively controlled through chemical pesticides, the main effect of switching to Bt crops is a reduction in chemical pesticide use. However, there are also situations where insect pests are not effectively controlled by chemical means. In those situations, Bt can help reduce crop damage and thus increase effective yield. Table 4.5.1 shows that both pesticide-reducing and yield-increasing effects of Bt crops can be observed internationally. Lower chemical pesticide use is also associated with environmental and health advantages.

Profit effects of Bt technologies are also shown in Table 4.5.1. In spite of higher prices for seeds, Bt-adopting farmers benefit financially. On average, the benefits are higher for Bt cotton than for Bt maize, and they are also higher in developing than in developed countries. The latter is due to higher insect pest pressure in the tropics and weaker IPR protection (and thus lower seed prices) in developing countries.

Especially in China, India, and South Africa, Bt cotton is often grown by small-scale farmers with land areas of less than 5 ha. In South Africa, many smallholders grow Bt white maize as their staple food. Several studies show that Bt technology advantages for small-scale farmers are in a similar magnitude as for large-scale producers, in some cases even higher. In a study for Bt cotton in India, Subramanian and Qaim (2010) revealed sizeable employment-generating and poverty-reducing effects in the small farm sector.

One concern with Bt crops is that insect pest species may develop resistance or that non-Bt target insects (secondary pests) may gain in importance. If this were to happen, the benefits of Bt crops would decrease over time, which could create social problems. Recent research in India has shown that the benefits of Bt cotton have been sustainable until now (Krishna and Qaim 2012). While long-term effects are uncertain, these results mitigate the concern that GM crops would soon become ineffective in smallholder environments.

Potential impact of future GM crops

Improved agronomic traits

While Bt technology so far has mainly been used in maize and cotton, there are also other Bt crops, such as Bt rice or different Bt vegetables, that are likely to be commercialized soon,

Table 4.5.1 Average effects of Bt crops

Country	*Pesticide reduction (%)*	*Increase in effective yield (%)*	*Increase in profit (US$/ha)*
	Bt cotton		
Argentina	47	33	23
Australia	48	0	66
China	65	24	470
India	41	37	135
Mexico	77	9	295
South Africa	33	22	91
USA	36	10	58
	Bt maize		
Argentina	0	9	20
Philippines	5	34	53
South Africa	10	11	42
Spain	63	6	70
USA	8	5	12

Source: Qaim (2009, p. 672)

with similar projected impacts. Also for other pest-resistant GM traits that are being developed in different crops – such as fungal, virus, nematode, or bacterial resistance – pesticide-reducing and yield-increasing effects can be expected (Qaim 2009).

Impacts of GM crops with tolerance to abiotic stress will be situation-specific. A drought-tolerant GM variety, for instance, can lead to substantially higher yields under water stress, whereas the effect may be small when sufficient water is available. Especially in the semi-arid tropics, many small-scale farmers are operating under drought-prone conditions, so that the benefits of drought tolerance may be sizeable. Biotech researchers are also working on tolerance to heat, salinity, flood, or coldness. Climate change is associated with more frequent weather extremes, so that more tolerant crops can help reduce the risks of crop failures. Furthermore, research is underway to develop crops with higher nutrient (e.g. nitrogen, phosphorous) efficiency. Nutrient-efficient crops may reduce chemical fertilizer use and associated environmental externalities in intensive agriculture, while they may contribute to yield gains in regions where fertilizers are currently underused, as is the case in large parts of Africa. Some of these traits are genetically complex, so that commercialization may only be expected in the medium run.

Improved nutritional quality

Nutritionally enhanced GM crops with particular relevance for developing countries include food crops bred for higher contents of vitamins and minerals (so-called biofortification). Especially poorer population segments often suffer from micronutrient deficiencies with severe negative health consequences. A well-known example of a GM biofortified crop is Golden Rice, which contains high amounts of beta-carotene to control vitamin A deficiency. Golden Rice could soon become commercially available in some Asian countries. A recent study suggests that widespread consumption of Golden Rice could lessen vitamin A-related health problems by up to 60 per cent (Qaim 2009). Positive health effects can also be expected for biofortified crops that contain higher amounts of iron, zinc, folate, and other micronutrients.

Institutional and policy issues

Most GM crops available so far were commercialized by private companies. The evidence demonstrates that proprietary GM crops can have positive development effects. Nonetheless, the small farm sector will hardly be served comprehensively by private multinationals alone. One concern is related to patents and seed prices. Most of the existing GM crops are not patented in developing countries, but this may potentially change in the future. Strengthening IPRs in developing countries can have advantages and disadvantages. Especially in the least-developed countries, it could entail undesirable social consequences, because higher seed prices would reduce technology accessibility for smallholders.

Beyond seed prices, the dominance of multinationals also has implications for the type of GM crops that emerge. The private sector develops technologies primarily for big lucrative markets. While technically feasible, it is unlikely that multinationals will commercialize GM innovations for niche markets in the least-developed world where market failures are commonplace. Such research gaps will have to be addressed by the public sector if GM crop developments are not to bypass the poor.

But also when suitable GM crops are being developed and commercialized, benefits for poor farmers and consumers will not occur automatically. A conducive institutional environment is important to promote wide and equitable access. Well-functioning input and output markets will spur the process of innovation adoption. Unfortunately, in many poor countries such conditions first need to be established, so that the GM crop impacts observed so far in India, China, Argentina, and other more advanced developing countries cannot simply be extrapolated. Like any agricultural technology, GM crops are not a substitute but a complement to much needed institutional change in developing countries.

Conclusion

The evidence suggests that the risks of GM crops are often overrated in the public debate, while the benefits are underrated. Impact studies of commercialized GM crops show that there are sizeable economic and environmental benefits. Pest-resistant crops in particular can have positive social effects in the small farm sector and contribute to poverty reduction. Farmers in developing countries sometimes benefit more than farmers in developed countries. But income distribution effects also depend on the wider institutional setting, including farmers' access to suitable seeds, credit, information, and other input and output markets. More public support will be needed to realize the benefits for the poor on a larger scale.

GM technologies in the research pipeline include crops that are tolerant to abiotic stresses and crops that contain higher amounts of micronutrients. The benefits of such future applications could be much bigger than the ones already observed. Against the background of a dwindling natural resource base and growing demand for agricultural products, GM crops could contribute significantly to food security and sustainable development.

However, GM crops are no panacea. Technical solutions cannot cure problems of lacking infrastructure, weak institutions, low education, or bad governance. Nor should GM technologies be considered as isolated solutions to increase agricultural productivity. GM techniques are a powerful tool in crop breeding. Yet, sustainable productivity growth in agriculture requires a smart combination of many different things, including breeding, improved agronomy, post-harvest handling, and various other approaches.

References

Krishna, V. V., M. Qaim (2012). Bt Cotton and Sustainability of Pesticide Reductions in India. *Agricultural Systems*, 107, 1, 47–55.
Qaim, M. (2009). The Economics of Genetically Modified Crops. *Annual Review of Resource Economics*, 1, 665–693.
Subramanian, A., M. Qaim (2010). The Impact of Bt Cotton on Poor Households in Rural India. *Journal of Development Studies*, 46, 2, 295–311.

Further reading

Carter, C., G. Moschini, I. Sheldon, (eds) (2011). *Genetically Modified Food and Global Welfare*. Emerald Publishing, Bingley, UK.
James, C. (2011). Global Status of Commercialized Biotech/GM Crops: 2011. *ISAAA Briefs* 43, International Service for the Acquisition of Agri-biotech Applications, Ithaca, NY.
Qaim, M., S. Kouser (2013). Genetically Modified Crops and Food Security. *PLOS ONE*, 8, 6, e64879.

4.6

Rural cooperatives

A new millennium?

Deborah R. Sick, Baburao S. Baviskar and Donald W. Attwood

The cooperative movement originated in nineteenth-century Europe. From the early twentieth century, cooperatives have been promoted for small producers in the developing world. In theory, economies of scale should enable organized small producers to compete with larger ones, while democratic governing structures based on 'one member, one vote' (regardless of a member's assets) should result in equitable decisions and benefits. Cooperatives have also been promoted for the non-economic benefits they can provide, such as social services and a sense of empowerment for small producers.

These characteristics fitted notions of participatory development which arose in the 1960s and 1970s. Developing nations, aid agencies, NGOs, and churches promoted cooperatives with high hopes as an alternative to private corporations or state enterprises. However, the institutional designs of cooperatives vary widely, as do the political agendas and rationales for their creation and their impact on the poor.

This chapter examines the role of cooperatives in rural development. Starting with examples from India, we discuss 'formal' cooperatives (established under official regulatory frameworks), 'informal' cooperation (customary methods of pooling labour, savings, etc.), and the importance of cooperatives in the 'new social economy'. What factors cause these varied organizations to succeed or fail, and what benefits accrue to the poor if they succeed?

Notable cooperatives by region

India

Cooperatives for rural credit and marketing were started under British administration in the early 1900s. After independence in 1947, cooperatives of many kinds were promoted for rural development. However, most proved ineffective. Few became economically viable, many became moribund, and few operated under democratic control by their members. Most have been managed by government officials.

Success in sugar and milk processing

There have been some genuine, member-controlled cooperatives. Notable examples include sugar factories in Maharashtra and dairies in Gujarat. India is the world's second-largest sugar producer; cooperatives in Maharashtra state make about 30 per cent of the total. Compared with private factories, these cooperatives became more efficient, enabling them to pay higher prices for sugarcane – a vital benefit for their members, most of whom are small farmers. These cooperatives were quite successful from the 1950s to the 1980s. Many expanded and diversified into ancillary enterprises. They built schools, colleges, clinics, and hospitals. Their success resulted from management by members and elected leaders, assisted by hired managers and technicians. But recently, many sugar cooperatives have declined.

As Maharashtra led in sugar cooperatives, Gujarat led in dairies. The renowned Kheda District Co-operative Milk Producers' Union had an annual turnover of more than Rs3 billion (roughly US$100 million). The Gujarat Co-operative Milk Marketing Federation has a distribution network all over India, with an annual turnover of more than Rs150 billion (Candler and Kumar, 1998). The main benefit for members, as dairy farmers, is access to distant urban markets. Because milk is perishable, such access is impossible without industrial processing and transport, as provided by the cooperatives.

Such cooperatives serve members with economic assets. The very poor benefit indirectly when production and the demand for labour increase. Women in India seldom own land, but there are grassroots organizations specifically for women. One of the first and most innovative is SEWA, the Self-Employed Women's Association, founded as a trade union for women working in the informal sector. Among its many activities, SEWA established a cooperative bank in 1974.

Africa

African cooperatives resemble India's in several respects. They were first established by colonial rulers and managed by bureaucrats. After independence, some regimes drew on socialist models to promote development, expecting to harness customary patterns of informal cooperation to build modern enterprises. The much-publicized *ujamaa* programme in Tanzania provided a classic example. National leaders thought collectivization would promote equality and productivity, yet the programme failed due to faulty assumptions about rural society coupled with top-down management by state officials (Hyden, 1988). Absence of established democratic traditions made it hard for member-controlled cooperatives to emerge in Africa, yet cooperatives have played a role in rural development. For example, farmer cooperatives have been key to the Ethiopian government's recent efforts to increase agricultural production and are 'a preferred mechanism through which many donor and non-governmental organizations are organizing their rural development and poverty reduction interventions' (Bernard and Spielman, 2009: 61). The results have been mixed. As elsewhere, the very poorest tend to be excluded, though they may benefit indirectly (2009: 67).

Latin America

In Latin America during the 1960s and 1970s, cooperatives were widely promoted by the United States as a means to raise rural incomes and thus counter the influence of communism. As in Africa, many regimes were hostile to organizations not under state control, yet member-controlled cooperatives exist. In Costa Rica, the state supports cooperatives (e.g. via tax concessions) without managing them directly. Under member control, many have failed due to poor management, but others have done well. To better compete in global markets, one coffee-processing cooperative decided to improve its product by more careful harvesting; it also used the internet to bypass middlemen and market directly abroad (Sick, 2008a). These strategies brought higher prices for small coffee farmers. To reduce reliance on a single volatile commodity, this cooperative has also diversified: it owns a sugar factory, supermarkets, agricultural supply stores, and gas stations. While these activities have helped it weather economic downturns and become the region's largest employer (Sick, 2008b), many members complain that the cooperative is now run more like a corporation, with little care for the problems of the small farmer.

Informal cooperation, credit, and savings

Many people use small, 'informal' institutions to pursue common interests. Almost everywhere, rural producers cooperate by pooling or exchanging labour. Informal savings groups are found in villages and cities around the world. They pool savings collected in small increments from their members, each member gaining access to the pool by rotation – as with some women's groups in Kenya (Thomas-Slayter and Rocheleau, 1995).

Formal credit cooperatives offer bigger loans but often have problems. India has poured vast sums into state-run credit 'co-ops', whose assets never consist of members' savings. Loans go mostly to male landowners, and most are never repaid. Thus, state-sponsored 'cooperative' credit is neither self-supporting nor directly beneficial to the very poor, including women.

NGOs sometimes try to fill the gap between informal savings groups and dysfunctional, state-run 'cooperatives'. In 1976, the Grameen Bank was established in Bangladesh. The bank makes small loans to poor women organized in local groups. As with informal savings groups, social discipline ensures repayment. Similar micro-credit programmes have been set up in many parts of the world. In India's Andhra Pradesh state, an NGO launched Women's Thrift Cooperatives (WTCs) in 1990. WTCs raise funds solely through small, regular contributions from their members, who earn interest on savings at 1 per cent per month. (For loans they pay 2 per cent.) A village WTC may consist of 200 to 500 women, divided into groups of 10 to 50. In less than a decade, over 33,000 women formed 101 WTCs, and their combined savings totalled Rs26 million, with no external grants or loans. The NGO provided training, but the WTCs soon become self-sufficient and self-managing. About half the members and leaders came from landless households (Biswas and Mahajan, 1997).

With its focus on disciplined repayment and targeting of poor people, especially women, micro-credit appeals to NGOs and donor agencies. Yet, the impacts have been disputed (Kabeer, 2001). Some research finds that micro-credit strengthens women's positions in their households and communities (Hashemi *et al.*, 1996; Holvoet, 2005). But a study in Bangladesh found that men often end up controlling loans that women bring into their households (Goetz and Sen Gupta, 1996), while a study in Cameroon found that using established networks for group formation can exacerbate inequalities among rural women (Mayoux, 2001). Overall, micro-credit is favoured for rural development, while the potential of 'micro-thrift', which is far less dependent on external donors, seems to be overlooked. Micro-thrift resembles more the age-old patterns of informal savings and group discipline.

Cooperatives and the new social economy

The crisis of the welfare state, plus growing discontent over neoliberal development policies, has led to a renewed interest in 'third sector' organizations as a means for socially responsible development. In 2001, UN Secretary General Kofi Annan called on governments to support autonomous cooperatives to help achieve millennium development goals. The enthusiasm for cooperatives 'to pursue both economic viability and social responsibility' has grown such that the United Nations declared 2012 'The International Year of Cooperatives' (UN, 2012).

This renewed interest in cooperatives for development takes various forms. Many developing nations, such as South Africa, have enacted policies to encourage the formation of cooperatives. Cooperatives have also been touted as a means of achieving 'fair globalization' when partnered with socially responsible corporations (Burke, 2010). Whether these new efforts will result in effective member-controlled organizations remains to be seen.

In today's increasingly competitive global economy, a new generation of cooperatives are emerging, focused on value-added processing and niche marketing (Harris *et al.*, 1996). Many producer organizations and alternative trade organizations are building on this trend to help small producers survive in competitive and volatile commodity markets. For example, the Fair Trade movement targets socially and environmentally conscious consumers who will pay more for 'fairly traded' products. To promote equity and empowerment, Fair Trade (FT) organizations deal only with cooperatives or other producer associations.

As in the 1960s, this revival of faith in cooperatives as economically sustainable and inherently democratic raises questions regarding how well cooperatives are meeting the needs of small producers. In a comparative study of FT and conventional coffee cooperatives in Nicaragua, Bacon (2010) found that not only did members of FT cooperatives fare better economically than their non-FT counterparts, they also had a greater sense of empowerment. Some argue that while financial benefits appear significant in the short run, empowerment and capacity building via fair trade will prove more important in the long run (Raynolds *et al.*, 2004: 1119).

Yet, while some cooperatives find that Fair Trade provides access to new markets and higher prices, limited demand means that most can sell only a small portion of their coffee through FT channels. In Costa Rica, long-established coffee cooperatives were slow to form Fair Trade partnerships, though since Fair Trade certification requirements have been streamlined, many are now doing so. However, farmers in FT-certified cooperatives often sell their better quality beans outside the cooperatives when world prices rise (Sick, 2008; Rice, 2001). Other problems include the high cost of FT certification (Bacon *et al.*, 2008), burdensome regulations (Moberg, 2005), heavy debt burdens incurred by Fair

Trade cooperatives (Utting-Chamorro, 2005), and issues of gender inequality (Bacon, 2010; Lyon, 2008).

Explaining success and failure

So many types of cooperatives have been attempted that it should be possible to explain patterns of success and failure. There are two broad approaches. *Institutional design analysis* looks inside organizations for factors promoting or impeding efficiency (Shah, 1996). For example, cooperative sugar factories in Maharashtra pay no dividends. Farmer-shareholders benefit from supplying cane to the factory. The directors (themselves cane farmers) try to maximize efficiency so that the factory can offer high cane prices and thus ensure a steady supply of cane from the members. (Reliable cane supply is vital to efficiency.)

Political economy analysis looks beyond internal design to external context. 'Cooperative' sugar factories in northern India superficially resemble those in Maharashtra yet are highly inefficient. Their main problems stem from management by state officials, who have no real stake in raising efficiency. When efficiency declines, so do cane prices; farmers look elsewhere to market their crop, and efficiency declines even more. Different policy approaches to cooperatives in these two regions make the difference.

However, shifts in party politics at the national and state levels have recently encouraged massive corruption in Maharashtra. In 2005, of 190 sugar cooperatives in the state, only 70 turned a profit; 77 had closed down, and others were in dubious financial condition (Vasantdada Sugar Institute, 2005). National policy changes now allow new private enterprises into this sector. This might have led to healthy competition, but corrupt leaders have been selling financially weakened cooperatives at a fraction of their asset value. Buyers and sellers collude to make profits, while shareholders suffer massive losses.

In general, cooperatives perform better with cautious government support; direct control or unlimited subsidization is lethal. Likewise, cooperatives that are grassroots-initiated tend to have greater success than those imposed by governments or created at the behest of NGOs. Nevertheless, member-initiated and member-controlled cooperatives are not without their problems. In terms of equity, people without land or capital – notably women – often are not eligible to join or are discouraged from participating in decisions. In terms of financial viability, lack of experience and management skills has in many cases led to bankruptcy. Finally, there is the question of leadership accountability. To cope with competition in global markets, some cooperatives are becoming more corporate in nature; decision-making can become less democratic (Corby, 2010: 31), and some members then feel their interests are overlooked.

If cooperatives are to play a renewed role in rural development, their long history of successes and failures suggests that we approach them with a critical eye, paying attention to past lessons as well as new ways forward.

References

Bacon, C. (2010) 'A spot of coffee in crisis: Nicaraguan smallholder cooperatives, fair trade networks, and gendered empowerment'. *Latin American Perspectives* 171(37): 50–71.

Bacon, C. M., Méndez, V. E., Flores Gómez, M. E., Stuart, D. and Diaz Flores, S. R. (2008) 'Are sustainable coffee certifications enough to secure farmer livelihoods? The millennium development goals and Nicaragua's Fair Trade cooperatives'. *Globalizations* 5(2): 259–274.

Bernard, T. and Spielman, D. J. (2009) 'Reaching the rural poor through rural producer organizations? A study of agricultural marketing cooperatives in Ethiopia'. *Food Policy* 34(1): 60–69.

Biswas, A. and Mahajan, V. (1997) 'Sustainable banking with the poor: A case study on women's thrift co-operative system in Warangal and Karimnagar Districts of Andhra Pradesh' (project report). Hyderabad: Co-operative Development Foundation.
Burke, B. (2010) 'Cooperatives for "fair globalization"? Indigenous people, cooperatives, and corporate social responsibility in the Brazilian Amazon'. *Latin American Perspectives* 37(6): 30–52.
Candler, W. and Kumar, N. (1998) *India: The Dairy Revolution. The Impact of Dairy Development in India and the World Bank's Contribution.* Washington DC: World Bank.
Corby, J. (2010) 'For members and markets: Neoliberalism and cooperativism in Mendoza's wine industry'. *Journal of Latin American Geography* 9(2): 27–47.
Goetz, A. and Sen Gupta, R. (1996) 'Who takes the credit? Gender, power, and control over loan use in rural credit programmes in Bangladesh'. *World Development* 24(1): 45–63.
Harris, A. B., Stefanson, B. and Fulton, M. (1996) 'New generation cooperatives and cooperative theory'. *Journal of Cooperatives* 11: 15–28.
Hashemi, S. M., Schuler, S. R. and Riley, A. P. (1996) 'Rural credit programs and women's empowerment in Bangladesh'. *World Development* 24(4): 635–653.
Holvoet, N. (2005). 'The impact of microfinance on decision-making agency: Evidence from South India'. *Development and Change* 36(1): 75–102.
Hyden, G. (1988) 'Approaches to co-operative development: Blueprint versus greenhouse', in D. W. Attwood and B. S. Baviskar (eds) *Who Shares? Cooperatives and Rural Development.* Delhi: Oxford University Press.
Kabeer, N. (2001) 'Conflicts over credit: Re-evaluating the empowerment potential of loans to women in rural Bangladesh'. *World Development* 29(1): 63–84.
Lyon, S. (2008) 'We want to be equal to them: Fair Trade coffee certification and gender equity within organizations'. *Human Organization* 67(3): 258–268.
Mayoux, L. (2001) 'Tackling the down-side: Social capital, women's empowerment, and microfinance in Cameroon'. *Development and Change* 32: 435–464.
Moberg, M. (2005) 'Fair Trade and eastern Caribbean banana farmers: Rhetoric and reality in the anti-globalization movement'. *Human Organization* 64(1): 4–15.
Raynolds, L. T., Murray, D. and Taylor, P. L. (2004) 'Fair Trade coffee: Building producer capacity via global networks'. *Journal of International Development* 16(8): 1109–1121.
Rice, R. R. (2001) 'Noble goals and challenging terrain: Organic and fair trade coffee movements in the global marketplace'. *Journal of Agricultural and Environmental Ethics* 14: 39–66.
Shah, T. (1996) *Catalysing Co-operation: Design of Self-Governing Organizations.* New Delhi: Sage.
Sick, D. (2008a) *Farmers of the Golden Bean: Costa Rican Households, Global Coffee, and Fair Trade.* DeKalb: Northern Illinois University Press.
Sick, D. (2008b) 'Coffee, farming families, and fair trade in Costa Rica: New markets, same old problems?' *Latin American Research Review* 43(3): 193–208.
Thomas-Slayter, B. and Rocheleau, D. (1995) *Gender, Environment, and Development in Kenya: A Grassroots Perspective.* London: Lynne Rienner.
United Nations (2012) Statement by Secretary General Ban-Ki Moon. Available online at http://social.un.org/coopsyear/
Utting-Chamorro, K. (2005) 'Does Fair Trade make a difference? The case of small coffee producers in Nicaragua'. *Development in Practice* 15(3–4): 584–599.
Vasantdada Sugar Institute. (2005) *Financial Performance of Co-operative Sugar Factories in Maharashtra, Financial Year 2004–2005.* Manjari (BK), Dist. Pune: Vasantdada Sugar Institute.

Further reading

Attwood, D. W. (1992) *Raising Cane: The Political Economy of Sugar in Western India.* London: Westview Press. (Social history of sugar cooperatives in Maharashtra.)
Baviskar, B. S. and Attwood, D. W. (1995) *Finding the Middle Path: The Political Economy of Co-operation in Rural India.* London: Westview Press. (Case studies, analysis of patterns of success and failure.)
Sick, D. R. (2008) *Farmers of the Golden Bean: Costa Rican Households, Global Coffee, and Fair Trade.* DeKalb: Northern Illinois University Press. (Case study of a coffee cooperative competing with private firms.)

4.7
Land reform

Ruth Hall, Saturnino M. Borras Jr and Ben White

Introduction

Land policies and particularly land reform have gained, lost and regained prominence in development strategies and debates since the Second World War. Land is an important topic in development theories and debates, and a major arena of rural social and political movements and agrarian conflict. In the early history of development thought it was widely recognized that, because of the predominance of agricultural production and employment in developing economies, access to land is crucial to poor people's capacity to construct viable livelihoods and overcome rural poverty. It was also widely understood that existing land tenure systems are often obstacles to improving farm productivity, and that landholding and power are intimately related in poor agrarian economies (Warriner 1969; Lipton 2010). Contemporary debates centre on redistributive versus 'market-led' agrarian reforms, gender and generational issues in land policy, the contemporary corporate 'land rush', and new ideas of food and land sovereignty driving today's agrarian movements.

Figure 4.7.1 Painting by Boy Domingues (Philippines) 'The new enclosures' (2012)
Copyright: Saturnino M. Borras Jr

Land reforms and agrarian reforms

The generally accepted definition of 'land reform' is the redistribution of property or rights in land. Normally definitions also include some notion of the intended beneficiaries (small peasant farmers, tenants and sharecroppers, landless farm workers, labour tenants, the 'rural poor'). Land reforms are efforts to correct historical distortions in the allocation of land ownership and use rights resulting from colonial land grabbing and dispossessions, enclosures, landlordism, or – as with post-socialist de-collectivization – previous reforms themselves. These reforms have often aimed to establish or consolidate a class of market-oriented and surplus-producing 'middle peasants' (Ghose 1983). They also have a broader macroeconomic objective of enhancing farm productivity; reformed land tenure structures are usually expected to promote agrarian transition (to capitalist, modernized smallholder, or collective systems).

It is common to distinguish 'land reforms' from more comprehensive 'agrarian reforms', which also aim to promote landholders' access to the inputs (knowledge, credit, markets) needed to increase productivity and enhance sustainable livelihoods. However, any successful 'land reform' is necessarily accompanied by such supporting measures, rendering the distinction redundant. The distinction is also untranslatable in many world languages – 'land reform' in French is *réforme agraire*, and in Spanish *reforma agraria*. Yet the conception of 'agrarian reform' may serve the purpose of emphasizing the inadequacy of redistributing land and securing tenure rights alone in bringing about lasting, structural change in rural economy and society.

The emergence, decline and resurgence of land reform policies in development agendas

Land reforms and the cold war, 1950s–1980s

From the 1940s to the 1960s, various models of land and agrarian reforms coexisted, reflecting different models of 'development' from purely capitalist to purely socialist, plus a variety in between. Land reform became 'an instrument of the cold war, used by Russia and China to promote communism and by America to prevent it' (Warriner 1969: 37). The decolonization era of the 1950s and 1960s saw extensive reforms in sub-Saharan Africa and Asia. The United States backed redistributive 'land-to-the-tiller' models in Japan, South Korea and Taiwan, involving confiscation of excess holdings and their redistribution to smallholders, and revisions in tenancy laws enabling tenants and sharecroppers to become owners of their plots. From the late 1960s onwards, international support shifted from land reforms to 'green revolution' small-farm modernization through the use of new high-yielding varieties of maize and rice and purchased inputs. Yet some US-backed land-to-the-tiller reforms aimed to forestall rural communist revolutions in the Philippines, South Vietnam and El Salvador, while China, North Vietnam and Cuba embarked on collectivization. By the 1980s, land reform had virtually disappeared on the agenda of international development and financial institutions and donors.

Market-based land reforms, 1990s–2010s

The end of the cold war led to decollectivization and a new wave of privatizing land reforms in many post-socialist countries (Spoor 1997). Since 2000, many new land projects of the World Bank and USAID have focused on 'land administration', focusing on improving the legal, technical and institutional framework for land ownership to enhance 'tenure security' and the more efficient working of land markets. The prime advocate of 'market-assisted land reform' has been the World Bank, whose approach has four key elements:

- Promotion of owner-operated family farms on both efficiency and equity grounds.
- Secure property rights to land so as to enhance owners' effort and investment and facilitate land transactions.
- A policy and regulatory environment that promotes transfers to more efficient land uses.
- A role for egalitarian asset distribution and redistributive land reform where the agrarian structure is highly dualistic, characterized by very large and very small holdings (Deininger and Binswanger 1999: 2).

To title or not to title?

The market-based approach promotes privatization through registration or titling of land held in communal tenure or under customary law. Hernando de Soto's influential book *The Mystery of Capital* advocates land titling to bring poor rural people with 'informal' tenure into the formal system, so they can acquire working capital through bank loans (De Soto 2000). These notions have influenced land titling projects in various Latin American, African and Asian countries. While proponents argue that customary tenure systems 'evolve' towards formalization, evidence from Asia, Africa and Latin America shows that, despite growing population pressure and competition for resources, property systems remain plagued by legal uncertainty and damaging conflicts over resources. Land titling programmes have often favoured elites, the older generation, and men, and may ultimately create not more but less security of tenure (Platteau 1996). Titling is also unnecessary according to Nobel laureate Elinor Ostrom, who showed that common pool resources (grazing lands, forests, water, rivers and marine resources) are central to the livelihoods of most of the world's rural population, and the customary property regimes that govern them can be durable and equitable. Instead of privatizing these resources through titling, collective self-governance by users themselves can and should be supported (Ostrom 1990).

Small or large farms?

Underlying arguments for and against land reforms is the enduring debate about the respective merits of large- and small-scale agriculture, reflecting opposing standpoints on the viability of small farms and their future place in globalized food (and fuel, feed and fibre) regimes. Rural social movements like La Via Campesina, the global alliance of peasant movements, claim that 'small-scale farmers can feed the world', and do this in environmentally friendly ways, while generating more employment than large-scale capital- and energy-intensive farming (La Via Campesina 2010). Authoritative scientific support comes from the International Assessment of Agricultural Science and Technology for Development report, *Agriculture at a Crossroads*, which concludes that industrial, large-scale monoculture farming is unsustainable and must be reconsidered in favour of small-scale mixed crop farming agro-ecosystems that conserve water, preserve biodiversity and improve the livelihoods of the poor (IAASTD 2009). Nevertheless, many national governments and international development institutions continue to promote large-scale agriculture.

Gender, generation and land reforms

The UN CEDAW Convention firmly established women's right to 'equal treatment in land and agrarian reform as well as in land resettlement schemes' (CEDAW 1979, Article 14(g)), but gender discrimination persists in land policy and practice, and land reforms have themselves often affected women detrimentally. State-codified, individual and 'household'-based forms of land allocation, resettlement and contract-farming schemes often annihilate women's customary rights to land (Moser 1993; Jacobs 2010). In many agrarian societies parents and/or community elders retain control of land as long as possible. Access to land is narrowing for young people, through land concentration, land sales and allocations by older generations to outsiders, including commercial investors, rather than to the next generation. It is not surprising if today's young rural men and women are reluctant to wait until they are 40 or 50 years old to be farmers, and decide to move to the city, a trend which now extends to all social classes and (in most countries) both genders. The issue of intergenerational transfer of land rights, or 'intergenerational dispossession', deserves more attention, especially if small-farm based alternatives to industrial capitalist agriculture are to be promoted (White 2012).

New enclosures? The debate on corporate land deals

There is currently much debate about corporate 'land grabs'. Large-scale, government-supported corporate acquisition of contested lands and common lands, and the accompanying dispossession of local farmers, pastoralists and forest users is occurring on an unprecedented scale in Africa, Asia, Latin America, and the former Soviet Union. Large land deals are typically shrouded in secrecy, and 'guesstimates' of the total area of large land deals worldwide in the period 2009 to 2011 range widely, from less than 50 million to more than 220 million hectares; there is reasonable consistency across all reports that at least 60 percent of total acquisitions are in sub-Saharan Africa.

Large-scale land deals are often premised on the vision of a 'dualistic' agrarian economy, with large-scale farms engaged in capitalist production mainly for export, while smallholder farms gradually disappear or are incorporated as part of contract-farming arrangements, with the former peasantry proletarianized, providing low paid labour to the new estates and plantations. This development path is supported by the World Bank's report on *Agriculture for Development* (World Bank 2007), arguing that only large farms or smallholder out-growers hooked into large agribusiness can compete in globally integrated value chains. These arguments have been taken on by national governments, investors and donor agencies alike. Many countries' land and agricultural policies are now ambivalent on issues of corporate land grabbing, on the one hand promoting large private sector-driven investments while at the same time employing a discourse of support for smallholder farming.

All major development and donor agencies recognize that the recent spate of land deals has involved major abuses, but there is disagreement on how to address them. The World Bank proposes corporate codes of conduct through its seven 'Principles for Responsible Agro-Investment': respecting land and resource rights; ensuring food security; ensuring transparency, good governance, and a proper enabling environment; consultation and participation; responsible agro-enterprise investing; social sustainability; and environmental sustainability (World Bank 2010: x, 68–91). Others are sceptical about the efficacy of codes of conduct (Borras and Franco 2010). The UN Special Rapporteur on the Right to Food argues for a broader vision: not regulating land grabbing as if this were inevitable, but developing alternative programmes for agricultural investment that do not involve corporate land acquisition and help to realize the human right to adequate food. Land investments that imply changes to local people's land rights 'should represent the last and least desirable option, acceptable only if no other investment model can achieve a similar contribution to local development' (De Schutter 2010: 20). In between these two positions are the FAO Voluntary Guidelines for the Responsible Governance of Tenure of Land, Fisheries and Forests in the Context of National Food Security. Negotiations for these voluntary guidelines have been politically complex and, without complementary action from outside, implementation by states and investors alike is likely to be much weaker than hoped for by civil society groups.

Concluding reflections

At least 1.5 billion people today have some farmland as a result of land reform, and are less poor, or not poor, as a result. But huge and inefficient land inequalities remain in many countries (Lipton 2010: 8). Despite decades of land and agrarian reforms, political debate and contestation over land appears to be on the rise, and old models and policies are proving inadequate in the face of the current trend of corporate 'land grabbing' across much of the developing world. An emerging, wider concept of 'land sovereignty' is defined as the right of working peoples to have effective access to, use of, and control over land and the benefits of its use and occupation (Borras and Franco 2012: 1). If embraced as the focus of struggles by social movements, indigenous people and rural labourers, this concept may become the centrepiece of a call to action against corporate attempts to enclose the commons, and in favour of redistributive reforms and the land rights of the poor.

References

Borras, Saturnino and Jennifer Franco (2010) 'From threat to opportunity? Problems with the idea of a "Code of Conduct" for land-grabbing'. *Yale Human Rights and Development Law Journal* 13: 508–523.

Borras, Saturnino and Jennifer Franco (2012). 'A "land sovereignty" alternative? Towards a people's counter-enclosure campaign'. Agrarian Justice Program Discussion Paper July 2012. Amsterdam: Transnational Institute (TNI).

CEDAW (1979) *Convention on Elimination of All Forms of Discriminaton Against Women*. Available online at www.cedaw2010.org/index.php/about-cedaw/cedaw-at-a-glance (accessed 9 July 2012).

Deininger, Klaus and Hans Binswanger (1999) 'The evolution of the World Bank's land policy: Principles, experiences and future challenges'. *World Bank Research Observer* 4(2): 247–276.

De Schutter, O. (2010) *Report of the Special Rapporteur on the Right to Food*. New York: UN General Assembly.

De Soto, Hernando (2000) *The Mystery of Capital*. New York: Basic Books.

Ghose, A. K. (ed.) (1983) *Agrarian Reform in Contemporary Developing Countries*. Geneva: International Labour Organization.

IAASTD (2009) *Synthesis Report: Agriculture at a Crossroads: International Assessment of Agricultural Science and Technology for Development*. Washington, DC: Island Press.

Jacobs, Susie (2010) *Gender and Agrarian Reforms*. London: Routledge.

La Via Campesina (2010) 'Sustainable peasant and family farm agriculture can feed the world'. Jakarta: Via Campesina Views. Available online at http://viacampesina.org/downloads/pdf/en/paper6-EN-FINAL.pdf (accessed 12 July 2012).

Lipton, Michael (2010) *Land Reform in Developing Countries: Property Rights and Property Wrongs*. London: Routledge.

Moser, C. (1993) *Gender Planning and Development: Theory, Practice and Training*. London: Routledge.

Ostrom, Elinor (1990) *Governing the Commons: The Evolution of Institutions for Collective Action*. Cambridge: Cambridge University Press.

Platteau, Jean-Philippe (1996) 'The evolutionary theory of land rights as applied to sub-Saharan Africa: A critical assessment'. *Development and Change* 27(1): 29–86.

Spoor, Max (ed.) (1997) *The 'Market Panacea': Agrarian Transformations in Developing Countries and Former Socialist Economies*. London: Intermediate Technology Publications.

Warriner, Doreen (1969) *Land Reform in Principle and Practice*. Oxford: Clarendon Press.
White, Ben (2012) 'Agriculture and the generation problem: Rural youth, employment and the future of farming'. *IDS Bulletin* 43(6): 9–19.
World Bank (2007) *Agriculture for Development: World Development Report 2008*. Washington, DC: The World Bank.
World Bank (2010) *Rising Global Interest in Farmland: Can it Yield Sustainable and Equitable Benefits?* Washington, DC: The World Bank.

Further reading

Deininger, Klaus (2003) *Land Policies for Growth and Poverty Reduction*. Washington, DC: The World Bank and Oxford: Oxford University Press. The World Bank's influential land policy report, setting out the arguments for 'market-led agrarian reform'.
De Schutter, O. (2011) 'How not to think of land-grabbing: Three critiques of large-scale investments in farmland'. *The Journal of Peasant Studies* 38(2): 249–279.
Lipton, Michael (2010) *Land Reform in Developing Countries: Property Rights and Property Wrongs*. London and New York: Routledge. A comprehensive overview of historical and contemporary land reform experiences.
White, Ben, Ruth Hall and Wendy Wolford (eds) (2012) *The New Enclosures: Critical Perspectives on Corporate Land Deals*. London: Routledge. Also published as special issue, *Journal of Peasant Studies* 39(3–4).
World Bank: The World Bank's influential land policy report, setting out the arguments for 'market-led agrarian reform'.

Websites

Land Deal Politics Initiative: www.iss.nl/ldpi – LDPI supports engaged research and debate on the global 'land grab'.
Land Resources Action Network: www.lran.org – a network of researchers and social movements committed to the promotion and advancement of the fundamental rights of individuals and communities to land, and to equitable access to the resources necessary for life with human dignity.
Land Rights in Africa: www.mokoro.co.uk/other-resources/land-rights-in-africa – online research and information on land rights, especially women's land rights, in Africa.

4.8

Gender, agriculture and land rights

Susie Jacobs

Introduction

Farmers are commonly assumed to be male in most parts of the world. This image, however, does not correspond with agrarian realities. Within smallholder households, women usually play crucial roles in food processing as well as in domestic labour and caring work. Women work to raise crops for consumption and for sale, and in associated small livestock husbandry. The global trend is for women to take increased responsibility within agricultural production (FAO, 2005).

The refusal to acknowledge women's contribution is part of a common pattern of disempowerment. Women often lack land rights as well as rights to other productive resources, and have little decision-making power over agricultural production, despite their active roles. They may be disadvantaged through lack of legal entitlements – or more often, through customary, or everyday practices such as (in south Asia), taboos that stop women handling ploughs.

This chapter discusses two policy directions concerning land rights; the first is the case of agrarian reforms, or redistribution of land through state-backed programmes; and the second concerns the contemporary trend for land titling and related debates about gender rights. The next section briefly reviews what other types of support might be needed to make farming successful for women smallholders. Last, the paper discusses some ways forward, including movements to gain land rights for women.

Land and agrarian reforms

Agrarian reforms, here defined as widespread *redistribution* of land as part of state programmes to the landless or land-hungry, aim to raise household incomes, improve food security and reduce rural poverty. They also aim to give more democratic rights to the rural poor. Realisation of these objectives, however, remains highly gendered.

Comparative studies on women, gender relations and agrarian reform across 19 countries in Asia, Africa, Eastern Europe and Latin America (Jacobs, 2010) show decidedly mixed results. A positive step is that many land reform programmes have allowed widows or divorcees with dependents to hold land or land permits. Many redistributionist land reforms raise incomes within households and increase food security (El-Ghonemy, 1990) and this often benefits wives. This was the case, for instance, in Ethiopia and in Andhra Pradesh, India, where women underlined the import of greater food security with land redistribution (Jacobs, 2010).

Less beneficial outcomes are, unfortunately, more numerous. Many stem from the near-universal award of land titles or permits to the 'head of household'. Negative outcomes include loss of women's existing land rights. Another common effect is the diminution of their *own* incomes and income sources. This takes place for a number of reasons including loss of trading niches where the family moves or is resettled. Or, because the land is perceived to be the husbands', wives must spend more time working on the land, but without controlling the proceeds.

Agrarian reforms usually affect family relations. Husbands often gain more control and power, both because they hold land rights and because they are more constantly present to monitor their wives. In the Chilean agrarian reforms enacted under Frei and then under Salvador Allende, wives' dependence increased as they worked less outside the home. Men's authority in the home, previously denied them as subjects of the landlord, increased (Tinsman, 2002). Men as husbands often gain power and control at the wife's expense – despite the democratic intent of land reforms.

Redistributionist agrarian reform programmes need to 'mainstream' gender so that policies will not be to marginalise women, as has happened so often. Equitable agrarian reforms should continue to be a priority for sustainable agriculture. In practice, however, neoliberal economic policies have favoured the retreat of state programmes and have encouraged privatisation or land 'titling' and more recently, land grabs. (Indeed, many now understand 'land reform' to mean parcellisation and titling rather than redistribution.)

Land titling and debates about gender and communal tenure

Land titling has often been seen as a way to 'secure' rights for smallholders. The World Bank, for instance, has advocated moves to title or to privatise land and women's land rights have sometimes been linked discursively to titling initiatives.

Here the issue of differing property regimes is significant. In much of sub-Saharan Africa land is held communally although is worked individually, and women are the main cultivators. In other parts of the world, most land is privately owned. Within both private and communal systems, women are marginalised in terms of landholding and control over land for a variety of reasons. In patrilineal systems traditionally women do not inherit land as this passes through the male line. Even where no patrilineages exist (e.g. Latin America) women may still be marginalised from landholding. In other systems, sharia law gives daughters the right to inherit property, but not in equal portion with sons. Social norms indicating that it is 'shameful' or a sign of low status for women to be seen doing agricultural labour operate in a number of societies; or women's rightful place may be seen to be in the home rather than in fields. For instance, in China women were traditionally seen as 'of the inside' [the home] and men, 'of the outside'. Where land is privatised, campaigns for women's landholding seek to equalise women's position by placing title deeds (or else land permits) in women's names or else in joint names of spouses. Without legal reforms such as inheritance rights, listing of women's names on legal documents and legal safeguards preventing sale of land without the wife's (and children's) consent, it is argued that women will remain subordinate and unable to use and control land effectively.

Bina Agarwal (1994) has summarised a strong case for the potential for legal land rights to transform women's lives, deriving initially from her work in India (see also Deere and León, 2001, on Latin America). The first argument is in terms of justice: women as the main, or important, 'tillers' should receive and be able to control land and proceeds from agriculture. A second is in terms of efficiency. It is now often acknowledged that rural women's lack of decision-making power often lowers agricultural output (FAO, 2011). Women's land rights would, third, improve family welfare: a number of studies indicate that women's enterprise and incomes are often more explicitly oriented to food security than are men's. Fourth, landholding might permit women to take advantage of economic opportunities. In some cases women have been able to seize opportunities either as entrepreneurs individually, or in other cases, collectively. Fifth, access to land is likely to increase women's social and economic status within households and communities and to enhance their decision-making powers (Wanyeki, 2003: 2).

Thus, strong arguments can be made for women's land rights. Must these, however, be accompanied by, or part of, initiatives for titling, demarcation and privatisation of land? This question is particularly significant in the sub-Saharan African context. Some authors have argued that customary authorities attend to the claims of the poorest, including women and that African customary law is flexible and open to complex, overlapping tenure claims. Others feel that customary tenure in patrilineal systems so marginalises women that tenure must be individualised in order to obtain equity (Wanyeki, 2003). Many lawyers and women's groups feel that women's land rights necessarily entail individual tenure. For instance, research with key informants in South Africa by the author indicated that most NGO workers and 'grassroots' activists interviewed dismissed the possibility that even joint title with husbands could benefit poor women. It was argued that without sole title wives would always be subject to male household power.

On the other hand, a number of authors (e.g. Meinzin-Dick and Mwangi, 2009) have warned against linking women's rights too firmly to privatisation and titling. The poorest usually lose out in marketised systems and this is especially so for women. It is quite possible that women who gain rights through titling may lose their land within the globalised marketplace.

The question of what type(s) of tenure are 'best' for women's rights is complex and likely to be context-specific. However, this should not obscure the need for gender-equitable land rights.

Other factors

There exists some evidence that women are gaining effective control over as well as access to land, despite many impediments (Budlender and Alma, 2011). This raises the question of what other supports women would need to be successful farmers. Landholding alone, even where land is available, is not usually enough to farm successfully. Smallholders require access to, for instance, capital for inputs such as fertilisers or seeds; to buy or rent ploughs and other implements; and command over labour – often a difficulty for women within communities. Women are, additionally, often marginalised from community groups that make decisions. Small farmers need the backup of agricultural extension services and training, but these are often not available for women, or they may not be easily able to take advice from male extension workers. Thus, there exists much need for training of women extension advisors. Most of the factors listed above depend on effective state provision or other forms of backup, for example, through NGOs. Access to credit is of particular importance. As noted, this is needed in a variety of contexts including survival of smallholdings or for expansion of enterprises. However, if land is used as collateral – as in many titling programmes – then it can be lost. Credit can be double-edged, placing farmers in situations of unmanageable debt, a situation particularly likely to affect poorer women. Last, provision of infrastructure helps to combat male bias, as shortfall in provision of schools, clinics and roads affects women disproportionately.

What is to be done?

Some issues can be addressed through legal or political change and within agrarian reform and titling programmes themselves. Where women face legal discrimination, the law should be amended to give them land rights equal with men's. Where land is held communally it may be that mechanisms other than privatisation of land can be found: there are some signs that communal tenure can be reformed in a 'gender-friendly' manner (Budlender and Alma, 2011) – but there is also much evidence of resistance. It is crucial that women be able to claim the same kinds of property rights as men in their own groupings (i.e. communities, kin groups, households). In contemporary situations in which many corporate and state 'land grabs' are taking place, it is easy to forget that women as well as men smallholders suffer from loss of lands and livelihoods.

Historically, women have made most gains within agricultural programmes where there exists state backing for gender-equitable measures – for example, rights for female-headed households, practical advice, and mechanisms for adjudication of disputes between husband

and wife. Reforms to family law connected with property are also important: one of the most contentious areas concerns wives' rights to remain on land in case of divorce. It is common for wives to be ejected from land and to lose their livelihoods along with their marriages. This points to the key factor that changes in legislation are insufficient in themselves. To be meaningful, these must be enforced (literally) 'on the ground' – through, for instance, educational campaigns to alter beliefs about women and landholding; through equitable legislation and through actual enforcement of laws. Women also need legal recourse when they are subjected to violence, which is frequently precipitated as a backlash to land claims. Most such social changes can be made only through political organisation.

Some campaigns for women's land rights take the form of direct action. For instance, the World March of Women participated in a world forum on food sovereignty in 2007; marches for women's land rights have taken place in India and Brazil and smaller actions have taken place in South Africa. However, many initiatives are on a small scale: for example, discussions with community elders; formation of women's cooperatives and many initiatives to list wives' names on permits or title deeds. The importance of women's land rights is beginning to be recognised globally, not least by United Nations institutions, state aid agencies, many NGOs as well as by local movements. Nonetheless, the struggle for gender-equitable land rights remains at an early stage: perhaps the disquiet arising from women's land struggles indicates their great importance.

References

Agarwal, Bina (1994) *A Field of One's Own: Gender and Land Rights in South Asia*, Cambridge: Cambridge University Press.

Budlender, Debbie and Eileen Alma (2011) *Women and Land*, Ottawa: IDRC.

Deere, Carmen Diana and Magdalena León (2001) *Empowering Women: Land and Property Rights in Latin America*, Pittsburgh: Pittsburgh University Press.

El-Ghonemy, M. Riad (1990) *The Political Economy of Rural Poverty: The Case for Land Reform*, London: Macmillan.

FAO (2005) *Gender and Land Compendium of Country Studies*, Geneva: Food and Agriculture Organisation of the UN. Available online at ftp.fao.org/docrep/fao/008/a0297e/a0297e00.pdf

FAO (2011) *The State of Food and Agriculture 2010–2011: Women in Agriculture, Closing the Gender Gap for Development*, Rome: Food and Agriculture Organisation. Available online at www.fao.org/docrep/013/i2050e/i2050e00.htm

Jacobs, Susie (2010) *Gender and Agrarian Reforms*, New York and London: Routledge.

Meinzin-Dick, Ruth and Esther Mwangi (2009) 'Cutting the web of interests: Pitfalls of formalizing property rights', *Land Use Policy* 26(1): 36–43.

Tinsman, Heidi (2002) *Partners in Conflict: The Politics of Gender, Sexuality and Labor in the Chilean Agrarian Reform, 1950–73*, Durham, NC: Duke University Press.

Wanyeki, L. Muthoni (2003) 'Introduction', in L. M. Wanyeki (ed) *Women and Land in Africa*, London: Zed.

4.9

The sustainable intensification of agriculture

Jules Pretty

All commentators now agree that food production worldwide will have to increase substantially in the coming years and decades (World Bank, 2008; IAASTD, 2009; Royal Society, 2009; Godfray *et al.*, 2010). But there remain very different views about how this should best be achieved. Some still say agriculture will have to expand into new lands, but the competition for land from other human activities makes this an increasingly unlikely and costly solution, particularly if protecting biodiversity and the public goods provided by natural ecosystems are given higher priority (MEA, 2005). Others say food production growth must come through redoubled efforts to repeat the approaches of the Green Revolution; or that agricultural systems should embrace only biotechnology or become solely organic. What is clear is that more will need to be made of existing agricultural land. Agriculture will, in short, have to be intensified. Traditionally agricultural intensification has been defined in three different ways: increasing yields per hectare, increasing cropping intensity (i.e. two or more crops) per unit of land or other inputs (water), and changing land use from low-value crops or commodities to those that receive higher market prices.

It is now understood that agriculture can negatively affect the environment through overuse of natural resources as inputs or through their use as a sink for waste and pollution (Dobbs and Pretty, 2004). What has also become clear is that the apparent success of some modern agricultural systems has masked significant negative externalities, with environmental and health problems documented and recently costed for some countries (Norse *et al.*, 2001; Tegtmeier and Duffy, 2004; Pretty *et al.*, 2005; Sherwood *et al.*, 2005). These environmental costs shift conclusions about which agricultural systems are the most efficient, and suggest that alternative practices and systems which reduce negative externalities should be sought.

Sustainable agricultural intensification is defined as producing more output from the same area of land while reducing the negative environmental impacts and at the same time increasing contributions to natural capital and the flow of environmental services (Pretty, 2008; Royal Society, 2009; Godfray *et al.*, 2010; Conway and Waage, 2010).

A sustainable production system would thus exhibit most or all of the following attributes:

1. Utilizing crop varieties and livestock breeds with a high ratio of productivity to use of externally and internally derived inputs;
2. Avoiding the unnecessary use of external inputs;
3. Harnessing agroecological processes such as nutrient cycling, biological nitrogen fixation, allelopathy, predation, and parasitism;
4. Minimizing use of technologies or practices that have adverse impacts on the environment and human health;

5 Making productive use of human capital in the form of knowledge and capacity to adapt and innovate and social capital to resolve common landscape-scale problems;
6 Quantifying and minimizing the impacts of system management on externalities such as greenhouse gas emissions, clean water, carbon sequestration, biodiversity, and dispersal of pests, pathogens, and weeds.

As both agricultural and environmental outcomes are pre-eminent under sustainable intensification, such sustainable agricultural systems cannot be defined by the acceptability of any particular technologies or practices (there are no blueprints). If a technology assists in efficient conversion of solar energy without adverse ecological consequences, then it is likely to contribute to the system's sustainability. Sustainable agricultural systems also contribute to the delivery and maintenance of a range of valued public goods, such as clean water, carbon sequestration, flood protection, groundwater recharge, and landscape amenity value. By definition, sustainable agricultural systems are less vulnerable to shocks and stresses. In terms of technologies, therefore, productive and sustainable agricultural systems make the best of *both* crop varieties and livestock breeds *and* their agroecological and agronomic management.

This suggests that sustainable intensification will very often involve more complex mixes of domesticated plant and animal species and associated management techniques, requiring greater skills and knowledge by farmers. To increase production efficiently and sustainably, farmers need to understand under what conditions agricultural inputs (seeds, fertilizers, and pesticides) can either complement or contradict biological processes and ecosystem services that inherently support agriculture (Royal Society, 2009; Settle and Hama Garba, 2011). In all cases farmers need to see for themselves that added complexity and increased efforts can result in substantial net benefits to productivity, but they need also to be assured that increasing production actually leads to increases in income. Too many successful efforts in raising production yields have ended in failure when farmers were unable to market the increased outputs. Understanding how to access rural credit, and especially how to sell any increased output, becomes as important as learning how to maximize input efficiencies or build fertile soils. Equally, the creation of a social infrastructure of relations of trust, connections, and norms is critical to spread innovation.

Food output and environmental improvements through sustainable intensification in Africa

Farmers have been able to increase food outputs by sustainable intensification in two ways. The first is *multiplicative* – by which yields per hectare have increased by combining use of new and improved varieties with changes to agronomic-agroecological management. The UK government's Foresight programme commissioned reviews and analyses from 40 projects from 20 countries of Africa where sustainable intensification has been developed or practiced in the 2000s (Pretty *et al.*, 2011). The cases comprised crop improvements, agroforestry and soil conservation, conservation agriculture, integrated pest management, horticulture, livestock and fodder crops, aquaculture, and novel policies and partnerships. By early 2010, these projects had documented benefits for 10.4 million farmers and their families and improvements on approximately 12.75 million hectares.

Across the projects, yields of crops rose on average by a factor of 2.13 (i.e. slightly more than doubled). The timescale for these improvements varied from 3 to 10 years. It was estimated that this resulted in an increase in aggregate food production of 5.79 million tons per year, equivalent to 557 kg per farming household (in all the projects). This does not include the additive benefits to yield production.

Many projects also improved food outputs by *additive* means – by which diversification of farms resulted in the emergence of a range of new crops, livestock, or fish that added to the existing staples or vegetables already being cultivated. These system enterprises included:

1 aquaculture for fish raising;
2 small patches of land used for raised beds and vegetable cultivation;
3 rehabilitation of formerly degraded land;
4 fodder grasses and shrubs that provide food for livestock (and increase milk productivity);
5 raising of chickens, and zero-grazed sheep and goats;
6 new crops or trees brought into rotations with staple (e.g. maize, sorghum) yields not affected, such as pigeonpea, soyabean, indigenous trees;
7 adoption of short-maturing varieties (e.g sweet potato, cassava) that permit the cultivation of two crops per year instead of one.

The environmental side effects or externalities have been shown to be highly positive in a number of cases. Carbon content of soils is improved where legumes and shrubs are used, and where conservation agriculture increases the return of organic residues to the soil. Legumes also help fix nitrogen in soils, thereby reducing the need for inorganic fertilizer on subsequent crops. In IPM-based projects, most have seen reductions in synthetic pesticide use (e.g. in cotton and vegetables in Mali, pesticide use fell to an average of 0.25 l/ha from 4.5 l/ha; Settle and Hama Garba, 2011). In some cases, biological control agents have been introduced where pesticides were not being used at all. The greater diversity of trees, crops (e.g. beans, fodder shrubs, grasses), and non-cropped habitats has generally helped to reduce run-off and soil erosion, and thus increased groundwater reserves.

A key constraint across Africa is nutrient supply. Many African soils are nutrient-poor, and fertilizer use is low across the continent compared with other regions. The average use of mineral fertilizers in SSA does not surpass a very low 6–7 kg of NPK ha^{-1}, against a middle and low income country average of nearly 100 kg/ha^{-1}, on land of generally low and declining inherent fertility (Reij and Smaling, 2008). As yields increase, so the net export of nutrients also increases (unless nutrient cycles are closed). Thus, farms in many contexts will need to import or fix nutrients. Many approaches have been used in these projects, including inorganic fertilizers, organics, composts, legumes, and fertilizer trees and shrubs. The Malawi fertilizer subsidy programme is a rare example of a national policy that has led to substantial changes in farm use of fertilizers and the rapid shift of the country from food deficit to food exporter (Dorward and Chirwa, 2011).

A common objection made about many agronomic-agroecological approaches is their perceived need for increased labour (Tripp, 2005). However, this is highly site-specific. In some contexts, labour is highly limiting, especially where HIV-AIDS has removed a large proportion of the active population; in other contexts, there is plentiful labour available as there are few other employment opportunities in the economy. Successful projects of sustainable intensification by definition fit solutions to local needs and contexts, and so thus take account of labour availability. In Kenya, for example, female owners of raised beds for vegetable production employ local people to work on the vegetable cultivation and marketing.

Labour for crop and livestock management is thus not necessarily a constraint on new technologies. In Burkina Faso, work groups of young men have emerged for soil conservation. Tassas and zai planting pits are best suited to landholdings where family labour is available, or where farm hands can be hired. The technique has spawned a network of young day labourers who have mastered this technique. Owing to the success of land rehabilitation, farmers are increasingly buying degraded land for improvement, and paying these labourers to dig the zai pits and construct the rock walls and half moon structures, which can transform yields. This is one of the reasons why more than three million hectares of land are now rehabilitated and productive.

Conclusions

These projects of sustainable intensification show that where there is a political and economic domestic recognition that 'agriculture matters', then food outputs can be increased not only without harm to the environment but also in many cases to increase the flow of beneficial environmental services. Such improvements then contribute to national domestic food budgets, foster new social infrastructure and cultural relations, help the emergence of new businesses and so drive local economic growth, and ultimately improve well-being of both rural and urban populations.

These projects contained many different technologies and practices, yet had similar approaches to working with farmers, involving agricultural research, building social infrastructure, working in novel partnerships, and developing new private sectors options. Only in some of the cases were national policies directly influential. These projects indicated that there were seven key requirements for such scaling up of sustainable intensification to larger numbers of farmers:

1. Scientific and farmer input into technologies and practices that combine crops and animals with appropriate agroecological and agronomic management.
2. Creation of novel social infrastructure that results in both flows of information and builds trust amongst individuals and agencies.
3. Improvement of farmer knowledge and capacity through the use of farmer field schools, videos, and modern ICTs.
4. Engagement with the private sector to supply goods and services (e.g. veterinary services, manufacturers of implements, seed multipliers, milk and tea collectors) and development of farmers' capacity to add value through their own business development.
5. A focus particularly on women's educational, microfinance and agricultural technology needs, and building of their unique forms of social capital.
6. Ensuring that microfinance and rural banking is available to farmers' groups.
7. Ensure public sector support to lever up the necessary public goods in the form of innovative and capable research systems, dense social infrastructure, appropriate economic incentives (subsidies, price signals), legal status for land ownership, and improved access to markets through transport infrastructure.

References

Conway, G. R. and Waage, J. 2010. *Science and Innovation for Development*. UKCDS, London.

Dobbs, T. and Pretty, J. N. 2004. Agri-environmental stewardship schemes and 'multifunctionality'. *Review of Agricultural Economics* 26(2), 220–237.

Dorward, A. and Chirwa, E. 2011. The Malawi agricultural input subsidy programme: 2005–6 to 2008–9. *Int. Journal of Agric Sust* 9(1), 232–247.

Godfray, C., Beddington, J. R., Crute, I. R., Haddad, L., Lawrence, D., Muir, J. F., Pretty, J., Robinson, S., Thomas, S. M., and Toulmin, C. 2010. Food security: The challenge of feeding 9 billion people. *Science* 327, 812–818.

IAASTD. 2009. *Agriculture at a Crossroads*. Island Press, Washington, DC.
Millennium Ecosystem Assessment (MEA). 2005. *Ecosystems and Well-being*. Island Press, Washington, DC.
Norse, D., Li Ji, Jin Leshan and Zhang Zheng. 2001. *Environmental Costs of Rice Production in China*. Aileen Press, Bethesda.
Pretty, J. 2008. Agricultural sustainability: Concepts, principles and evidence. *Phil Trans Royal Society of London* B 363(1491), 447–466.
Pretty, J., Lang, T., Ball, A., and Morison, J. 2005. Farm costs and food miles. *Food Policy* 30(1), 1–20.
Pretty, J., Toulmin, C., and Williams, S. 2011. Sustainable intensification in African agriculture. *Internat Journ Agric Sust* 9(1), 5–24.
Reij, C. P. and Smaling, E. M. A. 2008. Analyzing successes in agriculture and land management in sub-Saharan Africa. *Land Use Policy* 25, 410–420.
Royal Society. 2009. *Reaping the Benefits: Science and the Sustainable Intensification of Global Agriculture*. Royal Society, London.
Settle, W. and Hama Garba, M. 2011. The FAO integrated production and pest management programme in the Senegal and Niger River basins of francophone West Africa. *Int. Journal of Agric Sust* 9(1), 171–185.
Sherwood, S., Cole, D., Crissman, C., and Paredes, M. 2005. Transforming potato systems in the Andes. In J. Pretty (ed.) *The Pesticide Detox*. Earthscan, London.
Tegtmeier, E. M. and Duffy, M. D. 2004. External costs of agricultural production in the US. *Int. J. Agric. Sust.* 2(1), 1–20.
Tripp, R. 2005. *Self-Sufficient Agriculture: Labour and Knowledge in Small-scale Farming*. Earthscan, London.
World Bank. 2008. Agriculture for development: World Development Report 2008. Washington, DC.

Part 5
Urbanization and development

Editorial introduction

Urban places in poor- and middle-income nations currently house close to three-quarters of the world's total urban population. Although we frequently associate this with the growth of very large cities – those with a million or more inhabitants – and indeed large urban agglomerations have shown great growth; in fact, in much of the world, the majority of urban residents are living in cities considerably smaller than one million. It is in these areas of a range of sizes that jobs will be needed, homes built, schools and social services provided and environments suitably maintained in a context of good overall governance. Stated in these terms the challenges faced are revealed as considerable.

To put this overview into historical context, it was the Second United Nations Conference on Human Settlements, customarily referred to as 'Habitat II', held in Istanbul, Turkey, in 1996 that attested to the continuing importance of the urbanization process in developing societies. In the period since 1950, rapid urbanization has become one of the principal hallmarks of developing nations. It is the magnitude of the changes that are occurring which underscores the salience of such urban processes, for it is generally accepted that, on average, the conditions to be found in the rural areas of developing countries are much poorer than those that are to be encountered in the towns and cities.

It is now well established that in the contemporary world, for every urban dweller living in the affluent developed world, two exist in the poorer cities and towns of the developing world. In fact, by the end of the first quarter of the twenty-first century, this ratio is set to have risen to three to one in favour of urban residents in the developing world. This rapid rise in both *urbanization* (the proportion of the population living in urban places), and *urban growth* (the physical expansion of cities on the ground), is exemplified in a number of different ways.

Globally, these include increases in the number of large cities, as well as increases in the size of the largest cities themselves. It is also associated with the ever-increasing number of cities that have reached the million population level. By 1990, the average population of the world's 100 largest cities was in excess of 5 million. In 1800, the equivalent statistic had stood at fewer than 200,000 inhabitants. Further, by 1990, there were 12 'mega cities' with over

10 million inhabitants; and, most notably, seven were to be found in Asia, three in Latin America and two in the United States of America. But as we have noted, globally the growth of smaller urban places is just as important.

In the past, many analyses have treated the urban and the rural, physically, socially and economically, as being essentially separate, although in reality they are, of course, functionally closely interrelated. For example, food is required for growing city populations, and is often produced at low procurement prices, giving rise to one aspect of what is frequently described as urban bias in patterns and processes of national and international development. The implications of such issues in respect of policy prescriptions for future urban change are closely debated.

Over the last twenty years, the concept of the 'world' or 'global city' has come to prominence. This approach stresses that cities have to be seen as key points in the articulation of the global economic system, and are dominated by transnational corporations (TNCs) and flows of transnational capital. Overall, the world city paradigm stresses the economic importance of very large cities, and it shows their mounting salience in the global South (for example, São Paolo, Mexico City, Johannesburg, Dubai, Mumbai, Jakarta, etc.). World cities have through time been intimately associated with the generation of uneven development and global inequalities, as shown by classical dependency theory.

However, as noted above, recent trends point to the fact that in many regions it is medium-sized and small cities that are now showing the fastest overall rates of growth. In short, over the developing world as a whole, big and small cities are growing and exhibiting great dynamism and are a focus for future pure and policy-related research.

These observations stress the need to look at cities and processes of urban development from a strongly comparative methodological perspective. Generalizations based on long-standing notions of urban development based on the experiences of West European and North American cities in the past are likely to prove myopic. We need to factor in both diversity and commonality among cities. Think here of climate change where cities in poor nations have generally generated few of the pollutants so far, but all too often many face the environmental consequences in terms of rising sea level and the like.

Despite offering higher incomes on average, urban areas customarily show pockets of great disadvantage, poverty and inequality. Within fast-growing urban settlements, poor housing is perhaps the most conspicuous manifestation of generalized poverty. Possibly the second most overt sign of the stresses and strains of rapid urban change are witnessed in the urgent need for sound and effective environmental management. While more urban residents die due to preventable diseases than as the result of disasters, it is earthquakes, storms and floods that all too often receive far greater media coverage on a day-to-day basis.

Housing and environmental conditions are major areas calling for good governance in low-income and medium-income cities, and perhaps more than anything else, the political will to improve matters on a broad social and economic front is pressingly required. This is a vital area when one considers that the very existence of urban areas is predicated on economies of scale, which should therefore allow environmental and health hazards to be tackled more effectively than in rural areas. One great environmental issue is how to provide for more equitable transport conditions under rapid urban growth and social flux. What in particular is the desirable balance between private motorized transport and public transport? How can motorized transport be squared with the more friendly environmental policies needed given climate change?

Finally, crime in urban areas is often seen as a key dimension of development and is strongly associated with urban inequalities and the associated injustices noted previously. But we should avoid thinking in simple Western terms in this connection too. All too often, the urban poor can be criminalized through the everyday practices they are forced to be involved in to survive. For example, poor urban residents living in informal housing may be seen as criminal merely because their homes do not meet required planning legislation, they may not have ownership or rights to their plot, and they may access services such as electricity illegally. In these cases, the failure of cities to provide adequate housing for the growing urban poor can all too easily enforce so-called 'criminal' practices.

5.1

Urbanization in low- and middle-income nations in Africa, Asia and Latin America

David Satterthwaite

Urban trends

Urban centres in low- and middle-income nations now have close to two-fifths of the world's total population and close to three-quarters of its urban population. They contain most of the economic activities in these nations and most of the new jobs created over the last few decades. They are also likely to house most of the world's growth in population in the next few decades (United Nations 2012). Thus, how they are governed and what provisions are made to serve their expanding populations has very large implications for economic and social development – and also for environmental quality and whether dangerous climate change can be avoided (see Chapter 5.7).

Table 5.1.1 shows the scale of the growth in the urban population in what the UN terms 'less developed regions' from 304 million to over 2.6 billion between 1950 and 2010. This is almost the same as the population in low- and middle-income nations in Africa, Asia and Latin America. In 2011, all independent nations in Latin America and in Africa except for Equatorial Guinea were within the low- and middle-income categories. So too were the nations with most of the population of the Caribbean and Asia – the main exceptions being some oil exporters in the Middle East, Japan, Republic of Korea, Israel, Singapore and Hong Kong (China).

There has also been a shift from most of the world's urban population being in high-income nations to being in low- and middle-income nations. Although Asia and Africa still have more rural than urban dwellers, they both have very large and rapidly growing urban populations. Asia alone has more than half the world's urban population and more than half of this is within just two countries, China and India. Africa now has a larger urban population than northern America; so too does Latin America and the Caribbean which also has more than three-quarters of its population living in urban centres. In 1950, the countries now classified as within Europe had 37.6 per cent of the world's urban population; in 2010, they had 15.1 per cent.

Growth of large cities

Two aspects of this rapid growth in urban population have been the increase in the number of large cities and the historically unprecedented size of the largest cities (see Table 5.1.2).

Table 5.1.1 The distribution of the world's urban population by region, 1950–2010 with projections to 2030 and 2050

Urban population (millions of inhabitants) *Major area, region, country or area*	*1950*	*1970*	*1990*	*2010*	*Projected for 2030*	*Projected for 2050*
World	745	1,352	2,281	3,559	4,984	6,252
More developed regions	442	671	827	957	1,064	1,127
Less developed regions	304	682	1,454	2,601	3,920	5,125
Least developed countries	15	41	107	234	477	860
Sub-Saharan Africa	20	56	139	298	596	1,069
Northern Africa	13	31	64	102	149	196
Asia	245	506	1,032	1,848	2,703	3,310
China	65	142	303	660	958	1,002
India	63	109	223	379	606	875
Europe	281	412	503	537	573	591
Latin America and the Caribbean	69	163	312	465	585	650
Northern America	110	171	212	282	344	396
Oceania	8	14	19	26	34	40
Per cent of the population in urban areas						
World	29.4	36.6	43.0	51.6	59.9	67.2
More developed regions	54.5	66.6	72.3	77.5	82.1	85.9
Less developed regions	17.6	25.3	34.9	46.0	55.8	64.1
Least developed countries	7.4	13.0	21.0	28.1	38.0	49.8
Sub-Saharan Africa	11.2	19.5	28.2	36.3	45.7	56.5
Northern Africa	25.8	37.2	45.6	51.2	57.5	65.3
Asia	17.5	23.7	32.3	44.4	55.5	64.4
China	11.8	17.4	26.4	49.2	68.7	77.3
India	17.0	19.8	25.5	30.9	39.8	51.7
Europe	51.3	62.8	69.8	72.7	77.4	82.2
Latin America and the Caribbean	41.4	57.1	70.3	78.8	83.4	86.6
Northern America	63.9	73.8	75.4	82.0	85.8	88.6
Oceania	62.4	71.2	70.7	70.7	71.4	73.0
Per cent of the world's urban population						
World	100.0	100.0	100.0	100.0	100.0	100.0
More developed regions	59.3	49.6	36.3	26.9	21.4	18.0
Less developed regions	40.7	50.4	63.7	73.1	78.6	82.0
Least developed countries	2.0	3.0	4.7	6.6	9.6	13.8
Sub-Saharan Africa	2.7	4.1	6.1	8.4	11.9	17.1
Northern Africa	1.7	2.3	2.8	2.9	3.0	3.1
Asia	32.9	37.4	45.2	51.9	54.2	52.9
China	8.7	10.5	13.3	18.6	19.2	16.0
India	8.5	8.1	9.8	10.6	12.2	14.0
Europe	37.6	30.5	22.0	15.1	11.5	9.5
Latin America and the Caribbean	9.3	12.1	13.7	13.1	11.7	10.4
Northern America	14.7	12.6	9.3	7.9	6.9	6.3
Oceania	1.1	1.0	0.8	0.7	0.7	0.6

Source: Derived from statistics in United Nations 2012. There is a group of countries (mostly in sub-Saharan Africa) for which there are only 1–3 censuses from 1950 to the present, including some with no recent census so figures for their urban (and rural) populations are based on estimates and projections.

Just two centuries ago, there were only two 'million-cities' worldwide (i.e. cities with one million or more inhabitants): London and Beijing (then called Peking). By 2010, there were 449, three-quarters of which were in low- and middle-income nations. Seventy had populations that had grown more than tenfold from 1960 to 2010 – and many had populations that had grown more than twentyfold in these five decades including many cities in China and Abidjan, Abuja, Dar es Salaam, Dhaka, Karaj, Luanda, Ouagadougou, Sana'a and Yaoundé.

Table 5.1.2 also highlights how the size of the world's largest cities has grown dramatically. In 1800, the average size of the world's 100 largest cities was less than 200,000 inhabitants, but by 2010 it was 7.8 million.

Key characteristics of urban change

These statistics may give the impression of very rapid urbanization focused on large cities. But some care is needed when making generalizations, because there is such diversity in the scale and nature of urban change between nations and also, within each nation, over time.

Table 5.1.2 The distribution of the world's largest cities by region over time

Region	*1800*	*1900*	*1950*	*2010*
Number of 'million cities'				
World	2	16	75	449
Africa	0	0	2	50
Asia	1	3	26	226
China	1	1	9	95
India	–	1	5	43
Europe	1	9	23	54
Latin America and the Caribbean	0	0*	8	63
Northern America	0	4	14	50
USA	–	4	12	44
Oceania	0	0	2	6
Regional distribution of the world's largest 100 cities				
World	100	100	100	100
Africa	5	2	3	10
Asia	63	22	34	50
China	23	13	10	21
India	18	4	6	9
Europe	28	54	33	7
Latin America and the Caribbean	3	5	8	16
Northern America	0	15	20	15
USA	0	15	18	13
Oceania	0	2	2	2
Average size of the world's 100 largest cities	187,520	728,270	2.0 m	7.8 m

* Some estimates suggest that Rio de Janeiro had reached one million inhabitants by 1900 while other sources suggest it has just under one million.

Notes: Cities that have changed their country-classifications and nations that have changed regions are considered to be in the country or region that they are currently in for this whole period. For instance, Hong Kong is counted as being in China for all the above years while the Russian Federation is considered part of Europe.

Source: Satterthwaite 2007 but revised with data from United Nations 2012; for 1800 and 1900, the statistics are derived primarily from city populations in Chandler 1987.

Also, for the larger and more populous nations, there is such diversity in urban trends between different regions. However, certain general points can be highlighted.

The relative importance and spatial distribution of urban centres is in effect a map of where private enterprises have chosen to concentrate or to expand. Cities concentrate producers and consumers, buyers and sellers, firms and workers (World Bank 2008). An increasingly large and complex network of cities spanning high-, middle- and some low-income nations are the backbone of the globalized economy (Sassen 2006), and it is mostly cities that have greatly expanded their role in this that have grown most rapidly.

There is an economic logic underpinning urbanization and the location of large cities

All the world's wealthiest nations are predominantly urban and the nations with the largest increase in their economies over the last few decades are generally those that have urbanized most. There is also an economic logic to the location of most large cities. In 2010, the world's five largest economies (United States of America, China, Japan, India and Germany) had 11 of the world's 23 mega-cities (cities with 10 million plus inhabitants) and more than two-fifths of its 'million-cities'. This concentration of mega-cities is unsurprising since these can only develop in countries with large non-agricultural economies and large national populations; most nations have too small a population and too weak an urban-based economy to support a mega-city.

Within low- and middle-income countries, the largest cities were concentrated in the largest economies (which also tend to be among the most populous countries): Brazil, Mexico and Argentina in Latin America; China, India, Indonesia and the Republic of Korea in Asia. Although much of the rapid growth in the major cities in low- and middle-income nations in the 1950s and 1960s was associated with political change, such as the achievement of political independence and the expansion in government, economic change is now a more powerful and important factor in where major cities grow (or decline) in most nations and this is unlikely to change.

Less urbanized populations, smaller cities

In all regions of the world, most of the urban population live in urban areas with less than one million inhabitants. By 2010, mega-cities had 5.1 per cent of the world's population and most had had slow population growth rates during the 1990s and the first decade of the twenty-first century while several had more people moving out than in for these two decades – for instance, Kolkata, Buenos Aires, São Paulo, Rio de Janeiro and Mexico City. One reason for this is that most mega-cities are being challenged by a new generation of smaller cities that compete with them for new investment – for instance, in Mexico, Brazil, China and India; this may also reflect some of the dis-economies of inadequately managed large agglomerations. In addition, in various nations that have had effective decentralization, urban authorities in smaller cities have more resources and capacity to compete for new investment. There are interesting parallels to this in the USA as new large cities in the south came to draw new investment away from the long-established large cities in the northeast. In Brazil, cities such as Curitiba and Porto Alegre have attracted new investments away from the mega-cities of São Paulo and Rio de Janeiro. In India, cities such as Hyderabad, Bangalore, Surat and Pune have attracted new investment away from long-established large cities such as Kolkata, Mumbai and Chennai. However, some of these new cities will themselves become mega-cities. Trade liberalization and a greater emphasis on exports from the 1980s also increased the comparative advantage of many smaller cities. Meanwhile, improvements in transport and communications infrastructure in many nations lessened the advantages for businesses of concentrating in the largest cities. Another factor evident in many cities, was lower rates of natural increase, as fertility rates came down.

For much of Africa and some Latin American nations, the slower growth rates for major cities over the last two decades is linked to slow economic growth (or economic decline), so fewer people moved there (see Potts 2009 for a discussion of this for sub-Saharan Africa). In many cities in sub-Saharan Africa and some in Asia, population growth rates remain high because of high rates of natural increase, not high rates of net in-migration. In some cities, mostly in sub-Saharan Africa, the population growth has been much boosted by the movement there of people displaced by wars, civil strife or drought, but this is often largely a temporary movement, not a permanent one.

Although the size of the world's largest cities is historically unprecedented, most of the largest cities are significantly smaller than had been predicted. For instance, Mexico City had 20.1 million people in 2010 when it had been predicted to reach 31 million people by 2000 in 1980. Kolkata had 14.3 million inhabitants in 2010, not the 40–50 million people predicted during the 1970s. However, there are also large cities whose population growth rates have remained high during recent decades – for instance, Dhaka (Bangladesh) and many cities in India, China, Mexico and Brazil. Some very large cities have grown to sizes that were not predicted, including Dhaka with around 15 million inhabitants and Shenzhen with around 10 million.

More than half the world's population does not live in cities

More than half the world's population lives in urban centres but the proportion living in cities is considerably lower since a significant proportion of the urban population lives in urban centres too small to be called cities. There are no estimates for the proportion in cities because there is no agreement on how cities are defined but a considerable proportion of the world's urban population live in settlements that lack the size and the economic, administrative or political status that being a city implies. The economic and demographic importance of small urban centres is often overlooked; for instance, many nations have more than 10 per cent of their national population in urban centres with less than 20,000 inhabitants (Satterthwaite 2006).

The need for a better understanding of the economic, social, political and demographic influences on urbanization

We learn about this issue from a few detailed historically rooted analyses of urban change in particular nations and these remind us how complicated and varied is such change. Also, how much it can vary within a nation (and over time) and the complex mix of local, regional and international economic, social and political influences on it – along with the influences brought by demographic changes (see Hasan and Raza 2002; Martine and McGranahan 2010; UNCHS 1996).

Beyond a rural–urban division

Perhaps too much emphasis is given to the fact that the world's population became more than half urban in 2008, because of the imprecision in defining 'urban' and 'rural' populations, and the large differences between countries in the criteria used to define urban centres. These differences limit the validity of inter-country comparisons. For instance, it is not comparing like with like if we compare the 'level of urbanization' (the percentage of population in urban centres) of a nation that defines urban centres as all settlements with 20,000 or more inhabitants with another that defines urban centres as all settlements with more than 1,000 inhabitants. The comparison is particularly inaccurate if a large section of the population lives in settlements of between 1,000 and 19,999 inhabitants (which is the case in most nations). The proportion of the world's population living in urban areas could be increased or decreased by several percentage points simply by China, India or a few of the other most populous nations changing their definition of what is an urban centre. Thus, the proportion of the world's population currently living in urban centres is best considered not as a precise percentage (i.e. 51.6 per cent in 2010) but as being between 45 and 60 per cent, depending on the criteria used to define what is an 'urban centre'. But what is beyond doubt is the fact that the world is becoming increasingly urban, as most of the world's economic production and most new investment is now concentrated in urban areas – as has been the case for many decades.

There is a tendency in the discussions of urban change to concentrate too much on changes in urban populations or in levels of urbanization, and too little on the economic and political transformations that have underpinned urbanization – at a global scale, the very large increase in the size of the world's economy and the changes in the relative importance of different sectors and of international trade – and how this has changed the spatial distribution of economic activities, and the social and spatial distribution of incomes. The distinction between 'rural' and 'urban' populations has some utility in highlighting differences in economic structure, population concentration and political status (as virtually all local governments are located in urban centres) but it is not a precise distinction. First of all, large sections of the 'rural' population work in non-agricultural activities or derive some of their income from such activities, or commute to urban areas. Many rural households also derive some of their income from remittances from family members working in urban areas. Distinctions between rural and urban areas are also becoming almost obsolete around many major cities as economic activity spreads outwards – for instance, around Jakarta (McGee 1987), around Bangkok and within Thailand's Eastern Seaboard, around Mumbai (and the corridor linking it to Pune), around Delhi, in the Pearl River Delta in China and in the Red River Delta in Vietnam (World Bank 1999). Conversely, large sections of the 'urban' population work in agriculture or in urban enterprises that serve rural demand. In addition, discussing 'rural' and 'urban' areas separately can ignore the multiple flows between them in terms of migration movements and the flow of goods, income, capital and information (Tacoli 1998, 2006). Many low-income households draw goods or income from urban and rural sources.

An uncertain urban future

Most publications discussing urban change assume that all nations will continue to urbanize far into the future until virtually all their economy and workforce are based in urban areas. Such assumptions should be viewed with caution. Given the historic association between economic growth and urbanization, a steady increase in the level of urbanization in any low-income nation is only likely if it also has a steadily growing economy. While stronger and more buoyant economies for the world's lower-income nations should be a key development goal, the current prospects for many such nations are hardly encouraging, within the current world economic system. Many of the lowest-income nations have serious problems with political instability or civil war, and most have no obvious 'comparative advantage' on which to build an economy that prospers and thus urbanizes within the globalized economy.

There are also grounds for doubting whether a large proportion of the world's urban population will come to live in very large cities. As noted above, many of the largest cities have slow population growth rates and much new investment is going to particular medium-sized cities well located in relation to the largest cities and to transport and communications systems. In addition, in prosperous regions with advanced transport and communications systems, rural inhabitants and enterprises can enjoy standards of infrastructure and services, and access to information, that historically have only been available in urban areas. Thus, both low-income and high-income nations may have smaller than expected increases in their urban populations, although for very different reasons.

References

This chapter draws on Satterthwaite, David (2007), *The Scale of Urban Change Worldwide 1950–2000 and its Underpinnings*, Working Paper, London: IIED, that can be downloaded at no charge from www.iied.org/pubs/; and on United Nations (2012), *World Urbanisation Prospects: The 2011 Revision*, CD-ROM edition, New York: Department of Economic and Social Affairs, Population Division.

Chandler, Tertius (1987), *Four Thousand Years of Urban Growth: An Historical Census*, Lampeter, UK: Edwin Mellen Press.

Hasan, Arif and Mansoor Raza (2002), 'Urban change in Pakistan', Urban Change Working Paper 6, London: IIED.

Martine, George and Gordon McGranahan (2010), *Brazil's Early Urban Transition: What Can It Teach Urbanising Countries*? Urbanisation and Emerging Population Issues Paper 7, London and New York: IIED and UNFPA.

McGee, T. G. (1987), 'Urbanization or Kotadesasi: the emergence of new regions of economic interaction in Asia', working paper, Honolulu: East West Center, June.

Potts, Deborah (2009), 'The slowing of sub-Saharan Africa's urbanization: Evidence and implications for urban livelihoods', *Environment and Urbanization*, 21(1): 253–259.

Sassen, Saskia (2006), *Cities in a World Economy*, Thousand Oaks, CA: Pine Forge Press.

Satterthwaite, David (2006), 'Outside the large cities; the demographic importance of small urban centres and large villages in Africa, Asia and Latin America', Human Settlements Discussion Papers, London: IIED.

Tacoli, C. (1998), 'Rural–urban interactions: A guide to the literature', *Environment and Urbanization* 10(1): 147–166.

Tacoli, C. (ed.) (2006), *The Earthscan Reader in Rural-Urban Linkages*, London: Earthscan Publications.

UNCHS (1996), *An Urbanising World: Global Report on Human Settlements*, Oxford and New York: Oxford University Press.

United Nations (2012), *World Urbanisation Prospects: The 2011 Revision*, New York: Department of Economic and Social Affairs, Population Division.

World Bank (1999), *Entering the 21st Century: World Development Report 1999/2000*, Oxford and New York: Oxford University Press.

World Bank (2008), *World Development Report 2009: Reshaping Economic Geography*, Washington, DC: World Bank.

Further reading

Bayat, Asef and Eric Denis (2000), 'Who is afraid of Ashwaiyyat: Urban change and politics in Egypt', *Environment and Urbanization*, 12(2): 185–199.

Bryceson, Deborah F. and Deborah Potts (eds) (2006), *African Urban Economies: Viability, Vitality or Vexation?* Basingstoke and New York: Palgrave Macmillan.

Douglass, Mike (2002), 'From global intercity competition to cooperation for livable cities and economic resilience in Pacific Asia', *Environment and Urbanization*, 14(1), 53–68.

Montgomery, Mark R., Richard Stren, Barney Cohen and Holly E. Reed (eds) (2003), *Cities Transformed: Demographic Change and its Implications in the Developing World*, Washington, DC: The National Academy Press (North America)/Earthscan (Europe).

5.2
Urban bias

Gareth A. Jones and Stuart Corbridge

Students of development and development practitioners have long concerned themselves with urban–rural relationships. A familiar assumption has been that ambitious people will move to urban areas to improve their lot. In the 1950s and 1960s an apparent bias in favour of urban-industrial models of development was justified by three key ideas. W. Arthur Lewis (1954) proposed that there is disguised unemployment or underemployment in rural areas of poorer countries, where the marginal productivity of labour is often very low. Men and women move to the city to find more productive work and to pull their families out of poverty. Hans Singer (1950) and Raoul Prebisch (1950) further argued there is a long-run tendency for the terms of trade, the ratio of export commodity prices to import commodity prices, to move against primary commodities like foodstuffs and raw materials. Import-substitution was commended partly on this basis. Urban planners, meanwhile, argued that goods and services would most efficiently be diffused from major cities to smaller cities and rural areas. Cities benefited from, and generated, economies of scale. This, after all, had been the experience of most Western countries.

All of these arguments have been disputed. There is no reason why manufacturing or service-sector jobs must be based exclusively or even mainly in urban areas. China and Taiwan have each generated large numbers of industrial jobs in rural areas over the past fifty years. In addition, countries like Norway and Canada became rich largely as exporters of food and raw materials. In practice, however, for both economic reasons (including the benefits of industrial clustering), and political reasons (including a disposition to think of the urban as modern), most non-agricultural jobs *have* been based in urban areas of developing countries. The question is whether this matters, and, if so, why and for whom and in what terms?

One answer to this question is that concentrations of urban-industrial power damage the 'authenticity' of a country. Gandhi (1997[1908]) believed that the soul of India was to be found in its villages. For him, large-scale industrialization was a form of social evil. Some deep ecologists also think in these terms, which for the most part are recognizably normative (that is, they advance a value judgement). In the 1960s Michael Lipton began to develop a more positive (or testable) account of urban bias in the process of world development. His urban bias thesis (UBT) was formally presented in his book *Why Poor People Stay Poor: A Study of Urban Bias in World Development* (Lipton 1977). The UBT proposes that urban classes in poorer countries use their social power to bias (distort) a range of public policies against members of the rural classes. Lipton argued that urban bias 'involves (a) an allocation, to persons or organizations located in towns, of shares of resources so large as to be inefficient and inequitable, or (b) a disposition among the powerful [urban classes] to allocate resources in this way' (Lipton 2005: 724, summarizing Lipton 1977).

The urban bias thesis and its critics

In its first iteration, the UBT made five main claims.

1. Rural areas of developing countries suffer from too little spending on education and health care (relative to their population figures).
2. These inequalities, combined with excessively 'urban' forms of teaching and curriculum development, pull bright young people to the cities (when a more neutral incentive regime would keep them in the countryside).
3. People in rural areas are forced to pay a higher share of national taxes than is fair.
4. Urban bias is further and most damagingly evident in a series of government-imposed price twists that causes inputs into rural areas to be overpriced when compared with a market norm, and which causes outputs from rural areas to be correspondingly underpriced.
5. This combination of *distributional urban bias* (1, 2 and 3) and price-twisting is not only unfair but inefficient: at the margin, Lipton maintained, a given sum of government money will earn higher returns in the small agriculture sector or in rural off-farm employment creation than it will in cities or large-scale urban-based industries.

Two principal critiques were levelled at this initial iteration of Lipton's UBT. A first line of criticism was that the UBT lacked empirical validity and was overly generalized. The UBT failed to prove that the inter-sectoral terms of trade moved everywhere against rural areas. Critics also said that Lipton was inattentive to the issue of urban poverty; that most of the rural poor in South Asia were sellers of labour who benefited from cheap food; and that the UBT neglected the power of rural elites (see Byres 1979). A second line of critique was theoretical: it challenged Lipton's accounts of class, power and policy formation. Keith Griffin (1977) took exception to Lipton's attempt to account for intra-sectoral differences in wealth and power by counting members of the rural elite as members of the urban class, and members of the urban poor as part of the rural class. This seemed like sophistry to Griffin.

Early support for the UBT came from the political scientist Robert Bates, and from Elliott Berg at the World Bank. Bates (1981, 1988) argued that food production problems in sub-Saharan Africa were mainly the result of non-democratic governments using urban-biased policies to discriminate against smallholding agriculturalists. Governments used marketing boards and over-valued exchange rates to procure food cheaply from the countryside or overseas. They also made it difficult for farmers to sell food and other crops to private merchants, and spent large sums on industrial protection. Bates suggested that many ruling regimes in sub-Saharan Africa imposed urban-biased policies to keep the lid on unrest in towns and cities. The urban working class was bought off with cheap food, not least in nominally socialist countries. City-based bureaucrats also extracted large rents from systems of licenses and quotas. In the medium-term, however, rational farmers responded by producing less food for sale. This caused precisely those surges in food prices that triggered the urban food riots or even coups that Africa's ruling coalitions were keen to avoid.

Bates argued that international donors should exploit moments of political or economic crisis in sub-Saharan Africa to push more farmer-friendly policies. This view found support in the World Bank's 1981 report on *Accelerated Development in sub-Saharan Africa*, the so-called Berg report, and later became standard World Bank policy in sub-Saharan Africa when the debt crisis opened the way for structural adjustment loans and various forms of donor conditionality. Michael Lipton (2005), for his part, accepts that many structural adjustment programmes (SAPs) improved the inter-sectoral terms of trade in sub-Saharan Africa: price twists became less damaging to the countryside, although they remain significant at the global scale, given EU, Japanese and US farm-support policies. Lipton believes, however, that the focus on price twists alone caused the World Bank to ignore the more deeply rooted inequalities in social and spatial power that promote urban bias. Indeed, he has recently made a strong case for an old argument; greater equality of land holding and secure property rights as the most reliable means to lift small-scale producers out of poverty (Lipton 2009).

The revised urban bias thesis and some new criticisms

With Robert Eastwood, Lipton claimed that distributional urban bias has increased precisely at the same time as a series of successful neoliberal assaults on the urban–rural terms of trade. Eastwood and Lipton (2000) claim: (a) that overall within-country inequality increased significantly after 1980–85, following adjustment policies; (b) that these increases have not been offset by declining rural – urban inequality; (c) that this absence of offset, save for in a few countries in Latin America, must be accounted for by a rise in distributional urban bias at a time of reduced price twists against the countryside. Urban bias still exists and with it the case for off-setting measures to improve rural livelihoods.

There is considerable support for the view that urban/rural welfare ratios are not yet falling towards unity, and may even be diverging in some countries that are enjoying rapid economic growth (notably China). Nevertheless, we can identify at least four contemporary challenges to the UBT in its original and modified versions.

First, the UBT underestimates the scale and recent relative growth of urban poverty. It is estimated that about 550 million people in the cities of the global South live in absolute poverty with nutritional and health conditions close to rural areas, and over 920 million live in 'slums' with limited access to the resource allocations supposedly biased in favour of cities. Second, the UBT is beset by definitional and measurement problems. Definitions of urban and rural are not stable. A large village in Bangladesh might count as a town in Peru. It is also difficult to carry out a detailed audit of all transfers of goods and services across the urban–rural divide. Many subsidies are hidden, as when developers are not required to pay the full costs of infrastructure provision, which might lead to an undercount of UBT. At the same time, mayors of many cities can rightly claim to be hard done by from net allocations; cities such as Bogotá and Cape Town, for example, contribute far more to gross domestic product than they receive from national fiscal allocations.

Third, it has been argued that the UBT overestimates the possibilities for *in situ* improvement in agricultural livelihoods, especially in sub-Saharan Africa. In comparison to the arguments demonstrating the productivity of small-scale agrarian producers (see Schultz 1964), research on rural livelihoods has identified constraints to farmer productivity beyond price twists or the concentration of services in large towns. Under conditions of historically high commodity prices, the FAO (2011) has revealed that food security may worsen in the short term as input costs rise, risk adverse farmers delay investment in prediction of price falls and real incomes suffer as the poorest buy more food than they produce. The thesis also underestimates (even ignores) the benefits of migration and ceaseless circulation and the construction of livelihoods that move across the urban and rural divide (see Tacoli 2006). Boundaries are blurring, and where they are not blurring it is because of unhelpful government obstacles to mobility, notably in the case of pass laws or registration systems such as the *hukou* in China. Development agencies should be pushing hard for the removal or reduction of these restrictions (see World Bank 2005).

Fourth, there is a new challenge from a body of work inspired by the work of Paul Krugman called the 'new economic geography'. Research suggests that many non-primate cities in the developing world have grown less because of rent-seeking and distorted patterns of political access (urban bias in Lipton's terms, in decline following SAPs) than because of the returns to scale and spill-over effects that are associated with the clustering of innovative economic activities. In the case of many public goods, Overman and Venables (2005: 6) suggest, an 'urban bias in public expenditure and provision may [also] be an efficient allocation of resources'. Biases in outcome are not always the result of wilful distortion (of markets by states) and even if 70 per cent of the world's poorest people reside in rural areas, it is not always and everywhere rational that at least 70 per cent of development spending should be spent in rural areas. Indeed, the 2009 World Development Report went a step further, arguing that governments should abandon any pretence to spatially equitable investment, which will simply waste resources and reduce growth, and target efforts according to city scale and function (World Bank 2009).

'Urban bias' and public policy

Where does all this leave us? It makes little sense to choose between the UBT and what might be called its opposite. There is much we still do not know, as Lipton readily concedes. Non-economic forms of urban bias are under-researched. Many rural people are stereotyped as backward or ignorant and are treated by government officials on that basis. They might experience 'urban bias' in terms other than those set out in Lipton's UBT, but which are nonetheless consistent with his view that key development actors are disposed to equate the urban with the modern or developed (what he calls 'dispositional urban bias'). It is thus prudent to retain the provocation set out by Lipton's model (location matters for welfare), while avoiding reference to urban bias either as a social fact or as a pathology that always needs correction.

To the extent that the work of Lipton, Bates and Berg helped to authorize a change of policy within the World Bank and other development agencies in the 1980s, they rightly deserve acknowledgement and even some applause. Whether some of these agencies and governments have since moved to an implicit or explicit anti-urban bias is not clear, although this has been the claim of some critics. What is clear is that the charge of distributional urban bias has been widely and often properly subjected to criticism. There are good reasons for thinking that many non-primate cities in developing countries are growing on the back of a strong commitment to economic innovation and the production of dynamic growth clusters, and not on the backs of rural people or as a result of the activities of rent-seeking politicians. Not all 'bias' is bad, and many people in the countryside will rightly want to make their way to the city. In any case, many of the policies that Lipton and others want to see enacted – including better provision of primary education and health care in the countryside – can be argued for without resort to a generalized model of the exploitation of the countryside by the city. In this specific respect the UBT can be unhelpful.

References

Bates, R. H., 1981, *Markets and States in Tropical Africa*, Berkeley: University of California Press.

Bates, R. H., 1988, *Toward a Political Economy of Development*, Berkeley: University of California Press.

Byres, T. J., 1979, 'Of neo-populist pipe-dreams: Daedalus in the Third World and the myth of urban bias', *Journal of Peasant Studies* 6: 210–44.

Eastwood, R. and Lipton, M., 2000, 'Pro-poor growth and pro-growth poverty reduction: Meaning, evidence, and policy implications', *Asian Development Review* 18(2): 22–58.

Food and Agriculture Organisation (FAO), 2011, *The State of Food Insecurity in the World*, Rome: FAO.

Gandhi, M. K., 1997[1908], *Hind Swaraj* (published with an Editorial Introduction by Anthony Parel), Cambridge: Cambridge University Press.

Griffin, K., 1977, 'Review of *Why Poor People Stay Poor*', *Journal of Development Studies* 14: 108–9.

Lewis, W. A., 1954, 'Economic development with unlimited supplies of labour', *Manchester School of Economics and Social Studies* 22: 139–91.

Lipton, M., 1977, *Why Poor People Stay Poor: A Study of Urban Bias in World Development*, London: Temple Smith.

Lipton, M., 2005, 'Urban bias', in Forsyth, T. (ed.) *Encyclopedia of International Development*, London: Routledge.

Lipton, M., 2009, *Land reform in developing countries: Property rights and property wrongs*, London: Routledge.

Overman, H. G. and Venables, A. G., 2005, *Cities in the Developing World*, Centre for Economic Performance Working Paper No 695, LSE. Available online at http://cep.lse.ac.uk/pubs/download/dp0695.pdf

Prebisch, R., 1950, *The Economic Development of Latin America and Its Principal Problems*, New York: UN-ECLA.

Schultz, T. W., 1964, *Transforming Traditional Agriculture*, New Haven: Yale University Press.

Singer, H., 1950, 'The distribution of gains between investing and borrowing countries', *American Economic Review* 40: 473–85.

Tacoli, C. (ed.), 2006, *Rural-Urban Linkages*, London: Earthscan.

World Bank, 1981, *Accelerated Development in sub-Saharan Africa*, Washington DC: World Bank.

World Bank, 2005, *Beyond the City: The Rural Contribution to Development*, Washington DC: World Bank.

World Bank, 2009, *World Development Report 2009: Reshaping economic geography*, Washington DC: World Bank.

Further reading

Jones, G. A. and Corbridge, S. E., 2010, 'The continuing debate about urban bias: The thesis, its critics, its influence, and its implications for poverty reduction strategies', *Progress in Development Studies* 10(1): 1–18. A broad-based critical review of the urban bias thesis from inception to contemporary relevance.

Lipton, M., 1977, *Why Poor People Stay Poor: A Study of Urban Bias in World Development*, London: Temple Smith. The classic text, argued as both an eloquent narrative and through economic theory.

Tacoli, C. (ed.), 2006, *Rural-Urban Linkages*, London: Earthscan. Fifteen chapters, mostly based on original field research in South East Asia and sub-Saharan Africa demonstrating the inter-dependence of livelihoods across the rural–urban divide.

5.3

Global cities and the production of uneven development

Christof Parnreiter

Introduction

Global or world city research is vital for development studies for two main reasons. First, it provides critical insights on how economic globalization is organized and thereby contributes to a more comprehensive understanding of the processes through which uneven development is produced and maintained. Second, the global city paradigm's nuanced account of the geographies of globalization and, in particular, of global economic governance, helps to correct any simple notion of a clear-cut North–South power divide.

The global city paradigm

The global city paradigm emerged with the writings of John Friedmann and Saskia Sassen, whose principal interests were to explore and to theorize the new geography of the world economy, reflecting globalization processes. The approach denotes two key functions of world or global cities: they provide *connectivity* for geographically dispersed production units and are therefore 'the "basing points" in the spatial organization and articulation of production and markets' (Friedmann 1986: 71); and they serve as locations for the *governance* of cross-border economic activities, therefore they are 'highly concentrated command points' (Sassen 2001: 3), from where the world economy is managed and controlled.

A main innovation in global city research is the shift of attention from formal power functions as exercised by corporate headquarters to 'the production of those inputs that constitute the capability for global control' (Sassen 2001: 6). Drawing on the literature on producer services, Sassen maintains that a global city's function in the management and the governance of global operations of firms derives from the existence of sizeable clusters of (globalized) producer service firms, because these services accommodate a demand that emerges from the organizational needs of companies with global operations. From this it follows that each global city constitutes, as Brown *et al.* (2010) and Parnreiter (2010) maintain, a node for numerous commodity chains, while all commodity chains run through global cities seeking core-labour processes such as producer services.

A German corporation, for example, that invests in Mexico to produce for the US market needs the support of financial, law or accountancy professionals in order to get things right in a foreign environment. Because access to tacit knowledge is critical in getting acquainted with local modi operandi, service providers such as Deutsche Bank, Baker & McKenzie or KPMG cannot serve the global operations of their clients from headquarters in Frankfurt, Chicago or Toronto. They have created a global network of offices to support their clients, because, as the chief knowledge officer of a global accountancy firm in Mexico City puts it, 'you have to be here, you have to have the relations with the entrepreneurs, you have to be in the chambers, . . . you have to be at the cocktails'. The local offices of global producer service firms are, however, not only indispensable in making the global production line run, they also influence how commodity chains are governed. Given the complexity of doing business in different markets involving different jurisdictions, budgets, languages and cultures, the global headquarters of TNCs rely in their decision-making on the counselling of both their subsidiaries' regional headquarters and of producer service firms. A lawyer at the Mexico City office of a global law firm affirms that 'the partners of the [law] firm, they have conferences with the clients to plan a deal, to structure a deal. . . . I do believe that the one who makes the strategy, it's the partners of the law firm' (both quotes in Parnreiter 2010: 45f).

In short, because cross-border business needs worldwide networks of producer service firms in order to be conducted smoothly, the places where these firms establish their offices become highly central to the world economy. Moreover, since global city formation is a result of the emergence of intra-firm and inter-city networks of producer service firms, the global city itself is a function of a network: 'there is no such entity as a single global city' (Sassen 2001: 348). Yet, one of the most powerful critiques of the early global city research was that this conceptualization was not backed by relational data concerning the supposed city-to-city flows. In order to overcome this flaw and to substantiate the notion of a 'world city network' (Taylor 2004; Derudder and Witlox 2010), in the late 1990s the Globalization and World Cities research network (GaWC) was founded.

The world according to GaWC

Initially using data relating to 100 global producer service firms with offices in 315 cities around the world, and expanding the database recently to 2,000 firms in 525 cities (Taylor *et al.* 2010), GaWC developed an 'interlocking network model' to come to terms with the intra-firm and inter-city flows that establish the world city network. Acting on the assumption that the larger and the more important an office in a particular city is, the more flows of information, capital, people, and so on, to other cities it will generate, the 'interlocking network model' allows for estimating service flows between cities, which are then expressed as their connectivity values. As shown in Figure 1, the most recent account sees London and New York as the best connected cities, followed by Hong Kong, Paris, Singapore, Tokyo, Shanghai, Chicago and Dubai.

Two things are striking about the world city network. First, it is much bigger than the network of headquarter cities that emerges from the locations of the main offices of the world's biggest corporations. While about half of the revenues of the Fortune 'Global 500' companies are made by corporations headquartered in only 20 cities, GaWC's list of the 'Alpha world cities', which are characterized as 'very important world cities that link major economic regions and states into the world economy', comprised 47 cities (GaWC 2011). From this difference in the network size and from the assumption that producer service firms exercise command functions for global commodity chain follows, as does second, a more *decentralized geography of economic governance.* Figure 5.3.1 shows that many 'Third World' cities are on the map of 'Alpha world cities'. In fact, a third of these best-connected global cities are located in middle-income countries, with Asia being particularly well linked. Considering the complete list of Alpha, Beta and Gamma world cities, all 'Third World' megacities

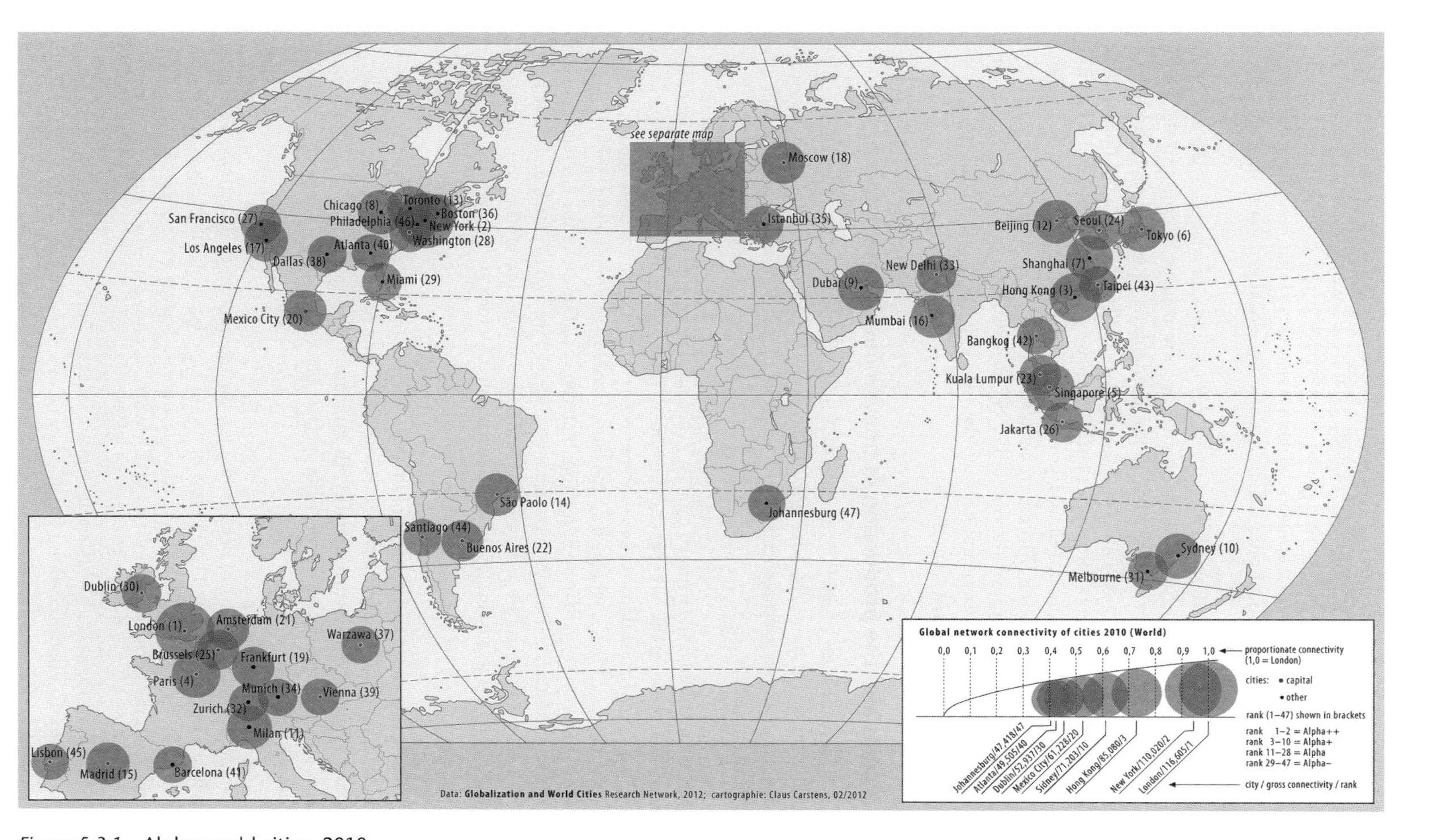

Figure 5.3.1 Alpha world cities, 2010

(except Dhaka) are included (GaWC 2011), contradicting the notion that (big) cities in poorer countries are economically insignificant or even parasitic. This finding also challenges post-colonial urban studies' complaints that global city research drops cities in poorer countries 'off the map' of urban studies (Robinson 2006). Mexico City, Johannesburg or Mumbai are *on* the map, not only for global production, but also for global economic management and governance.

Thus, the global city paradigm allows for a more inclusive appreciation of the role of cities in the global South within the world economy, because it helps us to grasp the multiple hinges or intermediaries between the few global headquarter cities and the countless cities where production for the world market is carried out. Globalization is neither like an oil slick that indifferently covers the whole world, nor is it exclusively controlled from a handful of 'supercities'. Rather, it is organized and governed from multiple places, including many global cities in poorer countries. Moreover, because all cities are unavoidably integrated into commodity chains, and because all commodity chains run through global cities, it can be contended that the ramifications upon which the world city network is built reach cities in the world economy's remotest hinterlands.

Global cities and the production of uneven development

In his conceptualization of the 'development of underdevelopment', Andre Gunder Frank suggested a *spatialized* model of how uneven development is organized. He delineated the Latin American city as a bridgehead for the interests of the dominant centres of the world economy, arguing that

> [j]ust as the colonial and national capital . . . become the satellite of the Iberian (and later of other) metropoles of the world economic system, this satellite immediately becomes a colonial and then a national metropolis with respect to the productive sectors and population of the interior. . . . Thus, a whole chain of constellations of metropoles and satellites relates all parts of the whole system from its metropolitan center in Europe or the United States to the farthest outpost in the Latin American countryside. . . . we find that each of the satellites . . . serves as an instrument to suck capital or economic surplus out of its own satellites and to channel part of this surplus to the world metropolis of which all are satellites.
>
> *(Frank 1969: 6)*

Obviously, this very lucid account of the role of cities in the making of uneven development matches up perfectly with key ideas of the global city paradigm developed two decades later. Frank's depiction also corresponds with Immanuel Wallerstein's (1983: 30) notion that 'core-ness' in the world system results from specific places' capability to attract many global commodity chains:

> Commodity chains have not been random in their geographical directions . . . they have been centripetal in form. Their points of origin have been manifold, but their points of destination have tended to converge in a few areas. That is to say, they have tended to move from the peripheries of the capitalist world-economy to the centers or cores.

Thus, while both Frank and Wallerstein suggest that the transfer of resources from peripheries to the cores is organized along commodity chains, Frank additionally emphasizes that this transfer of resources is managed from cities along these chains. These ideas have recently led some authors to suggest an integration of global city and global commodity chains research (Derudder and Witlox 2010) because, as Brown *et al.* (2010: 29) suggest, the 'creation and (unequal) distribution of value along commodity chains is organized in and governed from world cities'.

The production of uneven development is, therefore, based on innumerable connections between cities with different functions in commodity chains. While in many places production for the world market is carried out by people who are by and large restricted to being rule keepers, there are also places from which the rule makers – or the 'masters of the universe', as Tom Wolfe has called the professionals of the finance industries in his novel *Bonfire of the Vanities* – operate. It is a main merit of the global city paradigm that it emphasizes this distinction, rather than blurring it by suggesting a shift of attention to 'how "global" economic processes affect all cities' (Robinson 2006: 102). Concerned with the 'the *practice* of global control' (Sassen 2001: 6), global city research provides insights into the geography of the *making* of power asymmetries. These insights allow for a comprehension of uneven development that goes beyond overgeneralized notions of the powerful 'global North' versus the powerless 'global South'. Cities in poorer countries are not only 'affected' by uneven globalization, they are involved in its production. Mexico City, for example, is – through hosting many global producer service firms – a critical node for the functioning of myriad global commodity chains and thus is a node from whence the 'development of underdevelopment' is managed and governed.

References

Brown, E., Derudder, B., Pelupessy, W., Taylor, P. and Witlox F. (2010): World City Networks and Global Commodity Chains: Towards a World-systems' Integration. *Global Networks* 10, 1, 12–34.
Derudder, B. and Witlox, F. (eds) (2010): *Commodity Chains and World Cities*. New York: John Wiley & Sons.
Frank, A. G. (1969): *Latin America: Underdevelopment or Revolution: Essays on the Development of Underdevelopment and the Immediate Enemy*. New York: Monthly Review Press.
Friedmann, J. (1986): The World City Hypothesis. *Development and Change* 17, 69–83.
GaWC (2011): The World According to GaWC 2010. Available online at www.lboro.ac.uk/gawc/world2010t.html
Parnreiter, C. (2010): Global Cities in Global Commodity Chains: Exploring the Role of Mexico City in the Geography of Global Economic Governance. *Global Networks* 10, 1, 35–53.
Robinson, J. (2006): *Ordinary Cities: Between Modernity and Development*. London: Routledge.
Sassen, S. (2001): *The Global City: New York, London, Tokyo*. Princeton: Princeton University Press.
Taylor, P. (2004): *World City Network: A Global Urban Analysis*. London: Routledge.
Taylor, P., Ni, P., Derudder, B., Hoyler, M., Huang, J. and Witlox, F. (eds) (2010): *Global Urban Analysis: A Survey of Cities in Globalization*. London: Earthscan.
Wallerstein, I. (1983): *Historical Capitalism*. London: Verso.

Further reading

A useful book to get a comprehensive idea of Sassen's conceptualization of global cities is her *Cities in a World Economy*, published in a fourth edition in 2011 by Pine Forge Press. *The Global Cities Reader*, edited by Neil Brenner and Roger Keil and published in 2006 by Routledge will broaden your understanding of how the global city paradigm was developed, of local transformations associated with global city formation, and of critical contributions to the debate. Finally, the GaWC website at www.lboro.ac.uk/gawc/ is an indispensable resource for all interested in the relation between globalization and cities. There are more than 400 research bulletins online, many of which have later been published in peer-reviewed journals. Also, the GaWC website offers data on and visualizations of the world city network.

5.4
Studies in comparative urbanism

Colin McFarlane

Cities have always been understood comparatively. While comparison may appear as a prosaic set of methodological questions around case studies, in practice it is a critical part of how understanding, theory and research about cities are produced and contested. Recent years have witnessed not just a resurgence of comparison, but a new experimentalism with comparative thinking and methodologies. This is in part a response to the globalisation of urban policy, planning, economies, cultures and ecologies, but it is also an attempt to internationalise urban geography and development by thinking across intellectual and imaginative divides that have traditionally separated out the cities of the global North from those of the global South. Our inherited conceptions of the city are often premised on the experiences and theoretical work based upon cities in Western Europe and North America, accompanied by the often implicit slippage between claims about certain cities (e.g. New York, Los Angeles, Barcelona, Berlin, Paris or London) and claims about 'the city' as an abstract, generalised category. Part of the revival of comparison has been to widen the range of urbanisms that constitute urban theory.

This upsurge in comparative research has been multifaceted, and includes – and this is by no means exhaustive – efforts to compare one city with several (Nijman, 2007a, on Miami); comparing two cities (Huchzermeyer, 2007, on the production of informal settlements in São Paulo and Cape Town); exploring how specific processes or features recur or diverge in different cities (Gulger, 2004, on world cities in the South, or Roy, 2005, on planning and citizenship across North and South); research outlining frameworks for comparative urban research (Brenner, 2001; Kantor and Savitch, 2005); and work developing an explicitly postcolonial (Robinson, 2006, 2011; McFarlane, 2010) or relational (Ward, 2010) revisioning of comparative urbanism. In short, comparison is firmly on the agenda of urban studies, whether as a way of experimenting with the diversity of cities across and beyond inheritances of global North/South or global city/megacity, as a means for thinking through the relations between case studies and wider processes, or – increasingly – as a resource for locating difference rather than similarity (McFarlane and Robinson, 2012).

The emphasis in much of these debates has been on comparison as a set of practical, methodological and typological questions, including Abu-Lughod's (1999) variation-finding comparison that seeks to explain differences across New York, Chicago and Los Angeles as global cities; Nijman's (2007b) 'multiple-oriented comparative approach', which aims to better understand a particular case through individual comparisons with other (multiple) selected cases by seeking out both idiosyncrasies and analogies; and Ward's (2010) positioning of comparison as embedded in urban networks and flows rather than discrete or self-enclosed cities. These debates have also included important efforts to consider the assumptions through which objects of urban comparison are arrived at and pursued. Drawing on postcolonial and development scholarship, three closely interrelated sets of emerging concerns are crucial here: *theory culture*, *learning*, and *ethico-politics*. Theory culture, following Mufti (2005: 475), is 'the *habitus* that regulates "theory" as a discrete set of practices' within and sometimes between specialisms and regions (and see Connell, 2007, on *Southern Theory*). Comparative research *across* theory cultures prompts reflection not just on contrasting spaces or processes, but on the ontological and epistemological framings that inform how the world is being debated, how knowledge is being produced and questioned, and about the purpose of knowledge, research and theory. This means also considering the role of a whole range of institutional actors, from journals to forms and patterns of citation to modes of writing and dissemination that co-constitute theory cultures. By exploring how different theory cultures debate, for example, the city, politics, infrastructure, modernity, or globalisation, there is potential to develop more pluralised understandings through comparison.

In relation to learning, a key tension in comparative thinking, as Mbembe (2004: 375) has argued in relation to Johannesburg, is the 'temptation of mimicry' – the desire to copy, to learn directly from another urban experience. In contrast, thinking of urban comparison as learning-through-difference positions comparison as uncertain and provisional, because in widening the discursive field of cities the occurrence of unlikely translations involves the increased traversing of unfamiliar and unpredictable terrain. There is an important challenge here in trying to contextualise and understand knowledge from places unfamiliar to the researcher – texts that emerge from theory cultures that may require, for example, greater effort for an outsider to grasp (Connell, 2007).

Finally, efforts to compare between theory cultures raise ethical and political considerations (Jazeel and McFarlane, 2010). There is an ongoing challenge to engage, on as close a level playing field as possible, with the work of thinkers in different places: 'If a cosmopolitan urban theory is to emerge' writes Robinson (2002: 549–550), 'scholars in privileged western environments will need to find responsible and ethical ways to engage with, learn from and promote the ideas of intellectuals in less privileged places'. This requires a critical epistemic interrogation and reworking, such as that found in Appadurai's (2000) formulation of 'strong internationalisation'. He writes:

> ['Strong internationalization'] is to imagine and invite a conversation about research in which . . . the very elements of this ethic could be the subjects of debate. Scholars from other societies and traditions of inquiry could bring to this debate their own ideas about what counts as new knowledge and what communities of judgement and accountability they might judge to be central in the pursuit of such knowledge.
>
> *(Appadurai, 2000: 14)*

The demand here is an ethical commitment to learning and unlearning comparatively through different theory cultures. Drawing on Spivak (1993), McEwan (2003: 384) argues that 'unlearning' involves working hard to gain knowledge of others who occupy those spaces most closed to our privileged view through open-ended conversations. Part of this unlearning involves articulating the Western intellectual's participation in the formation of categories like 'Third World city'. This requires an ongoing sensitivity to the relationship between power, authority, positionality and knowledge, but it is a set of problems that cannot simply be acknowledged away; the positions, privileges and ways of seeing that help shape the comparisons we make cannot be stepped around.

The notion of strong internationalisation involves a particular and reflexive engagement with, for example, different regimes of academic knowledge production formulated through distinct patterns of collection, citation or judgement. There is a challenge here for academics to connect more closely and more frequently with the worlds, languages and vocabularies of disparate scholars (Desbiends and Ruddick, 2006). These ethico-political questions point to a range of practical challenges in working across different theory cultures, including negotiating new forms of collaboration for comparative research through journals, refereeing, and editorships, or in supporting scholarship and writing from different contexts. Other examples of strong internationalisation might include developing personal contacts and resources, such as a fund for translations to deal with language barriers, universities and departments investing in graduate language skills, creating funds for exchanges, and encouraging more collaborative postgraduate programmes.

There is, then, an emerging debate around thinking comparison not just as a research method – as crucial as that is – but as a mode of thought and as a *strategy* for international urban studies. If we are interested in a more international or postcolonial conception of the city – a conception that attempts to grapple with the multiplicity of different cities and ways of knowing the city across the global North–South divide – then it is inevitable that we examine what our implicit objects of reference are when we write urbanisms, and that we consider how we might bring other urban experiences, knowledges and theories into a more horizontal comparative field. What this opens is an expansive reading of comparison where a key question at stake is: what might be the implications for urban theory when we take comparison not just as a method, but as a mode of thought and set of institutional practices that inform how urban theory is constituted? Comparative thinking can be a strategy for, first, revealing the assumptions, limits and distinctiveness of particular theoretical or empirical claims, and second for formulating new lines of inquiry and more situated accounts.

The three overlapping areas of theory culture, learning and ethico-politics matter if comparison is to assist in producing research that reflects a more global understanding of urbanism. This outline of comparison as a strategy can be part of a wider effort to foster the collaborative formation of research projects, where the effort is to avoid – under ongoing constraints of history, positionality and unequal power relations – a privileging of one context over another, and that seeks not to assume that one theory culture represents a norm or standard of knowledge over others. It seeks to offer a route to alternative forms of comparative thinking and research that expands the field of inquiry to contribute to new ways of understanding contemporary urbanism.

References

Abu-Lughod, J. (1999) *New York, Chicago, Los Angeles: America's Global Cities*. Minneapolis: University of Minnesota Press.

Appadurai, A. (2000) 'Grassroots globalization and the research imagination'. *Public Culture*, 12(1), 1–9.

Brenner, N. (2001) 'World city theory, globalization, and the comparative-historical method: Reflections on Janet Abu-Lughod's interpretation of contemporary urban restructuring'. *Urban Affairs Review*, September, 124–147.

Connell, R. (2007) *Southern Theory: The Global Dynamics of Knowledge in Social Science*. Sydney: Allen and Unwin.

Desbiends, C. and Ruddick, S. (2006) 'Guest editorial. Speaking of geography: Language, power, and the spaces of Anglo-Saxon hegemony'. *Environment and Planning D: Society and Space*, 24, 1–8.

Gulger, J. (2004) *World Cities beyond the West: Globalization, Development and Inequality*. Cambridge: Cambridge University Press.

Huchzermeyer, M. (2007) 'Tenement city: The emergence of multi-storey districts through large scale private landlordism in Nairobi'. *International Journal of Urban and Regional Research*, 31, 4, 714–732.

Jazeel, T. and McFarlane, C. (2010) 'The limits of responsibility: A postcolonial politics of academic knowledge production'. *Transactions of the Institute of British Geographers*, 35, 109–124.

Kantor, P. and Savitch, H. V. (2005) 'How to study comparative urban development politics: A research note'. *International Journal of Urban and Regional Research*, 29, 1, 135–151.

Mbembe, A. (2004) 'Aesthetics of superfluity'. *Public Culture*, 16, 3, 373–405.

McEwan, C. (2003) 'Material geographies and postcolonialism'. *Singapore Journal of Tropical Geography*, 24, 3, 340–355.

McFarlane, C. (2010) 'The comparative city: Knowledge, learning, urbanism'. *International Journal of Urban and Regional Research*, 34, 4, 725–742.
McFarlane, C. and Robinson, J. (forthcoming) 'Rethinking comparative urbanism'. *Urban Geography*.
Mufti, A. (2005) 'Global comparativism'. *Critical Inquiry*, 31, 427–489.
Nijman, J. (2007a) 'Place-particularity and "deep analogies": A comparative essay on Miami's rise as a world city'. *Urban Geography*, 28, 92–107.
Nijman, J. (2007b) 'Introduction: Comparative urbanism'. *Urban Geography*, 28, 1–6.
Robinson, J. (2002) 'Global and world cities: A View from off the map'. *International Journal of Urban and Regional Change* 26, 3, 513–554.
Robinson, J. (2006). *Ordinary Cities: Between Modernity and Development*. London: Routledge.
Robinson, J. (2011) 'Cities in a world of cities: The comparative gesture'. *International Journal of Urban and Regional Research*, 35, 1, 1–23.
Roy, A. (2005) 'Urban informality: Towards an epistemology of planning'. *Journal of the American Planning Association*, 71, 2, 147–158.
Spivak, G. C. (1993) *Outside in the Teaching Machine*. London: Routledge.
Ward, K. (2010) 'Towards a relational comparative approach to the study of cities'. *Progress in Human Geography*, 34, 4, 471–487.

Further reading

King, A. D. (2004) *Spaces of Global Cultures: Architecture, Urbanism, Identity*. London: Routledge.
McFarlane, C. (2010) 'The comparative city: Knowledge, learning, urbanism'. *International Journal of Urban and Regional Research*, 34, 4, 725–742.
Nijman, J. (2007) 'Place-particularity and "deep analogies": A comparative essay on Miami's rise as a world city'. *Urban Geography*, 28, 92–107.
Robinson, J. (2011) 'Cities in a world of cities: The comparative gesture'. *International Journal of Urban and Regional Research*, 35, 1, 1–23.
Roy, A. (2005) 'Urban informality: Towards an epistemology of planning'. *Journal of the American Planning Association*, 71, 2, 147–158.
Ward, K. (2010) 'Towards a relational comparative approach to the study of cities'. *Progress in Human Geography*, 34, 4, 471–487.

5.5

Prosperity or poverty?

Wealth, inequality and deprivation in urban areas

Carole Rakodi

Cities and towns hold out great promise, but also can be unforgiving environments; opportunities abound, but the risks are great; the successful live well, but life for those who do not succeed is a struggle, marked by poverty, ill-health and insecurity.

Prosperity or poverty?

The correlation between urbanization and economic development, statistics which show the disproportionate share of GDP generated in urban areas, and the modern buildings and ostentatious wealth of central business districts, industrial estates, shopping malls and high-income residential areas combine to reinforce a view of cities as 'engines of economic growth' and sites for wealth generation. Many do not regard urban poverty as a serious problem. In earlier models, the path to development was expected to be industrialization and urbanization. For a time, progress appeared to be promising. Protected industrialization and the expansion of public-sector activities resulted in increased formal-sector wage employment, mostly in urban areas.

Not all entrants to the urban labour force could obtain full-time employment or earn sufficient wages to support a family. However, the good prospects that were thought to encourage high rates of rural–urban migration despite rising unemployment, led to a perception of urban areas as favoured environments. By the 1970s, it was clear that the basic needs of many, especially in Africa and South Asia, were not being met: trickle down was not working as expected, especially for those living in rural areas. An influential explanation was Michael Lipton's 'urban bias' thesis. In 1977, Lipton asserted that the major mistake in development policy was the 'urban bias' in expenditure and pricing policies, and he has subsequently reiterated and refined his arguments (for example Eastwood and Lipton, 2000). In 1977, he argued that failure to recognize the necessity of increased productivity in peasant agriculture for both sustained economic growth and increased prosperity for the majority of the population had led to a disproportionate emphasis on industrialization and thus concentration of investment in urban infrastructure. The political importance of concentrated urban populations reinforced this pattern and helped to explain the widespread adoption of cheap food policies. Subsidized food was paid for by low producer prices, which constrained the production and marketing of food crops and maintained small farmers in poverty, further fuelling out-migration.

The extent to which urban bias was a valid and sufficient explanation of development failure (especially persistent rural poverty) was questioned from the outset. It was noted that not all countries had an anti-rural policy bias; that other identities and political interests (ethnic, religious, class) cut across the rural–urban divide; and that rural/urban boundaries are arbitrary (Varshney, 1993). Jones and Corbridge (2010) revisited the debate, concluding that Lipton's more recent arguments can still be challenged. Thus, although there is some truth in his analysis, which figures showing higher average household income/expenditure and a lower incidence of poverty in urban than rural areas appear to bear out, the 'urban bias' thesis and the use of averages also serve to conceal a more complex reality.

Wealth, inequality and poverty

While until the 1960s some urban residents were able to secure employment, housing and access to services, many were not. Increasingly, wage employment, the formal housing development process and public provision of infrastructure failed to keep up with population growth. An increasing proportion of workers were forced, in the absence of social security systems, to seek economic opportunities in the so-called informal sector. Residents unable to rent or buy in the formal housing sector were forced to become house owners or tenants in informal settlements, with a variety of insecure tenure arrangements. Although physical infrastructure and social facilities were available in urban centres, many poor households could not access them and had to rely on self-provision (wells, pit latrines), the purchase of relatively costly private-sector services (water from vendors) or illegal tapping of publicly provided services (electricity, water).

The urban bias thesis, which labelled urban areas as 'wealthy' and rural areas as 'poor', failed to recognize high degrees of urban inequality and the exclusion of a large proportion of residents from the wealth, opportunities and good living conditions supposedly typical of urban areas. Nevertheless, it was widely accepted, and the economic reforms of the 1980s had, as a result, a strong 'rural bias'. The need to address trade and budget deficits led, in countries subject to structural adjustment policies, to trade and financial liberalization, the abolition of price controls and subsidies, and the commercialization of physical and social services. The effect, deliberately, was to remove 'urban bias' and other economic 'distortions' in order to encourage agricultural production and exports of all kinds. The results, in urban areas, included falling real incomes and job losses from all formal employment sectors. Thus much of the brunt of typical structural adjustment policies was borne by the urban poor and the gap between average rural and urban incomes was, more or less, eliminated, undermining the continued validity of the 'urban bias' explanation for underdevelopment (Varshney, 1993).

Before examining some evidence on the extent of inequality and poverty, and trends since the 1980s, a few methodological difficulties must be mentioned. The most common measure of poverty is household consumption or expenditure. An absolute poverty line represents the cost of a basket of necessities, including food (the food poverty line) and other needs. However, money-metric measures of poverty have limitations, many of which adversely affect comparisons of the incidence of poverty in urban and rural areas. In particular, prices and patterns of consumption vary between regions. However, some poverty assessments do not allow for differences in prices between urban and rural areas, and many fail to allow for differences in consumption bundles, especially the need for the urban poor to pay for housing and services which may not be monetized in rural areas, as well as the higher costs of transportation (especially for journeys to work). Further, the poorest urban residents often live in temporary accommodation or on the streets and may not be captured in sample surveys. Lastly, changes in, and the arbitrary nature of, urban boundaries affect estimates of poverty incidence in urban and rural areas, either excluding large numbers of residents in informal settlements beyond urban boundaries or including rural households living within them (Rakodi, 2002; Satterthwaite, 2004; UN-Habitat, 2006). As a result, official figures often underestimate the extent of urban monetary poverty.

Inequality is generally greater in urban than rural areas, although it varies between countries and cities (UN-Habitat, 2010), with Johannesburg and Bogota being amongst the most unequal cities (Gini co-efficients of 0.75 and 0.61 in 2005) and Chinese cities the least unequal (Gini co-efficients mostly 0.37 or less in 2004/5). However, in many of the least unequal cities, people are equally poor, and in Chinese cities inequality appears to have increased recently (He *et al.*, 2010). The incidence of poverty also varies, but is almost always lower in urban areas: in a sample of fifty countries (data from 1998 to 2007), 49 per cent of people on average were below the rural poverty line compared to 30 per cent below the urban poverty line (UN-Habitat, 2010: 22). Economic growth is associated with urbanization and declines in both urban and rural poverty, while both long-term economic decline and sudden economic crisis result in increases. For example, the economic crisis of the late 1990s in the industrializing countries of Asia precipitated a dramatic, if temporary, increase, especially in urban poverty, as formal-sector jobs were lost, as have soaring food prices since 2007/8, which have hit poor urban people, who spend a large proportion of their meagre incomes on food, particularly hard (UN-Habitat, 2010: 104).

The contribution of urban poverty to total poverty depends on both the incidence of poverty in urban and rural areas and the extent of urbanization. World Bank estimates, drawing on over 200 household surveys for 90 countries, and allowing for the higher cost of living in urban areas, show that in 2002 it varied from 25 per cent in Southern Asia (where 28 per cent of the population was urban) to 59 per cent in Latin America (where 76 per cent of the population was urban) (Ravallion *et al.*, 2007: 38). While overall, three quarters of the world's poor people still live in rural areas, the proportion living in urban areas is rising (from 19 per cent in 1993 to 24 per cent in 2002) (Ravallion, 2007). Average consumption figures and estimates of poverty incidence thus appear to show a lesser problem of poverty in urban than in rural areas, but high inequality and trends showing a shift in the locus of poverty demonstrate that these indicators conceal much of the reality.

Deprivation, vulnerability and insecurity

All headcount measures of poverty have been challenged on the basis that, even if refined, they do not recognize the life-cycle trajectories of households and therefore do not distinguish between permanent and transient poverty. Moreover, a consumption-based conceptualization of poverty may not coincide with the perceptions of the poor themselves, who define poverty to encompass not merely low incomes, but also deprivation and insecurity.

In high-density urban environments, housing and utilities of adequate standard are critical to health. Their availability is an outcome of the interaction between private provision and public policy. In the monetized economies of towns and cities, access to housing, utilities and social services is determined not just by availability, however, but also by household financial resources. To the poor, good-quality accommodation, education and healthcare are unaffordable. Their access to secure tenure, services and social facilities may, moreover, be constrained by social and political discrimination, affecting groups differentially by gender, ethnicity, caste, religion and so on.

Some basic health indicators, such as childhood mortality, are generally lower in urban areas. However, in some countries the reverse is true for other indicators, such as morbidity rates from infectious diseases (Haddad *et al.*, 1999). Higher population densities, combined with absent or inadequate piped water, drains, sanitation and refuse collection, mean that urban populations are more at risk from faecal contamination and other environmental hazards, and the high incidence of infectious diseases is associated with acute malnutrition. In addition, HIV/AIDS prevalence rates are invariably higher in urban areas, affecting the poor disproportionately and exacerbating poverty and economic recession (UN-Habitat, 2006).

Moreover, there is greater variation in access to basic environmental services, rates of mortality and morbidity and nutritional status *amongst* urban than rural populations, consonant with the greater inequality referred to above. Estimates based on demographic and health surveys show that 46 per cent of the urban poor lack access to three or more of the key services compared to 22 per cent of the non-poor, and women and girls generally continue to be more disadvantaged than men and boys (Montgomery *et al.*, 2004: 175; see also UN-Habitat, 2010). Although overall levels of morbidity and mortality due to disease or injuries and accidents vary between cities, all are greater in poor areas and in some cases as high or higher in slum as in rural areas. Finally, the proportion of malnourished children in slum areas is often double that in non-slum areas (54 and 21 per cent in India, 51 and 24 per cent in Bangladesh, 32 and 15 per cent in Bolivia, 48 and 11 per cent in Ethiopia) and similar to that in rural areas (for example, 37 per cent in Bolivia, 48 per cent in Ethiopia)

(UN-Habitat, 2010: 103). Driven partly by the adoption in 2000 of a Millennium Development Goal to improve the lives of at least 100 million 'slum' dwellers by 2020 and the subsequent focus on collecting data for non-slum and slum populations in cities (the latter defined as households lacking one or more of durable housing, sufficient living area, access to improved water and sanitation, and secure tenure; UN-Habitat, 2010: 33), there is an increasing tendency to equate poverty and deprivation with slum residence. This may be misleading: while it does recognize some of the non-monetary dimensions of poverty, not only is the definition of 'slum' as arbitrary as that of 'poverty', but also not all the residents of slums are poor and not all poor people live in slums.

Insecurity is related to vulnerability: the sensitivity of well-being to a changing environment and households' ability to respond to negative changes. Households have assets that may be drawn down in times of need and built up in better times to provide defences against shocks and stresses and to improve well-being. Assets of particular importance to urban households may be physical (housing, equipment for economic activities), human (labour power, skills, good health), financial (savings, credit), social (membership of formal or informal social organizations, which provide information, contacts and support) or political (channels of representation and influence) (Moser, 1998; Rakodi, 1999). Households which lack assets that they can mobilize in the face of hardship are more vulnerable to impoverishment. The assets available to households are determined in part by their characteristics and strategies. However, the potential for accessing and building up assets is strongly influenced by wider circumstances, including the operation of labour, land and housing markets; levels of crime and violence; arrangements for infrastructure and service provision; and the regulatory regimes governing urban activities, which may discriminate against self-help housing construction or informal-sector economic activities. Newer conceptualizations of household livelihoods emphasize the importance of structural factors in determining livelihood opportunities, especially power and politics, social relations and cultural identities, stressing that processes of social exclusion constrain the agency of individual households (Van Dijk, 2011; see also Devas *et al.*, 2004).

Access to labour market opportunities is a key element in the livelihood strategies of urban households. Wider economic circumstances, trade and industrial policies, availability of land and infrastructure, and the characteristics of local labour markets influence the opportunities available. Access to them depends on the resources available to and constraints on particular households: education and skills, dependency ratios, health status, relative location of affordable residential areas and employment centres, and access to public transport. Poor households with adults in work tend to depend on informal-sector activities, especially services, or on casual employment. They are, however, characterized by high unemployment rates amongst adults and also include households with limited labour resources: young households with children, especially single mothers; and those containing disabled, sick or elderly adults.

Much of the discussion so far has referred to households as though consumption is equally distributed amongst all their members, but there is extensive evidence to show that this is not the case: not only is access to resources and opportunities amongst the urban population as a whole differentiated by social characteristics such as gender and age, but also some individuals within households may be disadvantaged, including the economically inactive (some elderly people, women and children) and unpaid family workers (especially domestic servants).

In times of wider economic difficulties or in the face of household-level shocks, such as bereavement, illness or retrenchment, some households are better equipped than others to defend themselves against impoverishment. The key strategy seems to be diversification, so that households with multiple adults of working age, more than one source of income, an asset such as a house which can generate income, urban or rural land for food production, and urban or rural social networks that can be called upon for support, cope better with economic crisis (Rakodi, 1999). For some, impoverishment may be temporary, and improved well-being may follow national economic recovery or the development of alternative income sources. Others, however, may be forced to sell physical assets, move into inferior accommodation, send children out to work, reduce the quality and quantity of food consumed, postpone medical treatment, and/or withdraw from reciprocity arrangements, such as rotating savings and credit associations. The chronic poor are unable to take advantage of the opportunities offered in cities and become trapped in a vicious circle of poverty and deprivation.

Conclusion: Policy responses

The most powerful policies with a direct or indirect impact on urban poverty are national economic and social policies. Their effects are mediated through markets and through the activities of governments, which have varying capacity to adopt pro-poor policies, especially at the local level. The urban poor depend heavily on their assets of labour and human capital, tying them strongly into the money economy and labour markets. Access to housing and capital assets for informal economic activities is also important to their livelihoods. The role of social networks and the ways in which local political processes determine access by the poor to resources and assets are beginning to be better understood (Devas *et al.*, 2004). Increased understanding of poverty, vulnerability and social exclusion, and identification of constraints on the ability of individuals, households and communities to access key assets and services provide pointers to appropriate policy interventions.

References

Devas, N. with Amis, P., Beall, J., Grant, U., Mitlin, D., Nunan, F. and Rakodi, C. (2004) *Urban Governance, Voice and Poverty in the Developing World*, London: Earthscan.

Eastwood, R. and Lipton, M. (2000) Pro-poor growth and pro-growth poverty reduction: Meaning, evidence and policy implications, *Asian Development Review*, 18: 22–58.

Haddad, L., Ruel, M. T. and Garrett, J. (1999) Are urban poverty and undernutrition growing? Some newly assembled evidence, *World Development*, 27(11): 1891–904.

He, S., Wu, F., Webster, C. and Liu, Y. (2010) Poverty concentration and determinants in China's urban low-income neighborhoods and social groups, *International Journal of Urban and Regional Research*, 34(2): 328–49.

Jones, G. A. and Corbridge, S. (2010) The continuing debate about urban bias: The thesis, its critics, its influence and its implications for poverty-reduction strategies, *Progress in Development Studies*, 19(1): 1–18.

Lipton, M. (1977) *Why Poor People Stay Poor: Urban Bias in World Development*, London: Temple Smith.

Montgomery, M. R., Stren, R., Cohen, B. and Reed, H. E. (2004) *Cities Transformed: Demographic Change and its Implications in the Developing World*. London: Earthscan.

Moser, C. O. N. (1998) The asset vulnerability framework: Reassessing urban poverty reduction strategies, *World Development*, 26(1): 1–19.

Rakodi, C. (1999) A capital assets framework for analysing household livelihood strategies, *Development Policy Review*, 17(3): 315–42.

Rakodi, C. (2002) Economic development, urbanization and poverty. In Rakodi, C. with Lloyd-Jones, T. (eds) *Urban Livelihoods: A People-Centred Approach to Reducing Poverty*, London: Earthscan, pp. 23–34.

Ravallion, M. (2007) Urban poverty, *Finance and Development* 44(3): 15–17.

Ravallion, M., Chen, S. and Sangraula, P. (2007) *New Evidence on the Urbanization of Global Poverty*, Washington, DC: World Bank Policy Research Working Paper no. 4199, available online at http://econ.worldbank.org/docsearch

Satterthwaite, D. (2004) *The Under-Estimation of Urban Poverty in Low and Middle-Income Nations*, London: International Institute for Environment and Development, Poverty Reduction in Urban Areas Series Working Paper 14.

UN-Habitat (2006) *The State of the World's Cities 2006/7: The Millennium Development Goals and Urban Sustainability – 30 Years of Shaping the Habitat Agenda*, London: Earthscan.

UN-Habitat (2010) *State of the World's Cities 2010/11: Bridging the Urban Divide*, London: Earthscan.

Van Dijk, T. (2011) Livelihoods, capitals and livelihood trajectories: A more sociological conceptualisation, *Progress in Development Studies*, 11(2): 101–17.

Varshney, A. (1993) Introduction: Urban bias in perspective. In Varshney, A. (ed.) *Beyond Urban Bias*, London: Frank Cass, pp. 1–22.

Further reading

Relevant material published during the last ten years may be found in both special issues of journals and edited or multi-authored books: *International Planning Studies* 10(1) (2005); Rakodi, C. with Lloyd-Jones, T. (eds) *Urban Livelihoods: A People-Centred Approach to Reducing Poverty*, London: Earthscan; Devas *et al.* (2004); Montgomery *et al.* (2004); Martine, G., McGranahan, G., Montgomery, M. and Fernández-Castilla, R. (eds) *The New Global Frontier: Urbanization, Poverty and Environment in the 21st Century*, London: Earthscan. Extensive work has sought to improve the conceptualization and measurement of poverty, some of which refers explicitly to urban areas – see Addison, T., Hulme, D. and Kanbur, R. (eds) (2008) *Poverty Dynamics: Interdisciplinary Perspectives*, Oxford: Oxford University Press; and Stewart, F., Saith, R. and Harriss-White, B. (eds) (2007) *Defining Poverty in the Developing World*, Basingstoke: Palgrave Macmillan; Mitlin, D. and Satterthwaite, D. (2013) *Urban Poverty in the global South: Scale and Nature*, London: Routledge; Satterthwaite, D. and Mitlin, D. (2013) *Reducing Urban Poverty in the Global South*, London: Routledge. In addition UN-Habitat's alternating annual flagship reports – the Global Reports on Human Settlements and the State of the World Cities reports – are both readable and empirically rich.

Websites

UN-Habitat's website www.unhabitat.org contains a wide range of information on urban characteristics and programmes, as well as publications, some of them downloadable.

Reports on the Demographic and Health Surveys in many countries, most of which disaggregate rural and urban data, can be obtained at www.measuredhs.com/aboutsurveys/dhs/start.cfm

The World Bank Urban Development website covers a variety of topics, including health, providing services to the poor and strategies for local economic development. It contains links to data and reports, many of them downloadable: http://web.worldbank.org/WBSITE/EXTERNAL/TOPICS/EXTURBANDEVELOPMENT/0,,menuPK:337184~pagePK:149018~piPK:149093~theSitePK:337178,00.html

The International Labour Organization's Job Creation and Enterprise Development Department has programmes relevant to urban areas – www.ilo.org/empent/areas/ – and downloadable reports on urban employment and poverty. Data on informal-sector employment and informal-sector survey methodologies can be obtained in machine readable form through a clickable link – www.ilo.org/public/english/bureau/stat/info/dbases.htm

The International Institute for Environment and Development has a Human Settlements Programme which includes work on urban poverty reduction. Both reports and back issues of the journal *Environment and Urbanization* can be downloaded at www.iied.org/group/human-settlements

5.6
Housing the urban poor

Alan Gilbert

Millions of families in the cities in Africa, Asia and Latin America live in adequate accommodation and a few even live in luxury. Unfortunately, most of the poor do not. They live in homes built of flimsy materials, with an irregular supply of electricity and water, and lacking sanitation. Millions of others live in more solid and serviced accommodation but in overcrowded conditions. Millions more would claim to have a housing problem. They live in houses that do not match their hopes and needs: they have difficulty paying their rent or mortgage, they have a long journey to work, their home is too small, they wish to own a house rather than rent. The housing problem in the South, therefore, is enormous (UN-Habitat, 2003a; Davis, 2006).

Of course, poor housing is not just a matter of physical standards. As John Turner (1968) demonstrated long ago, there is little point providing a poor family with a fully serviced, three-bedroom house if the family cannot afford the rent or mortgage payment. The most suitable shelter for such a family may be something rather flimsy. In the short term, a poor family can survive in inadequate shelter whereas it cannot survive without food or water. Adequate accommodation is that which fits the circumstances of the family rather than being determined on purely physical grounds.

As such, the housing problem is not something that can be solved by architects and planners alone. It is a multifaceted problem that can only be helped through raising living standards, improving employment opportunities and applying sensible urban regulations. Unfortunately, policymakers and particularly politicians often push simplistic solutions, something that rarely resolve complex problems.

Variations in the nature of the housing problem

The housing problem differs considerably between countries both in form and severity. In general, housing conditions are linked to the level of economic development: African and Indian shelter standards are far below those of Latin America. Homelessness, however, does not follow the same pattern. The United States has relatively more homeless people than most countries in the South (UN-Habitat, 2003a).

Housing conditions also vary greatly even within the same country. The quality of shelter in the countryside is almost always far worse than that in the cities. Most rural families live in shacks and relatively few have access to infrastructure and services. Of course, this does not mean that millions of city dwellers live in adequate shelter, and in much of sub-Saharan Africa and the Indian subcontinent a majority live in what can only be regarded as appalling conditions. However, cities are diverse and the nature of their housing problem varies according to their size, the nature of the land market, the state of the economy, the ability of governments to provide services, and the local climate and topography. One city may have poor infrastructure and services although low density settlements offer plenty of space. In another city, families may suffer from severe overcrowding but have access to more services. Cities also vary greatly in terms of housing tenure even within the same country. In Ecuador, 64 per cent of Guayaquil's population own their homes whereas in Quito, where informal land access is much more difficult, only 48 per cent are homeowners.

Housing policy up to the 1980s

Over the years, the housing policy has changed. During the first half of the twentieth century, the authorities concentrated on improving hygiene in a bid to stop outbreaks of infections diseases like cholera, measles and the plague. Eviction and demolition of 'unhealthy' settlements was common and the central areas of some cities were almost completely rebuilt. A parallel approach was to apply inappropriate forms of rent control to prevent landlords exploiting the poor. While this represented a sensible political strategy at a time when the vast majority of urban families rented a home, it contributed to a decline in investment in rental housing.

Gradually governments supplemented these policies by building housing for the poor. Unfortunately, this strategy was undermined by the limited resources that governments dedicated to the task compared with the enormous numbers of families demanding a home. Too often, the outcome was long housing queues, corrupt allocation systems and the creation of poor quality shelter – the consequence of attempts to reduce the cost of construction.

Recognising the insufficiency of these approaches, most governments resorted to a more covert strategy. Informal self-help suburbs spread rapidly as politicians and officials encouraged land invasions or turned a blind eye to illegal subdivisions. In Brazil, self-help housing accounted for approximately three-quarters of all housing production between 1964 and 1986 (UN-Habitat, 2011a: 8).

Housing policy since the 1980s

It was clear that current shelter policies were not working but a strong stimulus was needed to persuade governments to change direction. The debt crisis that began in Latin America in 1982 and hit other countries like the Philippines, South Korea and Thailand was the spark that brought a significant change in approach.

The rules of the Washington Consensus were applied in many countries after Chile had seemingly demonstrated that neoliberal approaches could bring economic growth and reduce poverty. The IMF and the World Bank pushed a policy mix of fiscal discipline, public expenditure priorities, tax reform, financial liberalisation, exchange rates, trade liberalisation, foreign direct investment, deregulation and property rights. New shelter policies were devised to fit the new regime; one of the key motivations being to reduce government spending on housing. As a result, many public housing agencies were closed or their remit reduced and more responsibility was given to the private sector.

By the early 1990s, it was becoming clear that the new free-market approach had aggravated the problems of poverty and inequality. In response, the development banks were increasingly pushing poverty alleviation programmes and initiatives to improve urban governance. Shelter policies were also changing and, with the establishment of the Cities Alliance, UNDP's embrace of the Millennium Development Goals and the strengthening of UN-Habitat the enabling shelter approach was developed (Garau *et al.*, 2005; Buckley and Kalarickal, 2005; Roberts and Kanaley, 2006).

The new approach was based on nine principles: political endorsement and support; participation; needs driven; people-centred; pro-poor; results-oriented; comprehensive; based on partnership; and sustainability (UN-Habitat, 2011c: 47). It accepted that better shelter conditions could only be achieved as part of a more comprehensive urban programme. There was little point in dealing with the poor if the middle class lack adequate housing because that would merely cause down-grading. Sector-wide investment was required and that meant improving access to housing finance, providing services and infrastructure and organising urban growth. The public sector should limit itself to a facilitating role. Given the limits of government resources and the extent of the shelter problem, the main responsibility for solving housing problems had to lie with the private sector and with civil society. This required a new range of measures: improving mortgage finance, offering tax relief on interest payments, providing subsidies for the poor and selling off the public housing stock.

But something also had to be done about the existing stock of 'slums'. Gradually, most governments began to recognise that in-situ housing improvement was far superior to demolishing self-help housing and shifting poor families to formally constructed homes. Upgrading settlements would maintain existing social and economic networks, was relatively cheap and any spending would reach the poor. The perceived success of projects like *Favela-Bairro* in Rio de Janeiro seems to have convinced officials that such an approach should be replicated worldwide. Insofar as there were reservations about the approach it concerned how best to finance it.

There can be little doubt that the new approach has been more effective than earlier policies. Nevertheless, it contains a number of serious flaws. First, it relies too much on the delivery of property titles. Ideologues like Hernando de Soto (2000) argue that giving families legal tenure will transform their lives, an idea that was adopted by many governments and the development banks. 'Secure property rights facilitate economic transactions, ensure efficient and sustainable resource use, allow for the evolution of effective credit markets, improve business climate and investment opportunities, and ensure economic accountability and transparency' (Garau *et al.*, 2005: 64). In practice, millions of titles have been given out, a policy that proved popular with the recipients and with the politicians who gained votes from distributing them. However, there is little evidence that it has had much effect on improving shelter conditions because the poor are simply too poor to invest much in their housing (Gilbert, 2002).

Second, there is far too much emphasis on encouraging the poor to take on loans – something that the sub-prime crisis in the United States shows is potentially dangerous. Banks were encouraged to lend to poorer families and new micro-credit facilities developed to deliver mortgages to them. However, in practice few families in Africa, Asia and Latin America could prove that they earned enough for the banks to issue them with a loan.

Third, the new approach to subsidies involved targeting the deserving poor and offering them funds to contibute to the cost of buying a home built by private developers. While this was not an unsuccessful approach it often failed to reach the very poor. And, where it did, very poor families often failed to make full use of the subsidy because their budgets were too limited to pay their taxes and service charges. Many also found that getting to work from the distant estates that had been developed for social housing was too expensive. Many rented or sold their homes or gave up their subsidised houses and fled (Gilbert, 2004; Huchzermeyer, 2003).

Fourth, the enabling approach ignored the needs of a major group – the billion or so people who rent or share accommodation in the cities of the South (UN-Habitat, 2011b; Gilbert, 2008). Most governments are obsessed with the dream of creating nations of home owners.

While this constitutes excellent propaganda insofar as most families aspire to home ownership, there will always be a large group of people who need to rent: those planning short stays in the city, those who like living close to the city centre, those without family responsibilities and those who do not want the complications that home ownership brings. And, even if rates of homeownership are growing rapidly in many cities, tenants are increasing in number virtually everywhere. This is particularly true in large cities where the cost of acquiring land and housing is very high and in those cities where poor transport facilities force people to live close to their jobs. Policymakers in poor countries seem not to have noticed that the majority of inhabitants in extremely rich cities like Berlin, Los Angeles, New York and Zurich are tenants. Perhaps they should develop a policy for rental housing to cater for that sector of demand?

Finally, the new approach assumed that the quality of governance would improve. A mixture of decentralisation and private sector participation would improve housing delivery and bring faster delivery of services. But, while there are examples of greatly improved governance in cities like Bogotá and Curitiba, too many national and local authorities have not been up to the task. Cities have grown faster than servicing capacity. Too many governments have been inefficient and some have proved to be corrupt. Most rural areas are neglected.

Conclusion

There are no easy solutions to the shelter problems of the Third World because inadequate housing is merely one manifestation of generalized poverty. At the same time, sensible policies can help to reduce shelter problems. Governments should try to reduce the number of people living at densities of more than 1.5 persons per room, increase access to electricity and potable water, improve sanitary facilities, prevent families moving into areas that are physically unsafe, and encourage households to improve the quality of their accommodation (Buckley and Kalarickal, 2005). We now know enough to design sensible shelter policies. Unfortunately, it has rarely been the lack of knowledge that has stopped us improving living conditions. More often the critical barrier has been the inability of governments to introduce the necessary policies. Their reluctance to adopt new legislation or to modify inappropriate policies have been determined by their perception of political reality. At root, therefore, the Third World housing problem is about politics and economics. Without political reforms and sustained economic growth, housing conditions will not get much better. Appropriate action would at least help a substantial minority of Third World citizens to improve their shelter situation.

Bibliography

Buckley, R. M. and Kalarickal, J. (2005) 'Housing policy in developing countries: Conjectures and refutations', *The World Bank Research Observer* 20: 233–257.

Davis, M. (2006) *Planet of slums*, New York: Verso.

De Soto, H. (2000) *The mystery of capital*, New York: Basic Books.

Garau, P., Sclar, E. D. and Carolini, G. Y. (2005) *A home in the city: UN Millennium Project: Task Force on improving the lives of slum dwellers*, London: Earthscan.

Gilbert, A. G. (1998) *The Latin American city*, London: Latin America Bureau and New York: Monthly Review Press.

Gilbert, A. G. (2002) 'On the mystery of capital and the myths of Hernando de Soto: What difference does legal title make?', *International Development Planning Review* 24: 1–20.

Gilbert, A. G. (2004) 'Helping the poor through housing subsidies: Lessons from Chile, Colombia and South Africa', *Habitat International* 28: 13–40.

Gilbert, A. G. (2008) 'Slums, tenants and homeownership: On blindness to the obvious', *International Development Planning Review* 30, i–x.
Huchzermeyer, M. (2003) 'A legacy of control? The capital subsidy and informal settlement intervention in South Africa', *International Journal of Urban and Regional Research* 27: 591–612.
Roberts, B. and Kanaley, T. (eds) (2006) *Urbanization and sustainability in Asia: Case studies of good practice*, Manilla: Asian Development Bank.
Turner, J. F. C. (1968) 'The squatter settlement: An architecture that works', *Architectural Design* 38: 357–360.
UN-Habitat (2003a) *The challenge of slums: Global report on human settlements, 2003*, Nairobi: UN-Habitat.
UN-Habitat (2003b) *Rental housing: An essential option for the urban poor in developing countries*, Nairobi: UN-Habitat.
UN-Habitat (2008) *State of the world's cities, 2008/9 – harmonious cities*, Nairobi: UN-Habitat.
UN-Habitat (2011a) *Affordable land and housing in Latin America and the Caribbean*, Nairobi: UN-Habitat.
UN-Habitat (2011b) *A policy guide to rental housing in developing countries*, Nairobi: UN-Habitat.
UN-Habitat (2011c) *Enabling shelter strategies: Design and implementation guide for policymakers*, Nairobi: UN-Habitat.

5.7

Urbanization and environment in low- and middle-income nations

David Satterthwaite

Introduction

As described in Chapter 5.1, most of the world's urban population and most of its large cities are now in low- and middle-income nations in Latin America, Asia and Africa. The quality of environmental management in these urban centres (and of the governance structures within which this management occurs) has very significant implications for development. Despite great diversity in the size and economic base of urban centres, they all share certain characteristics. All combine concentrations of human populations and a range of economic activities. Their environment is much influenced by the scale and nature of these activities and in the resources they use and the pollution and wastes they generate. Urban environments are also much influenced by the quality and extent of provision for supplying fresh water and for collecting and disposing of solid and liquid wastes from households, enterprises and institutions.

In most countries in sub-Saharan Africa and many in Asia, North Africa and Latin America, the expansion in the urban population has occurred without the needed expansion in the services and facilities essential to a healthy urban environment, especially provision for water, sanitation, drainage and solid waste management. It has often occurred with little or no effective pollution control and very inadequate provision for health care. The deficits in provision for basic infrastructure are particularly serious in many sub-Saharan African cities where most of the population lack water piped to their home, sewers and drains and electricity and where it is common for 40–70 per cent of the population to live in informal settlements.

As well as these primarily local problems, there is the global problem of human-induced climate change and a high proportion of greenhouse gas emissions arise from the high-consumption lifestyles of middle- and upper-income urban dwellers. Businesses and populations in high-income nations are currently the greatest global threat because they contributed most to greenhouse gases already in the atmosphere and their populations remain the largest

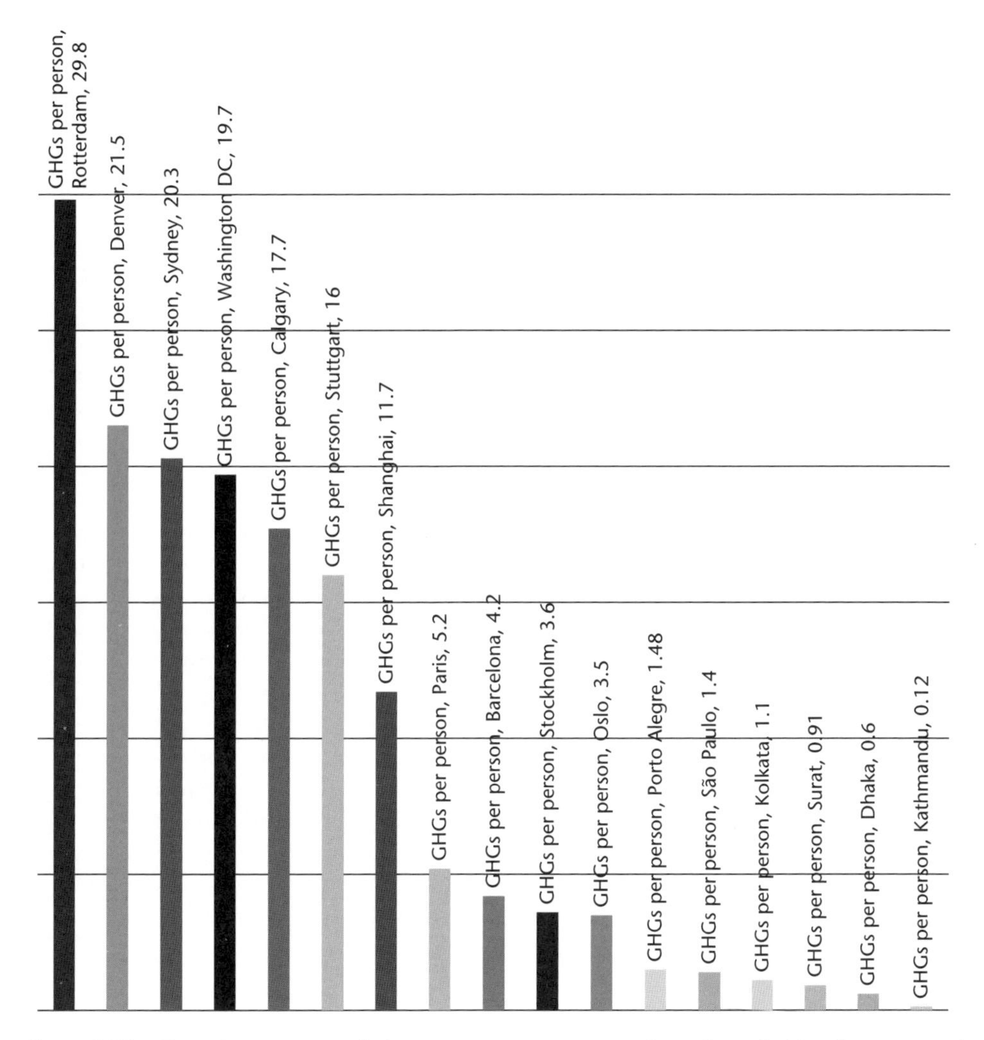

Figure 5.7.1 Greenhouse gas emissions per person per year for selected cities (in tonnes of CO_2 equivalent)

Source: Hoornweg *et al* 2011. The authors caution against making such direct comparisons between cities because of differences in the methodologies used to calculate and assign emissions and the different years for which data were available.

emitters per person. Emissions per person in cities in North America and Australia can be 25–50 or more times higher than those in cities in low-income nations (see Figure 5.7.1).

But urban centres in low- and middle-income nations in Africa, Asia and Latin America are likely to house most of the growth in the world's population, production and greenhouse gas emissions over the next ten to twenty years (United Nations 2012). In addition, they have a large and growing proportion of the world's population most at risk from climate change-related impacts including increased severity and frequency of extreme weather events and sea-level rise. Thus, what is done in their urban centres will influence strongly whether the current trend of increasing risks for the whole planet from climate change will be reduced or stopped.

Urban opportunities and disadvantages

The fact that large urban centres have high concentrations of people, enterprises and motor vehicles – and their wastes – can make them very hazardous places in which to live and work. With inadequate or no environmental management, environmental hazards become the main causes of ill-health, injury and premature death. The urban poor face the greatest risks as their homes and neighbourhoods generally have the least adequate provision for water supplies, sanitation, drainage, garbage collection and health care. Within the tenements and informal settlements where much of the urban population live, it is common for a fifth of all children to die before the age of five; also for a high proportion of children to suffer from malnutrition (Mitlin and Satterthwaite 2012).

But urban centres also provide many environmental opportunities. High densities and large population concentrations lower the costs per household and per enterprise for the provision of most kinds of infrastructure and most services. The concentration of industries should reduce the unit cost of making regular checks on plant and equipment safety, as well as on occupational health and safety, pollution control and the management of hazardous wastes. There are also economies of scale or proximity for reducing risks from most disasters. There is generally a greater capacity among city dwellers to help pay for such measures, if they are made aware of the risks and all efforts are made to keep down costs. With good environmental management, cities can achieve life expectancies that compare favourably with those in Europe and North America – as demonstrated by well-governed cities such as Porto Alegre in Brazil (Menegat 2002).

Cities also have many potential advantages for reducing resource use and waste. For instance, the close proximity of so many water consumers gives greater scope for recycling or directly reusing waste waters. In regard to transport, cities have great potential for limiting the use of motor vehicles (and thus also the fossil fuels, air pollution and greenhouse gases that their use implies). This might sound contradictory, since most large cities have problems with congestion and motor-vehicle-generated air pollution. But compact cities allow many more trips to be made through walking or bicycling. Cities make a greater use of public transport and a high-quality service more feasible (Kenworthy 2006).

Good environmental management can also limit the tendency for cities to transfer environmental costs to rural areas, for example, through:

- enforcing pollution control to protect water quality in nearby water bodies, safeguarding those who draw water from them, and also fisheries;
- an emphasis on 'waste reduction, reuse, recycle' to reduce the volume of wastes that are disposed of in the area around cities;

- comprehensive storm and surface drains, and garbage collection systems which reduce non-point sources of water pollution.

More resource-conserving, waste-minimizing cities can also contribute much to addressing global environmental problems, including de-linking a high quality of life from increased resource use, waste generation and greenhouse gas emissions (Satterthwaite 2011). Figure 5.7.1 shows how cities such as Oslo, Stockholm and Barcelona have relatively low average emissions per person yet these are also cities with a high quality of life. However, realizing these potential advantages of urban concentrations requires competent city governments that are accountable to their populations and guided by national urban policies that ensure they also meet their regional and global environmental responsibilities.

The importance of local governance

All urban areas require some form of government to ensure adequate quality 'environments' for their inhabitants. Governments must also act to protect key resources and ecosystems in and around cities – for instance, to regulate land use while ensuring sufficient supplies can be developed for housing, to protect watersheds, to control pollution and the disposal of wastes, and to ensure adequate provision of environmental infrastructure and services. Ensuring good quality environments becomes increasingly complex the larger the population (and the scale and range of their daily movements) and the more industrial the production base.

In all cities, environmental management is an intensely political task, as different interests compete for the most advantageous locations, for the ownership or use of resources and waste sinks, and for publicly provided infrastructure and services. The most serious environmental problems in urban areas are largely the result of inadequate governance and inadequate investment. Two concerns are particularly pressing:

- reducing environmental hazards;
- reducing the loss of natural resources and the damage or disruption of ecosystems.

The emphasis on 'governance' rather than 'government' is because good environmental management needs city authorities and politicians to be accountable to their citizens and to work with them. The term governance is understood to include not only the political and administrative institutions of government, but also the relationships between government and civil society (McCarney 1996). Much of the innovation in urban environmental policies in the last two decades has been driven or much influenced by community-based organizations and local NGOs and has taken place in nations where mayors and city governments are elected (see, for example, Velasquez 1998; Cabannes 2004; Almansi 2009).

The scale and range of environmental problems

Infectious and parasitic diseases

Many of the most serious diseases in cities are 'environmental' because they are transmitted through disease-causing agents (pathogens) in the air, water, soil or food, or through insect or animal disease vectors. Many diseases and disease vectors (for instance, the mosquitoes that transmit malaria or dengue fever) thrive when provision for water, sanitation, drainage, garbage collection and healthcare is inadequate.

Around half the urban population in low- and middle-income nations is suffering from one or more of the diseases associated with inadequate provision for water and sanitation (WHO 1999). Around half the urban population in Africa and Asia lack provision for water and sanitation to a standard that is healthy and convenient. For Latin America and the Caribbean, more than a quarter lack such provision (UN-Habitat 2003), although deficits in provision have significantly come down in many cities in this region. But in Africa and Asia, tens of millions of urban dwellers have no toilet in their home. Where there are public toilets, these are often poorly maintained, costly and difficult to access because of long queues. Consequently, very large numbers of urban dwellers have to defecate in the open or into plastic bags or waste materials. Good provision for water (with supplies piped into homes) and for toilets can bring great benefits in terms of improved health, reduced expenditures (on water vendors, public toilets and treatment from diseases) and much reduced burdens in time and physical effort (especially for those who collect and carry water from standpipes or other sources far from their shelters). In the early 1990s, many international agencies supported private-sector provision for water and sanitation in the hope that this would increase efficiency, expand provision and bring in new capital but the results of this have been very disappointing (Budds and McGranahan 2003; UN-Habitat 2003).

Airborne infections remain among the world's leading causes of premature death. For many, their transmission is aided by the overcrowding and inadequate ventilation that is common in the tenements, boarding houses or small shacks in which most low-income urban dwellers live. While improving housing and other environmental conditions can reduce their incidence (and by reducing other diseases also strengthen people's defences against these) medical interventions such as immunization or good quality accessible health care that provides rapid treatment are more important for reducing their health impact.

Chemical and physical hazards

The scale and severity of many chemical and physical hazards increases rapidly with urbanization and industrialization. While controlling most infectious diseases centres on provision of infrastructure and services (whether through public, private, NGO or community provision), reducing chemical and physical hazards is largely achieved by regulation – for instance, of enterprises, builders and motor-vehicle users.

A great range of chemical pollutants affect human health. Controlling exposure to chemicals in the workplace ('occupational exposure') is particularly important, with action needed from large factories down to small 'backstreet' workshops. In most cities, there is an urgent need for measures to promote healthy and safe working practices and to penalize employers who contravene them. In many urban areas, domestic indoor air pollution from open fires or poorly vented stoves that use coal or biomass fuels has serious health impacts. Lower-income households are affected most as people tend to shift to cleaner, safer, more convenient fuels when incomes rise.

There is also a need for more effective control of outdoor (ambient) air pollution, most of which is from industries and motor vehicles. World Health Organisation estimates for 2008 suggest that over a million premature deaths would be avoided if its air quality guideline values were implemented. Urban air pollution problems are particularly pressing in many Indian and Chinese cities.

Accidents in the home are often among the most serious causes of injury and premature death, where much of the population live in overcrowded accommodation made from temporary materials and use open fires or unsafe stoves and candles or kerosene lights. In 2005, 700 million urban dwellers still lacked access to clean fuels and 279 million lacked electricity (Legros *et al.* 2009).

Traffic management which protects pedestrians and minimizes the risk of motor-vehicle accidents is also important. Motor-vehicle accidents have become an increasingly significant contributor to premature deaths and injuries in many cities. Of the 1.3 million people who die each year from road traffic accidents, over 90 per cent occur in low- and middle-income nations.

Achieving a high-quality city environment

Attention should be given not only to reducing environmental hazards but also to ensuring provision or protection of those facilities that make urban environments more pleasant, safe and valued by their inhabitants, including parks, public squares/plazas and provision for children's play and for sport/recreation. This should be integrated with a concern to protect each city's natural landscapes with important ecological and aesthetic value, for instance wetland areas, river banks or coasts. This also has particular importance for cities where high temperatures are common (and likely to be exacerbated by climate change).

Managing wastes

Municipal agencies are generally responsible for providing waste collection services, although this is often contracted out to private businesses. Many urban governments lack the technical knowledge, institutional competence and funding base to ensure these responsibilities are met. In many urban centres, more than a third of the solid wastes generated are not collected and it is usually the low-income districts that have the least adequate collection service (or no collection service). This means that wastes accumulate on open spaces and streets, clogging drains and attracting disease vectors and pests (rats, mosquitoes, flies).

Many industrial and some institutional wastes (for instance, from hospitals) are categorized as 'hazardous' because of the special care needed to dispose of them – to ensure they are isolated from contact with humans and the natural environment. In most urban centres in low- and middle-income nations, governments do little to monitor their production, collection, treatment and disposal. Since managing hazardous wastes properly is expensive, businesses have large incentives to avoid meeting official regulations on how these wastes should be managed and generally little risk of prosecution.

Reducing the impact of disasters

Disasters are considered to be exceptional events that suddenly result in large numbers of people being killed or injured, or large economic losses. As such, they are distinguished from the environmental hazards discussed above. This distinction has its limitations, since far more urban dwellers die of easily prevented illnesses arising from environmental hazards in their food, water or air than from 'disasters', yet the death toll from disasters gets more media attention. However, recent studies have shown that in many cities, 'small' disasters that do not get included in national or international disaster records contribute much to premature death, injury, damage to homes, schools and health care centres and loss of assets especially for low-income groups (United Nations 2011).

Cyclones/high winds/storms have probably caused more deaths in urban areas than other 'natural' disasters in recent decades. Earthquakes (including those that cause tsunamis) have caused many of the biggest urban disasters. Flood disasters affect many more people than cyclones and earthquakes, but generally kill fewer people. Landslides, fires, epidemics and industrial accidents are among the other urban disasters that need attention. These may also take place as secondary disasters, after a flood, storm or earthquake.

Global warming will increase the frequency and severity of extreme weather events that can cause disasters in most urban areas. For instance, the rise in sea level will increase the risk of flooding for many port and tourist cities and low- and middle-income nations have a higher proportion of their urban population in low-elevation coastal zones than high-income nations (McGranahan *et al.* 2007). Sea-level rise will also disrupt sewers and drains and may bring seawater intrusion into freshwater aquifers. Changes in rainfall regimes caused by climate change are likely to reduce the availability of freshwater resources in many nations and cities or bring increased risk of floods as the intensity of storms increases. More frequent heat waves and higher temperatures are likely in many cities.

Increasingly, urban authorities recognize the need to integrate 'disaster prevention' within 'environmental hazard prevention'. Even where disasters have natural triggers that cannot be prevented, their impact can generally be greatly reduced by understanding who within the city population is vulnerable and acting to reduce this vulnerability before the disaster occurs. There are also important overlaps between 'the culture of prevention' for everyday hazards and for disasters. Now, it is important to integrate a concern for climate change adaptation into risk reduction and to build resilience to present and likely future climate change impacts into urban infrastructure, institutions and land use management.

Developing more sustainable interactions with nature

Cities transform natural landscapes both within and around them:

- The expansion of cities reshapes land surfaces and water flows. In the absence of effective land-use management, urbanization can have serious ecological impacts such as soil erosion, deforestation, and the loss of agricultural land and of sites that provide invaluable ecosystem services (Roberts *et al.* 2012).
- The 'export' of solid, liquid and airborne pollutants and wastes often brings serious environmental impacts to regions around cities, including damage to fisheries by untreated liquid wastes, land and groundwater pollution from inadequately designed and managed solid waste dumps and, for many of the larger and more industrial cities, acid rain.
- Freshwater resources are being depleted. Many cities have outgrown the capacity of their locality to provide fresh water or have overused or mismanaged local sources so these are no longer usable. Increasingly distant and costly water sources have to be used, often to the detriment of the regions from which these are drawn.

However, a focus only on such regional damages can obscure the positive rural–urban links. Urban consumers and enterprises provide the main market for rural produce (especially higher value-added products) while rural inhabitants and enterprises draw on urban enterprises for goods and services. A substantial proportion of rural households in low- and middle-income nations draw some of their income from working in urban areas or from remittances from family members working in urban areas. These enhance rural incomes and may also provide the basis for rural investments in better environmental management.

Global impacts

Over the last few decades, the demands concentrated in larger and wealthier cities for food, fuel and raw materials have increasingly been met by imports from distant ecosystems. For these cities, this avoids environmental costs as all energy, water and pollution intensive goods are imported, with their high environmental costs generated at their point of production. This makes it easier to maintain high environmental standards within and around wealthy cities, including preserving natural landscapes, but this is transferring costs to people and ecosystems in other regions or countries. Other cost transfers are pushed into the future. For instance, air pollution may have been cut in some wealthy cities but emissions of carbon dioxide (the main greenhouse gas) remain high in all wealthy cities and may continue to rise. This expanding ecological footprint of cities (Rees 1992) is also transferring costs to the future through the human and ecological costs that human-induced global warming will bring. As yet, there is no sign of the needed global agreements to cut greenhouse gas emissions so that dangerous climate change can be avoided. A large part of avoiding dangerous climate change depends on cities (and wealthy city inhabitants) that act to delink a high quality of life from high consumption.

All cities have to develop a more sustainable interaction with the ecocycles on whose continued functioning we all depend. But to promote a sustainable development-oriented urban policy needs a coherent and supportive national policy. It is difficult for city governments to act on this because they are accountable to the populations living within their boundaries, not to those living in distant ecosystems on whose productivity and carbon sinks the city producers or consumers may draw. It is also difficult for city authorities to take account of the needs and rights of future generations and of other species without a supportive national sustainable development framework.

There is also the issue of the profound unfairness globally in the likely impacts of global warming because the people, cities and nations most at risk are not those that are responsible for most greenhouse gas emissions. The very survival of some small-island nations and low-income nations is in doubt as much of their land area is at risk from sea-level rise (McGranahan, Balk and Anderson 2007), yet their contributions to global greenhouse emissions have been very small. There are also tens of millions of low-income urban dwellers in Asia and Africa whose homes and livelihoods are at risk from sea-level rise and storms yet they have contributed very little to greenhouse gas emissions. One wonders what will happen to global relations as the number and scale of urban disasters associated with climate change increases in low-income nations, within a global recognition that this is driven by the high-consumption lifestyles of people, most of whom live in high-income nations.

References

This article draws on Hardoy, Jorge E., Mitlin, Diana and Satterthwaite, David (2001), *Environmental Problems in an Urbanizing World*, London: Earthscan Publications; and on Satterthwaite, David (2011), 'How urban societies can adapt to resource shortage and climate change', *Philosophical Transactions of the Royal Society A*, 369, 1762–1783 that can be downloaded at no charge at http://pubs.iied.org/pdfs/G03104.pdf

Almansi, Florencia (2009), 'Rosario's development: Interview with Miguel Lifschitz, mayor of Rosario, Argentina', *Environment and Urbanization*, 21, 1, 19–35.

Budds, Jessica and McGranahan, Gordon (2003), 'Are the debates on water privatization missing the point? Experiences from Africa, Asia and Latin America', *Environment and Urbanization*, 15, 2, 87–114.

Cabannes, Yves (2004), 'Participatory budgeting: A significant contribution to participatory democracy', *Environment and Urbanization*, 16, 1, 27–46.

Kenworthy, Jeffrey R. (2006), 'The eco-city: Ten key transport and planning dimensions for sustainable city development', *Environment and Urbanization*, 18, 1, 67–86.

Legros, G., Havet, I., Bruce, N. and Bonjour, S. (2009), *The Energy Access Situation in Developing Countries: A Review Focusing on the Least Developed Countries and sub-Saharan Africa*, New York: UNDP.

McCarney, Patricia L. (ed.) (1996), *Cities and Governance: New Directions in Latin America, Asia and Africa*, Toronto: Centre for Urban and Community Studies, University of Toronto.

McGranahan, Gordon, Balk, Deborah and Anderson, Bridget (2007), 'Populations at risk: The proportion of the world's population living on low-elevation coastal zones', *Environment and Urbanization*, 19, 1.

Menegat, Rualdo (2002), 'Participatory democracy and sustainable development: Integrated urban environmental management in Porto Alegre, Brazil', *Environment and Urbanization*, 14, 2, 181–206.

Mitlin, Diana and Satterthwaite, David (2012), *Urban Poverty in the global South: Scale and Nature*, London: Routledge.

Rees, William E. (1992), 'Ecological footprints and appropriated carrying capacity', *Environment and Urbanization*, 4, 2, October, 121–130.

Roberts, Debra, Boon, Richard, Diederichs, Nicci, Douwes, Errol, Govender, Natasha, McInnes, Alistair, McLean, Cameron, O'Donoghue, Sean and Spires, Meggan (2012), 'Exploring ecosystem-based climate change adaptation in Durban, South Africa: "Learning-by-doing"', *Environment and Urbanization*, 24, 1.

UN-Habitat (2003), *Water and Sanitation in the World's Cities: Local Action for Global Goals*, London: Earthscan Publications.

United Nations (2011), *Revealing Risk, Redefining Development: The 2011 Global Assessment Report on Disaster Risk Reduction*, Geneva: United Nations International Strategy for Disaster Reduction.

United Nations (2012), *World Urbanisation Prospects: The 2011 Revision*, CD-ROM Edition, New York: Department of Economic and Social Affairs, Population Division.

Velasquez, Luz Stella (1998), 'Agenda 21: A form of joint environmental management in Manizales, Colombia', *Environment and Urbanization*, 10, 2, 9–36.

WHO (1999), 'Creating healthy cities in the 21st Century', in David Satterthwaite (ed.), *The Earthscan Reader on Sustainable Cities*, London: Earthscan Publications, pp. 137–172.

Further reading

Hardoy, Mitlin and Satterthwaite 2001 and Satterthwaite 2011 see above. There are also dozens of relevant papers in *Environment and Urbanization* and all but the papers published in the last two years can be accessed online at no charge at http://eau.sagepub.com/. Since 2007, almost all issues have had two or three papers on cities and climate change. The two issues in 2006 were on ecological urbanization; the first issue in 2011 (23, 1) was on health and the city.

Hoornweg, Daniel, Sugar, Lorraine and Trejos Gomez, Claudia Lorena (2011), 'Cities and greenhouse gas emissions: Moving forward', *Environment and Urbanization*, 23, 1, 207–227.

McGranahan, G., Jacobi, P., Songsore, J., Surjadi, C. and Kjellén, M. (2001) *Citizens at Risk: From Urban Sanitation to Sustainable Cities*, London: Earthscan Publications. For discussions on the transfer of cities' environmental costs.

Satterthwaite, David (ed.) (1999), *The Earthscan Reader on Sustainable Cities*, London: Earthscan Publications. For a discussion of sustainable development and cities.

www.who.int/topics/en/ – For data on health burdens and air pollution, see an index of relevant topics which also have fact sheets.

5.8

Transport and urban development

Eduardo Alcantara Vasconcellos

In urban environments, transport is vital for social and economic life and may be provided by several means. The specific modes used in a city are highly correlated to its size, to the quality and cost of available modes and to the economic and social characteristics of its inhabitants. As a general rule, in smaller cities walking and cycling are the dominant modes, while in larger cities, motorized means become essential, be they public (bus, train, metro) or private (automobiles and motorcycles). As transport implies the consumption of natural resources and the generation of some negative impacts such as accidents, pollution and congestion, the use of any particular mode and the overall balance of modes in a particular city have to be analyzed carefully.

Transport modal choices

When we observe a large number of societies, encompassing all income and wealth variations, we find people using a large variety of transport means, from walking to motorized private means such as the automobile. The transport modes people choose to use are constrained by several social, cultural and economic factors. Family or individual income is the most important factor, determining the possibility of paying to have access to motorized means (including the bus). Age also influences mobility because it involves mental or physical limits to the use of roads as a pedestrian and to drive vehicles, as in the case of children, youngsters and elderly people. Gender poses constraints that arise from the division of tasks among household members, which vary according to different societies. Finally, cultural or religious beliefs may limit access to some sort of vehicles or place incentives on using others.

Mobility – measured as the number of trips per person per day – generally increases as income increases, is higher among males and for those between 10 and 40 years old, who are either studying or working. Overall, the average individual mobility rates (trips/day) vary from 1 (for very poor people) to 4 (wealthy people). A parallel phenomenon is that of increasing use of motorized means with increasing incomes, either public or private, depending on the level of income and the availability of transport means. Figure 5.8.1 summarizes data from São Paulo in 2007. It may be seen that as income increases, walking decreases and travelling by public transport also decreases (for the higher income levels). In parallel, the use of private means increases across all income levels. The pattern revealed by Figure 5.8.1 may change according to the specific characteristics of every society and of transport alternatives. For instance, cycling is very important in Asia and in wealthy European cities and may replace trips made on public or motorized private modes; more importantly, the high quality of public transport may 'soften' the decrease in its use as income increases, as happens in

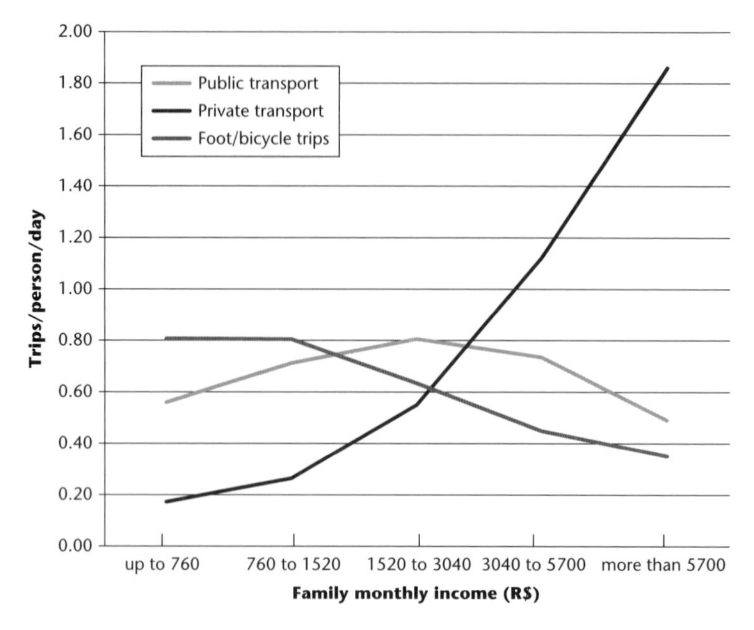

Figure 5.8.1 Income and modal split, São Paulo, 2007

Source: CMSP, 2008

Europe. The same effect may occur if the costs (fuel, parking) of using automobiles are high (such as in Europe and wealthy Asian cities).

Therefore, although the increase in the use of private motorized means is a worldwide phenomenon, the final pattern of modal share will be directly influenced by individual characteristics and by transport policies, concerning cost, availability and convenience of each transport mode.

Characteristics of various transport modes

Accessibility

The first important outcome of a transport system is the accessibility it promotes so people may reach desired destinations. The essential difference in accessibility is that provided by non-motorized and motorized transport means. On the one hand, both walking and cycling are universal transport means. However, they have natural limits, related to the time and distance a person may cover with them. On the other hand, the use of any form of motorized transport overcomes the distance barrier, however, it dramatically increases the consumption of money, energy and road space, and generates or increases negative externalities such as air pollution and accidents (Vasconcellos, 2001).

The overall accessibility provided by any transport means depends on several components. The first is made of walking, waiting and transferring times while trying to use a vehicle. Another essential component is actual speed inside vehicles (fluidity conditions). In non-congested cities, automobile and motorcycle speeds may be triple that of buses (60 km/h as compared to 20 km/h) in more congested cities – as most large cities of contemporary developing countries – speed of cars may still be double that of buses. Finally, the comfort of using sidewalks and public transport vehicles is an issue.

Family and personal transport costs

As distances increase and more motorized means are used, people have to spend more money to travel. In developing economies, a large number of people have no money to pay for decent transport. The most pressing problems occur with those who cannot pay public transport fares, forcing some of them to change to slower modes or to give up travelling.

Capacity

Transport modes have different characteristics concerning cost, capacity, fuel consumption, comfort and safety. Wright (1992) notes that no mode gets perfect grades on all proposed attributes: cycling may be too dangerous in some places, walking long distances is not feasible, transit may be inflexible and the car is highly polluting. The capacity of different transport modes is better translated by the quantity of people that can be transported per hour, per length of road (Table 5.8.1).

The social and environmental impacts of different transport choices

Use of energy

Energy spent by a particular transport mode, per passenger, depends on vehicle and operating characteristics and on the number of people using the mode. Public transport means are always less energy-intensive than private modes. Relations between buses and cars fall in the 1 : 5 range in most developing economies, however, it may be even higher should bus occupancy be very high. On a city or country basis, the average (and total) energy consumption is highly dependent on the transport modal share. Considering that private vehicles consume much more energy per passenger-kilometre, the higher their share, the higher the energy spent.

Use of space

The first impact of transport is the land required to build roads and provide parking spaces. On a grid-type road system with roads located 100 metres apart from each other, roads alone will occupy about 25 per cent of the city's area. If parking spaces are considered, this figure may grow up to 30 per cent or more, as happens in sprawling US cities such as Los Angeles and Houston. A second important issue is the road space required by any single mode. When

Table 5.8.1 Transport mode carrying capacity (interrupted flow)

Mode	*Speed range (km/h)*	*Capacity range (passenger/hr/metre)*
Walkway	3–4	3,609
Bicycle	10–16	1,000–1,750
Cycle-rickshaw	6–12	500–1,500
Motorcycle (1 person)	15–40	500–1,000
Car (1.5 person)	16–24	140–250
Regular bus	10–25	1,000–6,000 (busways)
Tram	10–30	1,000–6,000

Source: Wright, 1992

we compare different modes, the most demanding mode is the automobile, which consumes 30 times more area per passenger than a bus and about five times that of a two-wheeler. This also explains why the increase in the number of automobiles may rapidly exhaust existing road spaces and increase congestion.

Pollutant emissions

In most large cities, transport is the main source of pollution, especially private transport. Motor vehicles are responsible for most CO_2, almost all CO emissions, and most of HC and NO_x emissions. Diesel trucks and buses contribute to most of the PM, especially when low-quality diesel with a high percentage of sulphur is used. Two-stroke motorcycles are usually highly polluting. For policy analysis, the key factor is emission per passenger-kilometre, which depends on vehicle occupancies, favouring public modes. In a particular city or region, actual modal share will define the emission pattern. When comparing different motorized means the environmental advantages of increasing the use of public transport are clear. In the case of São Paulo in 2010, motorcycles and automobiles emitted, respectively, 11 and 5.5 times more local pollutants per passenger-kilometre than diesel buses (ANTP, 2010).

Traffic accidents

It is estimated that every year there are 1.2 million road traffic deaths and about 50 million injured in the world, with costs varying from 1 to 2 per cent of the GNP of each country, adding up to about US$500 billion a year worldwide (WHO, 2004). In developing economies the same report estimates that the number of traffic fatalities in 2000 (613,000) will rise almost 100 per cent by 2020 amounting to 1.2 million fatalities. The severity of the problem is related to the mix of transport modes, with public transport modes being safer than private modes. Safety conditions may become critical where intense economic development occurs, with the corresponding change in the quantity of vehicles and the mix of transport modes. Negative results are more severe when 'soft' modes such as walking and cycling are replaced with 'harder' modes, such as the automobile, trucks, buses and the motorcycle.

The 'barrier effect' and urban disruption

Motorized traffic and especially automobiles affect social relations occurring in space by reducing social interaction and the use of public spaces (Appleyard, 1981) and requiring people to redefine strategies for lowering the risk of accidents. Negative impacts may be attributed to all motorized means, according to their specific use. The most damaging effect is that of automobiles, when we consider their number and their need to consume space to survive, often at the expense of the urban physical tissue and its architectural and cultural heritage.

Congestion

The significance of congestion stems from two sorts of concerns. The first is equity, as people using the roads causes delay to others. Secondly, environmental factors are important. Congestion caused by motorized vehicles implies higher energy consumption and pollutant emissions. Contrary to automobile-centred cities such as those of the USA, the analysis of congestion in developing economy cities has to face other traffic conditions, especially the extensive use of non-motorized and public transport modes. A particular analysis of congestion that is relevant for developing countries is that of how bus traffic is affected by automobile traffic. If congestion lowers bus speed from 20 km/h to 12 km/h, every bus passenger will take an extra 20 minutes for a 10-kilometre trip.

Conclusion

The fundamental question for the future of cities in developing economies is how to accommodate and direct growth to ensure higher social equity and economic efficiency on the one hand, and more equitable transport conditions on the other. With the former, solutions rely mostly beyond the transport agenda – income distribution, better housing, educational and health conditions – although transport policies and social policies interact with one another. With respect to the latter, urban, transport and traffic policies may be worked out to change current conditions and help achieve the objectives. The variety of existing conditions requires a variety of solutions, even considering that a few basic solutions may be devised to apply to a large number of cities. Three fundamental questions may be proposed:

1 What is the desirable balance between private motorized transport and public transport?
2 How can a shift from motorized means to more environmentally friendly ones be promoted? The question divides into two others, relating to transferring trips to non-motorized or to public transport means.
3 Even if private motorization is supported by public policies, how can the needs of the large number of people that will still be dependent on walking, cycling and using public modes be met?

The relevance of these questions relates directly to the environmental, economic and social consequences of urban and transport policy decisions. There are no reason to believe that developing economies are condemned to automobile dominance, congestion and pollution. As Cervero (1998) and Newman and Kenworthy (1999) have pointed out, there are important lessons to draw from large cities that have managed to keep or create good public transport systems, despite tendencies of the increasing use of private transport.

References

ANTP (Associação Nacional de Transportes Públicos) (2010) *Sistema de informação da mobilidade urbana*, São Paulo (www.antp.org.br).
Appleyard, D. (1981) *Livable streets*, Berkeley: University of California Press.
Cervero, R. (1998) *The transit metropolis: A global enquiry*, Washington, DC: Island Press.
CMSP (2008) *Pesquisa origem-destino 2007 na RMSP*, São Paulo: CMSP.
Newman, P. and Kenworthy, J. (1999) *Sustainability and cities: Overcoming automobile dependence*, Washington, DC: Island Press.
Vasconcellos, E. A. (2001) *Urban transport, environment and equity: The case for developing countries*, London: Earthscan.
WHO (World Health Organisation) (2004) *World report on road traffic injury prevention*, Geneva: WHO.
Wright, C. (1992) *Fast wheels, slow traffic: Urban transport choices*, Philadelphia, PA: Temple University Press.

Further reading

Illich, I. (1974) *Energy and equity*, New York: Harper & Row.
Whitelegg, J. (1997) *Critical Mass: Transport, environment and society in the twenty-first century*, London: Pluto Press.

5.9

Cities, crime and development

Paula Meth

Introduction

Crime is a dominant social issue in many cities across the global South, and it can exacerbate key development challenges such as poverty and inequality, and undermine residents' sense of well-being and welfare. Crime is also tied to the realities of cities in complex ways. It is, in part, shaped by globalisation and rapid urbanisation, as well as levels of unemployment, forms of welfare regimes and the effectiveness of urban governance and policing. This chapter argues that crime within cities is a development issue, as it is fundamentally tied to urban inequalities and injustices (often constructed through unequal political and economic trajectories). Crime must be understood in an historical context: political strategies such as military dictatorships or authoritarian rule; and economic practices such as tax concessions for multinational companies or the destruction of informal economies, often have long term impacts on city residents and their ways of making a living as well as their sense of justice. In such contexts residents may commit crimes as a result of their economic, political or social marginalisation. However, this is not to suggest that the crimes of the poor are the only, or even the most significant crimes within cities, nor that all crimes can be explained so easily. Elite and state crimes are also significant, although their relationship to development studies is less clearly understood. This chapter explores the experiences of, and trends in, crime across cities in the global South and then considers ways of defining, theorising or explaining crime. It discusses the intersections between cities and crime, and concludes with a consideration of crime as a development concern.

Experiences of crime and crime trends

Much of the available research on crime tends to focus on analyses of crime statistics. This work is valuable because it illustrates the scale, extent and transitions or consistencies in crime over time and place, however, it often does not reveal the more qualitative experiences of crime as detailed by residents and nor does it point to the consequences of crime. Crime can exacerbate poverty and inequality, and this can occur through the loss of earnings and resources, as well as the loss of the ability to earn or to engage in livelihood practices. Theft, corruption, homicide and assault may all extend the poverty of an individual or household. Given the high crime levels facing particular cities (see Table 5.9.1), dealing with crime is part of everyday life for many residents who will adopt a range of strategies to try and minimise criminal experiences, including the ways in which they use and travel through the city and the role of community to provide protection. There is, as yet, inadequate research on fear and how this shapes the lives of urban residents, but a growing sensitivity to the emotional geographies of residents across cities of the global South will arguably point to fear (and coping with crime) as a key source of anxiety. Crime also impacts on cities more broadly. Nearly 30 per cent of firms in Latin America, Africa and Asia point to crime as the key concern for their business (World Bank, 2011) and thus as a development burden.

Analysing crime trends across cities is very difficult because the reliability and comparability of crime statistics is limited. A useful example of these difficulties is the recording and comparison of drug related crimes (per police recording), which reveals that in 2008 Mexico had 68.3 such crimes per 100,000 of the population, whereas Switzerland, in the same year had 619.0 (UNOCD, 2012). Here issues around crime classification, recording procedures, criminal justice systems, and policing practices all contribute to this final figure, making effective comparison futile. Indeed, the process of recording crime within countries is politically determined and allegations of strategic under-reporting are not uncommon. The UNOCD (United Nations Office on Crime and Drugs) provides some insight into urban comparison through listing particular crimes per most populous city, and these can be considered in relation to the Gini coefficients (as a measure of inequality) for those countries (see Table 5.9.1).

Table 5.9.1 reveals the extreme variety of homicide rates across cities, as well as tentative (but not conclusive) evidence of the links between inequality and crime. India's low prevalence raises important questions about the necessary relationship between poverty and the incidence of more violent crimes, such as homicide.

Ways of defining and theorising crime

Defining what is, and what is not, a crime is a complex political and social issue and this makes understanding crime in cities even more challenging. The notion of crime is contested and whether particular practices are criminalised (defined and treated as criminal) or not tells us much about power relations within a society (examples of these issues will be given below in relation to informality and also gender violence). All states through their legislative powers define crime and the UNOCD lists the following categories of crime under which their statistics are organised: homicide, assaults, sexual violence, robbery, kidnapping, theft, motor vehicle theft, burglary and drug related crimes (UNOCD, 2012). It is clear that crimes such

Table 5.9.1 Homicide rates per 100,000 of the population in most populous city by country (alongside Gini coefficients measuring the extent of income distribution)

Country	*City*	*2008*	*2009*	*National Gini coefficient**
Kenya	Nairobi	4.0	–	47.7 (2005)
Lesotho	Maseru	59.7	61.0	52.5 (2003)
Jamaica	Kingston	29	26.8	45.5 (2004)
Venezuela	Caracas	127.0	122	44.8 (2006)
Mongolia	Ulan Bator	11.5	10.2	36.5 (2008)
Philippines	Quezon City	2.0	5.3	43.0 (2009)
India	Mumbai	1.3	1.3	33.4 (2005)

*data in this column adapted from World Bank, 2012
Source: Adapted from UNOCD, 2012

as corruption and fraud and police violence are not recorded in such categories, yet these are significant urban crimes in many cities in the global South.

The poor within cities are frequently both criminalised as well as suffering as victims of crime and this is a significant development concern, and it is also starkly gendered. Globally men, particularly young men, suffer from violent crime disproportionately (World Bank, 2011), although women suffer extensively from particular crimes, such as sexual violence. But the poor are not simply the victims of crime in cities. Caldeira's work in Brazil reveals how young black men in particular are viewed as criminal (2000) and this impacts on their human rights, their access to justice and their ability to live safely within cities. In both Brazil and South Africa, state-directed violence against young black men within cities is high and more worryingly the wider citizenry do not always perceive this as a problem.

More broadly the urban poor are criminalised through their everyday practices of survival (housing, employment and access to justice). Poor urban residents living in informal housing (in many cities this accounts for over 80% of housing) may technically be deemed criminal because their building structures do not meet required planning legislation, they may not have ownership or rights to their plot, and they may access services such as electricity illegally. In these cases, the failure of cities to provide adequate housing for the growing urban poor enforces their 'criminal' practices. In many cities authorities turn a blind eye to such housing, but there are now multiple cases of mass evictions of slum housing which use the illegality of such structures to enforce the law (see Mumbai, India and Abuja, Nigeria, for instance). A similar argument can be made in relation to the informal economies across cities in the global South, where informal traders frequently suffer the consequences of police raids, imprisonment and eviction (see Kamete's (2010) work on street traders in Harare, Zimbabwe). The politics of the criminalisation of the informal practices of the urban poor is all the more significant when compared with the treatment of the middle classes, who also adopt informal and illegal practices, but which the state often condones. In Delhi, the state often condones middle-class water users' practices compared with the urban poor, despite the fact that middle-class residents are blatantly 'breaking the law' (Truelove and Mawdsley, 2011).

A final example of the difficulties of defining crime relate to gender based violence against women. Crimes such as domestic violence and rape are not fully recognised in all states across the world and hence are not deemed a crime. In addition, even when they are legally deemed a crime, multiple forces work to silence such crimes so that women are forced to suffer without protection from the law or health care professionals. In countries in the Middle East, for example, a combination of religion, culture and patriarchal structures means that gender-based violence is often viewed as a domestic private issue (see Douki *et al.*, 2003) and not a concern for those focusing on crimes within cities. Given that rates of domestic violence are 1 in 3 women across particular Arab states (Douki *et al.*, 2003), arguably such crimes are core to an understanding of urban development challenges.

Cities and crime

Globally, levels of crime tend to be higher in cities than in rural areas, but this is not always the case, and criminal practices can cross-cut the rural–urban divide (such as the trade in drugs). Nonetheless crime is often associated with city life and many cities in the global South are increasingly labelled in relation to their purported high crime levels. One key reason for the urban phenomenon of crime is simply population size and density, whereby the very concentration of populations within urban centres accelerates the occurrence of crime within a particular location. Cities also provide a market for criminals in that they host a large population of potential victims often living in very close proximity. Rapid urbanisation and the inability of city authorities to manage such urbanisation and population growth have meant that cities often lack the governance structures necessary to cope with crime. Policing services across most of the global South are often ineffective and stretched, and criminal justice institutions (courts and prisons) are over-burdened. This has led to varying responses to crime above and beyond 'legal' policing measures, including the dominance of gangs and/or vigilante groups, as well as privatised security in some cities. Dealing with crime, and being seen to deal with it effectively is a key political issue for city governance structures. Crime is a core election issue in many cities, at times even the core issue. As a result of this, crime management can be the focus of multiple governance structures and institutions including partnership organisations working between the police and communities, local political parties, local authority structures, and so on. In South Africa, the dominance of crime on the agenda of local governance structures can be described as the 'criminalisation of governance' whereby city governance structures concentrate their attention on crime management rather than on wider governance issues (such as participation or delivery of other services) (Meth, 2011). Crime then is a development issue because of its absorption of resources. Despite comments above about the ineffectiveness of police more broadly, urban police forces are key urban actors across cities of the global South. Their roles are often criticised, but they can and do play important roles as mediators and protectors. Styles of policing shift over time but more militarised combative policing is common in contexts where crime is high and the politics of crime are significant. Policing within the cities of South Africa and Brazil are again good examples here. Finally, the physical designs of cities are increasingly aimed at managing crime and the rise of gated communities, walled enclaves (see Caldeira's analysis of this in Brazil), privatised spaces and the securitisation of elements of the urban are evidence of this.

To conclude, crime within cities is a core development issue. It impacts on the social, political, physical, economic and emotional lives of residents and responses to crime can also be a developmental concern. Crime is also directly affected by changes in wider development indicators particularly employment and governance.

References

Caldeira, T. (2000) *City of Walls: Crime, Segregation and Citizenship in São Paulo*, Berkeley: University of California Press.

Douki, S., Nacef, F., Belhadj, A., Bouasker, A. and Ghachem, R. (2003) Violence against women in Arab and Islamic countries, *Archives of Women's Mental Health*, 6(3), 165–171.

Kamete, A. Y. (2010) Defending illicit livelihoods: Youth resistance in Harare's contested spaces, *International Journal of Urban and Regional Research*, 34(1), 55–75.

Meth, P. (2011) Crime management and urban governance: Everyday experiences within urban South Africa, *Environment and Planning A*, 43(3), 742–760.

Truelove, Y. and Mawdsley, E. (2011) Class and water: Discourses of citizenship and criminality in clean, green Delhi. In I. Clark-Decès (ed.) *A Companion to the Anthropology of India*. Malden, MA: Princeton University Press.

UNOCD (2012) United Nations Office on Crime and Drugs, available online at www.unodc.org (accessed 18 April 2012).

World Bank (2011) *Conflict, Security and Development, World Development Report 2011*, Washington, DC: The World Bank.

World Bank (2012) Gini Index, The World Bank, available online at http://data.worldbank.org/indicator/SI.POV.GINI (accessed 19 April 2012).

Further reading

Caldeira, T. (2000) *City of Walls: Crime, Segregation and Citizenship in São Paulo*, Berkeley: University of California Press. This text provides an excellent anthropological account of how crime, criminals and place are constructed in urban Brazil. It points to inequalities in the city and the ways in which approaches to crime shape the city.

Koonings, C. G. and Kruijt, D. A. N. M. (2007) *Fractured Cities: Social Exclusion, Urban Violence and Contested Spaces in Latin America*. London: Zed Books. This edited collection explores urban violence alongside analyses of six Latin American cities revealing the multidimensional character of urban crime and violence.

Part 6
Environment and development

Editorial introduction

Over the last twenty years or so, the environment has become a major dimension of development thinking. In the future, it is increasingly clear that it needs to become an even more important component of development practice. In the past, undoubtedly, too much attention has been paid to economics in the development equation, and far too little emphasis has been placed on the environment–development interface. However, since the Brundtland Commission in 1987, attention has increasingly focused on the concept of sustainable development. Although there are many discourses on sustainability, reflecting the interests of many different groups, most adopt the Commission's working definition of sustainability as meeting the needs of the present, without compromising the ability of future generations to meet their needs. But it has to be recognised that the concept of sustainable development is complex and contradictory: for example, it is generally harder for the poor to operationalise, as it is tempting for them to 'discount' the future, in order to provide for the ever-pressing needs of the present.

The United Nations Conference on Environment and Development (UNCED) held in Rio de Janeiro in 1992, and generally referred to as the 'Rio Earth Summit', was designed as a major effort to catalyse a more sustainable approach to development. In the summer of 2012, the follow up, 'Rio plus 20', was held in Brazil. However, some commentators have argued that the needs of national governments and business lobbies have been too dominant in the various discussions that have taken place as part of such conferences. Another major issue has been that during such proceedings, the nations broadly making up the North and the South seemed to have adopted and stressed quite different objectives. The nations of the North seem intent to argue that environmental problems can be cured by the application of technology. In contrast, the governments of the South have emphasised the need for real structural reforms in order to change the international economy, so as to impact directly on the effects of debt, poverty, inequality and the like.

Such circumstances make all too clear the need for collective international environmental agreements and action. These define how countries should manage the environment and resources. Thus, the 1997 Kyoto Protocol set out clear targets for greenhouse gases across certain countries, through to 2012. International agreements now exist in areas such as trans-boundary air pollution, freshwater resources, trade in hazardous substances, endangered species and the management of the international seas. But much remains to be achieved and such agreements have shown that they are not necessarily effective in changing behaviour in as comprehensive terms as is required.

A major threat to the environment is posed by global climatic change, and it is now generally accepted that climate is changing, in large part due to human activity. Once again, the concomitants of global warming, such as land degradation, desertification and flooding, threaten to impact on the poor disproportionately, rather than the rich. Such environmental circumstances are also expressed in terms of problems of food security. In this connection, urban environmental circumstances are also salient, not least as witnessed in waste dumping in watersheds, which causes major pollution problems downstream, where such waters may then be used for irrigation and domestic purposes, thereby leading to a variety of health risks.

Some pundits maintain that climate change represents one of the major challenges now facing African development. Of course, both historically and currently, Africa's contribution to climate change has been low compared to other world regions. Thus, issues of global justice and equity are centrally involved in the consideration of African climate change and development. But the challenge presented by this equation also affords unique opportunities, for example, to follow low carbon growth strategies and to build more resilient economies overall. The need is to explore the potential for the better exploitation of renewable energy, more sustainable cities and transport systems and more efficient land use and forest management systems.

Another major area of enquiry has re-examined the nature of disasters and vulnerability. Here views have been changing quite dramatically, in so far as there has been a move to recognise that even natural events need not turn into disasters, and that steps may be taken to avoid the harmful consequences of such events. The vulnerability and entitlement explanatory models are of particular salience in this connection.

Another major focus has been on the developmental issues associated with what are referred to as 'ecosystem services' (ES). ES can be defined as the free gifts of nature, such as the air and sea. Unlike many goods and services that are sold on open markets, they may, therefore, be undervalued and can be overlooked in the decision-making process. Thus the apparently economically rational degradation of ecosystems can result in significant losses in human welfare. The ES concept draws attention to the essential benefits that are provided to people and economies for free by such natural ecosystems, and aims to improve the efficient use of ecosystems to support human wellbeing. Turning to the wider employment aspects of ES, for more than 1.3 billion people worldwide, natural resources in the form of fisheries, forests and agriculture provide almost half of all jobs, particularly in rural environments. The future aim needs to be to promote the use of locally owned resources and solutions to natural resource management, which are participatory, democratic and non-hierarchical.

Only one per cent of the world's water is available to humans and the challenge is to supply water to the estimated 780 million people, mainly in Africa, Asia and Latin American, who lack access to safe water. Issues of access to water boil down to issues of power, as exemplified by low- and high-income access to water under privatisation, and are thereby essentially political. Turning to the important sector of energy, this has been closely related to human progress and thereby the quest for sustainable development. In the past, there was the well-marked general tendency for energy consumption to increase steeply as per capita incomes increased. What is pressingly needed in the future is more efficient technologies, cleaner resources, more effective demand management and better governance.

Tourism has become a major plank of development in many developing countries since the 1960s, bringing in valuable foreign exchange. But the social and cultural outcomes are clearly recognised in terms of dependency and demonstration effects. A further major set of issues relates to the fact that tourism effectively represents selling the environment, and can have major consequences, for example, the demand for water, the destruction of wetlands, coastal ecosystems and the like.

Looking toward the future, transport is potentially one of the most significant sectors in need of decarbonisation, with many individuals either still aspiring to, or already being largely reliant on, car use on a day-to-day basis. So, sustainability remains a major goal and quest, needing imaginative urban planning, effective traffic demand management, the enhancement of public transport systems, the use of low emission vehicles and alternative fuels, as well as the maintenance of efficient systems of walking and cycling.

6.1 Sustainable development

Michael Redclift

Discourses of sustainable development

The expression 'sustainable development' has been used in a variety of ways, particularly within the context of development studies. Today we are confronted with several different discourses of 'sustainable development', some of which are mutually exclusive. For example, campaigners for greater global equality between nations, huge international corporations and local housing associations have all had recourse to the term 'sustainable development' to justify their actions.

Sustainable development was defined by the Brundtland Commission in the following way: 'development that meets the needs of the present without compromising the ability of future generations to meet their own needs' (Brundtland Commission, 1987). This definition has been brought into service in the absence of agreement about a process which almost everybody thinks is desirable. However, the simplicity of this approach is deceptive, and obscures underlying complexities and contradictions. It is worth pausing to examine the apparent consensus that reigns over sustainable development.

First, following the Brundtland definition, it is clear that 'needs' themselves change, so it is unlikely (as the definition implies), that those of future generations will be the same as those of the present generation. The question then is, where does 'development' come into the picture? Obviously development itself contributes to 'needs', helping to define them differently for each generation and for different cultures.

This raises the second question, not adequately covered by the definition, of how needs are defined in different cultures. Most of the 'consensus' surrounding sustainable development has involved a syllogism: sustainable development is necessary for all of us, but it may be defined differently in terms of each and every culture. Furthermore, how do we establish which course of action is *more* sustainable? Recourse to the view that societies must decide for themselves is not very helpful. (Who decides? On what basis are the decisions made?) At the same time there are problems in ignoring culturally specific definitions of what is sustainable in the interest of a more inclusive system of knowledge. There is also considerable confusion surrounding *what* is to be sustained. One of the reasons why there are so many contradictory approaches to sustainable development (although not the only reason) is that different people identify the objects of sustainability differently.

What is to be sustained?

For those whose primary interest is in ecological systems and the conservation of natural resources, it is the natural resource base which needs to be sustained. The key question usually posed is the following: how can development activities be designed which help to maintain ecological processes, such as soil fertility, the assimilation of wastes, and water and nutrient recycling? Another, related, issue is the conservation of genetic materials, both in themselves and (perhaps more importantly) as part of complex and vulnerable systems of biodiversity. The natural resource base needs to be conserved because of its intrinsic value.

There are other approaches, however. Some environmental economists argue that the natural stock of resources, or 'critical natural capital', needs to be given priority over the flows of income which depend upon it (Pearce, 1991). They make the point that human-made capital cannot be an effective substitute for natural capital. If our objective is the sustainable yield of renewable resources, then sustainable development implies the management of these resources in the interest of the natural capital stock. This raises a number of issues which are both political and distributive: who owns and controls genetic materials, and who manages the environment? At what point does the conservation of natural capital unnecessarily inhibit the sustainable flows of resources?

Second, according to what principles are the social institutions governing the use of resources organized? What systems of tenure dictate the ownership and management of the natural resource base? What institutions do we bequeath, together with the environment, to future generations? Far from taking us away from issues of distributive politics, and political economy, a concern with sustainable development inevitably raises such issues more forcefully than ever (Redclift, 1987; Redclift and Sage, 1999).

The question 'what is to be sustained?' can also be answered in another way. Some writers argue that it is present (or future) levels of production (or consumption) that need to be sustained. The argument is that the growth of global population will lead to increased demands on the environment, and our definition of sustainable development should incorporate this fact. At the same time, the consumption practices of individuals will change too. Given the choice, most people in India or China might want a television or an automobile of their own, like households in the industrialized North. What prevents them from acquiring one is their poverty, their inability to consume, and the relatively 'undeveloped' infrastructure of poor countries.

Is there anything inherently unsustainable in broadening the market for TV sets or cars? The different discourses of 'sustainable development' have different answers to this question. Many of those who favour the sustainable development of goods and services that we receive through the market, and businesses, would argue that we should broaden the basis of consumption. Others would argue that the production of most of these goods and services today is inherently unsustainable – that we need to 'downsize', or shift our patterns of consumption. In both developed and, increasingly, developing countries, it is frequently suggested that it is impossible to function effectively without computerized information or access to private transport.

The different ways in which 'sustainability' is approached, then, reflects quite different underlying 'social commitments', that is, the patterns of everyday behaviour that are seldom questioned. People define their 'needs' in ways which effectively exclude others from meeting theirs and, in the process, can increase the long-term risks for the sustainability of other people's livelihoods. Most important, however, the process through which we enlarge our choices, and reduce those of others, is largely invisible to people in their daily lives.

Unless these processes are made more visible, 'sustainable development' discourses beg the question of whether, or how, environmental costs are passed on from one group of people to another, both within societies and between them. The North dumps much of its toxic waste and 'dirty' technology on poorer countries, and sources many of its 'needs', for energy, food and minerals, from the South. At the same time, the elevated lifestyles of many rich and middle-class people in developing countries are dependent on the way in which natural resources are dedicated to meeting their needs. Finally, of course, the inequalities are also intergenerational, as well as intragenerational: we despoil the present at great cost to the future. Discounting the future (as economists call it), valuing the present above the future, is much easier to do in materially poor societies, where survival itself may be at stake for many people.

The final element in the redesign of 'sustainable development' policy was the creation of the 'consumer-citizen', the idea that the individual could best express their preferences for goods and services through their own (and their household's) personal consumption. Parallel with the development of cleaner technology, and carbon markets, came the concern with sustainable consumption. Partially as a result of their insufficient understanding of the link between social structures and consumer habits, and the awkward politics of wealth redistribution, governments came to favour consumer encouragement to live more sustainably and to reduce household 'footprints'. This implied the design of new 'lighter' consumer goods, evocations to act in more environmentally responsible ways, and an accent on 'lifestyle' and the consumer, at the expense of livelihoods and citizenship.

From the perspective of those most critical of market-based environmental valuation, the conjunction of newly 'liberated' markets and environmental concern was a necessary contradiction of capitalism seeking a resolution, and could with hindsight be seen as a 'managed senescence', if we continue with the biological metaphors of 'development' (Redclift, 2009; Bellamy-Foster, 2010). A more mainstream view, however, would be that they addressed system failures, and could even lead to a rejuvenated, if scarcely recognizable, type of materials 'light' capitalism (Lovins and Lovins, 2000).

Human rights, democracy and sustainable development

Finally, since the various discourses of sustainable development began to flourish, it has become evident that another dimension to the problem of diminished sustainability needs to be considered. This is the extent to which, at the beginning of the twenty-first century, we need to refer to processes of democracy and governance in the context of sustainable development. The Brundtland Report took a highly normative view of both the environment and development, as did the Earth Summit deliberations in 1992. Today questions of sustainability are linked, intellectually and politically, to other issues, such as human rights and 'identity', with which they are connected. But notions of 'rights' and 'identity' are themselves changing. In the era of genetic engineering, the genetic modification of humans, as well as plants and animals, is shaping new senses of 'identity'. As individuals change, so do the groups to which they belong. In the era of globalization it is sometimes argued that 'sustainable development' may be more difficult to achieve, as economies converge towards shared economic objectives. At the same time it may prove impossible to achieve 'sustainable development' (if it *is* achievable) without acknowledging quite distinctive accounts of human rights in nature, and even the rights *of* nature, which were hitherto ignored. The concept of 'sustainable development', as understood by different people, is contradictory, obscure and illuminating at the same time.

Bibliography

Bellamy-Foster, John (2010) 'Marx's ecology and its historical significance', in Michael R. Redclift and Graham Woodgate (eds) *The International Handbook of Environmental Sociology*, Second Edition, Cheltenham, UK: Edward Elgar.

Brundtland Commission (World Commission on Environment and Development) (1987) *Our Common Future*, Oxford: Oxford University Press. This report led directly to the term 'sustainable development' passing into common use. It was also the first overview of the globe which considered the environmental aspects of development from an economic, social and political perspective; cf. Man and the Biosphere (MAB) almost a decade earlier. Among the principal omissions was detailed consideration of non-human species, and their 'rights'. The 'Brundtland Report' (named after its chairperson, the Norwegian prime minister at the time) also opened the way for non-governmental organizations (NGOs) to be considered a serious element in environment and development issues.

Habermas, J. (1991) 'What does socialism mean today?' in R. Blackburn (ed.) *After the Fall*, London: Verso.

Lovins, L. H. and Lovins, A. B. (2000) 'Pathway to sustainability', *Forum for Applied Research and Public Policy* 15(4): 13–22.

Meadows, D. H., Meadows, D. L., Randers, J. and Behrens, W. (1972) *The Limits to Growth*, London: Pan Books. This book was a milestone in thinking about the environment in the 1970s. It identified the main problem for development as the shortage of natural resources – these were the 'limits' to growth. This view changed in the 1970s and 1980s, largely because the price of oil rose very quickly and alternative substitutes for many natural resources were exploited in their place. In addition, of course, the 'Green Revolution' in staple cereal crops, which was largely undertaken in the 1970s and 1980s, appeared to show that the same land base could produce much more food.

Pearce, D. (1991) *Blueprint 2: Greening the World Economy*, London: Earthscan. Following on from *Blueprint 1*, David Pearce illustrated the applications of economic analysis to environmental problems, in a way that particularly interested policymakers.

Redclift, M. R. (1987) *Sustainable Development: Exploring the Contradictions*, London: Routledge. This was the first, and probably the best, treatment of 'sustainable development'. The case studies and ethnographic illustrations, combined with the accessible intellectual discussion, make this a 'classic'.

Redclift, M. R. (2009) 'The environment and carbon dependence: Landscapes of sustainability and materiality', *Current Sociology* 57(3): 369–387.

Redclift, M. R. and Sage, C. L. (1999) 'Resources, environmental degradation and inequality', in Andrew Hurrell and Ngaire Woods (eds) *Inequality, Globalisation and World Politics*, Oxford: Oxford University Press. A good general overview of the relevance for development of inequality in resource endowments.

Smith, N. (2007) 'Nature as accumulation strategy', in Leo Panitch and Colin Leys (eds), *Socialist Register 2007: Coming to Terms With Nature* 43.

6.2

International regulation and the environment

Giles Atkinson

Introduction

It is now a number of years since a comprehensive report on the economics of climate change concluded that, in terms of policy proposals to stabilise the stock of greenhouse gases, such as carbon dioxide, in the global atmosphere: 'the benefits of strong and early action far outweigh the economic costs of not acting' (Stern *et al.*, 2006, p. xiii). These conclusions should also be viewed in the context of mounting evidence from the Intergovernmental Panel on Climate Change (e.g. IPCC, 2007), which has continued to build the scientific foundations that underpin concern about climate change.

However, it is clear that almost all countries are still far from anything like taking the strong action demanded by the Stern Review. This is only partly due to lingering and inevitable uncertainties about the climate change problem and its likely future impacts (Barrett, 2007). Such uncertainties remain relevant of course along with the fact that the costs of acting are borne now, whereas the (expected) benefits of these actions are enjoyed in the (far-off) future. A further crucial consideration is that any country acting unilaterally to cut back its emissions incurs all of the costs of these actions but only a small portion of the overall benefits. The rest of these benefits accrue to the global community. Not surprisingly, left to its own devices, any individual country typically concludes that it has little incentive to contribute to global public goods in this way.

What is required to realign these incentives is some degree of formal international cooperation. Needless to say, achieving such cooperation is far from straightforward and critically will depend on the characteristics of the underlying environmental challenge to be addressed. In the case of climate change, for example, the evidence thus far is that genuine cooperation has been exceptionally difficult to achieve and so far has certainly militated against the fulfilment of bold objectives. Yet, in instances where cooperation has been brokered, the formal 'glue' that holds the international environmental regime is typically some form of international agreement. This chapter provides a brief overview of some of the main issues that have characterised ongoing efforts to construct international responses to, in particular, global environmental problems.

International environmental problems

International environmental agreements (IEAs) – defined as legal documents between states to manage a natural or environmental resource[1] – are the outcomes of typically lengthy negotiation processes and any single IEA may itself be one treaty within a framework of such agreements that characterise an 'environmental regime'.[2] For example, the Framework Convention on Climate Change (FCCC) initially set out the broad aspiration: '(t)o achieve . . . stabilisation of greenhouse gas concentrations in the atmosphere at a level that would prevent dangerous anthropogenic interference with the climate system' (United Nations, 1992, p. 4). However, it was the 1997 Kyoto Protocol (KP) that set out the first legally binding targets for GHGs (greenhouse gases) across certain countries. Given that the first commitment period of the KP ends in 2012, attention is now on what (if anything) will succeed it.

International agreements now exist across a whole range of environmental issues including transboundary air pollution, shared freshwater resources, trade in both hazardous substances and endangered species and the management of the international seas. An increasing number of these problems can be viewed as global problems in terms of the sheer numbers of countries affected, albeit to differing degrees (Pearce, 1999). Problems arise as a result of natural and environmental resources being depleted or degraded that have no owner; that is, they are 'open access'. Put simply, these resources are prone to over-use. For example, the global atmosphere acts as a sink for many of the by-products of economic activity including carbon dioxide (CO_2) and other greenhouse gases. The fact of no ownership means that no one has an incentive to limit their contribution to the increasing stock GHGs in the global atmosphere. Increasing mean global temperature, and associated adverse socioeconomic impacts, is one consequence of this (see, for a discussion, Helm, 2005). Other global resources have an owner in that they are located within a sovereign country. The world's tropical forests which act as a significant store of biological diversity would be one such example. The problem here is that each 'owner' may have insufficient incentives to continue to provide these services when the value of alternative uses of the land, on which there is currently standing forest, 'out-competes' the conservation option (at least from the owner's standpoint).

In each of these cases, these incentives mean there is too little provision of globally valued resources. Changing these incentives is thus the foremost challenge facing the international community in seeking a collective response to shared environmental problems (Barrett, 2003). Moreover, it is not enough that an agreement simply confers an overall 'global' gain but also that each state perceives that it is better off being part of the agreement than remaining outside of it. The last notion has particular significance, however, given there is no 'global government' that can impose environmental objectives, no matter how worthy, on states. Matters are complicated further by the need to deter 'free-riding'. To take the example of climate change again, those outside of an agreement to limit emissions of GHGs cannot be excluded from enjoying the global benefits of the actions of those countries within the agreement. The free-rider problem is not just confined to climate but is ubiquitous in other challenges such as managing the oceans and stratospheric ozone protection.

Not surprisingly, significant attention has been devoted to questions about the enforcement of IEAs both in terms of encouraging participation (i.e. being a party to an agreement) and compliance (i.e. fulfilling the obligations that one is party to). Enforcement methods might refer to rewards ('carrots') for becoming a party to or complying with an IEA or punishments ('sticks') for 'anti-social' behaviour. Most famously of all, perhaps, the 1987 Montreal Protocol (MP) has within it the potential to ban trade in products which contain prohibited stratospheric ozone-depleting chemicals. Whether analogous provisions are needed or could be rolled out across new international environmental challenges is debatable. One anxiety is that IEAs, which make use of such mechanisms, may fall foul of international trade regime rules. Whether this would apply, for example, to proposals for border taxation of the CO_2 content of goods imported into a country, arguably depends on the specific details of the policy enacted (see, for a discussion, Atkinson *et al.*, 2010; Helm *et al.*, 2012). In general, the economics and global politics of using the trade regime to reach environmental protection goals are fraught. Yet, if the credible threat of punishment such as trade restrictions exists then there is a case in principle for it being available to negotiators or to those implementing commitments. In practice, this case depends on the ability to shape a consensus that any punishment is 'just' or (less grandly) politically acceptable and, this in turn, may well depend on its being used in judicious combination with rewards such as financial and technological inducements (Barrett, 2003).

A related puzzle, however, is that there are numerous examples of IEAs which appear to operate with no tangible provisions for enforcing the obligations placed on countries. This has been interpreted by some as evidence that a change in state behaviour, and so meaningful cooperation, can be triggered without countries having to sign up to coercive provisions (see, for example, Chayes and Chayes, 1995; Young, 1994). This emphasis on meaningful cooperation is important as a rather different perspective is that many IEAs in reality might only obligate countries to do little more than they would have done in the absence of the agreement. This is articulated by Barrett (2003) in his assertion that 'most treaties . . . fail to alter state behaviour appreciably' (p. xii). And the reason for this is often the very complexity of trying to achieve deeper cooperation. Of course, whether or not IEAs actually achieve something more is an empirical question; albeit a challenging one, given the need to speculate about the counterfactual: that is, what would have happened without an agreement (see, for a review of evidence on IEA effectiveness, Mitchell, 2003).

One mechanism whereby treaties might end up being watered down is when unanimity is focal to the agreement. Froyn (2007), for example, notes that the fact that the KP which was finally entered into in 2005 was a far softer version of what was initially negotiated in 1997, and was primarily due to those countries favouring this lighter touch being able to 'exploit' the US withdrawal in 2001 and the treaty's minimum country participation rule (i.e. if these remaining sceptical countries did not ratify, the KP would not enter into force). This 'law of the least ambitious programme' (Underdal, 1980) has been observed in other environmental regimes as well. More broadly still, climate negotiations involve discussions between many nations on different stages along development paths. In this context, negotiations must reconcile debates about, amongst other things, responsibility for climate change, and the economic wealth built on that environmental liability (see, for example, World Bank, 2010).

Financing international cooperation

The implementation of IEAs typically involves the introduction of domestic policies within participating countries. Increasingly, however, there is an international dimension to implementation possibly involving substantial financial flows by creating markets in environmental services. The pre-eminent example of this – *carbon trading* – is concerned with cost-effectiveness in meeting climate change goals. One impetus for carbon trading is the KP. Under the KP, a number of countries (which include the OECD countries and 'economies in transition') have GHG targets defined with reference to 1990 levels of emissions. A feature of these targets is that GHG cuts do not necessarily all have to be achieved domestically. That is, a country, or perhaps an enterprise in that country, can purchase GHG reductions which have been achieved elsewhere in the world. The incentive for the buyer to make this trade occurs when mitigation carried out elsewhere is cheaper than that achieved domestically. Similarly, a seller will be willing to trade if it is adequately compensated. Most of these trades are currently being made via the European Union Emissions Trading Scheme (EU ETS) and the Clean Development Mechanism (CDM) (Hepburn, 2007).

In principle, carbon trading makes GHGs (reduction) akin to any other internationally traded good. Yet challenges abound. Some of these are philosophical in flavour and reflect debates about 'buying' one's way out of domestic responsibilities and whether sellers are getting, in turn, a fair deal (Wiener, 1999). Those discussions are not entirely unrelated to equally important technical issues about monitoring and verification of *additional* GHGs saved as a result of a trade (see, for a review, Tietenberg, 2005). Many hope that *global* carbon trading will be a centrepiece of any future (post KP) climate agreement. However, an emerging irony is that the prospect of 'scaling-up' existing trading regimes depends on expectations about the climate regime that lie beyond the KP. Uncertainty about this future translates into uncertainty in carbon markets. For example, buyers will not want to hold carbon credits that may turn out to be of little value if the outcome of climate change negotiations involves significant back-pedalling from even the modest GHG reduction targets in the KP.

Carbon trading involves a 'polluter' paying for GHG reduction elsewhere. For other problems, however, it might be the polluter who has a property right underpinning their current behaviour perhaps because a threatened biological resource of international importance is sovereign property. For example, the 1992 Convention on Biodiversity (CBD) commits parties to the conservation and sustainable use of biological resources but designates protection as an issue of national sovereignty to be determined within national boundaries. To the extent that others, elsewhere in the world, place a value on conserving biodiversity, it is in the interests of these beneficiaries to pay for conservation if biodiversity otherwise will be under threat.

By attracting money payments in excess of the cash flows associated with an alternative use, an opportunity arises for conservation to 'pay its way' and so be on a sustainable financial footing (see, for a recent review, Pattanayak *et al.*, 2010). An estimate by Parker and Cranford (2010) finds that the amount of finance directed towards protecting global biodiversity was just over USD50 billion in 2010. Nevertheless, the authors of that study warn that this total is considerably less than is reckoned to be required to save the world's biodiversity and is geographically uneven in its spread. Moreover, novel payment schemes, such as those described above, are currently only a very modest proportion of this total. There is optimism over schemes for protecting forest carbon such as REDD+ (Reducing Emissions from Deforestation and Degradation). Scaling up those schemes is a challenge but could have the secondary benefit of protecting (some) biodiversity as well. Of course, uncertainty about future climate change negotiations will influence the market for forest carbon just as it has done in other carbon markets.

Concluding remarks

A glance at the number of existing agreements across an array of international environmental issues might lend substantial encouragement as to future prospects for facing new challenges. This is not the end of the story unfortunately. Not all past agreements necessarily have been effective in changing countries' behaviour despite giving every impression of having achieved that end. This is not to say that meaningful environmental agreement is not possible; rather, it cannot be presumed. Much of the international community's problem is one of balancing of distributive considerations, in terms, for example, of dividing the gains from cooperation amongst parties. And while the creation of environmental markets, through carbon trading, for example, has emerged as an important bridge between competing interests, this significance does not deflect from deeper arguments about the distribution of the burdens of taking action to mitigate problems such as climate change.

Notes

1 See Mitchell (2003) for a discussion of definitional issues.
2 The stages of an IEA are numerous (see, for a discussion, Barrett, 2003). Initially processes of agenda-setting will evolve a negotiating text and subsequent deliberation. The end result is a final treaty that countries sign up to (or otherwise). Nestling between a country becoming a signatory to and implementing (i.e. complying with) an agreement is ratification by the country's national parliament. When there is significant disagreement within a country about the merits of what has been negotiated, ratification may not be a simple rubber-stamping of the earlier decision to sign (Froyn, 2007).

Bibliography

Atkinson, G., Hamilton, K., Ruta, G. and Van der Menbrugghe, D. (2010) 'Trade in "Virtual Carbon": Empirical Results and Implications for Policy', *Global Environmental Change*, 21(2), 563–574.
Barrett, S. (2003) *Environment and Statecraft*, Oxford University Press, Oxford.
Barrett, S. (2007) *Why Cooperate? The Incentive to Supply Global Public Goods*, Oxford University Press, Oxford.
Chayes, A. and Chayes, A. (1995) *The New Sovereignty*, Harvard University Press, Cambridge, MA.
Froyn, C. B. (2007) 'International Environmental Cooperation: The Role of Political Feasibility', in Atkinson, G., Dietz, S. and Neumayer, E. (eds) *Handbook of Sustainable Development*, Edward Elgar, Cheltenham, UK.
Helm, D. (ed.) (2005) *Climate-Change Policy*, Oxford University Press, Oxford.
Helm, D., Hepburn, C. and Ruta, G. (2012) 'Trade, Climate Change and the Political Game Theory of Border Carbon Adjustments', Grantham Research Institute on Climate Change and the Environment Working Paper No. 80, London School of Economics.

Hepburn, C. (2007) 'Carbon Trading: A Review of the Kyoto Mechanisms', *Annual Review of Energy and Environment*, 32, 375–393.
IPCC (Intergovernmental Panel on Climate Change) (1995) *Climate Change 1995: Economic and Social Dimensions of Climate Change*, Cambridge University Press, Cambridge.
IPCC (Intergovernmental Panel on Climate Change) (2007) *Climate Change 2007: The Physical Science Basis – Summary for Policymakers*, IPCC, Geneva.
Mitchell, R. B. (2003) 'International Environmental Agreements: A Survey of their Features, Formation and Effects', *Annual Review of Energy and Environment*, 28: 429–461.
Parker, C. and Cranford, M. (2010) *The Little Biodiversity Finance Book: A Guide to Proactive Investment in Natural Capital (PINC)*, Global Canopy Programme, Oxford.
Pattanayak, S. K., Wunder, S. and Ferraro, P. J. (2010) 'Show Me the Money: Do Payments Supply Environmental Services in Developing Countries?', *Review of Environmental Economics and Policy*, 4(2): 254–274.
Pearce, D. W. (1999) 'Economic Analysis of Global Environmental Issues', in Van den Bergh, C. J. M. (ed.) *Handbook of Environmental and Resource Economics*, Edward Elgar, Cheltenham.
Stern, N. *et al.* (2006) *Stern Review: The Economics of Climate Change*, Cambridge University Press, Cambridge.
Tietenberg, T. (2005) 'The Tradable-permits Approach to Protecting the Commons: Lessons for Climate Change', in Helm, D. (ed.) *Climate-Change Policy*, Oxford University Press, Oxford, pp. 167–193.
Underdal, A. (1980) *The Politics of International Fisheries Management: The Case of the North-East Atlantic*, Scandanavian University Press, Olso.
United Nations (1992) *United Nations Framework Convention on Climate Change*, United Nations, New York.
Wiener, J. B. (1999) 'Global Trade in Greenhouse Gas Control: Market Merits and Critics' Concerns' in Oates, W. (ed.) *The RFF Reader in Environmental and Resource Management*, Resources for the Future Press, Washington, DC, pp. 243–248.
World Bank (2010) *The Changing Wealth of Nations*, World Bank, Washington, DC.
Young, O. R. (1994) *International Governance: Protecting the Environment in a Stateless Society*, Cornell University Press, Ithaca and London.

6.3

Climate change and development

Emily Boyd

Introduction

Global climate change is a 'wicked' problem that is unprecedented (Rockström *et al.*, 2009). Climate change will incur impacts at a scale and speed that is unparalleled (Boyd and Juhola, 2009). One can think of the problem in terms of a set of 'planetary boundaries'. If boundaries are crossed then that may lead to irreversible disasters. Important boundaries include overstepping 350 parts per million (ppm) of carbon dioxide in the atmosphere, enhancing stratospheric ozone depletion and increasing ocean acidification. The outcomes of crossing these boundaries may lead to impacts such as sea level increases of several meters, a collapse of agricultural systems in dry regions, a total loss of coral reefs and fishing resources, and the dehydration of the Amazonas (Rockström *et al.*, 2009).

Many of the world's poorest people are exposed to the risks of climate change (IPCC, 2007). The 'faces' of poverty include the 800 million people who go to bed hungry every night and 200 million women that are at risk of maternal mortality. Climate change stands to exacerbate existing interlinked problems of social inequality, ecological degradation, and conflict over natural resources. The challenge of climate change is 'messy' and not easy to solve with conventional methods of scientific inquiry (Funtowicz and Ravetz, 1993). A greater focus on how people on the ground build resilience through existing social institutions and how they relate to environmental risk is needed urgently. In this regard, development studies can inform scientists about how human institutions such as families and kinship groups have evolved over time, how these social institutions work most effectively and what the barriers are that they face in adapting to a changing climate.

While it is important to consider ways in which societies can tackle greenhouse gas emissions (mitigation), this chapter focuses on the adaptation measures instead. This is because whatever mitigation actions society takes now there will be a time lag when societies will have to prepare to live with a changing climate. In this regard, the chapter examines the linkage between climate change and development and identifies key barriers to adaptation measures. The chapter discusses in brief the role of development in meeting these challenges.

Climate change and development

The majority view in the world now is that climate change is happening. Although scientists remain unsure about by how much and where the change is occurring what is unequivocal is that some parts of the world will get warmer and others will get cooler. Phenomena such as El Niño are likely to get more severe. The Intergovernmental Panel on Climate Change scenario projections show temperature rising between two and six degrees by 2099. The consequences of this warming will have risks for the developing world and these risks will vary, but nevertheless will have implications for millions of people.

Climate change and development are intricately linked in that climate change is a result of human-induced actions. Global warming is a result of industrialisation and the burning of fossil fuels. Industrialised development has therefore contributed to the problematic of climate change and the warming of the global mean surface temperate, which lead to changes in rainfall patterns and precipitation across the planet. Other changes in the biophysical system include prolonged periods of drought, rising sea levels, increased intensity of storms, and more erratic monsoon patterns with potentially devastating consequences.

Climate change risks

Climate change poses a challenge to existing development problems. For example, climate change is expected to exacerbate the stress of food insecurity such as the recurrent and persistent drought conditions in the West African Sahel region. The impact of climate change on precipitation and rainfall is of major concern in the global South. For example, Africa's development is closely linked to rain-fed agriculture rather than irrigation. Most rural communities rely on rainfall patterns for their crops. These links have implications for health. Most diseases are associated with poor water quality and parasites thrive when it floods, which can result in cholera and malaria. Mosquitoes are able to survive in some highlands with warming temperatures. Most cities in the world are in low lying coastal zones and are particularly vulnerable to flooding and others risk the impacts of storm surges. What is important here are the early warning systems to prepare people for crises, and the ability to continue to recover after crises have hit, and to build long-term capacity to overcome climate change risks.

The 'faces' of development

The United Nations Millennium Development Goals (MDGs) are development targets that are to be met by 2015. Generally, the trend suggests that there are very few locations where the MDG targets have been met to date. The majority of countries fall into the category of no progress or not expected to be met. The least progress on the MDGs has been made in sub-Saharan Africa. More than 50 percent of Africans still suffer from water-related diseases such as cholera and infant diarrhea. Over 2.6 billion people do not have basic sanitation, and more than one billion people still use unsafe sources of drinking water. Declining soil fertility, land degradation and the AIDS pandemic have led to a 23 percent decrease in food production per capita in the last 25 years, even though population has increased dramatically. More than 80 per cent of farmers in Africa are women.

One area of critical concern is the Horn of Africa, which is one of the world's most food-insecure regions. It includes eight countries: Djibouti, Ethiopia, Eritrea, Kenya, Somalia, Sudan, South Sudan and Uganda and has a combined population of 160 million people. Of that 70 million live in areas prone to extreme food shortages (UNEP, 2011). Scientists say that between the years 1970 and 2000 the threats of famine occurred every 1 in 10 years. Yet, it is anticipated that future climate risk, with growing populations and declining per capita agricultural capacity, will lead to further threats. Because there is limited capacity to respond to drought or food crises there is need to build long-term resilience, addressing root causes of the region's vulnerability (UNEP, 2011).

Links between development and adaptation

The definition of adaptation is the 'adjustment in natural or human systems in response to actual or expected climatic stimuli or their effects, which moderates harm or exploits beneficial opportunities' (IPCC, 2007). The emphasis in this definition is on vulnerability, to climate change impacts, as the trigger for adaptation action. Vulnerability is thought to consist of exposure of a particular system to climate impacts, of sensitivity of that particular system, and the adaptive capacity of that system (IPCC, 2007). There have been many debates about how the concept of adaptation relates to development (Smit and Wandel, 2006; Schipper, 2007). At the core, both concepts deal with social processes and issues of vulnerability, change, and human agency. Development addresses the root causes of poverty to reduce the vulnerability of the poor, while adaptation also takes into account the biophysical impacts of climate change that are likely to exacerbate development problems. The challenge, in reality, is how to measure the additional impact of climate change on development, given that the climate is variable and there are a lot of uncertainties in the existing climate models.

There are four adaptation measures: planned, autonomous, anticipatory and reactive. Planned adaptations are best documented and include actions such as building seawalls to protect coastal areas from storm surges and investments in agricultural technologies to combat drought stress. In response to drought and famine in the Horn of Africa, there has been action to avoid mass loss of life and the malnourishment of millions of people. This includes efforts to improve short-term food access, with long-term livelihood interventions such as agricultural development and diversification, integrated land management, market failures and food production (UNEP, 2011). More is required, however, to sustain the livelihoods of the 600 million people in the region.

Adaptation barriers and bridges

A brief history of adaptation policy

Since the 1970s the United Nations has been a forum for discussing the problem of climate change, identifying the problem and ways to address it. Essentially, the issue of climate change is focused on the problem of greenhouse gas emissions, in particular, carbon dioxide and its release into the atmosphere from burning fossil fuels, which lead to a rise in mean global temperature. The way to fix this has been thought of as assigning individual countries an emissions reduction target to not transcend carbon dioxide emissions above 450 parts per million. The United Nations Framework Convention on Climate Change (UNFCCC) was agreed in 1994 following major efforts at the United Nations Conference in Rio 1992. The convention created an overall framework for intergovernmental efforts to tackle climate change and included mention of adaptation to climate change and policy measures to support developing countries. Still, progress on adaptation was slow and not much happened until 2002 when new finance for adaptation was agreed at the United Nations climate conference in New Delhi, India. In 2006, governments adopted a five-year plan to support climate change adaptation by developing countries, and also agreed on the rules and processes for the Adaptation Fund. Another negotiation milestone was the Bali Road Map laid out by the UN negotiations in 2007, which is a plan for the international climate negotiations following on from the Kyoto Protocol in 2012. In 2009, the international community met again in Copenhagen to deliberate emissions targets for the future and to pledge funds for climate change adaptation. Developed countries pledged US$30 billion to poor countries between 2010 and 2012, and to rise to US$100 billion a year by 2020. Still, it is anticipated that much more will be required to fund adaptation projects and programmes.

International–national adaptation

At the international level adaptation has never received much attention even though developing countries have raised their concerns about climate change since the early days of negotiations. The UN negotiations were entrenched from the onset because of questions of whether climate change was caused by humans or a natural phenomenon, who was responsible for it and who should pay for the damages caused by it, and how payments should be designed, as loans or compensation. The focus on stabilising greenhouse gases was a priority issue. Since 2007, IPCC scientists agreed on the 90 per cent likelihood that global warming was human-induced. The emergence of a policy debate around adaptation in the context of extreme weather-related events (e.g. Katrina in 2005) confirmed that things were moving from climate impacts to resilience, that is, the need for building responses to short-term shocks (resilience) in tandem with building strategies for long-term change (sustainability). Although managing for resilience still 'suffers' from the way that trade-offs are conceptualised between the ability to recover from short-term shocks such as coping with seasonal food deficits and sustainability, which is the long-term planning for adaptation capacity under a changing climate. While a global consensus on tackling greenhouse gas emissions is needed. Governments are faced with the challenge of providing adaptation measures in the face of recurring and potentially irreversible crises. This will require more partnerships between international and national levels. In terms of enabling adaptation, more needs to be understood about the role the state plays in provisioning adaptation measures and public expectations.

National–local adaptation

At the national and local levels there has been quite extensive research into barriers to, and enablers of, adaptation amongst a wide variety of institutions (see Gifford *et al.*, 2011; Jones and Boyd, 2011; Moser and Ekstrom, 2010). In general, this research has demonstrated that there are a series of structural and psychological barriers to adaptation. Jones and Boyd (2011), for example, identified caste (structure) as a constraint to community engagement on adaptation measures, while Moser and Ekstrom (2010) presented a framework to diagnose barriers to adaptation, which identified barriers separately in the understanding, planning and managing phases of adaptation. Some of these barriers are structural or institutional, but many relate to individual cognitive and normative factors. In this regard, more needs to be understood about the individual and household level of engagement in adaptation, what kinds of public support and incentives exist for particular adaptation measures, and what do people consider is an acceptable risk and in this regard what are people's expectations? Development studies can help to explore the entrenched structural causes and manifestations of poverty that act as barriers to adaptation. To have extensive impact on the deep-rooted structures of poverty in the context of climate change adaptation, it will be important to seek approaches that are truly participatory and experimental to empower people.

Conclusions

The chapter has described linkages between climate change and development and identified some key barriers and bridges in adapting to climate change across levels of decision making. Development studies as a discipline can help to expand our understanding of how to address the challenges of climate change. In particular, development can offer insights, analytical tools and approaches that lend themselves to critical, rich and context-based understandings of adaptation.

References

Boyd, E. and Juhola, S. (2009) Stepping up to the climate change: Opportunities and barriers in reconceptualisation development futures. *Journal of International Development* 21(6): 792–804.

Funtowicz, S. and Ravetz, R. (1993) Science for the post-normal age. *Futures* 25(7): 739–755.

Gifford, R., Kormos, C. and McIntyre, A. (2011) Behavioral dimensions of climate change: Drivers, responses, barriers, and interventions. *Wiley Interdisciplinary Reviews–Climate Change* 2: 801–827.

IPCC (2007) *Assessment of Adaptation Practices, Options, Constraints and Capacity.* Cambridge: Cambridge University Press.

Jones, L. and Boyd, E. (2011) Exploring social barriers to adaptation: Insights from Western Nepal. *Global Environmental Change–Human and Policy Dimensions* 21: 1262–1274.

Moser, S. C. and Ekstrom, J. A. (2010) A framework to diagnose barriers to climate change adaptation. *Proceedings of the National Academy of Sciences of the United States of America* 107: 22026–22031.

Rockström, J., Steffen, W., Noone, K., Persson, Å., Chapin, S., Lambin, E. F., Lenton, T. M., Scheffer, M., Folke, C., Schnellnhuber, H. J., Nykvist, B., De Wit, C. A., Hughes, T., Van der Leeuw, S.,

Rodhe, H., Sörlin, S., Snyder, P. K., Costanza, R., Svedin, U., Falkenmark, M., Karlberg, L., Corell, R. W., Farby, V. J., Hanse, J., Walker, B., Liverman, D., Richardson, K., Crutzen, P. and Fole, J. A. (2009) A safe operating space for humanity. *Nature* 461: 472–475.
Schipper, E. L. F. (2007) *Climate change adaptation and development: Exploring the linkages*. Tyndall Working Paper No. 107. University of East Anglia, Norwich.
Smit, B. and Wandel, J. (2006) Adaptation, adaptive capacity and vulnerability. *Global Environmental Change* 16(3): 282–292.
UNEP (2011) Food security in the Horn of Africa: The implications of a drier, hotter and more crowded future. *Global Environmental Alert Services* (November). Available online at www.unep.org/GEAS/

6.4

A changing climate and African development

Chukwumerije Okereke

Introduction

Climate change is arguably the greatest development challenge facing Africa in the twenty-first century. As the continent most susceptible to climate variability and vulnerability, Africa faces a threat that can reverse many decades of national and international development effort and plunge large sections into severe poverty. Vulnerability to climate change is compounded and exacerbated by multiple stressors such as low level of development, rapid population growth, weak institutions and low adaptive capacity. Both historically and currently, Africa's contribution to climate change is very low compared to other world regions. This makes the changing climate in Africa and associated negative consequences very much an issue of global justice and equity. However, while climate change poses profound challenges to African development, it also offers unique opportunities to pursue low carbon growth and build more resilient economies. Whether or not climate change will impede African development or be turned into an opportunity for sustainable development will depend on how governments and other stakeholders at different scales – from the local through national to international level – choose to respond.

Climate types and trends in Africa

Africa is a large and geographically diverse continent. The diversity in geography and climate regimes is determined by the location, size and shape of the continent, which is the second largest in the world. A number of climatic zones can be distinguished. These include the rain forest climate in the central portion of the continent, the equatorial climate in the Guinea coast, a large tropical savannah zone which encompasses about one-fifth of the continent and the semi-temperate zone found in the extreme northwest and southwest of Africa. There is also a very large expanse of arid or desert climate zone comprising the Sahara (in the north), the Horn (in the east), and the Kalahari and Namib deserts (in the southwest).

According to IPCC (Metz *et al.*, 2007) the climate of Africa is controlled by complex maritime and terrestrial factors. The key forces include the El Niño-southern oscillation (ENSO) which influences mostly eastern and southern Africa and the North Atlantic oscillation (NAO) which affects mostly northern Africa. The Atlantic Ocean current together with the African easterly jet (AEJ) and the tropical easterly jet (TEJ) have significant influence on western and central Africa. Southern African climate is influenced by the combination of the Southern Oscillation Index (SOI), the Indian Ocean sea-surface temperature (SST), the intertropical convergence zone (ICTZ) and also ENSO. Terrestrial vegetation covers mostly in the rainforest regions and the mineral dust in the Sahel region also have an impact on climate systems. The complex interactions of these physical forces and human factors such as deforestation, land use and migration account for the extremely high climate variability of Africa. And precise knowledge of the mode of interaction and effect of these various forces is severely limited by paucity of data and a lack of modelling tools.

Observed temperatures indicate a greater warming trend since 1960 and the IPCC suggests that all of Africa is very likely to warm during the twenty-first century and at a rate that is larger than the global, annual mean warming throughout all seasons. Future and historical warming is not uniform within the continent but varies according to regions. For example, the observed temperature rise over the last three decades has ranged from 0.1°C in South Africa (Kruger and Shongwe, 2004) to 0.29°C in the tropical forest region (Malhi and Wright, 2004).

The current and future picture for rainfall looks more complex but the trend clearly suggests high levels of variability in mean seasonal and decadal rainfall over the century. In large parts of West Africa, there has been a 4 per cent decline in mean annual precipitation with a decrease of up to 40 per cent noted between 1968 and 1990 (Nicholson, 2001). In contrast, an average of 10 per cent increase in rainfall in the last 30 years has been reported along the Guinea coast. Southern Africa does not show a significant long term trend, although increased inter-annual variability has been observed (Metz *et al.*, 2007). Overall, Africa has witnessed increased anomalies in weather pattern, and more extreme weather events including droughts and floods. These changes have major implications on African lives and economy in ways that are more significant than many populations in other continents.

Key dimensions of impact

Climate exerts significant influence on economic activities in Africa from the national to household level. This is primarily because the African economy is still heavily resource-based with a very large portion of the population depending on rain-fed subsistence agriculture for their livelihood. Accordingly, even slight to medium variability in seasonal climate can have immediate and direct impact on yield and livelihood. Climate variability affects the degree and pattern of precipitation, temperature and water availability. Climatic change is also associated with increased floods, droughts, cyclones, typhoons and frost. These events on their own or in combination can result in increased risks of crop failure with devastating effect on food security. The effects of these weather events are more difficult to mitigate given the general lack of mechanisation, improved seed varieties and irrigation systems. Indeed sub-Saharan Africa is the world's poorest and most rain-dependent region. The result is that several millions in this region have been exposed to the risk of hunger, malnutrition and death associated with drought and crop failure in the last three decades. For example, it has been estimated that up to 8.7 million where affected by the drought that beset Ethiopia in 1994. This drought resulted in the death of about 1 million people and 1.5 million livestock (FAO, 2000). More recently millions are facing malnutrition, famine and death with the East African drought of 2009–2011 proving to be one of the worst that the Horn of Africa has faced in 60 years. Southern Africa experienced an over 50 per cent drop in cereal harvest with more than 17 million exposed to the risk of starvation at the beginning of the 1990s (FAO, 2000). Both the West African and the Sahelian regions have experienced drought of unprecedented severity and frequency in recorded history since 1960 with millions facing famine and death (Zeng, 2003).

Predictions about possible future scenarios do not offer much comfort. For example, it is suggested that by 2010, an estimated 32 per cent of sub-Saharan Africa's total population of 400–500 million will suffer from malnutrition, compared with 4–12 per cent in other developing countries (Baro and Deubel, 2006). Furthermore Arnell (2004) estimates that between 75 million and 250 million people in sub-Saharan Africa could have their livelihoods compromised by 2020 due to climate, while the IPCC report (Metz *et al.*, 2007) suggests that climate change could result in about a 50 per cent drop in agricultural production in Africa by 2030.

The situation in agriculture and food security mirrors the situation in other sectors, including water, health, ecosystem, energy, infrastructure and human settlement. About 25 per cent of Africans experience acute water stress and of the 69 per cent that live in areas of relative abundance, a significant majority does not have access to clean water. These figures are very much likely to increase in the future as a result of climate change-induced factors such as flash flooding, siltation of river basins and degradation of land water sheds.

Rising temperatures and flooding will affect pathogen life cycles and rate of infections. The health consequences of climate change are likely to be felt far more in Africa than in any other continent. For example, the loss of healthy life years as a result of climate change is predicted to be 500 times greater in poor African populations than in European populations (McMichael *et al.*, 2006). Climate change will affect the spread of malaria, meningitis, neglected tropical diseases and a host of other water and air borne diseases. Infections and deaths are likely to be compounded by a decrease in immunity due to malnutrition and famine. Indeed, malnutrition is considered the most important health risk globally as it accounts for an estimated 15 per cent of total disease burden in disability adjusted life years (DALYs). At present, undernutrition causes 1.7 million deaths per year in Africa and is currently estimated to be the largest contributor to climate change-related mortality around the world (Patz *et al.*, 2005).

Many ecosystems will collapse due to a combination of drought, increase in population and land use changes. In the absence of effective adaptation measures, this will likely result in the exacerbation of drought, hunger, as well as a loss of earnings from tourism. Many zones, countries and entire sub-regions are so vulnerable to sea-level rise and flooding that climate change becomes a major threat. These pressures will affect critical infrastructure such as roads, housing and energy installations. Ultimately, the result will be large-scale human displacement followed by unprecedented levels of internal and cross-border migrations. It is evident then that African development hangs on a balance due to climate change and that effective climate adaptation is one of the most urgent needs facing Africa.

Africa and climate justice

Historically and currently, Africa's contribution to climate change is very low compared to other world regions. This makes the changing climate in Africa and associated negative consequences very much an issue of global justice and equity. On the basis of global surface temperature mean increases in 2000, it has been calculated that the average contribution of Africa to climate change, based on a start date of 1890 is a mere 7 per cent, a statistic which pales in comparison to the 40 per cent contribution from OECD countries and 24 per cent contribution from Asia (Den Elzen *et al.*, 2005; Höhne and Blok, 2005) (see Figure 6.4.1). The picture looking forward does not tell a significantly different story. Höhne and Blok calculate that by 2050 the OECD will be responsible for about 41.7 per cent of global average surface temperature increase due to fossil CO_2 while Africa and Latin America combined would be responsible for just 17.05 per cent (Höhne and Blok, 2005). Given the huge disparity in both current and historical contribution and the negative consequences involved, climate change is essentially a case of the rich imposing their burden on the poor.

Although nearly all the other continents of the world will face climate change-induced challenges, the effect in magnitude does not compare with what is envisaged in Africa. Moreover many of these economies have adequate financial, technical and institutional capabilities to cope. African countries and many environmental organisations have therefore been making the case that justice should be placed at the centre of the international effort to tackle climate change (Okereke, 2010). Making justice the cornerstone of international climate cooperation requires at least four things (Okereke and Schroeder, 2009). First is that developed countries should take urgent steps to cut their carbon and other greenhouse gas emissions

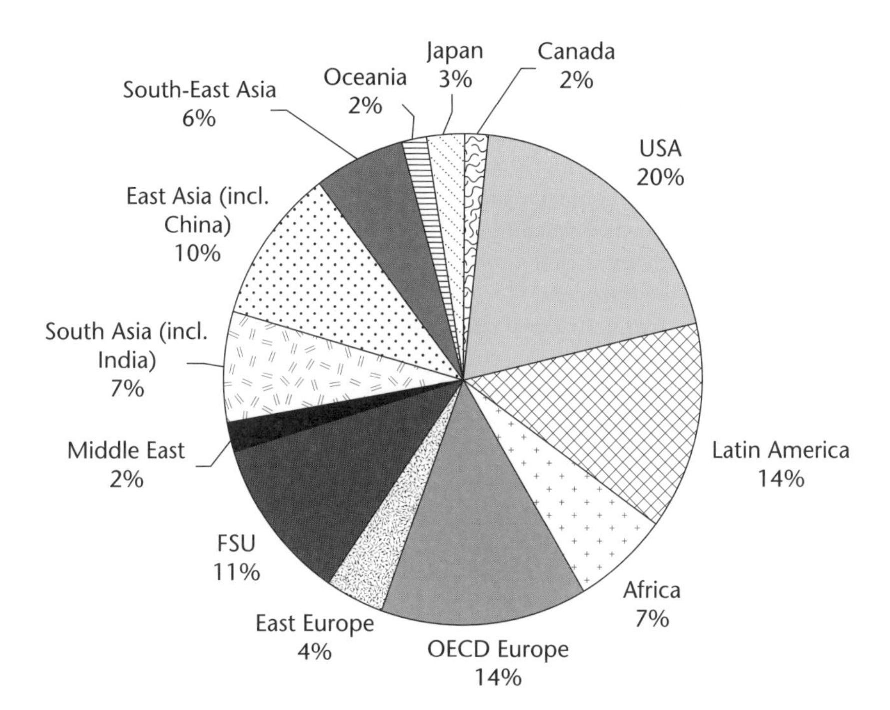

Figure 6.4.1 Percentage contribution of greenhouse gas according to regions

Source: http://www.match-info.net

responsible for global warming and climate change. Second, developed countries must provide Africa with adequate and long-term technical and financial assistance; it needs to help adaptation to climate change and to develop in a sustainable manner. Third, there should be greater commitment to procedural equity with more spaces provided for Africa to be effectively represented in climate negotiations. And fourth, the global community needs to dismantle background structures of inequality such as unfavourable terms of international trade which work to exacerbate poverty in Africa (Okereke, 2011).

African governments themselves also have a lot to do to increase resilience against climate change. Effort should not be placed solely on achieving climate justice at the international level but also by pursuing more equitable policies within borders. Relevant institutional reforms would have to be embraced to ensure that financial and technical assistance received from the developed countries are put to best use. And critically there should be greater commitment to more effective governance, internal capacity building and homegrown economic development. The emphasis should be on turning climate change into opportunities by pursing a different type of economic development – the so-called climate resilient, climate smart or climate compatible economic development.

Climate compatible development options for Africa

It has been indicated that while climate change poses profound challenges to African development, it also offers unique opportunities to pursue low carbon growth and build more resilient economies. Notable aspects of such a 'green development' approach would include greater exploitation of renewable energy, construction of sustainable cities and transportation systems and more effective land use and forest management.

Low carbon development has the potential to help some African countries reduce economic vulnerability associated with dependence on oil, especially since about one-third of African countries are landlocked and pay huge amounts to import oil. Many African countries have access to abundant renewable energy such as solar wind and hydro power. Although gas is a finite energy source it is much cleaner than oil and coal. There is a vast reserve of proven oil in many African countries such as Nigeria, Congo and more recently Ghana. African governments and their development partners would need to concentrate their efforts on how best to harness these vast renewable resources for economic growth and increased energy security within the continent.

The finance available for international climate change is still a far cry from what is needed. But public and private climate funding is getting significant and likely to grow even more in the coming years. The Green Climate Fund established in Cancun for assisting poor countries is expected to be worth USD100 billion a year by 2020 and frontline donor agencies such as the World Bank will likely be committing considerable money to climate development. African governments could find creative ways to leverage these and related funding mechanisms in an effort to build climate-smart economies. Individual African countries would need to prepare detailed climate-resilient growth plans which identify strategies that are sensitive to their unique circumstances. Such plans would be needed to attract the much-needed funding from international donor agencies but also provide a clear road map for pursuing economic growth that will not exacerbate climate change and related vulnerabilities. In Africa and elsewhere climate change also warrants serious questions about what exactly development is all about and how best to achieve the desired objectives. Given the finiteness of earth resources and ever rapidly increasing population, there is a need for a thorough examination of prevailing thinking which ties development too strongly with economic growth, mass production, and mass consumption.

Bibliography

Arnell, N. W. (2004) 'Climate Change and Global Water Resources: SRES Emissions and Socio-Economic Scenarios', *Global Environmental Change*, 14: 31–52.
Baro, M. and Deubel, T. F. (2006) 'Persistent Hunger: Perspectives on Vulnerability, Famine, and Food Security in sub-Saharan Africa', *Annu. Rev. Anthropol.*, 35: 521–538.
Den Elzen, M., Fuglestvedt, J. Höhne, N. Trudinger, C., Lowe, J., Matthews, B., Romstad, B., Pires de Campos, C. and Andronova, N. (2005) 'Analysing countries' contribution to climate change: Scientific and policy-related choices', *Environmental Science & Policy*, 8(6): 614–636.
FAO (2000) *FAO Statistics Database*. Rome: United Nations Food and Agriculture Organisation.
Höhne, N., and Blok, K. (2005) 'Calculating Historical Contributions to Climate Change: Discussing the "Brazilian Proposal"', *Climatic Change*, 71: 141–173.
Kruger, A. C. and Shongwe, S. (2004) 'Temperature Trends in South Africa: 1960–2003', *Int. J. Climatol.*, 24(15): 1929–1945.
Malhi, Y. and Wright, J. (2004) 'Spatial Patterns and Recent Trends in the Climate of Tropical Rainforest Regions', *Philos. T. Roy. Soc. B*, 359: 311–329.
McMichael, A. J., Woodruff, R. E. and Hales, S. (2006) 'Climate Change and Human Health: Present and Future Risks.' *Lancet* 367: 859–869.
Metz, B., Davidson, O. R., Bosch, P. R., Dave, R. and Meyer, L. (2007) *Climate Change 2007: Mitigation of Climate Change* (IPCC Fourth Assessment Report). Contribution of Working Group III to the Fourth Assessment Report of the Intergovernmental Panel on Climate Change, 2007. Cambridge: Cambridge University Press.
Nicholson, S. E. (2001) 'Climatic and Environmental Change in Africa During the Last Two Centuries', *Climate Res.*, 17: 123–144.
Okereke, C. (2010) 'Climate Justice and the International Regime', *WIREs Climate Change*, 1: 462–474.
Okereke, C. (2011) 'Moral Foundations for Global Environmental and Climate Justice', *Royal Institute of Philosophy Supplement*, 69: 117–135.
Okereke, C. and Schroeder, H. (2009) 'How can the Objectives of Justice, Development and Climate Change Mitigation be Reconciled in the Treatment of Developing Countries in a Post-Kyoto Settlement?' *Climate and Development*, 1(1): 10–15.
Patz, J., Campbell-Lendrum, D., Holloway, T. and Foley, J. (2005) 'Impact of Regional Climate Change on Human Health', *Nature*, 438: 310–317.
Zeng, N. (2003) 'Drought in the Sahel', *Atmospheric Science*, 302(5647): 999–1000.

6.5

Vulnerability and disasters

Terry Cannon

Environment as risk and provider for production

The key to understanding natural hazards and disasters is that the environment has a dual character: it is both a source of production opportunities (farming, fisheries, forests, pastures, energy sources, etc.) *and* of hazards. Many of the locations that provide for livelihoods are also at risk from geophysical and meteorological hazards. In much of the world, people are willing – or forced by poverty – to risk being exposed to hazards in order to reap the everyday production benefits that happen to be in dangerous places: flood plains provide farmers with fertile soils, and flat land for settlement, transport links and production activities; the slopes of volcanoes normally give rise to very fertile soils for farmers; tsunami and storm-prone coasts are often suited to agriculture, cities and trade, and even active geological faults that trigger earthquakes channel water to the surface and provide the basis for life in the desert.

In recent decades, the idea that extreme environmental events can be called 'natural disasters' has been challenged: they are not 'natural' since hazards can only be disastrous when they affect people who are vulnerable and exposed to them. Disasters should be considered as the socially constructed outcome of environmental trigger events visited upon human systems. An earthquake in an unoccupied area is merely a natural phenomenon; it only leads to a disaster when it affects humans and people get in its way. Disasters therefore happen only because of socioeconomic reasons: various factors encourage or compel people to live in dangerous places (often even when there is general awareness of the dangers). This means that it is vital to understand disasters not only in terms of the environmental 'trigger', but most importantly in regard to the social construction: why are they in places that *expose* them to the risk, what factors make them *vulnerable*, and what cultural and psychological *attitudes to risk* lead people to put themselves in harm's way. This section is focused on *vulnerability* as the key factor related to social, economic and political processes that determine how some groups of people suffer more in disasters than others.

At the simplest level, people can be 'vulnerable' to the hazard impact itself because they have occupied places that are hazardous (river valleys in Pakistan devastated in the 2010 floods, low-lying deltaic land in Myanmar/Burma in the 2009 cyclone Nargis, the Indian Ocean tsunami in 2004, New Orleans). However, it is also evident that in many (probably most) disasters, the hazard's impact is socially constructed in a much more significant way: some groups of people are much more badly affected than others. In this deeper sense, *vulnerability* is a characteristic not simply of being human and in a dangerous location (*exposure*), but of being part of a socioeconomic system that allocates risk unequally between different social groups (such as those based on class, gender, ethnicity, age). This is often much more evident in developing countries, where there is a very high correlation between being affected in a disaster and being poor (though logically this effect would have to take account of the high proportion of poor people). But as seemed evident in the aftermath of Hurricane Katrina in 2004, such socially constructed vulnerability can also be a feature of rich countries.

Disasters and development

It is only in recent years that donors and governments (of both rich and poor countries) have begun to accept that there are significant connections between disasters and development (e.g. IEG 2006; UNDP 2004; DFID 2004). Disasters triggered by natural hazards destroy and delay development: not just structures like schools, bridges and factories, but especially jobs and livelihoods. Worse still, many projects and other development activities are still carried out without the simple precaution of checking the potential impact of known risks in the area concerned (e.g. many village water pumps in parts of Asia installed in development projects that are covered when it floods). And there was no references to disasters in the Millennium Development Goals, despite the need to protect poverty reduction and development gains from hazards.

The lack of proper connections between disasters and development is made all the more complicated by the effects of global warming. There is evidence for an increase in number and/or intensity of hydro-meteorological events (floods, drought, storms and hurricanes, heat-waves and related wildfires, landslides related to heavy rainfall), and it is highly likely that these are linked to climate change. Of all natural hazards, floods and drought are responsible for the vast majority of deaths and disrupted lives, and it is therefore likely that the number of disasters related to such events will increase unless vulnerability can be reduced. But climate change is also (through the negative impacts of changes in temperature and rainfall on agriculture, fisheries, health) increasing the number of people who will be vulnerable to all types of hazards. Climate change is therefore producing new vulnerability to hazards, and undermining the progress of development. In part, the UNFCCC (UN Framework Convention on Climate Change) acknowledges the damage done to the economy of poor countries, through a framework of funding for adaptation and (at the Cancun COP in 2010) for loss and damage. However, these are not properly integrated with programming for development, and do not work through concepts of vulnerability.

The Hyogo Framework for Action (HFA) on disaster risk reduction, which aims to be 'Building the Resilience of Nations and Communities to Disasters' paid little attention to climate change. Running from 2005–2015, the HFA mentions the need to reduce vulnerability, but does not have a strong development context and has little to say on poverty reduction or livelihoods protection (or climate change). The international agreements for the HFA, the MDGs and the Kyoto protocol of the UNFCCC all come to an end in 2015, bringing an opportunity to create an international framework that combines work on development and poverty, disaster risk reduction and climate change. However, given the failure of the Rio+20 meeting in 2012, it is doubtful that there will be much progress in this.

New thinking about disaster–development linkages has been a result of several factors. One is the devastating impact of a number of disasters in recent decades that have highlighted their significance in creating poverty and undermining development. The Pakistan floods of 2010 caused billions of dollars of losses, which added to the $5 billion costs of the 2005 earthquake, equivalent to the total of all development assistance in the previous three years (IEG 2006, p. xx). Hurricane Mitch devastated several central American countries in 1998, and Carlos Flores (the then president of Honduras) said: 'We lost in 72 hours what we have taken more than 50 years to build, bit by bit.' More than a decade later, it is widely considered that many poor people in the countries affected are still struggling to catch up.

Vulnerability analysis

The use of the concept of vulnerability in analysing disasters has become much more widespread in the past two decades. It is part of the challenge to the notion that disasters are 'natural'. It has shifted the argument for dealing with disasters more to issues of prevention (including precautions and reductions of vulnerability) rather than the usual focus on emergency response or technical fixes for hazard mitigation: there is still an overwhelming imbalance in national and international spending on emergency relief as compared with preventive measures. Also the term vulnerability is now widely abused, without precision or clarity, much as the word sustainable has lost its meaning.

'Hard' hazard mitigation measures are usually expensive, difficult or inappropriate (e.g. flood defences), and early warnings are not possible for all hazards. But if disasters are a consequence of vulnerability, then the effects of hazards can be reduced or avoided by changing people's vulnerability: 'soft' preparedness should be the focus for disaster prevention (through development and poverty reduction). Because vulnerability is inherently connected to people's livelihoods, level of poverty, and their ability to protect themselves (with the right type of house) or live in a safe place, it is also strongly linked to development.

Despite its widespread use (e.g. by NGOs and increasingly by international organizations), prospects for the success of vulnerability analysis in the actual reduction of disasters still seem remote. This is largely because the reduction of vulnerability requires changes in economic and political systems similar to those required to reduce poverty across the world. In other words, the barriers to reduction of disasters (and global warming) are similar to those that prevent or constrain the improvements in people's lives through development. But there is an additional layer to contend with in disasters: even where poverty has been reduced, development is not a guarantee that people will be safe from hazards and disasters: their homes and locations must still be safe in relation to specific known risks, and people's *attitude to risk* needs to be shifted to take hazards seriously.

Making significant changes in people and institutions in relation to risk behaviour is difficult. People are often willing to take risks, especially when the risk is from a sudden-onset hazard that is infrequent and in some indeterminate future. There is no shortage of people willing to live in earthquake zones in California and hurricane-prone Florida, and in poorer countries there are also cultures of risk-taking that are difficult to change. Vulnerability analysis also typically finds economic and political causes of exposure to risks that are greater for some groups than for others. Remedies are therefore likely to upset existing power structures.

Vulnerability is not the same as poverty, marginalization, or other conceptualizations that identify sections of the population who are deemed to be disadvantaged, at risk, or in other ways needy. *Vulnerability* must involve a *predictive* quality that is specifically related to the relevant hazard(s) that may affect people (vulnerability to one type of hazard will differ from another). Precisely because it is predictive, the concept of vulnerability should be capable of directing disaster prevention. It should do this by seeking ways to protect and enhance people's livelihoods, assist vulnerable people in their own self-protection, and support institutions in their role of disaster prevention (Cannon 2008). Unfortunately the word vulnerability has been used in many different ways and it has generated almost as much confusion as insight. In recent years it is also being eclipsed by the use of the term *resilience*, which some consider as a dangerous move towards a depoliticizing of risk analysis (Cannon & Muller-Mahn 2010). A widely accepted definition is that '[vulnerability is] the characteristics of a person or group and their situation that influences their capacity to anticipate, cope with, resist and recover from the impact of a natural hazard' (Wisner *et al.* 2003, p. 11).

The vulnerability conditions are themselves determined by processes and factors that often involve the power relationships and institutional factors operating in the wider political and economic context. These conditions can be analysed as five specific components of vulnerability:

- people's base-line status;
- the strength and hazard-proof status of their livelihoods;
- their ability *and willingness* to protect themselves from known hazards (even when they have money, many people choose to expose themselves to hazards as a trade-off with the benefits they get in that location);
- adequate social protection (against specific hazards) by other institutions (such as local government, NGOs, religious groups and so on) – for instance, through early warning systems and earthquake building codes;
- the quality of governance – the power systems that determine how good the social protection is, and how resources are allocated in the society to determine income and welfare between different groups of people.

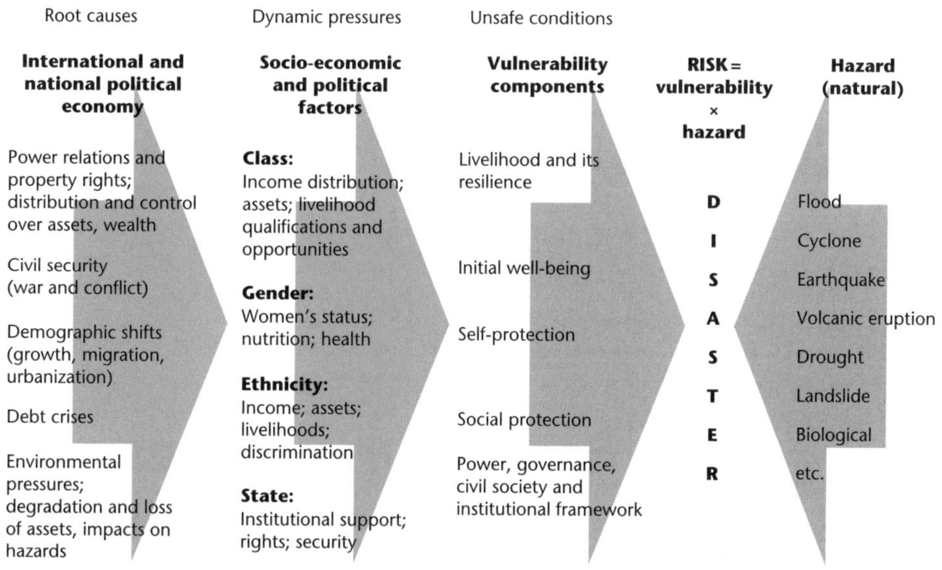

Root causes

International and national political economy

Power relations and property rights; distribution and control over assets, wealth

Civil security (war and conflict)

Demographic shifts (growth, migration, urbanization)

Debt crises

Environmental pressures; degradation and loss of assets, impacts on hazards

Dynamic pressures

Socio-economic and political factors

Class: Income distribution; assets; livelihood qualifications and opportunities

Gender: Women's status; nutrition; health

Ethnicity: Income; assets; livelihoods; discrimination

State: Institutional support; rights; security

Unsafe conditions

Vulnerability components

Livelihood and its resilience

Initial well-being

Self-protection

Social protection

Power, governance, civil society and institutional framework

RISK = vulnerability × hazard

D I S A S T E R

Hazard (natural)

Flood

Cyclone

Earthquake

Volcanic eruption

Drought

Landslide

Biological

etc.

Figure 6.5.1 The PAR model

Source: Adapted from Wisner *et al.* (2003)

People's vulnerability to disasters is a function of this wider political-economic environment. The causes of vulnerability are inherently connected with people's livelihoods (vulnerability is likely to be reduced when livelihoods are adequate, diversified and robust) and their position in society and its systems of power and governance (which determine access to livelihood resources and adequacy of social protection). Understanding livelihoods and power relations, and the pattern of assets and incomes, is therefore crucial to understanding much of how vulnerability differs between various groups of people.

It is therefore essential to put patterns of vulnerability in a wider social context. One attempt to do this is the 'pressure and release' (PAR) model (Figure 6.5.1). This identifies a web of causative factors, from the international to the local, which can be identified as 'root causes' and 'dynamic pressures', which lead to the 'unsafe conditions' of people's vulnerability. In effect, this PAR model is a graphic representation of the commonly used 'equation':

$$\text{Risk} = \text{Vulnerability} \times \text{Hazard}$$

In the diagram, the risk of being harmed by a hazard is represented by the 'pressure' between the hazard on one side and the vulnerability conditions on the other. To 'release' people from risk involves reducing the hazard and/or reducing their vulnerability, turning the arrows around so that they discharge the pressure of risk.

References

Cannon, T. (2008) 'Reducing people's vulnerability to natural hazards: Communities and resilience', *WIDER Research Paper 34*. Available online at www.wider.unu.edu/publications/working-papers/research-papers/2008/en_GB/rp2008-34/

Cannon, T. & Muller-Mahn, D. (2010) 'Vulnerability, resilience and development discourses in context of climate change', *Natural Hazards* 55: 621–635.

DFID (2004) *Disaster risk reduction: A development concern – a scoping study on links between disaster risk reduction, poverty and development*, London: Department for International Development.

IEG (Independent Evaluation Group) (2006) *Hazards of nature, risks to development: An IEG evaluation of World Bank assistance for natural disasters*, Washington DC: World Bank.
UNDP (United Nations Development Programme) (2004) *Reducing disaster risk: A challenge for development*, New York: UNDP Bureau for Crisis Prevention and Recovery.
Wisner, B., Blaikie, P. M., Cannon, T. and Davis, I. (2003) *At Risk: Natural Hazards, People's Vulnerability and Disasters*, 2nd edition, London: Routledge. The first three chapters are available free on the internet at www.preventionweb.net/files/670_72351.pdf

6.6

Ecosystem services and development

Tim Daw

Natural ecosystem processes support and enrich human lives and contribute to economies in a wide range of ways, conceptualised as 'ecosystem services' (ES). Unlike many goods and services that are sold on markets, these benefits are provided for free, with no prices to indicate their value. They are, therefore, undervalued by conventional economics and can be overlooked in decision making. Thus, apparently economically rational degradation of ecosystems, can result in significant losses in human welfare. The ES concept draws attention to the essential benefits that are provided to people and economies for free by natural ecosystems, makes the economic case for conserving nature and aims to improve the efficient use of ecosystems to support human well-being.

This chapter provides an overview of the concept and definition of ES, their valuation, their relevance for poverty alleviation and their application through payment schemes. The future challenges of ES research are then discussed.

The Millennium Ecosystem Assessment (MA) (2005) drew attention to the profound modification of ecosystems around the world, and the changing portfolio of ES that they produce. While some ES, such as food production, have increased dramatically through agriculture, most others are declining as natural ecosystems continue to be degraded or converted. This has created an impetus for the assessment and valuation of ES and their communication to policy (e.g. The Economics of Ecosystems and Biodiversity, the Intergovernmental Platform on Biodiversity and Ecosystem Services – see list of websites).

The growing academic literature on ES includes conceptual development, identification and modelling of how ES are generated from ecosystems, investigation into how ES benefit different social groups under different conditions, and use of economic theory to represent the value of ES so that they can be considered in development decisions, or national accounts. Another evolving field is the design and appraisal of payments for ecosystem services (PES), in which ecosystem stewards are financially incentivised to maintain flows of ES by conservation interests, governments or parties who can benefit from the ES.

Examples and definitions of ecosystem services

The case of mangrove forests offers some classic examples of ES. These tropical coastal ecosystems have limited obvious development value, and thus have frequently been cleared for more lucrative land uses such as prawn farming. However, mangroves generate a wide range of ecosystem services. They provide a range of products for local communities including wood, shellfish and medicinal plants and protect coastlines and communities from storm surges, and erosion. Their root systems consolidate sediments that might otherwise smother coral reefs, and serve as feeding and nursery habitats for fish, thus supporting fisheries productivity in adjacent waters. Mangroves may also be valued by traditional cultures, or for ecotourism that can support local livelihoods. Fully accounting for these multiple ES may demonstrate that the costs of losing mangroves exceed the benefits from replacing them with productive industries (Barbier *et al.* 2008).

Various definitions of ecosystems services have been proposed, but all reflect a human-centred view of ecology. The MA (2005) used a generalised definition of 'the benefits people obtain from ecosystems' and identified four categories of (a) provisioning services such as food production; (b) regulating services that maintain a benevolent environment and protect against environmental disturbance, such as flood defence; (c) cultural services that are reflected in religious, recreational or cultural values and practices; and (d) supporting services comprising the underlying natural framework and processes (such as sediment consolidation) on which other services rely (Figure 6.6.1). The MA definition and categories have been very successful as a communication tool but are somewhat vague for the purposes of valuation or understanding how ES contribute to the well-being of specific people. Various authors have pointed out the need for more specific distinction between ecological processes, intermediate services, final potential services, goods (which may incorporate human inputs such as labour or capital) and actual benefits to people in terms of improvements in well-being. These distinctions are needed to avoid double counting during valuation (Fisher *et al.* 2009) and for conceptual clarity when assessing the contribution of ES to well-being. Ultimately, the definition of services and benefits varies according to the question of interest and the perspective of the beneficiary. Thus, from a poverty alleviation perspective, ES should be appraised in terms of how poor people's lives are actually improved, whether by food, shelter or employment (Daw *et al.* 2011).

Valuing ecosystem services

Much of the ES literature has been concerned with deriving values for ES using a range of economic valuation tools. Where provisioning services produce goods that are sold, market prices are assumed to reflect people's willingness to pay for goods, and thus their value to society. For non-marketed ES, willingness to pay may be inferred from people's behaviour (called revealed preferences), such as spending on travel to visit a natural area, or by directly asking them in surveys (stated preferences). Where the extensive data required for these methods are not available, previously calculated values for similar ecosystems elsewhere are sometimes used. Considerable debate and uncertainty revolves around these methods and valuation is a developing field (Bateman *et al.* 2011).

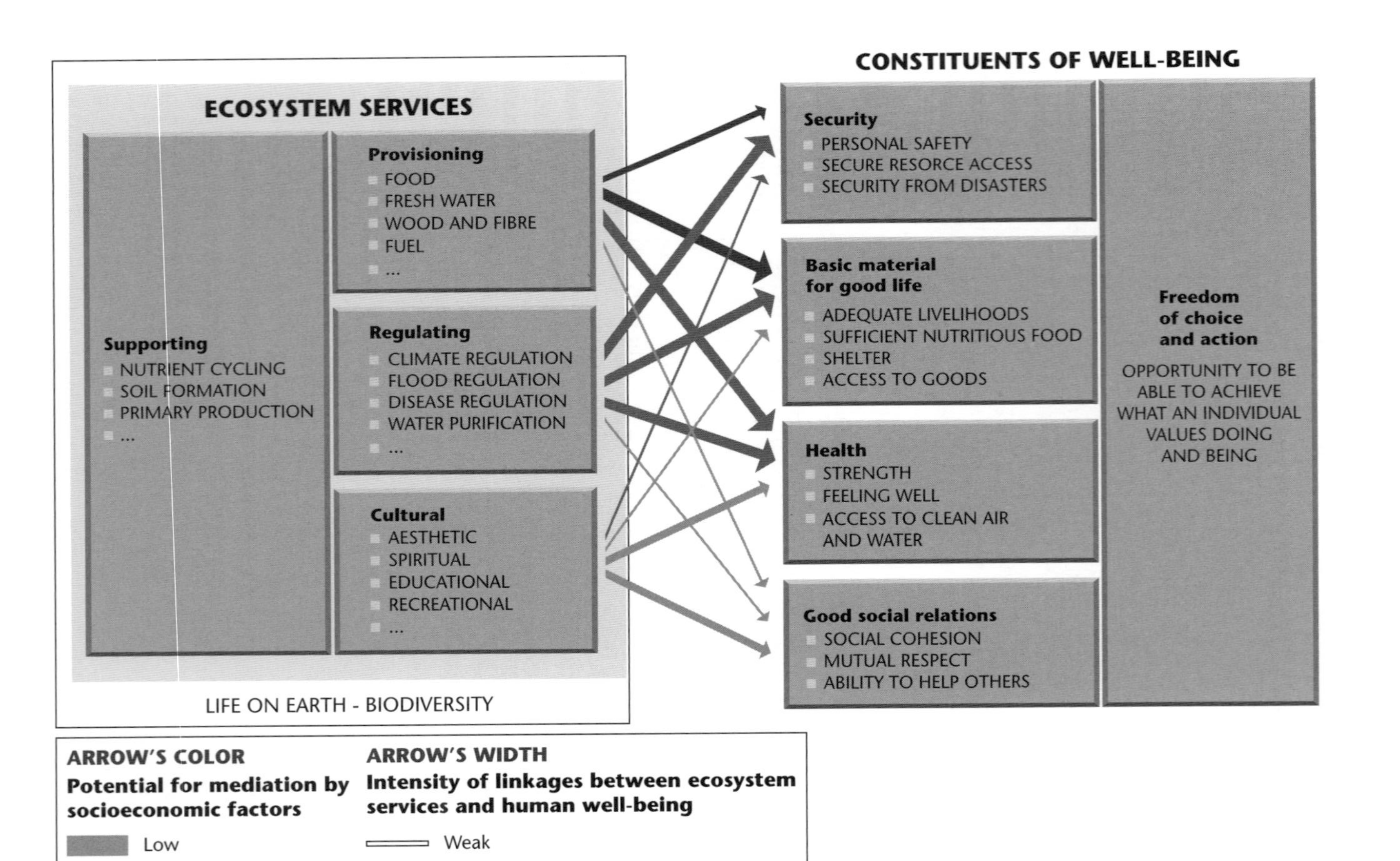

Figure 6.6.1 Linkages between ES and human wellbeing identified by the Millennium Ecosystem Assessment (2005)

Source: Millenium Ecosystem Assessment (2005)

Ecosystems services, human well-being and poverty alleviation

The MA (2005) applied the concept of ecosystem services specifically to 'human well-being and poverty alleviation', highlighting how ES can contribute to multiple dimensions of well-being including security, basic material for a good life, health, good social relations and freedom of choice and action (Figure 6.6.1).

While economic valuations of ES aim to capture the aggregate societal value, in reality different individuals and groups benefit from different ES to different extents, so that changes in ES inevitably create winners and losers. The poor are typically more sensitive to impacts on ES due to their greater reliance on natural resource-based livelihoods, and their vulnerability to natural hazards. This also applies globally, as poorer nations have suffered more costs and reaped fewer rewards from environmental changes (Srinivasan *et al.* 2008).

The degree to which an individual benefits from ecosystems depends on a range of mechanisms of access including social relationships, institutions, capabilities, rights and various capitals. Thus, the best opportunities to improve the poor's well-being may be through more secure and equitable access to ES rather than through increasing flows of ES.

ES contributions to well-being also depend on the perspectives and circumstances of ES beneficiaries. For example, earnings from an ES has a greater well-being impact for a poor individual than the same benefit would for a wealthier individual with multiple earning opportunities. Likewise the well-being impact of food or flood protection depends on how hungry someone is or how exposed their property is to flooding.

While valuation is useful for communication, advocacy and aggregate cost–benefit decisions, it is problematic for addressing the links between ES, well-being and poverty alleviation. Economic values are derived from people's willingness to pay, but (a) people may be ignorant about ES (e.g. they may be unaware they are benefiting from regulating services) and (b) poor people have less ability to affect prices. Thus valuation is skewed towards ES of interest to the wealthy (e.g. tourism) valuing and prioritising them over values of the poor (e.g. food security) (Hicks *et al.* 2009).

Payments for ecosystem services

PES represent an exchange between people who value an ES (such as downstream water users) and those who have agency to provide them by maintaining ecosystems (such as upstream landowners). PES are possible where ES can be measured, and functioning markets can be created between a 'provider' and a 'beneficiary'. Great interest has arisen around the potential for possible 'triple wins' from PES; economically efficient maintenance of ES, financial incentives for conservation, and raising the incomes of poor local service providers. However, experience so far suggests mixed outcomes. Poor groups may benefit from PES, but schemes cannot be optimised for both conservation and poverty alleviation (Wunder 2008). Furthermore PES may suffer from elite capture or corruption, excluding the poor from benefits or exacerbating existing inequalities (Kosoy and Corbera 2010). PES for reducing emissions from deforestation and forest degradation (REDD) are currently under development through UN agreements and could see payments to developing countries for safeguarding the carbon storage ES of forests that could run to billions of dollars.

Critiques of ES

Critics of ES approaches reject the neoliberal commodification of nature, particularly PES, pointing out that the reduction of complex bundles of ecological processes to single market prices is unlikely to reflect the true value of ES to people (Kosoy and Corbera 2010). Some commentators fear that the expansion of capitalist markets into ecology will be to the detriment of local and traditional users. For example, where land tenure is weak, and power inequalities not addressed, PES can lead to displacement of the poor from commonly held resources so that more powerful interests can benefit from payments (Kosoy and Corbera 2010).

While some conservationists have embraced ES as a rationale and source of resources for conservation, others fear that such a utilitarian perspective can 'crowd out' intrinsic values of nature and leave conservation vulnerable to changing market conditions or technology. For example, the valuation of pollination services from a forest reserve in Costa Rica fell from US$60,000 to nil as local producers switched from coffee to pineapples, which do not require pollination (McCauley 2006).

Future challenges and priorities for ecosystem services

The ecosystem service approach has moved from relative obscurity to mainstream policy within little more than a decade. Growing awareness of the impacts of development on natural ecosystems lends urgency to understand and value ecosystem services, as hard choices are faced between growing human populations and economies and shrinking reserves of natural assets and biodiversity. The new discipline is evolving fast but still faces significant conceptual, scientific and institutional challenges. The complex relationship between natural capital, ecosystem services and human well-being needs further exploration and understanding to allow ecosystems to be used to promote well-being and poverty alleviation. Institutional arrangements need to be developed that mitigate the risks of commodification of ES for the poor. Valuation methodologies need to account for unequal wealth distribution to support just decision making. Finally, tools are needed to understand and represent trade-offs in space and time, and to resolve them in a just and equitable way.

References

Barbier, E. B., Koch, E. W., Silliman, B. R., Hacker, S. D., Wolanski, E., Primavera, J., Granek, E. F., Polasky, S., Aswani, S., Cramer, L. A., Stoms, D. M., Kennedy, C. J., Bael, D., Kappel, C. V., Perillo, G. M. E. and Reed, D. J., 2008. Coastal Ecosystem-Based Management with Nonlinear Ecological Functions and Values. *Science*, 319(5861), 321–323.

Bateman, I. J., Mace, G. M., Fezzi, C., Atkinson, G. and Turner, K., 2011. Economic Analysis for Ecosystem Service Assessments. *Environmental and Resource Economics*, 48(2), 177–218.

Daw, T., Brown, K., Rosendo, S. and Pomeroy, R., 2011. Applying the Ecosystem Services Concept to Poverty Alleviation: The Need to Disaggregate Human Well-Being. *Environmental Conservation*, 38(04), 370–379.

Fisher, B., Turner, R. K. and Morling, P., 2009. Defining and Classifying Ecosystem Services for Decision Making. *Ecological Economics*, 68(3), 643–653.

Hicks, C. C., McClanahan, T. R., Cinner, J. E. and Hills, J. M., 2009. Trade-Offs in Values Assigned to Ecological Goods and Services Associated with Different Coral Reef Management Strategies. *Ecology and Society*, 14(1).

Kosoy, N. and Corbera, E., 2010. Payments for Ecosystem Services as Commodity Fetishism. *Ecological Economics*, 69(6), 1228–1236.

McCauley, D. J., 2006. Selling Out on Nature. *Nature*, 443(7107), 27–28.

Millennium Ecosystem Assessment (MA), 2005. *Ecosystems and Human Well-Being: Synthesis*. Washington, DC: Island Press.

Srinivasan, U. T., Carey, S. P., Hallstein, E., Higgins, P. A. T., Kerr, A. C., Koteen, L. E., Smith, A. B., Watson, R., Harte, J. and Norgaard, R. B., 2008. The Debt of Nations and the Distribution of Ecological Impacts from Human Activities. *Proceedings of the National Academy of Sciences of the United States of America*, 105(5), 1768–1773.
Wunder, S., 2008. Payments for Environmental Services and the Poor: Concepts and Preliminary Evidence. *Environment and Development Economics*, 13(03), 279–297.

Websites

The Economics of Ecosystems and Biodiversity: www.teebweb.org
The Intergovernmental Platform on Biodiversity and Ecosystem Services: www.ipbes.net
The Millennium Ecosystem Assessment: www.maweb.org
United Nations Collaborative Programme on REDD in Developing Countries: www.un-redd.org

6.7

Natural resource management

A critical appraisal

Jayalaxshmi Mistry

Introduction

Natural resources play a critical role in our lives, providing us with, for example, food, shelter and energy. For more than 1.3 billion people around the world, natural resources in the form of fisheries, forests and agriculture provide close to half of all jobs worldwide, with rural people, in particular, directly depending on natural capital compared to other parts of the population. In Africa, more than seven in ten poor people live in rural regions, and most are engaged in resource-dependent activities such as subsistence farming, livestock production, fishing, hunting, artisanal mining and logging. In these areas, natural resources provide the primary sources of income (financial and otherwise), but also the safety nets when crops fail, droughts hit and employment opportunities are few and far between.

Current thinking in the management of natural sources has linked natural and social systems into integrated and nested social-ecological systems, implying that effective management and governance of natural resources requires an understanding of the multiple, networked and dynamic interrelationships between socio-ecological systems at different scales (Pierre and Peters, 2009). Yet, to date many development policies and actions have supported a command-and-control approach which lacks the ability to manage and adapt to unexpected and rapid change. In addition, these management solutions/policy interventions have come from higher-scale structures, for example, national governments, which are not always compatible with the realities and perspectives of smaller-scale units, such as isolated rural communities and their associated natural resources. Even at the community scale, issues emerge with regard to how distinct groupings, for example, women, youth, the disabled and the elderly are engaged in natural resource management. The challenge of future policy evolution, therefore, is to link more explicitly socio-ecological systems at different scales with governance approaches that are reflexive in the way that they deal with multiple perspectives, interests and values of stakeholders (Voss *et al.*, 2006).

Current approaches to natural resource management

Although within the theoretical debates there seems to be growing advocacy for holistic, social-ecological, adaptive and participatory approaches to natural resource management, current policies seem to have hijacked both protectionist or 'fortress' and community-based approaches, and rolled them into one, where a clear distinction between the two can be hard to identify in practice. The way this has happened has been through the dominance of neo-liberal thinking leading to the increasing commodification of nature, where natural resource 'goods' and 'services' are transformed into 'objects' meant for trading as commodities (Castree, 2008). Notions closely associated with the commodification of nature include the separation of humans from the environment (e.g. narrowing down an ecological function to the level of an ecosystem service, hence separating the latter from the whole ecosystem), the establishment of a monetary value of nature (where price becomes paramount over other metrics of worth) and the separation of the resource user from the manager and the rise of the managerial class (the 'provider' and 'consumer' of resources set up a supply–demand relationship in market or market-like exchanges).

In this paradigm, and under the label of 'green capitalism' or 'market environmentalism', the creation and capture of market value for the services provided for humans by the non-human world is considered the most efficient and sustainable means of mitigating global environmental problems, such as climate change, while maintaining and even enhancing economic growth (Arsel and Büscher, 2012). Nowhere is this seen more clearly than in the arena of tropical forest management, where new forces are acting to modify and/or repartition access to, and exploitation of, natural resources. Sources of development financing are being made available to developing countries to maintain their forest stands, especially through assistance for climate change mitigation and adaptation through, for example, Reducing Emissions from Deforestation and Forest Degradation (REDD+), Payments for Ecosystem Services (PES), Multilateral Environmental Agreements (MEAs) and direct investments from private and non-governmental initiatives (Ghazoul *et al.*, 2010). However, little is known about the potential impacts of these new funding initiatives on forest conservation and to what extent they recognise, if not ensure, inclusion of divergent values, participation in political decision-making and equitable distribution of benefits, as determined by ethnicity, gender, age, income distribution and other differentiating factors (Sikor *et al.*, 2010). It has been argued that these new funding streams could potentially escalate inequalities, with some critics arguing that these initiatives may potentially diminish the power of local communities to control the management of their own natural resources. For example, national governments could apply carbon reduction strategies by disincentivising carbon intensive livelihoods within rural settings, such as traditional slash-and-burn agriculture, while supporting development projects, such as large-scale hydroelectric dams, that do not always directly and fully compensate for these lost livelihoods.

Guyana, for example, has been actively seeking ways to sustainably manage its extensive tracts of natural forest resources. Two particular policies championed by the government of Guyana are the adoption of the Low Carbon Development Strategy (LCDS) and REDD+ (Mistry *et al.*, 2009). The LCDS in particular calls for a reorientation of the economy from a resource extraction development paradigm to a supplier of environmental services, with low carbon and climate resilience as the focus. Out of this initiative emerged a memorandum of understanding (MoU) and a concept note agreed between Guyana and Norway for interim bilateral funding for a five-year period to support forest carbon and ecosystem service management through PES schemes. In addition, there is the growing involvement of the private sector in forest resource management. For example, the investment firm Canopy Capital and the related environmental alliance known as the Global Canopy Programme (GCP) signed agreements to help finance protected areas in return for 'ownership' of Guyanese forest ecosystem services, including carbon retention, rainfall generation and climate regulation, and a claim in any future profits. Incorporating environmental externalities into the market economy reflects a belief that the dynamic behaviour of the capitalist system changes from one of unlimited growth to one where environmental limits feed back to balance growth within sustainable limits. However, significant questions still remain on how one can value environmental externalities such as biodiversity loss, the effects of greenhouse gas emissions and ecosystem services.

Although there has been considerable international and national coverage of REDD+ (e.g. see the REDD Monitor website at www.redd-monitor.org), how these schemes will be implemented and the implications for the most vulnerable and marginalised members of society are unclear (McAfee, 2012). Although there is evidence that local communities have been consulted and/or participated in many PES and REDD+ processes, whether they understand and fully accept the commodification in situ of their natural resources and the potential implications is still very unclear. For many, entering a market economy and/or placing monetary value on their resources are also radically new cultural practices and little is known of the potential impacts on local societies and cultures, especially because these often do not have the governance structures and practices able to cope with such global systems.

Devolution of natural resource control to local communities has occurred in participatory community-based approaches as well as in the form of integrated conservation and development projects (ICDPs). However, under the neoliberal agenda, true local control of resources has been limited. In some instances, corporations and international agencies, such as large environmental/development NGOs, have increased their influence over local resource use through the decentralised governance structures. In many, economic activities linked to the market, such as ecotourism or involvement in PES schemes, are promoted as livelihood options, which could further disfranchise marginal communities.

Collaborative adaptive management

Today, in many parts of the world, particularly intertropical biodiversity-rich regions, we are dealing with increasingly complex, dynamic and unpredictable social-ecological systems, made even more so principally as a result of direct and indirect human interventions at a range of scales. Many natural resource management problems seem intractable and often involve the convergence of multiple crises: unpredictable and dynamic development financing; food insecurity; unplanned urbanisation; climate change; escalating organised crime; environmental degradation and biodiversity extinction; unemployment; life-threatening pandemics such as AIDS and avian flu, corruption, and the list could go on and on.

Three distinct approaches to managing these complex situations have been identified: authoritarian, market-led and collaborative. The authoritarian approach has the appearance of rapid implementation. However, the limited number of people involved in the decision-making process runs the risk of missing out crucial issues, and prescriptive implementation can be derailed by a range of disenfranchised stakeholders thus requiring significant investments in enforcement authorities. The market-led approach can generate a range of creative and innovative solutions, but frequently runs the risk of allowing the most financially successful approach to dominate, which principally benefits the established economic elite rather than addressing wider issues of social justice and ecological sustainability. The majority of scholars researching the management of complex problems have instead encouraged a clear shift towards the adoption of a transdisciplinary, multi-scalar, participatory and adaptive approach to problem-solving which is led and owned by local communities (see, for example, Project Cobra at www.projectcobra.org). However, this shift in rhetoric has run well ahead of knowledge about how adaptive management might work in practice.

Looking forward, the aim should be to champion locally owned solutions to natural resource management which are genuinely participatory, democratic and non-hierarchical. There should be more focus on learning from and with local communities on how they have developed their own processes of adaptive management, and how other organisations can support these communities in the development process. For example, recent research on community capacity building for adaptive management has identified a range of skills as well as a knowledge base on which it needs to be implemented (Mistry *et al.*, 2011). Key skills needed for adaptive management include risk management, initiative, critical and systemic awareness and reflective practice, and as the number of and/or mistrust between stakeholders involved in management processes increases, skills of negotiation, communication, conflict management and empathy become increasingly important. As there have been calls for a significant shift away from the top down, expert-led (usually foreign) decision-making, strengthening local and institutional capacity for adaptive management through enhancement of these skills can, as Heyman and Stronza (2011) point out, help people in biodiversity-rich developing countries to take the lead in finding long-term sustainable solutions to their own natural resource management and conservation/poverty dilemmas.

What is required is an alternative framework to the mainstream models of how nature is viewed and resources are managed as an alternative to the neoliberal globalising agenda which increasingly places control over natural resource management in the hands of multinational corporations and transnational bodies such as the World Bank. New paradigms need to build on localised relationships between communities and the ecology within which they are nested. This approach does not advocate protectionism and isolation, but rather the sharing of experiences and resources amongst communities so as to maximise benefits for all rather than the elite minority.

References

Arsel, M. and Büscher, B. (2012). Nature™ Inc.: Changes and continuities in neoliberal conservation and market-based environmental policy. *Development and Change*, 43: 53–78.

Castree, N. (2008). Neoliberalising nature: The logics of deregulation and reregulation. *Environment and Planning A*, 40: 131–152.

Ghazoul, J., Butler, R. A., Mateo-Vega, J. and Pin Koh, L. (2010). REDD: A reckoning of environment and development implications. *Trends in Ecology and Environment*, 25: 396–402.

Heyman, W. D. and Stronza, A. (2011). South–south exchanges enhance resource management and biodiversity conservation at various scales. *Conservation and Society*, 9: 146–158.
McAfee, K. (2012). The contradictory logic of global ecosystem services markets. *Development and Change*, 43: 105–131.
Mistry, J., Berardi, A. and McGregor, D. (2009). Natural resource management and development discourses in the Caribbean: Reflections on the Guyanese and Jamaican experience. *Third World Quarterly*, 30: 969–989.
Mistry, J., Berardi, A., Roopsind, I., Davis, O., Haynes, L., Davis, O. and Simpson, M. (2011). Capacity building for adaptive management: A problem-based learning approach. *Development in Practice*, 21: 190–204.
Pierre, J. and Peters, G. B. (2009). *Governing complex societies*. Palgrave McMillian, Basingstoke, UK.
Sikor, T., Stahl, J., Enters, T., Ribot, J. C. and Singh, N. (2010). REDD-plus, forest people's rights and nested climate governance. *Global Environmental Change*, 20: 423–425.
Voss, J. P., Bauknecht, D. and Kemp, R. (2006). *Reflexive governance for sustainable development*. Edward Elgar Publishing, Cheltenham, UK.

6.8

Water and hydropolitics

Jessica Budds and Alex Loftus

Water and politics

Managing water is an important global challenge. Only about 2.5 per cent of the world's water is freshwater, less than 1 per cent of which is accessible by humans. Given this finite supply, managing demand for water and controlling pollution are crucial. However, perhaps the most pressing challenge is to extend water supply to the estimated 780 million people, overwhelmingly concentrated in Africa, Asia and Latin America, who still lack access to sufficient and safe water, despite claims of substantial progress achieved under the Millennium Development Goals (WHO/UNICEF, 2012).

Why do so many of the world's poor remain excluded from water in the twenty-first century? We aim to show how this failure is a political issue. We will argue that an analysis of the power relations underpinning water issues and challenges is key to explaining how, why and by whom such challenges become defined and addressed in different ways.

Our approach to hydropolitics is based on the idea that water is more than just a material substance (H_2O): water is physically manipulated by humans in multiple ways (for example, by dams that modify watercourses) and seen in particular ways by different people (such as spiritual meanings attached to water). These changes and attitudes, in turn, are shaped by the vested interests and particular goals of social groups who endeavour to influence water provision. As such, water embeds and reflects power relations: water flows, patterns of allocation, types of infrastructure, design of policies and dominant arguments will all be configured by particular motives and perspectives (Swyngedouw, 2004).

Whose scarcity?

The 2006 Human Development Report (UNDP, 2006) examined the underlying causes of the lack of access to water among the world's poor. It questioned the relationship between available water resources and provision of water services, and strongly argued that deficient provision was not due to physical scarcity, to population pressure or to technical incapacity, but rather to the unequal distribution of water among wealthy and poor groups.

Exclusion from water supply among low-income groups is more closely related to the ways in which water services are organised, and how access to them is determined. The administration and allocation of water services are determined by the particular stances and decisions taken by those in charge. For instance, city authorities or water companies in developing countries have often refused to serve low-income settlements, even if they could be easily connected, due to illegal land occupation, low political priority and/or concerns over non-payment. In contrast, higher-income neighbourhoods typically enjoy reliable, unlimited and often inexpensive supplies piped to their homes. Water provision is thus not a neutral matter, but one that is defined and addressed in certain ways, which in turn determine who receives water and who does not.

Water privatisation

During the 1990s, international financial institutions promoted the increased participation of formal private enterprises in water utilities to improve and extend water services in Africa, Asia and Latin America. Privatisation was often a condition of structural adjustment policies that aimed to reduce indebted governments' public spending and investment. Privatisation was also supported by bilateral development agencies, which realised that development assistance could not fund the large-scale improvements needed in low-income contexts (Winpenny, 2003).

Water service deficiencies were attributed to the failure of public utilities to organise water services effectively, and to recover the full costs of water provision, which in turn limited their ability to improve and extend water networks (Bakker, 2010). Privatisation was justified on the basis that public utilities:

- were inefficient, as they have no incentive to improve;
- had weak capacity, leading to poor performance;
- were often corrupt; and
- lacked funding and access to finance.

In contrast, private enterprises were deemed more efficient because they operate according to commercial incentives, which encourage them to maximise efficiency and expand their business in order to generate profit. Unserved households were thus seen as a huge and untapped customer base that would be willing to pay for formal water services to replace expensive and unsafe water from informal vendors (Brocklehurst, 2002).

Despite the vigorous promotion of privatisation, and the expectation that it would significantly improve and extend services to low-income groups, privatisation has remained limited in the global South, for two main reasons (Budds and McGranahan, 2003). First, private companies 'cherry-picked' contracts in larger and wealthier cities in middle-income countries (such as Argentina or South Africa), which had comparably large customer bases, large proportions of middle and upper class inhabitants, good and extensive existing infrastructure, and political stability. Second, water provision proved to be more complex and less profitable than companies had expected, and some contracts were cancelled due to economic crises or social protests. In many contexts, people accepted the logic to engage private providers for some utilities, but specifically rejected privatisation for water, due to its status as a basic human need. The improvement and expansion of water provision among low-income groups has been particularly disappointing, as many private operators considered these groups commercially unviable.

If privatisation has made so little difference to delivering water to the poor, why was it proposed, and who has benefited? Private sector participation was not based on successful experiences within the water sector, but rather the ideologies and interests of the institutions that promoted it. For instance, the World Bank enforced structural adjustment and promoted free-market ideals, and, as Schulpen and Gibbon (2002) suggest, bilateral development agencies endorsed privatisation to create new market opportunities for water companies from their own countries. Privatisation also reconfigures power relations over water provision, by passing control from the state to the private sector. Thus, multinational companies' contracts may be backed by international financial institutions, while developing country regulators have often had little leverage over private operators that miss targets or retract obligations.

Despite the arguments made, extending water provision to the poor could never simply be resolved by engaging private operators, because the underlying causes of deficiencies are not due to the absence of commercial intervention, but to factors including indebtedness and exclusion of informal settlements from infrastructure and services.

Water infrastructure

If water privatisation has slowed, emphasis on full-cost recovery has intensified. Whether publicly or privately operated, the dominant orthodoxy suggests that the sustainability of water providers is dependent on costs (and profit, under private contracts) being recovered through user charges for operation, maintenance and often also construction. Individualised full-cost recovery, whereby each household pays the full cost of connection, has been particularly contentious: households already served had not paid for their infrastructure, whereas (low-income) unserved residences were expected to bear the full cost. Some authors have argued that the focus on full-cost recovery is part of a broader shift towards the corporatisation of public providers (McDonald and Ruiters, 2005). In the process, water users become defined as consumers of a commercial service, rather than citizens entitled to universal provision of a basic service. Public-private partnerships, bringing together national or municipal water providers with private companies (or civil society organisations), further illustrate this changing relationship (Bakker, 2010).

The shift towards achieving full-cost recovery in previously highly indebted as well as heavily subsidised services has required an assault on what is often considered to be a culture of non-payment. In many cases, this assault has diverted attention from the inability of those with low and unstable incomes to pay user charges. An array of new and old technologies has been enlisted to ensure that household water consumption can be measured and billed accordingly. In part, the emphasis on water metering emerges from the replacement of supply side policies with demand management strategies, due to growing consciousness over the over-consumption of scare resources (Bakker, 2010). Nevertheless, within the global South, water meters have often become material symbols of low-income citizens' inability to access a resource to which they feel they are entitled. Although banned in some countries of the global North, including England and Wales, prepayment meters have been introduced in cities of the global South. Flow restrictors and flow limiters, also banned in England and Wales, have often accompanied these meters, and both public and private companies use disconnection to ensure bills are paid. In a curious paradox, the challenge for some water service providers becomes how best to limit the supplies of poor households rather than how best to extend the network (Loftus, 2009).

If water meters have become symbols of the barriers poor communities now face, they also demonstrate the politicisation of water infrastructure. The ordinary technologies that mediate everyday life embody and express the politics of the societies of which they are part. The tap and the water meter thus capture and reflect the shifting political economy of water provision. At the same time, water meters play a role in disciplining individuals into being the economically efficient subjects that the technologies assume them to be (Ekers and Loftus, 2008).

Aside from technologies expressing political changes, the post-war ideal of providing water to all through a comprehensive network has given way to an unequal, and generally inadequate, system of splintered water networks (Graham and Marvin, 2001; Bakker, 2010). Although moves off-grid can be part of a radical challenge to the supply side approach to water provision, fragmented networks are typically a result of the power of some, and the powerlessness of others, to be able to pay for piped supplies. The splintered networks of many developing country cities now suggest that progressive ideals have given way to an intensification of clientelistic models benefiting the few.

Domestic water technologies provide a bridge between the high politics of statecraft and the intimacies of the home and can, in turn, serve to consolidate societal shifts, such as the intensification of commoditisation in the global South. The making of water networks can thereby be seen as the making of history under particular conditions. With new technologies being designed to ensure monetary returns for service providers rather than improvements in water provision to low-income groups, more politicised responses are required in order to envision radical and necessary alternatives.

Conclusion

Water serves as one of the most powerful reminders of the stark inequalities in today's world: that at least 780 million people in the global South lack access to sufficient and safe water is still a damning indictment of development efforts. Yet, these deficiencies are actively produced through the ways in which water infrastructure and services are organised, how access to them is determined, and how new institutional and technical responses are promoted and justified. Policies such as water privatisation, and changes to the provision and regulation of water infrastructure, are thus not merely technical, but inherently political. Continuing to address such inequalities thus requires fundamentally political interventions.

Bibliography

Bakker, K., 2010, *Privatizing Water: Governance Failure and the World's Urban Water Crisis*, Cornell University Press: Ithaca, NY.

Brocklehurst, C. (ed.), 2002, 'New designs for water and sanitation transactions: Making private sector participation work for the poor', World Bank: Washington, DC.

Budds, J. and McGranahan, G., 2003, 'Are the debates on water privatization missing the point? Experiences from Africa, Asia and Latin America', *Environment and Urbanization* 15(2): 87–113.

Ekers, M. and Loftus, A., 2008, 'The power of water: Developing dialogues between Foucault and Gramsci', *Environment and Planning D* 26: 698–718.

Graham S. and Marvin, S., 2001, *Splintering Urbanism: Networked Infrastructures, Technological Mobilities and the Urban Condition*, Routledge: Abingdon, UK.
Loftus, A., 2009, 'Rethinking political ecologies of water', *Third World Quarterly* 30(5): 953–968.
McDonald, D. and Ruiters, G., 2005, *The Age of Commodity: Water Privatization in Southern Africa*, Earthscan: London.
Schulpen, L. and Gibbon, P., 2002, 'Private sector development: Policies, practices and problems', *World Development* 30(1): 1–15.
Swyngedouw, E., 2004, *Social Power and the Urbanisation of Water: Flows of Power*, Oxford University Press: Oxford.
UNDP, 2006, *Human Development Report 2006 – Beyond Scarcity: Power, Poverty and the Global Water Crisis*, Palgrave Macmillan: London.
WHO/UNICEF, 2012, *Progress on Drinking Water and Sanitation: 2012 Update*, UNICEF: New York.
Winpenny, J., 2003, 'Financing water for all', World Water Council, Third World Water Forum and Global Water Partnership: Marseille.

6.9

Energy and development

Subhes C. Bhattacharyya

Introduction

The critical role played by energy in achieving sustainable development (used in the sense of the Brundtland Report) is now well recognised. Energy is an essential input to economic activities and for meeting the basic human needs of food and shelter. The modern lifestyle cannot be imagined without adequate, reliable and affordable energy supply. While the 'energy haves' are concerned about ensuring their future, billions of others are struggling to get access to affordable, reliable, and acceptable energy services. At the same time, activities related to supply and consumption of energy created various environmental damages at local, regional and global dimensions. More importantly, the symptoms of unsustainable energy practices, such as unprecedented demand growth, skewed regional demand distribution, profound divergence in consumption level, and environmental effects of energy use, could continue to haunt us in the future unless corrective actions are taken.

This chapter provides an overview of historical energy use and presents the energy–development link. In the process it highlights the unsustainable practices, and suggests remedial actions. The chapter is organised as follows: the evolution of energy demand is first considered where the grand energy transitions and regional demand patterns and disparities are presented. This is followed by a section on energy–development links that considers issues like intensification of energy use, energy access problems and environmental consequences of energy use. The final section briefly indicates some ideas to remedy the challenges to promote sustainable energy futures.

Historical evolution of energy use

Grand transitions

Energy use patterns have undergone significant changes since mankind started using energy. Prior to the Industrial Revolution, energy was mostly derived from the natural energy flows and human and animal power, and energy usage was limited. Since then, two grand transitions have shaped energy systems at all levels. The first transition involved a gradual shift towards coal from traditional energies in the nineteenth century leading to a coal-dominated energy system in the early twentieth century. The second involved a decline of coal and the emergence of petroleum as the dominant fuel in the middle of the last century. The arrival of steam engines powered by coal was the main driver in the first transition and it gained acceptance because of its higher energy density, flexibility and mobility. For the second transition, the innovations of internal combustion engines and electricity conversion technologies proved crucial. Oil emerged as the dominant energy form by virtue of its ease of use arising from fluidity, versatility, relative cleanliness compared to coal and universal appeal.

These transitions brought far-reaching structural changes in the economies: agriculture was displaced by industry and manufacturing. Industrialisation brought urbanised style of living, which in turn changed lifestyles, social values, and led to relocation from rural to urban areas. These changes affected energy demand and supply, leading to a shift towards commercial energies away from traditional energies. Demand for flexible, convenient and cleaner energy forms increased. See Bhattacharyya (2011) for further details.

Global demand trend

As a consequence of economic growth and lifestyle changes, the demand for commercial energies has seen a rapid growth. In 1965, the global primary energy demand was 3,767 Mtoe, 82 per cent of which (or about 3,100 Mtoe) arose from North America and Europe (including Eurasia). In 2010, the global primary energy demand has increased to 12,002 Mtoe, registering an annual average growth rate of 2.6 per cent over this period. The share of North America and Europe (including Eurasia) has reduced significantly as well – just about 48 per cent of the global demand in 2010 due to the rise of a third demand centre in the form of Asia-Pacific (see Figure 6.9.1).

Fossil fuels continue to play an important role in the global energy mix. In 2010, 33.5 per cent of global primary energy came from oil, followed by coal (about 30%), natural gas (almost 24%), and thereby ensuring fossil fuel contribution of 87 per cent in the energy mix. Coal plays a dominant role in Asia, while oil and natural gas contribute more significantly in other regions. Despite a growing interest in modern renewable energies (solar energy, wind, tidal, etc.), their share remains insignificant in the global energy picture in 2010.

Regional disparity

Despite the growth in energy consumption, the extreme level of disparity prevalent across regions is clearly visible. For example, a person in Africa consumes just 20 per cent of global average consumption while Asia and Latin America also consume between 68 and 76 per cent of the world average. Moreover, a large section of the population in the developing world relies on traditional energies such as firewood or agricultural residue and does not have access to modern energies. As the global population is projected to rise to 9 billion by 2050 with most of the increase in global population residing in Asia and Africa, there will be increasing pressure on energy from the developing world.

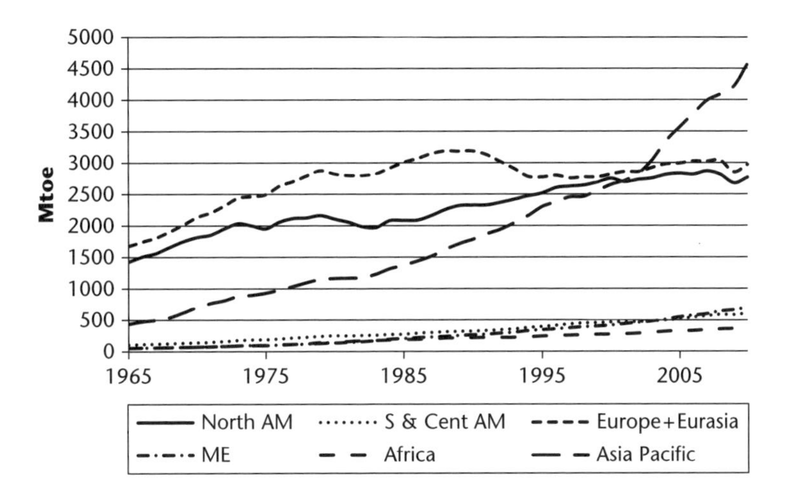

Figure 6.9.1 Global primary energy demand trend

Data source: BP Statistical Review of World Energy (2011)

Energy and economic development

Intensification of energy use

The positive correlation between energy and economic development can be easily verified from a cross-sectional scatter plot of per capita GDP and per capita annual energy consumption information (see Fig 6.9.2). As expected, the need for energy tends to increase with higher per capita income. As energy demand is a derived demand that arises for satisfying the needs through the use of appliances, the demand for energy depends on economic activities, services and choice of energy using processes or devices. Developing economies transit from traditional energies to modern energies as they climb up the income ladder and as consumers follow more energy-intensive lifestyles at higher incomes, energy need increases. Further, the changing economic structure due to industrialisation and consequent rapid urbanisation of these economies results in a growing urban sector and growing energy demand.

It is generally presumed that an economy will first industrialise from an agrarian economy and then move to service-dominated activities. Accordingly, the energy demand tends to increase first to support an energy-intensive mode of industrial development and then slows down or declines as the economy matures and moves towards a service-oriented economy. In addition to the structural change, energy demand is directly related to efficiency of energy use in an economy. The lower the energy intensity (ratio of energy demand and economic output), the more efficient the energy use is in an economy. In the developed economies, the energy intensity has fallen drastically indicating that these economies are deriving their outputs from less energy-intensive activities and that there is a decoupling of energy use and the economic development taking place through structural change, fuel substitution and energy efficient systems.

However, the development trajectory of developing countries can be very different from that of the developed economies. A growing urban sector is found to coexist with a predominantly rural economy in most cases and the nature of economic activities as well as opportunities differs significantly between urban and rural areas due to the prevalence of informal economic activities in rural and semi-urban areas where poverty and inequity is more acute. In addition, the structure of their energy sector can also be very different due to specific characteristics such as high reliance on traditional energies, an inefficient energy

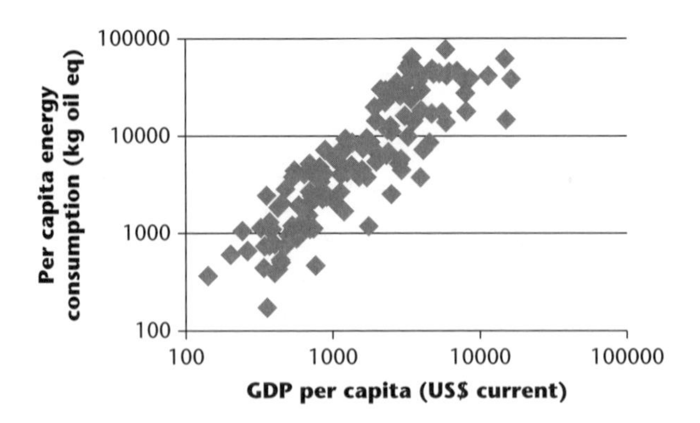

Figure 6.9.2 Energy–GDP link

Data source: World Bank Development Indicator Database

sector characterised by supply shortages and poor performance of energy utilities, and existence of multiple social and economic barriers to capital flow and slow technology diffusion. Consequently, developing economies offer a contrasting picture, where a growing commercial energy sector coexists with the traditional energy systems. Their energy intensity tends to be higher due to inefficient energy use and depending on whether commercial and traditional energies are considered, the picture can be different.

In general, economic development of developing economies will have a direct impact on future energy demand. IEA (2011) indicates that 90 per cent of the projected demand growth over the next 25 years originates from non-OECD economies and that fossil fuels are likely to play a significant role in this development. As these resources are finite in size and are unevenly distributed in the world, the concern for security of energy supply to fuel future development remains a major challenge.

Energy access and development

A related issue facing the developing world is the issue of energy access, where more than 1.4 billion people in the world did not have access to electricity and more than 2.7 billion did not have access to clean cooking energies in 2009 (IEA, 2011). Two regions stand out in this picture: South Asia had 614 million while sub-Saharan Africa had 587 million without access to electricity in 2009. Another 195 million people without access to electricity are found in East Asia. In terms of cooking energy access, 72 per cent of those lacking clean cooking energy access live in Asia and a majority of the rest live in sub-Saharan Africa.

Lack of energy access clearly affects the economic development of countries, as human capital development is greatly affected by lack of energy services. Countries with poor electricity or cooking energy access perform relatively poorly on the Human Development Indicator ranking. High reliance on traditional energies has significant social impacts in terms of air pollution and health effects on women and children. UNDP-WHO (2009) reported that about 2 million deaths per year and 40 million disability-adjusted life years can be attributed to solid fuel use in developing countries. See www.worldenergyoutlook.org/resources/energydevelopment/ and www.oasyssouthasia.info/ for further details on this issue.

Serious thoughts in terms of policy analysis and implementations will be required to address these issues. Sporadic or ad hoc international support and non-governmental organisation involvement alone will not resolve the problem.

Adverse environmental consequences

Another adverse consequence of high fossil-fuel dependence for economic development is the environmental degradation and the threat of climate change. As Figure 6.9.3 indicates, the energy-related carbon emissions have significantly increased: emissions in 2010 were 30 per cent higher than 2000 levels and 46 per cent higher compared to 1990 levels. A dramatic increase in emissions from the Asia-Pacific region is clearly noticeable as well. Rapid economic growth in the region fuelled by coal to a large extent is responsible for such a change in the regional emission balance. According to the Global Environmental Outlook 4 (UNEP, 2007), emissions of SOx and NOx have declined in affluent countries of Europe and North America but have increased in developing countries. In fact, the air pollution level in many cities of the developing world exceeds the World Health Organisation guidelines. This in turn causes public health concerns.

The way forward

Clearly, the above discussion points to an unsustainable path of an energy development link that was followed in the past. High reliance on fossil fuels and inefficient technologies cannot be sustained in the long run. At the same time, the social aspects of energy use cannot be ignored. Providing affordable, reliable and clean energies to the entire population will remain a challenge for the sustainable energy future of a country. Improving energy supply security and reducing environmental damages arising from energy use will also assume greater importance in a changing world.

To achieve energy sustainability, a clear strategy is required. Developing countries have the possibility of leapfrogging and learning from others' experiences (Urban *et al.*, 2007) instead of following a development path that steeply raises energy intensity as per capita income increases, as exemplified by the industrialised countries. Simultaneously, the emphasis on efficient technologies, clean resources, demand management practices and good governance will be essential to ensure a transition to a sustainable path. However, there is no readymade template to follow, as no country has resolved the problem yet. Therefore, each country would have to find its own solution through trial and experimentation.

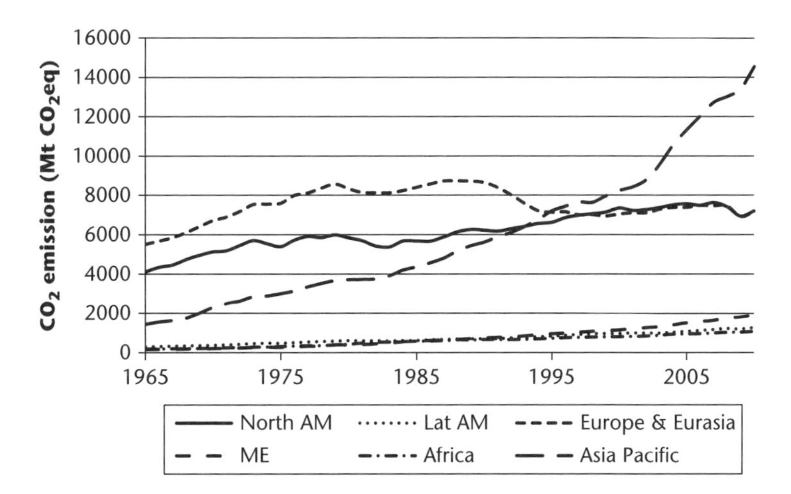

Figure 6.9.3 Energy-related global CO_2 emissions

Data source: BP Statistical Review of Energy (2011)

References

Bhattacharyya, S. C., 2011, *Energy economics – concepts, issues, markets and governance*, Springer-Verlag, London. UK.
IEA, 2011, *Energy for all: Financing access for the poor*, (special early excerpt of the World Energy Outlook 2011), International Energy Agency, Paris.
UNDP-WHO, 2009, The energy access situation in developing countries: A review focusing on the least-developed countries and sub-Saharan Africa, United Nations Development Programme, New York.
UNEP, 2007, Global Environmental Outlook 4, UNEP, Nairobi, Kenya.
Urban, F., Benders, R. J. M. and Moll, H. C., 2007, Modelling energy systems for developing countries, *Energy Policy*, 35, 3473–3482.

6.10

Tourism and environment

Matthew Louis Bishop

Tourism is the fastest growing industry in the world today, and, since the 1960s, many developing countries have sought to place it at the centre of their development strategies. There are a number of reasons for this. On the demand side, reductions in the relative cost of air travel and growth in disposable incomes in the West permitted the emergence of newer and more 'exotic' destinations. On the supply side, developing countries have viewed tourism as a way of diversifying out of those industries characterised by a precarious primary product dependency and declining terms of trade such as sugar, cocoa and other commodities. It is also considered a good generator of foreign exchange, a means to attract foreign investment, infrastructural development and connections to the outside world, and, for those countries – such as the many remote tropical islands scattered across the planet – that enjoy the plentiful sun, sea and sand coveted by tourists there have been precious few alternative development options.

Underpinning the expansion of tourism in the developing world has been a concerted effort on the part of development agencies, the Washington-based International Financial Institutions (IFIs) and, especially, the UN World Tourism Organisation to promote its benefits almost as a developmental panacea (Ferguson 2007). Intellectually, this has been buttressed by the notion of 'sustainable tourism development' which has guided policymaking, and sought to reconcile the effects of what is a notoriously unsustainable industry with the imperatives of engendering development in a genuinely sustainable fashion. However, the very idea of sustainable tourism development has tended to be poorly conceptualised, subject to myriad competing interpretations, and has come under sustained and 'vociferous criticism' (Sharpley 2009: 57–58).

In this chapter we explore some of these issues by looking at, first, the often deleterious impact of tourism on both the physical and social environment. Then, we reflect on whether and how the industry can ever be sustainable, drawing on emerging ideas and examples of greener approaches to its development. Finally, we offer some concluding thoughts on the relationship between sustainability, tourism, development and the environment.

The impact of tourism on the physical environment

The thorny tensions that exist within the notion of sustainable tourism development have been far from easy to resolve in practice. At its most basic, there is an inherent contradiction between an industry which depends fundamentally on a pristine environment, yet which often involves the degradation of that environment. As Cater (1995: 22) put it some years ago, 'unless the environment is safeguarded, tourism is in danger of being a self-destructive process, destroying the very resources upon which it is based'. More broadly, and for all of the revenue that it undoubtedly generates, tourism provokes a number of problematic trade-offs with often dire environmental consequences.

For many countries, approaches to tourism have generally been predicated on encouraging as many visitors as possible to come, and this in turn has placed heavy pressure on limited natural resources. So-called 'mass' tourism is especially destructive since it often generates sustained and continued degradation, as described in Butler's (1999) 'tourism life cycle'. As more tourists arrive and the associated paraphernalia such as hotels, restaurants, bars, casinos, tourist shops, beach and marine activities and so on proliferate, the destination loses its sheen of exclusivity, the environment gradually becomes blighted by excessive construction and intensive overuse, and arrivals begin to decline. In order to stave this off, ever-more desperate attempts are made to entice people to an increasingly less appealing location, resulting in lower prices, lower revenues, and the place becomes locked into a generalised pattern of decline from which it can neither emerge refreshed nor return to its pre-tourism idyll. Today, examples as disparate as parts of the Costa Del Sol in Spain, some of the Thai islands, and even Blackpool in northern England epitomise this sad story.

Although the major destinations in the world – in terms of absolute arrival numbers – remain large and resilient countries like France, the USA, China and the UK, tourism development has been relatively more penetrating in much smaller, poorer and more vulnerable places. Indeed, the parts of the world in which tourism development has been most pervasive are those which are ecologically the most fragile. The delicate ecosystems in the Caribbean, the Indian Ocean and the Pacific contain vast numbers of rare and distinctive species of flora and fauna which are highly vulnerable to being degraded by human pressure. There are many examples, including: effluent from cruise ships and beachfront property which is spewed into the sea, killing marine life; over-exploitation of shellfish, crustaceans and other native species by hunters, fishermen and poachers to serve the tastes of wealthy visitors; and the seizing of arable land or the destruction of forests to facilitate the construction of hotels, golf courses and other attractions, many of which place an inordinate amount of stress on water, energy and other limited resources.

There is, of course, also an intrinsic contradiction in the notion of sustainability in an industry which relies heavily on long-haul aviation and which has therefore contributed significantly to climate change. The Caribbean, for example, is the most heavily tourism-penetrated region in the world (Pattullo 2005). As a consequence, the coastlines of the many islands and territories that comprise the region are dense with hotels, air and seaports, shops, restaurants, homes, and other critical infrastructure. Yet as climate change intensifies, these remain acutely vulnerable to sea-level rise and extreme weather patterns such as more frequent and violent hurricanes (Bishop and Payne 2012). Already many Caribbean countries have reported severe environmental degradation, much of which is caused either directly or indirectly by tourism: such as the bleaching of coral reefs, saltwater intrusion into freshwater resources, and worrying declines in biodiversity.

The Maldives, a multi-island nation in the Indian Ocean, and a destination prized by wealthy tourists, lies just 1.5 metres above sea level and provides a stark example of the paradoxes embedded in the relationship between tourism and the environment. On the one hand, its own tourism depends heavily on aviation and has contributed – albeit relatively insignificantly – to global warming. Yet on the other, the country stands to be rendered uninhabitable by climate change as the sea level rises, destroying not only the tourism industry, but the territory itself. The Maldives government has been steadily investing money in order to purchase land to resettle the entire state and population, should that dreadful day ever arrive (Adger 2010).

Although perhaps less dramatic, there are numerous other deep-seated negative implications of tourism for developing societies. First, tourism plant often tends to be owned and controlled by foreign capital, profits are usually repatriated to metropolitan countries, and investment is facilitated by both tax holidays and publicly funded infrastructure. Aside from (often insecure) employment, local people receive few benefits, and arguably bear a greater overall share of the costs of tourism development. In the Caribbean, there is great anger in many places at the way in which huge stretches of pristine beaches are lost to foreign-owned all-inclusive hotels from which local people are subsequently excluded, and out of which tourists often do not venture to spend their money locally (Bishop 2010). Second, resident populations tend also to be exoticised, with their art, culture and even their bodies increasingly packaged and commodified for external consumption (Crick 1989). Indeed, in many places, the cycle of decline is such that integral to the tourism 'product' are drugs, prostitution and even the exploitation of children. Third, the structure of local social traditions and customs is irreversibly changed by tourism, and new patterns of inequality and exclusion are created and reproduced (Ferguson 2011).

Is sustainable tourism possible?

There have been many attempts – both intellectual and practical – to develop genuinely sustainable approaches to tourism (Sharpley 2009). Notions such as 'green' tourism or 'eco' tourism have peppered the literature, and they have influenced policy. Such ways of thinking have tended to be problematic when their central concepts are poorly theorised, and where policymakers have applied them unthinkingly or misguidedly. For example, the construction of 'eco-lodges' in tropical rainforests to host river-based activities, hiking, nature-watching and so on are often seen as intrinsically 'green'. However, there is no innate reason why visitors will be any less destructive in these environments than in a traditional mass tourism location, particularly given the acutely delicate ecological and social balance that exists in plant, animal and human communities in such remote places. Recent reports from Vang Vieng, a village deep in the interior of Laos, provide a salutary example: in just a few years, a sleepy river community has been turned into a destination where backpackers outnumber locals by as many as fifteen to one, leading to disorder, prostitution, widespread drug abuse, along with confusion and turmoil amongst the local population as the town becomes rapidly and dramatically changed beyond recognition (Haworth 2012).

Elsewhere, though, sustainable approaches to tourism have gained some traction, and it is increasingly the case that many tourists themselves are deliberately seeking to travel in a way that has a mutually beneficial impact upon local societies. At its most simple, this can involve eschewing multinational firms for local providers of accommodation and other services in order to ensure that money does not leak from the receiving economy and stays in the pockets of local people. Yet many are going well beyond this by engaging in all manner of cultural activities, staying with local people in their homes, and even deliberately using their holidays to invest their skills and money in local development projects. For example, Harrison (2007) has described how, on the rural northeast coast of Trinidad the village of Grand-Rivière witnessed the decline of its cocoa industry just as it became an important and globally renowned nesting site for endangered leatherback turtles. Although not without its problems, this transition has been managed sensitively: as more tourists have arrived, villagers have been able to participate at every level of the new industry and it has been tightly integrated into, rather than compromising and uprooting, existing social structures. Interactions between tourists and local people are considerably more authentic and mutually rewarding than is often the case. Many of the jobs provided, moreover, involve the protection of the turtles and other aspects of conservation, leading to the perpetuation of a positive cycle in terms of sustainability.

Concluding thoughts

In summary, modes of sustainable tourism development are to some extent possible. Yet the major problem with them is that they currently remain a relative minority interest: for every nature-conscious turtle-watcher in Trinidad, there are probably two hundred people on the beaches of its sister island, Tobago, enjoying the sun, sea, sand, and, at times, sex that their relative wealth and power permit them to commodify and consume. In narrow economic development terms, this does not have to be considered intrinsically problematic. Mass tourism generates the kind of revenue which developing countries often desperately require, and which more sustainable variants simply cannot provide.

However, if we are more interested in the sustainability part of the equation, and particularly if we consider sustainability to be a *sine qua non* of development, then mass tourism carries with it many troubling implications. The kinds of greener approaches which we have briefly discussed here may not generate large amounts of foreign exchange. Yet when developed thoughtfully, they can certainly lead to positive-sum development outcomes, such as conservation, more fruitful engagements between tourists and local people, and a wider distribution of the income and other benefits which subsequently accrue. Over time, they will also surely lead to far less depletion – and potentially even protection and renewal – of the natural capital on which tourism, of whatever kind, depends.

Ultimately, contemporary modes of tourism cannot ever be entirely sustainable. Much human economic activity in our post-industrial age involves the application of one kind of capital to augment, alter, or destroy another kind of capital. But it can be rendered *more* sustainable than is presently the case. If the concept of sustainable tourism development is ever to signify anything beyond a nice phrase, institutionalising such an eventuality is the challenge its theorists and practitioners must meet in the years and decades ahead.

References

Adger, W. N. (2010). 'Climate Change, Human Well-Being and Insecurity'. *New Political Economy* 15(2): 275–292.

Bishop, M. L. (2010). 'Tourism as a Small State Development Strategy: Pier Pressure in the Eastern Caribbean?' *Progress in Development Studies* 10(2): 99–114.

Bishop, M. L. and A. Payne (2012). 'Climate Change and the Future of Caribbean Development'. *The Journal of Development Studies* 48(10): 1536–1553.

Butler, R. (1999). 'Sustainable Tourism: A State-of-the-Art Review'. *Tourism Geographies* 1(1): 7–25.

Cater, E. (1995). 'Environmental Contradictions in Sustainable Tourism'. *The Geographical Journal* 161(1): 21–28.
Crick, M. (1989). 'Representations of International Tourism in the Social Sciences: Sun, Sex, Sights, Savings and Servility'. *Annual Review of Anthropology* 18: 307–344.
Ferguson, L. (2007). 'Global Monitor: The United Nations World Tourism Organisation'. *New Political Economy* 12(4): 557–568.
Ferguson, L. (2011). 'Tourism, Consumption and Inequality in Central America'. *New Political Economy* 16(3): 347–371.
Harrison, D. (2007). 'Cocoa, Conservation and Tourism Grande Riviere, Trinidad'. *Annals of Tourism Research* 34(4): 919–942.
Haworth, A. (2012). 'Vang Vieng, Laos: The World's Most Unlikely Party Town'. *The Guardian*. London, 7 April.
Pattullo, P. (2005). *Last Resorts: The Cost of Tourism in the Caribbean*. New York, New York University Press, 2nd Revised Edition.
Sharpley, R. (2009). *Tourism Development and the Environment: Beyond Sustainability?* New York: Earthscan/James & James.

Further reading

Crick's (1989) article gives a detailed overview of tourism in the social sciences literature, written when it was still a comparatively novel area of study. This paper remains highly relevant today.
Pattullo's (2005) account of the many tensions inherent in every aspect of Caribbean tourism is the most definitive – and highly readable – book on the subject available.
Sharpley, R. and D. J. Telfer (2008). *Tourism and Development in the Developing World*. London: Routledge. Provides an excellent overview of tourism in the developing world.
Sharpley (2009) offers a detailed, critical and insightful analysis of the notion of 'sustainable tourism development' with reference to a wide range of interesting examples.

6.11

Transport and sustainability

Developmental pathways

Robin Hickman

Unsustainable transport

Great challenges are facing transport planning in cities in the developing (and developed) world as economic development is sought in a manner that is also consistent with environmental and social goals. Transport is a critical part of the design of the sustainable city, and plays a major role in people's everyday lives. Yet the transportation system is in crisis as transport energy usage and carbon dioxide (CO_2) emissions rise and the impact of traffic and its associated infrastructure on the city fabric often becomes untenable.

At the international level, the globalisation of trade and movement, and the widespread aspiration towards motorisation, means that there are rapid rises in travel distance, including in private car travel, international aviation and shipping. Global air traffic has, for example, increased by nearly 9 per cent per year since 1960, and is projected to grow by nearly 5 per cent per annum to 2015. In policy terms, international travel demand is allowed to grow almost unrestrained. Taking the UK as an example, assuming the national CO_2 reduction target is met (an 80% reduction by 2050 on 1990 levels, Climate Change Act, 2008), then continued growth in the air sector would mean that air accounts for all of the allowed emissions by 2050 (Bows and Anderson, 2007). A similar trend is found in many other countries. Clearly this position of unrestrained growth in international travel, supporting the ideal of free trade and globalisation, is incompatible with aspirations to reduce CO_2 emissions at the national and city levels.

The aspiration for motorisation

Within many urban areas there are often excessive and rising levels of motor traffic congestion and too much space devoted to the motor car. Cities in Asia and South America are commonly built around infrastructure for the car, with sprawling urban areas. The rapidly rising demand for motorisation in the developing world, alongside that already found in the developed world, poses serious problems for transport and city planners (Dimitriou, 1992; Hickman *et al.*, 2011a). Transport accounts for 62 per cent of oil consumption in 2008 (International Energy Agency, 2010; Hickman *et al.*, 2011b) and there are already very high traffic casualty rates, with over 1.2 million deaths per year internationally in traffic accidents; and rising levels of energy use and carbon dioxide (CO_2) emissions. The policy aspiration – at the international, national and city levels, and also with the international funding institutions – is usually to move towards more sustainable transport systems and behaviours; but often the financing levels, the structures of governance, and local capacity levels are seriously inadequate.

The aspirations of the public, with few exceptions, are to own a private motor car, at least in the future as incomes rise, with perceived and often realised benefits in terms of convenience, 'freedom' and status – hence there are instrumental, symbolic and affective dimensions to car use (Steg, 2005; Anable and Gatersleben, 2005). The motor car, still largely driven using the late nineteenth-century technology of the four stroke engine, and fuelled by petrol or diesel, remains the dominant mode of travel for many. There is little availability of the new, emerging technologies such as hybrid-electric and electric vehicles. Where they are available, vehicles are expensive relative to the petrol or diesel comparator. The history of the Tata Nano is perhaps very instructive in illustrating the importance of status and the 'sign value' of the car. The Nano was billed as the first mass production, cheap and affordable car; priced to be available for the lower income segments within the car market in India, and ultimately the rest of Asia and wider. Initially made available for just US$2,000 (one lakh, or 100,000 rupees), the Nano was expected to develop a large market share and transform the vehicle market, particularly for those who could not previously afford a car. Individuals would suddenly be able to leave the two or three wheeler or crowded bus behind and purchase their first vehicle. Many commentators saw this as a defining moment for mass car ownership in Asia. The *Financial Times* (Pilling, 2008), for example, reported:

> If ever there were a symbol of India's ambitions to become a modern nation, it would surely be the Nano, the tiny car with the even tinier price-tag. A triumph of homegrown engineering, the . . . Nano encapsulates the dream of millions of Indians groping for a shot at urban prosperity.

Purchasing levels have, however, remained surprisingly low. There have been problems with the vehicle design, prices have risen, and importantly the public have not identified with the low-cost vehicle; instead wanting something 'more special', signifying their rising status in the world. The sign value within consumption remains critical, perhaps increasingly so, and consumers have tended to buy second hand or new, larger cars. There are important lessons here – the aspirations of the public are critical and need to be understood as part of the design of future transport policies and investments.

Developmental pathways

Developing countries often follow divergent pathways in terms of transport (Figure 6.11.1), moving from the walking and cycling cities of the pre-1970s to the current traffic saturated bus and metro cities. Many suffer from a situation of entrenched traffic saturation and congestion. Some cities have managed to develop more extensive public transport networks and much less reliance on the car (such as Hong Kong, Singapore and Shanghai), but these are few and far between. There has been a recent investment in bus rapid transit in Asia, following the lead from the 1970s in South America (in Curitiba and Bogotá). There is also a developing and extensive high speed rail network in China, linking many of the major urban areas with quick connections. As part of this, the urban form is being shaped to support the transport investments, with high density new city quarters being developed around the key interchanges.

Towards sustainability in transport

There are many policy options available to reduce energy consumption and emissions in transport, including infrastructural, technological and behavioural interventions. Some of the key policy areas are described below.

- *Urban planning*: the form and layout of land use can be designed to help support public transport and reduce the length of trips. Development can be located in the larger urban areas, higher-density development clustered around an upgraded public transport system, and urban areas planned to vastly improve their urban design quality and attractiveness for living and working.
- *Traffic demand management (TDM)*: TDM measures cover road pricing, high occupancy lanes, traffic priority measures and pedestrianisation schemes, and are aimed at reducing the attractiveness of motor car travel. The price of fuel varies markedly by country, with many of the oil producing nations tending to opt for low taxation levels and petrol prices (e.g. Saudi Arabia, Qatar, Kuwait, Malaysia, USA, Indonesia – below US$0.80 per litre), and the non-oil producers tending towards higher petrol prices (e.g. Thailand, Singapore, Japan, China, much of Europe – above US$1.50 per litre). Road pricing can make a substantial difference to travel, by increasing the price of car travel relative to other modes, whether this is operated nationally or within cities and on the motorways. There are, however, very few pricing schemes in operation internationally, mainly due to the political difficulties of implementation, with the exceptions including Singapore, Stockholm and London.
- *Public transit*: options range widely in themselves, including mass rapid transit (MRT), to light rapid transit (LRT), bus rapid transit (BRT), ultra-light transit and demand responsive transit (DRT). There are over 140 MRT systems internationally, known

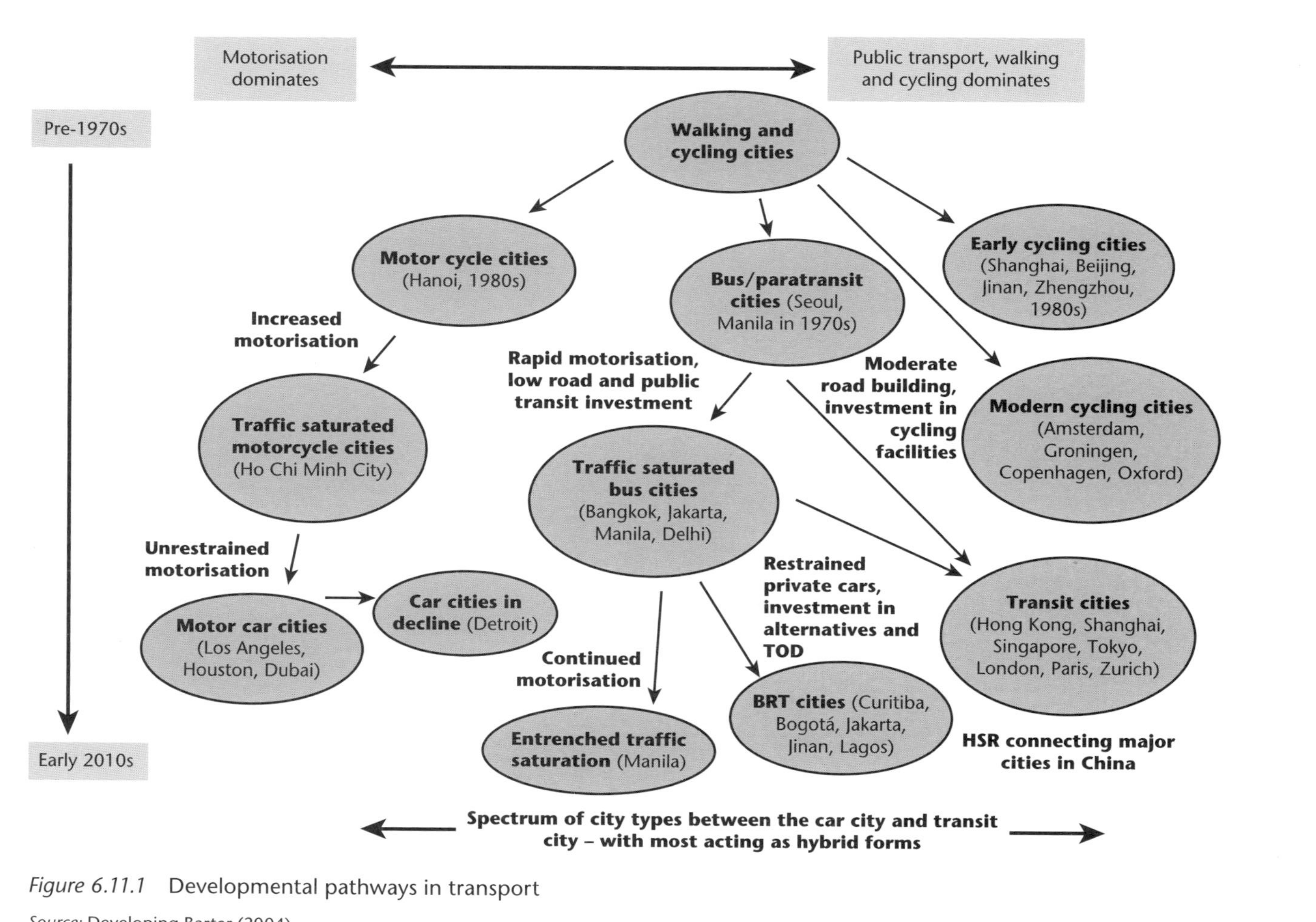

Figure 6.11.1 Developmental pathways in transport

Source: Developing Barter (2004)

as subways or metros, offering high capacity services to the core of major urban areas, including Beijing, Bangkok, Delhi, Kuala Lumpur, London, Paris, Shanghai, Singapore and Seoul. LRT generally has lower capacity and speed than heavy rail and MRT systems, but higher capacity than bus-based systems. There are examples in Hong Kong, Manila, Singapore, and many in Europe. BRT is experiencing a renaissance in developing countries, with many new schemes being planned and built. Curitiba's Rede Integrada de Transporte and Bogotá's TransMilenio systems have successfully moved large numbers of people, including the urban poor, for over 30 years. Similar schemes have been developed in Ahmedabad, Bangkok, Delhi, Jakarta; and many in China including Guangzhou, Jinan and Zhengzhou.

- *Walking and cycling*: both are extremely valuable means of travel, and virtually zero in CO_2 emissions. For both, there are also large gains in terms of supporting an active lifestyle. Despite this, the quality of the walking and cycling environment is often poor in the developing world. Improved networks and facilities are required in almost all cities, with segregated networks, cycle parking facilities, and vehicle speeds reduced, drawing on experience found in the Netherlands and Denmark.
- *Low emission vehicles and alternative fuels*: much reduced CO_2 emissions can be achieved if the vehicle fleet comprises of a large proportion of the new emerging technologies such as hybrid petrol, diesel and electric engines. The current low emission vehicles can emit less than $100gCO_2/km$, such as the Toyota Prius and Volkswagen Polo BlueMotion at $89gCO_2/km$. There is typically a very poor availability of these types of vehicles in the developing world, though there are plans for some hybrids in India and China. In Brazil, flex-fuel vehicles (FFV) account for nearly 80 per cent of the motor car market. They have an internal combustion engine but are designed to run on more than one fuel, usually gasoline blended with ethanol or methanol fuel. Both fuels are stored in the same common tank. Two and three wheelers can also be converted to use alternative fuels; for example, compressed natural gas is used in three wheelers in Delhi and electric two wheelers are popular in China.

There is often a concentration in policy implementation on the technological options (Hickman and Banister 2007; Sperling and Gordon, 2009; Gilbert and Perl, 2010). Governments seem to feel more comfortable in encouraging mobility growth, and to rely on the use of clean vehicles to reduce emissions. But governmental influence in changing the motor vehicle fleet is limited, to voluntary or rarely mandatory emission standards, and the end result is that the increase in vehicle distance travelled offsets any gains made with cleaner vehicles.

A wider approach to achieving sustainable mobility requires a full range of policy measures, tailored to the particular context and the normative vision of 'what is desired', this includes the infrastructural measures, the physical dimension in urban planning, the vehicle technologies and also the behavioural options which can be used to influence people's attitudes to travel (Newman and Kenworthy, 1999; Hickman and Banister, 2007; Banister, 2008; Hickman *et al.*, 2011a). The social and cultural aspects behind travel, including how car use is often central to everyday life, are frequently misunderstood (Urry, 2007). A better understanding of the psychological and sociological dimensions behind travel may help tailor policies more to the requirements, constraints and aspirations of individuals.

Hence there are some very unpalatable decisions to be made in the transition to sustainable transport. The current neoliberal 'governing at a distance' stance – of limited public transit, walking and cycling investments, poor use of urban planning and traffic demand management measures, limited penetration of low emission vehicles; and, instead, continued road building, sprawling development patterns and growth in international travel – means that transport CO_2 emissions continue to rise in most cities in the developing world and internationally. Transport is potentially also the most difficult sector to decarbonise in social and cultural terms, with many individuals either aspiring to or already reliant on car use in their everyday lives. As yet, the transition to greater sustainability in transport remains little more than conjecture.

References

Anable, J. and Gatersleben, B. (2005) All work and no play? The role of instrumental and affective factors in work and leisure journeys by different travel modes. *Transportation Research Part A: Policy and Practice*, 39, 163–181.

Banister, D. (2008) The sustainable mobility paradigm. *Transport Policy*, 15, 73–80.

Barter, P. (2004) A broad perspective on policy integration for low emission urban transport in developing Asian cities. Conference paper for International Workshop on Policy Integration towards Sustainable Energy Use for Asian Cities. Kanawaga, Japan.

Bows, A. and Anderson, K. (2007) Policy clash: Can projected aviation growth be reconciled with the UK Government's 60% carbon-reduction target? *Transport Policy*, 14, 103–110.

Dimitriou, H. (1992) *Urban Transport Planning: A Developmental Approach*. London: Routledge.

Gilbert, R. and Perl, A. (2010) *Transport Revolutions: Moving People and Freight Without Oil*. Philadelphia, PA: New Society.

Hickman, R. and Banister, D. (2007) Looking over the horizon: Transport and reduced CO_2 emissions in the UK by 2030. *Transport Policy*, 14, 377–387.

Hickman, R., Ashiru, O. and Banister, D. (2011a) Transitions to low carbon transport futures: strategic conversations from London and Delhi. *Journal of Transport Geography*, 19, 1553–1562.

Hickman, R., Fremer, P., Breithaupt, M. and Saxena, S. (2011b) *Changing Course in Sustainable Urban Transport: An Illustrated Guide*. Manila: Asian Development Bank.

International Energy Agency (2010) *Key World Energy Statistics*. Paris: IEA.

Newman, P. and Kenworthy, J. (1999) *Sustainability and Cities: Overcoming Automobile Dependence*. Washington, DC: Island Press.

Pilling, D. (2008) India hits a bottleneck on its way to prosperity. *Financial Times*, 25 September.

Sperling, D. and Gordon, D. (2009) *Two Billion Cars: Driving Toward Sustainability*. New York: Oxford University Press.

Steg, L. (2005) Car use: Lust and must. Instrumental, symbolic and affective motives for car use. *Transportation Research A*, 39, 147–162.

Urry, J. (2007) *Mobilities*. Cambridge: Polity.

Part 7

Gender and development

Editorial introduction

It is widely accepted that development must be informed by gender analysis. All major development agencies include a mandatory framework for all activities to check that gender is considered, even in neutral projects. Some might even argue that the way gender is integrated into development thinking and practice indicates a high degree of co-option or neutralisation of feminist objectives, rather than their success in transforming the development agenda. Recent processes of globalisation have questioned the state, institutional rules and processes to determine what open space is available to women to claim gender rights and justice. The important question is how do women engage the state in policymaking and hold it accountable. Increasingly, the political participation of women has become crucial to access the male-dominated world of policymaking. Empowerment involves challenging existing power relations and gaining greater control over sources of power especially challenging patriarchy and inequality. It is a process by which women redefine and extend what is possible for them on an individual basis to bring about transformation. Women's interest can only be served when both the formal and informal spaces are brought together and negotiated by the women's movement.

This section argues that gender remains central to development and offers a lens through which to deconstruct gender-related processes and issues. It enhances understanding of women's lives and the gendered nature of economic, social and political processes.

UNICEF has identified women, or rather mothers and young children, as the groups most vulnerable to economic adjustment programmes. There has been a very rapid growth of women's employment, from which women have gained self-esteem and provided visible contribution to household income. Women's willingness to work is partly determined by the extent to which they have control over income generated in the household. This relates to gender bias in the household, and the gendered nature of negotiation and exchange within households. In addition, there is extensive evidence that women's income is almost exclusively used to meet household needs, whereas men tend to retain a considerable portion of their income for personal spending. Women with access to better education and employment opportunities rely less on their children for economic security and support in later life. Gender remains a problematic issue for post-structural adjustment policies as local constructions are not taken into account when deriving policies in developing countries.

Technological change and globalisation raise many gender issues and have led to the disruption of cultural norms in households and traditional societies. Conflicts within households, within generations and between spouses have become quite prevalent. Female employment has led to the breakdown of joint families and nucleation of households; this has implications for the elderly, particularly in the context of demographic transition in relative numbers of children (due to falling fertility rates), working age population and older population (improvements in health, leading to higher life expectancy) in developed as well as developing countries. Feminist critiques from the South have consistently pointed to the similarities that exist between gender issues (e.g. domestic or sexual violence) in the global South and global North and criticised development agencies from the North to having a very Eurocentric view.

Global capitalism has had immense effects on women's lives all over the world, it is shaped by mass migration and economic exchange and women are moving around the globe as never before. Every year, many leave Eastern Europe, Mexico, Philippines and Sri Lanka, to work in the homes, nurseries and brothels of developed countries in the northern hemisphere creating a 'care deficit' in their own country to ease the 'care deficit' in developed countries.

Today as in the past struggles and achievements in the domains of gender equality, reproductive autonomy and sexual pluralism are marked by tensions and deep controversies. By the 1990s a diverse and lively political agenda around sexual orientation and gender identity has also become globalised. Sexual rights activists from the global South are engaged in a struggle against broader issues of justice, elimination of all forms of discrimination and exclusion and its economic, political, social and psychological effects and help in creating space for sexual diversity.

Family planning programmes introduced over the last forty years in many developing countries have implanted the value of a small family among people with aspirations to rise up the social ladder. But the mindset for a small family has not erased the deeper cultural roots of gender disparity.

7.1 Demographic change and gender

Tiziana Leone and Ernestina Coast

Introduction

The demographic transition theory studies the progression from a regime of high mortality and high fertility to low mortality and low fertility through various stages. These are linked to wealth and industrial changes in countries which are often linked to changes in the gender systems such as an increasing female autonomy which led to increasing contraceptive use. In the last three decades demographers have become more and more interested in the gender dimension of demographic change and they have tried to account for it in the analyses as well as in the theoretical work. More studies than ever have highlighted the need to come up with methodological innovations in order to capture in some instances the impact of gender systems on demographic change. This chapter will consider key points (without being exhaustive) of demographic changes which have been linked to the broader term of gender systems.

Demographic change can refer to change over time in a population's growth or territorial distribution, or to change in the major components of population growth, especially fertility (the number of births per woman) and mortality (the risks of dying at a particular age) and to a lesser extent to migration (people moving from their place of usual residence).

Mason (1997) defines gender systems as socially constructed expectations for male and female behaviour that are found in every known human society. For the sake of simplicity given the wide range of issues this terminology refers to, this paper will refer loosely to this concept as well as just showing changing trends between men and women. In this chapter reference is made to both demographic dynamics for both men and women (e.g. differential mortality) and to demographic changes as a result of gender systems' influence.

Population structure and gender

A key value in demography is the sex ratio at birth (male births per female ones) which is set to 1.05/1.06 men per woman. This value is often used to analyse demographic phenomena which might be influenced by the social construct of gender (e.g.: sex-selective abortion or infanticide). Along with key biological determinants of survival/mortality as we will see in the mortality section below, it does influence the key population structure which is the fundamental starting point for any demographic change analysis.

Changes in the gender imbalance, whether behavioural or cultural or in general as part of the gender systems, do have an impact on the overall population structure in more general terms and as we will see more in detail in the three demographic components (fertility, mortality and migration). Figure 7.1.1 shows that while the age structure has a more obvious progress throughout the various estimates/projections, the gender structure might not be as straightforward. While fertility decline leads to ageing populations, it could also lead to a gender imbalance towards men as in many Asian countries. Similarly behavioural differences between men and women lead to different levels of survival and life expectancy at birth as in many eastern European countries where the mortality gap is greater. Gender adds an extra layer of complexity to the projections of the population structure in the future which is rather challenging to account for and often neglected. The issue becomes even more challenging if we consider a region or country where migration should be included.

The sections that will follow will attempt to briefly touch on these points.

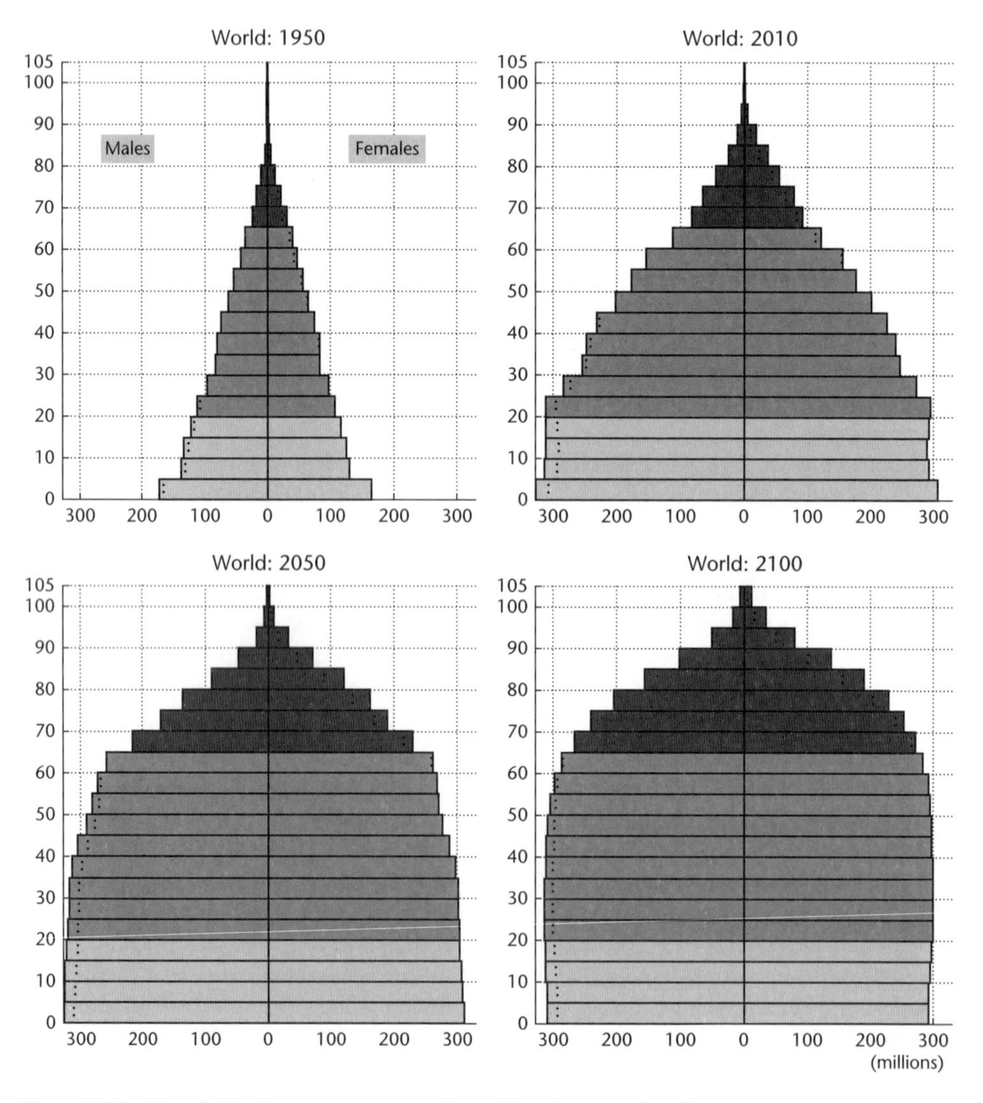

Figure 7.1.1 Population by age groups and sex (absolute numbers)

Source: UN Population Division, DESA 2012

Fertility and gender

The relationship between fertility trends and gender systems is not as straightforward as expected. Declining fertility trends have been associated on one hand with increasing female autonomy, on the other hand to have contributed in many Asian countries to an exacerbated gender discrimination. Paradoxically low gender equity within family institutions has had a detrimental effect on the level of fertility (McDonald, 2000). When considering the fertility transition, changing family morality, religious values as well as declining mortality are equally relevant (for more information on the first and second demographic transition, see Van de Kaa, 1987).

Through the proximate determinants one can disentangle fertility changes down to the key components such as contraception, marital unions, abortions or postpartum amenorrhea (Mason, 1997). Through these components we have been able to look at the impact that female autonomy through increasing education and female employment have had on increasing contraception use, shortened reproductive span (later age at first childbearing), and so on. At the individual level, for example, Presser (1971) has suggested that the timing of marriage and motherhood can have strong effects on women's subsequent achievements and well-being. In general, several experimental studies showed that increasing autonomy through credit schemes, increasing female employment and education (Mason, 1997) will eventually decrease the level of fertility.

While the causality is usually analysed as the impact that gender systems have had on demographic changes, the direction of the relationship is not always straightforward, as Mason (1997) points out, and might often be the result of shortcomings in the statistical models used by demographers. The demographic transition usually considers that increasing women's autonomy leads to a decline in the level of fertility. The opposite could also be true, therefore, women could aim for a lower number of children in order to increase their level of autonomy or at least their dependence on her.

An example of how the causality might go in the other direction is the demographic transition in eastern and southern Asia. The impact of son preference, however moderate, must be viewed in the context of fertility transition in Asia: the higher the overall level of fertility, the weaker the effect of sex preference on the TFR. For example, if a population strictly applies a stopping rule after the birth of a boy, son preference would have a higher impact on parity progression at parity one than at parity two (Arnold *et al.*, 1998; Mutharayappa *et al.*, 1997).

There is plenty of evidence that once the transition of the fertility goes further the sex ratio at birth (Figure 7.1.2) is strongly influenced. Studies conducted in India suggest that sex preference influences not only discrimination and excess mortality, but also demographic transition (Leone *et al.*, 2003). Through the postponement of stopping behaviour until the birth of a son, sex preference may exert its strongest effect on fertility during the intermediate stage of demographic transition, thereby slowing fertility decline. However, in Korea and China, where aggregate fertility levels are low, gender bias has persisted through the demographic transition – no longer as sex-determined stopping behaviour, but as sex-selective abortion.

Even where there is a convergence in gender interests in relation to fertility goals within a society, gender concerns will remain if such goals incorporate and reproduce the wider devaluation of women by the society in which such goals are formed. This devaluation manifests itself in the discriminatory provision of critical expenditures in the human capital, productivity and life options of daughters as well as in more extreme forms of discrimination, including female foeticide, infanticide and life-threatening neglect of daughters. These phenomena have unsettling implications for population structure, marriage, the labour market and women's status in these two countries.

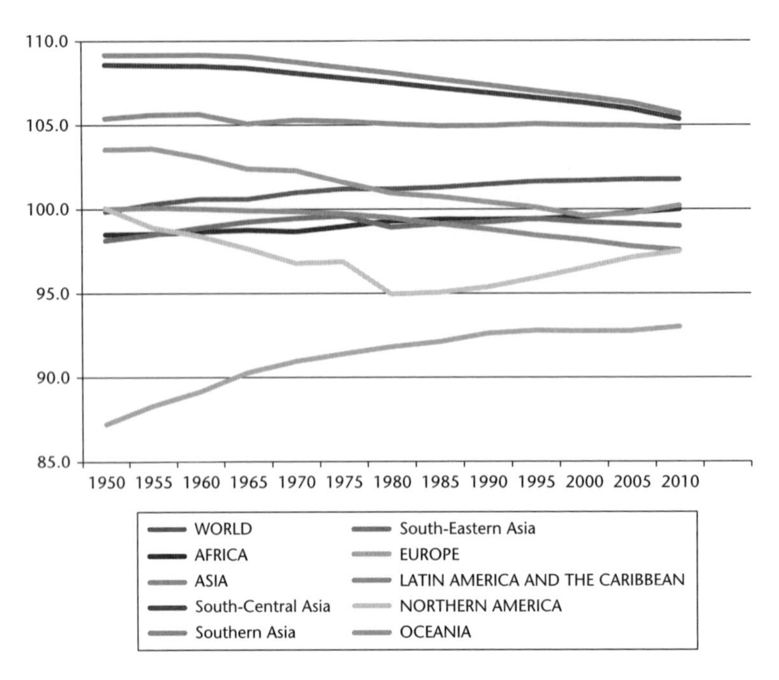

Figure 7.1.2 Sex ratio at birth

Source: UN Population Division, DESA 2012

Mortality and gender

Mortality changes are influenced by gender systems at different levels and at different ages across both developed and developing countries. Different causes are reported in gender differentials for child and adult mortality.

A key indicator of the effect of gender discrimination on children is the sex-specific child mortality rate, after correction for sex-determined disease susceptibility (girls are generally more resistant to disease than are boys) and for other gender-related differences (e.g. accidental deaths are more common among boys than among girls). A mortality rate among girls that is higher than that among boys indicates excess female child mortality and hence female disadvantage. In this respect, South Asia is again prominent, for although overall childhood mortality continues to decline in this region, the relative plight of girls has worsened. For example, between the 1970s and the 1980s, excess mortality among females aged 1–5 years increased from 11 per cent to 28 per cent in India and from 8 per cent to 22 per cent in Nepal (Leone *et al.*, 2003). Son preference may be a direct contributing factor, through shortened birth intervals after the birth of girls and the neglect of younger daughters in large families of girls in which parents are still hoping for a boy. Son preference has also promoted more extreme forms of discrimination, such as feticide or infanticide of females (Arnold *et al.*, 1998; Mutharayappa *et al.*, 1997).

While education is a strong predictor of female autonomy and also of improved child survival, previous research in Asia shows that education is not always a strong predictor of improved child preference for boys (Leone *et al.*, 2003). As for fertility, culture and religious values could be key in explaining gender discrimination for babies. At adult level biological as well as behavioural differences are important in explaining the differences in life expectancy at birth for men and women (Figure 7.1.3).

There is a vast literature on the determinants of gender differences in mortality (Vallin, 2001). They include biological factors such as hormonal influences on physiology and behaviour, and environmental factors, such as cultural influences on gender differences in health behaviours. Higher male hospitalizations and mortality is often linked to higher rates of smoking. Changing health behaviours and increasing numbers of female smokers at a time when men's numbers are declining, have slowly narrowed the gap (Gjonça *et al.*, 2005).

On the biological side different factors have been proposed in order to explain why women live longer than men. In general, higher male mortality rates are found among infants and children, a time when only small differences in behaviour are present. Women's sex hormones reduce the risk of ischaemic heart disease, in part by the favourable effects they have on serum lipids. Oestrogens help remove damaging blood fats and prevent blood clots that lead to heart disease. Additionally oestrogens help protect brain cells and prevent degenerative diseases associated with ageing (Vallin *et al.*, 2001). In contrast, men's higher testosterone levels have unfavourable effects on serum lipids, contributing to increased risk of ischaemic disease.

The behavioural explanation of the differential mortality which has been fundamental in understanding the mortality changes over the last few decades in relationship both with the demographic transition and economic changes. Many studies have focused on behavioural and cultural factors that can explain men's higher mortality (Gjonça *et al.*, 2005). Men are more likely to partake in risky jobs like mining or heavy manufacturing and behaviours like smoking and drinking, are more likely not to have a healthy diet, and are less likely to use health services for prevention of diseases. It also may be that men suffer on average greater levels of social stress linked to their professional life and their social status. Risky behaviours tend to be more associated with lower social status. Changes in mortality in later life are more likely to affect the overall sex gap in mortality (Gjonça *et al.*, 2005).

The gender gap in survival is a constant feature of European data over the last century. Women do show higher life expectancy at birth than men across all countries. This is generally a worldwide phenomenon with a few exceptions in pockets of developing countries such as India or to a lesser extent in China where often a gender imbalance is caused by female infanticide or sex-selective abortion (Arnold *et al.*, 1998; Mutharayappa *et al.*, 1997). The Baltic countries in

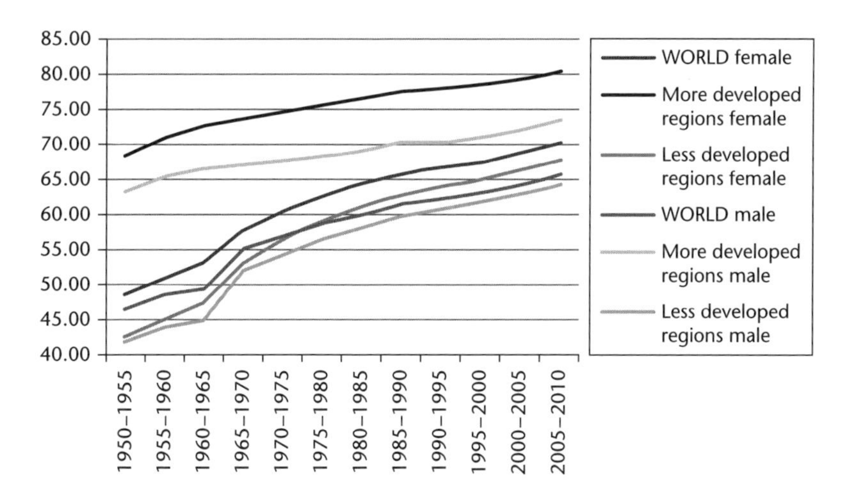

Figure 7.1.3 Life expectancy

Source: UN Population Division, DESA 2012

particular have the widest gap. This is due to health lifestyles of men which are usually characterised by higher levels of smoking and binge drinking (Vallin, 2001). Around 40 per cent of the gender gap in life expectancy could be due to traditional risk factors, such as smoking, obesity, hypertension, high plasma total, and low high-density lipoprotein cholesterol. In general, women are reported to cope better at times of crisis and to deal with adversities with strategies rather than avoidance, such as admitting being ill or depressed and getting treated.

Migration and gender

Migration is the component of demographic dynamics possibly least studied in connection with gender. In particular, the key criticism is that migration theories do not incorporate gender. Despite common perceptions 49 per cent of migrants worldwide are women and migration trends by gender have not changed over time (UN Population Division, 2011). What is changing is the role of women in migration rather than the extent of it. While before they were mainly involved in family reunifications, now they move particularly to find a new job and to be the breadwinner of the family.

Until the early 1980s, most migration studies focused on economics. Male migrants were seen as the main economic players and women were often seen as passive followers – the guardians of tradition and stability at home. But perceptions are changing and now migrant women are increasingly migrating to find jobs as individuals, although many still migrate as dependants (Zlotnik, 1995).

Women often migrate for reunifications but are also increasingly becoming key players when finding a job. This is not to say though that gender roles are still not strongly rooted in many migration flows. Migrant women are important economic actors and their participation in economic activity is closely related to the needs of their families, so that the choices that migrant women make regarding work cannot be understood without taking into account the situation of their families and women's roles within them. Any analysis pertaining to migration flows and understanding of causes and consequences of migration will need to take into account the key gender factors that coexist to create different experiences all along the migration spectrum.

Conclusion

Far from being exhaustive in considering the key issues of demographic change and gender, this chapter gave an overview of some of the matters that connect demographic change and gender. Fertility, mortality and migration are all intertwined within gender systems. The relationship between the single components and gender is not as straightforward as one might expect. In sum, quantitative research into the question of how demographic change influences gender systems is largely absent. It is feasible to assume that the demographic transition/change can be a precursor of 'gender transition' (Mason, 1997) in many parts of the world. Demographic change alone will not revolutionise a society's gender stratification system. It might, though, produce changes to the conditions that can change the balance within gender systems such as an increase in female education and employment.

References

Arnold, F., Choe, M. K. and Roy, T. K. (1998) 'Son preference, the family building process and child mortality in India'. *Population Studies*, 52(3): 301–315.

Gjonça, A., Tomassini, C., Toson, B. and Smallwood, S. (2005) 'Sex differences in mortality, a comparison of the United Kingdom and other developed countries'. *Health Statistics Quarterly*, 26(Summer): 6–16.

Leone, T., Matthews, Z., Dalla Zuanna, G. (2003) 'Impact and determinants of sex preference in Nepal'. *International Family Planning Perspectives*, 29: 69–75.
Mason, K. O. (1997) 'Gender and demographic change: What do we know?' In G. W. Jones, R. M. Douglas, J. C. Caldwell and R. M. D'Sousa (eds), *The Continuing Demographic Transition*. Oxford: Clarendon Press, pp. 158–182.
McDonald, P. (2000) 'Gender equity in theories of fertility transition'. *Population and Development Review*, 26(3): 427–140.
Mutharayappa, R., Choe, M. K., Arnold, F. and Roy, T. K. (1997) 'Son preference and its effect on fertility in India'. *National Family Health Survey Subject Reports*, 3.
Presser, H. (1971) 'The timing of the first birth, female roles and black fertility'. *Milbank Memorial Fund Quarterly*, 49: 329–361.
UN Population Division (2012) 'Trends in international migrant stock: Migrants by age and sex'. United Nations database, POP/DB/MIG/Stock/Rev.2010. New York: Department of Economic and Social Affairs, UN.
Vallin, J., Meslé, F. and Valkonen, T. (2001) *Trends in Mortality and Differential Mortality*. Strasbourg: Council of Europe Publications.
Van de Kaa, D. (1987) 'Europe's second demographic transition'. *Population Bulletin*, 42(1): 1–59.
Zlotnik H. (1995) 'Migration and the family: The female perspective'. *Asian Pacific Migration Journal*, 4(2–3): 253–271.

7.2

Women and the state

Kathleen Staudt

Until the last two decades, the state has been relatively neglected in political studies. Analysts had long referred to the 'nation-state', but were much more taken with the nation: the growth of nationalism, national and cultural values, political participation, popular attitudes toward government and society. The state, defined in its Weberian sense as the exercise of sovereign authority within territorial boundaries, was the empty box that structural-functional theorists drew in their political systems graphics, seemingly relegating that box to the sometimes tedious studies of public administration.

Even women and gender analysts succumbed to these tendencies. They studied 'inputs' to the box with research on social movements, revolutions, public opinion and political participation. They studied 'outputs' from the box in their analyses of public politics and laws that 'developed' women and men differently in various class, cultural and geographic contexts.

State analysis winds up

With the publication of *Bringing the State Back In* (Evans *et al.*, 1985) comparative political theorists put the state on the analytic agenda. With historical and comparative perspectives, they ended the pretence (so common in studies of the USA) of an irrelevant or minimalist state, as in the classic liberal ideal. This focus led people to examine government institutions more carefully for the way they opened or closed doors to people and policy debates.

The study of institutions experienced some revival, with analysts attentive to the grand institutions and rules that enveloped the political scheme of legislative bodies, administrative agencies and electoral systems. As far back as 1955, Maurice Duverger compared the single-member and proportional representation electoral systems for their impacts on geographical or ideological politics and two- or multi-party systems. Soon analysts of female under-representation in high-level political decision-making positions would embrace these grand-level approaches, asking whether proportional representation and/or parliamentary systems would seat more elected women and, most importantly, what difference would that make in the gendered decisions and outcomes of the political process. Policy analysts searched for ways to understand the connections between policies, institutions and decision-makers for their resistance to gender justice and attention to deep inequalities.

These connections began to be made with attention to the state. Statist critique and analysis became attached to grand explanatory narratives, from pluralism to Marxism, but with attention to women and gender. States were conceptualized as historical and institutional shells that protected and advanced male privilege. Theorists challenged existing conceptions of the state as neutral umpire between competing interests (the pluralist view ingrained in US politics), or as the instrument of the dominant economic class (the Marxist view). The institutionalization of male interests reached beyond capitalist profiteering rationales. Moreover, although socialist models were few and flawed, gender inequality persisted in seemingly intractable ways.

During the 1980s, the wind-up period on women and the state analyses, the state and its ideology were tied to development policies. Debates existed among 'feminists': should feminists work with the state or against the state? In contrast to the few states with a positive track record on women and gender justice, many states seemed doomed to perpetuate gender hierarchy.

Several collections emerged on the state–development–women connections in the 1980s. Charlton *et al.* (1989) reviewed literature on statism, calling for analysts to examine state officials and their gender ideologies, state policies and institutions, and state definitions of the parameters of politics. The area-studies chapters in their collection made theoretical use of the public–private divide that emerged in Western theories. Deere and León (1987) set their collection in terms that linked agricultural and land policies in Latin America to national economic policies and the global debt problem. Afshar's collection (1987) focused on nation-state case studies of policies in Asia and Africa. In their edited collection, Parpart and Staudt (1989) drew on critical theoretical perspectives of dependency and mode of production analysis to examine the origins of the state in Africa, women's access to the state, and state management of resources. These collections traced state origins to Europe in centuries past, with the transnational spread of state structures and ideologies through world capitalism, colonization and imperialism. Yet even before the rise of the modern state, it appeared that men dominated women, drawing in part on public, institutional authority.

A proliferation of edited collections emerged using the nation-state as a unit of analysis. It was refreshing and long overdue that country case studies emerged, for comparative politics had rarely integrated women and gender into analyses. Yet this flurry of country cases made it clear that analysts could not easily generalize about states in the nearly 200 countries worldwide.

State analysis winds down

Analysis on women and the state wound down for a variety of reasons. Related analyses re-emerged with new conceptual language such as democracy, governance, political representation and accountability. Why the wind-down? Some of the debates pointed toward new analytic trends.

First, states began to be differentiated, not only in their strength and weakness as agents of control, authority and power, but also in the degrees to which they opened space to women's claims to be active and heard, for space within the state, and for justice in policy and legal terms. States were never all alike, and some of the most overdrawn feminist analyses treated them as monolithic, with men and women as monolithic inhabitants within them. All but the most simplistic research acknowledges the diversity among women and men by class, nationality, culture and geographic space, among other factors (Mohanty *et al.*, 1991). Further, international financial institutions began structural adjustment programmes in the 1980s that aimed to reduce the size of states, make them more efficient and expose more economic resources to market forces. It was an analytic mistake to assume that states were static.

Second, weak states and even some strong states never had full power, authority and agency to envelop the societies they claim (Migdal, 1988). Important theorizing recognized that another part of power was the ability of inhabitants to resist, sabotage and ignore state machinery. In parallel fashion, women exercise power in resistance to men and to the state. It was an analytic mistake to accept theoretical or state claims of omnipotence.

Third, states are not monolithic, but rather bundles of contradictions that do not work in perfect harmony. Some of the most overdrawn statist analysis emerged from the US with over-generalized, almost biological notions of women as victims and men as sexual aggressors (MacKinnon, 1989). The welfare state began to emerge as a category worth analysis and engagement. Canada and the Scandinavian states emerged as models for ways that rights and justice agendas might be consolidated within (Vickers *et al.*, 1993; Gelb, 1989; Hernes, 1987).

States revisited: Democracy, governance and politics

With democratic space and process, along with healthy judicial systems, people make use of contradictions within the state to achieve gains. In the 1990s, with the so-called transitions to democracy, new contexts led to greater analytic attention to the kinds of institutions and democracies which permit democratic openings to women and to gender justice agendas. A mammoth collection with 43 country case studies focused on political institutions (Nelson and Chowdhury, 1994). Collections on women engaging state and international bureaucracy institutions offered insights into openings for women and gender justice agendas (Staudt, 1997; McBride and Mazur, 2010 on 13 European countries in the Research Network on Gender, Politics and the State). Goetz (1997) aptly titled a collection *Getting Institutions Right for Women in Development*.

With globalization increasing in full force, in a context lacking accountable global governance, claims for gender rights and justice are operable primarily in existing nation-state governance and transnational organizations. Thus, analysts have put institutional rules and processes on the agenda to determine which open space to women and new policy agendas. In the United Nations Development Programme, which prepares annual *Human Development Reports*, readers can compare nation-state 'Human Development Index' (HDI) rankings, how gender-disaggregated (GDI) data reduce rankings, and what difference the gender empower measure (GEM) might make. Although useful, such measures are not sensitive to the intersection of gender with class, ethnicity and race.

At the World Conference on Women in Beijing, in 1995, participants resolved to follow up its elaborate Platform for Action. Accountability is now a key concept for organization and for research. Currently, debates are less likely to be anti-state or pro-state, but rather: how do people engage the state and public affairs for accountability, not only on traditional women's policy issues but also on mainstream policy issues (Staudt, 1998)? One vehicle of engagement is through transnational non-government organization (NGO) activism, whether analysed at global, well established NGO levels (Keck & Sikkink, 1998; Moghadam, 2005) or at grassroots levels (Staudt, 2008; Jaquette, 2009; Basu, 2010), particularly in left–center political coalitions that resonate with feminist agendas (Waylen, 2007).

So analysts are back to institutions once again, but in ways focused more closely on strategies to increase women's representation, such as legislative quotas that exist in nearly a hundred countries or to strengthen policy responses on such issues as violence against women (Weldon, 2002).

Conclusion

The rise and fall of 'women and the state' analysis parallels the contemporary challenge to meta-analysis and grand narratives. The state as all-powerful agent of male control is one of those grand, but wobbly scaffolds that, with a critical eye, falls with its own flaws. 'Women and the state' analysts now pursue various paths that examine national and global governance with a wide variety of institutional rules and policies. Debates on women and the state moved analysts beyond the pro- or anti-state stance, useful in applications to women's activism. Analysts also moved beyond the notions of states as static and the same. What emerged after statist analyses was a comparative approach that examined institutions and accountability strategies. Yet a common thread of masculinist ideologist and male control lingers in the institutions, policies and laws in countries worldwide, part of a growing global economy. Hopefully, comparative studies of women in politics will not regress to a pluralist approach, for the analysis convincingly demonstrated that few, if any, states have operated as neutral umpires in gender terms.

Bibliography

Afshar, Haleh (ed.) (1987) *Women, State and Ideology: Studies from Africa and Asia*, Albany, NY: SUNY/Albany Press.

Basu, Amrita (ed.) (2010) *Women's Movements in the Global Era: The Power of Local Feminisms*, Boulder, CO: Westview.

Charlton, Sue Ellen M., Everett, Jana and Staudt, Kathleen (eds) (1989) *Women, the State, and Development*, Albany, NY: SUNY/Albany Press.

Deere, Carmen Diana and León, Magdalena (eds) (1987) *Rural Women and State Policy: Feminist Perspectives on Latin American Agricultural Development*, Boulder, CO: Westview.

Duverger, Maurice (1955) *The Political Role of Women*, Paris: UNESCO.

Evans, Peter B., Rueschemeyer, Dietrich and Skocpol, Theda (eds) (1985) *Bringing the State Back In*, Cambridge: Cambridge University Press.

Gelb, Joyce (1989) *Feminism and Politics: A Comparative Perspective*, Berkeley and Los Angeles: University of California Press.

Goetz, Anne Marie (ed.) (1997) *Getting Institutions Right for Women in Development*, London: Zed Books.

Hernes, Helga Maria (1987) *Welfare State and Woman Power*, Oslo: Norwegian University Press.

Jaquette, Jane (ed.) (2009) *Feminist Agendas and Democracy in Latin America*, Durham, NC: Duke University Press.

Keck, Margaret and Sikkink, K. (1998) *Activists Across Borders*, Ithaca, NY: Cornell University Press.

Krook, Mona Lena (2009) *Quotas for Women in Politics: Gender and Candidate Selection Reform Worldwide*, New York: Oxford University Press.
MacKinnon, Catharine (1989) *Toward a Feminist Theory of the State*, Cambridge, MA: Harvard University Press.
McBride, Dorothy and Mazur, Amy (2010) *The Politics of State Feminism: Innovation in Comparative Research*, Philadelphia, PA: Temple University Press.
Migdal, Joel S. (1988) *Strong Societies and Weak States: State-Society Relations and State Capabilities in the Third World*, Princeton, NJ: Princeton University Press.
Moghadam, V. M. (2005) *Globalizing Women: Transnational Feminist Networks*, Baltimore, MD: Johns Hopkins University Press.
Mohanty, Chandra Talpade, Russo, Ann and Torres, Lourdes (eds) (1991) *Third World Women and the Politics of Feminism*, Bloomington: Indiana University Press.
Nelson, Barbara and Chowdhury, Najma (eds) (1994) *Women and Politics Worldwide*, New Haven, CT: Yale University Press.
Parpart, Jane and Staudt, Kathleen (eds) (1989) *Women and the State in Africa*, Boulder, CO: Lynne Rienner Press.
Staudt, Kathleen (ed.) (1997) *Women, International Development and Politics: The Bureaucratic Mire* (2nd edn), Philadelphia, PA: Temple University Press.
Staudt, Kathleen (1998) *Policy, Politics & Gender: Women Gaining Ground*, West Hartford, CT: Kumarian Press.
Staudt, Kathleen (2008) *Violence and Activism at the Border: Gender, Fear and Everyday Life in Ciudad Juárez*, Austin: University of Texas Press.
United Nations Development Programme *Human Development Reports*, New York: Oxford University Press. Available online at www.hdr.undp.org/en/reports/
Vickers, Jill, Rankin, Pauline and Appelle, Christine (1993) *Politics as if Women Mattered: A Political Analysis of the National Action Committee on the Status of Women*, Toronto: University of Toronto Press.
Waylen, Georgina (2007) *Engendering Transitions: Women's Mobilization, Institutions, and Gender Outcomes*, New York: Oxford University Press.
Weldon, S. Laurel (2002) *Protest, Policy, and the Problem of Violence Against Women: A Cross-National Comparison*, Pittsburgh, PA: University of Pittsburgh Press.

7.3

Gender, families and households

Ann Varley

Family and household as cultural constructs

The distinction between 'the family' and 'families' is a key fault line in contemporary social and political thought. Using the term 'families' signifies rejection of the nuclear family household as universal norm. Insistence on the definite article signals support for an ideal family form or repudiation of 'Western' influences allegedly undermining local traditions. 'Family' is often used as a touchstone by which communities differentiate themselves from each other or gauge how they have changed over time. Changes in family life may therefore symbolize the effects of development or modernization processes, for better or for worse.

Debates about the family are often, in reality, debates about women's behaviour. The figure of the young woman leaving the family home to find work in the city has often served as a symbol of modernity displacing tradition. Women may be blamed for a range of social ills, whilst men's behaviour too often remains unquestioned. There is also a danger in isolating families as a subject for discussion. Families may be blamed for the consequences of structural forces because it is convenient to the state to divert attention from its responsibility to address poverty.

Families are clearly an emotive subject. The term 'household' is often seen as more neutral. Households are usually defined in functional terms (what they do), whereas families are kinship units that need not be identified with a particular site. Confusion arises because household membership is also generally based on kinship. Although households may consist of single individuals, friends or homosexual partners, members are usually recruited through heterosexual partnerships, childbearing or the incorporation of other kin. Many discussions about 'the family' thus refer to the family household.

Household functions include co-residence, economic cooperation, reproductive activities such as food preparation and consumption, and socialization of children. It is mistaken, however, to regard 'household' as unproblematically descriptive by comparison with the value-laden 'family'. Households are also cultural constructs. We should not assume that our society's understanding of 'household' is shared by others. For example, in parts of sub-Saharan Africa the functions mentioned above may be spread across several residential units.

Functional accounts of the household unit run the risk of portraying it as isolated from the rest of society. In practice, members' survival and well-being is also influenced by social network connections with other households. The importance of social networks was demonstrated by Larissa Lomnitz's classic (1977) study of a Mexican 'shanty town'.

An emphasis on households as bounded units can also lead us to mistake them for individual social agents, and it is then easy to confuse the group with a specific human being. The 'head of household' concept has been much criticized for ignoring conflicts of interest between members and assuming that the head acts altruistically and equitably. Feminist scholars have challenged this assumption that family life is based on consensus and cooperation, arguing that households are characterized by inequality structured around the axes of gender, age and generation.

Gender, headship and household relations

The theme of gender and households is often narrowly interpreted as referring to 'female-headed households'. It is often assumed that the proportion of households headed by a woman is increasing dramatically, but the evidence is not clear. Many countries show increases in recent years, but others report the opposite (United Nations, 2000, 2006). Many discussions of the subject are, however, seriously flawed, statistically and conceptually.

Statistically, problems of definition and meaning are daunting. An economic definition of headship may be preferred but is not practical for census purposes. Many census agencies ask respondents to say who heads their household or exclude the possibility of a woman being the head unless there is no adult male present. Other agencies have replaced the head with a 'reference person', who may simply be the person who filled in the form. This may account, in part at least, for increases recorded in UN data for countries such as the United States or New Zealand (with some of the highest figures globally).

Many sources nonetheless still confidently assert that one-third of the world's households are headed by a woman. This figure comes from an 'educated guess' by participants in the 1975 International Women's Year conference (Irene Tinker, personal communication, 1995). It is not supported by census data (Varley, 1996). Although the UN data on headship rates by country suffer from the problems noted and are now rather old, it is still worth reviewing them to challenge the persistent myth about the extent of female headship. Table 7.3.1 suggests that, excluding Europe, the United States, Canada and Oceania, one in six households is headed by a woman. Including all countries reporting a figure, the proportion is one in five.

Why has the 'one-third' figure proved so attractive? As Cecile Jackson (1996: 492) suggests, there has been an 'inclination . . . to "talk up" the numbers of women-headed households, and their poverty, to justify GAD [Gender and Development approaches] in numerical terms'. 'One in three' grabs the attention, especially if these households are supposedly the poorest of the poor.

There are dangers in prioritizing impact over accuracy. There is no clear and consistent link between poverty and headship (United Nations, 2010). In addition, the 'feminization of poverty' debate can distract attention from issues that may be more important, such as intra-household or 'secondary' poverty – in particular, the denial to women and children of full access to men's income (Chant, 2007; Jackson, 1996). This may be why the *World Development Report 2012:*

Table 7.3.1 Percentage of household heads who are women (1995–2003)

Region	*Percentage*	*No. of countries for which data are available*
Africa	23.8	37
Northern Africa	12.9	2
Southern Africa	42.2	3
Rest of sub-Saharan Africa	23.5	32
Asia and Oceania	13.4	27
Eastern Asia	20.0	5
South-eastern Asia	15.4	5
Southern Asia	9.6	5
Central Asia	27.6	5
Western Asia	10.8	5
Oceania	54.1	2
Latin America and the Caribbean	23.9	21
Caribbean	33.5	7
Central America	21.2	6
South America	24.2	8
Canada and United States	45.9	2
Europe	30.5	22
Eastern Europe	33.1	7
Western Europe	29.4	15
All	21.4	109
Excluding Canada, USA, Europe and Oceania	16.3	83

Source: Author's calculations, from United Nations (2006), Table A6. Source of data *Demographic Yearbook* database, plus United Nations (1999).

Notes: No data for China (except Hong Kong and Macau) or the Russian Federation. Weighted regional totals calculated using estimated number of households for each country obtained by dividing population size by mean household size. For the need for caution in interpreting these figures, see text and Varley (1996).

Gender Equality and Development highlights headship only when observing that female-headed households in rural areas own less land and have poorer access to credit and productive inputs (World Bank, 2011: 224–9). The latest (2010) edition of the United Nations publication series on *The World's Women* no longer provides headship figures by country.

The focus on female-headed households has other dangers. First, there is a temptation to conflate households headed by women and those headed by men with, respectively, 'women' and 'men'. This renders women who are not heads of household invisible. The problems with such an approach can be seen, for example, in programmes providing security of tenure that seek to attain gender equity by including women without a partner, but ignore those who do currently have a partner (Varley, 2007). What will happen to such women should their relationship subsequently break down? Will they be able to keep a roof over their heads? Second, there is a tendency to over- or under-play the agency of women who head their household. Portraying them as opting out of patriarchal relationships, for example, overlooks the question of widowhood. Table 7.3.2 suggests that, in many regions, the highest proportion of woman-headed households occurs where the head is aged 60 or over. Different life expectancies for men and women, together with a tendency for women to marry partners older than themselves, mean that demographic ageing may increase the percentage of female-headed households.

Underplaying women's agency makes female heads the victims of male desertion. Although it is more likely to be women who are left caring for children, it is not always men who leave. Recent interest in masculinities and development questions the stereotype of male 'irresponsibility' (Bannon and Correia, 2006). In addition, intra-household conflict is not confined to conflict between spouses. There is a history of conflict between women in extended households, particularly in southern and eastern Asia and in Latin America. Where brides have traditionally moved in with their husband's parents, there has long been a problem of mothers-in-law teaching them to 'know their place' in a sometimes brutal fashion. Where greater prosperity and female employment has made young couples less dependent, the unhappiness created by such arrangements encourages household nucleation.

Such changes have implications for the welfare of elderly women, who have been marginalized by the literature's emphasis on 'lone mother' households (Varley, 1996). Developing countries are currently ageing much faster than developed countries. Fertility decline implies that older people will have fewer children to support them, although not all children may have supported their parents in the past anyway. In addition, 'the growing number of young married couples who live on their own . . . means that care by a daughter-in-law is no longer automatic' (Wilson, 2000: 120). We should not assume that people who *do* live with younger relatives have been 'taken in'. Inter-generational support does not necessarily flow 'upwards', especially when economic difficulties facing younger people make even meagre pensions attractive to other family members (Lloyd-Sherlock, 2010). Older people can find themselves sharing their accommodation, pensions, or limited incomes from informal employment with the younger generation far longer than they may have anticipated or desired.

Elderly people can find themselves in a vulnerable position – in extreme cases, as victims of abuse – when they live with relatives: for example, when the balance of power in the mother-in-law/daughter-in-law relationship shifts towards the younger woman. Yet older women may be in a better position than older men. Women's association with housework and childcare means they can continue to 'make themselves useful'. By contrast, older men who are no longer the breadwinner risk being seen as 'useless' or blamed for earlier failures to provide adequate financial or emotional support for their families (Varley and Blasco, 2000).

The significance of gender in relation to families and households is, in short, far broader than the persistent emphasis on female-headed households would suggest.

Table 7.3.2 Percentage of household heads who are women, by region and age, and percentage aged 60+, by region and sex (1997–2010)

Region (no. of countries)	*Percentage of heads of household who are women, by age of head:*				*Percentage of heads aged 60+ years:*	
	< 25 years	*25–44 years*	*45–49 years*	*60+ years*	*women*	*men*
North America (2)	53	39	33	40	30.5	27.6
Latin America + Caribbean (7)	23	21	28	37	29.8	18.1
Europe (21)	50	34	35	51	41.7	28.0
Africa:						
- East (1)	18	18	29	36	31.7	17.3
- West (1)	14	9	12	14	18.5	14.6
- Southern (3)	45	38	41	53	24.8	16.1
Asia:						
- East (4)	41	18	18	28	40.4	30.0
- West (6)	22	11	17	30	37.9	18.5
- South (2)	9	6	11	22	38.5	16.1
- South East (3)	28	12	20	31	27.2	13.7
Oceania (2)	59	46	36	40	26.9	29.0
All (52)	35	24	27	40	36.7	23.5

Source: Author's calculations, from United Nations (2012): households by type of household, age and sex of head of household or other reference member.

Notes: Countries with data available (mostly from 2000–2006): North America: Bermuda, Canada; Latin America and Caribbean: Brazil, Chile, Colombia, Cuba, Dominican Republic, Ecuador, Mexico.

Europe: Austria, Bulgaria, Croatia, Czech Republic, Estonia, Greece, Hungary, Ireland, Latvia, Lithuania, Malta, Norway, Poland, Portugal, Romania, Russian Federation, Serbia, Slovakia, Slovenia, Spain, Switzerland.

East Africa: Uganda; West Africa: Burkina Faso; Southern Africa: Botswana, Lesotho, South Africa.

East Asia: Japan, North Korea, South Korea; West Asia: Armenia, Azerbaijan, Cyprus, Israel, Occupied Palestinian Territory, Turkey; South Asia: Iran, Nepal; South East Asia: Cambodia, China (Hong Kong SAR), Malaysia, Timor-Leste; Oceania: Australia, New Zealand.

Age ranges differ for Chile (15–29; 30–44), Cyprus (45–64; 65+) and Uganda (45–54; 55+).

References

Bannon, Ian and Correia, Maria C. (eds) (2006) *The Other Half of Gender: Men's Issues in Development*, Washington, DC: World Bank.

Chant, Sylvia (2007) *Gender, Generation and Poverty: Exploring the 'Feminisation of Poverty' in Africa, Asia and Latin America*, Cheltenham, UK: Edward Elgar.

Jackson, C. (1996) 'Rescuing gender from the poverty trap', *World Development*, 24, 3, 489–504.

Lloyd-Sherlock, Peter (2010) *Population Ageing and International Development: From Generalisation to Evidence*, Bristol, UK: Policy Press.

Lomnitz, L. (1977) *Networks and Marginality: Life in a Mexican Shanty Town*, London: Academic Press.

United Nations (1999) *Women's Indicators and Statistics Database (Wistat)*, Version 4 CD-Rom, New York: UN.

United Nations (2000) *The World's Women 2000: Trends and Statistics*, New York: UN.

United Nations (2006) *The World's Women 2005: Progress in Statistics*, New York: UN.

United Nations (2010) *The World's Women 2010: Trends and Statistics*, New York: UN.

United Nations (2012) *United Nations Statistics Division Demographic Statistics*, New York: UN. Available online at http://data.un.org/Explorer.aspx (accessed 28 Aug 2012).

Varley, A. (1996) 'Women heading households: Some more equal than others?' *World Development*, 24, 3, 506–20.

Varley, A. (2007) 'Gender and property formalization: Conventional and alternative approaches', *World Development*, 35, 10, 1739–53.

Varley, A. and M. Blasco (2000) 'Intact or in tatters? Family care of older women and men in urban Mexico', *Gender and Development*, 8, 2, 47–55.
Wilson, G. (2000) *Understanding Old Age: Critical and Global Perspectives*, London: Sage.
World Bank (2011) *World Development Report 2012: Gender Equality and Development*, Washington, DC: World Bank.

Further reading

Aboderin, I. (2006) *Intergenerational Support and Old Age in Africa*, New Brunswick, NJ: Transaction Publishers. Demonstrates the lack of evidence for the widespread belief that vulnerable older people will be supported by their children.
Bannon, Ian and Correia, Maria C. (eds) (2006) *The Other Half of Gender: Men's Issues in Development*, Washington, DC: World Bank. Reviews the connections between normative manhood and development in Latin America and Africa.
Chant, Sylvia (2007) *Gender, Generation and Poverty: Exploring the 'Feminisation of Poverty' in Africa, Asia and Latin America*, Cheltenham: Edward Elgar. Questions the 'feminization of poverty' argument and proposes the existence of a 'feminization of responsibility and obligation'.
Lloyd-Sherlock, Peter (2010) *Population Ageing and International Development: From Generalisation to Evidence*, Bristol: Policy Press. Examines family based and formal provision of social protection for older people across a range of different development contexts.
United Nations (2010) *The World's Women 2010: Trends and Statistics*, New York: UN. Reviews the evidence on gender, poverty and household structure and headship.

7.4

Feminism and feminist issues in the South

A critique of the "development" paradigm

Madhu Purnima Kishwar

I would like to clarify at the outset that feminism and commitment to strengthening women's rights in society through concrete, demonstrable actions are not always synonymous in the countries of the South. However, the public domain on women's issues is dominated by the discourse made fashionable by international development agencies of developed countries.

Taking the example of India, there is a long tradition of male social reformers dedicating their entire life to battling prejudices against equal participation of women in public life and strengthening their rights within the family. The nineteenth century social reform movements and the twentieth century freedom movement led by Mahatma Gandhi brought women's rights to the core of these movements. Even in contemporary India, many radical mass-based women's rights struggles have been led by men. Several outstanding women have emerged in leadership roles in rural movements. But the development analysts invariably present a unidimensional picture of patriarchal oppression.

Many of the rights for which valiant feminists in the West fought long drawn-out battles in the face of hostile attacks from men of their societies, especially those in the political establishment, came to Indian women through the efforts of male social reformers and freedom fighters who took the lead in fighting battles along with women. They braved forces of resistance in order to make dignified space for women in the public and political life of India. Unlike in the Western democracies, rights of women to education at par with men, to vote, to hold political office, enter male dominated professions, came without a lonely and harsh battle. The right to contraception and abortion have never evoked religious opposition or hostility in India as it did in countries where the Church holds the veto power on these issues (see Kishwar 1985).

Eurocentric view of development

To illustrate the limitations and inbuilt biases, I would like to use the World Development Report 2012 entitled *Gender Equality and Development (GED)*. The very first chapter of GED, "A Wave of Progress" starts by saying,

> Despite the hardships many women endure in their daily lives, things have changed for the better and at a speed that would not have been expected even two decades ago . . . Improvements that took 100 years in wealthier countries took just 40 years in some low- and middle-income countries.

It further reminds us that

> Japan's Equal Employment Opportunity Act of 1985 obliged employers merely to endeavour to treat men and women equally during job recruitment, assignment, and promotion. The mandate for equal treatment came about in 1997. The first domestic violence law was passed in 2001.

This is meant to hammer the point that even a "developed" Asian country like Japan is decades behind Europe and America – in comparison to "underdeveloped" countries of Africa and Asia! We are then reminded that it is "the ratification of CEDAW (Committee on the Elimination of Discrimination against Women) and other international treaties" – all of which emanated from Western countries – that "established a comprehensive framework to promote equality for women".

The GED report is supposed to provide us with a global overview of the challenges faced by women the world over. However, there is hardly any mention of the problems faced by women in developed countries. The assumption is that they have solved all their problems and have provided benchmarks and a road map for countries of the South.

That is why in almost all the charts and graphs presented in the report, countries of the South are presented as not yet measuring up to "developed" countries. Countries of the North are no doubt ahead of most countries of the South in matters of education and employment opportunities for women. But such reports completely overlook the fact that domestic violence is fairly rampant in developed countries, which have very high participation of women in the workforce as well as high education levels for females. Sexual violence is also very common in these societies. What does the flourishing pornographic industry and sex trafficking say about the status of women in developed countries? Why are these not used as important indicators for evaluating the status of women?

In this Eurocentric worldview, problems of women in developed countries are erased out of existence and the non-European world projected as politically and culturally backward and, therefore, in need of perpetual guidance by the West. This is not the only "development report" that displays a total lack of awareness about the fact that many of the "underdeveloped" countries of the South, such as India, China and other Asian countries were highly prosperous economies and societies during the period Europe describes as its "Medieval Dark Ages". Even in the eighteenth century, India and China were world leaders in trade, manufacturing, education, medicine, architecture, arts, and even engineering.

The era of "progress" for Europe was the era of colonization for many of the countries of the South. While Europe prospered, the colonized societies were politically subjugated through brute force and wrecked – economically, socially and culturally. Apart from their legendary wealth, many of the Asian countries had matrilineal family structures. Even in the eighteenth century, India's large regions in the south and east followed matrilineal inheritance. In patrilineal family structures too, Indian women had inalienable right to family property in the form of *stridhan* (totally different from modern day dowries) which passed from mother to daughter.

But because in Britain and US the law began changing in favour of women only in 1857 and 1858 respectively, it is assumed that the world over women had lived like helpless dependents. This worldview wants us to erase memories of colonial rule in the South which disinherited women by enacting laws which transformed community owned assets and joint family owned property into individual property vested only in male hands. For the Victorian minded British, patriarchal control over property and women was the "natural" God-given order of things. Matrilineal communities in India were condemned for encouraging "promiscuous and immoral" conduct among women. Women of matrilineal families were described with the disdain reserved for prostitutes because they could change their partners at will and divorce was not an issue in the way it was in the Christian world. They were legally pushed into giving up matrilineal inheritance systems in favour of patriarchal controls. For examples see Schneider & Gough (1961) and Saradamoni, K (1999). Today, the same societies are being delivered sermons against the sexual repression of women and advised to be more sexually liberated!

Obsession with 50 percentism

Another problem with the "Women and Development" framework is its obsession with measuring women's well-being through the "50 per-cent' benchmark. The parameters crafted by international development agencies for evaluating women's status in different societies are all centred on counting the percentage of women in various fields of life such as education, professions, labour force and politics. The march to "equality" means only one thing – how far are women from occupying 50 per cent of the spaces in all domains including jobs. Wherever they are less than 50 per cent it is assumed that they are deprived and excluded, and therefore oppressed.

"Percentism" in and of itself makes good sense in some areas such as sex ratio figures since the ratio at birth for males and females is near equal or in favour of baby girls. However, in China and most South Asian countries the growing culture of son preference as well as neglect of women's health and nutritional needs has resulted in an alarming deficit of females in the overall population of these countries. Therefore, "50 percentism" is a valid criterion in this domain. Nevertheless, in many other domains, it does not tell the whole story or can lead to erroneous conclusions. For example, there is a great deal of excitement about equal participation of women in parliaments and state legislatures as a means of political empowerment. However, feminists and development experts have paid far less attention to the nature of representative democracy in countries of the South as well as North.

When political parties are themselves under the grip of money and muscle power and institutions of governance are steeped in crime and corruption because they lack accountability and transparency, can the presence of a certain percentage of women by itself make our polity and governance machinery more citizen friendly? India, Pakistan, Bangladesh and the Philippines have seen women rise to the top echelons of power only to outperform the worst of men in crime and corruption. Did American foreign policy become less hegemonic or militarist when Condoleezza Rice or Hillary Clinton became Secretary of State? Did they try to curb the arms manufacturing lobby any more than the macho males who assumed those positions of power? It is not enough to demand that women be given a 50 per cent share of the available pie without first examining whether the pie is worth eating at all. And if it is rotten, women should know how to bake a new and healthy pie which has enough portions for all. All this is not to undermine the importance of enabling women to have a decisive say in public policy and political affairs but merely to point out that this "ladies compartment" approach to women's issues which demands that women should have reserved compartments in every train, never mind whether the train is headed for doom and destruction because its engineering is intrinsically faulty and disaster prone.

Similarly, much is made of school enrolment figures for girls as an indicator of high or low status. But the percentage of girls' enrolment in school tells a partial story about the future prospects of these girls or for that matter that of boys. The quality of education being imparted to them has far greater implications on what doors open or remain closed for them in life. In countries like India that have adopted English as the language of elite education, administration and access to job opportunities – the quality of education in most government schools for the poor is abysmal.

There is pride associated with being a master craftswoman or -man. There is no pride in a man becoming a street vendor, low paid assembly line worker, a peon or lowly paid clerk after some years of shoddy schooling. The economic and social status of supposedly "illiterate" women in weaver or potter households was far better and respect worthy than that of a woman who has received poor quality high school or even a bachelor's degree which does not equip her for anything better than a lowly clerical or sales girl job involving soulless drudgery. In a traditional weaver or potter household, women were an indispensable part of a very creative production process involving specialized skills. Even if they did not get a separate pay cheque they were not treated as dispensable. If today women of such households come out low on the Human Development Index, it is primarily on account of state policies aimed at marginalizing and destroying these traditional crafts and technologists despite the fact that these artisanal groups produce high quality aesthetic products which have ready buyers in the national and international markets. We need policy interventions that enable these traditional home-based production units to earn dignified incomes – not necessarily make factory workers or clerks out of them, unless of course they abandon these occupations out of choice.

Ideological preference for employment in the organized sector

The "development" paradigm has a strong bias against employment in the self-organized sector. Any country that shows a higher percentage of people employed in the organized sectors of the economy is considered "developed" and "advanced" and those with a high percentage in the self-organized sectors are considered "backward" and "under developed".

Low percentage of women in the organized sector is also held up as an example of discriminatory policies against women. However, in countries like India, the self-organized sectors of the economy have proved to be a boon and provide far greater avenues of economic advancement and upward mobility than most jobs in the organized sectors of the economy. In most developed economies, over 90 per cent of people are in the organized sector and a relatively small proportion is self-employed. Therefore, the moment there is an economic crisis it leads to job cuts and the retrenchment of workers. However, in countries like India where over 92 per cent are in the self-organized sectors of the economy (with no more than 3 per cent in corporate sector and 3 per cent in government employment) most people do not even notice economic downturns. This is an important reason why the economic crisis that hit Europe and US in the last decade did not produce any cataclysmic changes for countries like India or Bangladesh. These economies kept growing through the period of global economic crisis.

Many in the self-employed sector prosper fast *except* when government policies are outright hostile to them. Take the example of street vendors – many woman hawkers earn much more than a factory worker or an office clerk because profit margins in retail are usually high and business in street hawking brisk since they sell low value goods. They experience rapid upward mobility because they get an opportunity to develop entrepreneurial skills and explore new opportunities. By contrast, a bank cashier or an office clerk may have relatively greater security but avenues of upward mobility are limited.

If the vast majority of street vendors and other self-employed groups remain trapped in low income ghettos it is mainly because most governments treat them as illegal encroachers and an unwanted nuisance. They are forced to survive by seeking protection from political mafias and police who siphon off large parts of their incomes by way of bribes. The routine violence during clearance operations unleashed on vendors makes it doubly risky for women to survive in this occupation. That is why in countries that have hostile policies towards street vendors far fewer women are found in this trade. But in vendor-friendly countries like Thailand women thrive and dominate the markets and make good money. Most important of all, it provides free business training to the younger generation with young boys and girls joining their mothers as helping hands after school. No organized sector job allows such hands-on training for family members with flexible timings – all for free.

Similarly, women who set up small tailoring shops or beauty parlours at home are able to combine housework with their small enterprises. Many expand their businesses and start earning good money within no time minus the guilt of neglecting their family responsibilities. But "development" experts would rather put all these small entrepreneurs in the organized sectors.

Devaluation of women as homemakers

Apart from undervaluing the importance of home-based enterprises, development experts have a strong ideological preference for seeing women "gainfully employed" in paid jobs outside their homes. The flip side of this ideological position is systematic devaluation of women's role as mothers and as homemakers and as nurturers of future generations. But the dominant feminist discourse within the "development" framework treats women who focus on this role as "unproductive" workforce which is not contributing to national wealth and economic growth.

We need to respect the fact that a lot of women would rather not have dual jobs – as homemakers and wage earners. For many going out to labour outside the house is a distress response. Many prefer to devote their full attention to family and children. For many women withdrawal from wage labour comes as a big relief. Many prefer that their husbands earn enough to support the whole family in comfort because worldwide experience shows that the best of paid childcare does not compare favourably with parental and family care. If both husband and wife are in high pressure jobs, it is difficult to do justice to children, unless one has the support of extended family – grandparents, uncles and aunts. But joint family is also an anathema for most feminists. They only see it as the site of restrictions and oppression for women. This despite the fact that there is plentiful evidence in countries like India that women who live in supportive extended families where grandparents and others share a large part of the responsibility for childcare, rise high in their professions because they do not have to take mid-career breaks and can devote time on their professions without worrying about neglecting family.

It is noteworthy that the early battles waged by the working class in Europe were for family wage so that women did not have to work under compulsion. The job of women's rights activists should be limited to ensuring that every woman who wishes to take up a paid job or profession outside the house is not prevented from doing so due to lack of opportunities or discriminatory social practices. But it is presumptuous to insist that every woman must work for a wage and those who do not are socially "unproductive" and culturally "backward". Feminism should be about respecting women's choices not imposing a pre-set, ideologically determined road map for all. To conclude, a one-size fits all approach advocated by development experts does not take into account the diverse needs and aspirations of women.

References

Kishwar, Madhu (1985) "Gandhi and Women" in *Economic and Political Weekly*, XX, 40–41.

Saradamoni, K. (1999) *Matriliny transformed: Family, law, and ideology in twentieth century Travancore*, Sage Publications, New Delhi.

Schneider, David M. & Gough, Kathleen (eds) (1961) *Matrilineal kinship*, University of California Press, Berkeley, CA.

World Development Report (2012) *Gender Equality and Development (GED)*, World Bank, Washington, DC.

7.5

Rethinking gender and empowerment

Jane Parpart

Empowerment, particularly for women and the poor, has become a central issue for development institutions such as the UNDP, the World Bank, Oxfam, government agencies and many smaller non-governmental organizations (NGOs). Initially seen as a key element for challenging and transforming unequal political, economic and social structures, empowerment was regarded as a weapon for the weak – best wielded through participatory, grassroots community-based NGOs. However, empowerment is a flexible concept and by the mid-1990s, mainstream development agencies had begun to adopt the term. The language of participation, partnership and empowerment increasingly entered mainstream development discourse (World Bank, 1995; Elson and Keklik, 2002). While the wording remained the same, meanings varied, and mainstream institutions and their practitioners for the most part envisioned empowerment as a means for enhancing efficiency and productivity within the status quo rather than as a mechanism for social transformation (Parpart *et al.*, 2002).

The issue of empowerment first surfaced in gender and development debates in the work of Caroline Moser (1993) and Gita Sen and Caren Grown (1987). These writings reflected a growing concern that gender equality would not be achieved unless women could challenge patriarchy and global inequality. To achieve this end, Moser argued that women needed to gain self-reliance and internal strength in order 'to determine choices in life and to influence the direction of change, through the ability to gain control over crucial material and non-material resources' (Moser, 1993: 74–5). Sen and Grown emphasized the need for a collective vision, a set of strategies and new methods for mobilizing political will and empowering women (and men) to transform society (Sen and Grown, 1987: 87).

Scholar/activists from the global South raised crucial questions. Srilatha Batliwala (1994) warned that 'empowerment', which had virtually replaced terms such as poverty alleviation, welfare and community participation, was in danger of losing its transformative edge. She called for a more precise understanding of power and empowerment, asserting that empowerment must be seen as 'the process of challenging existing power relations and of gaining greater control over the sources of power' (Batliwala, 1994: 130). It requires political and collective action against cultural as well as national and community power structures that oppress women and some men, and consequently, transformative political action.

Naila Kabeer (1994), who has played a key role in operationalizing empowerment as development practice, has placed empowerment at the centre of efforts to achieve gender equality. Criticizing the liberal and Marxist emphasis on *power over* resources, institutions and decision making, Kabeer adopted a more feminist position, which emphasizes the transformative potential of *power within*. This power is rooted in self-understanding that can inspire women (and some men) to recognize and challenge gender inequality in the home and the community (Kabeer, 1994: 224–9). Like Batliwala, she highlights collective, grassroots participatory action – the *power to* work *with* others 'to control resources, determine agendas and to make decisions' (Kabeer, 1994: 229). Kabeer is particularly concerned with enhancing women's ability to exercise choice (associated with access and claims on resources, agency and achievements) (1999: 437).

Jo Rowlands argues that 'empowerment is more than participation in decision-making; it must also include the processes that lead people to perceive themselves as able and entitled to make decisions' (Rowlands, 1997: 14). It is personal, relational and collective, and 'involves moving from insight to action' (Rowlands, 1997: 15). Drawing on in-depth research in Honduras, she points to the crucial role played by social, political and economic contexts, warning that consciousness and agency are always context specific. Building on these debates, Sarah Mosedale (2005: 252) suggests that women's empowerment is best seen as 'the process by which women redefine and extend what is possible for them to be and do in situations where they have been restricted compared to men'. For her the issue is not simply enhancing choice, but extending the limits of the possible.

Increasingly, particularly as top-down approaches to development failed to alleviate poverty in the 1990s, especially among women, empowerment became a central focus for mainstream women and development programs. For example, the *Beijing Platform*, produced at the 1995 UN global women's conference, stated unequivocally that women's empowerment is 'fundamental for the achievement of equality, development and peace' (United Nations, 1995: para. 13). Official development institutions picked up the language of empowerment, gender equality and gender mainstreaming. The Canadian International Development Agency's (CIDA) *Policy on Gender Equality* declared women's empowerment one of its guiding principles for development policy (CIDA, 1999). While generally framed within neoliberal discourses of productivity and efficiency, the official commitment to empowerment and gender equality continues (CIDA, 2010; UN Women and UN Global Compact Office, 2010).

However, the adoption of empowerment as a basis for gender and development policy and practice has obscured the difficulties facing those trying to understand, implement and measure women's empowerment projects. Empowerment has been framed as a doable, reachable and measureable goal. Yet it is highly context specific, fluid and messy – all factors which affect attempts to operationalize and evaluate empowerment projects. Kabeer has critiqued empowerment projects for assuming that 'we can somehow predict the nature and direction that change is going to assume'. As she points out, 'In actual fact, human agency is indeterminate and hence unpredictable in a way that is antithetical to requirements of measurement' (Kabeer, 1999: 462). These critiques have been picked up by many others. Indeed, even internal evaluations often prove the difficulties facing agencies that have promised to empower women and ensure gender equality (Cornwall *et al.*, 2007; Parpart, 2013).

Critical voices have begun to question whether the gender and empowerment approach can produce fundamental change. Parpart *et al.* (2002) warned that critical thinking about power and empowerment must be incorporated into empowerment projects and policies. Women's empowerment cannot be regarded simply as the need to bring women into established power structures. This approach ignores the deeply held resistances facing marginalized groups around the world, and the subtle attitudinal and structural impediments to collective action (*power with*) and generative *power to* support gender equality. Empowerment requires attention to language and meanings, identities and cultural practices as well as the forces that enhance *power to* act *with* others to fight for change, often in hostile and difficult environments. It also requires moving away from the limits of a preoccupation with difference, which as Rai (2007) points out, has undermined efforts to support feminist activism across borders and cultures.

The shift towards the local has enriched knowledge of local conditions, and highlighted the plight of small communities and the gendered crises faced by many women (and some men) in these communities. However, while empowerment is often a local affair, the local is also embedded in the global and the national, and vice versa. Power (and empowerment) can only be understood within this fluid, complex and messy context. It requires attention to local struggles and their intersection with broader forces, particularly the increasingly global inequality and its gendered nature. As Sharma points out, women's struggles for empowerment and governance require analysis of local complexities within their regional, national and global context (Sharma, 2008). Understanding local processes requires attention to larger forces as well. Only then will local empowerment projects be able to address local inequalities and injustices.

Empowerment projects and advocates must also pay more attention to the ways institutional structures, material and discursive frameworks shape the possibilities and limits of individual and group agency and choices. This does not undercut the importance of local participation and consultation. It does, however, point to the need to situate individual and group action/agency within the material, political and discursive structures in which they operate. This requires careful, historically situated analyses of women's struggles to gain power in a world often neither of their own making or choosing (Berik *et al.*, 2009).

Finally, empowerment is both a process and an outcome. At times the two are indistinguishable, at others they merge, and sometimes the process is the outcome. While recognizing that specific outcomes should (and often can) be measured, measuring empowerment remains elusive. Many subtle and often unexpected strategies have the potential, but not the certainty, of empowerment (Kabeer, 1999). Others, such as international covenants and gender-sensitive laws, seem to guarantee empowerment but fail due to patriarchal cultural practices and structures. Thus, while attempts to measure outcomes can focus the mind and encourage new thinking, an obsession with outcomes and measurement can endanger the very processes most apt to nurture women's empowerment, even if not apparent at the time.

These critiques offer some guidelines for trying to ensure that women's empowerment is more than simply a 'motherhood' term for development agencies. They offer ways of making both the concept and practice of empowerment more rigorous, effective and nuanced. Empowerment, particularly as part of the struggle for gender equality around the world, continues to be a development concern, both for policy and praxis. The inclusion of men and masculinity (Cornwall *et al.*, 2011) as well as attention to cultural differences and economic empowerment is a welcome expansion of the term and key to addressing gender empowerment in an increasingly complex, global and still very patriarchal world.

References

Batliwala, S. (1994) 'The Meaning of Women's Empowerment: New Concepts from Action', in G. Sen, A. Germain and L. C. Chen (eds) *Population Policies Reconsidered*, Boston: Harvard University Press.

Berik, G., Meulen Rogers, Y. and Zammit, A. (eds) (2009) *Social Justice and Gender Equality: Rethinking Development Strategies and Macroeconomic Policies*, New York: Routledge.

Canadian International Development Agency (CIDA) (1999) *Policy on Gender Equality*, Ottawa: CIDA.

CIDA (2010) *Gender Equality*, Ottawa: CIDA.

Cornwall, A., Edstrom, J. and Greig, A., (eds) (2011) *Men and Development: Politicising Masculinities*, London: Zed Books.

Cornwall, A., Harrison, E. and Whitehead, A. (2007) *Feminisms in Development*, London: Zed.

Elson, D. and Keklik, H. (2002) *Progress of the World's Women: Gender Equality and the Millennium Development Goals*, New York: UNIFEM.

Kabeer, N. (1994) *Reversed Realities: Gender Hierarchies in Development Thought*, London: Verso.

Kabeer, N. (1999) 'Resources, Agency, Achievements: Reflections on the Measurement of Women's Empowerment', *Development and Change* 30(3): 435–64.

Mosedale, Sarah (2005) 'Assessing Women's Empowerment', *Journal of International Development* 17: 243–57.

Moser, C. (1993) *Gender Planning and Development*, London: Routledge.

Parpart, J. (2013) 'Exploring the Transformative Potential of Gender Mainstreaming', *Journal of International Development* 8(8).

Parpart, J., Rai, S. and Staudt, K. (eds) (2002) *Rethinking Empowerment: Gender and Development in a Global/Local World*, London: Routledge.

Rai, S. (2007) *The Gender Politics of Development*, London: Zed.

Rowlands, J. (1997) *Questioning Empowerment: Working with Women in Honduras*, Oxford: Oxfam Publications.

Sen, G. and Grown, C. (1987) *Development, Crises, and Alternative Visions*, New York: Monthly Review Press.

Sharma, A. (2008) 'Crossbreeding Institutions, Breeding Struggle: Women's Empowerment, Neoliberal Governmentality, and State (Re)Formation in India', *Cultural Anthropology* 21(1): 60–95.
United Nations (1995) *Beijing Platform for Action*, New York: United Nations.
UN Women and UN Global Compact Office (2010) *Women's Partnership Principles*, New York: United Nations.
World Bank (1995) *World Bank Participation Source Book*, Washington, DC: World Bank Environment Department Papers.

Websites

http://web.worldbank.org: see Ruth Alsop, Mette, F. Bertelsen and J. Holland (2005) *Empowerment in Practice*, Washington, DC: World Bank.
www.undp.org/women: see the Resource Guide for Gender Theme Groups, January 2005.
www.unifem.org or http://unwomen.org

7.6

Gender and globalisation

Harriot Beazley and Vandana Desai

Over the past three decades, processes of globalisation resulted in mixed consequences for men and women. This chapter examines the consequences afforded to women in the developing world as a result of new opportunities presented by globalisation. Analysis begins with an overview of how production processes have sourced cheap and compliant labour in export-oriented processing zones (EPZs) of different countries, resulting in new employment opportunities for women. Some negative consequences of neoliberal globalisation, including exploitation and abuse of women working in EPZs will then be explored. The chapter considers how international migration of women has rapidly increased as a result of new opportunities, and how the feminisation of labour has contributed to shifting gender relations and notions of femininity and masculinity. The chapter describes some diverse and 'subtle strategies' employed by women to negotiate a sense of place within patriarchal societies in response to challenges posed by neoliberal policies. Here it is seen how globalisation influences young women to develop new forms of feminism.

Globalisation

In recent years the term 'globalisation' has gained prevalence, with varied meanings. In the twenty-first century it is usually understood as the rapid international mobility of labour, goods, service delivery and capital, facilitated by the liberalisation of economies, privatisation and 'deregulation' of labour markets. In many developing countries, neoliberal policies have been adopted through World Bank and International Monetary Fund (IMF) supported packages of structural adjustment programmes (SAPs) (Elson, 1995). SAPs aim to strengthen the role of markets and reduce the role of the state in trade and industry. These goals were considered critical for realising economic growth and poverty reduction. Such programmes lead many governments and NGOs involved in development in the North and South to argue that globalisation processes are consolidating a new kind of colonialism. This is because power resources are increasingly being held by a relatively small number of global players who remain unaccountable to the vast population experiencing poverty in the global South (see Kerr and Sweetman, 2003: 5)

For over 25 years transnational corporations (TNCs) and multinational companies (MNCs) have been building factories and producing goods off shore, where labour is cheaper and workers have fewer unions or rights. These investors are usually from the minority, and generate enormous profits through employment and exploitation of cheap, predominantly female, labour from developing countries. This practice has led to what is known as 'the feminisation of labour' particularly within the service and manufacturing industries. As a result: 'unequal gender relations have been shaped by and, in turn, shape globalisation' (Çağatay and Ertürk, 2004: 1).

Feminisation of labour

At face value globalisation presents many new opportunities to young women in the developing world, and TNCs often promote women as 'winners' in these new arrangements. In efforts to strive for economic development and industrialisation most countries see private foreign investment from these corporations as the answer to their economic woes. Many countries focus on export-oriented industrialisation, through setting up EPZs, where factories produce garments, shoes, electronics, toys and other goods for export. MNCs are also setting up offshore service centres, in countries such as India, Philippines and Indonesia. These data processing and call-centres include telecommunication, banking and car rental companies. Host countries need a 'comparative advantage', in order to appear attractive to foreign investors. Such 'advantage' is dependent on the abundant supply of low-paid, passive, flexible, acquiescent, young female labour who are suited to mundane repetitive tasks, thereby reinforcing sexist and stereotypical images of women (Elson, 1995).

In a study of young female factory workers in Java, Indonesia, Diane Wolf (1992) described them as 'Factory Daughters', due to the patriarchal gender relations in the factories where they worked. In this research young women were poorly educated and escaping poverty and strict Muslim restrictions on young unmarried girls. Their work experiences and disposable incomes delivered new-found freedoms, including ability to consume products of capitalism. This resulted in their increased status in the household, society, and self-esteem. These young women, however, also endured exploitative working conditions, including shift work and excessive overtime, non-existent worker's rights and job insecurity. The patriarchal conditions in factories, were enforced by male managers, who dispensed discipline to those who complained. There was a significant neglect of health and safety regulations, and an ease in which women were dismissed if they got married, became pregnant, complained or contracted work-related illnesses (Wolf, 1992). More recent research with female factory workers in EPZ in Sri Lanka has identified social stigma and negative opinions of society towards women who work in factories, resulting in experiences of public humiliation and sexual harassment (Hancock, 2009).

Gender and transnational migration

Transnational labour flows are a result of economic globalisation. More than 120 million migrants have left their home countries in search of economic opportunities abroad (Piper, 2004). In Southeast Asia in 2000, the majority of migrant workers were women and numbers are growing (Piper, 2004). This demand for workers overseas has led women from Asia to migrate overseas as domestic workers, live in maids, health professionals, caregivers, sex workers and other care services (Pyle, 2011; Law, 2002). With these opportunities, risks prevail as they are predominantly informal, unprotected low-paid and low-skilled positions (Pyle, 2011). Saudi Arabia, is increasingly becoming a destination for Indonesian and Indian women, who go to work as domestic or sex workers. Female migrants from the Philippines go to Hong Kong, Canada, and Singapore, to work as housemaids (Law, 2002). For women from poor, uneducated backgrounds, migration can often only take place illegally, via a globalised network of human traffickers. In this way women take an extraordinary gamble to bridge the gap between poverty and gaining a livelihood to prosperity.

While the economic benefits of remittances from migration are well documented, studies have also began to examine the social costs of migration on migrants themselves, including health concerns, labour conditions and treatment by employers (Piper, 2004; Silvey, 2006). Like factory workers, migrant labourers often face adverse conditions in overseas employment. A combination of racial discrimination, employer abuse, lack of health services, and vulnerable legal status confront women seeking low-paid jobs abroad (Piper, 2004). Cases of torture, rape, sexual assault, overwork and non-payment of wages abound (Silvey, 2006). Increasing sexual harassment cases and human rights violations against female workers, particularly in Singapore, Hong Kong and the Middle East have sparked major protests from labour unions and women's organisations globally (Law, 2002) and bring forth the question of the accountability of the state in protecting citizens who work abroad.

Implications for gender relations

The sociocultural impacts of migration on families left behind, including children, has also been documented. Further, migration of men has led to an increase in female-headed households. In this context migration has led to the breakdown of extended family networks. In this new division of labour there have been changes in men's roles at the household level. As men struggle to fulfil traditional gender-ascribed roles as main providers, they face increased levels of depression, alcoholism, and violence. The longer term consequences of this 'crisis of masculinity' in developing countries has yet to be explored sufficiently (Kanji and Menon-Sen, 2001).

The extended family is particularly pressured in cities where nuclear families are common, as they must manage work and home without traditional extended family support networks. As well as daily struggles to catch up with the increasing cost of living, aspirations and attitudes change with exposure to fresh ideas and values, which oblige them to revaluate social identities, roles and undergo processes of psychological adjustment (Desai, 2006). Married women experience the double pressure of performing at work and adhering to traditional gender expectations at home. This creates conflicts within families, between husband and wife, and between wives and mother-in-laws, resulting in increased familial breakdown.

Young women must also manage expectations for girls in the developing world to take on family obligations at an early age and their increased exposure to different ideas of childhood and femininity presented by forces of globalisation. In India and Indonesia, for example, young unmarried girls have restricted mobility and are expected to be 'good girls', perform domestic duties and be good daughters and wives who do not stray far from home (Chakraborty, 2009; Beazley, 2007). Transnational migration then, becomes one way for girls to negotiate new subjectivities. Going to work in factories or overseas offers many attractions and opportunities, which are often perceived as a source of emancipation (Beazley, 2007), despite stories of exploitation and abuse. The burning desire young women have to migrate for work for the perceived benefits continues to compel them to navigate such risks, in order to earn money, gain independence, and escape the restrictive roles in the family and village (Beazley, 2007; Silvey, 2006). These young women can be understood as active agents who are organising their lives and negotiating the gendered power relations around them (Punch, 2007). So through the production of 'diverse and subtle strategies' (Desai, 2006), new forms of feminism are emerging as women seek to negotiate traditional religious and cultural values and traditions, while embracing a more global culture.

Within this context the gendered geographies of emotions have also changed as a result of globalisation, especially regarding the emotion of love. How love is perceived in these transformed societies remains an unexplored subject (see Morrison *et al.*, 2012). Povinelli (2006: 175) notes, 'Love is not merely an interpersonal event, nor is it merely the site at which politics has its effects. Love is a political event'. It is important to reflect upon how globalisation impacts upon emotions such as love. Chakraborty (2012), for example, has noted how over 80 per cent of all marriages in the slums where she conducted her research in Kolkata were 'love marriages', a significant trend away from the arranged marriage systems which were typically the norm in this community a generation ago. More research is needed that focuses on the forms, spaces and politics of love and the manner in which love constitutes people, places and subjectivities.

Gender, cultural change and globalisation

As social, technological, lifestyle and consumer choices change in these societies, identities based on familiar social class or caste hierarchies are often challenged, resulting in multiple, fragmented and shifting identities. For example, young girls will avidly watch TV and communicate via instant messaging and social network sites which can compensate for their social and geographical exclusion. Also they are fascinated by soap operas and movies and consume media images of glamorous 'Western' lifestyles of free and independent women (see Ansell, 2005). Research exploring media consumption and cultural production among young women in Indonesia, India and Vietnam during times of political transition and globalisation, reveals transformation in beliefs, values, spatial practices and identities of women in these countries (Beazley, 2007; Chakraborty, 2009). In these ways young women in the global South have been able to construct alternative geographies to the dominant cultural models of femininity which persist in their societies. Mass media is far more persuasive than state constructions of femininity, and much more appealing as agents of socialisation. Such circumstances can also give rise to fundamentalism, which is connected to globalisation, as politico-religious movements react to Western-imposed cultural domination. These movements promote unique interpretations of 'tradition' and 'culture', often at the expense of women's rights (Kerr and Sweetman, 2003: 7).

Globalisation has not only influenced identity constructions of Western women and men, but has resulted in increased media consumption, cultural production and identity formation among men and women in the developing world. It is important to note that although transnational media provides a plethora of new ideas for identity constructions, access to transcultural media is not necessarily synonymous with increased gender equity. Following Connell (1998), Derne (2002: 145) argues that the emerging global gender order is not a 'hotbed of gender progressivism'. On the contrary, the reality is that global forces that challenge men's identity and power often lead men to 'reaffirm local gender orthodoxies and hierarchies' (Derne, 2002: 145). On researching Indian men living abroad, Derne (2002) concludes that migrant men assert their Indianness by celebrating and being attracted to Indian women who adhere to traditional women's roles and dress, and while attracted to the lifestyles of the rich and famous in Hollywood, 'root their Indianness in women's adherence to oppressive roles' (Derne, 2002: 146). In this way transnational influences can include new modes of subordination, while simultaneously enhancing existing local modes of gender division. In this way 'the effects of transnational cultural flows . . . expand men's choices, while generating anxieties about identity that push men to limit the choices of women' (Derne, 2002: 146).

Conclusion

This chapter has demonstrated that although millions of women are participating in globalised networks of production and consumption in the developing world, they are not necessarily the beneficiaries of this interaction. This is because in many cases globalisation compounds their poverty, increasing marginalisation and heightened gender inequalities. Contemporary research into how globalisation impacts upon gender relations and subjectivities in developing countries – via neoliberal economic policies including SAPs, the rapid mobility of labour and capital – suggests that benefits of globalisation have not been equitably distributed. In fact, women have been – and still are – the 'shock absorbers' of the global economic system (Elson, 1995: 11).

Despite their marginalisation, however, women are not necessarily accepting their position passively. Mainstream Western culture has been influential in shaping a new form of feminism in the developing world, and this feminism is articulated through young women's involvement in global cultures. Young women are participating in a global society, by drawing upon international images of young people reflected in magazines, television, the Internet, and in established youth cultures. The 'traditional' concept of women, and their role in developing countries is, in fact, being challenged by newly emerging identities and changing expectations of globally aware young women. Implicit in all the above is that action on the part of the international women's movement and gender and development policymakers and practitioners is needed more than ever (Desai, 2005). The key is how do we turn the gender analysis of globalisation into action for social change?

References and Guide to Further Reading

Ansell, N. (2005) *Children, Youth and Development*, Routledge, London.

Beazley, H. (2007) 'The Malaysian Orphans of Lombok: Children's Livelihood Responses to Out-Migration in Eastern Indonesia', in R. Panelli, R. Punch and E. Robson (eds), *Young Rural Lives: Global Perspectives on Rural Childhood and Youth*, Routledge, London, pp. 107–120.

Çağatay, N. and Ertürk, K. (2004) 'Gender and Globalization: A Macroeconomic Perspective', Working Paper No. 19, Policy Integration Department, World Commission on the Social Dimension of Globalization, International Labour Organization, Geneva.

Chakraborty, K. (2009) '"The Good Muslim Girl": Conducting Qualitative Participatory Research to Understand the Lives of Young Muslim Women in the Bustees of Kolkata', *Children's Geographies*, 7(4): 421–434.

Chakraborty, K. (2012) 'Young Married Muslim Couples Negotiating their Sexual Lives in the Urban Slums of Kolkata, India', *Intersections: Gender and Sexuality in Asia and the Pacific*, 28, March. Available online at http://intersections.anu.edu.au/issue28/chakraborty.htm#n1 (accessed 4 June 2013).

Connell, R. W. (1998) 'Masculinities and Globalization', *Men and Masculinities*, 1(1): 3–23.

Derne, S. (2002) 'Globalization and the Reconstitution of Local Gender Arrangements', *Men and Masculinities*, 5(2): 144–164.

Desai, V. (2005) 'NGOs, Gender Mainstreaming, and Urban Poor Communities in Mumbai', *Gender and Development*, 13(2): 90–98.

Desai, V. (2006) 'Women's Social Transformation, NGOs and Globalisation in Urban India' in Saraswati Raju, Satish Kumar and Stuart Corbridge (eds), *Colonial and Postcolonial Geographies of India*, Sage Publications, New Delhi, Thousand Oaks, London, pp. 120–140.
Elson, D. (1995) 'Male Bias in Macro-Economics: The Case of Structural Adjustment', in D. Elson, (ed.), *Male Bias in the Development Process*, Second Edition, Manchester University Press, Manchester.
Hancock, P. (2009) 'Violence, Women, Work and Empowerment: Narratives from Factory Women in Sri Lanka's Export Processing Zones', *Journal of Developing Societies*, 25: 393–420.
Kanji, N. and Menon-Sen, K. (2001) 'What does Feminisation of Labour Mean for Sustainable Livelihoods?' London, IIED.
Kerr, J. and Sweetman, C. (eds) (2003) *Women Reinventing Globalisation*, Oxfam, Oxford.
Law, L. (2002) 'Sites of Transnational Activism: Filipino Non-Government Organisations in Hong Kong', in Brenda S. A. Yeoh, Peggy Teo and Shirlena Huang (eds), *Gender Politics in the Asia-Pacific Region*, London: Routledge.
Morrison, C. A, Johnston, L., and Longhurst, R. (2012) 'Critical Geographies of Love as Spatial, Relational and Political', *Progress in Human Geography* (published online 15 November 2012), DOI: 10.1177/0309132512462513: 1–17.
Piper, N. (2004) 'Rights of Foreign Workers and the Politics of Migration in Southeast and East Asia', *International Migration*, 42(5): 71–97.
Povinelli, E. A. (2006) *The Empire of Love: Toward a Theory of Intimacy, Genealogy, and Carnality*, Duke University Press, Durham, NC.
Punch, S. (2007) 'Negotiating Migrant Identities: Young People in Bolivia and Argentina'. *Children's Geographies*, 5(1): 95–112.
Pyle, J. (2011) 'Globalisation and the Increase in Transnational Care Work: The Flip Side', *Globalizations*, 3(3): 297–315.
Silvey, R. (2006) 'Consuming the Transnational Family: Indonesian Migrant Domestic Workers to Saudi Arabia', *Global Networks*, 6(1): 23–40.
Wolf, D. L. (1992) *Factory Daughters: Gender, Household Dynamics, and Rural Industrialization in Java*, University of California Press, Berkeley, CA.

7.7

Migrant women in the new economy

Understanding the gender-migration-care nexus

Kavita Datta

Introduction

The predominance of migrant women in care migrations from the global South to advanced economies is increasingly recognised and attributed to the emergence of a 'new economy' on a global scale (Yeates, 2012). Initially focusing upon the migration of poorer women to undertake child care for richer women, academic enquiry has since broadened to reflect that care comprises of a complex mixture of unpaid and paid, unskilled and skilled, private and public occupations (Hochschild, 2000). Correspondingly, research on the provision of institutionalised care, elderly care and skilled care has burgeoned (Datta *et al.*, 2010). Furthermore, while the significance of care migrations to the advanced economies of the USA, UK, Canada and Australia is well established, commentators have also noted South to South care migrations particularly to the Middle East, Japan and Singapore.

This chapter is arranged in three sections. The first explores the meanings of care, the factors leading to the concentration of migrant women in 'care chains' and the theorisation of the gender-migration-care nexus. The second section focuses on the impact of care migrations upon sending countries in the global South. The chapter ends with a consideration of the challenges migrant women face in host societies and the strategies they deploy in order to survive.

Theorising care and the role of migrant women in care chains

While definitions of care are contested, there is some agreement that it is 'simultaneously a construction, a process and an achievement that includes family care, homemaking and health care, but also encompasses physical activity as well as emotional work.' Implicit in this definition are two overlapping yet distinctive interpretations related to 'reproductive labour' and 'nurturance' (Duffy, 2005). Rooted within Marxist thinking, 'reproductive labour' emphasises the importance of women's unpaid domestic labour (referred to as the 'other economy' by feminist economists) in supporting and maintaining the world of work via non-relational tasks such as cleaning and cooking which are necessary for the reproduction of the workforce. In contrast, 'nurturance' is associated with relational and interdependent work characterised by responsive action and underwritten by emotional connections. While set up in an oppositional framework, it is increasingly accepted that care involves both reproductive labour and nurturance thus incorporating visible and invisible, relational and non-relational, physical and emotional work (Duffy, 2005).

In turn, the gendering of care work is apparent globally, and is attributable to gender ideologies which depict it as a biologically determined task, a 'natural' extension of female roles related to motherhood and imbued with 'feminine' attributes such as nurturing and caring for others. Further, this gendered division of household labour plays a central role in the subordination of women in labour markets such that care is shaped by, and in turn determines, gender relations and well-being. While research has privileged gender in illustrating the nature and politics of care work, commentators have stressed the importance of adopting an intersectional approach (Nakano Glenn, 1992). The intertwining of gender with ethnicity, race and nationality is particularly evident in determining who does what care work with women of specific ethnicities/nationalities sought after as housemaids, nannies and cooks (Duffy, 2005).

Care migrations have been predominantly understood by a theoretical approach termed 'global circuits of care' (GCC) which is conceptualised as a 'series of personal links across national boundaries between people engaged in paid or unpaid work of caring for others' (Hochschild, 2000: 131). Involving the transfer of physical and emotional labour from the global South to advanced and emerging economies, these care chains are shaped by neoliberal restructuring, demographic change and the persistence of uneven development (Teo and Piper, 2009). Taking these in turn, the 'new' global service economy has increased pressures *and* opportunities for women to work in both the global North and South. Yet, men's take up of unpaid reproductive work has been limited. In the global North, these changes have occurred alongside declining public provisioning of care as well as ageing populations leading to a 'care deficit' as women struggle to combine their paid labour with their gender ascribed roles of being primary carers. Further, pressures on women to work in the global South have grown apace due to global restructuring and falling wage earning potential among men which has meant that women's employment has gained significance in terms of providing for the family. However, opportunities to work are not necessarily found locally leading to the feminisation of transnational migration as women migrate to provide for their families while leaving their own dependants to the care of others (Yeoh *et al.*, 2005).

While contributing substantially to understandings of the gender-migration-care nexus, GCCs have been particularly critiqued for an insufficient consideration of the role of the state – in both sending *and* receiving countries – in configuring care chains (Yeates, 2009). Correspondingly, research on 'care diamonds' has extended the focus beyond households and the market to consider how the state shapes care chains which cut across three regimes: namely gender, care and migration (Razavi, 2007). Empirical research illustrates that the gendered immigration and emigration policies adopted by states are crucially important in determining who can migrate for what length of time and under what conditions (Kofman and Raghuram, 2009).

Long distance caring: Impacts of care migrations on sending countries

Women who migrate as care providers leave behind their own care responsibilities, the implications of which have been explored at a range of scales from the household to the state. Considering these in turn, transnational migration has resulted in the formation of a variety of new diverse types of transnational households as intimate relations have been stretched across transnational space (Yeoh *et al.*, 2005). Transnational mothering has been a particular focus of attention whereby mothers leave their children in home countries while they migrate in order to undertake paid care work (Hondagneu-Sotelo and Avila, 1997). Female out-migration results in significant adjustments in care giving which are both gendered and class specific. Research reports on the significance of 'other mothers' (grandmothers, aunts and older daughters) in caring for migrant women's children with limited – if geographically varied – involvement of non-migrant fathers. Further, while middle-class families may employ paid carers, their poorer counterparts are more likely to utilise unpaid family labour (Parreñas, 2012).

These care arrangements are often overseen by migrant women who strive to parent from afar through the practice of 'long distance mothering' which is facilitated by the telecommunications revolution including instant messaging, cheap international calls as well as new visual communications such as Skype. Yet, while these developments support virtual-cyber relationships and illustrate the simultaneity of transnational living, not all migrant women and/or their children have access to these technologies. Furthermore, long periods of physical separation can result in resentment as migrant women struggle to return back to their home countries or bring their children to live with them. It can also lead to a shift in inter-generational power as children assume (financial) responsibility from a young age while non-parental carers gain greater importance vis-à-vis transnational mothers (Parreñas, 2012).

At a societal level, care migrations can compound 'care deficits' in sending countries. The feminisation of migration often prompts fears about a 'crisis of care' given the ascription of parental and broader caring responsibilities to women. Further, increased migration from the global South is occurring alongside ageing populations, heightened childcare burdens in the aftermath of the HIV/AIDS epidemic, commercialisation of health care and the greater availability of education and work opportunities for girls and women (Kofman and Raghuram, 2009). In other instances the 'care' and 'brain' drain intertwine as illustrated by the migration of skilled health personnel including nurses particularly from the Philippines but also Indonesia, Sri Lanka, Bangladesh, Vietnam, China, Pakistan, Thailand and India (Yeates, 2009).

Thinking through the broader implications of GCCs, sending country states have been accused of seeking temporary solutions to complex structural problems such as unemployment through the export of people. In turn, host countries offshore the costs of reproducing a migrant labour force to sending countries who assume responsibility for caring of migrants' children and/or elderly parents, an arrangement which Silvey (2008) identifies as a 'direct development subsidy' from poorer states to the economies of wealthy countries.

Care and migrant women

The welfare of migrant women often eludes attention partly due to perceptions that migrants return home when they themselves need care. Migrant women confront significant challenges in service/care sector occupations in host countries both within the public but particularly in the private sphere with empirical research illustrating the vulnerabilities associated with live in care work (Ehrenreich and Hochschild, 2002). These vulnerabilities are further heightened where such work is undertaken on an irregular basis, as well as immigration policies which bind domestic workers to specific employers. While the conditions associated with institutional care work are relatively better, low wages and few employment benefits can lead to a combination of long working hours with migrant women working two or more jobs as they attempt to make a living while meeting their transnational caring responsibilities through the sending of remittances (Datta *et al.*, 2010). Furthermore, the devaluing of care work which involves dealing with 'dirty' and 'polluted' bodily fluids is also problematic as evidenced in McGregor's (2007) work on Zimbabwean carers in the UK who are stigmatised as 'BBCs' or 'British Bottom Cleaners'. Beyond the world of work, migrant women who bring dependent children with them often struggle to organise their own child care. For example, in the UK, migrants have little or no recourse to the public provision of care while being separated from extended family who might otherwise have been able to provide care. Migrant women cope with these challenges in myriad ways through relying upon their partners and/or husbands to share caring responsibilities while also drawing upon migrant and faith networks to arrange informal care for their dependants (Datta *et al.*, 2010).

Migrant civic organisations have taken the lead in articulating the need to organise around migrants' labour rights including those of domestic workers (www.kalayaan.org.uk). Yet, it is important to recognise that it is difficult to ensure migrants' labour rights in situations where the rights of native workers may also be eroded which is especially evident during a time of global recession. Further, while there is some recognition of the importance of ensuring migrants rights in the face of employer abuse, there is less foresight of global structural inequalities which render this kind of work necessary.

Conclusion

This chapter has focused upon the relationship between gender, migration and care, and the increased importance of migrant women in care chains. Shaped by the ongoing neoliberal restructuring of economies, states and society, these care migrations reflect individual and privatised solutions to the retrenchment of the welfare state in the global North, the reinforcement of gendered inequalities between men and women as well as the entrenchment of uneven development between the global North and South.

References

Datta, K., McIlwaine, C. J., Evans, Y., Herbert, J., May, J. and Wills, J. (2010) Towards a migrant ethic of care? Negotiating care and caring among migrant workers in London's low pay economy, *Feminist Review*, 94(1): 93–116.

Duffy, M. (2005) Reproducing labour inequalities: Challenges for feminists conceptualising care at the intersections of gender, race and class, *Gender and Society*, 19(1): 66–82.

Ehrenreich, B. and Hochschild, A. R. (eds) (2002) *Global Women: Nannies, Maids and Sex Workers in the New Economy*, New York: Owl Books.

Hochschild, A. (2000) Global care chains and emotional surplus value, in W. Hutton and A. Giddens (eds) *On The Edge: Living with Global Capitalism*, London: Jonathan Cape.

Hondagneu-Sotelo, P. and Avila, E. (1997) 'I'm here, but I'm there': The meanings of Latina Transnational Motherhood, *Gender & Society*, 11(5): 548–571.

Kofman, E. and Raghuram, P. (2009) *The Implications of Migration for Gender and Care Regimes in the South*, Social Policy and Development Programme Paper 41, UN Research Institute for Social Development (UNRISD), Geneva.

McGregor, J (2007) 'Joining the BBC (British Bottom Cleaners)': Zimbabwean migrants and the UK Care Industry, *Journal of Ethnic and Migration Studies*, 33(5): 801–824.

Nakano Glenn, E. (1992) From servitude to service work: Historical continuities in the racial division of paid reproductive labor, *Signs: Journal of Women in Culture and Society*, 18: 1–43.

Parreñas, R. S. (2012) The reproductive labour of migrant workers, *Global Networks*, 12(2): 269–275.

Razavi, S. (2007) *The Political and Social Economy of Care in a Development Context: Conceptual Issues, Research Questions and Policy Options*, Gender and Development Programme Paper 3, United Nations Research Institute for Social Development (UNRISD), Geneva.

Silvey, R. (2008) Development and geography: Anxious times, anaemic geographies and migration, *Progress in Human Geography*, 33(4): 507–515.

Teo, Y. and Piper, N. (2009) Foreigners in our homes: Linking migration and family policies in Singapore, *Population, Space and Place*, 15(2): 147–159.

Yeates, N. (2009) Production for export: The role of the state in the development and operation of global care chains, *Population, Space and Place*, 15(2): 175–187.

Yeates, N. (2012) Global care chains: A state-of-the-art review and future directions in care transnationalization research, *Global Networks*, 12(2): 135–154.

Yeoh, B. S. A., Huang, S. and Lam, T. (2005) Transnationalizing the 'Asian' family: Imaginaries, intimacies and strategic intents, *Global Networks* (Special issue on Asian Transnational Families), 5(4): 307–315.

7.8

Women and political representation

Shirin M. Rai

Why is representation an important issue for development? Development policies are highly politically charged trade-offs between diverse interests and value choices. 'The political nature of these policies is frequently made behind the closed door of bureaucracy or among tiny groups of men in a non-transparent political structure' (Staudt, 1991: 65). In 2002, 94.5 per cent of the World Bank Board of Governors were men as were 91.7 per cent of its Board of Directors. Of the 24 IMF Directors in 2012 three were women, all of them appointed rather than elected. The question then arises, how are women to access this world of policy-making so dominated by men? The answers that have been explored within women's movements have been diverse – political mobilization of women, lobbying political parties, moving the courts and legal establishments, constitutional reform, mobilization and participation in social movements such as the environmental movement, and civil liberties campaigns at local and global levels.

Representation has been the focus of reformist, inclusionary strategies in public policy in many political systems. Of itself, the concept is not such that it can bear the burden of close scrutiny – there are too many caveats that have to be taken on board for it to work. It is attractive largely in contrast to other political arrangements. Lack of representation is perceived as a problem, and citizens largely accept democratic institutions as important to the expansion of possibilities of political participation. Further, exclusion generates political resentment, adversely affecting not only the political system but also social relations within a polity; no individual or group likes being regarded as part of an excluded, and therefore disempowered, group.

Representation as a concept makes certain assumptions that are problematic. These affect the policymaking and functioning of political institutions. The first set of assumptions are related to representation of interests – that there are identifiable (women's) interests, that (women) can represent. For women this raises issues about what are women's interests when they are being constantly disturbed by categories of race, ethnicity and class, and whether women can be homogenized in terms of their sex/gender without regard to their race, ethnic and class positionings. Martha Nussbaum has emphasized the importance of equality and human rights as the irreducible minimum rights for women in the development of their capabilities (Nussbaum, 1995). Iris M. Young has addressed these issues through her exploration of the idea of group interests (1990). She suggests that group interests can be formulated by groups meeting as groups supported by public resources; that the interests thus formulated would be more reflective of group concerns, securing greater legitimacy within the group. As such these interests should be made part of policy, and policymakers asked to justify their exclusion from processes of policymaking. As a final protection of interests, Young suggests that these more cohesive groups should have power of veto over policy affecting them. This analysis would presumably take account of class-based groups such as trade unions, though most research which has taken on board Young's framework has focused on issues of race, ethnicity, sexuality and gender.

A second set of assumptions regarding representation is about appropriate forms of representative politics, and the levels of government and policymaking. Despite an enduring interest in direct participatory politics, representative government operates largely through a party political system. Political parties thus form an important constraint for individual representatives, especially if the representative seeks to support certain group interests. Group interests have often been regarded as too particularistic; 'general interests' or more often 'national interests' take precedence in party political rhetoric. Political parties also perform 'gatekeeping' functions – interests are given recognition through the agenda-setting of political parties. My study of Indian women parliamentarians suggests that institutional constraints, and systems of organizational incentives and disincentives are important explanations of the limited role that women can play in advancing the agenda of gender justice through party-based political work (Rai, 1996, 1997).

Representation is a key concept in liberal understandings of governance – it focuses on institutions, organization and practices. Representation is also central to the concept of citizenship – in both its normative sense, participation in politics as a prerequisite for the subjective public self, and in the practical, the inclusion within state boundaries where particular development agendas take shape. The framework of citizenship has recently been used by corporate-sector engagements with the policy world, and civil society organizations – 'corporate citizenship' is an emergent field of enquiry in development studies as well as business studies. Representation is also important in terms of accountability – of both public and private organizations. Increasingly we see a critique of not only state-based institutions, but also of INGOs and NGOs in terms of an accountability deficit and how this might influence interest articulations and formulations and implementation of policy. As it signifies consent, representation is also an essential element of good and therefore legitimate government which is increasingly becoming the focus of the delivery network institutions of development (World Bank, 1992: 6; 2000). Representation is, however, used as a universal and an undifferentiated concept which does not take into account particular positionings, needs and claims of groups constituting a particular civil and political society.

Women in political institutions

A headcount of the officers of the state in all sectors – legislature, executive and the judiciary – in most countries of the world reveals an overwhelming male bias despite many mobilizations furthering women's presence at both national and global levels. An Inter-Parliamentary Union Study found that the number of sovereign states with a parliament increased sevenfold between 1945 and 1995, while the percentage of women MPs worldwide increased fourfold (Pintat in Karam, 1998: 163); 19.6 per cent of parliamentarians are women, with the Nordic states leading the way with 42 per cent and the Arab states with 11.7 per cent (IPU, 2012). Furthermore, data shows that there is no easy positive correlation between economic indicators and the presence of women in public bodies. While in Europe (without the Nordic countries) women were 20.9 per cent of the total number of MPs, the figure in sub-Saharan Africa was 19.7 per cent. In recognition of the painfully slow improvement in women's representation in national parliaments, enhancing women's presence within state bodies is now being pursued as a goal both by women's movements as well as international institutions. This suggests that an engagement with state structures is now considered an appropriate means of bringing about shifts in public policy.

Feminist debates on representation

Why are debates on women's representation important? First, women have recognized that interests need to be articulated through participation and then represented in the arena of politics. Within the women's movements there has been a significant shift in the 1980s towards engaging positively with state feminism as a strategy that could be effective in furthering the cause(s) of women. The argument about women's presence in representative politics, as Anna Jonasdottir (1988) has pointed out, concerns both the *form* of politics and its *content*. The question of form includes the demand to be among the decision-makers, the demand for participation and a share in control over public affairs. In terms of content, it includes being able to articulate the needs, wishes and demands of various groups of women. The interest in citizenship has also been prompted by the shift in women's movements, in the 1980s, from the earlier insistence upon direct participation to a recognition of the importance of representative politics and the consequences of women's exclusion from it (Lovenduski and Norris, 1993; McBride Stetson and Mazur, 1995; Rai, 2000; Lovenduski, 2006). It is here that politics – public and private, practical and strategic – begins to formalize within the contours of the state.

Gender and representation is also an important issue in the international sphere. Good governance has become important to the political discourses of world aid agencies and financial institutions. After having concentrated attention on the processes and practices of structural adjustment programmes (SAPs) in the Third World countries during the 1970s and 1980s, the World Bank and other aid agencies are faced with decelerating growth in many countries accompanied by a rise in social tensions, in many cases the consequence of SAPs. Issues of governance have therefore become important and the focus has shifted to supporting processes of institution-building that would help manage the social and political fallout of economic policies supported by international aid agencies and Western governments (World Bank, 2000). The UN has also played an important role in placing the issue of women's exclusion from political processes on the international agenda. The Platform for Action agreed upon in Beijing in September 1995 emphasized both the need to increase the levels of women's participation in politics, and also the need for the development of national machineries for the advancement of women (Rai, 2001).

While we cannot assume that more women in public offices would mean a better deal for women in general, there are important reasons for demanding greater representation of women in political life. First, is the intuitive one – the greater the number of women in public office, articulating interests, and seen to be wielding power, the more disrupted the gender hierarchy in public life could become. Without a sufficiently visible, if not proportionate, presence in the political system – 'threshold representation' (Kymlicka, 1995) – a group's ability to influence either policymaking or, indeed, the political culture framing the representative system is limited. Further, the fact that these women are largely elite women might mean that the impact that they have on public consciousness might be disproportionately bigger than their numbers would suggest.

Second, and more important, we could explore the strategies that women employ to access the public sphere in the context of a patriarchal sociopolitical system. These women have been successful in subverting the boundaries of gender, and in operating in a very aggressive male-dominated sphere. Could other women learn from this cohort? The problem here is, of course, precisely that women in national legislatures are an elite. The class from which most of these women come is perhaps the most important factor in their successful inclusion into the political system. We can, however, examine whether women's representation at the local levels of governance and in sociopolitical movements provide opportunities for women to use certain strategies that might be able to subvert the gender hierarchy in politics. Finally, we can explore the dynamic between institutional and grassroots politics. The 'politicization of gender' in the Indian political system, for example, is due largely to the success of the women's movement. Women representatives have thus benefited from this success of the women's movement. There has, however, been limited interaction between women representatives and the women's movement. This, perhaps, is the issue that the women's movements need to address.

Strategies for increasing women's representation

Many strategies have been discussed for increasing women's representation in politics – party lists in South Africa and the UK, and quotas at the local governance levels in India are two important experiments. There has also been discussion of parity representation for women and men, especially in the Nordic states (Karam, 1998; Dahlerup, 2005). All three are contested strategies, as the following discussion of quotas demonstrates. The arguments for quotas for women in representative institutions are fairly well rehearsed. Women's groups are now arguing that quotas for women are needed to compensate for the social barriers that have prevented women from participating in politics and thus making their voices heard: that in order for women to be more than 'tokens' in political institutions, a level of presence that cannot be overlooked – 'critical mass' – by political parties is required, hence the demand for a 33 per cent quota: where the quota system acknowledges that it is the recruitment process, organized through political parties and supported by a framework of patriarchal values, that needs to carry the burden of change, rather than individual women. The alternative, then, is that there should be an acknowledgement of the historical social exclusion of women from politics, a compensatory regime (quotas) established, and 'institutionalized . . . for the explicit recognition and representation of oppressed groups' (Young, 1990: 183–91). However, there is some unease felt by many women's groups with elite politics and elite women, and the role that quotas might play in consolidating rather than shifting power relationships.

Conclusions

While political representation is crucial to women's empowerment, there are also some critical issues that we need to reflect upon. State institutions cannot be the only focus of women's political struggles. Both spaces – the informal and formalized networks of power – need to be negotiated by the women's movements in order to best serve women's interests (Rai, 1996). This would include the work of civil society associations, increased representation of women in state and political bodies that allows for a wide-ranging set of interests to be represented by women, and also a discursive shift in the way in which politics is thought about to enable women to function effectively in politics. Women's participation in representative institutions can be effective only in the context of such continuing negotiations and struggles.

Bibliography

Dahlerup, D. (2005) *Women Quotas and Politics*, London: Routledge.

Goetz, A. M. (ed.) (1997) *Getting Institutions Right for Women in Development*, London: Zed Books.

IMF (2012) IMF Executive Directors and Voting Power. Available online at www.imf.org/external/np/sec/memdir/eds.aspx (accessed 13 May 2012).

IPU (Inter-Parliamentary Union) (2012) 'Women in national parliaments'. Available online at www.ipu.org/wmn-e/world.htm (accessed 13 May 2012).

Jonasdottir, A. G. (1988) *The Political Interests of Gender: Developing Theory and Research with a Feminist Face*, London: Sage.

Karam, A. (ed.) (1998) *Women in Parliament: Beyond Numbers*, Stockholm: IDEA.

Kymlicka, W. (1995) *Multicultural Citizenship*, Oxford: Oxford University Press.

Lovenduski, J. (2006) *Feminising Politics*, Cambridge, Polity Press.

Lovenduski, J. and Norris, P. (1993) *Gender and Party Politics*, London: Sage.

McBride Stetson, D. and Mazur, A. (1995) *Comparative State Feminism*, London: Sage.

Nussbaum, M. (1995) 'Human capabilities, female human beings', in M. Nussbaum and J. Glover (eds) *Women, Culture and Development*, Oxford: Clarendon Press.

Rai, S. M. (1996) 'Women and the state: Some issues for debate', in S. M. Rai and G. Lievesley (eds) *Women and the State: International Perspectives*, London: Taylor and Francis.

Rai, S. M. (1997) 'Women in the Indian Parliament', in A. M. Goetz (ed.) *Getting Institutions Right for Women in Development*, London: Zed Books.

Rai, S. M. (ed.) (2000) *International Perspectives on Gender and Democratisation*, Basingstoke, UK: Macmillan.
Rai, S. M. (ed.) (2001) *Mainstreaming Gender, Democratizing the State? National Machineries for the Advancement of Women*, Manchester, UK: Manchester University Press.
Staudt, K. (1991) *Managing Development*, London: Sage.
World Bank (1992) *Good Governance*, Washington: World Bank.
World Bank (2000) *World Development Report: Attacking Poverty*, Washington: World Bank.
Young, I. M. (1990) *Justice and the Politics of Difference*, Princeton, NJ: Princeton University Press.

7.9
Sexuality and development

Andrea Cornwall

What would the world be like if we really did have the right to choose our sexuality and pleasurable sexual relations?

(*Karin Ronge, Women for Women's Human Rights, Turkey*)

Sexuality has had a place in international development from the early days of colonial and missionary intervention in the countries of the global South. Late nineteenth-century European sexual mores were transported to the colonies as part of imperial expansion. With it came an ordering of the world into rigid gender and sexual binaries, with the naturalization of the categories 'women' and 'men', and the late nineteenth-century European categories 'heterosexual' and 'homosexual'. Development came to be cast as a *moral* as well as economic endeavour, famously described by Gayatri Spivak (1988), for India, as 'white men rescuing brown women from brown men'. Laws and social policies sought to encourage the formation of nuclear families based on the – still relatively novel in the Victorian period – European model of heterosexual monogamy. Practices considered 'barbaric', such as child marriage and female genital cutting, became a prime focus of colonialism's 'civilising' project. Education for girls and women sought to produce domesticated 'good women' who would maintain hygienic homes and respect their husbands (Hunt, 1990); the contradictions of colonial extraction also fostered 'bad women', who left rural homes for lives as sex workers in the city (Jeater, 1993; White, 1990).

Colonial Christian moralities produced a normative categorization of 'good' and 'bad' women (Hunt *et al.*, 1997), and a notion of the perversity of certain sexual acts that came to be constitutive of certain types of persons. In many countries of the global South, colonial governments put laws in place that regulated sexuality (Stoler, 1989), proscribing non-reproductive sexual expressions as 'carnal intercourse against the order of nature' (Baudh, 2008) and reshaping the contours of marital practices. Many of these laws remain in place, notoriously the laws on 'unnatural practices' that have been used to persecute men who have sex with men, described by Judge Michael Kirby (2011) as 'England's least lovely criminal law export'.

Traditions of rescue and the uses of law, social and economic policy and education in the institutionalization of normative heterosexuality – with all its contradictions – connect the colonial past and a significant part of contemporary development practice. As the development industry came to take institutional form in the post-World War II period, the overt concern of the colonial powers with sexual morality receded as the modernization agenda brought its technocratic rationality to development practice. The control of reproduction came to take centre stage, as efforts were made to curb rapid population growth in the global South. Sex came to be represented within international development as a cause of unwanted pregnancies, disease, harm and hazard, rather than a source of pleasure, joy, intimacy and happiness. And social and economic development policies came to be premised on a particular form of normative heterosexuality, exemplified in the hegemonic concept of the 'household' with its assumptions about male breadwinners and female dependents, that shunted all other forms of economic, sexual and domestic arrangements out of the frame.

Feminist critiques of mainstream development's 'gender blindness' often tended to reproduce rather than challenge the normative assumptions about sexuality implicit in development policy. The first wave of Women in Development projects naturalized women's roles as wives and mothers. The next wave of feminist engagement with development, Gender and Development (GAD), sought interventions in policy discourse that inscribed an asymmetrical power relation between women and men as constitutive of 'gender relations'. Premised on a very specific 'coital and conjugal' (Ogundipe-Leslie, 1994) relation, it came to eclipse other kinds of relations of gender and power. Powerful gender myths cast women as heroines or victims, and men as shadowy, oppressive figures (Cornwall, 2000). A curious prurience hung over feminist engagement with development that relegated questions of sexuality to the margins. Amidst much talk about paying closer attention to women's lived experiences, when it came to sexuality a guarded silence often prevailed. Attempts to break that silence were met with such reactions as a concern that sexuality was 'frivolous' compared to the *real* issues – as if sexuality had nothing to do with poverty, or indeed any other core concern in international development – or plain 'embarrassing' (Cornwall and Jolly, 2006).

Sex and sexuality remained a 'health issue'; and in striking contrast to mainstream social and economic development, an explosion of interest in sex and sexuality was provoked by the AIDS epidemic. Epidemiological research revealed patterns of sexual networking that profoundly challenged normative beliefs about sexuality. The AIDS response made visible a rich seam of sexual practices that departed from – and flouted – these norms. Categories created for epidemiological analysis and HIV prevention efforts, such as men who have sex with men (MSM), came to take on a life of their own as they were appropriated as identities, and used to claim rights (Boyce and Khanna, 2011).

Anthropologist James Scott explores in his book *Seeing Like a State* (1998) the ways in which people come to be made legible to the state through forms of discursive ordering – counting, registering, surveying, categorizing. Foucault drew attention to how discourses produce subjects; the very categories through which we come to know ourselves are not natural or fixed but products of particular historical and cultural frames. So too the 'homosexual', a category produced in late nineteenth-century Europe in a very particular cultural context (Plummer, 1981). Its corollary, which came into being a few years later, 'heterosexual', corresponded with a stage in the growth of capitalism that came to depend on the very regulation of sexuality and production of nuclear families with the heterosexual couple at their heart. Like MSM or WSW, the categories 'gay' and 'lesbian' – and indeed that of 'heterosexual' – are, then, less ontological categories with a universal existence than the products of a particular period in Euro-American history. 'Transgender' is of more recent provenance still, originating in the United States in the 1990s (Stryker, 2008).

The circulation of a globalized narrative on sexual identity amongst metropolitan elites has, some would argue, rendered as fixed identities local sexual expressions and arrangements that were practices rather than identifications. Writing of the Arab world, Joseph Massad argues that what he calls 'the Gay International' – Western LGBT activist organizations engaged in constituency-building in the global South – operate in a way 'that both produces homosexuals, as well as gays and lesbians, where they do not exist, and represses same-sex desires and practices that refuse to be assimilated into its sexual epistemology' (Massad, 2002: 363). This, he contends, is '*heterosexualizing* a world that is being forced to be fixed by a Western binary' by 'inciting discourse about homosexuals where none existed before' (2002: 383). Yet, as Oliver Phillips argues, in a paper on the emergence of the 'global gay' written some years before Massad's provocative piece, this process is made more complex by the ways in which these global categories come to be taken up and used:

> These 'new' identities are merged into local histories and contexts, so that the end product has signifiers of local significance while simultaneously providing a strategy for either laying claim to international human rights agreements, or enabling more effective AIDS/HIV preventative work, or simply buying into an expanding market of western signifiers of 'modern' and bourgeois status, or serving all of these purposes. *What is clear is that these identities are not simply imposed through an imperialistic cultural discourse or economic dominance, but they are actively assumed and proclaimed from below, by those marginalised in these hegemonic formations.*
>
> *(Phillips, 2000: 17–18, my emphasis)*

Through the taking up of identities that Phillips describes, we see activists in the global South coalescing around such labels as MSM, and using them politically to claim rights and recognition; we also see spaces being created across and within social movements, including the feminist movement, for greater recognition of gender and sexual diversity.

Yet it does not seem to be this openness to diversity of gender expressions, of sexualities that is being manifested in the embrace of LGBT rights by US State Department and the UK's Foreign Office. Rather, what we are seeing it seems is 'homonormativity', described by Lisa Duggan as:

> A politics that does not contest dominant heteronormative assumptions and institutions, but upholds and sustains them, while promising the possibility of a demobilized gay constituency and a privatized, depoliticized gay culture anchored in domesticity and consumption.
>
> *(2002)*

Indeed, some would charge, this 'recognition' and defence of LGBT rights by prominent Western governments represents not only homonormativity, but 'homonationalism' (Puar, 2007): a process through which selective acts of recognition and incorporation of gay and lesbians within normative liberal institutions work to deflect attention from the persistent injustice experienced within – *and perpetrated by* – the very societies that construct themselves as liberal and progressive. Agathangelou and colleagues ask:

> What bodies, desires, and longings must be criminalized and annihilated to produce the good queer subjects, politics, and desires that are being solidified with the emergence of homonormativity?
>
> *(Agathangelou et al., 2005: 124)*

Those who are most marginalized by this are typically those whose departure from dominant gender, sexual and social norms makes them the most visible dissenters from normative gender and sexual categories. Giuseppe Campuzano (2008) uses Robert Chambers' matrix of poverty to highlight the multiple forms of exclusion experienced by Peruvian *travestis* whose transits of gender flout normative expectations of sexual and gender binaries.[1] From bullying and marginalization at school making it difficult to remain in education, difficulties gaining and remaining in employment because of pervasive prejudices, prejudice making it hard to find housing, the violence of stigma affecting access to health and social services, Campuzano spells out myriad effects of discrimination.

That the 'global gay' has become an object of interest to the development industry, and with it categories of persons produced by discourses on sexual risk and hazard such as MSM, has arguably had the effect of reducing the discussion on sexuality in development to one about political rights and deflecting attention from the glare of social and economic injustice. After all, arguably, what Peruvian *travestis*, Indian *hijras*,[2] South African *tommy boys*[3] and Brazilian *putas*[4] have in common is not membership of a category of persons – the 'sexually marginalized' or 'sexual minorities' – but marginalization and vulnerability to violence that must be understood in intersectional terms, exposing the race and class dimensions of these effects of discrimination and difference. Armas (2007) and Jolly (2010a) explore the far-reaching dimensions of these forms of exclusion and discrimination on the basis of sexuality, and highlight how international development's disregard for sexual and gender diversity can injure, impoverish and produce ill-being.

It is engagement with precisely these broader issues of justice that animates the struggle for the elimination of all forms of discrimination on the basis of gender and sexuality in which so many sexual rights activists in the global South are engaged. This is a struggle that is about creating space for sexual diversity and coming together in common struggle against exclusion and its economic, political, social and psychological effects. And it is about establishing connections that, by reclaiming pleasure as a right, reframe the relationship between sexuality and development. Such work involves opening a space for dialogue about pleasure (Jolly, 2010b) that can confront the internalized societal prejudices that may otherwise prevent people from exploring what they might have in common, as people experiencing sexual exclusion. An inspiring example of this can be found in the work of Chinese NGO Pink Space, who facilitate conversations between the wives of gay men, female sex workers and lesbians that shatter prejudices and create opportunities for these women to explore their own sexualities, rights and pleasures (He, 2013). Another example comes from the human rights training courses run by Turkish organization Women for Women's Human Rights, who teach rural women sexual pleasure as a human right (Seral, 2013). And such work involves confronting society with its prejudices: like the Indian sex workers' collective VAMP who mobilize with the mission of 'changing society', confronting stigma and asserting the right to have rights and, whose slogan 'Save us from Saviours' sends a powerful message to the development industry

(Seshu, 2013). It is in these sites and spaces that a very different kind of 'development' is happening, one that is about people taking charge of their own 'liberation' and pursuing their own struggles, in their own ways.

Notes

1 Born as male, living in a gender identity that is neither man nor woman and that is constructed from exaggerated femininities, with the use of silicone and plastic surgery to reshape hips, buttocks and breasts.
2 Born as male or intersex, some removing genitalia to become 'eunuchs' and using hormones to develop breasts, living in a gender identity that is neither man nor woman, in feminine clothes.
3 'Masculine' women-loving-women, who adopt masculine styles of dress and expression.
4 'Prostitutes' – as the leader of the Brazilian Association of Prostitutes put it, 'don't sanitize us with your labels and call us sex workers, we're prostitutes, call us "prostitutes"'.

Bibliography

Agathangelou, Anna, Bassichis, Daniel and Spira, Tamara (2005) 'Intimate Investments: Homonormativity, Global Lockdown, and the Seductions of Empire', *Radical History Review*, 100: 120–143.
Armas, Henry (2007) *Whose Sexuality Counts? Poverty, Participation and Sexual Rights*, Working Paper 294, Brighton: IDS.
Baudh, Sumit (2008) *Human rights and the criminalization of consensual same-sex sexual acts in the Commonwealth, South and Southeast Asia*. New Delhi: South and Southeast Asia Resource Centre on Sexuality.
Boyce, P. and Khanna, A. (2011) 'Rights and Representations: Querying the Male-to-Male Sexual Subject in India', *Culture, Health & Sexuality*, 13: 89–100.
Campuzano, Giuseppe (2008) *Building Identity While Managing Disadvantage: Peruvian Transgender Issues*, IDS Working Paper 310, Brighton: IDS.
Cornwall, Andrea (2000) 'Missing Men: Reflections on Men, Masculinities and Gender in GAD', *IDS Bulletin*, 31 (2): 18–27.
Cornwall, A. and Jolly, S. (2006) 'Sexuality Matters', *IDS Bulletin*, 37 (5): 1–11.
Cornwall, Andrea, Corrêa, Sonia and Jolly, S. (2008) *Development with a Body: Sexuality, Development and Human Rights*, London: Zed Books.
Duggan, Lisa (2002) 'The New Homonormativity: The Sexual Politics of Neoliberalism', in Russ Castronovo and Dana D. Nelson (eds) *Materializing Democracy: Toward a Revitalized Cultural Politics*, Durham, NC: Duke University Press.
He, Xiaopei (2013) 'Building a Movement for Sexual Rights and Pleasure in China', in Susie Jolly, Andrea Cornwall and Kate Hawkins (eds) *Women, Sexuality and the Political Power of Pleasure*, London: Zed Books.
Hunt, Nancy Rose (1990) 'Domesticity and Colonialism in Belgian Africa: Usumbura's Foyer Social, 1946–1960', *Signs: Jounal of Women in Culture and Society*, 15 (3): 447–474.
Hunt, Nancy Rose, Liu, T. and Quataert, T. (eds) (1997) *Gendered Colonialisms in African History*, London: Blackwell.
Jeater, Diana (1993) *Marriage, Prostitution and Power: Divorce in Colonial Southern Rhodesia*, Oxford: Oxford University Press.
Jolly, Susie (2010a) *Poverty and Sexuality: What are the Connections*, Stockholm: Sida.
Jolly, Susie (2010b) 'Why the Development Industry should Get Over Its Obsession with Bad Sex and Start to Think about Pleasure', in Suzanne Bergeron and Amy Lind (eds) *Development, Sexual Rights and Global Governance: Resisting Global Power*, New York: Routledge.
Kirby, Michael (2011) 'The Sodomy Offence: England's Least Lovely Criminal Law Export?' *Journal of Commonwealth Criminal Law*, 22–43.
Massad, Joseph (2002) 'Re-Orienting Desire: The Gay International and the Arab World', *Public Culture*, 14(2): 361–385.
Ogundipe-Leslie, Molara (1994) *Recreating Ourselves: African Women and Critical Transformation*, Trenton, NJ: Africa World Press.

Phillips, Oliver (2000) 'Constituting the Global Gay: Issues of Individual Subjectivity and Sexuality in Southern Africa', in D. Herman and C. Stychin (eds) *Sexuality in the Legal Arena*, London: Athlone Press. (Also, paper presented to the Queering Development seminar series, IDS, Autumn 1999).
Plummer, Ken (ed.) (1981) *The Making of the Modern Homosexual*, London: Hutchinson.
Puar, Jasbir (2007) *Terrorist Assemblages: Homonationalism in Queer Times*, Durham, NC: Duke University Press.
Scott, James (1998) *Seeing Like a State: How Certain Schemes to Improve the Human Condition Have Failed.* New Haven, CT: Yale University Press.
Seral Aksakal, Gulsah (2013) 'Sexual Pleasure as a Woman's Human Right: Experiences from a Human Rights Training Programme for Women in Turkey', in Susie Jolly, Andrea Cornwall and Kate Hawkins (eds) *Women, Sexuality and the Political Power of Pleasure*, London: Zed Books.
Seshu, Meena (2013) 'Sex, Work and Citizenship: The VAMP Sex Workers' Collective in Maharashtra', in Naila Kabeer, Ratna Sudarshan and Kirsty Milward (eds) *Organizing Women Workers in the Informal Economy: Beyond the Weapons of the Weak*, London: Zed.
Spivak, Gayatri (1988) 'Can the Subaltern Speak?' in C. Nelson and L. Grossberg (eds) *Marxism and the Interpretation of Culture*, Chicago: University of Illinois Press, pp. 271–313.
Stoler, Ann (1989) 'Making Empire Respectable: The Politics of Race and Sexual Morality in 20th-Century Colonial Cultures', *American Ethnologist*, 16(4): 634–660.
Stryker, Susan (2008) 'Transgender History, Homonormativity and Disciplinarity', *Radical History Review*, 100: 145–157.
White, Luise (1990) *The Comforts of Home: Prostitution in Colonial Nairobi*, Chicago: Chicago University Press.

7.10 Indigenous fertility control

Tulsi Patel

Introduction

Demographic studies have generally seen the world as divided into two broad parts. The less developed societies with high fertility, under the spell of religious superstition and tradition and absence of any conspicuous attempt to limit fertility were termed as 'natural fertility' populations (Henry 1961). Couples in the developed world are seen as capable of rational choice and use contraception to control population. Except for anthropological studies, little recognition was found for populations that practiced fertility regulation through customary institutional practices influencing longer intervals between menarche and first birth, interval between births and prescriptions for when to begin and cease having babies and how many to have. Reported evidence and methodological individualism overlooked reflective social agents working out their fertility through personal, familial and community considerations. Anthropological (Handwerker 1990; Rapp and Ginsburg 1991; Patel 1994; Greenhalgh 1995) and historical (Harris and Ross 1987; Hufton 1995) studies convince that population and fertility is situated amidst cultural, gendered, and political-economic considerations. These studies show that, though barrenness was regarded as a curse, an over-abundance of living children was seen as something less than a blessing.

Acknowledging that rationality is not instrumental rationality alone, this chapter focuses on indigenous fertility control in two ways: mortality control on the one hand, and fertility on the other. It also shows how colonial powers refused to recognize customary and contemporary attempts of the colonies and especially of women to control population growth as do the policies of postcolonial nation states.

Social mechanisms and population regulation

The prevalence of indigenous means of population control is reported in many societies. Freebeme (1964) reports for traditional China, Smith (1977: 142–3) for Japan, Polgar (1971) for several societies in the edited collection, Harris and Ross (1987) for societies ranging from Palaeolithic times to the present, in contemporary times Patel (1994, 1998) for Indian peasant and tribal society, Bledsoe (1997) for Gambia, and Ram and Jolly (1998) for Asia-Pacific during the colonial times. (See also female infanticide, widow deaths and prohibited marriage.) Ahluwalia's (2008) account of historical records analyses how the prevalent population control practices – written and oral – in Indian society were overlooked in favour of modern and Western practices by colonial powers in India. Both historical accounts as well as demographic studies in the Post-World War II era have denied agency to the erstwhile colonies and women under the pretext of their irrationality in not adopting modern contraception. Let us see how research reveals that 'non-modern' societies regulated population.

Limiting fertility within marriage

Matters related to marriage and fertility are intimate, gendered and elicit intense interest from surrounding kin, community, and state. The German ethnologist, Felix Speiser, attributes low birth rate to the precolonial sexual economy – the way in which older, less virile men monopolized women, the sterility in women 'used' for sexual purposes at a very early age.

Celibacy

Prayer to have self-control is known besides the New Testament praising virgins as Christ's brides or jewels. Hindu *sadhus* and *sadhvis*, and some wrestlers in India remain celibate for life. A strict and elaborate regimen is prescribed for wrestlers to control semen loss (Alter 1992: 133).

Unwelcome babies

The infamous infanticide, especially female infanticide has for centuries depressed the number of surviving children with fewer girls, exacerbated by colonial land policies. Vishwanath (2000) provides figures for reduced population of Patidars during 1820s and 1890s in Gujarat.

Douglas (1966) describes the customary beliefs and practices of eliminating babies among a few primitive communities. Female children's neglect is customary in south Asia characterized by son preference leading to millions of girls missing in China and India. It holds true for India even today. Though impossible to quantify them, accounts of female infanticide are available from autobiographical accounts (Dureau 1998).

Eugenics politics

Many colonizers and missionaries encouraged breeding among the middle and upper classes despite a mixed response from them (see Ram and Jolly 1998). This collection reveals the colonial politics of control of women's bodies to prevent contraception, in particular abortion and infant and child mortality with the dual aim to produce more labour power and introduce Western medical systems. Fijian mothers were under strict surveillance to increase the population, and were often accused of procuring abortions and were liable to inquests. Speiser (1990: 50) evokes this clearly from a mother, 'Why should we go on having children? Since the white man came they all die.'

Abortion

In the early stages of pregnancy, abortion was widely tolerated as one of the only dependable methods of fertility control. Even the Catholic Church views the foetus to be animated by the rational soul, making abortion a serious crime at 40 days after conception for a boy and 80 days for a girl (The Boston Women's Health Collective 1971: 216).

Among the Javanese (Alexander 1986) there are occasional reports of the occurrence of abortion. Treating delayed menstruation with herbs and massage is evidently common (see the special issue of *Social Science and Medicine* 1996). Herbs purchased at fairs and/or prepared from garden plants to bring on 'women's courses' before the pregnancy was far advanced are also reported for fifteenth- to seventeenth-century European society, their knowledge passed from one generation of women to another. Midwives were thought to be well versed in preparing these concoctions, as were prostitutes. Demographic studies ignored these customary traditions, enabling correlations of modern contraception with development.

Mechanisms of fertility regulation

Fertility control practices, though gender biased, are often related directly with the belief and value structures of a given society, and with political and economic structures. Population policies have targeted women for fertility control. But use of new technologies in reproduction – contraception, abortion and gestational screening – have boomeranged throwing population policies in a tizzy. Wealthy, urban, modern and other rational couples in India (also China) practice sex-selective abortion/female feticide despite legislation outlawing it (Patel 2007).

Abstinence

Handwerker (1986: 103) refers to an old Liberian farm woman stating that the foolproof method was to avoid your husband. A Rajasthani woman in her eighties had confided that she had borne only two children because she had sternly kept her husband at an arm's length. They are known to ingeniously introduce a barrier to reduce or prevent a possibility of congenial sex. A demographer from Philippines corroborated Patel's (1994) observation of enticing children to sleep in one's bed specifically to deter any sexual advances. 'Madame de Sevigne, in seventeenth-century France, urged upon her daughter after three pregnancies in quick succession, the desirability of having her maid sleep in the same room to depress the sexual urges of the Comte de Grigan' (Hufton 1995: 178). Predominantly the women's domain, fertility control knowledge passed on through women across generations.

Gandhi's exhortation to not separate sex from procreation is well known. Erstwhile colonies' ideologies of procreation differed from the Western one, though in common was women's secret knowledge and strategies to control the number of offspring. Ahluwalia shows Indians publishing literature on limiting coital frequency by age and fertility during colonial times, including avoiding a bedroom or at least the same bed (2008: 168).

Sleeping arrangements

The housing architecture usually provides for only one room, allowing a rather generous ventilation but disallowing privacy (Patel 1994, for privacy: 173, 188–9). Except for very young couples, others sleep in common open spaces like the courtyard. Streets are lined with cots at night and men talk themselves to sleep. Sex relations of the elderly do not remain a secret for long and become a subject for gossip (see also Dureau 1998: 244). In Mogra (the village of Patel's 1994 study), if elderly (parents of married children) display undue concern over sensual enjoyment they risk criticism and ridicule (cf. Caldwell 1982 for the Yoruba of Nigeria).

Social onomastics

The pattern of nomenclature signals the names of uncalled-for births after parents have had the socially optimum number of children. Examples of these are Madi (one who has barged in), Aichuki (enough of coming), Santi (peace/quiet), Santos (satisfaction/complacency) and Dhapuri (satisfied/full/complete) and more Sanskritic, Itishri (the end) and Sampurna (complete).

Lactation practices

Among the Asia-Pacific communities (Ram and Jolly 1998), children are ideally suckled for about four years. The association between lactation and anovulation is well known. In Mogra, mothers breastfeed their infants on demand. If an infant were to die, the older sibling is put to the breast with the clear purpose of avoiding a quicker conception. MacCormack reports for Sierra Leone, 'Sande women, with their secret knowledge, public laws, legitimate sanctions and hierarchical organization, bring women's biology under the most careful cultural control' (1977: 94).

Political and ritual considerations

Some African and Pacific women prolong breastfeeding and abstinence to have only one or two children. The Bhils in Rajasthan (Patel 1998) practice abstinence for a month during the Gavari worship festivities, and during fasting. Many men adorn the ritually processed copper ring (called a *gole*) which ensues frequent abstinence.

Besides, the popular practice of *pomanchar,* literally guesting, involves frequent reciprocal visits of kin, relatives and friends. Sleeping is especially gender-segregated, inhibiting sex.

Contraception

Coitus interruptus, or withdrawal, has been practised and passed on from generation to generation. The semiology of seed and soil (*beej-kshetra*) is the most common one in many parts of the world. Hufton records a case, 'When Isabella de Moerloose of the region of Ghent married a pastor in 1689, he tried by coitus interruptus and oral sex to avoid making Isabella pregnant' (1995: 174).

Conclusion

Prolonged disregard and illegitimization of customary population control mechanisms by colonial powers' civilizing mission projected their image as an irrational people. Policy discourse denies agency to peoples and societies that have below 'natural fertility'. The grand project of development found its mission in spreading rational development, population control being among the means. The taken-for-granted female passivity colours contemporary masculine-state population policies. Differential access to power (individual/state/colonial) determines how conflicts over reproduction/population are conducted and resolved (if they ever are resolved) and the argument for development is deployed in the discourse.

References

Ahluwalia, Sanjam (2008) *Reproductive Restraints: Birth Control in India; 1877–1947.* New Delhi: Permanent Black.

Alexander, P. (1986) 'Labour expropriation and fertility: Population growth in nineteenth century Java', in W. P. Handwerker (ed.) *Births and Power: Social Change and Politics of Reproduction.* Boulder, CO: Westview Press.

Alter, J. (1992) *The Wrestler's Body.* Berkeley: University of California Press.

Bledsoe, C. (1997) 'Reproduction and Gambia: African empirical challenges for the culture of Western science', Unpublished Manuscript.

The Boston Women's Health Collective (1971) *Our Bodies, Our Selves.* New York: Simon and Schuster.

Caldwell, J. C. (1982) *Theory of Fertility Decline.* London: Academic Press.

Douglas, M. (1966) *Purity and Danger: An Analysis of the Concepts of Pollution and Taboo.* London: Routledge and Kegan Paul.

Dureau, C. (1998) 'From sisters to wives: Changing contexts of maternity on Simbo, Western Solomon Islands', in K. Ram and M. Jolly (eds) *Maternities and Modernities: Colonial and Postcolonial Experiences in Asia and the Pacific.* Cambridge: Cambridge University Press.

Freebeme, M. (1964) 'Birth Control in China', *Population Studies* 18 (1): 5–16.

Greenhalgh, S. (ed.) (1995) *Situating Fertility: Anthropology and Demographic Inquiry.* Cambridge: Cambridge University Press.

Handwerker, W. P. (ed.) (1986) *Births and Power: Social Change and Politics of Reproduction* (pp. 1–33). Boulder: Westview Press.

Handwerker, W. P. (1990) *Culture and Reproduction.* Boulder, CO: Westview Press.

Harris, M. and Ross, E. (1987) *Death, Sex, and Fertility: Population Regulation in Preindustrial and Developing Societies.* New York: Columbia University Press.

Henry, Louis (1961) 'Some data on natural fertility', *Eugenics Quarterly* 8: 81–91.

Hufton, O. (1995) *The Prospect Before Her. Vol. 1: 1500–1800.* London: Harper Collins.

MacCormack, C. P. (1977) 'Biological events and cultural control', *Signs* 3 (1): 93–100.

Patel, Tulsi (1994) *Fertility Behaviour: Population and Society in a Rajasthan Village.* Delhi: Oxford University Press.

Patel, Tulsi (1998) 'Reproduction and tribal society: A study in southern Rajasthan'. Project Report submitted to the ICSSR, Delhi.

Patel, Tulsi (2007) 'Female feticide, family planning and state-society interface in India', in T. Patel (ed.) *Sex-selective Abortion in India: Gender Society and New Reproductive Technologies.* New Delhi: Sage.

Polgar, S. (ed.) (1971) *Culture and Population: A Collection of Current Studies.* Cambridge: Schenkman.

Ram, K. and Jolly, M. (eds) (1998) *Maternities and Modernities: Colonial and Postcolonial Experiences in Asia and the Pacific.* Cambridge: Cambridge University Press.

Rapp, R. and Ginsburg, F. (1991) 'The politics of reproduction', *Annual Review of Anthropology* 20: 311–43.

Smith, T. C. (1977) *Nakahara: Family Planning and Population in a Japanese Village, 1717–1830.* Stanford, CA: Stanford University Press.
Social Science and Medicine (1996) Special issue, 42(4) (for several articles in the special issue on abortion).
Speiser, F. (1990[1923]) *The Ethnology of Vanuatu: An Early Twentieth Century Study.* Bathurst: Crawford House.
Vishwanath, L. S. (2000) *Female Infanticide and Social Structure.* Delhi: Hindustan.

Further reading

Delaney, D. (1991) *The Seed and The Soil: Gender and Cosmology in Turkish Village Society.* Berkeley: University of California Press, is a graphic account of the suffering and anguish infertility can cause to men and especially to women in the patriarchal society of Turkey.
Patel, Tulsi (2007). 'Female feticide, family planning and state-society interface in India', in T. Patel (ed.) *Sex-selective Abortion in India: Gender Society and New Reproductive Technologies.* New Delhi: Sage. (See other articles in this book for more on female feticide, development, etc.)
Scheper-Hughes, Nancy (1992) *Death Without Weeping.* Berkeley: University of California Press, is an ethnographic account of cultural practices of dealing with child mortality, though not as a measure of fertility control.

Part 8
Health and education

Editorial introduction

Should poor countries be concerned about the implications for human welfare when they cannot afford to provide health and educational needs and when their economic resources are heavily constrained? There are three reasons for concern. First, the basic health, nutritional and educational needs of the most vulnerable groups (children, women and the elderly) are urgent and compelling. If these are neglected, they can set back the health and welfare of the whole future generation of a country, in addition to adding to present human and economic miseries. Second, there is considerable evidence that there are positive economic returns to interventions supporting basic nutrition, health and education; and third, human welfare and progress is the ultimate end of all development policies and so many of the UN Millennium Development Goals (MDGs) are related to health and education. However, as we approach 2015 MDG targets, there is evidence of much less improvement and we are unlikely to meet these targets, for example, of progress in improving maternal health and reducing maternal deaths. There is also recognition of the unevenness of progress within countries and regions and the severe inequalities that exist among populations, especially between rural and urban areas.

Levels of child malnutrition have fallen slowly during the last two decades despite significant economic growth and considerable expenditure on various programmes. The poor and the very young in low-income countries seem particularly vulnerable to disease and there is evidence to suggest that, in many countries, the position of women in society relative to that of men makes them more prone to disease. Women's nutritional status influences children's nutritional status in a variety of ways both during pregnancy and early childhood. Women who are malnourished are more likely to deliver smaller babies, who, in turn, are at increased risk of poor growth and development. Additionally, malnourished women may be less successful at breastfeeding their children, have lower energy levels, and have reduced cognitive abilities, all of which hamper their ability to adequately care for their young child. Prenatal care provides an opportunity for a number of preventative interventions, including various immunisations, prevention and treatment of anaemia and infections, and detection of high-risk pregnancies needing special delivery care. Many of those concerned with women's reproductive health argue that family planning services should be available not for population control but for birth control, and only as part of comprehensive health and welfare provisions. Preventative and curative measures to control infectious diseases in young children are also crucial to their nutritional status. Infectious diseases, many of which are relatively easy to cure, remain the major killers in low-income countries (e.g. malaria). The provision of contraceptive devices and nutritional supplements will not in themselves ensure a reduction in maternal morbidity. Poverty is clearly connected with vulnerability to infections, and undernutrition is often a major contributing factor. Lack of access to safe water and poor environmental sanitation due to unsanitary waste disposal are considered to be important causes of infectious diseases and disabilities among children. Mainstream schools do not include or have provisions for disabled students and hence disabled children do not have access to education. This, in turn, leads to high levels of illiteracy, reduced skills and employment opportunities for disabled people in adulthood, perpetuating the cycle of poverty from one generation to the next and becoming vulnerable to abuse and exclusion. These issues raise serious problems with the prevalent approaches to primary healthcare and accessibility for the poor to the health services.

Over the last four decades there has been great concern about the transmission of HIV/AIDS, its impact on communities, households, livelihoods, and access to treatment. Studies are now highlighting that, as countries are developing, health and education inequalities are widening between the rich and poor. These issues need to be put in the context of human rights and social justice within emerging societies of the developing world.

Rising income leads to changing consumption patterns and this has led to changing disease patterns. There are now 'diseases of the rich' such as diabetes, obesity, and heart disease which coexist with disease patterns of the poor such as malnutrition and micro-nutrient deficiencies, TB, hepatitis and malaria. This can potentially cause a shift in resources away from preventative, public and basic health for the poor. With the rising age expectancy leading to an ageing population, higher per capita health expenditure among older people further adds to the strains of caregiving. This has implications for the ageing population as demographic transition takes place to an increasingly older population generated by a rise in life expectancy and falling fertility rates.

Social protection programmes have become popular in development policy with the aim to promote more inclusive development within a framework of universal rights and broader concerns with poverty and vulnerability. It provides safety nets and the management of risk concerned with preventing, managing and overcoming situations that adversely affect people's well-being.

Adult literacy is a huge problem for human resource development especially amongst women. Much emphasis has been placed on the role of female education in achieving gender equality and that education stimulates a range of personal, economic and social benefits that well exceed their costs. It increases their ability to benefit from health information and to make good use of health services; it increases their access to income and enables them to live healthier lives.

Globalisation has resulted in the growth in size and proportion of the informal economy, as well as the fragmentation of the service and commodity chains in developing countries resulting in children entering the world of work in large numbers, in the service sector and as unpaid family labour. A large proportion of the workforce is engaged as casual labour, low paid salaried work or self-employed in micro enterprises with the help of family labour. These workers constitute a large and growing informal economy with very low skill levels and education and many emerging countries face the challenges of developing skills and technology absorption. Development of the 'knowledge economy' and high technology capital intensive industry requires an educated and skilled labour force. Children out of school and in work should be targeted to secure long-term development goals. The right to education is a fundamental universal human right and managing education systems at all levels and creating accessibility has become very crucial.

During the current austerity measures in the global North stronger links are needed with global civil society networks and social movements. There is a need for critical understandings of global processes taking place and of unequal interdependence. A more active engagement is needed through development education in poverty reduction in the South.

8.1

Nutritional problems, policies and intervention strategies in developing economies

Prakash Shetty

Introduction

A critical examination of the principal causes of mortality and morbidity worldwide indicates that malnutrition and infectious diseases continue to be significant contributors to the health burden in the developing world. Although reductions in the prevalence of undernutrition is evident in most regions, the numbers of individuals affected remain much the same or have even increased, largely the result of increases in the population in these countries. What is striking, however, is that the health burden due to non-communicable diseases (NCDs) such as heart disease, diabetes mellitus and cancer is dramatically increasing in some of these developing economies with modest per capita GNPs, particularly those that appear to be in some stage of rapid economic and developmental transition. Even modest increases in prosperity that accompany economic development seem to be associated with marked increases in the mortality and morbidity attributable to NCDs related to diet and nutrition. These dramatic changes in the disease burden of the population are probably mediated by changes in the dietary patterns and lifestyles which typify the acquisition of urbanized and affluent lifestyles. Countries in rapid developmental transition seem to bear a 'double burden' of undernutrition coupled with overnutrition concurrently.

The spectrum of malnutrition

The term 'malnutrition' refers to all deviations from adequate nutrition, including under- and overnutrition, and encompasses both inadequacy of food or excess of food relative to need. 'Undernutrition' is the result of insufficient food caused primarily by an inadequate intake of dietary energy, whether or not any other nutrient is an additional limiting factor. This emphasis on dietary energy as a measure of food adequacy is justified, since food energy derived from a habitual diet brings with it most other nutrients. Hence, increased dietary energy is a necessary condition for nutritional improvement, even if it is not always sufficient in itself.

Malnutrition also encompasses specific deficiencies of essential nutrients such as vitamins and minerals largely the result of diets based on the wrong types, quantities or proportions of foods. Goitre, scurvy, anaemia and xerophthalmia are forms of malnutrition, caused by inadequate intake of iodine, vitamin C, iron and vitamin A respectively. Conditions such as overweight and obesity, though not the result of inadequacy of food, are also included in the spectrum of malnutrition.

While it is possible to arrive at the prevalence and the numbers of individuals within a population manifesting signs of specific nutrient deficiencies (for instance anaemia), they are almost always associated with marginal or low food energy intakes. Thus 'malnutrition' is often used in the broader sense, referring to any physical condition implying ill health or the inability to maintain adequate growth, appropriate body weight and to sustain acceptable levels of economically necessary and socially desirable physical activities brought about by an inadequacy in food – both quantity and quality. Thus 'malnutrition' is generally used to characterize the nutritional problems of relatively poorer socioeconomic populations of developing countries.

The causes of malnutrition

To develop policies related to food and nutrition that affect the health of populations in developing countries, it is helpful to review the multidimensional factors that cause malnutrition. Its determinants include both food and non-food related factors, which often interact to form a complex web of biological, socioeconomic, cultural and environmental deprivations. Establishing a relationship between these variables, and the indicators of malnutrition do not necessarily imply causality, but they do demonstrate that in addition to food availability many social, cultural, health and environmental factors influence the prevalence of malnutrition. Although, people suffering from inadequacy of food are poor, not all the poor are undernourished. Even in households that are food secure, some members may be undernourished. Income fluctuations, seasonal disparities in food availability and demand for high levels of physical activity, and proximity and access to marketing facilities may singly or in combination influence the nutritional status of an individual or a household. For example, the transition from subsistence farming to commercial agriculture and cash crops may help improve nutrition in the long run, however, they may result in negative impacts over the short term unless accompanied by improvements in access to health services, environmental sanitation and other social investments. Rapid urbanization and rural to urban migration may lead to nutritional deprivation of segments of society. Cultural attitudes reflected in food preferences and food preparation practices, and women's time constraints including that available for child-rearing practices, influence the nutrition of the most vulnerable in societies. Inadequate housing and overcrowding, poor sanitation and lack of access to a protected water supply, through their links with infectious diseases and infestations, are potent environmental factors that influence biological food utilization and nutrition. Inadequate access to food, limited access to healthcare and a clean environment and insufficient access to educational opportunities are, in turn, determined by the economic and institutional structures as well as the political and governance superstructures within society.

Poor nutritional status affects physical growth, cognitive development and intelligence, behaviour and learning abilities of children and adolescents. It impacts on their physical and work performance and has been linked to impaired economic work productivity during adulthood. Inadequate nutrition predisposes them to infections and contributes to the negative downward spiral of malnutrition and infection. Good nutritional status, on the other hand, promotes optimal growth and development, contributes to better work performance, enhances adult economic productivity, increases levels of socially desirable activities and promotes better maternal birth outcomes. Good nutrition of a population reflects on the nutritional status of the individual in the community and contributes to an upward positive spiral that reflects improvement in the resources and human capital of societies.

Economic growth and prosperity can also contribute to the problem of malnutrition in a population by creating conditions that are conducive to the development of chronic diseases of adulthood which include heart disease, diabetes and cancer. Urbanization which accompanies economic development of societies, alters several environmental factors including the diet and lifestyles of individuals. While economic prosperity helps attain adequacy of food in quantitative terms for much of the population, it is, however, accompanied by qualitative changes in the diet with increased dietary energy from fat replacing the carbohydrates from staples or cereals. There is an increase in the consumption of animal source foods which has other ecological consequences. Consumption of salt and sugars also increases. Lifestyle changes, particularly with relation to the level of occupational and leisure time activities, predisposes to an increasingly sedentary lifestyle which consequently leads to the occurrence of overweight and obesity. These developmental changes in largely rural societies break down social support systems and networks, favour inequalities and increase stress levels of individuals. In addition, the deterioration of the physical environment, particularly the increase in levels of environmental pollution, contributes to the health burden of these developing economies.

Policies and interventions to promote good nutrition in developing economies

A proper understanding of the range, complexity and interplay of the factors that sustain the problem of malnutrition in developing societies is essential to help develop policies and interventions that meet the nutritional needs of populations and to reduce the burden of all forms of malnutrition.

Improving household food security is one of the stated objectives of all democratic societies and constitutes an important element of the human right to adequate food now endorsed by the international community. *Food security* is defined as the access by all people at all times to the food they need for an active and healthy life. The inclusion of the term household ensures that the dietary needs of all the members of the household are met throughout the year. The achievement of household food security requires an adequate supply of food to all members of the household, ensuring stability of supply all year round, and the access, both physical and economic, which underlines the importance of the entitlement to produce and procure food. The links between food security and malnutrition are evident as nutritional status is the outcome indicator. The presence of undernutrition is not only causally related to food insecurity at the household or at the individual level ensuring fair and appropriate intra-household food distribution, but is also determined by other health-related factors such as access to safe water, good sanitation and health care and the care practices that include proper breastfeeding and complementary feeding.

The pre-eminent determinant of household food insecurity is poverty in societies. Several policy measures undertaken by governments in developing countries are aimed at ensuring food supply and household food security. These include the following:

- Macroeconomic policies and economic development strategies that ensure both public-sector and private-sector investment in agriculture and food production.
- Appropriate policies to promote expansion and diversification of food availability and agricultural production in a stable and sustainable manner, and to regulate the import or export of foods and agricultural products to ensure food security.

- Policies that help create adequate employment opportunities for the rural poor and improve market efficiencies and opportunities.
- Policies that improve distribution and access to land, and to other resources such as credit as well as other agricultural inputs.
- Legislating for policies that deter discrimination and ensure equal status for women, and ensuring their effective implementation.
- Identification of good and culturally appropriate caring practices and policies that protect, support and promote good care and nutrition practices for children.
- Policies that enable public health measures to reduce the burden of infectious diseases and to ensure access to primary health care.

The two complementary approaches to addressing nutritional problems are direct nutrition-specific interventions and broader multi-sectoral approaches. Several direct nutritional intervention programmes have been initiated in developing countries to improve the nutritional situation. These interventions have immediate or short-term goals and the strategies used include the following:

- *Supplementation* of food or specific nutrients to meet the immediate deficits. Examples include food supplementation during acute food shortages such as famines or disasters and the mandatory provision of iron and folate supplements to all pregnant mothers attending antenatal clinics in primary healthcare centres.
- *Fortification* of food items in the daily diet is another successful strategy that has been adopted to deal with specific nutritional problems or nutrient deficiencies. A good example is the fortification of common salt with iodine (iodised salt) to address iodine deficiency and goitres; one of the most successful strategies that has helped reduce the burden of iodine deficiency globally.
- *Food-based approaches* include attempts to improve the nutrition of households by promoting kitchen gardens to enable families to produce and consume a diversified diet rich in vitamins and minerals. This has been promoted to reduce vitamin A and iron deficiency in developing countries.

The *Scaling Up Nutrition (SUN)* initiative led by the World Bank is an example of a multi-sectoral approach which aims to scale up evidence-based cost-effective interventions that includes integrating nutrition in related sectors and using indicators of undernutrition as the key measures of overall progress in these sectors at the country level.

Preventing malnutrition is just as important as solving the problem, and both goals require the need to assess the severity of the problem as well as the ability to predict its occurrence to make prevention a realistic goal. *Nutritional surveillance* plays an important role in assembling information to assist the development of policy and programme decisions and involves the regular and timely collection of nutrition-relevant data, as well as its analysis and dissemination. There are obvious advantages in the collection of nutrition-relevant information since the nutrition of a population is the outcome of social, economic and other factors, and is hence a good indicator of the overall development, and often a better indicator of the equitable development of societies. Nutritional surveillance systems may vary depending on the immediate objective for which they have been set up and some developing countries have more than one system in place. National governments of developing countries are also involved in *improving food quality* and *ensuring food safety*. Most countries have legislation to help ensure safety and quality of foods from production to retail sale. They also have established institutions or ministries to oversee food safety and quality.

There is wide consensus that strong economic growth has potential to result in significant improvements of nutrition. However, economic growth must be both 'nutrition-sensitive' and 'pro-poor'. Nutrition sensitivity denotes a comprehensive concept which aims at improving all dimensions of food and nutrition security and evolved out of the experience that agricultural development is essential, but insufficient in itself to improve nutrition. In order to be 'pro-poor', the process of growth must benefit disproportionately the low-income groups which in low-income countries is best achieved through agriculture-based growth, because agriculture is still the main direct or indirect source of livelihood for the majority of those at risk of poverty and hunger. There is a need for well-conceived policies for sustainable economic growth and social development that will benefit the poor and the undernourished. Development strategies need to be nutrition-sensitive by combining pro-poor investment in agriculture with social investments in health and nutrition, in particular in education, research and broad investment in rural infrastructure. The deleterious consequences of rapid growth and development need to be guarded against and policies need to be in place to prevent one problem of malnutrition replacing another in these societies. Given the complexity of factors that determine malnutrition of all forms, it is important that appropriate food and agricultural policies are developed to ensure household food security and that nutritional objectives are incorporated into development policies and programmes.

Further reading

ACC/SCN (2000) *Nutrition Through the Life Cycle*, Fourth Report on the World Nutrition Situation, Geneva: UN Sub-Committee on Nutrition.

Berg, A. (1987) *Malnutrition: What Can Be Done?* Baltimore, MD: Johns Hopkins University Press.

Brun, T. A. and Latham, M. C. (1990) *Maldevelopment and Malnutrition*, World Food Issues 2, Ithica: Cornell University.

FAO (1993) *Developing National Plans of Action for Nutrition*, Rome: Food and Agriculture Organization.

Gopalan, C. (1992) *Nutrition in Developmental Transition in South-East Asia*, New Delhi: World Health Organisation, SEARO.

Gopalan, C. and Kaur, H. (1993) *Towards Better Nutrition: Problems and Policies*, New Delhi: Nutrition Foundation of India.

Headey, D. (2011) *Turning Economic Growth into Nutrition-Sensitive Growth*, IFPRI 2020 Conference, Paper 6.

Latham, M. C. (1997) *Human Nutrition in the Developing World*, Rome: Food and Agriculture Organization.

Sen, A. (1981) *Poverty and Famines: An Essay on Entitlement and Deprivation*, Oxford: Clarendon Press.

Shetty, P. S. (2006) *The Boyd Orr Lecture: Achieving the Goal of Halving Global Hunger by 2015*, Proceedings of the Nutrition Society, 65: 7–18.

World Bank (1994) *Enriching Lives: Overcoming Vitamin and Mineral Malnutrition in Developing Countries*, Washington, DC: World Bank.

World Bank (2010) *Scaling up Nutrition: A Framework for Action*, Washington, DC: World Bank.

8.2

Motherhood, mortality and health care

Maya Unnithan-Kumar

Most resource-poor countries in South Asia and sub-Saharan Africa remain high-mortality settings particularly marked out by maternal and infant deaths.[1] Out of the 1000 women who die every day due to maternal complications, 570 live in sub-Saharan Africa, 300 in South Asia, as compared to five in high-income countries (Berer, 2010). Poverty, along with skewed gender ideologies, poor quality of health-care provision and negligibly supported health-seeking behaviour are the significant drivers of high mortality. The main underlying epidemiological conditions of maternal mortality,[2] such as high levels of anaemia or reproductive tract infections, sepsis and hypertensive disorders[3] are strongly correlated with the lack of adequate food, nutrition, absence of clean water and good sanitation, respectively. Other major causes of maternal death such as postpartum haemorrhage as well as infant death are connected with the kind of care received at the time of birthing and here the role of birth assistants and the public health infrastructure is critical.

What the epidemiological conditions underlying the mortality statistics reveal, however, is how preventable maternal and infant deaths are in terms of available drugs and medical procedures. Why then does becoming a mother and the period of infancy remain marked by the risk of survival? This chapter will situate an understanding of this question in the social and cultural contexts which frame motherhood as well as issues of 'structural violence' (Farmer, 2004).[4] Using the lens of structural violence enables an understanding of how systemic inequalities (such as those of poverty, skewed gender ideologies and health-care policies and programmes) contribute to the persistence of unjust conditions in which motherhood and infant health are experienced. We conclude with a consideration of the role of human rights frameworks in addressing maternal mortality.

When motherhood is precarious

> *Munga lived in a village on the outskirts of Jaipur, the capital city of Rajasthan in NW India. She was 20 years old and expecting her third child. She already had a daughter and a son. When her labour pains started, she called for Samina (her father's younger brother's wife) and Munaan an elderly relative, both known in the local area for their expertise in birthing babies. A boy was born at 6 p.m. that day. Samina and Munaan left shortly after this. Munga continued to bleed profusely and twice vomited the milk she was given to drink. Six hours later, at 1 a.m., she died.*
>
> *(Unnithan-Kumar, fieldwork notes, 2002)*

Why did Munga die? It could have been for any one, or all of the following reasons. Munga's parents were poor and she had little nutrition as a child and later suffered from tuberculosis.

When she married, her husband earned enough but never gave her any money to consult with doctors or pay for medicines, or arrange for transport to the government centres, which were far away, and, in any case, it was uncertain that the doctors would be there. Her husband may have also been reluctant to give her time off from her household and child-caring duties for any antenatal check-ups. Munga relied on the local midwives for any problems during pregnancy. The midwives, competent in delivering babies in normal circumstances, were anxious as Munga was weak and could not push hard enough during her contractions. As her labour was felt to be too extended in duration, family members decided she should receive an injection by a local doctor to speed up the birth (medically the unassessed provision of intra-muscular oxytocin greatly enhances the risk of rupture to the uterus). For Munga, like for a number of other young women in her village, childbearing and birth carried with it the potential of a serious risk to one's life.

Given the poverty-related nutritional deficiencies, poor access to emergency care, scarcity of water for washing, and sheer physical work that frame the contexts of their continuous biological reproduction, it is not surprising that childbearing is a physically debilitating and potentially life-threatening period for rural and urban poor women in India as described above. And yet, given the social pressure to bear children and the defining role that birth has in conferring a whole range of social privileges on women, from defining their position as adults to ensuring access to financial and economic resources, it is also a life-stage event that few women can deny or put off. There is often very little choice in undertaking motherhood even if it is associated with medical risk.

Understanding the persistence of maternal mortality

A study on women's access to emergency obstetric care in Koppal, the poorest district in the state of Karnataka, India, identifies 'systemic bias' in the routine delivery of health care to be the main cause of the persistence of maternal mortality in the region (George, 2007). The mechanisms which underlie such bias include: (a) weak information systems – where there is no record or proper review of maternal death; (b) discontinuity in care – where diagnoses takes place but there is no follow up; (c) unsupported health workers – where a lack of knowledge and training results in poor technical judgement and irrational care; (d) haphazard referral systems – where referral information is not exact or supported; and (e) distorted accountability mechanisms – with blame and defensive practices among health providers making it difficult to identify responsibilities among these health care professionals.[5]

These systemic failures become evident in the case of local health workers such as midwives. The presence of skilled midwives has been critical in rural settings, where most births take place at home rather than in institutional settings. However, as studies have shown most women and families only seek institutional assistance in crisis conditions. In addition, public health workers at the village level are ill equipped to deal with emergencies, which, in turn, curtail their importance as safe and reliable health providers in local perceptions.

Health-care workers often suffer from a lack of institutional infrastructural and technical support, which, if addressed, would enhance their engagement and rapport with the community. Where midwives are not trained or equipped to deal with obstetric complications, referral becomes a vital mechanism to ensure the survival and health of the birthing mother and infant, and yet one which often fails in practice as the study in Koppal demonstrates. In the absence of good and timely referral services, it is often most effective for midwives to work in close contact with obstetric and gynaecological professionals at the local health centre itself, where emergency care is most needed (Mavlankar *et al.*, 1999; Iyengar and Iyengar, 2009).

The provision of a high quality of maternal health care is related to an understanding of what constitutes good quality of care, on the one hand, and a commitment on the part of the state to invest resources towards this end. In their framework on the quality of maternal care, Hulton *et al.* (1999)[6] suggest two basic levels at which 'quality' must be addressed: first, at the level of the provision of care the focus should be on elements of care such as the use and quality of the personnel employed and the infrastructure available to provide the essential obstetric care as laid down by WHO regulations, an efficient referral system, universal emergency provisioning and the use of appropriate technologies and an awareness of internationally recognised good practice. The second level must address the experience of care where the important elements include dealing with the patient's cognition of the services, the respect received and emotional support during care.

Some of the important effective interventions which address the risks associated with motherhood are considered to be the distribution of contraceptives to space pregnancies and the provision of iron, folate and malarial prophylaxes to strengthen women and enhance their immunity. The focus on contraceptives are considered important in public health frameworks as they address the potential problems associated with closely spaced deliveries and an early initiation into childbearing (see, for example, McDonagh, 1996, for East Africa).

However, empirical observations of health-seeking behaviour in sub-Saharan Africa and South Asia show that the provision of contraceptive devices and nutritional supplements will not in themselves ensure a reduction in maternal morbidity (e.g. Jejeebhoy and Rao, 1998; Patel, 1994). There are both cultural reasons for this (gender ideologies make it difficult to discuss sexual matters across the sexes; the association of pills and injections with power and heat which threaten childbearing; the cultural expectations and value placed on women's sexual and reproductive availability to men, women's subservience to men in decisions relating to childbearing) as well as physiological reasons (the low cost contraceptive technologies available are not sustainable by weak, anaemic and nutritionally deficient women).

Anthropological studies in the Gambia and Cameroon show how women's perceptions of childbirth as associated with reproductive ageing or individual women's aspirations for education, respectively, are shaping their engagement with contraceptive technologies (Bledsoe, 2002; Johnson-Hanks, 2006). In other contexts where the high levels of reproductive tract infections and related sterility are linked to local anxieties to do with the ability to conceive, the supply of contraceptives is misplaced as it ignores more pressing local demands for infertility treatment. There is thus a disjunction between the contraceptive-oriented health services routinely provided through public health services in high population settings and the local demands for assistance in conception as well as for birthing assistance to be locally provided. The quality of health-care provision is a crucial factor to consider in this context as is the wider issue of how the contours of maternal and child well-being are shaped by the framing of the health systems themselves.

Biopolitics of the state

Maternal and child health programmes in resource-poor contexts in particular have been shaped by population-centred development ideologies, and in high population settings continue to focus on family planning programmes, especially with regard to fertility control.[7] The quality of public health service provision is significantly linked to the state financial allocations for the health sector. The economic reforms such as those brought in by structural adjustment programmes or changing conditions of aid packages place severe constraints on the allocation of public funds for health in these countries. The dependence on the vagaries of external funding have seen a deceleration in the real capita revenue and capital expenditure on health since the mid-1980s in India, for example (Seeta Prabhu, 1999: 121), which accorded with the World Bank's recommendations to confine the role of the public sector to providing preventive rather than curative health services.

However, the more recent global turn to rights-based models of development from the late 1990s has meant that national governments are signatories to a whole set of conventions (on reproductive rights and child rights such as CEDAW) which make them accountable to promoting and delivering people-centred rather than population-centred health programmes. The commitment to the Millennium Development Goals has further especially placed a direct requirement on states to address maternal and infant mortality.

Concluding comments: Motherhood and human rights

In the absence of appropriate obstetric and gynaecological expertise at the local level of the health-service structure, and where there exist exacting cultural expectations which devalue the health risks of motherhood, the reproductive phase will continue to be a period in which poor women and new-born infants in resource-poor as well as highly populated middle-income countries face high health risks. More generally, a rights-based commitment to good health care needs to develop within the service sector and in the community so that women receive adequate support and understanding during, before and after their childbearing years (UNFPA, 2008; Unnithan-Kumar, 2003).

The significance of the recent human rights-based approach to health lies in the connection it makes between the existing conditions of structural violence and mechanisms of accountability. The persistence of maternal mortality, such an approach argues, demonstrates a *failure to value* women's lives. Women have a *right to survive but also the right to experience* pregnancy and childbirth as safe life-course events. Accordingly, governments have an *obligation to ensure* the provision of pregnancy related care, no matter who the women are, where they live or what their economic status (International Maternal Mortality Initiative, 2010).

Notes

1 Maternal mortality figures remain high at 358,000 worldwide, despite the overall decline of 34 per cent in the number of women dying from pregnancy, childbirth and unsafe abortion, as noted in the 2008 interagency report by WHO, UNICEF, UNFPA and the World Bank (Berer, 2010). Some reasons for explaining this decline in countries such as India have been attributed to the introduction of policies such as the cash incentive scheme within the National Rural Health Mission, 2005, which encourage childbirth within health-care institutions.
2 Maternal death 'is defined as the death of a woman while pregnant or within 42 days of termination of pregnancy, irrespective of the duration and site of pregnancy, from any cause related to or aggravated by the pregnancy or its management but not from accidental or incidental causes'. (International classification of diseases, 10th revision, http://apps.who.int/classifications/icd10/browse/2010/en).
3 UNFPA, 2008. Also referred to as morbidity factors as they are life-threatening conditions which have the potential to result in death.
4 Farmer uses the concept to refer specifically to the ways in which violence or inequality is embedded in policies and programmes themselves.

5 The systemic approach draws on but moves beyond the three-delay model of Thaddeus and Maine, 1994.
6 They define the quality of care as, 'the degree to which maternal health services for individuals and populations increase the likelihood of timely and appropriate treatment for the purpose of achieving desired outcomes that are both consistent with current professional knowledge and uphold basic reproductive rights' (Hulton *et al.*, 1999: 4).
7 This continues to be the case in India despite sweeping health-sector reforms and rights-based health programmes such as the National Rural Health Mission.

Bibliography

Berer, M., 2010, Round up: Maternal health and maternal mortality. *Reproductive Health Matters* 18(36): 197–205.

Bledsoe, C., 2002, *Contingent Lives: Fertility, Time and Aging in West Africa*. Chicago: University of Chicago Press.

Farmer, Paul, 2004, An anthropology of structural violence. *Current Anthropology* 45(3): 305–325.

George, A., 2007, Persistence of high maternal mortality in Koppal District, Karnataka, India: Observed service delivery constraints. *Reproductive Health Matters* 15(30): 91–102.

Hulton, L., Matthews, Z. and Stones, R., 1999, *A framework for the evaluation of the quality of care in maternity services*. Southampton, UK: University of Southampton: Department of Social Statistics publication.

International Maternal Mortality Initiative, 2010, Human rights based approaches to maternal mortality reduction efforts. Available online at www.righttomaternalhealth.org

Iyengar, K. and Iyengar, S., 2009, Emergency obstetric care and referral: Experience of two midwife-led health centres in rural Rajasthan, India. *Reproductive Health Matters* 17(33): 9–20.

Jejeebhoy, S. and Rama Rao, S., 1998, Unsafe motherhood: A review of reproductive health, in M. Das Gupta, L. Chen and T. N. Krishnan (eds), *Women's Health in India: Risk and Vulnerability*. Oxford: Oxford University Press.

Johnson-Hanks, J., 2006, *Uncertain Honour: Modern Motherhood in an African Crisis*. Chicago: University of Chicago Press.

Mavlankar, D., Bang, A. and Bang, R., 1999, Quality reproductive health services in rural India, in *Services on Upscaling Innovations in Reproductive Health in Asia, No.2: India*. Selangor, Malaysia: The International Council on the Management of Population Programmes.

McDonagh, M., 1996, Is antenatal care effective in reducing maternal morbidity and mortality? *Health, Policy and Planning* 11(1): 1–15.

Patel, T., 1994, *Fertility Behaviour: Population and Society in Rajasthan*. Delhi: Oxford University Press.

Seeta Prabhu, K., 1999, Structural adjustment and the health sector in India, in Mohan Rao (ed.), *Disinvesting in Health* (pp. 120–129). New Delhi: Sage.

Thaddeus, S. and Maine, D., 1994, Too far to walk: Maternal mortality in context. *Soc. Sci. Med.* 38(8): 1091–1110.

UNFPA, 2008, *State of the World Population 2008: Reaching Common Ground- Culture, Gender and Human Rights*. New York: UNFPA.

Unnithan-Kumar, M., 2001, Emotion, agency and access to healthcare: Women's experiences of reproduction in Jaipur, in S. Tremayne (ed.), *Managing Reproductive Life*. Oxford: Berghahn.

Unnithan-Kumar, M., 2003, Reproduction, health, rights: Connections and disconnections, in J. Mitchell and R. Wilson (eds), *Human Rights in Global Perspective: Anthropology of Rights, Claims and Entitlements*. London: Routledge.

Unnithan-Kumar, M., 2010, Learning from infertility, in A. Doron and A. Broom (eds), *South Asian History and Culture* (special issue) 1(2): 315–328.

8.3
The development impacts of HIV/AIDS

Lora Sabin, Marion McNabb, and Mary Bachman DeSilva

HIV and AIDS

The human immunodeficiency virus (HIV) is a virus that attacks the human immune system. Once infected, a person's immune system deteriorates over a number of years, until it cannot resist infections. At this stage, clinically defined by a low CD4 cell count (below 200 cells/μL) or an illness the immune system cannot fight, the person has AIDS (acquired immunodeficiency syndrome). HIV is transmitted through the exchange of body fluids, including semen, vaginal secretions, blood and breast milk. Worldwide, most people contract HIV from sexual contact, although needle-sharing, blood transfusions and breastfeeding are responsible for many infections.

Without treatment, people living with HIV/AIDS (PLWHA) typically experience numerous symptoms and illnesses, including weight loss, declining energy and infections such as pneumonia, tuberculosis and meningitis. PLWHA also often experience stigma and the pain of leaving children and parents in need of economic and caregiving support.

Since the late 1980s, antiretroviral therapy (ART) has been used to boost the immune system and suppress the HIV virus. ART is now available in most countries through support from a wide range of international and domestic organizations, with PLWHA generally able to live 10 years or more after starting therapy. While some 22,000 facilities worldwide provided ART to 14.2 million people, 47 per cent of those in need, inadequate human resources, drug shortages and insufficient facilities continue to limit treatment access in low resource settings, where the epidemic is concentrated (WHO 2011).

Global trends in HIV and AIDS: The view from 2012

Worldwide, total HIV infections continue to rise, though fewer PLWHA die annually due to ART. In 2010, there were an estimated 34 million PLWHA globally, compared to 28.6 million in 2001. HIV/AIDS-related deaths numbered 1.8 million, down from the 2002 peak of 2 million; 2.7 million individuals were newly infected, continuing the downward trend that began in 2001 (3.1 million new infections). However, trends vary by region and country due to uneven progress in prevention and treatment efforts. Between 2002 and 2010, the number of new infections rose in the Middle East and North Africa, while AIDS-related deaths increased tenfold in Eastern Europe and Central Asia and more than doubled in East Asia (WHO 2011).

Sub-Saharan Africa (SSA) still bears the greatest HIV/AIDS burden, with 68 per cent of all PLWHA residing in a region containing 12 per cent of the global population. In 2010,

SSA was home to 22.9 million PLWHA, and new infections (1.9 million) accounted for 70 per cent of the global total. In 2009, South Africa had the largest number of PLWHA of any country (5.6 million). One in four adults lived with HIV in Botswana, Swaziland and Lesotho (WHO 2011; UNAIDS 2010).

Worldwide, the proportion of infections in women continues to rise. In 2010, 50 per cent of HIV infections were among women aged 15 years and older, compared to 42 per cent in 1996. However, the burden of HIV on women varies by region. In SSA, six of every ten infected people are women (WHO 2011).

HIV/AIDS and development

HIV/AIDS has particularly devastating impacts on development because it attacks society's most productive members in the prime of their lives. In countries with a generalized epidemic, many individuals are at risk of infection. Other countries have concentrated epidemics, meaning that most PLWHA are among most-at-risk populations (MARP), including injection drug users, sex workers and men who have sex with men. HIV/AIDS can change population structures, decrease the supply and productivity of labour and reduce household and business incomes. At the household, community and national levels, AIDS shifts resources from productive uses, such as investment in education and preventive health care, to care and treatment for PLWHA. As children are subjected to economic, social and psychosocial impacts, HIV/AIDS can reduce a society's future development potential.

Different impacts occur separately or simultaneously (Figure 8.3.1). As individuals perform various social functions, each role/function is affected by an illness or death. PLWHA may be less engaged as family or community members, contribute less labour and change

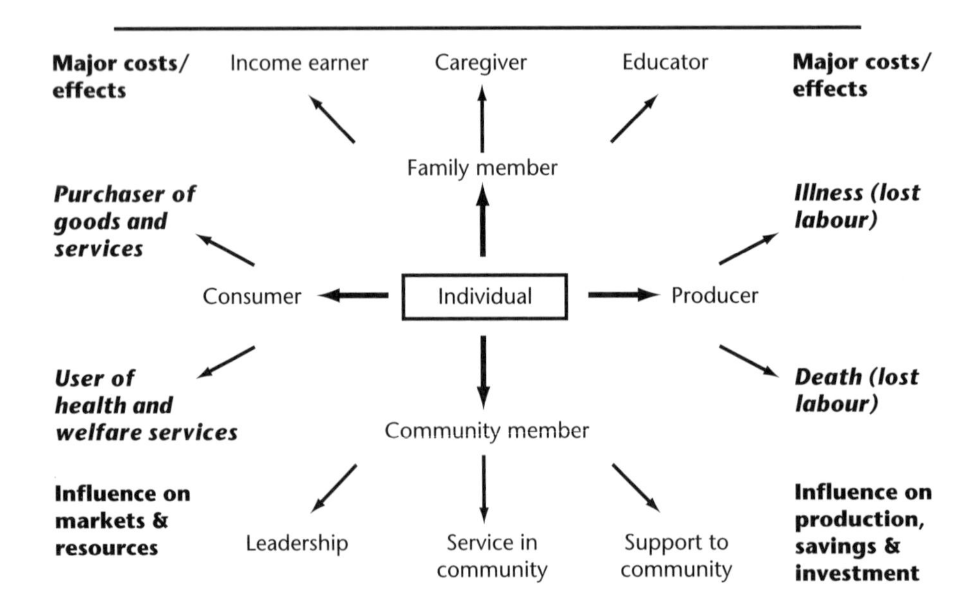

Figure 8.3.1 The radiating effects of HIV/AIDS

Source: Adapted from Barnett & Whiteside, 2002, p. 184

consumption patterns. As this scenario is multiplied throughout a community, country or region, the macro-level impacts may be severe.

The timing of development impacts

Development impacts may be felt as immediate and severe shocks or as complex, gradual, and long-term changes. Impacts emerge as total cumulative infections rise and AIDS deaths occur, and may linger long afterward (Figure 8.3.2).

Challenges with information and measurement

Nearly three decades into the epidemic, we are still studying the myriad ways in which HIV/AIDS impacts society. Compared to before the epidemic, national monitoring systems are far more capable of collecting key HIV/AIDS indicators, including the numbers of people tested for HIV, those infected, pregnant women who receive ART to prevent transmission to their children, and PLWHA on first and second line treatment. Although international recommendations can assist countries in designing HIV data collection and reporting systems as well as improving overall health surveillance, data evaluating prevention, treatment and care programs remain limited, underscoring the need for rigorous evaluations of program impact (Padian *et al.* 2011; Bryant *et al.* 2012).

Impacts at different levels: What we have learned

There is a growing body of evidence on the impact of HIV/AIDS at various levels, from children and households up to communities and countries. Impacts are complex and can vary considerably by geographic and socioeconomic context. Below is a brief summary of some established impacts.

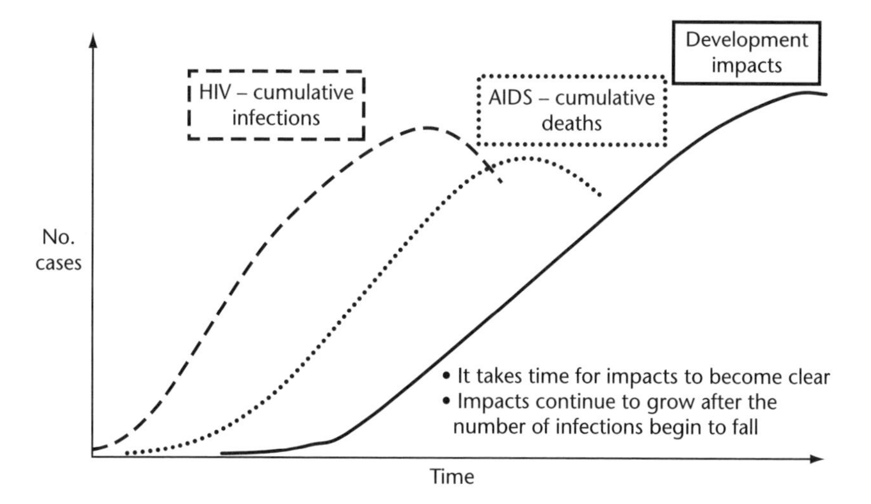

Figure 8.3.2 The timing of impacts

Source: UNAIDS, 2002, report on the global HIV/AIDS epidemic, UNAIDS, Geneva

Households

In many households, the death of the main breadwinner plunges the family into poverty and leaves grandparents to care for children. Where health care services are limited, care for the sick usually takes place at home, with responsibilities falling disproportionately on women. Consequently, grandparents and children absorb the greatest HIV/AIDS burden, earning an income and holding the family together, often while coping with social, psychosocial and health problems. To survive, families report borrowing money, selling assets, sending children to work, and reducing spending on food. Compromised food security is common in affected regions, potentially exacerbating the long-lasting health and productivity impacts of AIDS. In response, community-based organizations and programs now provide a variety of services for families, caregivers and PLWHA.

Children infected and affected by HIV/AIDS

At the end of 2010, 3.4 million children below 15 years of age were living with HIV. Each day over 1,000 children worldwide were infected, and more than one-half died from lack of ART. Over 16 million (15 million in SSA) were estimated to have lost one or both parents to AIDS (UNAIDS 2010). Millions more became vulnerable as a result of death or illness in their families. Other documented impacts include: over 10 per cent of children have been orphaned due to AIDS in Botswana, Kenya, Lesotho, Malawi, Uganda, Swaziland, Zambia and Zimbabwe (UNAIDS 2010); throughout SSA, households with orphans tend to have fewer assets, higher dependency ratios and thus fewer adults to care for children, and be headed by women and older persons compared to households without orphans (Bachman DeSilva *et al.* 2008); orphaned children are less likely to attend school than non-orphaned children (UNICEF 2006; Evans and Miguel 2007); orphans generally have elevated rates of depression and post-traumatic stress disorder compared to non-orphans, although these psychosocial effects may be mediated by poverty, gender, stigma and social support (Cluver and Gardner 2007).

Communities

In the hardest-hit countries, the social fabric that once knitted communities together has unraveled due to heavy AIDS-related economic and caregiving burdens. However, over the last few years, communities have taken on greater ownership of the local response, which also helps reduce stigma related to HIV, while providing important support for PLWHA. Local-level activities range from organizing support groups for PLWHA, to educating the public about HIV transmission, and conducting outreach testing campaigns (WHO 2011).

Sectors

HIV/AIDS affects businesses and the public sector through direct and indirect costs, particularly when HIV prevalence is high. Businesses and government offices may experience reduced productivity, increased absenteeism, and vacancies; they may also be compelled to pay for medical care. Despite these challenges, private and public sector partnerships are emerging as a source of support for workers, with HIV testing and treatment programs now offered at many factories, farms and other businesses.

National impacts and strategies

The leaders of many countries have acknowledged the threat that HIV/AIDS poses to their nations' development. Perhaps most notably, HIV/AIDS has caused demographic changes, lowering life expectancy by decades in southern Africa, negating fifty years of gains in some countries. Numerous nations realize the importance of fighting HIV and have committed funding, resources and revisions of national policies to address epidemics.

Global response

By the twenty-first century, the world community had developed international goals and targets to respond to the epidemic. The Millennium Development Declaration (2000) established goals to combat HIV/AIDS. The UN Declaration of Commitment on HIV/AIDS (2001) set targets for implementing policies and strategies, which were further articulated by The Political Declaration drafted at the 2006 United Nations General Assembly Special Session on HIV/AIDS. Funding for HIV/AIDS programs increased dramatically to US$6.9 billion in 2010 from US$300 million in 1996, largely from the Global Fund for AIDS, Tuberculosis, and Malaria, PEPFAR (US President's Emergency Plan for AIDS Relief), the Clinton Foundation, the Bill and Melinda Gates Foundation, and other international, bilateral, and non-government organizations (NGOs) (Henry J. Kaiser Foundation and UNAIDS 2011). The private sector, NGOs, and a broad spectrum of faith- and community-based groups have also contributed to the response.

In the early 2010s, the global focus shifted to increasing local efficiency and sustainability. In June 2010, UNAIDS announced 'Treatment 2.0', an initiative building on WHO's previous '3 by 5' plan (3 million people on ART by 2005), emphasizing several pillars of HIV prevention and care, including increased access to drugs and improved diagnostics, service delivery and community mobilization (WHO 2011). Additionally, PEPFAR began concentrating on long-term sustainability bolstered by partnership frameworks and local responsibility. At the community level, numerous programs offer a wide range of support to hospitals, clinics, households and individuals. Most focus on increasing uptake of prevention and treatment; others seek to increase wellbeing by improving food production, nutrition and shelter, particularly for orphans and vulnerable children.

Remaining challenges

The need for effective prevention, care and treatment continues to outstrip available resources, even as many international donors seek to shift responsibility to local actors. National and community NGOs bear the brunt of the pressure to support the most vulnerable, though their efforts often lack consistency, funding and an evidence base to guide activities. Major obstacles to an effective HIV/AIDS response persist, including uncoordinated responses; disagreement over best approaches and priorities for available resources; poor information regarding effective approaches to mitigating impacts; and HIV/AIDS-related stigma, which undermines testing, prevention, treatment and impact mitigation efforts.

As the global community transitions to a long-term and sustainable response to the epidemic, a continued global, national, and local commitment will be essential to ensure that support for prevention, treatment, and care reaches those most in need.

References

Bachman DeSilva, M., Beard, J., Cakwe, M., McCoy, K., Nkosi, B., Parikh, A., Quinlan, T., Skalicky, A., Tshabangu-Soko, S., Zhuwau, T. and Simon, J. 2008. Vulnerability of orphan caregivers vs. non-orphan caregivers in KwaZulu-Natal. *Vulnerable Children and Youth Studies* 3(2): 102–111.

Bryant, M., Beard, J., Sabin, L., Scott, N., Larson, B., Miller, C., Biemba, G., Brooks, M. and Simon, J. 2012. PEPFAR's support for orphans and vulnerable children: Some beneficial effects, but too little data, and programs spread thin. *Health Affairs* July.

Cluver, L. and Gardner, F. 2007. The mental health of children orphaned by AIDS: A review of international and southern African research. *Journal of Child & Adolescent Mental Health* 19(1): 1–17.

Evans, D. K. and Miguel, E. 2007. Orphans and schooling in Africa: A longitudinal analysis. *Demography* 44(1): 35–57.

Henry J. Kaiser Foundation and UNAIDS. 2011. Financing the response to AIDS in low- and middle-income countries: International assistance from donor governments in 2010. Available online at http://www.unaids.org/en/media/unaids/contentassets/documents/document/2011/08/20110816_Report_Financing_the_Response_to_AIDS.pdf

Padian, N. S., McCoy, S. I., Manian, S., Wilson, D., Schwartländer, B. and Bertozzi, S. M. 2011. Evaluation of large-scale combination HIV prevention programs: Essential issues. *J. Acquir. Immune Defic. Syndr.* 58(2): e23–28.

UNAIDS. 2010. Global Report: UNAIDS report on the global AIDS epidemic, 2010.

UNICEF. 2006. Africa's orphaned and vulnerable generations: Children affected by AIDS.

World Health Organization (WHO). 2011. Global HIV/AIDS response: Epidemic update and health sector progress towards universal access: progress report 2011.

Further reading

Mishra, V. and Bignami-Van Assche, S. 2009. Orphans and vulnerable children in high HIV-prevalence countries in sub-Saharan Africa. *DHS Analytical Studies* 15.

For updates on the HIV/AIDS epidemic, see:

Joint United Nations Programme on HIV/AIDS (UNAIDS) and World Health Organization (WHO), Global HIV/AIDS Response: Epidemic update and health sector progress towards Universal Access (2011). Available online at www.unaids.org

The Global Fund website: www.theglobalfund.org/en/

The PEPFAR website: www.pepfar.gov/

For guidance on HIV/AIDS-related monitoring and evaluation, see UNAIDS. 2012. Global AIDS response progress reporting: Guidelines, construction of core indicators for monitoring the 2011 political declaration on HIV/AIDS. Available online at www.unaids.org

For more on Treatment 2.0, see WHO, UNAIDS (2011), The Treatment 2.0 Framework for Action: Catalysing the next phase of treatment, care and support. Available online at www.who.int/hiv/pub/arv/treatment/en/index.html

Basic information on HIV/AIDS may be found at:

Essex, M., Mboup, S., Kanki, P. J., Marlink, R. G. and Tlou, S. D. 2002. *AIDS in Africa* (2nd Edition). New York: Kluwer Academic/Plenum Publishers.

AIDS Education and Research Trust. 2006. Available online at www.avert.org

Barnett, T. and Whiteside, A. 2006. *AIDS in the Twenty-First Century: Disease and Globalization* (2nd Edition). New York: Palgrave Macmillan.

Numerous World Bank publications, including: *Confronting AIDS: Public Priorities in a Global Epidemic.* New York and Oxford: Oxford University Press.

More specialized information is available through:

The International Labour Organization's publications, including: HIV/AIDS and work: Global estimates, impact on children and youth, and response (2006). Available online at www.ilo.org

The United Nation's Children's Funds publications, including: Africa's orphaned generations children affected by AIDS (2006). Available online at www.unicef.org/publications/index_35645.html?lpos=main&lid=unicef

8.4
Ageing and poverty

Vandana Desai

Development initiatives and population policies adopted in the 1960s and 1970s have helped to reduce fertility and mortality rates, with rising life expectancy due to improvements in health in many developing countries but mainly in the Asian region. As a consequence many countries are experiencing demographic transition from a young to an increasingly older population (Hussain *et al.* 2006; Dyson 2010). With this in mind it is important to understand how social and economic changes have affected or will affect the well-being and support situation of present or future older people, and how older people's needs and position in society are related to development issues and the consequences for policy. This has particular implications for developing countries that need to recognise ageing and development as part of poverty reduction strategies (see Desai and Tye 2009; Aboderin and Ferreira 2008).

Already two-thirds of the world's older population live in developing countries, with the absolute numbers of older people in these regions estimated to double to reach some 900 million within 25 years (Harper 2006: 1). From 1950 to 2005, the number of older people worldwide grew from 205 million (8% of the total global population) to 672 million (10%). By 2050 it is expected to reach almost 2 billion, representing 21 per cent of the world's total population (UNPD 2006). Population ageing mainly means the progressive increase in the numbers and thus the proportions, of older people, compared to working age adults and children in a total population. The UN definition of 'old age' is 60+ years. This definition is quite useful for comparing internationally, but poses questions on its appropriateness when applied in different societies and culture (UNPD 2006). One of the major differences in the context of Asian developing countries is the pace of population ageing and the sheer numbers of older people in the developing world, which far outstrip those in the developed world. This in itself makes ageing a development issue.

The reality of the global demographic situation is that some developing countries will face 'unprecedented ageing' without the social and economic infrastructure or public institutions to support those frail and dependent elderly people who are unable to support themselves economically. Developing countries will contain the most 'frail dependent elders' in more vulnerable situations (Harper 2006: 30). This end of the spectrum contains the pool of those living without the support of a 'healthy retirement', but who survive on the bare minimum social insurance and patchy health services. The time has come for pro-active planning of appropriate institutional frameworks and policies. The challenge is to create those 'safety nets' that will protect the present generation over the coming decades, but also provide guidance, support and care for the ageing populations. Questions related to ageing populations have secured very little attention. The elderly face economic and resource deprivation but also face identity-based discrimination. These multiple sources of deprivation create deep-rooted social exclusion which cannot be addressed solely through conventional income poverty reduction programmes. Younger rather than older generations are employed in higher proportions in various sectors. This has also been associated with unemployment of older people and marginalisation of certain age groups in metropolitan areas.

The emerging global trade pattern, in which the capital and/or technology intensive goods and services constitute the fastest rising segment of exports (e.g. electronics, garments and IT), tends to reduce employment per unit of exports. Thousands of jobs have been created for younger skilled workers in big metropolitan cities like Mumbai, especially in call centres, the IT sector, tourism and the garments sector. For example, economic liberalisation in 1991 in India created large scale retrenchments from no-longer competitive industries (e.g. textiles and printing industries) that left older urban poor workers unemployed in the forties and fifties age groups, with no prospect of re-employment in the new emerging sectors. They are also likely to experience age discrimination (Vera-Sanso 2004). There is an increasingly urgent concern to create conditions for inclusive citizenship and participation of those ageing people in wider society. Honouring older people's dignity and rights has received very little general attention and is a matter of social justice, human rights and an extension of rights to social protection (Lloyd-Sherlock 2004).

Older people are particularly vulnerable to multidimensional poverty, ill-health and social exclusion, particularly in the context of diminishing employment opportunities and barely any social service and security provision in many of the Asian developing and transitional countries. Elders living in absolute poverty are a considerable challenge for many developing countries.

Decline of family support

There is an assumption by policymakers that the elderly live with and be cared for by their children or extended family (e.g. Vietnam and India), but with shifts in family structure caused by industrialisation, urbanisation, increasingly different forms of migration and the increasing number of women entering the labour market and education the support and security for elderly family members has declined. Traditional three-generational households (comprised of grandparents, parents and children) are decreasing in number in many Asian countries. The rates of elderly living in their own homes separate from children are increasing. There is widespread concern over the decline of the family, its stability and 'effectiveness' to cater for the needs of the elderly. Fertility decline also implies that older people will have fewer children to support them (see Glaser *et al.* 2006). This does not mean that all children may have supported their parents in the past (Aboderin 2005). Even if parents had a large number of children, it does not necessarily correlate with having more support in old age.

Older people can also find themselves in a vulnerable position – in extreme cases, as victims of abuse – when the balance of power in the mother-in-law/daughter-in-law relationship shifts towards the younger woman. There is a history of conflict between women in extended households, particularly where brides have traditionally moved in with their husbands' parents (very common in Asian countries), which can generate conflict and impact on the welfare of elderly women (i.e. the mother-in-law). Elderly women in such situations are marginalised and can consequently live on their own. In addition, 'the growing number of young married couples who live on their own means that care by the daughter-in-law is no longer automatic' (Wilson 2000: 120).

Widowhood

Different life expectancies for men and women, together with a tendency for women to marry partners older than themselves, mean that demographic ageing may be increasing the percentage of woman-headed households. The highest proportion of women-headed households occurs where the householder is aged 60 or over, pointing to the significance of widowhood in this context (Varley and Blasco 2000). Women who are unmarried (widowed, divorced, or separated) are at greater risk of economic difficulty and have to keep working to earn a living. The demographic and fertility transitions take on particular significance for women in the Asian region which marks gender imbalances arising from deeply entrenched forms of discrimination such as those arising from son-preferences in parts of East and South Asia.

Intergenerational support and employment

Most of the cost of the support for the elderly falls on the family, which also largely bears the costs of educating children. This burden on young families may prove unsustainable in the context of high unemployment or under-employment, if job creation fails to keep up with labour force increases, or where incomes are low and uncertain. A trend towards nuclear households, with the elderly living separately from their offspring, may also undermine the family support system, as may out-migration of younger workers from the rural areas, leaving behind the elderly and infirm (as is increasingly occurring in China).

The economic contribution of older people in some form of employment either for their own economic security or for the collective economy of the household is not well understood. Old-age dependency ratios are based on the questionable assumption that older people do not work, ignoring the one-in-five over 65s who remain in employment worldwide, higher in low-income countries (see Lloyd-Sherlock 2010). There is no doubt, though, that there are high rates of economic participation and diverse strategies being adopted by older men and women until late in their lives mainly because of the limited degree of formal social security coverage in regions like South Asia. The majority of older workers – particularly those in the private and informal sector – continue to work until physical disability or sickness prevents them from further participation in the labour force. Greater still is the economic contribution made to the informal economy and there is rising evidence of 'income strategies' adopted by older urban people primarily to provide vital income.

Older people have naturally assumed roles such as tending to childcare, cooking and other household tasks and many part-time jobs in the informal sector. Younger adults are 'released' for employment purposes and the cumulative effect is beneficial to the whole household unit. If a longer time is spent in productive employment, ageing challenges can be offset without shifting unacceptable burdens onto the young. Increasing the level of economic activity in the economy provides a pro-active policy solution and increases the per capita income of the population. By reducing economic dependency, the state can expand social security protection too.

Old people can find themselves sharing their accommodation, pensions or limited incomes from informal employment with the younger generation for longer than they may have anticipated or desired. The vicious circles of poverty these households experience from generation to generation are very prominent in some parts of South Asia.

Conclusion

Though many older people are working beyond the age of 60, what should be a realistic retirement age has become the subject of debate in both developed as well as in developing countries. Is there a 'surplus' share of the population at 'productive' ages? What are older people's important 'productive' contributions to families and communities and what are the potential economic 'costs' of failing to cater for older people has become important questions that need further exploration. There is a need to reform social security arrangements particularly pension reforms (Willmore 2007) which are fairer and sustainable for longer life expectancy and the provision of basic minimum living standards in old age, especially as very limited state welfare systems (social security and health care) are in place for the elderly. The experience of ageing and the impact of ageing population on development are complex, difficult to predict and highly dependent on context. NGOs working on ageing are trying to simplify these complexities in an effort to catch the attention of the policymakers, but are accused of generalisation of impact of ageing on development, but the fact remains that there is almost total absence of discussion on ageing and poverty in the global development agenda. The knowledge base on well-being, poverty, and vulnerability among older poor and their households is very limited.

There is a need to adopt a critical perspective in understanding ageing and the preparedness of developing countries to address future challenges of ageing populations especially those living in poverty in later stages of their lives. How can opportunities be created for economically active ageing populations so that they are not dependent on either the state or their families in the later part of life cycles? Some Asian countries have less time to set in place the institutions needed to cope with population ageing, and are disadvantaged by fiscal constraints, poor governance and weak public agencies. It is important to create a society for all ages in which the changing demographic, economic and social changes will mean an evolution of policies catering for the elderly in a society which promotes support between generations and at the same time sustains economic growth and development.

References and Guide to Further Reading

Aboderin, I. (2005) 'Changing family relationships in developing nations', in M. L. Johnson (ed.) *The Cambridge Handbook of Age and Ageing*. Cambridge: Cambridge University Press, pp. 30–46.

Aboderin, I. (2006) *Intergenerational Support and Old Age in Africa*. New Brunswick, NJ: Transaction Publishers.

Aboderin, I. (2008) 'Ageing', in V. Desai and R. B. Potter (eds) *Companion to Development Studies*. London: Hodder Education, pp. 418–423.

Aboderin, I. and Ferreira, M. (2008) 'Linking ageing to development agendas in sub-Saharan Africa: Challenges and approaches'. *Journal of Population Ageing*, 1 (1): 51–73.

Barrientos, A. and Lloyd-Sherlock, P. (2002) 'Older and poorer? Ageing and poverty in the South'. *Journal of International Development*, 14: 1129–1131.

Croll, E. (2008) 'The intergenerational contract in the changing Asian family', in R. Goodman and S. Harper (eds) *Ageing in Asia*. Oxford: Routledge.

Desai, V. and Tye, M. (2009) 'Critically understanding Asian perspectives on ageing'. *Third World Quarterly*, 30 (5): 1007–1025. Available at: www.globalaging.org/elderrights/world/2009/asianperspective.pdf

Dyson, Tim (2010) *Population and Development: The Demographic Transition*. London: Zed Books.

Gender and Development (2009) Special Issue on Ageing, *Gender and Development*, 17 (3).

Glaser, K., Agree, E. M., Costenbader, E., Camargo, A., Trench, B., Natividad, J. and Chuang, Y. L. (2006) 'Fertility decline, family structure, and support for older persons in Latin America and Asia'. *Journal of Ageing and Health*, 18 (2): 259–291.

Gutiérrez-Robledo, L. M. (2002) 'Looking at the future of geriatric care in developing countries'. *Journal of Gerontology: Medical Sciences*, 57A (3): M162–M167.

Harper, S. (2006) *Ageing Societies: Myths, Challenges and Opportunities*. London: Hodder Arnold.

Hussain, A., Cassen, R. and Dyson, T. (2006) 'Demographic transition in Asia and its consequences'. *IDS Bulletin*, 37 (3): 79–87.

Lloyd-Sherlock, P. (2000) 'Population ageing in developed and developing regions: implications for health policy'. *Social Science and Medicine*, 51: 887–895.

Lloyd-Sherlock, P. (ed.) (2004) *Living Longer: Ageing, Development and Social Protection*. London: Zed Books.
Lloyd-Sherlock, P. (2010) *Population Ageing and International Development: From Generalization to Evidence*. Bristol, UK: Policy Press.
United Nations, Population Division (UNPD) (2006) *World Population Policies 2005*. New York: United Nations.
Varley, A. and Blasco, M. (2000) 'Intact or in tatters? Family care of older women and men in urban Mexico'. *Gender and Development*, 8 (2): 47–55.
Vera-Sanso, P. (2004) 'Modelling intergenerational relations in south India'. *Generations Review*, 14 (1): 21–23.
Willmore L, (2007) 'Universal Pensions for Developing Countries'. *World Development*, 35 (1): 24–51.
Wilson, G. (2000) *Understanding Old Age: Critical and Global Perspectives*. London: Sage.

Websites

Ageing in Developing Countries Network: http://adcnet.psc.isr.umich.edu/
HelpAge International: www.helpage.org/Home
UN Programme on Ageing: www.un.org/ageing/
WHO Ageing: www.who.int/ageing/en/
Global Action on Ageing: www.globalaging.org/
AARP American Association of Retired People: www.aarpinternational.org/ includes databases such as 'AgeStats Worldwide' and provides access to statistical data that compare the situation of older adults across countries or regions around a variety of issues areas. The most recent data and projections as far ahead as 2050 are provided – see AARP statistics: www.aarpinternational.org/resource-library/agesource-agestats

8.5

Health disparity

From 'health inequality' to 'health inequity' – The move to a moral paradigm in global health disparity

Hazel R. Barrett

Introduction

'Promoting and protecting health is essential to human welfare and sustained economic and social development' (WHO, 2012, p. ix). Improving the health of people living in developing countries has been at the top of the global development agenda since the UN Millennium Development Goals (MDGs), were agreed and adopted by 189 countries in September 2000.

Of the eight MDGs, three directly address health challenges: reducing child mortality (Goal 4); improving maternal health (Goal 5); and combating HIV/AIDS, malaria and other diseases (Goal 6). These are known as the health-related MDGs. The other five MDGs all indirectly impact on health: eradicating extreme poverty and hunger (Goal 1); achieving universal primary education (Goal 2); promoting gender equality and empowering women (Goal 3); ensuring environmental sustainability (Goal 7); and developing a global partnership for development (Goal 8). The eight MDGs are measured through 21 targets and 60 official indicators. Most of the MDG targets have a deadline of 2015, using 1990 as the baseline against which progress is measured (UN, 2012).

With only three years to go before the MDG deadline, the UN Millennium Development Goals Report for 2012 highlights some impressive progress: extreme poverty has fallen in every region; the target to halve the proportion of people without access to improved sources of water has been met; and parity has been achieved in primary education between boys and girls. Whilst other goals and targets have not yet been met there has been considerable progress. For example, the world is on track to achieve the target of halting and beginning to reverse the spread of tuberculosis; access to treatment for people living with HIV has increased in all regions; and global malaria deaths have declined (WHO, 2012). However a number of MDG targets have showed much less improvement and are unlikely to meet their 2015 target. For example, progress in improving maternal health and reducing maternal deaths; access to improved sources of water remain lower in rural than urban areas; and gender inequality persists, with women continuing to face societal discrimination, with violence against women undermining the efforts to achieve all MDGs (UN, 2012).

There can be no doubt that the implementation of the MDGs has saved many lives, has reduced human suffering and has improved living conditions for millions of people. However, the UNDP and international community recognise that advances are uneven, as identified by Ban Ki-moon (UN, 2012), 'We must also recognize the unevenness of progress within countries and regions, and the severe inequalities that exist among populations, especially between rural and urban areas' (p. 3). It is now accepted that average health achievements, such as those stated in the MDGs are not a sufficient indicator of a country's performance in health, rather the distribution of health and health disparity is paramount, with future challenges being posed by multiple crisis and large inequalities (WHO, 2012).

Thus whilst the MDGs have contributed to the delivery of many real health gains in developing countries and progress at the national level, socially determined health disparities still exist. As a result, the debate on health disparities has increasingly focused on the social determinants of health and issues of social justice. Thus the debate is moving into a moral/ethical dimension, where the term 'health inequality' is being replaced by the concept of 'health inequity' which has its basis in the principles of human rights and social justice.

Health inequality and the political economy approach to health disparities

Whilst there is much evidence to demonstrate the differences in health achievements between countries (WHO annual reports) there is no internationally accepted definition of 'health inequality'. Gakidou *et al.* (2000) define health inequality as 'variations in health status across individuals in a population' (p. 42) which allows cross-country comparisons to be made as well as the study of the determinants of health inequality. According to Shaw *et al.* (2002) 'where the chances of good or bad health are not evenly distributed among groups of people (defined by the area in which they live or work or some other common characteristic), we say that there is *health inequality*' (p. 126). Gwatkin (2000) states that when referring to health inequalities 'the principal objective is the reduction of poor-rich health differences' (p. 6).

There is not only a lack of a standard definition of health inequality there is also a dearth of measurement strategies and indicators. The traditional measures of health inequality, the Gini Coefficient and Concentration Index, are both taken from the field of economics. The Gini Coefficient ranges from 0 which represents perfect equality to 100 which is perfect inequality. This is often represented diagrammatically as a Lorenz Curve. The Gini Coefficient is often supplemented with the Concentration Index which indicates the extent to which a health outcome is unequally distributed across groups with ranges between –1 and +1. These two measures focus on health status as measured by morbidity and mortality. Due to sparse and unreliable data for adult mortality in developing countries, infant and under five mortality rates are usually used, often derived from demographic and health surveys.

But health inequalities can also be measured according to need, access, efficacy and the effectiveness of health systems. In 2000 *The World Health Report* (WHO, 2000) unveiled a new measure called the Health Inequality Index. Based on the child mortality data of 191 countries, this index measures the performance of different national health systems, based on fairness and achievement. The index, which is not based on any preconceptions about the dimensions along which mortality is unequally distributed, has been robustly criticised as it 'bears little relationship to socioeconomic inequalities in mortality' (Houweling *et al.*, 2001, p. 1672; Wolfson and Rowe, 2001). Whilst there is much merit in a measure of health inequality that focuses on health delivery systems, it tells us little about the health risk environment and the socioeconomic factors that result in health inequality.

Health inequity and the social justice approach to health disparities

The arguments and empirical evidence that social action can improve health disparities motivated the WHO setting up a Commission on Social Determinants of Health in 2005. The driving principle of the commission was 'social justice: to reduce unfair differences in health between social groups within a country and between countries' (Marmot, 2006, p. 2091). In line with the human development paradigm, the commission aimed to change social conditions to 'ensure that people have the freedom to lead lives they have reason to value' and which will 'lead to marked reductions in health inequalities' (Marmot, 2006, p. 2082). This philosophical position draws attention to the central role played by human freedoms in health and calls for social action to achieve the MDGs and reduce both inter- and intra-country health disparity.

The commission has highlighted that many of the inequalities in health between and within countries are avoidable and are the result of social injustice (Marmot, 2006) and the outputs from the commission have contributed to the ongoing debate as to how to best reduce disparities in health. As the debate has progressed the term 'health inequity' has been adopted to refer to health inequalities that are avoidable and unfair (Asada, 2005; Braveman and Gruskin, 2003) and which reflect social injustice (Braveman *et al.*, 2011). Fotso (2006) explains the difference between the two terms: '*health inequality* is a generic term used to designate differences and disparities in the health achievements of individuals and groups, whereas the term *health inequities* refers to inequalities that are unjust or unfair' (p. 9).

Marmot (2005, 2006) suggests that health inequalities can be explained by a lack of material conditions for good health, which includes a nutritionally balanced diet, clean water, sanitation and the provision of medical and public health services. On the other hand health inequity is the result of a lack of community and individual empowerment. He argues that health inequity is inextricably linked to issues of empowerment, which includes community social cohesion, depth of social capital and systems of governance, as well as individual empowerment with respect to social networks, community engagement, life style preferences, health perceptions and personal stresses. For him both health inequality and inequity are clear outcomes of social injustice.

Kahan (2012) confirms that populations which are disadvantaged because of low income, discrimination and other marginalising circumstances suffer poorer health and die younger than the mainstream population. She goes on to state that 'These differences in health status constitute a socially unjust state of health inequity. Identifying ethical processes and activities will direct us to the actions required to achieve optimal health for all, for example, by ensuring that evidence regarding the root causes of health inequities is acted on' (Kahan, 2012, p. 431). Thus all factors, including economic determinants, that contribute to health inequity must be addressed.

Adopting the argument that health inequalities and inequities are related in highly complex ways the WHO (2012) have adopted the following definition of health inequities as: 'unfair and avoidable differences in health and health service provision – that arise for example from socioeconomic factors, such as level of education, occupation, and household wealth or income, from geographical location, and from ethnicity and gender' (p. 145). The WHO has produced a table of health inequities (WHO, 2012, table 8) which disaggregates six health indicators (contraceptive prevalence; antenatal care; births attended by skilled health personnel; DTP3 immunisation coverage among one-year olds; children under five years of age that are stunted; and under five mortality) according to urban or rural residence, household wealth by quintile and maternal educational level. Whilst the choice of health indicator and social characteristic can be criticised, it does demonstrate evidence of the beginning of a rebalancing of health disparity approaches and the first step in comparing health outcomes according to different social factors/groups.

Equity in health can be summarised as the equal opportunity to be healthy for all population groups and the equalising of health outcomes of disadvantaged social groups. As a concept it therefore focuses attention on the distribution of resources and the social determinants of health (Braveman and Gruskin, 2003). As such equity in health is grounded in ethical and human rights principles and focuses on the health differences that reflect social injustice (Braveman and Gruskin, 2003; Braveman *et al.*, 2011). Asada (2005) summarises this as equating health inequity with the moral dimension of health distribution. Equity in health is thus an ethical and moral approach to reducing health disparities. The challenge for the international community is how social justice issues can be identified and addressed to reduce health disparities, both between and within countries.

Towards a new moral paradigm of health disparity?

Over the last decade the conceptualisation of health disparities and development has shifted from 'health inequality' which is heavily influenced by the political economy approach to development, towards 'health inequity' which uses the social determinants of health as its principle framework. This shift has resulted in human rights and social justice taking a much more central role in health and development discourse. It will be interesting to see how influential the health inequity arguments are in the renegotiations of the MDGs and how the international community proposes to deal with this new moral paradigm in health disparity which confronts the controversial and sensitive issues of social injustice which are impacting on the health of millions of people globally.

References

Asada, Y. (2005) A framework for measuring health inequity. *Journal of Epidemiology and Community Health*, 59: 700–705.
Braveman, P. and Gruskin, S. (2003) Defining equity in health. *Journal of Epidemiology and Community Health*, 57: 254–258.
Braveman, P. A., Kumanyika, S., Fielding, J., Laveist, T., Borrell, L. N., Manderscheid, R. and Troutman, A. (2011) Health disparities and health equity: The issue is justice. *American Journal of Public Health*, Suppl. 1: S149–55.
Fotso, J. (2006) Child health inequities in developing countries: differences across urban and rural areas. *International Journal for Equity in Health*, 5: 9. Available online at www.equityhealthj.com/content/5/1/9
Gakidou, E. E., Murray, C. J. and Frenk, J. (2000) Defining and measuring health inequality: An approach based on the distribution of health expectancy. *Bulletin of the World Health Organisation*, 78 (1): 42–54.
Gwatkin, D. R. (2000) Health inequalities and the health of the poor: What do we know? What can we do? *Bulletin of the World Health Organisation*, 78 (1): 3–18.
Houweling, T. A. J., Kunst, A. E. and Mackenbach, J. P. (2001) World Health Report 2000: Inequality index and socioeconomic inequalities in mortality. *The Lancet*, 357: 1671–1672.
Kahan, B. (2012) Using a comprehensive best practices approach to strengthen ethical health-related practice. *Health Promotion Practice*, 13 (4): 431–437.
Marmot, M. (2005) Social determinants of health inequalities. *The Lancet*, 365: 1099–1104.
Marmot, M. (2006) Health in an unequal world. *The Lancet*, 368: 2081–2094.
Shaw, M., Dorling, D., and Mitchell, R. (2002) *Health, Place and Society*. London: Pearson Education.
UN (2012) *The Millennium Development Goals Report 2012*. New York: UN.
Wolfson, M. and Rowe, G. (2001) On measuring inequalities in health. *Bulletin of the World Health Organisation*, 79 (6): 553–560.
World Health Organisation (WHO) (2000) *The World Health Report*. Geneva: WHO.
WHO (2012) *World Health Statistics 2012*. Geneva: WHO.

Further reading

Asada, Y. (2005) A framework for measuring health inequity. *Journal of Epidemiology and Community Health*, 59: 700–705. In this paper, Asada discusses why health inequity is a topic of moral concern and proposes a framework for measuring the moral or ethical dimension of health inequity.
Braveman, P. and Gruskin, S. (2003) Defining equity in health. *Journal of Epidemiology and Community Health*, 57: 254–258. This paper discusses the definition of health inequity, distinguishing it from health inequalities. It explores the consonance of health equity with ethical and human rights principles.
Gakidou, E. E., Murray, C. J. and Frenk, J. (2000) Defining and measuring health inequality: An approach based on the distribution of health expectancy. *Bulletin of the World Health Organisation*, 78 (1): 42–54. A seminal paper which raises pertinent issues about defining and measuring health inequalities.
Gatrell, A. C. (2002) *Geographies of Health: An Introduction*. Oxford: Blackwell Publishers.
Gwatkin, D. R. (2002) Reducing health inequalities in developing countries. In R. Detels, J. McEwen, R. Beaglehole and H. Tanaka (eds) *Oxford Textbook of Public Health, 4th Edition*. Chapter which traces the history of concern about health inequalities, explores the main concepts associated with poverty and inequality and discusses policies for reducing inequality.
Leon, D. and Walt, G. (2001) *Poverty, Inequality and Health: An International Perspective*. Oxford: Oxford University Press. This book contains 17 chapters written by leading experts and covers issues of measurement, economic and social determinants of health inequality, life-course approaches through to health-care systems and their roles in health inequality.
Marmot, M. (2006) Health in an unequal world. *The Lancet*, 368: 2081–2094. This paper makes the case for understanding the socioeconomic determinants, of health inequality and inequity. It particularly stresses the importance of social justice in reducing health inequalities. Examples are taken from both the developed and developing world.
World Health Organisation (WHO) (2010) *The World Health Report*. Geneva: WHO.
World Health Organisation (annual) *The World Health Report*. Geneva: WHO. Gives an annual review of the global health situation. Each report focuses on a different health issue.

8.6
Disability

Ruth Evans

Disability, poverty and development

The complex links between disability, poverty and development have been increasingly acknowledged within development discourses and research in recent years. Poverty is regarded as both a 'cause' and a 'consequence' of disability (DFID, 2000). Although reliable statistical data on the incidence of disability/impairment is not widely available and there is no agreed consensus on how disability should be defined, over a billion people worldwide (15% of the global population) live with disability (WHO, 2011). The majority of the global population of disabled people (80%) live in low- and middle-income countries, which often have limited resources available to meet their needs. Furthermore, disabled people are disproportionately represented among the numbers of people living in chronic poverty. The high rates of impairment and preventable illness in the global South are largely caused by malnutrition, poverty, lack of access to sanitation, safe drinking water, health care and other services, hazardous work, landmines and armed conflict (Yeo and Moore, 2003; McEwan and Butler, 2007).

Being poor also increases the likelihood of an individual experiencing ill health and becoming disabled (Yeo and Moore, 2003). Childhood impairment, for example, is often caused by preventable injuries and illnesses in homes and neighbourhoods that are related to poor living conditions, inadequate access to health care and sanitation and accidents among working children (McEwan and Butler, 2007). Many disabled children are denied access to education, due largely to the fact that education systems in the global South lack the resources and skills to include disabled students within mainstream educational settings and adequately meet the needs of disabled students. This in turn leads to high levels of illiteracy, reduced skills and employment opportunities for disabled people in adulthood, perpetuating the cycle of poverty from one generation to the next (Yeo and Moore, 2003). Disabled women and girls often experience multiple disadvantages, on the basis of their gender and disability, and are particularly vulnerable to abuse, chronic poverty and exclusion.

Improving health systems, infrastructure and the prevention and treatment of diseases are critically important in preventing and reducing the prevalence of disability in the global South. Fulfilling disabled people's rights and enhancing their well-being, however, also requires efforts to tackle poverty and the sociocultural, political and economic inequalities and structural violence that people experience.

The 'medical' and 'social' models of disability

Development approaches based on multidimensional understandings of poverty that aim to enhance human capacities and wellbeing have been influenced by disability politics and social theories of the body, health and disability that developed from the 1970s onwards. Disability activists in the global North rejected medical, rehabilitative models of disability which were based on assumptions that disabled people suffer primarily from physical and/or mental abnormalities that medicine can, and should treat, cure, or at least prevent (Oliver, 1990). Within the dominant 'medical model' approach, disability is perceived as an 'individual misfortune' or 'tragedy'. The disability movement in the global North (led by mainly UK-based activists and allies) instead developed a 'social model' of disability to focus attention on the sociocultural, economic, political and spatial barriers to participation that disabled people experience. An individual's 'impairment' was seen as separate from the social, attitudinal and environmental dimensions of 'disability' that exclude disabled people. The 'social model' thus focuses on changing society to facilitate the participation and inclusion of disabled people, rather than on efforts to 'rehabilitate' individuals and overcome biological constraints of the body.

These understandings of disability have been crucial to improving accessibility, achieving equality of opportunity and securing disabled people's rights within the public sphere in the global North, including in education, employment, health and social care. Commentators, however, have questioned the appropriateness of applying Western-centric social models of disability in the global South (McEwan and Butler, 2007). The wider macroeconomic context, resource constraints and limited availability of technical solutions to make environments more accessible constrain the implementation of social model approaches to disability in many low income countries. Chronic poverty, limited income earning options and restricted access to health and education that many disabled people experience mean that access to basic services is likely to represent a higher priority for disabled people, governments and policymakers, rather than issues of accessibility or assistive technology.

Debates about the need to reconcile both medical and social models of disability in the 1980s led to the establishment of 'community-based rehabilitation' approaches in the global South. Such approaches aimed to provide rehabilitation through medical intervention and care, as well as promoting the social inclusion and participation of disabled people within their communities. Community-based rehabilitation projects have been criticised, however, for being ill-conceived and lacking sensitivity to local cultures and practices, including overlooking the existing care and support that many families and communities provide for disabled people (McEwan and Butler, 2007). Community-based rehabilitation can be seen as reinforcing medical/charitable models of disability that were introduced in the colonial era, perpetuating ideas that disabled people are dependent and need to be supported by charitable fundraising and donations (McEwan and Butler, 2007). Community-based rehabilitation projects have sought to shift towards a more community development approach in recent years and aim to empower disabled people and facilitate their participation in the development process.

Although debates about disability and chronic illness within the social sciences have been largely dominated by urban, Anglophone and Western-centric concerns to date, there has been growing recognition of the need to investigate the interconnections between sociocultural representations of health, illness and disability and development processes at both a local and global scale (McEwan and Butler, 2007; Power, 2001). Understandings of bodies, disability and illness vary according to the economic, geopolitical, sociocultural and spatial context. While most cultures ascribe to notions of a 'normal' or 'ideal' body or mind, the meanings attached to different illnesses and impairments and the social responses that are deemed appropriate are not universal. In many sub-Saharan African countries, for example, disability in children is associated with maternal wrongdoing and witchcraft and in contexts of poverty, negative cultural attitudes and a lack of support, families may 'hide' or abandon disabled children who are considered 'abnormal' (Kabzems and Chimedza, 2002). However, impairment does not always lead to exclusion and many individuals are supported and included within their families and communities (Barnes and Mercer, 2003).

Research from the global South is increasingly challenging Northern framings of the disability debate (Connell, 2011). Social model approaches to disability have been criticised for failing to acknowledge the materiality of the body such as the effects of pain and impairment on people's everyday lives, and the impacts of structural violence, such as impairments resulting from processes of imperialism and colonisation. Meekosha and Soldatic (2011) argue that a politics of impairment is critical for understanding 'disability' in the global South. The authors point to the example of the Vietnamese Agent-Orange Movement's claims for redistributive justice against the US military. Such political mobilisations to achieve compensation for impairments caused by crimes committed as part of the colonisers' project draw heavily on medical science to make claims for a global resource transfer from the North to the South. Similarly, people living with HIV in Africa and others with chronic illness may identify with others on the basis of their biomedical diagnosis rather than according to a strategic notion of 'disability' and the focus of their activism may be on access to healthcare and medical treatment, which differs from the focus of the disability movement in the global North (Evans and Atim, 2011).

The separation of 'impairment' from 'disability' that underpins the social model can result in impairment being constructed as 'natural' (as opposed to disability which is viewed as 'social'). However, as Meekosha and Soldatic (2011: 1393) argue, 'impairment is not in fact always natural, but the outcome of deeply politicised processes of social dynamics *in* bodies that then become medicalised and then normalised through a raft of moral discursive and real practices'. They call for a 'politics of diversity within unity' as a central strategy of global mobilisation on disability and impairment.

Disability politics and rights

The adoption and rapid ratification of the United Nations Convention on the Rights of Persons with Disabilities (UNCRPD) by many countries from 2008 has resulted in a high level of state and civil society mobilisation around disability. This has been accompanied by a growing awareness that many Millennium Development Goals and other international development targets are unlikely to be met unless disabled people's views, needs and priorities are taken into account within efforts to alleviate poverty. Disability issues appear to be increasingly mainstreamed within the 'rights-based development' agenda, although the legal rhetoric is often very distant from the lived reality experienced by many disabled people in the global South (Meekosha and Soldatic, 2011).

Disabled people's organisations, led by disabled people in the global South, have played an important role in collective advocacy for the representation of disabled people in all stages of the development process at the national and international levels. For example, the National Union of Disabled People of Uganda (NUDIPU) lobbied for the inclusion of disabled people at all levels of political administration. As a result, disabled people have achieved a higher level of political representation in Uganda than in any other country (McEwan and Butler, 2007). International non-governmental organisations, coalitions and networks, such as Disabled People's International (established in 1981) have helped to strengthen national disabled people's organisations and facilitate collective advocacy for disabled people's rights at the global level.

The UN Convention on the Rights of Persons with Disabilities is broadly informed by the social model of disability. Disability and impairment are not explicitly defined, but 'persons with disabilities' include: 'those who have long-term physical, mental, intellectual or sensory impairments which in interaction with various barriers may hinder their full and effective participation in society on an equal basis with others' (UN, 2010). Key tenets of the convention are disabled people's rights to participation and inclusion, non-discrimination and accessibility. While the emphasis is on mainstreaming disability into all development activities, such as Poverty Reduction Strategy Papers and the MDGs, it is recognised that disability specific measures may be necessary to 'accelerate or achieve de facto equality of persons with disabilities' (UN, 2010).

While the 'mainstreaming' of disability within rights-based approaches to development has been broadly welcomed by advocates and activists, tensions remain. Concerns focus on the dangers of tokenistic involvement of disabled people and the neglect of their self-determination and equality, in addition to the lack of attention to global structural inequalities and the role of imperialism and colonialism as root causes of violations of human rights, famines, malnutrition, ecological degradation and growing impairment in the global South (Meekosha and Soldatic, 2011).

Note

This chapter draws on a more extensive discussion of health, disability and development in Potter, Conway, Evans and Lloyd-Evans's 2012 book, *Key Concepts in Development Geography*, London: Sage.

References

Barnes, C. and Mercer, G. (2003) *Disability*, Cambridge: Polity Press.

Connell, R. (2011) 'Southern bodies and disability: Rethinking concepts', *Third World Quarterly*, 32(8): 1369–1381.

Department for International Development (DFID) (2000) *Disability, Poverty and Development*, London: DFID.

Evans, R. and Atim, A. (2011) 'Care, Disability and HIV in Africa: Diverging or interconnected concepts and practices?', *Third World Quarterly*, 32(8): 1437–1454.

Kabzems, V. and Chimedza, R. (2002) 'Development assistance: Disability and education in Southern Africa', *Disability and Society*, 17(2): 147–157.

McEwan, C. and Butler, R. (2007) 'Disability and development: Different models, different places', *Geography Compass*, 1(3): 448–466.

Meekosha, H. and Soldatic, K. (2011) 'Human rights and the global South: The case of disability', *Third World Quarterly*, 32(8): 1383–1397.

Oliver, M. (1990) *The Politics of Disablement*, Basingstoke: Macmillan and St Martin's Press.

Power, M. (2001) 'Geographies of disability and development in Southern Africa', *Disability Studies Quarterly*, 21(4): 84–97.

UN (2010) 'Convention on the Rights of Persons with Disabilities and Optional Protocol', available online at www.un.org/disabilities/documents/convention/convoptprot-e.pdf (accessed 26 January 2011).

World Health Organisation (WHO) (2011) *World Report on Disability*, Geneva: WHO/The World Bank.

Yeo, R. and Moore, K. (2003) 'Including disabled people in poverty reduction work: "Nothing About Us, Without Us"', *World Development*, 31(3): 571–590.

Further reading

Barron, T. and Manombe Ncube, J. (eds) (2010) *Poverty and Disability*, London: Leonard Cheshire Disability. This edited collection provides analysis of the links between poverty and disability in the global South from policy, practice and academic perspectives.

Eide, A. and Ingstad, B. (eds) (2011) *Disability and Poverty: A Global Challenge*, Bristol: The Policy Press. This wide-ranging collection explores the social, cultural and political dimensions of disability and poverty in different contexts in the global South.
Groce, N., Kembhavi, G., Wirz, S., Lang, R., Trani, J.-F. and Kett, M. (2011) 'Poverty and disability – a critical review of the literature in low and middle income countries', Working Paper Series No. 16, Leonard Cheshire Disability and Inclusive Development Centre, University College London. This paper provides a useful review of the evidence regarding the links between poverty and disability in the global South.
Third World Quarterly (2011) 'Disability in the global South', special issue, 32(8). This collection of articles discusses recent conceptualisations of Southern bodies, disability, poverty and human rights.

Website

www.un.org/disabilities

8.7

Social protection in development context

Sarah Cook and Katja Hujo

What is social protection?

Social protection is concerned with preventing, managing and overcoming situations that adversely affect people's well-being. Social protection programmes aim to assist individuals and households in maintaining basic consumption and living standards when confronted by contingencies such as unemployment, illness, maternity, disability or old age, as well as economic crisis or natural disaster. In an increasingly volatile economic context associated with market liberalization and globalization, social protection has risen on the development policy agenda, initially with a narrow focus on safety nets and the management of risk (Holzmann and Jørgensen, 2000), but expanding to encompass broader concerns with poverty and vulnerability (Cook and Kabeer, 2010). Supported by a renewed global commitment to poverty reduction represented in the Millennium Development Goals, the social protection agenda has also embraced the promotion of livelihoods and more inclusive development within a framework of universal rights (UNRISD, 2010).

The history of social protection of course is not new: instruments for social protection go back to European welfare states and before. In more developed economies, they have evolved to deal primarily with temporary or foreseeable income shortfalls and transitory experiences of poverty in otherwise relatively stable life trajectories and living conditions. They have been enshrined in the global human rights framework, with the Universal Declaration of Human Rights of 1948 establishing the right to social security and numerous conventions of the International Labour Organization (ILO), in particular the Minimum Standards Convention No. 102 of 1952, setting international standards with regard to, for example, medical care, unemployment, old age, family and other benefits.

While significant debates exist over both the appropriate scope of social protection and the principles and design of interventions, commonly used definitions include the following elements:

- *social insurance*: generally employment-related programmes financed from contributions such as unemployment and health insurance and pensions;
- *social assistance*: non-contributory transfers (conditional or unconditional; in cash or kind) to those deemed eligible, whether on the basis of income, vulnerability status or rights as citizens or residents. Work-related interventions such as public employment or food for work programmes are also a form of social assistance;
- *labour market policies* that ensure basic standards and rights at work, including collective bargaining, minimum wage policies, unemployment insurance and prohibition of child labour.

Broader definitions of social protection include the activities of non-state actors; a wider range of programmes, such as tax credits, microfinance and microinsurance, subsidies for smallholder farmers, and so on; as well as the provision of basic public or social services (water, housing, education, health care, transport) to the needy.

New social protection programmes – towards a developmental social policy?

The new social protection agenda responds to failures of development policies during recent decades as well as the negative impacts of ongoing crises. It involves innovative mechanisms of social assistance which have emerged largely from the South, designed to fit contexts of widespread poverty and crisis and of informal employment relations which limit formal coverage, and to overcome obstacles to development particularly through facilitating investments in health and education.

Major new social assistance programmes now being widely promoted (see Table 8.7.1) include: conditional cash transfers (CCTs), employment and public works programmes and unconditional income transfers such as social pensions and child grants. CCT programmes that require recipient households to comply with specific conditions such as regular school attendance and health check-ups for children have been particularly successful, rapidly expanding in different country contexts and promoted by international organizations and donors. Cash transfers support household consumption and directly improve household welfare, although it is more difficult to trace their impact on broader poverty and inequality indicators, especially when programmes are targeted to the poorest households and are small in terms of coverage and fiscal investment. In order to be transformative and developmental, social assistance programmes need to be sufficiently ambitious in terms of scale and financial investments; they also need to be closely articulated with labour markets and the national social policy system, including social services and revenue policies (UNRISD, 2010).

Social assistance coverage has expanded over recent years with new programmes achieving high coverage rates: CCT programmes in Latin America cover 19.3 per cent of the total

Table 8.7.1 Social assistance in developing countries: Selected programmes, objectives and impacts

Instruments	*Programmes (start)*	*Objectives*	*Impacts*
Unconditional income transfers			
Income transfers targeted to poorest	Kalomo pilot social transfer scheme, Zambia (2004); Mchinji pilot social transfer, Malawi (2006), China *dibao* programme (1999)	Reduce poverty and vulnerability among poorest households without economic capacity and with children	
Categorical income transfers: social pensions and child transfers	Social pensions in Bangladesh, Bolivia, Botswana, Brazil, Chile, India, Lesotho, Mauritius, Namibia, Nepal, South Africa Child Support Grant (CSG), South Africa (2001), *Asignación Universal por Hijo*, Argentina (2009)	Reduce poverty and vulnerability among older people and their households Reduce poverty and facilitate investment in schooling, break poverty across generations	CSG: improved height-for-age scores for children and positive educational outcomes; decreased likelihood of child labour or risky health behaviours (DSD *et al.* 2012).
Income transfer conditional on work			
Public works, cash-for-work, employment guarantees	Employment Guarantee Scheme (NREGP), India (2006); *Jefas y Jefes*, Argentina (2002) Productive Safety Net Programme (PSNP), Ethiopia (2006)	In rural areas, to smooth seasonal income fluctuations. In urban areas, to reduce poverty caused by unemployment and underemployment	NREGP: significant job creation; improved rural infrastructure and agricultural productivity (UNDP 2010) PSNP: positive nutritional impact (Gilligan *et al.* 2009)
Income transfers conditional on human capital investment			
Human development–targeted conditional transfers	*Bolsa Família* (2001/2005), Brazil; *Oportunidades* (1997/2004), Mexico; *Familias en Acción*, Colombia (2001); *Bono de Desarrollo Humano*, Ecuador (2003); *Keluarga Harapan* Programme, Indonesia (2007); *Pantawid Pamilyang Philipino* Programme, Philippines (2008)	Improve consumption for poorest households; facilitate investment in nutrition, health and schooling; reduce intergenerational poverty	*Oportunidades*: higher birth weight among participating women (Barber and Gertler 2008). *Bolsa Familia*: positive impact on school attendance and drop-out rates, positive impact on equality (Veras Soares *et al.* 2007)
Integrated poverty reduction/eradication programmes targeting the extreme poor	Targeting the Ultra Poor, Bangladesh (2002) Chile Solidario, Chile (2004)	Stabilize consumption of poorest households; improve human and productive asset base Achieve minimum thresholds for: income, employment, housing, health, education, etc.	Bangladesh: increased school enrolment and reduced gender gaps in school enrolment (World Bank 2009); positive impact on nutrition

Source: Authors, based on Barrientos 2010

population, reaching almost half of the population in Ecuador and around 25 per cent in Brazil, Colombia and Mexico (Cecchini and Madariaga, 2011). Brazil and Mexico cover 12.7 million (2011) and 5.6 million households (2010) respectively; social pension programmes cover 5.9 million elderly in rural Brazil, 2 million in Mexico city, 5.6 million in Thailand, 2.7 million in South Africa and 2.5 million in Bangladesh (Helpage, 2012); the non-contributory child support grant in South Africa benefits 10.8 million children (2012) and 3.5 million in Argentina (2011); and the National Employment Guarantee Scheme in India benefits 46 million households (UNDP, 2010). Non-contributory programmes are likely to gain further importance in the context of the global *Social Protection Floor* initiative.[1]

Adequacy of benefits is equally important for poverty reduction and livelihood promotion. The Minimum Standards Convention of ILO (No. 102) sets clear minimum levels for social security benefits (for example, pensions at 40 per cent of the reference wage). Non-contributory benefits vary considerably between programmes and countries, often with very low benefit levels exacerbated by problems of access and quality of related social services such as health and education necessary to comply with conditionalities or achieve their objectives.

Financing and sustainability

Coverage and adequacy are closely linked to the availability and allocation of financial resources. Quantity and composition of social protection expenditures differ widely across countries and regions: developed and transition countries spend on average around 20 per cent of GDP on social security (including health), this figure is approximately 5 per cent for large parts of Asia and sub-Saharan Africa, albeit with significant differences between countries at similar income levels (ILO, 2011; Barrientos, 2010). In low-income countries financial constraints can be eased by international aid supporting the initial introduction or scaling up of social programmes. In many cases there is space to shift the financing mix within existing public expenditures – from subsidized schemes for formal workers (which tend to be regressive) and towards more progressive public expenditures; as well from less efficient individual solutions (such as out-of-pocket payments), and towards collective schemes with greater scope for risk pooling, economies of scale and redistribution. Ultimately, long term sustainability will require improvements in domestic resource mobilization, including taxation, but this can encounter strong domestic opposition particularly where programmes are targeted to the poor. Sustainability is further secured through political commitment, institutionalization of programmes based on national law, and participation of the non-poor in universal schemes.

Principles of eligibility

Extending coverage also depends on the basis on which people access social protection benefits – whether through contribution payments, on the basis of need or vulnerability/dependency status, or by right as citizens or residents. Eligibility according to need or status implies *targeting* whether categorical (by population group) or by income (means-testing). The latter is often argued for on the basis of containing expenditures and minimizing adverse incentive effects (changes in individual behaviour particularly with regard to consumption, savings and labour supply); but it entails high administrative costs, usually significant errors of inclusion or exclusion, while potentially stigmatizing beneficiaries. De-linking rights to social protection from citizenship through targeted schemes may also enhance the discretionary power of authorities to assign benefits, increasing the risk of clientelism and corruption, while targeted programmes may also foster segmentation in social service access and quality by class/income.

Group targeted programmes (social pensions, child benefits) are often politically more popular, and are not associated with adverse labour market incentives. They can also help overcome structural inequalities or vulnerability based on individual or collective characteristics (gender, disability, ethnicity) which may impede individual or group access to benefits. Such affirmative targeting of excluded groups may be essential to ensure universal coverage of social protection (Mkandawire, 2005).

Future challenges

Effective social protection is widely acknowledged as important for macroeconomic stability and economic development, for ensuring essential investments in human capacities and for creating more resilient and cohesive societies. Resolving issues raised above will involve attention to broader political-economy and policy contexts, including issues of accountability, transparency and participation in institutional and governance arrangements.

Additional challenges stand out for the future: one is the need to guarantee social protection grounded in values of social justice and basic human rights, while ensuring inclusion of the least secure groups in society. When rooted in principles of universalism, social protection programmes can do more than protect individual livelihoods; they also support the creation of more equitable, cohesive and democratic societies. A second challenge is to provide effective pathways out of the new social assistance programmes into employment-based or contributory social protection; this requires linking social protection to the creation of employment-intensive growth strategies that expand economic opportunities for the poor. Third, social protection needs to link with environmental and climate challenges that particularly affect the poor: innovative approaches are needed that harmonize poverty reduction and social protection with sustainable development goals.

Acknowledgement

The authors are grateful to Harald Braumann for excellent research assistance.

Note

1 The Social Protection Floor Recommendation (2012, No. 202) adopted at the International Labour Conference in June 2012 calls for countries to establish and maintain national social protection floors, which constitute a set of nationally defined basic social security guarantees, including effective access to essential health care and basic income security throughout the life course.

References

Barber, Sarah L. and Gertler, Paul J. (2008) The impact of Mexico's conditional cash transfer programme, Oportunidades, on birthweight, *Tropical Medicine and International Health*, 13: 11, 1405–1414.

Barrientos, A. (2010) Social protection and poverty. Social Policy and Development Paper No. 42, Geneva: UNRISD.

Cecchini, S. and Madariaga, A. (2011) *Conditional Cash Transfer Programmes. The Recent Experience in Latin America and the Carribbean*. Santiago, Chile: ECLAC.

Cook, S. and Kabeer, N. (eds) (2010) *Social Protection as Development Policy: Asian Perspectives*. New Delhi: Routledge.

DSD, SASSA and UNICEF (2012) *The South African Child Support Grant Impact Assessment: Evidence from a Survey of Children, Adolescents and Their Households*. Pretoria, South Africa: UNICEF.

Gilligan, Daniel O., Hoddinott, John, Alemayehu, Seyoum Taffesse (2009) The impact of Ethiopia's Productive Safety Net Programme and its linkages, *Journal of Development Studies*, 45: 10, 1684–1706.

Helpage (2012) Helpage International Pension Watch Database, Version 6 July 2012. Available online at www.pension-watch.net/pensions/about-social-pensions/about-social-pensions/social-pensions-database/
Holzmann, R. and Jørgensen, S. (2000) Social risk management: A new conceptual framework for social protection and beyond. Social Protection Discussion Paper no. 0006. World Bank.
International Labour Organization (ILO) (2011) *World Social Security Report*. Geneva: ILO.
Mkandawire, T. (2005) Targeting and universalism in poverty reduction. Social Policy and Development Paper No. 23. Geneva: UNRISD.
United Nations Development Programme (UNDP) (2010) *What Will It Take To Achieve The Millennium Development Goals?* New York: UNDP.
United Nations Research Institute for Social Development (UNRISD) (2010) *Combating Poverty and Inequality: Structural Change, Social Policy and Politics*. Geneva: UNRISD.
Veras Soares, F., Perez Ribas, R. and Guerreiro Osório, R. (2007) Evaluating the impact of Brazil's Bolsa Família: Cash transfer programmes in comparative perspective, Evaluating Note No. 1, UNDP International Poverty Centre.
World Bank (2009) *Conditional Cash Transfers: Reducing Present and Future Poverty*. Washington, DC: World Bank.

Websites

www.unrisd.org
www.worldbank.org/socialprotection
www.socialprotectionfloor.org
www.ids.ac.uk/go/csp

8.8

Female participation in education

Christopher Colclough

In the year 2000, both the Millennium Declaration and the Dakar Framework for Action committed all nations to universalising primary education, to eliminating gender disparities in education over the short-term to 2005 and to achieving gender equality in education by the year 2015. These commitments reflected a concern that no one, whether boys or girls, should be denied access to education – as affirmed in the human rights legislation enacted by most countries over the previous half century. They also implied that patterns of discrimination and inequality required purposive policy action if they were to be overturned.

At the turn of the century, the educational gap was substantial. Globally, the Gender Parity Index (the proportion of girls to boys enrolled) was about 94 per cent at primary level, and some 57 per cent of the out-of-school population were girls. Furthermore, almost two-thirds of the 860 million people estimated to be non-literate were women (UNESCO 2011: 309, 280).

Much was achieved over the first decade of the new century. The proportion of primary-aged children in school increased from around 82 per cent to 88 per cent, while the numbers not in school were reduced by about 40 million. Transition rates from primary to secondary school increased significantly for both boys and girls. Gender disparities in enrolments at all levels of education were greatly narrowed, with the GPI at primary level increasing from 94 per cent to more than 98 per cent and at secondary level from 91 per cent to 96 per cent over those years (UNESCO 2011: 309, 325).

Notwithstanding these considerable accomplishments, the objective of achieving gender parity of enrolments by 2005 was missed, and the high hopes for meeting the broader development goals for education by 2015 will turn out to have been too optimistic. Additionally, the overall enrolment tally hides many national and regional imbalances. Dropout rates from school remain high in sub-Saharan Africa (SSA) – often because children enrol late and the quality of schooling they receive is low, making it difficult for parents to justify their continuation. Furthermore, achieving gender parity of enrolments remains off target in almost 70 countries – particularly in those of SSA and south and west Asia, where GPIs are often less than 90 per cent. The absence of girls from school in these cases represents a huge loss to the individuals involved, and reduced benefits for the next generation.

Many of these imbalances are caused by primary systems being of such low quality that those enrolled do not learn enough, or do not learn quickly enough, to make staying on worthwhile. Yet leaving school after five or six years without having achieved basic literacy and numeracy is a tremendous waste of financial and human resources, and sets up losses for society that extend for years ahead. The gaps for girls tend to be higher in poorer countries, and in rural and poorer areas, where social norms continue to give preference to the education of boys.

It is now clear that the goal set for 2005 was overly ambitious. Gender parity at primary and secondary levels could not have been achieved within a five-year period without large numbers of out-of-school girls enrolling in (or rejoining) classes at levels well beyond primary grade 1 (Colclough 2008). Such 'mid-career' enrolment would also have been extensively required if secondary enrolment parity were to have been achieved within the five-year period – at least in those many school systems in which male pupils significantly outnumbered girls at all grade levels. This kind of enrolment behaviour would have been unsustainable over the medium term and, in most countries, it would not have been feasible in the first place. It should, therefore, have been predictable that the initial gender goals in education would not be achieved within the chosen five-year time frame. Nevertheless, the target for 2015 of achieving gender parity in enrolments throughout education remains realistic for many countries, even though the deeper goals of achieving equality in education, and in society more broadly, will remain challenging for most.

Priorities for policy change can best be informed by an understanding of the causes of existing inequalities. These are dependent on national contexts, history and social and economic conditions. However, there are some important common circumstances shared by many countries, and a discussion of their nature and extent can help to inform strategies for tackling them. This is the task for what follows, which draws upon evidence from a number of recent comprehensive reviews (Colclough *et al.* 2003; UNESCO 2003; Subrahmanian 2007).

The social context

Educational discrimination is both a cause and a consequence of much wider discrimination against girls and women in society. The most marked gender inequalities are found in countries and regions where women's roles are constrained by patrilineal principles of inheritance and descent, by early marriage, by family resources being controlled by the senior male household member and sometimes by traditional restrictions preventing women from participating in the public sphere.

Such traditions are often stronger in the poorest countries, which are also farthest from achieving universal schooling. It is striking that, in countries where enrolments are low, boys tend to be given preference in access to schooling (Figure 8.8.1). Creating an environment where equality between women and men is a guiding principle is essential to improving girls' educational chances and outcomes. The agenda involves reforming family law, granting women property and inheritance rights and implementing legislation to provide for equal opportunities in the labour market. Where they are absent, such changes are often controversial, but they provide a decisive counterweight to entrenched social norms that, inter alia, affect whether or not children go to school.

Reducing the costs of school attendance

Despite the almost universal ratification of human rights treaties which commit nations to the provision of 'free and compulsory education' at primary level, school fees or other charges are still found in almost 100 countries – including in many of those that are farthest from achieving the enrolment and gender parity goals. The average parental contribution differs between countries, but it appears typically to amount to around one-third of total annual unit costs at primary level. This can represent a significant proportion of household income, especially for the poorest households, and leads to under-enrolment of school-age children, particularly amongst girls. Thus, abolishing school fees at the primary level can have a major positive impact on enrolments, as the recent experiences of Malawi, Uganda, Kenya and other African countries have demonstrated. Bilateral aid agencies can help to support such initiatives – by compensating governments, or schools, for the loss of fee revenues and by providing resources to support the quality of schooling during its subsequent expansion – in ways which can have a crucial role in ensuring the success of school-fee reforms. For example, during the period 1994–2002, tuition fees at primary level were abolished in Malawi, Uganda and Tanzania. In each of these countries primary enrolments, net enrolment ratios and the proportion of public spending allocated to education more than doubled. In these same countries between 40 and 60 per cent of primary education budgets were covered by support from aid agencies, following the reforms (UNESCO 2003: 220).

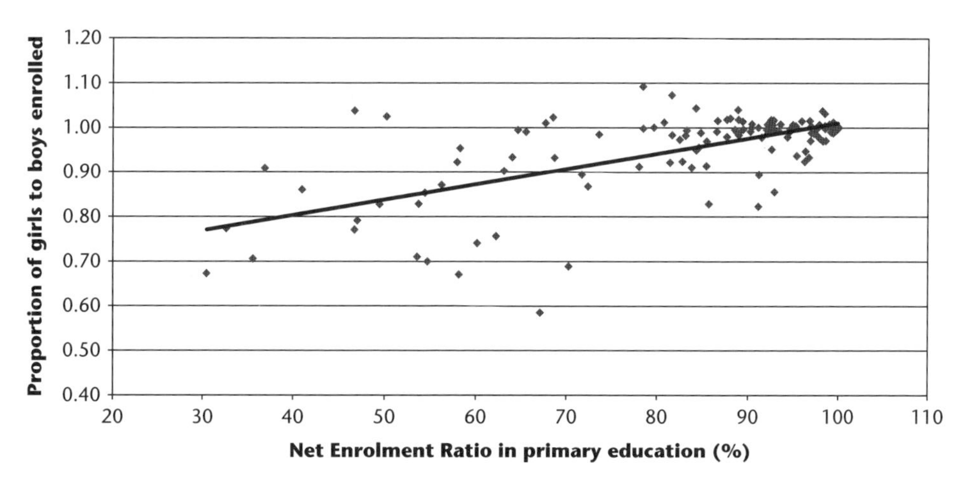

Figure 8.8.1 Gender inequality and enrolments at primary level in developing countries. Where enrolment rates are low, boys are given preference in most countries and most regions

Source: UNESCO data, various years

Removing the need for child labour

The provision of paid and unpaid child labour is a leading cause of under-enrolment and dropout, particularly amongst girls. Recent estimates suggest that more than 200 million children aged 5–14 years are in some form of paid work in developing countries, about half of whom are girls (ILO 2002; UNESCO 2003: 120). In addition, there are many more children – mainly girls – who spend considerable time each week in unpaid domestic work, in ways that affect their access to, or progress in, schooling. Parents are the main employers of these children – whether in the household or in the informal economy. Legislation can help to tackle this problem but, in many cases, incentives are needed to persuade households to forego the income benefits they derive from their school-age children. In Latin America, income support schemes that provide cash to poor families, conditional on school attendance by their children, have been successful in reducing the incidence of child labour. Food incentives, provided as meals at school or as dry food rations, can help to boost enrolments. Scholarships and bursaries can also strongly influence whether girls stay in school to the higher primary and secondary levels. The Food for Education Programme in Bangladesh has had a marked impact upon girls' enrolments. Scholarships at secondary level have also been shown to have major effects upon school attendance amongst girls in Bangladesh, Zambia, Zimbabwe and elsewhere in Africa.

Reforming classroom practice

Designing a gender-sensitive curriculum and training teachers to be 'gender-aware' are not simple processes. Indeed, the fact that girls are still underrepresented in science subjects in industrialised countries at secondary and tertiary levels points to enduring biases. In the developing world, although progress has often been made in removing from school texts the portrayal of women in stereotypical activities, regular gender auditing of curricular materials remains needed in many countries if the barriers are to be broken down.

Irrespective of curriculum design, the way in which it is interpreted by teachers has a crucial influence on students. Teachers can provide role models, a sense of direction and encouragement to both boys and girls. However, they cannot be expected to separate themselves easily from the powerful cultural and social norms in the context of which they themselves have grown up. In strongly patriarchal environments, female teachers face powerful obstacles in choosing their own careers and activities, and in such circumstances, gender sensitisation courses are seldom a standard part of teacher training programmes.

Notwithstanding these difficulties, hiring more female teachers and training them well are priorities for countries where gender disparities remain high. In sub-Saharan Africa, women hold only one-third, or less, of teaching posts, and in some countries their numbers are extremely small. There are strong correlations between girls' enrolments at primary level and the presence of female teachers (Figure 8.8.2) and, where the female/male teacher ratio is low, deliberate policy measures to increase it should be a leading component of any gendered strategy.

Promoting safety and women's empowerment

Schools are often not safe places for learning: violence against girls by other pupils and even by teachers helps to perpetuate the gender gap in education. A number of studies from Africa suggest that much gender violence in schools goes unreported because students fear victimisation, punishment or ridicule. Placing schools closer to homes is also important in many

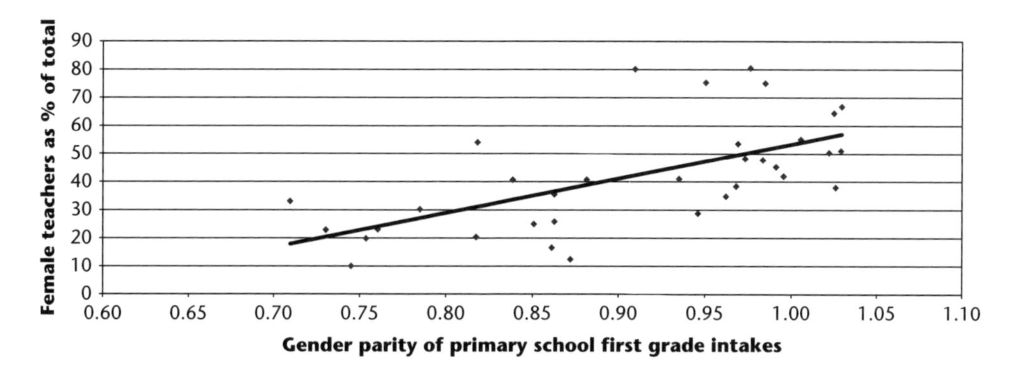

Figure 8.8.2 Sub-Saharan Africa: The proportion of female teachers at primary level, and the gender balance of first grade intakes (a value of 1.00 indicates gender parity in enrolments). Where female teachers are a small proportion of the total, many fewer girls than boys enrol in school

Source: UNESCO data, various years

countries – where the distance to school is large, the safety of children during the journey to school is particularly influential in affecting parental decisions to send their girls to school.

Girls have a much higher chance of attending school if their mother is educated – even if such education was acquired at a mature age. Literacy programmes tied to income-generating activities, and to health and human rights awareness campaigns have shown considerable potential to empower women. Linking such programmes to early childhood care and education generates strong benefits, often at low cost. Because of the strong links between attendance by children at pre-school programmes and their subsequent enrolment in primary school, an effective recognition of the links between different generations of learners paves the way towards greater equality, in education and beyond.

Conclusion

Although the causes of inequality are complex, policy reforms to improve women's rights in the household and the market place, to reduce the direct costs of schooling to households, and to improve school quality in gender-aware ways can do much to encourage and sustain increased enrolments amongst both girls and boys. The state has a critical role to play in creating an environment that promotes this through legislative and policy reform. There is a need to target resources for female education, and to introduce deliberate measures to reduce inequities. Arguments that equality cannot be afforded, or that such policy shifts would generate pressures that conflict with other, more pressing, development priorities, are largely misguided. The well-documented outcomes of educating girls (Schultz 1995; Colclough 2012) imply that achieving gender equality in education stimulates a range of personal, economic and social benefits that well exceed their costs. It is also foundational for achieving the equal rights and freedoms spelt out in the UN Charter on Human Rights more than half a century ago, and which, through the MDGs, the world has recommitted itself to achieve.

Bibliography

Aikman, S. and Unterhalter, E., 2005, *Beyond Access: Transforming Policy and Practice for Gender Equality in Education*, Oxfam, Oxford.

Colclough, C., 2008, 'Global Gender Goals and the Construction of Equality: Conceptual Dilemmas and Policy Practice', in S. Fennell and M. Arnot (eds), *Gender, Education and Equality in a Global Context: Conceptual Frameworks and Policy Perspectives*, Routledge, London: 51–66.
Colclough, C., (ed.) 2012, *Education Outcomes and Poverty – A Reassessment*, Routledge, London.
Colclough, C., Al-Samarrai, S., Rose, P., and Tembon, M., 2003, *Achieving Schooling for All in Africa – Costs, Commitment and Gender*, Ashgate, Aldershot, UK.
ILO, 2002, *A Future Without Child Labour*, Global Report under the Follow-up to the ILO Declaration of Fundamental Principles and Rights at Work, International Labour Conference, 90th session, Geneva.
Kabeer, N., 1994, *Reversed Realities: Gender Hierarchies in Development Thought*, Verso, London.
Schultz, T. P., (ed.) 1995, *Investment in Women's Human Capital*, University of Chicago Press, Chicago and London.
Subrahmanian, R., 2007, *Gender in Primary and Secondary Education: A Handbook for Policy-Makers and Other Stakeholders*, Commonwealth Secretariat, London.
UNESCO, 2003, *Gender and Education for All: The Leap to Equality*, EFA Global Monitoring Report 2003/4, UNESCO, Paris.
UNESCO, 2011, *The Hidden Crisis: Armed Conflict and Education*, EFA Global Monitoring Report 2011, UNESCO, Paris.

8.9

The challenge of skill formation and training

Jeemol Unni

Introduction

In the developing countries, a large proportion of the workforce is engaged as wage labour, in low paid salaried work or as self-employed generating their own incomes and employment by operating small or micro enterprises with the help of family labour alone. Much of these workers constitute a large and growing informal economy, with little legal regulation and social security benefits. In fast changing globalising economies labour and small enterprises struggle to survive, remain viable and grow in terms of income for labour and turnover or profits for enterprises. Education and skills are seen as routes out of poverty for labour and technology as a route to increase productivity in small enterprises. Skill training and access to technology are the two lifelines for them in developing economies. However, with low levels of basic education, acquiring skills for labour, and with low levels of investment, acquiring technology for small enterprises, is a challenge. On the supply side, provision of relevant skill training and technology is also a challenge for governments due to the low capacity to absorb such inputs. This chapter provides some insights into the challenges faced in skill formation and technology absorption in developing countries. We argue that skills and technology are closely linked for productivity and income enhancing effects.

Skills and technology in a globalising society

There is a rising skill premium in labour markets in these countries mainly due to the widespread entry of skill-based technology in industry. Solow's model of the growth theory brought out the importance of technical change by acknowledging the heterogeneity of labour in terms of observable and non-observable characteristics such as skills, dexterity, adaptability and motivation. The measurement issue with regard to skills of workers was acknowledged. The endogenous growth theory took this further and technology was considered an endogenous variable. That is, capital accumulation consisting of both physical and human capital occurred mainly through 'learning by doing'. Technology therefore clearly included accumulation of skills. In the globalising world the heterogeneity of labour inputs plays a decisive role. Besides the theoretical progress in the growth literature the real world was also taking note of the development of the 'knowledge economy'. Development of the 'knowledge economy' and high technology capital intensive industry implies that the education and skill requirements are high as well.

Measurement and definition of skills

The first challenge is to define skill in order to measure it. A macro definition is required for planning and to analyse shifts in demand. A micro definition is necessary at the enterprise level to understand the complementarities of technology with the skills of workers at an individual or organisational level and to articulate the demand for skills (Autor *et al.*, 2001).

Skills

Macro measure

There is no direct measure of the level of skill of the workers. The two ways in which to identify skill at a macro level is by the occupational classification and by the level of education. Both have their limitations. The occupational classification gives broad categories of non-production workers: professional, managerial and administrative, clerical, sales and service workers. And the rest are classified as production workers. The former, non-production workers, are assumed to be more skilled compared to the production workers. One can also consider managerial and professional workers only as skilled if one is interested in higher levels of skills.

The other method of identification of the level of skill is based on level of formal education. In the United States unskilled workers are defined as those with high school diplomas and skilled workers as those with college degrees. Most studies of other countries define skilled workers as high school graduates and above between the age of 15 to 64 years. The rest of the workers in this age group are treated as unskilled workers (Unni and Rani, 2008). Definition by level of education has its limitations. A large proportion of workers in developing countries are illiterate. However, these workers have various forms of skills acquired through informal training which are not captured in a measure using formal schooling only.

Micro measure

According to the World Employment Report, 1998, the term 'skill' refers to an acquired and practiced ability, or to a qualification needed to perform a job or certain task competently. It is a multidimensional concept as most jobs require a combination of skills for adequate performance, ranging from physical abilities like eye-hand coordination, dexterity and strength, to cognitive skills (analytical and synthetic reasoning, numerical and verbal abilities) and interpersonal (supervisory, leadership, social communicative) skills.

Autor *et al.* (2001) define five levels of skills using the Handbook of Analysing Jobs, Department of Labor, US. These can be used to arrive at micro measure of skill or identify skill level of workers:

1 Routine manual tasks: ability to move fingers and manipulate small objects with fingers rapidly and accurately.
2 Non-routine manual tasks: ability to move hand and foot coordinately with each other in accordance with visual stimuli.
3 Routine cognitive tasks: adaptability to situations requiring the precise attainment of set limits, tolerance or standards.
4 Non-routine cognitive/interactive tasks: adaptability to accepting responsibility for the direction, control or planning of an activity.
5 Non-routine cognitive/analytical tasks: general educational level and mathematics.

In a study of small enterprises in the auto component industry many dimensions of these broad indicators were observed in the definitions provided by the owner-manager of skilled, semi-skilled and unskilled workers among the production workers (Unni and Rani, 2008). These entrepreneurs and managers made a clear link between skills of workers and technology used.

Understanding needs of training

Skill insecurity

Skills become a necessity and a form of security to improve the employability of the workers in such enterprises. Skills are a method of improving human capital, which ensures income security to the workers, particularly for the poorly educated workers. A major insecurity faced by the informal workers particularly from poor households is lack of marketable skills. The inability to invest in skills and knowledge seems to be a major factor leading to skill insecurity.

In a micro study in the city of Ahmedabad, India one indicator of skill insecurity was defined as the ease with which anyone else could acquire the skill. Given the rudimentary nature of the skills among the informal workers, nearly two-thirds reported that their skills could be transferred easily. Easy transferability of skills was reported to be the highest among the salaried who had limited formal schooling. Two-thirds of piece rate, self-employed non-agricultural and casual workers also reported easy transferability of skills. Thus, while the salaried jobs were coveted in poor households as it provided a regular source of income, the low level of skills in these activities made them equally vulnerable to loss of jobs and low levels of income. This could also mean that the skill involved in their work was not very technical but simple with the use of less advanced production technologies. A large proportion of workers also felt that due to lack of education and skill it was difficult to move into other alternative jobs. This basically raises the need to focus on promotion strategies for increasing skills to improve the quality of employment, particularly among the vulnerable groups in the informal economy (Unni and Rani, 2002).

Micro and small enterprises

In the literature on 'flexible specialisation', in the 1980s, Piore and Sabel (1984) describe the growth of small-scale production units in developed countries, such as Italy, Germany and

Japan, with a combination of flexible technology and specialised production. The rapidly changing market demand and technological advances in electronics and computers led to this development.

This literature finds its echo in the developing countries in studies on sub-contracting, ancillarisation and industrial clustering. The human resources needs and skill training requirements of these small-scale enterprises are different and crucial for their sustainability and growth. These enterprises require not just technical skill training, but training in entrepreneurship and managerial skills.

Entrepreneurship training programmes

A relatively new class of training institutes and training programmes have come into existence known as entrepreneurship training. In India this is organised by the Ministry of Micro, Small and Medium Enterprises (MSME) and institutions funded by it to create a spirit of entrepreneurship among the youth and new entrants into the labour force. India has a strong tradition of self-employment and small informal enterprises in manufacturing and service activities. The African countries, also faced with unemployment, are now interested in fostering the spirit of entrepreneurship to create small enterprises.

Self-employed: Home-based workers

In Asian countries a large proportion of workers, mainly women, undertake production of garments, embroidery work, *bidi*-rolling (traditional cigarette), incense stick rolling, even assembling brass nuts and bolts and electrical switches, at home. The skill levels of some of these workers is very high, but are not recognised as such.

These industries follow the traditional way of acquiring skills, often transferred from mother to daughter, neighbour and so on. Non-government organisations that support such economic activities take up skill training of these workers as they try to improve the productivity and incomes of these workers. Formal systems sometimes do not work because these women have limited mobility out of their homes due to customs in the society. Very innovative methods have to be devised to provide skill training to these women and women interested in such activities.

A large proportion of these women also felt that due to lack of education and skills it was difficult for them to obtain alternative jobs (Unni, 2009). This raises the need to visualise ways to focus on promotion strategies for increasing skills, to improve quality of employment, particularly vulnerable groups in the informal sector who may remain outside the reach of the formal systems of skill training.

Delivering skills training

Skill-training models

A number of countries have traditionally designed their own skill-training models. An example of the successful German system of skills training, takes the famous '*dual apprenticeship system*', combining two basic models: centre-based training and enterprise-based training. The system is based on a longstanding tradition of apprenticeship that is firmly rooted in German corporate culture. Public vocational training centres provide the theoretical training and practical training is provided within the enterprise. Apprentices sign an employment contract with an enterprise, which gives them on average three and half years of formal training under the supervision of a certified master and receive an allowance fixed by a collective agreement. The graduates receive a nationally recognised diploma. Skill-training models that take into account the needs of the employers need to be designed. Two models are presented below.

Model 1: Private-public partnership within an on-the-job model

A major form of acquiring occupational skills in small businesses and informal enterprises in developing economies is through the traditional apprenticeship model. How best can the government and private sector cooperate to help upgrade the existing initiatives of this model? A system by which such a model can be partly formalised and the cost borne by the worker and the enterprise can be financed by an external agency perhaps in the form of a loan to be repaid later.

Model 2: Private-public partnership within 'skills training centres'

The service providers charged with the responsibility of training will design the curricula to suit the needs that have been established either through market scans and/or that can be justified based on the established skills gaps among the potential trainees. The objective of the curriculum should be to increase the employability of trainees in the local or regional context and/or raise the ability of the trainee to absorb higher levels of skills training available locally or at other locations. These will not be 'free' courses. A cost should be attached to each course based on overall expenditure (capital expenses amortised appropriately). Each trainee will be given a 'training loan' (without collateral) from a bank, which is to be repaid in small instalments after the trainee secures a job. The 'training loan' should include cost of the course plus the cost of commuting to the centre if the commute is long.

The first version of unregulated apprentice training has evolved in many developing countries such as India and sub-Saharan Africa. It is a written or oral contract between the master craftsman and the trainee with the objective of acquiring relevant practical skills. This is a form of on-the-job training and is the method through which traditional skills, such as weaving, pottery and leather work, are transmitted over the generations in developing countries. Nigeria has recognised this traditional system through its National Open Apprenticeship Scheme (NOAS). The scheme involves encouraging employers to take apprentices and train them (GOI, 2009). It officially exists in India as well, but has not been a success. However, the traditional system is the method used to train workers in most small enterprises without any incentive from the government. This traditional apprenticeship model is informally known as the '*ustad shagird*' model in India.

Various versions of the second model are being financed by the National Skill Development Corporation in India (www.nsdcindia.org). These two models of skill training take into account the skill-training needs of the local economy and of the trainees and are more likely to create a faster rate of absorption of the trainees into the workforce and growth of employment in the local economy.

References

Autor, D., Levy, F. and Mumane, R., 2001, 'The Skill Content of Recent Technological Change: An Empirical Exploration', Working Paper No. 8337, NBER Working Paper Series, Cambridge, www.nber.org/papers/w8337

GOI, 2009, *Skill Formation and Employment Assurance in the Unorganized Sector*, National Commission for Enterprises in the Unorganised Sector, Government of India.

Piore, Michael J. and Sabel, Charles, 1984, *The Second Industrial Divide: Possibilities for Prosperity*, Basic Books, New York.
Unni, Jeemol, 2009, 'The Unorganised Sector and Urban Poverty: Issues of Livelihood', India: Urban Poverty Report 2009, Ministry of Housing and Urban Poverty Alleviation, Government of India.
Unni, Jeemol and Rani, Uma, 2002, 'Insecurities of Informal Workers in Gujarat, India', SES Paper No. 30, In-Focus Programme on Socio-Economic Security, International Labour Organization, Geneva, September.
Unni, Jeemol and Rani, Uma, 2008, *Flexibility of Labour in Globalizing India: The Challenge of Skills and Technology*, Tulika Books, New Delhi.

Further reading

Chen, M., Vanek, J., Lund, F. and Heintz, J., 2005, *Progress of the World's Women: 2005*, UNIFEM, New York.
GOI, 2007, Report on the Condition of Work and Promotion of Livelihoods, National Commission for Enterprises in the Unorganised Sector, Government of India, New Delhi.
GOI, 2007, Report on Financing of Enterprises in the Unorganised Sector & Creation of a National fund for the Unorganised Sector (NAFUS), National Commission for Enterprises in the Unorganised Sector, Government of India, New Delhi.
Unni, Jeemol, 2011, 'Skilling the Workforce in India: Different models?' *Norrag News* 46, September, www.norrag.org/fr/publications/norrag-news/online-version/towards-a-new-global-world-of-skills-development-tvets-turn-to-make-its-mark/detail/skilling-the-workforce-in-india-different-models.html

Websites

www.wiego.org
www.nsdcindia.org
www.norrag.org

8.10

Development education, global citizenship and international volunteering

Matt Baillie Smith

Introduction

Popular communications of development are dominated by NGO fundraising activities and disaster focused media, producing ideas of aid, charity and development in which the global

North as 'giver' and South as 'receiver' are privileged over accounts which recognise the structural causes of global inequality and the agency of actors in the global South. Short-lived mass campaigns, such as Make Poverty History, have only effected temporary changes in those perceptions (Darnton and Kirk, 2011, p. 5). Whilst it has been marginalised in development scholarship and practice and remains a contested field of activity, development education works to foster more critical and reflexive understandings. It does this in the belief that these will produce more active global citizens and better informed engagements with the causes rather than effects of global poverty and injustice.

Defining development education

Defining development education (DE) is not straightforward. It emerged in the 1960s and 1970s, drawing particularly on the radical education ideas of Paulo Freire and sharing structural analyses of poverty with emerging development scholarship: 'educationalists talked about empowerment, structural causation, political change and social justice' (Cohen, 2001, p. 179), with dependency theory and theories of neo-colonialism being deployed by 'radical critics' to attack 'the traditional "starving child" appeals used by Oxfam and similar charities' (2001, p. 178). However, whilst there are now DE teams within some INGOs, a reducing number of smaller dedicated development education centres (DECs), a Development Education Research Centre (DERC), UN commitments to the importance of DE and European and national government policies and declarations on DE, what DE exactly is remains contested. Part of the problem with this lies in a mixing of the normative and positive. Some definitions are aspirational expressions of what *should* happen, and act as templates to guide practice. For example, the UK Development Education Association defined DE in 2005 as lifelong learning that:

- explores the links between people living in the 'developed' countries of the North with those of the 'developing' South, enabling people to understand the links between their own lives and those of people throughout the world;
- increases understanding of the economic, social, political and environmental forces which shape our lives;
- develops the skills, attitudes and values which enable people to work together to take action to bring about change and take control of their own lives;
- works towards achieving a more just and a more sustainable world in which power and resources are more equitably shared (DEA, 2005).

This does not tell us how DE works, who is doing it and what impacts it has. More empirically, DE can be understood as

> work done by a variety of organisations, including INGDOs, trade unions and schools, to educate constituencies in the North about development and global interdependence and global/local responsibilities. The emphasis is on critical reflection and, in the UK at least, is differentiated from more general awareness raising around development issues.
>
> *(Baillie Smith, 2008, p. 9)*

Andreotti has used a postcolonial framework to reimagine DE as emphasising unequal power relations rather than equal interdependence, reflexivity and an ethical relationship to difference (Andreotti, 2006), Baillie Smith has conceptualised it in terms of its relationship to global civil society and deliberative democracy (Baillie Smith, 2008) and Humble has analysed it in relation to ideas of the development encounter (Humble, 2012). However, the lack of a clear definition and limited research, particularly in development studies, has underscored the fragility of DE in funding and policy terms and undermined attempts to demonstrate its significance.

Development education policy: Mainstreaming and the emergence of 'global learning'

During the 1970s, 1980s and most of the 1990s, DE was principally funded and guided by international development NGOs. This was done through their own DE teams or units, as well as through funding other actors' DE work. A key tension this produced was between DE teams, committed to changing perceptions and fostering reflexivity, and the fundraising for overseas programme work, which often relied on images and ideas which DE teams actively wished to counter. A further difficulty lay in not being able to attract public donations to support DE when compared to overseas activity. More recently, there have been tensions with NGO campaigning, which is often premised on policy led and strategically defined goals which do not sit easily with DE's commitments to dialogue and reflexivity (Baillie Smith, 2008).

Government support for DE varies significantly between countries and over time, and can have significant impacts on its practice and politics. For example, reflecting an ideological opposition to DE's focus on structures of inequality, as well as broader trends within their education policies towards narratives of 'British' history, the UK Conservative governments of 1979 to 1997 provided no government funding for DE. Under New Labour, and connected with financial and political investment in development and different ideas of active British global citizenship, DE received a substantial injection of funding from the Department for International Development. A government White Paper, *Building Support for Development* (DFID, 1999) identified strategic target groups and partners, including faith groups and trade unions alongside the formal education sector, and underpinned the establishment of funding mechanisms. This process engaged new actors and accelerated processes of mainstreaming started when INGOS and DECs ensured the inclusion of 'development' in a prescribed National Curriculum introduced in 1989. From 2000, DE become interpreted through the lens of the 'global dimension' in schools, supported by curriculum guidance from the Department for Education and Skills and Qualifications and Curriculum Authority, emphasising eight concepts: global citizenship; sustainable development; conflict resolution; values and perceptions; diversity; human rights; social justice; interdependence (Bourne and Hunt, 2011). The change of government to a Conservative-led coalition in 2010 meant the dismantling of much of this support, despite commitments to maintain development spending, on the basis that DE does not 'give the taxpayer value for money' (DFID, 2010). This has meant the loss of UKDECs and DE practitioners, whilst INGOs have increasingly moved away from DE as they seek to maintain funding for overseas work during recession.

Paralleling wider arguments in development studies and beyond highlighting the impacts of state appropriation, neoliberal professionalisation and de-politicisation (Bondi and Laurie, 2005), the mainstreaming of DE has been strongly critiqued, as goals are aligned with government ones to gain funding (Selby and Kagawa, 2011, p. 25). Biccum argues that UK government funding of development awareness needs to be seen as part of a re-articulation of British cultural values through a focus on 'Britain's role in the institutions of global governance and global poverty reduction' with the result that it has become about producing young people as 'little developers' (Biccum, 2007, pp. 1113–1114). Another key debate centres on DE's transition from being a 'movement of NGOs' to a 'theory of learning' (Bourne, 2003) increasingly defined with reference to a pedagogy for exploring global connections. Increasingly, DE has been rebranded as 'global learning', the 'terminology used to articulate it or even promote it rarely uses the term "development"' (Bourne, 2008b, p. 4) and the UK DE Association has been re-named 'Think Global'. For some, these changes have de-politicised DE, its radical agenda for change through education diminished by moving away from the structural analyses of poverty and inequality it shared with early development studies (Baillie Smith, 2013, p. 409).

Development education and the changing development landscape

Changes to the aid and development landscape associated with the rise of the BRICS, austerity in the global North and the popularisation of development through celebritised mass campaigns, the growth of the 'gap year' and international volunteering and increases in 'fair trade' consumption, all provide changing and challenging opportunities and contexts in which to reinvigorate DE.

For example, international volunteering has become an increasingly popular practice, providing opportunities for engaging in and with development for growing numbers of people. Its connections to development have changed over time, with recent years seeing a move from an emphasis on benefits in the global South to the impacts on volunteers (Baillie Smith and Laurie, 2011). The encounters with development and the global South enabled by volunteering are seen as a means of creating global citizens, fostering social inclusion post-9/11 (Lewis, 2006), enhancing CVs and practising corporate social responsibility (Baillie Smith and Laurie, 2011). International volunteering has also been promoted by an increasingly diverse set of actors, from established NGOs to small private providers and the state, evidenced in government funded programmes me in a range of European and North American countries. In the UK, a New Labour government international volunteering initiative, *Platform 2*, was explicitly linked to DE agendas and social inclusion through the requirement of participants to share their learning on return and a targeting of participants from diverse and low income backgrounds. But whilst the claims for international volunteering to raise awareness and produce global citizens are seductive, evidence for this is less clear cut. Volunteering by individuals who do not have skills that address identified gaps in the global South runs the risk of reinforcing stereotypes of 'givers' and 'receivers' rather than fostering deeper understanding. There is also evidence that international volunteering provides opportunities to exercise existing understandings of citizenship as much as produce new forms (Baillie Smith *et al.*, 2013). Simpson highlights the need for accompanying high quality education and training to help foster critical reflexivity (Simpson, 2004, p. 690) suggesting that international volunteering should not, in itself, be understood as a form of DE.

Changing narratives of rich and poor, donor and recipient, produced by the rise of the BRICS and austerity in the global North, alongside expanding middle classes in the global South, provide a new set of histories, contexts and imaginaries for DE to engage with. To continue with an agenda for change rooted in fostering critical understandings of unequal interdependence and more active engagement in poverty reduction, DE will need to engage new actors and constituencies in new places. The changing aid landscape may even provide new opportunities for DE's challenging of established ideas of 'giver' and 'receiver'. This is particularly the case as austerity measures in the global North significantly reduce capacity, at least at a formal level. However, the lessons from the mainstreaming of DE in the UK and elsewhere may also point to the need to define new ways of practicing DE which are not reliant on state funding or INGO promotion, with attendant risks around de-politicisation and organisational marginalisation. Indeed, to return to its more radical ambitions may require stronger links with global civil society networks and social movements than INGOs and government departments.

Concluding thoughts: Development education and development studies

DE currently has a very limited relationship with development studies (Baillie Smith, 2013). With the exception of this chapter, there is almost no reference to it within other development studies texts; it is not mentioned in the index of Sumner and Tribe's *International Development Studies* (2008), nor is development studies included in the index to Bourne's edited collection, *Development Education. Debates and Dialogues* (Bourne, 2008a). As Andreotti comments in her doctoral thesis, 'there are no formal structures that connect development studies to development education' (Andreotti, 2007, p. 35). The move away from development in DE partly explains this, as does this history of development studies in anthropology and fieldwork in the global South. But if development studies is concerned with the problematics and possibilities of enhancing well-being and life chances for the world's poor, then DE's focus on the role of critical reflexivity and education in achieving this, wherever it takes place, should surely be treated as part of what we understand as development practice.

References

Andreotti, V (2006) Soft versus critical global citizenship education. *Policy and Practice. A Development Education Review* 9: 40–51.

Andreotti, V (2007) A Postcolonial Reading of Contemporary Discourses Related to the Global Dimension in Education in England. *Schools of Education and Critical Theory and Cultural Studies.* Nottingham: University of Nottingham.

Baillie Smith, M (2008) International non-governmental development organizations and their Northern constituencies: Development education, dialogue and democracy. *Journal of Global Ethics* 4: 5–18.

Baillie Smith, M (2013) Public imaginaries of development and complex subjectivities: The challenge for development studies. *Canadian Journal of Development Studies/Revue canadienne d'études du développement* 34(3): 400–415.

Baillie Smith, M and Laurie, N (2011) International volunteering and development: Global citizenship and neoliberal professionalisation today. *Transactions of the Institute of British Geographers* 36: 545–559.

Baillie Smith, M, Laurie, N, Hopkins, P and Olson, E (2013) International volunteering, faith and subjectivity: Negotiating cosmopolitanism, citizenship and development. *Geoforum* 45: 126–135.

Biccum, A (2007) Marketing Development: Live 8 and the Production of the Global Citizen. *Development and Change* 38: 1111–1126.

Bondi, L and Laurie, N (2005) Introduction. Special Issue: Working the Spaces of Neoliberalism *Antipode* 37: 393–401.

Bourne, D (2003) Towards a theory of development education. *The Development Education Journal* 10: 3–6.

Bourne, D (ed.) (2008a) *Development Education. Debates and Dialogues.* London: Institute of Education.

Bourne, D (2008b) Introduction. In Bourne, D (ed) *Development Education. Debates and Dialogues.* (pp 1–17). London: Institute of Education.

Bourne, D and Hunt, F (2011) Research Paper 1: Global Dimension in Secondary Schools. Development Education Research Centre, Institute of Education.

Cohen, S. (2001) *States of Denial: Knowing about Atrocities and Suffering.* Cambridge: Blackwell.

Darnton, A and Kirk, C (2011) *Finding frames: The ways to engage the UK public in global poverty.* London: BOND.

DEA (2005) Available online at www.dea.org.uk/dea/deved.html. Accessed: 21 November 2006.

DFID (1999) *Building Support for Development.* London: DFID.

DFID (2010) Freeze on UK-based development awareness projects. Available online at www.gov.uk/government/news/freeze-on-uk-based-awareness-projects (accessed July 25, 2012).

Humble, D (2012) 'This isn't getting easier': Valuing emotion in development research. *Emotion, Space and Society* 5: 78–85.

Lewis, D (2006) Globalization and International Service: A Development Perspective. *Voluntary Action* 7: 13–26.
Selby, D and Kagawa, F (2011) Development education and education for sustainable development: Are they striking a Faustian bargain? *Policy and Practice. A Development Education Review* 12: 15–31.
Simpson, K (2004) 'Doing development': The gap year, volunteer-tourists and a popular practice of development. *Journal of International Development* 16: 681–692.
Sumner, A and Tribe, M (2008) *International Development Studies: Theories and Methods in Research and Practice.* London: Sage.

Part 9

Political economy of violence and insecurity

Editorial introduction

For many people concerned about development, the most disturbing aspects are political instability, conflicts, and war, particularly civil war, as illustrated by images of turbulence, and distraught and destitute fleeing people. Some states are in 'complex emergency', characterized by protracted crisis and the collapse of state structures. These prolonged crises create fragility of state structures and weaken the capacities of the state. To eradicate poverty and generate development, effective states are crucial and hence an analysis of state fragility has become important and a relevant concept in recent times. Human rights violations resulting from current conflicts and rising violence are unprecedented. The costs are to be measured in deaths, broken lives, destroyed livelihoods, lost homes and increased vulnerability.

Total numbers of official refugees are difficult to measure, with official data on refugee flows never consistently collated and some very large displaced groups never formally registered as refugees. Many governments are increasingly unwilling to recognize and take responsibility for displaced populations. 'Displacement' is not just an economic transaction, substituting property with monetary compensation; it also involves 'resettlement' and requires true 'rehabilitation'. Rich countries attempt concertedly to restrict the arrival of asylum seekers on their territory and their exit from war zones.

In civil wars civilians tend to be the object of fighting and bear most consequences. Material resources and social networks, which made daily life possible, are destroyed. Longstanding arrangements of exchange between groups are often forcibly broken down. Levels of violence are less regulated in civil wars. It is important to recognize that contemporary internal conflicts have beneficiaries as well as victims. They include young men whose livelihoods become bound up with ongoing violence, and 'war lords' powerful because no authority can impose the rule of law. In some places, such as parts of Afghanistan and Sudan, civil wars are so prolonged that they have become the norm. Much of the population has had no experience of peace.

The international arms trade increased the availability of modern military equipment, strongly affecting the way in which civil wars break out and escalate. It is well known that industrialized countries compete with each other to secure lucrative export contracts for military hardware and know-how. Increased access to highly effective military equipment has been a major catalyst. Small arms have become profitable for powerful organizations and commercial companies prepared to engage in illegal trade. The availability of these modern weapons has indubitably facilitated escalating violence. Similarly, violence against women and children increasingly concerns development agencies as women challenge structures and practices of subordination. It is also nowadays used as a political tool of war. Most of the gender- and age-based violence occurs in private spaces in diverse forms and remains invisible in many societies of the South, as it is viewed acceptable and hence under-reported. Most of the age-based violence is against children and the elderly.

Economic deprivation is both cause and consequence of war. Social stratification linked to highly inequitable distributions of income is not new, and has often caused violent conflict. Recently, however, the exclusion of populations from benefits of economic growth has assumed a more overtly structured conflict of interest and has regional dimension.

Many commentators have suggested that the end of the cold war allowed scores of ethnic groups to compete ferociously for power and influence, encouraging demands for greater autonomy for populations within states and the restricting of central government controls. Development policies have become intertwined with group aspirations and shaped nationalism and multiple identities. These identities are subject to interpretations, change over time, interact, oppose and complement each other. National identities based around religion and ethnicities are always viewed with suspicion. It is important to understand the relationship between religion and society and faith traditions and development and its implications for poverty reduction in developing countries and how beliefs, practices and organizations change over a period of time especially in the context of economic development and globalization. Multilateral organizations such as the World Bank and the Department for International Development (UK) are keen to explore the potential of faith-based groups as agents of development.

The first Article of the United Nations Charter committed governments to maintain international peace and security, to take effective collective measures to prevent and remove threats to peace, and to suppress acts of aggression. Yet international response to conflict appears ever more inadequate. The United Nations operations have proved very costly, in terms both of the funding required and of political credibility.

International NGOs have gained considerable experience of providing relief aid to traumatized populations. Interventions not intended to be sustainable have created aid dependency (for example, in Mozambique). The possibilities of providing for human security do not seem to have been enhanced. Embarrassing failures have resulted; the effectiveness of humanitarian aid has been questioned especially in the context of an increase in natural disasters and a change in the nature of emergencies leading to substantial increase in humanitarian assistance. The "War on Terror" and the subsequent interventions in Iraq and Afghanistan post-11 September 2001 have created new challenges for the implementation of humanitarian assistance, human security, peace-keeping roles, partnership for post-conflict reconstruction, promotion of peace, good governance and the increasing securitisation of development and civil society in the field of international development which also got absorbed into global and national security agendas.

9.1
Gender- and age-based violence

Cathy McIlwaine

Violence is firmly established as an important development issue. This is due to global increases in everyday violence, the globalisation of crime and violence, and the recognition that violence undermines sustainable development (Moser and McIlwaine, 2006). An important dimension of these debates is gender- and age-based violence. While these types of violence are now widely recognised as restricting women's and young people's freedom of participation in society, impeding the efficiency of development interventions and eroding their human rights, the international community has been slow to respond. Indeed, only in 1989 were the rights of children to protection from various forms of violence recognised in the United Nations Convention on the Rights of the Child. Similarly, the elimination of violence against women was only formally called for in 1993 through a United Nations declaration. This relates to the invisible nature of much gender- and age-based violence, the fact it often occurs in the private sphere, is often deemed to be culturally acceptable and is associated with high levels of stigma.

Identifying gender-based violence

While definitions of violence are highly contested, it usually involves the use of physical force that causes hurt to others in order to impose a wish or desire. The primary motivating factor behind perpetration, either conscious or unconscious, is the gain and maintenance of power which may be political, economic or social (Moser and McIlwaine, 2006). All violence is inherently gendered although gender-based violence is distinguished where the gender of the victim of violence is directly related to the motive for the violence. One of the main definitional issues has been the shift from the term 'violence against women' to 'gender-based violence'. Although some conflate these, strictly, violence against women is one type of gender-based violence together with violence against men, boys and transgendered. Some argue that using the term gender-based violence de-politicises it and diverts attention from the reality that women and girls suffer disproportionately from violence at the hands of men.

Definitions of gender-based violence are contested. The benchmark is usually the 1993 UN Declaration of the Elimination of Violence against Women in Article 1: 'Any act of gender-based violence that results in, or is likely to result in, physical, sexual or psychological harm or suffering to women, including threats of such acts, coercion or arbitrary deprivations of liberty, whether occurring in public or in private life'. Article 2 continues that it may occur in the 'family, community, perpetrated or condoned by the state, wherever it occurs' and may refer to assault, sexual abuse, rape, female genital mutilation and other 'traditional' practices, as well as sexual harassment, trafficking in women, and forced prostitution. These have also been articulated further in the Beijing Platform for Action with the newly created UN Women (established in 2010) identifying 'Violence against Women' as one of its seven priority areas 'that are fundamental to women's equality, and that can unlock progress across the board'. Central to these definitions is that gender-based violence occurs not only in the private sphere thus challenging its invisibility and its associated impunity.

There is a huge diversity in types of gender-based violence, especially against women. Some types are also specific to particular cultures and countries. These may include acid-throwing in Bangladesh, involving men throwing acid on women's faces as a form of attack (Zaman, 1999) or female genital mutilation that is concentrated in Northern African and Middle Eastern societies (Rahman and Toubia, 2006), as well as the 'femicides' in Latin America (Prieto-Carrón *et al.*, 2007). Women also commit or collude with gender-based violence against other women in their role as mother-in-law, noted widely in South Asian societies.

One crucially important type of gender-based violence is the use of rape as a tool of war (Moser and Clark, 2001). The recognition of sexual violence against women as an act of conflict rather than individual act, with women being valued as peace-builders and decision-makers has been enshrined in the UN Security Council Resolution 1325 in 2000 (and subsequent resolutions 1820, 1888, 1889 and 1960). These resolutions also highlight high levels of violence against women by state and non-state armies and peace-keeping forces. 'Peace and Security' is another key priority area for UN Women. Although the focus here is on women as victims of armed conflict, they can also be perpetrators as combatants and soldiers. In addition, men can be victims of male-on-male violence and rape as detainees and prisoners of war.

Trends in gender-based violence

Under-reporting of gender-based violence is widespread despite calls to improve data collection. This is mainly due to the taboo attached to it in many cultures, as well as the sensitivity surrounding data collection. Yet, data sources have improved dramatically in recent years, although they remain inconsistent.

Despite this, there is a consensus that violence against women is more prevalent than first thought. As many as 71 per cent of women experienced physical or sexual assault by an intimate partner in rural Ethiopia (Table 9.1.1). There is also evidence that violence against women by male partners is less prevalent in cities than rural areas, while violence by a

Table 9.1.1 Proportion of adult women experiencing physical and/or sexual violence by intimate partner and non-partner by rural/urban residence

Country	*% of adult women ever experienced physical and/or sexual violence by intimate partner*		*% of adult women ever experienced physical and/or sexual violence by non-partner*	
	Rural	Urban	Rural	Urban
Bangladesh	62	–	10	–
Brazil	37	29	23	40
Ethiopia	71	–	5	–
Namibia	–	36	–	23
Peru	69	51	18	31
Tanzania	56	41	19	34
Thailand	47	41	14	20

Source: Adapted from WHO (2005: 6, 13); McIlwaine (2008: 446–447)

Table 9.1.2 Women's attitudes towards wife beating

	Husband justified in hitting or beating wife (% of women who agree)									
	Burns food		*Argues with him*		*Goes out without telling him*		*Neglects children*		*Refuses to have sex with him*	
	Rural	Urban	Rural	Urban	Rural	Urban	Rural	Urban	Rural	Urban
Benin 2001	35.3	20.2	45.5	30.6	50.6	34.3	59.3	38.9	19.4	13.5
Cameroon 2004	25.8	13.8	33.3	21.1	40	28.2	48.7	42	27.1	14.8
Ethiopia 2005	67.5	30.8	63.9	34.6	69.1	41.5	69	44.2	49.6	19.8
Nigeria 2003	35	22.7	47.8	35.4	58.4	42.2	53.8	41.1	42.5	28.1
Egypt 2005	25	10.4	47	23.8	49.4	27.4	48.6	27.1	43	20.1
Nepal 2001	5.1	3.9	8.8	8	12.1	13.2	24.8	29	3.1	2.7
Philippines 2003	4.3	2.2	7.1	3.7	12.6	6	25	17.3	4.5	2.4
Bolivia 2003	9.2	3.6	9.9	4.9	12.9	7.2	19.9	15.3	4.5	1.9
Haiti 2000	14.4	6.6	14	6.4	34.4	23.4	33.2	21.6	16.9	10.2
Nicaragua 2001	7.9	2.6	7.5	2.6	9.9	3.4	13.4	7.2	4.8	1.7

Source: Macro International Inc (2010): http://www.measuredhs.com/(accessed 10 January 2012)

non-partner is higher in urban areas (see Table 9.1.1). These patterns also prevail in relation to beliefs surrounding violence against women by intimate partners. Women are more likely to report that husbands are justified in beating or hitting their wives in rural areas compared with cities, especially when they went out without telling him or when they neglected their children (see Table 9.1.2). While even less is known about the prevalence of violence against men, a WHO (2005: 20) study found that only in Thailand did more than 15 per cent of physically assaulted women report having initiated violence against their male partner more than twice in their life.

Causes of gender-based violence

There are multiple causes of gender-based violence. In general, it is rooted in ideological differences between women and men related to the concentration of power in men's hands and the construction of gender identities, especially hegemonic masculinities (Barker, 2005).

One set of explanations focuses on accepting male violence as 'natural' and rooted in biological differences, making it difficult to change. The second, shaped by feminist scholarship, relates male violence to social constructions of patriarchal forces which may be prevented (O'Toole and Schiffman, 1997). It may also be linked with psychological factors in that men with 'impaired masculinity' may abuse women, influenced by socialisation processes involving witnessing violence. Socioeconomic factors such as increased poverty may exacerbate violence, especially when masculinity is undermined. Gender-based violence is also aggravated by a range of risk factors that may be 'triggered' by infidelity, male alcohol or drug abuse (see Plate 9.1.1), an absent or rejecting father, low educational level and high rates of employment or HIV status among women (Morrison *et al.*, 2007).

Identifying age-based violence

Age-based violence involves the abuse of those who are vulnerable in terms of their age. This may 'involve physical or mental violence, injury and abuse, neglect or negligent treatment,

(a)

(b)

Figure 9.1.1 Police officers

Source: (a) Michael Keith and (b) Cathy McIlwaine

maltreatment or exploitation, including sexual abuse' (Pinheiro, 2006: 4). While this is based on the UN Convention on the Rights of the Child Article 1, it can equally be applied to elderly people. However, while most analysis of age-based violence in the South relates to violence involving children and youth rather than elder abuse, the latter has emerged as important not least because the world's population of people aged 60 years and older will more than double by 2025 with the majority being women. Although most elder abuse is against older women, older men can also be at risk of abuse on the part of spouses, adult children and other relatives. There are also specific types of elder abuse that can be culturally and socially sanctioned in some societies. This can include mourning rites for widows that can entail sexual violence, force levirate marriages (where a woman is forced to marry her brother-in-law when she is childless), as well as violence against women accused of practising witchcraft. For example, in Tanzania, around 500 older women accused of being witches are murdered every year with many more being ostracised as a result of accusations and their land being taken from them (Krug *et al.*, 2002). Socioeconomic changes are often identified as the main reasons for elder abuse in developing societies linked with migration weakening extended family networks, and an erosion of respect for the elderly. It has also been suggested that older people are especially at risk because they are often vulnerable economically and more likely to be in ill health than younger people. This also means that the elderly are less able to cope with such abuse and to escape from risky situations.

Turning to violence against young people, Pinheiro (2006) usefully identifies the main places where violence against children occurs as the home and family, schools, care and justice systems, workplaces and communities. Although the home and family is where children can be protected against violence, it is also where they can suffer most, especially younger children aged under one year. Here, physical and sexual assault occur, together with traditional practices such as female infanticide, psychological abuse and neglect. In schools, physical and psychological violence against children may be perpetrated by teachers or by other children through bullying, playground fighting and gangs. Abused and neglected children may end-up in care homes, where their situation is likely to worsen; this particularly affects AIDS orphans (globally there were 15 million in 2006). The workplace can be dangerous for children who may suffer at the hands of employers, co-workers and customers. Child slavery is also widespread especially among domestic and sex workers (Pinheiro, 2006).

Most is known about violence against children in the community because of its visibility. Street children have received the most attention in this sphere, and are associated with petty theft. This is reflected in some of the slang words used for street children such as 'the plague' (Colombia) or 'vermin' (Ethiopia). Because of their association with crime, they are often targeted and killed. Gang activity is also important and growing. Largely urban-based, there is a huge diversity of gangs, most being male-dominated. They have different generic names, such as *Maras Salvatrucha* in Central America, as well as names such as the Area Boys of Lagos, Nigeria. Gang membership often reflects an attempt by young men to assert their identity within dysfunctional families and as a response to high rates of unemployment (Moser and McIlwaine, 2004). Children also suffer severe psycho-social effects as witnesses of war, as well as committing violence as child soldiers.

Trends and causes of violence against young people

Although it is often assumed that young women and girls are more likely to experience abuse, there has been a steep increase in rates of victimisation and perpetration among boys aged 15 and over. Indeed, globally 77 per cent of all homicides are male, with most aged 15–29 (Krug *et al.*, 2002: 10). Non-fatal violence is also higher among boys than girls. However, girls are more likely to experience sexual violence; an estimated 150 million girls and 73 million boys have experienced sexual violence (Pinheiro, 2006).

The reasons for aged-based violence also revolve around power. Their unequal position in the home and society makes children (and the elderly) vulnerable targets for those stronger than themselves. Economic development is also important in that the rate of homicide of children in 2002 in low-income countries was twice that of high-income countries (Krug *et al.*, 2002). However, besides those who are forced, committing violence can be empowering for children. There is often a continuum of causality that runs over generations in that those experiencing or witnessing violence in the home may become perpetrators in later life (Moser and McIlwaine, 2004).

In terms of looking to the future, the scale of gender- and aged-based violence must be recognised, as well as the fact that it undermines sustainable development. The Millennium Development Goals highlight both gender- and age-based inequalities and recognise that these groups must be protected. Yet, concrete policies and campaigns must be developed further. The groundwork has already been laid by organisations such as UN Women and UNICEF, but much remains to be done.

References

Barker, G. T. (2005) *Dying to be Men: Youth and Masculinity and Social Exclusion*, London: Routledge.

Krug, E. G., Dahlberg, L. L, Mercy, J. A., Zwi, A. B. and Lozano, R. (eds) (2002) *World Report on Violence and Health*, Geneva: World Health Organisation.

Macro International Inc (2010) http://www.measuredhs.com/ (accessed on 10/01/12).

Morrison, A., Ellsberg, M. and Bott, S. (2007) Addressing gender-based violence: A critical review of interventions, *The World Bank Observer*, 22(1): 25–51.

Moser, C. O. N. and Clark, F. (eds) (2001) *Victims, Perpetrators or Actors? Gender, Armed Conflict and Political Violence*, London: Zed.

Moser, C. O. N. and McIlwaine, C. (2004) *Encounters with Violence in Latin America*, Routledge: London.

Moser, C. O. N. and McIlwaine, C. (2006) Latin American urban violence as a development concern: Towards a framework for violence reduction, *World Development*, 34(1): 89–112.

O'Toole, L. L. and Schiffman, J. R. (eds) (1997) *Gender Violence: Interdisciplinary Perspectives*, New York and London: New York University Press.

Pinheiro, P. S. (2006) *World Report on Violence Against Children*, United Nations Secretary General's Study on Violence Against Children, Geneva: United Nations.

Prieto-Carrón, M., Thomson, M. and Macdonald, M. (2007) No more killings! Women respond to femicides in Central America, *Gender and Development* 15(1): 25–40.

Rahman, A. and Toubia, N. (eds) (2006) *Female Genital Mutilation*, London: Zed.

WHO (2005) *Multi-country Study on Women's Health and Domestic Violence against Women*, Geneva: World Health Organisation.

Zaman, H. (1999) Violence against women in Bangladesh: Issues and responses, *Women's Studies International Forum* 22(1): 37–48.

Further reading

Gender and Development (2007) Special issue on 'gender-based violence', 15(1). Considers gender-based violence as a development issue and includes case studies from a range of countries.

Krug, E. G., Dahlberg, L. L, Mercy, J. A., Zwi, A. B. and Lozano, R. (eds) (2002) *World Report on Violence and Health*, Geneva: World Health Organisation. A global report on violence from a health perspective.

Pickup, F., with Williams, S. and Sweetman, C. (2001) *Ending Violence Against Women: A Challenge for Development and Humanitarian work*, Oxford: Oxfam. Summarises violence against women from research and policy perspectives.

Pinheiro, P. S. (2006) *World Report on Violence Against Children*, United Nations Secretary General's Study on Violence against Children, Geneva. Provides an overview of the incidence, causes and consequences of violence against children (available to download: http://www.unviolencestudy.org/).

UN Women (2011) The United Nations trust fund to end violence against women, New York: UN Women. Outlines recent projects that have been funded to end violence against women (available to download: http://www.unwomen.org/wpcontent/uploads/2011/03/UNTF_AnnualReport2010_en.pdf).

9.2
Fragile states

Tom Goodfellow

The concept of a 'fragile state' is rooted in the end of the cold war and has been evolving constantly since; it is now used in such differing ways by a growing range of development agencies and academics that it is increasingly difficult to pin down. Nevertheless, the term reflects important ongoing efforts in recent decades to understand the relationship between patterns of violent conflict and the nature of the state in less developed countries. Today even market-oriented donor agencies such as the World Bank acknowledge that effective states fundamentally matter for development. The civil conflicts ravaging many parts of the globe not only cripple the capacities of states, but according to some analyses are caused by the weakness or fragility of state structures and institutions. State fragility is therefore likely to remain a relevant concept; however, clearer specification and more rigorous analysis are necessary to maximise its utility for development researchers and practitioners.

Emergence of the 'failed states' paradigm

The idea of 'state fragility' was preceded by that of 'state failure', which was essentially born after the fall of the Berlin Wall. During the cold war, superpower rivalry allowed little scope for states to collapse completely. Civil conflicts were rife, as were failures on the part of many governments to deliver basic services to their citizens; but the United States and Soviet Union propped up client regimes and intervened militarily in countries where spiralling conflict threatened their geopolitical interests. With the demise of the Soviet empire in the early 1990s, however, states imploded in places as diverse as Yugoslavia to Somalia. The proliferation of extremely violent civil conflicts around this time, often characterised as 'new wars' and linked to processes of globalisation and international criminal networks (Kaldor 1999), represented state failure in a very fundamental sense (Helman and Ratner 1992). At its most basic, state failure can thus be seen as the loss of the monopoly of violence by the central government within a given territory, reflecting a classically Weberian conception of the state (Ignatieff 2002).

However, notwithstanding the tumultuous events in Eastern Europe and parts of Africa in the 1990s, it was September 11, 2001 that really projected the idea of failed states centre stage. Anarchy in developing countries was increasingly perceived as fostering international terrorism, so the capacity of far-flung states to govern their own territories became viewed as an urgent concern for developed countries in the global North (Beall *et al.* 2006). Rotberg aptly captured this mood when he noted that after the 2001 attacks the problem of failed states acquired 'an immediacy and an importance that transcends its previous humanitarian dimension' (Rotberg 2002: 127). At the same time, research in the early 2000s claimed that civil wars were in part caused by 'weak' statehood (Fearon and Laitin 2003), further bolstering the idea that building the capacities of states was critical for global security.

This renewed concern with state capacity was welcome, and a necessary rebalancing after the late twentieth-century's overenthusiastic heralding of the 'retreat of the state' (Strange 1996). Nevertheless, the 'failed states' discourse was highly problematic. The term was often used ideologically in reference to states deemed undesirable by the world's remaining superpower, the United States, even though cases such as Iran (or indeed pre-2003 Iraq) were hardly 'failed states' even if their governing regimes were considered abhorrent. The concept of 'failed states' thus came to have intrinsically normative undertones, begging the question of how one should define its opposite, a 'successful' state. Some authors propose broad lists of core functions to 'fix' failed states that go far beyond the concern with conflict from which the idea arose, and which effectively stigmatise almost all developing countries as 'failing' (Ghani and Lockhart 2008).

Let them fail? Echoes of European history

A related problem with the idea of state failure is that it raises the question of when the international community should intervene to bolster statehood or 'save' populations from their failing state. This is a pertinent question in view of the history of nation-states in more developed parts of the world: Tilly, for example, has argued that the process of state-building in medieval Europe was essentially a by-product of war and the need to raise taxes to fund it (Tilly 1992). This idea that historically 'war made the state and the state made war' has been influential in contemporary debates on state failure, implying that it may be through the intrinsically violent efforts of 'coercion-wielders' to establish dominance over territory that capable states are consolidated. Some scholars effectively argue in favour of 'giving war a chance': if certain *de jure* states cannot provide security within a given territory then the international community should 'let them fail', enabling capable de facto states to emerge rather than protecting pseudo-states that exist in name only (Herbst 2004).

To some degree the international community has indeed been 'allowing' failure in places such as Somalia and the Democratic Republic of Congo (DRC). However, even these virtually stateless places are treated as sovereign states in international law – a crucial difference from the period of European history discussed by Tilly. No matter how much a state today may be 'failing', its borders are reified by the international system, which some scholars argue undermines processes that could lead to successful statehood (Englebert 2009). Yet also unlike in medieval Europe, in our globalised world of 24-hour media coverage the idea of completely abandoning nations to their fate is simply unacceptable to most people, even if it might produce more capable states in the long run. We therefore face a situation in which states with very little capacity do exist, but letting them collapse and vanish from the map is unpalatable, while simply labelling them 'failed' is unhelpful. The category of 'fragile states' represents an acknowledgement of this inescapable part of the contemporary international landscape.

From 'failure' to 'fragility'

The state fragility concept represents a more nuanced approach the enduring problems faced by states in conflict-affected parts of the world, recognising degrees of state weakness and not stigmatising states as 'failed' if they underperform. As the 2000s wore on, donor organisations including the World Bank, OECD and the UK Department for International Development adopted the language of fragility. In many respects this discursive shift is a positive move, resulting in more aid going to the countries that need it most (Putzel 2010) as well as shedding some ideological baggage of 'state failure' and refocusing more keenly on the specific issue of vulnerability to conflict (Naudé *et al.* 2011). The propensity to conflict is critical to defining fragile states; yet specifying exactly what makes states vulnerable in this way has proved problematic. Focusing solely on the monopoly of legitimate violence within a given territory is insufficient and renders the concept of fragility somewhat tautological.

The key problem in operationalising state fragility is therefore defining which capacities states have to be deficient in to be considered fragile, and how deficient they have to be. The definitions used by many development agencies are extremely broad. A typical example is that used by the OECD, which defines fragile states as countries 'where there is a lack of political commitment and insufficient capacity to develop and implement pro-poor policies'. As Putzel notes, this definition fails to distinguish between states that are particularly prone to violent conflict and those characterised by poverty and underdevelopment generally. In other words, such definitions render fragile states equivalent to all developing countries, making the concept somewhat redundant (Putzel 2010).

To clarify the nature of fragility it is helpful to consider what exactly constitutes the opposite of a fragile state. Just as the idea of 'failed states' intrinsically implied a normative ideal of 'successful' ones, it is difficult to conceive of fragility without defining its counterpart. Some commentators conceptualise the opposite of a fragile state as a 'developmental state' (Naudé *et al.* 2011: 5), but again this is problematic, not least because of the association of this term with the East Asian 'tigers', suggesting that any less successful states (i.e. all developing countries) are fragile. In fact, it is critically important to distinguish between underdeveloped countries that are particularly prone to conflict and those that are not. For example, while Tanzania is still very poor and would hardly qualify as a 'developmental state', it is clearly not fragile in the way that the neighbouring DRC is. Having experienced no large-scale internal conflict, it can be thought of as resilient. Indeed, the logical opposite of a fragile state is a not a developmental one but, as Putzel (2010) notes, a resilient state.

Understanding exactly what makes states resilient to conflict (rather than 'developmental', which may depend on different factors), is therefore the central problem of fragility. The work of the Crisis States Research Centre at the London School of Economics provides some insights in this regard, focusing on four fundamental characteristics of fragility. As well as the monopoly of the legitimate use of force, they highlight the degree to which (a) states possess bureaucratic capacity, as represented by a monopoly over taxation; (b) institutions (or rules) of the state 'trump' those of non-state organisations (such as rebel groups or traditional authorities); and (c) states attain territorial 'reach' within their borders (Putzel 2010). Ultimately, the extent to which a state achieves these characteristics of resilience depends on the nature of the political settlement within a given territory, particularly in terms of the 'bargains' struck among elites (Di John and Putzel 2009). Resilience is thus a profoundly political matter, about far more than enhancing a state's technical capacities.

Conclusion

Fragile states are a historically specific feature of the contemporary global order, reflecting the proliferation of *de jure* states since decolonisation alongside the limited willingness and ability of the rich world to intervene and bolster their resilience. Despite the very real changes wrought by globalisation, the significance of the state is clearly not in retreat; without resilient and effective states people in many parts of the developing world will continue to suffer from chronic insecurity, with state fragility and violent conflict perpetuating one another in a vicious cycle. Breaking this cycle means not only attempting to end violence but also supporting politically contentious processes of state building. This is difficult terrain for donors, who are wary of engaging explicitly in politics; but it may ultimately be the only substitute for the processes that built states historically in developed parts of the world, which often involved letting violent conflict itself play out.

References

Beall, Jo, Thomas Goodfellow, and James Putzel. 2006. Introductory article: On the discourse of terrorism, security and development. *Journal of International Development* 18 (1): 51–67.

Di John, Jonathan, and James Putzel. 2009. *Political Settlements: Issues Paper.* Birmingham: Governance and Social Development Resource Centre.

Englebert, Pierre. 2009. *Africa: Unity, Sovereignty, and Sorrow.* Boulder, CO: Lynne Rienner.

Fearon, James D., and David D. Laitin. 2003. Ethnicity, insurgency, and civil war. *The American Political Science Review* 97 (1): 75–90.

Ghani, Ashraf, and Clare Lockhart. 2008. *Fixing Failed States: A Framework for Rebuilding a Fractured World.* Oxford: Oxford University Press.

Helman, Gerald B., and Steven R. Ratner. 1992. Saving failed states. *Foreign Policy* 89: 3–20.

Herbst, Jeffrey. 2004. Let them fail: State failure in theory and practice. In Robert Rotberg, (ed.) *When States Fail: Causes and Consequences.* Princeton, NJ: Princeton University Press. 302–318.

Ignatieff, Michael. 2002. Intervention and state failure. *Dissent,* Winter.

Kaldor, Mary. 1999. *New and Old Wars: Organized Violence in a Global Era.* Stanford: Stanford University Press.

Naudé, Wim, Amelia U. Santos-Paulino, and Mark McGillivray. 2011. *Fragile States: Causes, Costs, and Responses.* Oxford: Oxford University Press.

Putzel, James. 2010. Why development actors need a better definition of state fragility. *Policy Directions,* Crisis States Research Centre (September).

Rotberg, Robert I. 2002. Failed states in a world of terror. *Foreign Affairs,* July/August.

Strange, Susan. 1996. *The Retreat of the State: The Diffusion of Power in the World Economy.* Cambridge: Cambridge University Press.

Tilly, Charles. 1992. *Coercion, Capital and European States, AD 990–1992.* Oxford: Blackwell Publishing.

Further reading

Di John, Jonathan. 2010. The concept, causes and consequences of failed states: A critical review of the literature and agenda for research with specific reference to sub-Saharan Africa. *European Journal of Development Research* 22 (10–30). This article provides an overview of some of the main themes in the academic literature on failed and fragile states.

Gutiérrez, Francisco, D. Buitrago, A. González, and C. Lozano. (2011). *Measuring Poor State Performance: Problems, Perspectives and Paths Ahead.* London: London School of Economics and Political Science. A thorough analysis of the problems involved in measuring aspects of state fragility, and critique of existing fragile states indices. Available online at http://www.dfid.gov.uk/r4d/Output/186126/Default.aspx

Naudé, Wim, Amelia U. Santos-Paulino, and Mark McGillivray. (2011). *Fragile States: Causes, Costs, and Responses.* Oxford: Oxford University Press. An edited collection with a range of perspectives encompassing some of the latest thinking on state fragility.

Putzel, James. 2010. Why development actors need a better definition of state fragility. *Policy Directions,* Crisis States Research Centre (September). A short policy paper that provides a succinct overview of some of the main findings of the Crisis States Research Centre. Available online at http://eprints.lse.ac.uk/41300/1/StateFragilityPD.pdf

Rotberg, Robert I. (2004). *When States Fail: Causes and Consequences.* Princeton: Princeton University Press. An edited collection containing many important perspectives on the idea of state failure, including the above-mentioned chapter from Jeffrey Herbst.

Websites

Crisis States Research Network homepage: http://www2.lse.ac.uk/internationalDevelopment/research/crisisStates/Home.aspx. Many publications from this 10-year project funded by UK DFID are available for download here.

Fund for Peace homepage: http://www.fundforpeace.org/global/.This organisation publishes an interactive 'failed states index' every year, which can be arranged in accordance with the different criteria considered constitutive of 'failure'.

World Bank's 2011 report on Conflict, Security and Development: http://wdr2011.worldbank.org/fulltext. The report is available for download and contains a wide range of statistics and perspectives on conflict and fragility in the twenty-first century.

9.3 Refugees

Richard Black and Ceri Oeppen

Introduction

The relevance of a chapter on refugees in a compendium on development may not be immediately obvious. The plight of refugees might be seen, at best, as irrelevant to the goals of international development agencies. There is no 'Millennium Development Goal' relating to refugees or directly to the causes of refugee flight. Indeed, the production of refugees might be seen as the *opposite* of development, especially as development has come to be defined more in terms of human rights, democracy and the rule of law.

Yet refugees are more than simply symptoms of the absence of development. Analysis of refugee flows both within the developing world and from 'South' to 'North' throws up discourses, policies and socioeconomic processes that directly parallel wider development issues. In this context, interesting opportunities exist for enhanced learning and understanding in both directions between 'refugee studies' and 'development studies'.

Defining refugees

Refugees exist in both North and South. In the North, they are commonly defined (by states) according to the 1951 Geneva Convention on refugees, as those who have a well-founded fear of being persecuted for reasons of race, religion, nationality, membership of a particular social group or political opinion. In contrast, in many Southern countries a wider definition is often employed, as in Africa where 'victims of war or external aggression' are included, and accordingly far greater numbers proportionally are held to fall within the purview of 'refugee policy'. Based mainly on government figures, which involve a mixture of these two definitions, the United Nations High Commissioner for Refugees (UNHCR) estimated that at the end of 2011 there were some 15.2 million refugees worldwide. Of these, 45 per cent were living in countries with a GDP below US$3,000 per capita. There were also a further 876,100 asylum seekers – those who had claimed refugee status but whose claims were not yet recognized. Most of these were in Europe and North America but South Africa received 107,000 applications (UNHCR, 2012).

Such geographical variations in the definition of who is a refugee stand as an initial warning of the lack of robustness of the category, and the potential for confusion. Indeed, there have been attempts both to narrow and to widen the definition of refugees. States in both the North and, increasingly, the South, have sought to justify more restrictive immigration policies, by cracking down on those they label as 'bogus' asylum seekers. The result has been a progressive narrowing of practical applications of the refugee definition, with a parallel reduction in the number of refugees reported by UNHCR compared to the 1990s.

At the same time, though, moves have come from both academic and policy quarters to widen the refugee definition – or at least to focus on the broader issue of 'forced migration' rather than a narrow definition of 'refugees'. As a result, terms such as 'internally displaced person' (IDP), 'environmental refugee', 'relocatee' and even 'economic refugee' have entered the literature. For example, some 15.5 million IDPs were either protected or assisted by the UNHCR in 2011 – in effect, treated as if they were refugees – while estimates of the number of 'environmental refugees' reach over 200 million, though such estimates – and indeed the concept itself – have been criticized widely. Although far from unproblematic in themselves, these additional categories of forced migration testify to the overlapping causes and consequences of migration. These range from situations of forced displacement across international borders as a result of essentially political events, to instances where state borders are not crossed, or where environmental degradation, large-scale development schemes or economic crisis play a significant role in producing migration.

A major part of this difficulty in defining refugees is the dual context in which the term is used. In the policy world, the term 'refugee' is first and foremost a legal category. It confers rights to protection, but also to assistance, and possibly the right to stay in a country that offers far better conditions in terms of its economic opportunities or human rights record (Shacknove 1985). Yet the transplanting of the term to social science as a term with any explanatory power or analytical worth requires at the very least a consistent application of the term by policymakers, and preferably a conceptual basis that remains independent of its policy implementation. It is far from clear that either exists. Even a simple division of migration into 'forced' (refugees) and 'voluntary' (not refugees) is problematic, as people who are displaced have varying degrees of choice over when, where and how to move, even in conditions of extreme crisis.

Explaining forced migration

If the term 'refugee' itself is not self-explanatory, the next step is to explore different explanations of forced migration, to see whether any more consistency can be found. One view is that increasing numbers of refugees and forced migrants are the result of a breakdown of order in a post-cold war world. During the cold war, there was a perception in the West that refugees were mainly those fleeing persecution under communist regimes, even if some observers recognized that proxy wars being fought between the USA and the Soviet bloc had also produced mass flows of refugees in Africa and Asia. In contrast, in the post-cold war world, the 'security' of superpower rivalry is argued to have given way to a much more localized and bloody series of wars and internal conflicts. In these new wars, 'tribal' or 'ethnic' hatreds are seen as driving an ever-expanding number of refugees out of their homes. This is linked to environmental degradation caused by poverty, conflicts over increasingly scarce natural resources and a declining commitment to development aid among the world's leading donor nations.

Such a perspective, at the very least, suggests that moves to draw up typologies of refugees and forced migrants, and to identify the distinguishing features of different categories, may be misplaced. From this standpoint, the wars that cause displacement increasingly mix political, economic, social and environmental factors. The result is a new vicious circle of poverty, violence and displacement that makes a rigid categorization of refugees as 'political', 'economic' or 'environmental' highly problematic. Meanwhile, as poverty, violence and displacement are interlinked, the promotion of development – especially a pro-poor, pro-human rights and pro-good governance version of development – might be seen as a potential solution to the world's 'refugee crisis'. Yet much as it is compelling, such a perspective arguably misses both the main causes of forced migration and the importance of the links between refugees and development.

A first clue is provided by the one form of forced migration that does not fit neatly into the downward spiral hypothesis of poverty, violence and displacement – namely, so-called 'development-induced displacement'. Those who have been displaced by dams and other major public works bear many of the hallmarks of the refugee – they are forced to leave their homes and land, often at short notice and with little or no compensation. Yet by definition their flight is caused not by a lack of development, but by 'development' itself, albeit a particular vision of what development should be.

Second, a number of major wars that have produced refugees in the last decade can be seen not as conflicts of poverty, but as wars driven by political elites seeking power over resources, territory or the state (or all three). Recent conflicts in the Middle East, including Iraq, Libya and Syria have all produced significant flows of people across borders to neighbouring countries, with an estimated 2.8 million Iraqis still displaced (UNHCR, 2012) as a result of intervention by a US-led coalition force. Elsewhere, war was taking place between rival factions, often fighting for control of territory and access to lucrative business opportunities, while appealing to ethnic, religious and other divisions. Where environmental factors are relevant, it is as often abundance, rather than lack of natural resources that leads to conflict. Thus in countries as diverse as Libya, Congo, and Angola, access to major deposits of diamonds, oil and other minerals has both caused and sustained long-running conflicts which have displaced millions of people. In these countries also, external intervention has often been neither absent nor benign, as Western companies (and even, in some cases, aid agencies and 'peace-keeping' forces) have played a partisan role in bolstering the position of one or more of the warring factions.

Refugees and the discourse of development

The above discussion suggests that there are links between development and the production of forced migration, even if the latter is not the simple consequence of the former, or of its absence or failure. Meanwhile, such links do not stop at the causes of refugee flight. In their responses to the humanitarian crises created by massive refugee flows, international aid agencies and donor governments have mirrored the policy shifts that have occurred in the development field. They have also used similar representations of the 'other' as they try to promote what Silk has described as 'caring at a distance' (Silk 1997). This leads us to probe further the connections between refugee and development discourse.

In the 1980s, discussion of the need to promote a 'relief to development continuum' was laudable in principle. Yet it led to grandiose projects for refugee resettlement based on development ideas that were already well out of date. In a review of over 100 such schemes in Africa since 1962, Kibreab (1989) found that just nine had reached the point of self-sufficiency by 1982. In the 1990s, attention shifted to the basic needs of refugees – food, water, shelter and health care – before moving on to a growing concern with environmental awareness and ensuring the sustainability of emergency interventions (Black 1998). Yet, at each stage, external interventions have been dogged by bureaucratic, technical and political problems, linked in grand part to a failure to heed the warnings derived from development experience.

Yet another, and arguably more important, element of these problems of refugee (and development) assistance has been the characterization of both the intended recipients of aid (passive, powerless, poor) and the problem itself (technical, unicausal, easily defined). Scholars have highlighted examples of refugees supporting themselves and their relatives rather than being solely passive recipients of assistance. This has been, in part, influenced by a wider trend that has looked at the development impact of migrants' transnational networks. Although much of this literature focusses on 'voluntary migrants', many refugees are also transnational actors (Koser, 2007).

Although development and refugee agencies have grappled with how to promote the participation of 'beneficiaries' in assistance programmes, it has proven difficult to break a pattern that Harrell-Bond (1986) describes as simply 'imposing aid'. Indeed, international humanitarian agencies have their own political and bureaucratic motives for involvement in refugee assistance, which may be far from benign or disinterested.

In this sense, refugee situations provide us with similar dilemmas and challenges to the broader development scene – how to deal with complex situations, in which poverty and need are not simply technical issues, but also political ones. Humanitarian aid may be driven as much by supply (excess food stocks, defence of strategic interests, guilt) as by demand (i.e. hunger or poverty). Increasing amounts of aid are provided by non-governmental organizations (NGOs), whose numbers have mushroomed. According to the Ministry of Economy, there were almost 300 international and 1,700 local NGOs operating in Afghanistan in 2012; the potential for problems of coordination and duplication are immense. The proliferation of agencies is encouraged by donor policies that favour the channelling of funds through NGOs.

It is important to hold up to scrutiny the role and motivations of these aid givers, as much as those who receive assistance. For example, while there has been some progress towards the definition of minimum standards for humanitarian assistance agencies, there is little to prevent NGOs from moving into refugee emergencies, whatever their objectives or capabilities. The results can be disastrous, as was seen in the 1994–96 emergency in eastern Zaire, when uncontrolled international agency activity may well have contributed to suffering, rather than relieving it. At the same time, whether we are concerned with emergencies or not, the task of engaging with recipients of aid requires us to understand people's own strategies and constraints, and resist the temptation to categorize and stereotype.

References

Black, R. (1998) *Refugees, Environment and Development*, Harlow: Longman.

Harrell-Bond, B. H. (1986) *Imposing Aid: Emergency Assistance to Refugees*, Oxford: Oxford University Press.

Kibreab, G. (1989) 'Local settlements in Africa: A misconceived option?', *Journal of Refugee Studies*, 2(4): 468–90.

Koser, K. (2007) 'Refugees, transnationalism and the state', *Journal of Ethnic and Migration Studies*, 33(2): 233–54.

Shacknove, A. (1985) 'Who is a refugee?', *Ethics*, 95(2): 274–84.

Silk, J. (1997) *The Roles of Place, Interaction and Community in Caring for Distant Others*, Geographical Papers No. 121, University of Reading, Department of Geography.

UNHCR (2012) *2011 Global Refugee Trends*, Geneva: United Nations High Commission for Refugees (available at http://www.unhcr.org/4fd6f87f9.html).

Further reading

Ager, A. (ed.) (1999) *Refugees: Perspectives on the Experience of Forced Migration*, London: Cassell Academic. This edited collection contains papers from a range of anthropological, sociological and psychosocial perspectives.
Black, R. (1998) *Refugees, Environment and Development*, London: Longman. Provides a critical analysis of both the growing literature on environmental causes of migration and attempts to make refugee assistance more closely linked to sustainable development priorities.
Harrell-Bond, B. H. (1986) *Imposing Aid: Emergency Assistance to Refugees*, Oxford: Oxford University Press. A classic study, by the scholar credited with initiating the field of 'refugee studies'. It provides a detailed anthropological account of the situation of Ugandan refugees in southern Sudan in the mid-1980s.
Horst, C. (2007) *Transnational Nomads: How Somalis Cope with Refugee Life in the Dadaab Camps of Kenya*, Oxford: Berghahn. An anthropological account of the transnational livelihood strategies of those living in a protracted refugee situation.
Jacobsen, K. (2005) *The Economic Life of Refugees*, Bloomfield: Kumarian Press. Explores the role of refugees as economic actors and suggests how refugee assistance could support a livelihoods and local development approach.
Mehta, L. (2009) *Displaced by Development: Confronting Marginalisation and Gender Injustice*, New Delhi: Sage. An edited collection that takes a gendered approach to the impacts of development-induced displacement in India.
UNHCR (2012) *The State of the World's Refugees 2012: In Search of Solidarity*, Oxford: Oxford University Press. Sets out basic facts and figures about refugees worldwide, and looks at key trends in refugee protection.
Zolberg, A. and Benda, P. (2001) *Global Migrants, Global Refugees: Problems and Solutions*, Oxford: Berghahn Books. This edited collection places the study of refugees within the wider context of international migration.

Websites

Forced migration online: www.forcedmigration.org
UNHCR: www.unhcr.org
Relief Web: www.reliefweb.int

9.4 Humanitarian aid

Phil O'Keefe and Joanne Rose

Background

Humanitarian aid is the assistance given to people in distress by individuals, organisations or governments with the core purpose of preventing and alleviating human suffering.

The principles of humanitarian intervention are impartiality, neutrality and independence. Impartiality means no discrimination on the basis of nationality, race, religious beliefs, class, gender or political opinions: humanitarian interventions are guided by needs. Neutrality demands that humanitarian agencies do not take sides in either hostilities or ideological controversy. Independence requires that humanitarian agencies retain their autonomy of action. These principles, originally drawn up for war and consolidated in humanitarian law expressed in the Geneva Convention of 1949, underlie response to conflict-related and natural disasters.

The international humanitarian system consists, principally, of four sets of actors: donor governments, including the European Commission Humanitarian Office (ECHO), the United Nations, the International Red Cross and Red Crescent Movement (ICRC) and international non-governmental organisations (INGOs). Local NGOs and beneficiaries have little voice in the system.

Since the early 1990s, there has been both an increase in the number of disasters and a change in the nature of emergencies leading to a substantial increase in humanitarian assistance. Natural disasters have grown in number largely due to increased climate variability, especially the increase in extremes. Complex emergencies (conflict) have increased, particularly since the end of the cold war, and are now characterised by high levels of civilian casualty, deliberate destruction of livelihoods and welfare systems, collapse of the rule of law and large numbers of displaced people (15.4 million refugees, 27.5 million internally displaced persons (IDPs) and a further 840,000 people waiting to be given refugee status as of 2011; UNHCR, 2011).

Emergencies have changed in nature from predominantly natural disasters, dominated by flood and drought, to complex emergencies and technological disasters. The 'War on Terror', following events on 11 September 2001 and the subsequent interventions in Afghanistan and Iraq, have created new challenges for the implementation of humanitarian assistance, not least the use of military to guarantee humanitarian space (i.e. to ensure the delivery of emergency aid) if not the military themselves delivering aid.

The increasing frequency and changing face of emergencies has caused humanitarian expenditures to soar. The overall humanitarian expenditure of Organisation for Economic Co-operation and Development (OECD) Development Assistance Committee (DAC) member governments – the major contributors to crises – increased from US$11.2 billion in 2009 to US$11.8 billion in 2010. This substantial increase, however, was made by just three donors (the United States, Japan and Canada) and does not reveal the reductions made by most other OECD DAC members (Development Initiatives, 2011). While the overall response to humanitarian crises illustrates an upward trend, many governments are under increasing pressure to justify existing levels of aid spending. These figures do not account for charitable donations from individuals or groups, such as churches and they do not capture non-Western assistance such as that provided by Islamic entities.

The United Nations as lead agency has an effective division of labour that sees the Office for the Coordination of Humanitarian Assistance (OCHA) in charge of the policy and planning framework; World Food Programme (WFP) responsible for emergency food delivery and logistics; the United Nations High Commission for Refugees (UNHCR) responsible for shelter; the United Nations Children Fund (UNICEF) responsible for nutrition and water, sanitation and hygiene; and the Food and Agricultural Organisation (FAO) is responsible for emergency agriculture, which if successful should mark the end of the emergency and the diminution of the role of WFP. Medical interventions are largely left to international NGOs and the Red Cross.

Three key challenges to humanitarian assistance since the cold war

Three challenges dominate discussions of humanitarian assistance:

1. Where does global leadership in the planning and implementation of humanitarian assistance lie?

The short answer is with the United Nations system through the Office for the Coordination of Humanitarian Assistance (OCHA). In 1991, the United Nations General Assembly adopted Resolution 46/182 which led to the establishment of the Department of Humanitarian Affairs: OCHA had its origins in this system which is designed for the 'prompt and smooth delivery of relief assistance'. The two critical mechanisms for this coordination are the sharing of a Common Humanitarian Action Plan (CHAP) by all humanitarian actors, including those such as ICRC, which are not under the United Nations humanitarian effort, and the Consolidated Appeals Process (CAP). The latter is the critical fundraising effort for humanitarian aid although donor response, particularly in protracted African crises, tends to be poor undermining delivery according to need. The 2011 CAP of global humanitarian contributions indicated that 27 per cent of funds were allocated for food; 10.2 per cent for health; 7.4 per cent for coordination and support services; 4.4 per cent for agriculture; 4.1 per cent for water and sanitation; and 1.3 per cent for education (33 per cent of funds were not specified) (OCHA 2012). In general, however, appeals remain underfunded with 60 per cent response for emergency food, 40 per cent for shelter and less than 20 per cent for water and sanitation.

Although OCHA, on behalf of the United Nations, provides leadership, it is leadership as a coordinating function and not as a 'Command and Control' structure in most emergency situations. To do this requires building shared platforms of information from the different parts of the humanitarian system: evidence of these platforms can be sourced through www.reliefweb. Figure 9.4.1 outlines these shared platforms and, by implication, maps the flow of funds. While the United Nations, as the lead global humanitarian actor, is frequently heavily criticised for poor performance, there is no doubt that for global reach it remains the key player. If it did not exist, it would have to be invented. To improve delivery of humanitarian assistance the United Nations, with lead NGOs has developed a 'cluster' approach to the delivery of emergency services: evaluations of the 'clusters' are ongoing.

2. How does humanitarian assistance link across the broader issues of development?

The transition from relief to development, also known as the 'gap issue' or 'grey area', has been debated internationally for the last fifteen years, initially in the hope that a smooth 'continuum' linking the three phases of relief, rehabilitation and development could be achieved. By the mid-1990s it was widely accepted that no such continuum was feasible and that elements of all three phases could best be implemented simultaneously.

Humanitarian aid remains organised around short-term, largely project-based, funding cycles and the concept that emergencies are temporary interruptions of normal processes. It is true humanitarian aid has been provided for long periods to populations of countries characterised by chronic conflicts, however, the humanitarian system is essentially ill equipped to engage with long-term crises, which can continue for decades and whose effects may stretch across an entire country and/or countries within a region. The humanitarian system is arguably already fully stretched.

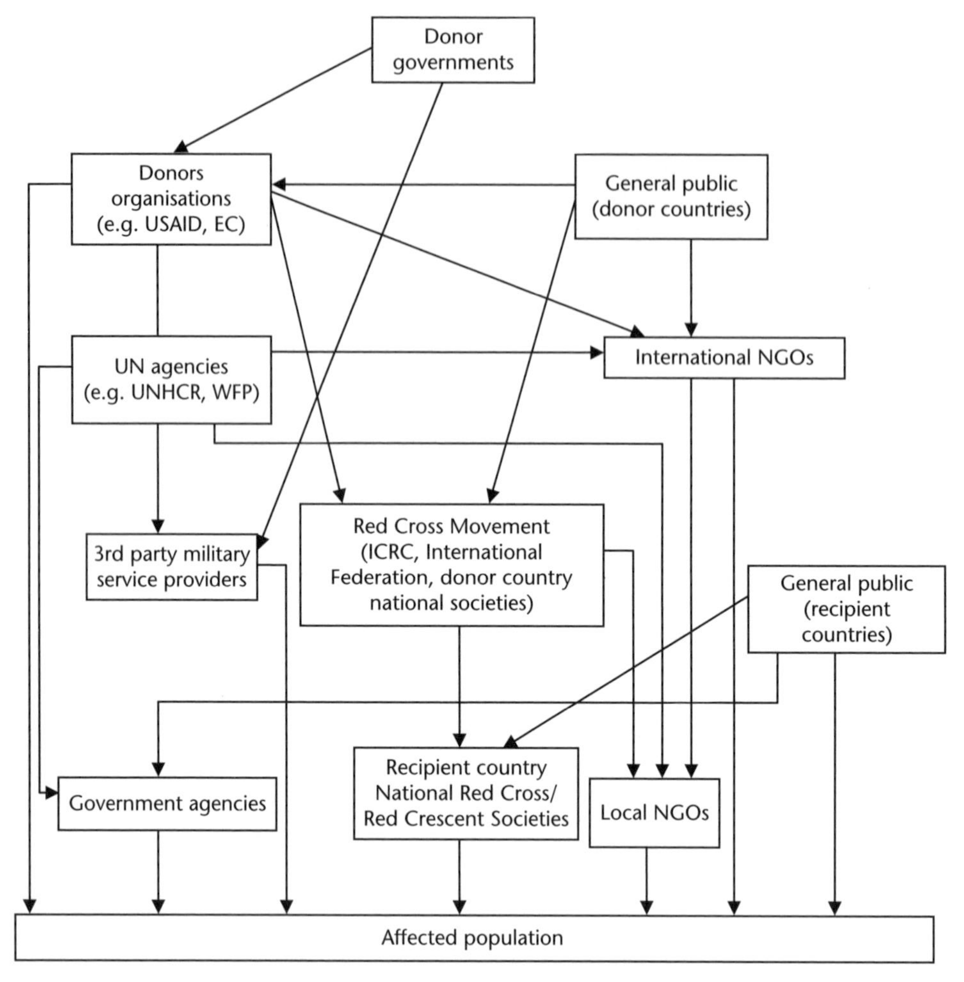

Figure 9.4.1 Resource flows within the international relief system

Source: Hallam, 1998

In striving to close the gap between relief, rehabilitation and development assistance, several donors, like Sweden, have taken the view that the distinction between development and humanitarian relief obstructs recovery. More widely, the view is that careful coordination at and amongst all levels is the appropriate framework. One workable version transpired from the late 1990s to become accepted and is referred to as 'humanitarian aid plus'.

'Humanitarian aid plus' was developed in realisation that basic needs alone were not sufficient to secure a durable beneficial outcome from an emergency. It seeks to support progress to more developmental activities. It was hoped through linking relief to rehabilitation and development the likelihood of further humanitarian need would be reduced: development would prevent conflict and humanitarian need. Humanitarian agencies looked to 'developmental relief' as a route towards this aim. This was obstructed by four obstacles: (a) humanitarian agencies have little influence; (b) there was little commitment to a radical redesign of the development and humanitarian assistance components of organisations; (c) protracted emergencies became even more intractable; and (d) donors sometimes used humanitarian aid to avoid engagement with repressive or undemocratic states (Macrae and Harmer, 2005).

3. How does humanitarian action, especially in complex emergencies, relate to military intervention?

The most significant development in humanitarian action since the cold war has been the increase in violent conflict. On average 52,000 people are directly killed each year as a result of armed conflicts (this is a conservative estimate including only recorded deaths, the actual total may be much greater), which although high, is not that different from the 60,000 people that die each year as a result of natural disasters (Geneva Declaration, 2008; Kenny, 2009). However, a further 22,000 people die indirectly each year as a result of these conflicts (Geneva Declaration, 2008).

Over the last two decades, military interventions have become common place within humanitarian operations. This encroachment of 'humanitarian space' led to significant questions of principle and policy and subsequently resulted in the Oslo Guidelines (OCHA, 1994) outlining the use of civilian and military assets in natural and technological disasters. The problem remains, however, that military intervention threatens the neutrality of humanitarian action and the humanitarian workers themselves.

Key principles for military involvement are complementarity, which implies that the military will not be used if civilian assets are available; control of the military in support of humanitarian action must be the responsibility of civil authority; no costs associated with the military can be charged to the affected population; and finally, the military must withdraw at the earliest possible moment.

The use of the military raises wider issues in humanitarian aid, especially whether humanitarian assistance is independent of foreign policy. It is how humanitarian practice unfolds that allows a typology of donors and agencies. Broadly speaking, they are classified as Wilsonian (after Woodrow Wilson) and Dunantist (after Jean Heenri Dunant). The former are dependent on, and cooperative with, government while the latter are more independent of, and oppositional towards, government: the former emphasise delivery, the latter advocacy (Overseas Development Institute, 2003).

Recurrent themes

Beyond the four key challenges a number of recurrent themes exist throughout humanitarian assistance, including the impact of HIV/AIDS on humanitarian action, delivery of aid to internally displaced persons (IDPs) and the ability and right to protection. These themes highlight the difficulties in delivering appropriate humanitarian response despite good intentions.

Conclusion

Despite progress, there are still well-known problematic areas in the delivery of humanitarian assistance especially over coordination of interventions and connectedness to development activities once the emergency period has passed. There remains a lack of attention to preparedness and pre-disaster planning (currently referred to as disaster risk reduction) in general and there is limited attention to indigenous coping strategies. Targeting of humanitarian assistance, particularly around issues of gender, remains problematic. It seems that these problems are somewhat intractable because ultimately no one has responsibility for the management of humanitarian assistance and since essentially it is assumed to be 'doing good'.

Bibliography

Borton, J., Buchanan-Smith, M. and Otto, R. (2005) Support to Internally Displaced Persons – Learning from Evaluations, Swedish International Development Cooperation Agency.

Deng, F. (1998) *Guiding Principles on Internal Displacement*. United Nations, New York.

Development Initiatives (2005) *Global Humanitarian Assistance Update 2004–2005*. Global Humanitarian Assistance and Good Humanitarian Donorship Project, UK.

Development Initiatives (2011) Global Humanitarian Assistance 2011. Development Initiatives, Somerset, UK.

Geneva Declaration (2008) *Global Burden of Armed Violence*. Geneva Declaration Secretariat, Geneva.

Hallam, A. (1998) Evaluating Humanitarian Assistance Programmes in Complex Emergencies. *Good Practice Review*, No. 7, p. 102.

Harmer, A. Cotterrell, L. and Stoddard, A. (2004) From Stockholm to Ottawa: A progress review of the Good Humanitarian Donorship initiative *Humanitarian Policy Group Research Briefing*. No. 18, October.

International Federation of Red Cross and Red Crescent Societies (2001) World Disasters Report. IFRC.

Kenny, C. (2009) *Why Do People Die in Earthquakes? The Costs, Benefits and Institutions of Disaster Risk Reduction in Developing Countries*. The World Bank Sustainable Development Network. Available online at http://gfdrr.org/docs/WPS4823.pdf (accessed 05 May 2012).

Macrae, J. and Harmer, A. (2005) Re-thinking Aid Policy in Protracted Crises. *Opinions*. September 2005. ODI, London.

OCHA (1994) *OSLO Guidelines*. OCHA, New York.

OCHA (2012) *Global Humanitarian Contributions in 2011: Totals per Sector*. Financial Tracking Service. Available online at http://fts.unocha.org/pageloader.aspx?page=emerg-globalOverview&Year=2011 (accessed 08 May 2012).

Overseas Development Institute (2003) *Humanitarian NGOs: Challenges and Trends*. Humanitarian Policy Group, Briefing Paper, no. 12, July.

Steering Committee of the Joint Evaluation of Emergency Assistance to Rwanda (1996) *The International Response to Conflict and Genocide: the Rwanda Experience*. Steering Committee, Copenhagen.

UNAIDS, UNFPA and UNIFEM (2004) *Women and HIV/AIDS: Confronting the Crisis*. UNAIDS, UNFPA and UNIFEM, New York.

UNHCR (2011) *World Refugee Day: UNHCR Report Finds 80 Per Cent of World's Refugees in Developing Countries*. UNHCR, Geneva.

Waal, A. and Tumushabe, J. (2003). *HIV/AIDS and Food Security in Africa*. A report for the Department for International Development (DFID), UK.

Further reading

The Sphere Project (2004) *Sphere Handbook*. The Sphere Project, Geneva.

United Nations Central Emergency Response Fund (CERF) (2006) *What is the CERF?* http://ochaonline2.un.org/Default.aspx?tabid=7480: United Nations Central Emergency Response Fund.

United Nations High Commission for Refugees (UNHCR) (2000) *Handbook for Emergencies*. Geneva, UNHCR.

United Nations High Commission for Refugees (UNHCR) (2006) *2005 Global Refugee Trends*. Geneva, UNHCR.

9.5

Global war on terror, development and civil society

Jude Howell

Following the Al Qaeda attacks on the Twin Towers on 11 September 2001, US President George Bush launched a global 'War on Terror'. This was to be won not only through military might but also through extra-ordinary legal and regulatory measures and ideological means. Human rights lawyers and activists campaigned vociferously against these new measures that in the name of security served to reduce civil liberties. The impact of these measures was not, however, confined to the field of human rights. The global War on Terror has led in turn to the deepening securitisation of development and civil society.

This contribution sets out to demonstrate the increasing securitisation of development and civil society since 9/11. By securitisation we refer to the processes through which development and civil society actors become socially constructed in terms of security. The chapter begins by outlining the increasing convergence of development, security and civil society during the 1990s. In the second section it analyses how the agendas, resources, and activities of development and security agencies became more deeply entwined after 9/11. The third part explores the ways in which civil society actors operating in the field of international development become absorbed into global and national security agendas, and depicted simultaneously as potential sources of terrorist threat and as adjutants in national security agendas. The conclusion draws together key findings and reflects on their implications for desecuritisation.

Closer encounters: Development, security and civil society

With the end of the cold war and the triumph of liberal democracy and capitalism, aid agencies were poised to place poverty reduction and democratisation at the forefront of their agendas and to involve civil society more directly. Closer cooperation between development agencies and NGOs had already begun during the 1980s when issues of basic needs were coming onto the agenda. By the end of the decade development organisations were channelling increasing amounts of funding towards NGOs. With the rise of the 'good governance' agenda in the 1990s, donor agencies began to adopt the language of civil society and draw civil society organisations more closely into their programming, though still working predominantly with NGOs (Howell and Pearce 2001).

At the same time the post-cold war scenario led to closer linkages between development and security. In fact, aid, military and foreign policy had always been inextricably intertwined (Duffield 2007). The changing role of the UN in peace-keeping and post-conflict reconstruction created new openings for development work and NGOs. Reconciliation, mediation, re-integration of soldiers and post-conflict reconstruction became part of the development portfolio. With the focus on improving governance, donors embarked on promoting security sector reforms and sought to involve civil society organisations in these processes. This was a volte-face both for security agencies and civil society organisations that had long regarded each other with mutual hostility (Pearce 2006). In the context of the Bosnia, Sierra Leone and Kosovo conflicts, the military began to engage more in development, provoking tensions with humanitarian workers about this blurring of boundaries.

By the turn of the millennium development agencies and civil society organisations had accumulated considerable experience of working together. Against the backdrop of a more general reassessment of aid, donors also began to take stock of their engagement with civil society organisations. Of concern were the increasing transaction costs of dealing with multiple NGOs, issues around representation, legitimacy, probity and fraud, and the challenges of scaling-up. The launch of the War on Terror marked a watershed moment when these mounting concerns around aid and civil society came to the fore. In the next section we examine how the War on Terror has affected aid policy.

Impact of War on Terror on aid policy

The security framework introduced through the War on Terror has affected aid policy and practice in several ways and at multiple levels. At the global and national level political leaders underlined the links between poverty, alienation and terrorism, thereby paving the way for closer policy integration between defence, diplomacy and development. For example, the US National Security Strategy of 2002 listed development, diplomacy and defence as three cornerstones of national security strategy, an approach known as the '3Ds'. UN leaders and multilateral institutions propagated common approaches such as 'whole-of-government approach' or an increased focus on 'fragile' states.

Whilst the validity of these links between poverty, extremism and terrorism remains unproven, development institutions at the meso-level have in turn emphasised in their mission statements the importance of aid to global and national security. As the then Director General of AusAID, Bruce Davis, stated in 2005, 'security is a pre-requisite for development . . . It is also recognised that underdevelopment is itself a security threat' (Davis 2005). Similarly, as stated in the UK Department for International Development 2005 paper on security and development, 'we need better collaboration between development, defence and diplomatic communities to achieve our respective and complementary aims' (DFID 2005). In this spirit, development agencies have participated in the 'whole-of-government' approach to terrorism, pooling resources, and establishing joint strategy groups with defence and foreign policy.

The growing prominence of security in development agendas has also filtered into programming. This is reflected in the increased flows of aid to countries at the frontline of the War on Terror such as Iraq, Pakistan, Ethiopia, Yemen and Afghanistan; in the expansion of security sector reform programmes; and in the increasing prominence of 'fragile' states as a category for intervention. This, in turn, becomes evident at the micro-level of operations, where aid agencies have established specific projects and programmes aimed at countering radicalisation. A prime example of this is a 'Peace, Development and Security' project set up by the Danish development agency, DANIDA in Kenya as a counter-terrorism initiative. This provides small grants for community projects addressing issues of social justice and conflict, focussing particularly on young Muslims. Donors have also supported inter-faith dialogues and sought to engage more with 'moderate' Muslim clerics as part of a broader strategy to counter terrorism. In conflict contexts such as Afghanistan, militaries have played an increasing role in development, drawing on the rationale that 'winning hearts and minds' through quick-impact projects is a vital prong in counter-insurgency. This has intensified a debate, already underway since the 1990s, about the risks this poses to humanitarian workers who rely on being perceived as 'neutral and impartial' to operate in conflict situations.

Impact of War on Terror on civil society

Since 9/11, Western governments have adopted a dual strategy of control and engagement towards civil society. Governments across the world tightened up national legislation relating to NGOs, increasing government scrutiny of their operations, requiring greater accountability and transparency, and demanding commitments to ensure they have no association with terrorist organisations (Bloodgood and Tremblay-Boire 2010). In this way humanitarian and developmental NGOs, particularly in Muslim-majority countries or in conflict contexts have simultaneously been treated as sources of potential threat and as adjutants in pursuing security agendas. Furthermore, they have themselves become securitisers in relation to the state and insurgents, both investing in security systems and distinguishing their activities from the military.

This gradual securitisation of civil society can be observed at multiple levels. Statements by national leaders that NGOs were a cause for concern contrasted starkly with politicians' enthusiastic embrace of civil society in the golden era of the 1990s. For example, then Chancellor of the Exchequer, Gordon Brown, stated in October 2006 that charities were one of the 'three of the most dangerous sources of terrorist finance'. Counter-terrorist measures have directly and indirectly affected the operations of international and local development NGOs. In the USA Muslim charities were the only charities to be added to the designated lists of terrorist organisations. The introduction of Special Recommendation VIII by the global Financial Action Task Force in 2002 required banks to monitor the international financial transactions of charities. The US Patriot Act of 2001 made it an offence to provide material support to terrorist organisations, creating significant problems for humanitarian organisations rendering assistance in conflict-torn regions and Muslim-majority countries (Guinane and Sazawal 2010). Development institutions, too, have introduced new regulations which make the work of local and international development NGOs more burdensome and difficult. For example, the US Agency for International Development requires all NGOs receiving its funds to sign an anti-terrorist certificate.

The War on Terror has also deployed ideological strategies aimed at 'winning hearts and minds'. New civil-military structures such as the provincial reconstruction teams (PRTs) in Afghanistan and Iraq seek to involve NGOs in development projects to retain military hold over occupied areas and persuade local populations of the benefits of cooperation (Perito 2007; Gordon 2010). Developmental intervention by militaries, often using civilian vehicles and clothing, have along with the PRTs created considerable problems for humanitarian agencies (McHugh and Gostelow 2004). These have become increasingly subject to attacks by insurgents who do not see them as neutral actors. This in turn has led international NGOs to find ways of strengthening their own security, such as employing security officers and remote programme management, thereby themselves becoming securitisers. The raft of 'hard' regulatory measures and 'soft' ideological strategies deployed since 9/11 have provoked mixed responses from civil society, with human rights organisations, Muslim groups and international NGOs at the forefront of challenging the justification and effects of these.

Conclusion

The War on Terror has shaped the contours of aid and civil society in the new millennium. Not all of this has been negative; some marginalised groups in society have gained increased access to resources and government, whilst development agencies can use security discourses to justify continuing budgetary support. Nevertheless, it does raise deeper questions about the purposes of aid and the desirability of the convergence of aid, civil society and security. Though the new US administration led by President Obama has renounced the language of the War on Terror and taken steps to revoke waterboarding and close the detention facility at Guantanamo Bay, the bulk of policies, legislation and practices introduced since 9/11 continue to prevail. This raises a set of issues around the possibilities of desecuritisation. It would imply at the very least de-coupling aid and civil society from security agendas. More significantly it would require rolling back those counter-terrorist measures that have impinged detrimentally on the activities of aid and civil society.

Bibliography

Bloodgood, Elizabeth, A. and Joannie Tremblay-Boire, 2010, 'Counter-Terrorism and Civil Society', *The International Journal of Not-For-Profit Law*, 12, 4, November, pp. 1–9, available online at www.icnl.org (accessed in April 2012).

Davis, Bruce, 2005, 'Aid and Security after the Tsunami', archived speech delivered on 27th October, Australia.

DFID, 2005, 'Fighting Poverty to Build a Safer World. A Strategy for Security and Development', UK Department for International Development, London.

Duffield, Mark, 2007, *Development, Security and Unending War. Governing the World of Peoples*, Polity Press, Cambridge.

Gordon, Stuart, 2010, 'Civil Society, the "New Humanitarianism", and the Stabilisation Debate: Judging the Impact of the Afghan War', in Jude Howell and Jeremy Lind (eds), *Civil Society Under Strain: Counter-Terrorism Policy, Civil Society and Aid Post-9/11*, Kumarian Press, Sterling, pp. 109–126.

Guinane, Kay and Suraj K. Sazawal, 2010, 'Counter-terrorism Measures and the NGO Section in the United States: A Hostile Environment' in J. Howell and J. Lind (eds), *Civil Society Under Strain: Counter-Terrorism Policy, Civil Society and Aid Post-9/11*, Kumarian Press, Sterling, pp. 53–74.

Howell, Jude and Jeremy Lind, 2009, *Counter-terrorism, Aid and Civil Society. Before and After the War on Terror*, Palgrave Macmillan, Basingstoke.

Howell, Jude and Jenny Pearce, 2001, *Civil Society and Development: A Critical Exploration*, Lynne Rienner Publishers, Boulder and London.

McHugh, Gerard and Lola Gostelow, 2004, *Provincial Reconstruction Teams and Military-Humanitarian Relations in Afghanistan*, Save the Children Fund Report, London.

Pearce, J., 2006, Case Study of IDRC-supported Research on Security Sector Reform in Guatemala (final report), Department of Peace Studies, University of Bradford, UK.

Perito, Robert, 2007, *The US Experience with Provincial Reconstruction Teams in Iraq and Afghanistan*, Briefing and Congressional Testimony, US Institute of Peace, Washington, DC.

9.6

Peace-building partnerships and human security

Timothy M. Shaw

> *It is easy to fault UN peacekeeping missions around the world. 'Blue helmets' are frequently wasteful, toothless and lack a clear line of command . . .*
>
> Nonetheless, the eight UN missions in Africa have made the difference between a new descent back into civil war and a slow but hopeful climb towards stability.
>
> (*Economist 2012: 15*)

After almost two decades, 'human security' as concept and practice still seeks to privilege personal economic and social rather than 'national' strategic concerns, including ecological, educational, food, habitat and health priorities (UNDP 1994: 22–40). But, despite its relatively recent post-cold war definition and advocacy, human or individual security along with human development has become increasingly more problematic in the twenty-first century. In part, this reflects the interrelated impacts of a decade of the BRICs/BRICS and a series of 'global' economic crises around the turn of the second decade. In turn, debates about its definitions and elusiveness have proliferated and intensified as indicated in the second section, including the articulation of the notion of 'citizen security' from the global South (UNDP 2012).

Any post-bipolar 'peace dividend' was shattered after a short decade by the 'global' shocks of 9/11 in the US in late-2001 then 7/7 in the UK in mid-2005. Subsequent 'wars on terrorism' and interventions in Iraq and Afghanistan have complicated previously rather simplistic or idealistic notions of peacekeeping roles and partnerships – R2P – for post-conflict reconstruction as articulated in ICISS (2001).

This chapter for the third edition of this companion has been rewritten because the interrelated worlds of human security and peace building are quite different at the beginning of the second decade of the twenty-first century. These shifts were reflected in revisionist UN (2003 and 2004) and donor (DFID 2005) deliberations yet they remain resilient as indicated in a new overview of fragile states with a focus on a trio of cases: Afghanistan, DRC and Haiti (Brock *et al.* 2012). But they remain as important as ever for 'development studies' and for state and non-state policies as suggested ten years ago (Shaw 2002) especially now that the UNECA (2011) advocates a developmental state policy for Africa. Further, as indicated in the second section below, there is growing anxiety within INGO and related circles about the tendency towards the conflagration of development and security post-9/11.

This contribution has two interrelated themes. The first is that human security, however defined, remains as relevant as ever to global development even if it was momentarily displaced by an apparent return to the hegemony of 'national security' after 9/11. And the second is that peace building is likewise vital but it has now become much more dangerous or 'robust' than at the end of the twentieth century. In short, development and security are more intertwined and inseparable than ever, leading to controversial notions like the securitization of development as well as the privatization of security as indicated below, even the militarization of refugee communities and camps (Muggah 2006).

Human security before and after 9/11 and 7/7

Human security was articulated in the post-bipolar world as an antidote to established notions of national security and balance of power: the privileging of individual rather than collective security against the threat and practice of violence as the principal referent of security. Symptomatic of contemporary policy development, like parallel discourses on human development and human rights, human security has been defined by international agencies and think tanks. Such 'public diplomacy' reflects growing 'contracting-out' by national regimes to combinations of non-governmental organizations, multinational corporations and international institutions: the bases of 'new multilateralisms' of global mixed actor coalitions around 'new' security issues like landmines and blood diamonds, small arms and child soldiers. The apex of the first period of human security deliberations was the late-1990s Ottawa Process around landmines advanced by the 1400-member International Campaign to Ban Landmines (ICBL) and spearheaded by Lloyd Axworthy as Canadian foreign minister.

As conceived and advocated in the mid-1990s by the UNDP (1994: 22–40), human security includes interrelated community, economic, environmental, food, health, personal and political securities. But, by the middle of the first decade of the new century, uneven 'globalization' had served to proliferate security issues especially around the latest generation of 'new' states which has led towards +/– 200 today, South Sudan being the 193rd member of the UN. Any lingering idealism was shattered by the shocks of 9/11 and the US unilateralist response: its declaration of a 'war on terrorism'.

Symbolically, the development of a human security doctrine and elaboration of the related 'responsibility to protect' (R2P) those communities and countries where it was threatened was in process when 9/11 diverted attention. So the December 2001 report of the blue-ribbon Canadian-supported International Commission on Intervention and State Sovereignty (ICISS 2001) was overshadowed by the new preoccupation with international terrorism. Thus, while the notion of human security is now almost two decades old, its definition and realization are more problematic than ever (MacLean *et al.* 2006). The ICISS (2001: xi) intended to extend the notion of international law for the new millennium from state sovereignty to the protection of people:

> Where a population is suffering serious harm, as a result of internal war, insurgency, repression or state failure, and the state in question is unwilling or unable to halt or avert it, the principle of non-intervention yields to the international responsibility to protect.

However, as the war on terrorism has dragged on and its costs – human, financial, regional and so on – have become ever more apparent, so analysts have begun to rediscover human rather than other varieties, such as national, regional, or even global, security. Thus the 2003 report of the UN Commission on Human Security and subsequent reports on 'new' security threats and responses around the millennium summit in 2004–2005 served to both rehabilitate and refine the concept. The post-bipolar and post-9/11 pre-summit panel report on *A More Secure World: Our Shared Responsibility* recognized that 'The threats are from non-state actors as well as states, and to human security as well as state security. The central challenge for the twenty-first century is to fashion a new and broader understanding . . . of what collective security means' (UN 2004: 11). Amongst the half-dozen clusters of threats identified by the UN (2004: 12) panel were: 'Economic and social threats, including poverty, infectious diseases and environmental degradation . . . Internal conflict, including civil war, genocide and other large-scale atrocities . . . Transnational organized crime.'

I conclude this section by noting the continuation of an intense debate around human security in the second decade of the twenty-first century. Initially this revolved around narrower versus broader conceptualization: freedom from fear versus freedom from want (MacLean *et al.* 2006). Initial formulations arising from the human development genre around the UNDP favoured the latter. By contrast, more cautious or conservative analysts oriented towards traditional international relations (IR) favoured the former. Reflective of such revisionist inclinations, Neil MacFarlane and Yuen Foong Khong (2006: 228) lamented the 'conceptual overstretch' around 'freedom from want' preferring to limit such security threats to those 'against their physical integrity' or 'organised violence'. This standoff was previewed by 21 analysts in a special section of *Security Dialogue* in mid-decade (Burgess and Owen 2004). As indicated in the next part, it entailed more than conceptual disagreement; it profoundly affected data and policy. But at the start of the second decade the field is being subjected to conceptual development around 'citizen security' (UNDP 2012) as well as a radical critique of the humanitarian enterprise as being intended to advance and insulate global inequalities (Soderbaum and Sorensen 2012).

Threats to human security in theory and policy

This second part juxtaposes a set of overlapping discourses which increasingly impact both the definition and implementation of human security at the start of the second decade of the new century. In addition to freedom from fear/want, these include: uneven globalization, 'African' international relations (IR), and redefinitions of both development and security. Taken together, this trio of debates has profoundly complicated the conceptual and empirical relationships between development and security.

First, 'globalization' has clearly become more uneven than ever leading towards a new trio of 'worlds': (a) the OECD states, (b) the 'emerging economies' of BRICS and N-11, and (c) the approximately 50 fragile states (Brock *et al.* 2012). Such inequalities are both intra- and inter-state, encouraging transnational alliances amongst the rich and the poor to either defend or challenge the status quo. Hence the privatization of security on the one hand and the tendency towards political and/or religious radicalization on the other.

Second, a further overview of 'African' IR as part of a global trend towards the 'discovery' of perspectives from the global South (Cornelissen *et al.* 2012) has raised the issue of whether such relations can be limited to inter- rather than intra-state relations of cooperation and conflict. This 'academic' discussion poses profound implications for security policy; for example, do only deaths from classical inter-state wars count? Reflective of conservative, traditional inter-state definitions, the first *Human Security Report 2005* from UBC (2005) can claim an optimistic picture of a decline in conflicts and related deaths at century's turn. By contrast, data on 'internal' or intra-state wars and body-counts would be much less sanguine. Thus (Cornelissen *et al.* 2012: 12) note that a wide range of cross-border relations on the continent are unrecorded despite the several land-locked states: from economic and ecological, ethnic and criminal to disasters and viruses.

And third, as regional conflicts have continued in, say, Central Asia, the Great Lakes, Horn (from Darfur to Somalia), West Africa, and so on (Boas and Dunn 2007; Muggah 2006), so two divergent responses can be identified. First, an extended version of the optimistic perspective on 'new multilateralisms' around 'new' security issues like landmines, blood diamonds, child soldiers, small arms (GIIS 2011) and so on: further Ottawa and Kimberley Processes augmented by more robust peace building as in, say, Sierra Leone along with longer-term developmental innovations like the Diamond Development Initiative. But second, a more critical or sceptical response based on the recognition of 'greed' rather than 'grievance': the 'political economy of conflict'. This cautions that peacekeeping per se cannot be efficacious as conflict is over resources and revenues rather than principles, notwithstanding efforts around corporate codes of conduct, Extractive Industries Transparency Initiative (EITI) and so on (Boge *et al.* 2006): the 'new' extended peacekeeping partnerships.

Growing scepticism about the feasibility of a 'liberal peace' is reinforced by the trend towards the securitization of development and/or militarization of security (Soderbaum and Sorensen 2012). INGO policy analysts are increasingly apprehensive about this direction as it erodes the boundary between development and security, enabling national and private security organizations to claim to qualify for development funds under the DAC of the OECD.

Futures for peace building and human security?

Given the above analytic and empirical trends and debates, optimism about the future of peace-building partnerships and human security is a scarce commodity. While the world of peace building continues to expand given demand and supply, post-9/11 and 7/7 conflicts have eroded and compromised its niche. Continued enlightened new multilateralisms and informed public diplomacy around the human development/security nexus are to be encouraged (MacLean *et al.* 2006). But the context is now complicated by improved prospects for the securitization of development and the militarization of security (Muggah 2006) exacerbated by interventions in Iraq, Afghanistan and Syria. Moreover, the advocacy of citizen security in response to fear of crime and violence, especially around TOC in the Caribbean and Central America (UNDP 2012), may widen policy options but complicate decision making (Hanson *et al.* 2012)

References

Boas, Morten and Kevin C. Dunn (eds) (2007) *African Guerrillas: Raging Against The Machine* (Boulder: Lynne Rienner).

Boge, Volker, Christopher Fitzpatrick, Willem Jaspers and Wolf-Christian Paes (2006) 'Who's Minding the store? The business of private, public and civil actors in zones of conflict' (Bonn: BICC. Brief #32).

Brock, Lothar, Hans-Henrik Holm, Georg Sorenson and Michael Stohl (2012) *Fragile States* (Cambridge: Polity).

Burgess, J. Peter and Taylor Owen (2004) 'Special section: what is "human security"?' *Security Dialogue* 35(3), September: 345–387.

Cornelissen, Scarlett, Fantu Cheru and Timothy M. Shaw (eds) (2012) *Africa and International Relations in the 21st Century* (London: Palgrave Macmillan).

DFID (2005) 'Why we need to work more effectively in fragile states' (London, January).

Economist (2012) 'UN troops in Africa: Blue berets in the red' 9 June 403(8788): 15–16.

GIIS (2011) *Small Arms Survey 2011: States of Security* (Oxford: Oxford University Press).

Hanson, Kobena T., George Kararach and Timothy M. Shaw (eds) (2012) *Rethinking Development Challenges for Public Policy: Insights from Contemporary Africa* (London: Palgrave Macmillan).

ICISS (2001) 'The responsibility to protect: Report of the International Commission on Intervention and State Sovereignty' (Ottawa: IDRC, December).
MacFarlane, S. Neil and Yuen Foong Khong (2006) *Human Security and the UN: A Critical History* (Bloomington: Indiana University Press for UNIHP).
MacLean, Sandra M., David R. Black and Timothy M. Shaw (eds) (2006) *A Decade of Human Security: Prospects for Global Governance and New Multilateralisms* (Aldershot: Ashgate).
Muggah, Robert (ed.) (2006) *No Refuge: The Crisis of Refugee Militarization in Africa* (London: Zed for BICC and SAS).
Shaw, Timothy M. (2002) '9.6 Peace-building partnerships and human security' in Vandana Desai and Robert B. Potter (eds) *The Companion to Development Studies* (First edition) (London: Oxford University Press), 449–453.
Soderbaum, Fredrik and Jens Stilhoff Sorensen (eds) (2012) 'The End of the Development-Security Nexus? The rise of global disaster management' *Development Dialogue* 58, April: 1–176.
UBC (2005) *Human Security Report 2005* (Oxford: Oxford University Press).
UN (2003) 'Human Security – Now. Report of the Commission on Human Security' (New York, May).
UN (2004) 'A More Secure World: our shared responsibility. Report of the High-level Panel on Threats, Challenges and Change' (New York, December).
UNDP (1994) *Human Development Report 1994* (New York: Oxford University Press).
UNDP (2012) 'Caribbean Human Development Report 2012: Human development and the shift to better citizen security' (Port of Spain).
UNECA (2011) 'Economic Report on Africa 2011: governing development in Africa – the role of the state in economic transformation' (Addis Ababa).

Further reading

MacFarlane, S. Neil and Yuen Foong Khong (2006) *Human Security and the UN: A Critical History* (Bloomington: Indiana University Press for UNIHP). A comprehensive, informed but sceptical overview of the evolution of human security which advocates narrower freedom from fear rather than broader freedom from want.
MacLean, Sandra M, David R. Black and Timothy M. Shaw (eds) (2006) *A Decade of Human Security: Prospects for Global Governance and New Multilateralisms* (Aldershot: Ashgate). A reflective, somewhat revisionist history of the first decade of human security informed by Canadian perspectives.

9.7

Nationalism

Michel Seymour

The first definition of nationalism that could be offered comes from Ernest Gellner (1983). It is expressed in the nationalist principle according to which each nation should have its own state. It is the claim that the borders of the nation and those of the state should always coincide. This radical definition must be contrasted with the thesis that each nation has the right to have its own state, for this latter definition is compatible with the suggestion that, in many circumstances, the best solution is rather to accommodate a nation in some kind of multinational arrangement. In other words, the kind of nationalism involved in this latter approach is less radical, because it is not always a good thing to exercise the right to have its own state, for it is sometimes better for the nation to remain stateless if it is able to get some kind of recognition within a multination state.

Cultural and political nationalisms

Some will thus be tempted to distinguish between cultural nationalism and political nationalism. The former would be a kind of nationalism that seeks cultural recognition without trying to achieve full political sovereignty, while the other kind of nationalism would always be looking for political sovereignty on a particular territory. The distinction is, however, not very well founded, because there is perhaps always something political going on in cultural nationalism and always something cultural going on in political nationalism. Nevertheless, the distinction between the two kinds of nationalism can be captured by two different senses in which the right to self-determination can be exercised.

Internal and external self-determination

We should distinguish between the right to internal self-determination and the right to external self-determination. The former right is the right for a people to develop economically, socially and culturally and the right to determine its political status within a sovereign state, while the right to external self-determination is the right to have its own sovereign state. When the distinction between cultural and political nationalism is understood in accordance with these two sorts of self-determination, it is easier to understand why the distinction between them is not so neat. If the cultural nationalist asks for internal self-determination, he usually asks at the same time for some kind of political arrangement that will allow for political autonomy within the state. And if the political nationalist is in the process of trying to have her own sovereign state, it is often because the attempt to achieve some kind of cultural recognition within the state has failed. The distinction between internal and external self-determination is thus a very useful one, especially since it reveals the connection between the two kinds of nationalism.

There are many different ways of interpreting internal self-determination. In the weakest sense, it is a right of political representation within the state. If the stateless people is able to elect its own political representatives, if these representatives come from the people and if they play an important role in the political institutions of the encompassing state, this is a first kind of internal self-determination. In the canonical sense, internal self-determination amounts to some kind of self-government. In the more robust sense, it is the right to have a special constitutional status like, for example, a constitutionalized asymmetrical arrangement within a multinational federation. Similarly, there are many different ways of exercising the right to external self-determination. It can mean the right to own a state for a population that already has a state. It can mean secession, which implies the creation of a brand new state for the stateless people. Finally, it can mean the right that a people would have to violate the territorial integrity of an existing state in order to associate with the already existing state of another people.

Primary right and just cause theories of external self-determination

There are also many different theories concerning the way to exercise the right to external self-determination. The main opposition is between those who believe that a people should have a right to external self-determination without having to comply with certain moral principles. This would entail for instance that a sovereign people would have no obligation to meet concerning its own national minorities in order to have the right to keep its own sovereign state. It could also entail that a stateless people could have the right to secede even in the absence of injustice. These two examples illustrate the primary right to external self-determination. The opposing view is the just cause theory of external self-determination (Buchanan 2004). According to this view, a sovereign people must respect the collective rights of its own internal national minorities in order to have the right to own its own sovereign state. Conversely, for a stateless people, secession would be justified only on the basis of a past injustice. Buchanan thinks that the most important injustices are the violation of basic human rights (Kurds in Iraq), annexation of territory (Baltic states) and the systematic violation of intra-state autonomy agreements (Kosovo). But the most obvious injustice would be the failure on the part of the state to recognize the right to internal self-determination of the stateless people (through colonization, oppression or domination) (Seymour 2007).

The just cause theory of external self-determination illustrates the existence of a continuum between cultural and political nationalism. It shows that a cultural nationalist population that wishes to exercise its internal self-determination may be led to become more and more political when the state refuses to grant internal self-determination. Conversely, it shows that very often, a political form of nationalism may be explained by the failure to meet the expectations of cultural nationalism.

Nation-state building and state-nation building

This problem raises the question concerning the origins of nations and nationalism. Where do nations and nationalism come from? The first issue concerns the relationship between the two concepts. Some will argue that nations exist prior to nationalism while others will argue that nations are created by nationalism. The debate leads to two different ways of conceptualizing nationalism itself. The classic distinction between nation-state building policies and state-nation building policies illustrates the debate. The former is associated with ethnic nationalism, while the latter illustrates civic nationalism. The advantage of arguing for the primacy of nationalism is that nations are then conceived as constructions and not as given, fixed, objective kind of entities. The disadvantage is that nationalism becomes *sui generis*, that is, a phenomenon that cannot be explained by the failure to get recognition for its own nation.

The origin of nations

The debate also has consequences concerning the origin of nations. The debate opposes modernists and pre-modernists. One of the most famous modernists is Benedict Anderson (1965) who explained the origin of nations by the capacity to imagine the existence of a population far beyond the local community, a capacity that was induced by print capitalism. Ernest Gellner (1983) explained the origin of the nation by the influence of the state who forced a single education system with a single language on populations speaking diverse languages, turning peasants into full-blown citizens. Liah Greenfeld (1992) explains the origin of the nation as being the result of the influence in England of an aristocratic power over a whole population. Anthony D. Smith (1991) is perhaps the most articulate pre-modernist. In numerous works, he shows that there existed "*ethnies*" long before the presence of the modern nation. Another famous proponent of pre-modernism is Clifford Geertz (1973) who defends a perennial variant in which certain primordial traits are described as constitutive of nations and have been there since the dawn of humanity.

A pluralist conception

Trying to define the concept of nation is no easy task. Actually, for most philosophers, political scientists, sociologists and historians, it is an impossible task. But the reason may be that there are many different sorts of nations. The ethnic nation is thus composed of a population representing itself as sharing the same ancestral origin (some indigenous peoples). The cultural nation gathers a multi-ethnic community sharing the same language, culture and history but no political institutions (Roma, Acadians). The sociopolitical nation is a multi-ethnic, multicultural and sometimes also multinational group sharing the same common public identity and organized into a non-sovereign political government (Scotland, Wales, Catalonia, Basque Country, Quebec). The civic nation is a mono-national sovereign country (Portugal, Iceland, Korea). The multisocietal nation is also organized into a sovereign country, but it contains many different peoples (Spain, Belgium, Great Britain and Canada). The multiterritorial nation is located on a continuous territory overlapping the official boundaries of many different countries, with no absolute majority located on either of those countries (Kurdistan, Akwesasne). The diasporic nation is dispersed on many discontinuous territories, with no absolute majority on either of those territories and forming minorities on all those territories (the old Jewish diaspora before the creation of Israel).

This account dissolves or resolves most of the traditional debates. Concerning the debate between modernists and pre-modernists, it will be observed that some nations were there long before the modernity (ethnic, cultural, multiterritorial, diasporic) while some of them were created with the modern state (civic, multisocietal, sociopolitical). So the debate is somewhat dissolved. A pluralist conception can also explain why some nations seemed to be created by the state (French nationalism) while some were created before the state (German nationalism). In addition, the account serves to explain where nationalism comes from. It is explained by the failure to implement reciprocal recognition among nations.

References

Anderson, B. (1965), *Imagined Communities*, London: Verso.
Buchanan, A. (2004) *Justice, Legitimacy, and Self-determination*, Oxford: Oxford University Press.
Geertz, C. (1973) "The Integrative Revolution: Primordial Sentiments and Civil Politics in the New States", in *The Interpretation of Cultures: Selected Essays*, New York: Basic Books.
Gellner, E. (1983) *Nations and Nationalism*, Oxford: Blackwell.
Greenfeld, L. (1992) *Nationalism: Five Roads to Modernity*, Cambridge, MA: Harvard University Press.
Seymour, M. (2007) "Secession as a Remedial Right", *Inquiry*, Vol. 50, no 4, 395–423.
Smith, A. D. (1991) *National Identity*, Harmondsworth, UK: Penguin.

Further reading

Couture, J., Nielsen, K. and Seymour, M. (eds) (1998) *Rethinking Nationalism, Canadian Journal of Philosophy, Supplemental Volume 22.* The book contains seventeen essays and an extensive Introduction and Afterword by the editors. It contains some of the most innovative samples of present reflection on this contentious subject. Moreover, contributions are from a variety of disciplines, from different parts of the world, often reflecting very different ways of thinking about nationalism and sometimes reflecting very different methodologies, substantive beliefs, and underlying interests.

Hutchinson, J. and Smith, A. D. (1995) *Nationalism* (Oxford Readers), Oxford: Oxford University Press. This book provides a concise, accessible introduction to the concept of nationalism. It is an excellent collection of articles with good historical coverage and a useful bibliography.
McKim, R. and McMahan, J. (eds) (1997) *The Morality of Nationalism*, Oxford: Oxford University Press. An excellent anthology of high-quality philosophical papers on the morality of nationalism.
Miscevic, N. (2010) "Nationalism", *The Stanford Encyclopedia of Philosophy* (Summer 2010 Edition), Edward N. Zalta (ed.), available online at http://plato.stanford.edu/archives/sum2010/entries/nationalism/. A great and authoritative presentation of the concepts of nation and nationalism. It is the best available introduction to the topic.

9.8
Ethnic conflict and the state

Rajesh Venugopal

Introduction

The large majority of countries in the world are multi-ethnic, and ethnicity is one of the most important forms of collective identity of relevance in the competition for and the constitution of state power, particularly in developing countries. The state is the terrain over which ethnic conflict is fought, and it features significantly in the theorization of ethnic conflict at a variety of different assumptions and levels, both in the cause and consequences of conflict, and as the central element in its resolution. The classical liberal conception views the state as a neutral arbiter suspended above society and adjudicating competing demands from rival interest groups. However, empirical research on the political sociology of conflict frequently describes a very different picture: of the hegemonic control of the state by dominant ethnic groups; of strong states generating insecurity rather than security; and of the deployment of violence and disorder as instruments of control by state elites. The literature on ethnic conflict and the state dwells largely in this domain of tension between the reality of the embedded, ethnicized state and the useful fiction of the neutral, dis-embedded, liberal state. Following a brief overview of the conceptual terrain, this chapter explores the relationship of ethnic conflict to the state by drawing on each of the state's three putatively 'core' functions: welfare, representation, and security.

While the more widely known cases of ethnic conflict are those that have involved large-scale, protracted or demonstrative episodes of violence such as the Lebanese civil war (1975–1990), urban riots in Kenya (2007), or ethnic massacres in Rwanda (1994), it is important to note the conceptual distance between violence and conflict. Conflict is widespread, but violence is not. Most multi-ethnic countries in the developing and developed world experience ethnic tensions and conflicts of some form or another – but these are for the most part either negotiated peacefully through formal or informal institutional mechanisms, or they remain latent and suppressed.

The term 'ethnic group' entered contemporary academic discourse through social anthropology on the one hand (where it displaced the term 'tribe'), and North American sociology on the other, where it referred to recent immigrant communities. In common use, ethnicity is widely understood to imply a primordial attachment – an ancient, cultural-biological identity that is rigid, impenetrable, and inescapable. However, primordialism has been viewed with scepticism in the academic literature, and has been the subject of intense critical scrutiny by constructivist and instrumentalist scholars who have challenged its fundamental claims. As a result, ethnicity is now understood to be far less immutable, ancient, or rigid than primordial claims suggest, but more contingent, plastic, and open to manipulation and re-definition. Primordialism continues to have significance in the study of ethnic conflict because it forms the cognitive basis on which ethnic consciousness and powerful ethnic attachments are based. It helps to explain how ethnic groups cast themselves, frame belief structures of group self-consciousness, construct origin myths, cultivate group solidarity and enforce loyalty.

It is also important to consider that the usage of the term ethnic conflict has changed considerably since the 1970s, particularly with respect to its conceptual overlap to related terms such as race, religion, caste or nation. While some of these terms had specific disciplinary pedigrees, they are frequently used in an inconsistent or idiosyncratic way. For example, 'confessionalism' in Lebanon, 'sectarianism' in Northern Ireland and 'communalism' in India are all localized monikers for group conflicts based on religious identities. As Horowitz (2000) describes, these diverse concepts display a 'family resemblance', and in the context of their transformation into political identities, refer to a common set of problems relating to the management of inter-group conflicts in pluralistic societies.

With this in mind, many scholars working on this topic, particularly in comparative political research, have opted to use ethnic conflict as a convenient meta-concept that conceptually equates and conflates these different terms. As such, ethnic conflict has to some extent superseded theories and definitions of ethnicity, and now commonly describes a range of contentious and violence-prone socio-political inter-community engagements between ascriptively defined communities who cohabit the territory of a given state.

Representation

The quantum and quality of ethnic group representation in the structure of state power is the most important problem in the political resolution of ethnic conflict. At its core is the issue of power-sharing, and the design of constitutional provisions that structurally distribute power in an appropriate manner between ethnic groups.

The issue of representation traditionally refers to the aggregation of individual voters' choices into democratically elected representatives in a legislature. However, when the unit at stake is the ethnic group rather than the individual citizen, democracy and elections may not always be relevant or even appropriate. Liberal democracy can be counter-productive in ethnically divided societies if the principle of majority rule translates the crude demographic advantage of a majority community into the 'hegemonic control' by one ethnic group (McGarry and O'Leary 1994).

The Westminster system, which was widely adopted in many former British colonies is particularly problematic in this respect. Under its 'first past the post – winner takes all' electoral system, smaller, territorially dispersed ethnic groups are systematically under-represented in the legislature, and can be completely excluded from the executive. In contrast, the more recently designed electoral systems and constitutions in post-conflict societies such as

Afghanistan, Iraq, Guatemala, El Salvador, Northern Ireland, Bosnia and Herzegovina, and Timor Leste, have all adopted institutional features that moderate the principle of majority rule in order to ensure greater ethnic representation.

At the electoral level, this involves mechanisms such as proportional representation, multi-member constituencies or preferential voting that are calibrated to ensure that the legislature better represents all ethnic groups. The contemporary approach to ethnic power-sharing has been influenced by Arendt Lijphart's writings on consociationalism, which marries the principle of ethnic proportionality in the legislature and bureaucracy, with carefully engineered constitutional requirements for ethnic power sharing in the executive (Lijphart 1977). For example, under Lebanon's 1989 Ta'if Agreement, the post of prime minister is reserved for a Sunni Muslim, the presidency for a Maronite Christian, and the speaker of parliament for a Shi'a Muslim. In addition, the consociationalist approach also involves two other key principles intended to protect minority groups: each ethnic group retains the right to veto key legislation, and has a measure of self-government and internal autonomy over their community affairs.

The second major problem that relates ethnic conflict to democracy is that the process of democratization is frequently violent and destabilizing. In the historical literature on ethnicity and colonialism, democratization has been associated with the construction of hostile ethno-nationalist ideologies, ethnic violence and ethnic cleansing (Mann 2005). The more contemporary, policy-focused literature has found that democratization in poorly institutionalized post-conflict states is conflict-inducing and results in a reversion from 'ballots to bullets', or in the ascendancy of ethnic extremists and former warlords to power (Snyder 2000). As Ottaway (1995: 235) describes:

> Democratization . . . encourages the conflicts that exist in a collapsing state to manifest themselves freely, but without the restraint of the checks and balances and of agreement on the basic rules, that regulate conflict and make it manageable in a well-established democratic system.

Security

The application of realist international relations theory has explained ethnic conflict as the outcome of a Hobbesian 'security dilemma' faced by vulnerable ethnic groups within collapsing states. As Posen (1993) describes, 'the condition of anarchy makes security the first concern', and the weakening of centralized authority effectively gives rise to ethnicized militias and violence. Indeed, the evidence from the Caucasus, Yugoslavia, Somalia, Afghanistan, Sierra Leone, Liberia, or D. R. Congo demonstrates how the collapse of state authority is accompanied by the rise of ethnic militias, territorial fragmentation and protracted insecurity.

The most important prognosis of the security-centric view of ethnic conflict is that it persists not only until the re-establishment of physical security, but until this security is monopolized – that is, provided and guaranteed by a single hegemon. The implications are that conflict resolution requires either a decisive military victory by one side (rather than a negotiated settlement), or in its absence, external military intervention to provide firm and unilateral security guarantees. Indeed, the empirical evidence on civil war termination (Walter 1997) broadly supports this conclusion, as decisive military victories are found to be more durable in resolving civil war and less prone to reversal than negotiated settlements.

Nevertheless, the security-centric view of ethnic conflict has some significant shortcomings. First, it is a theory of violence rather than conflict. A decisive military victory by one group of combatants may well end a civil war and restore order, but is unlikely to address the political issues that animate the ethnic conflict such as power sharing, territory, or resources, in an equitable way. Indeed, as Licklider (1995) finds, one-sided military victories do not even necessarily restore security, as they are far more likely to be accompanied by acts of genocide.

Second, deadly ethnic violence is not just a function of absent or weak states, but is frequently observed in many strong states where dominant ethnic groups appropriate state power, and where the state is a partisan participant in the violence.

Third, it makes the assumption that violence is the default condition of inter-ethnic relations, and that peace is little more than an ephemeral, externally enforced ceasefire by an external agent. However, the work of diverse scholars such as David Keen, Will Reno, Paul Brass and Chabal and Daloz shows that violence is not necessarily reflective of an absence of state control, but can be instrumental to the 'normal' exercise of power by elites.

Welfare

The material basis of conflict is a well-established subject of research in the social sciences, although it has emerged in its present form into the study of contemporary ethnic conflict largely through development economics. Here too, there is a familiar division between research that identifies the problem as arising from the ethnically neutral incompetence of the state versus conscious ethnic favouritism.

Cross-country research into the sources of economic growth and poverty in the 1990s found robust evidence that the incidence of armed conflict correlates with poverty and underdevelopment, although there are three problems in interpreting this finding. First, the direction of causality is often difficult to establish, so that it is uncertain whether conflict was the cause or consequence of economic failure. Second, poverty and conflict are too heterogeneous as analytical categories, and more precise causal pathways need to be elaborated in order to make this plausible – in particular to explain why a general problem such as poverty leads to the specific form of conflict that is prevalent, which is between ethnic groups. Third, there is a need to account for the large number of poor countries that have escaped armed conflict.

In response to this, Paul Collier has advanced a radical instrumentalist argument that poverty is indeed a factor because it reduces the opportunity cost of rebellion by making it cheap to recruit insurgents. In common with earlier generations of economic reductionism, it holds that ethnicity is merely the form or the ruse through which economic incentives, primarily the pursuit of private economic self-interest, are manifest. In contrast, Frances Stewart argues that ethnicity and group-based grievances are indeed relevant. Rather than poverty, she finds that it is inequality, and specifically, inter-group or 'horizontal' inequalities in terms of economic welfare, public service provision, political access, and other such variables which are critical both in the cause of conflict and in its resolution.

Both lines of argument have important consequences for the role of the state. The former holds that ethnic conflict can be addressed by accelerating economic growth to escape the poverty trap, and by strengthening state institutions of governance, security and natural resource management. The latter argues for a far more ethnically conscious form of intervention: that the state needs to actively redress inter-group inequalities by expanding service provision to deprived regions, and target welfare, employment and education towards deprived ethnic groups.

Epilogue

For the purpose of policy-relevant analysis, the state is often deliberately de-ethnicized in order to evaluate it technocratically in terms of its putative functions as a provider of welfare, security and representation. This would suggest that ethnic conflict results from states that are incompetent in providing welfare, poorly designed to adjudicate conflict, or too weak to provide security. In contrast, by looking at each of these functions in terms of processes rather than outcomes, it becomes clear that not only is ethnic conflict shaped by the state, but that the process is two-way, such that as Tilly (1975: 42) describes of another context: 'war makes the state, and the state makes war'.

Bibliography

Horowitz, D. (2000), *Ethnic Groups in Conflict*. Berkeley: University of California Press.
Licklider, R. (1995), The Consequences of Negotiated Settlements in Civil Wars, 1945–1993. *American Political Science Review* 89(3), 681–690.
Lijphart, A. (1977), *Democracy in Plural Societies*. New Haven, CT: Yale University Press.
Mann, M. (2005), *The Dark Side of Democracy: Explaining Ethnic Cleansing*. Cambridge: Cambridge University Press.
McGarry, J. and O'Leary, B. (1994), The Political Regulation of National and Ethnic Conflict. *Parliamentary Affairs* 47(1), 94–115.
Ottaway, M. (1995), Democratization in Collapsed States, in Zartman, W. (ed.) *Collapsed States: The Disintegration and Restoration of Legitimate Authority*. Boulder, CO: Lynne Rienner.
Posen, B. (1993), The Security Dilemma and Ethnic Conflict. *Survival* 35(1), 27–47.
Snyder, J. (2000), *From Voting to Violence: Democratization and Nationalist Conflict*. New York: W. W. Norton.
Tilly, C. (1975), Reflections on the History of European State-Making, in Tilly, C. (ed.) *The Formation of National States in Western Europe*. Princeton, NJ: Princeton University Press.
Walter, B. (1997), The Critical Barrier to Civil War Settlement, *International Organization* 51(3), 335–364.

9.9

Religions and development

Emma Tomalin

Introduction

Religious traditions continue to exert a strong influence on the lives of many people in developing countries. Religious beliefs can shape values about key development concerns, such as education and economics, and they frequently have an impact on styles of social and political organisation. Moreover, many so-called 'faith-based organisations' (FBOs) are engaged in development related activities, from service delivery to humanitarian aid and disaster relief. While the work of the larger international FBOs, such as CAFOD, Christian Aid or Islamic Relief is well known, there are numerous smaller religious organisations as well as places of worship (e.g. churches, mosques and temples) working at the grassroots level that are not documented and are less likely to receive international funds. Much of the work carried out by these religious organisations and places of worship pre-dates the post-World War II donor-driven development project and the formal rise of the FBO in the 2000s.

Nevertheless, until recently, considerations of religion have tended to be ignored or marginalised by mainstream development research, policy and practice (Selinger 2004; Haynes 2007; Deneulin and Bano 2009). A survey of three major development studies journals between 1982 and 1998 revealed that religion and spirituality are 'conspicuously under-represented in development literature and in the policies and programmes of development organizations' (Ver Beek 2002: 68). More recently, however, there has been a 'turn to religion' within development research, policy and practice. For instance, the views of religious leaders have been invited on key initiatives such as the 'Millennium Development Goals', donors, such as the UK Government's Department for International Development (DFID), increasingly fund faith-based organisations and development studies journals are witnessing an upturn in the publication of research articles that deal with the relationships between religions and international development.

This short chapter will provide an overview of some of the debates that have emerged around the topic of religions and development in recent years. By 'development' I primarily mean the agenda of the (predominantly) Western driven international development agencies and NGOs that have emerged since the 1950s. First, the reasons for the marginalisation of religion in development research, policy and practice will be addressed. Second, I will examine the factors that account for the more recent 'turn to religion'. Finally, this chapter will briefly explore the significance and implications of the rise of the 'faith-based organisation' (FBO).

Understanding why faith traditions have been ignored

One reason for this neglect of religion is due to the idea, dominant in modernisation theory, that secularisation (the removal of the influence of religion from public or social life, and the separation of religion and the state) goes hand-in-hand with the transition to modernity and economic development. Thus, religion was expected to either disappear altogether or to become a purely private matter that would have a limited effect on public life. It was anticipated that people would give up their 'primitive superstitions' and 'backward religious worldviews' (particularly in their application to social matters) as their communities evolved into developed societies modelled on the West. However, this 'secularisation thesis' has come under attack given the significant role that religion continues to play in public life in many developed as well as developing contexts.

Another reason for this avoidance of religion concerns the difficulties that may face secular development agencies (governmental and non-governmental) if they are seen to be supporting or opposing particular religious traditions (which are often politically significant and may divide communities). However, it is arguable whether this means that such agencies should never support the work of religious groups, given that they have the alternative option of adopting a careful and informed approach when building alliances with religious organisations. There is a need for development agencies to better understand the controversial and complex links between religion, politics and social development and their implications for policy.

A further reason for the uneasy relationship between religions and development is the fact that religious values and practices may act against the interests of the development process. For instance, the rise of 'fundamentalist' or 'conservative' religious forms is often considered to have negative consequences for women's human rights. The Catholic Church's prohibition on condom use can increase women's chances of contracting HIV when their husbands have been unfaithful, and can lead to unwanted pregnancies, which, in Catholic contexts, often cannot be legally terminated. In some Muslim countries, Islamic law is interpreted to deny women any form of political participation, while in others women are virtually confined to the home due to religious attitudes towards maintaining their sexual modesty (Tomalin 2011). However, in such contexts many women's rights activists are now strongly committed to pursuing improved conditions for women from within a religious framework rather than relying upon the secular rhetoric of mainstream (Western) feminist discourse. They promote styles of 'religious feminism', which argue for re-interpretations of religious beliefs that are consistent with the 'core' values of the tradition as well as various types of feminist thinking (see special editions of *Gender and Development* 1999 and 2006).

The above discussion has provided reasons for explaining why mainstream development actors have largely avoided consideration of and engagement with religion in pursuit of their goals, and suggested that, to a large degree, this is grounded in a Western secularist bias. Over the past 5–10 years, however, things have started to shift, with development studies displaying an upsurge in research interest and publishing on the topic of religions and development. This reflects an increased engagement between development actors and religious organisations and leaders. In the next section of this chapter I will discuss the reasons for this 'turn to religion' within development.

Accounting for the 'turn to religion' in development research, policy and practice

A previous version of this chapter was published in the second edition of this volume at a time when it had not yet been possible to either fully observe or reflect upon the 'turn to religion' within development. However, the processes that can be suggested to account for the entry of religion into development research, policy and practice have been underway for several decades. One important factor has been the shift from the dominance of purely economic understandings of development in the 1950s to an emphasis on more human-centred approaches, including the impact of the 'capabilities approach' on human development and the emergence of participatory development approaches, demonstrated, for instance, in the ground-breaking World Bank *Voices of the Poor* study (Narayan *et al.* 2000). This shift opened up space for considerations relating to religion and culture in development.

Also of significance have been critiques of simplistic theories of secularisation and the recognition that rather than disappearing or diminishing in significance, religion continues to exist alongside modernising and globalising processes, often adapting and even intensifying in response to changing social, economic and political environments. Even in so-called 'developed' nations, religious belief and practice is an important aspect of many people's lives. Debates about the global 'resurgence of religion' had been building momentum since the 1980s, if not earlier. The Iranian Islamic revolution of 1979 was, in many ways, a wake-up call to those who considered religion to have lost its public influence, and many examples of religiously inspired social and political action were witnessed across the globe in the following decades – including the '*Satanic Verses*' controversy in the early 1990s, the influence of the

Catholic Church on reproductive rights issues, the rise of nationalist Hindu politics in South Asia, and the cataclysmic event of 9/11.

The 'turn to religion' within development research, policy and practice is part of a broader renewal of global political interest in religion, at least since 9/11. There has been increasing concern about the rise of so-called extremist religious identities amongst Western governments, and spurred by the religious right in the USA, we also find an increased focus on the role that 'faith-based organisations' can play in various social and welfare activities both domestically and internationally (Clarke and Jennings 2008).

The rise of the FBO

The most striking feature of the 'turn to religion' by mainstream development actors over the past decade has been the increasing trend to support and even favour the work of 'faith-based organisations' (FBOs). While religious organisations have been involved in various types of charitable, philanthropic, humanitarian and development work for far longer, the origins of the term 'faith-based organisation' can be traced to the era of the Republican government of George W. Bush in the USA (2001–2009), when there was an increased desire to draw religious organisations into public life around the agendas of social cohesion and service delivery (Tomalin 2012).

An understanding of these 'faith-based organisations' as having a distinctive role or even a comparative advantage over secular organisations has come to shape much discussion about the role of FBOs in development (Tyndale 2006; Marshall and Marsh 2003). Particular attention is drawn to the perception that '*many FBOs, even more than NGOs*:

- provide efficient development services
- reach the poorest
- are valued by the poorest
- provide an alternative to a secular theory of development
- ignite civil society advocacy
- motivate action' (James 2009: 2; italics added).

Although many have welcomed the increased focus on FBOs as an attempt to bring a hitherto neglected body of stakeholders into the development process, others are more cautious (Balchin 2011). They perceive that there has been rather a rush to engage with FBOs without appropriate assessment of their effectiveness in achieving development goals (for instance, those relating to gender), and argue that this is a product of a broader political culture generated in the USA that (over-)values the faith contribution to welfare and social services (Tomalin 2011). Moreover, it is by no means clear what counts as an FBO, where the definition is often restricted to only include formally registered organisations that resemble NGOs, thereby excluding much religiously inspired development work (e.g. that occurring within places of worship) with which donors might usefully engage. In addition to the issue of what to include in the definition of FBO, the very act of distinguishing between faith-based and secular organisations in contexts where religion permeates almost all aspects of life may render the term 'FBO' meaningless in some locations.

A final thing to consider is the claim by donor agencies that their engagement with FBOs is evidence of their acceptance of the importance of incorporating cultural and religious values and practices into development policy and practice. However, it can instead be suggested that religion is being brought into development in ways that reflect Western perceptions of not only the nature of development processes but also understandings of and attitudes towards religion. First, the concept and term FBO is, to a large extent, a product of the NGO-isation of development, which is ideologically as well as practically driven (often by international agencies) and influences who works together, potentially marginalising other actors (i.e. organisations that do not fit the accepted FBO model). Second, there is evidence that organisations are explicitly shaping their activities and discourses to reflect the emergence of the FBO and to benefit from new opportunities to engage with donors, including playing down or moderating their faith identities (Tomalin 2012).

References

Balchin, C. 2011: Religion and Development: A Practitioner's Perspective on Instrumentalisation, *IDS Bulletin*, 42(1): 15–20.

Clarke, G. and M. Jennings (eds) 2008: *Development, Civil Society and Faith-Based Organisations: Bridging the Sacred and the Secular*, Basingstoke and New York: Palgrave Macmillan.

Deneulin, S. with M. Bano 2009: *Religion in Development: Rewriting the Secular Script*, London: Zed Books.

Gender and Development 1999, 7(1) and 2006, 14(3), (special editions on religion, gender and development).

Haynes, J. 2007: *Religion and Development: Conflict or Cooperation?* Basingstoke and New York: Palgrave Macmillan.

James, R. 2009: *What is Distinctive about FBOs*? Oxford: INTRAC Praxis Paper 22, available online at www.intrac.org/data/files/resources/482/Praxis-Paper-22-What-is-Distinctive-About-FBOs.pdf (accessed on 27 April 2012).

Marshall, K. and R. Marsh (eds) 2003: *Millennium Challenges for Development and Faith Institutions*, Washington, DC: World Bank Publications.

Narayan, D. with R. Patel, K. Schafft, A. Rademacher and S. Koch-Schulte 2000: *Voices of the Poor: Can Anyone Hear Us?* Oxford and New York: Oxford University Press for the World Bank.

Selinger, L. 2004: The Forgotten Factor: The Uneasy Relationship Between Religion and Development. *Social Compass* 51(4): 523–543.

Tomalin, E. 2011: *Gender, Faith and Development*, Oxford: Oxfam and Rugby: Practical Action Publishing.

Tomalin, E. 2012: Thinking about Faith-based Organisations in Development: Where Have We Got to and What Next? *Development in Practice* 22(5–6): 698–703. (Part of a special edition on religion and development).

Tyndale, W. (ed.) 2006: *Visions of Development: Faith-based Initiatives*, Aldershot: Ashgate.

Ver Beek, K. A. 2002: Spirituality: A Development Taboo, pp. 60–77 in T. Verhelst and W. Tyndale (eds) *Development and Culture*, Oxford: Oxfam.

Websites

www.religionsanddevelopment.org/ – This is the website of the 'Religions and Development' (RaD) research programme, a five year project that was based at the University of Birmingham, UK, (2005–2010) and funded by the UK Government's Department for International Development (DFID).

http://berkleycenter.georgetown.edu/wfdd – This is the website of the World Faiths Development Dialogue, established in 2000 by George Carey, then the Archbishop of Canterbury, and James Wolfensohn, then the President of the World Bank.

Part 10
Governance and development

Editorial introduction

Development activity was for long virtually the monopoly of the state. However, the lack of alternatives did not mean the state was always a positive force for development. Moreover, in the late twentieth century, the state's claim to this monopoly weakened, while other agencies of development such as the World Bank, the IMF and non-governmental organisations (NGOs), gained a higher profile. Development has to be seen in the economic context of global capitalism, but also in the political context. The most crucial relationship is between the state and the economy: states participate directly in processes of productive capital formation (establishing a set of economic policies favourable to capitalist accumulation), provide infrastructure and affect private-sector resource allocation through monetary and fiscal policies. The state provides an enabling environment/structure for development by other agencies. The state is the network of government, quasi-government and non-government institutions that coordinates, regulates and monitors economic and social activities. The role of non-state actors seems destined to grow as the power of the nation-state declines and global economic activity intensifies. The narrative of the study of development has changed dramatically in the last decade with the changes occurring post global recession.

Total official development assistance (ODA) allocated by all major donors is low and declining. Even though some ODA appears altruistic, much is manifestly deployed to promote the political and economic concerns of donors. Increased emphasis on aid conditionality underlines this, which has been controversial. Now not only the World Bank and the IMF but also many bilateral ODA programmes require recipients to adhere to certain policies. Aid is viewed as a means of promoting donors' perceptions of 'good governance' and 'sound' economic practices, leading many analysts and politicians to become very critical of aid. Research has indicated that economic growth has taken place post structural adjustment and reforms. Economic growth has also led to emerging rising powers becoming important development donors and partners which has challenged the traditional donors to assess their aid strategies and reflects a radically changed development aid landscape. Governments and social and political elites have prospered, but questions have been raised regarding reaching the ordinary poor people. Recently, corruption has been highlighted as leading to inefficient and unfair distribution of scarce benefits. It undermines the purpose of public programs, encourages officials to create red tape, increases the cost of doing business, and lowers state legitimacy. Corruption is a feature of state/society relations that undermines the legitimacy of the state and leads to wasteful public policies and begs for increased transparency and accountability.

Good governance is defined as sound management of a country's economic and social resources for development. What is 'sound' for the World Bank and others holding the view that 'democratisation stimulates development' are a range of management techniques that are believed to work well within a standardised liberal democratic model. Critics contend that there are, and ought to be, different paths for development; they are not opposed to 'good governance' but urge that this is compatible with alternatives to liberal democracy in poor countries with different institutional contexts. Despite various campaigns led by various advocacy groups such as Jubilee 2000 and Make Poverty History urging for debt cancellations, progress on this front has been very slow. Since 1999 governments in the South are also required to develop pro-poor policies (poverty reduction strategies – PRSPs) by consulting local stakeholders, especially civil society and cultivating 'partnership'. The question remains if aid is effective in alleviating poverty and making a difference. What has become evident over the last few decades is that the two multilateral institutions and their sister agencies, the regional development banks occupy a dominant position in the global political economy, with increasing involvement in the whole range of cross-cutting themes and politicised agendas, making their role very difficult.

Throughout the 1990s the private sector's role in public service delivery has generated significant opposition and protest. It emphasises the renewed interest of the markets in the development process, in the context of the state failure and a possible solution to respond to the vast needs of the low-income citizens of the developing countries. Search continues for combining the assumed efficiency of the private sector and the democratic accountability of the public bodies.

The purpose of giving ODA and its deployment is more open to scrutiny. One response has been increased funding for NGOs, generally thought more able to reach the local grassroots level.

In the New Policy Agenda, combining market economics and liberal democratic politics, NGOs are simultaneously viewed as market-based actors and central components of civil society which is reformulating itself since the global recession with various social movements, revolutions and protests taking place around the world. NGOs fill gaps left by the privatisation of state services as part of a structural adjustment or donor-promoted reform package. Governments need NGOs to help ensure their programmes are effective, well targeted, and socially responsible and well understood, hence the importance of the NGO – state relationship, which exists in diverse forms in different countries.

There are various types of NGO. At the international level we can differentiate between campaigning and charitable or service-providing NGOs. Both of these are generally based in the North. International campaigning NGOs are epitomised by Greenpeace. Such NGOs will address development policy issues from a distance. Northern-based, service-providing NGOs include Save the Children, Oxfam, and Christian Aid and so on. These generally have branches in the Southern countries in which they work. Often they will run their own projects, sometimes setting up their own partnerships with NGOs in the South, effectively by-passing those of the state. In other circumstances they will fund and monitor local service-providing NGOs or membership organisations. In recent years the NGO donor relationship and its mutual trust has been tested mainly because of government's reactions against terrorism, insecurity and instability. This has led to a shift in governments' attitudes towards non-governmental donors and civil society at large.

At the national level many NGOs are public-interest research or campaigning organisations. Some are Western-style human rights or conservationist NGOs. Usually relatively few, they often represent the concerns of particular groups. Other types of NGO are indigenous, national (and provincial) service-providing NGOs – mostly concerned with welfare and development. Many international NGOs have moved from directly running projects to working through partnerships with such NGOs. Lastly, there are membership organisations, often called 'grassoots organisations', which exist to further their members' interests.

Recent findings have shown that NGOs are weak in contextual analysis of societies in which they work and that their approaches to monitoring and evaluation are rarely adequate. Further, certain key technical skills are frequently seen as lacking in their human resource base and many are more concerned with micro- than macro-context work. The practice of participation and innovation in project implementation can be poor too. Questions are asked about their accountability, legitimacy, performance and effectiveness. On the other hand, drawing on the new opportunities afforded by growth in their numbers and size and in the resources available to them, and facilitated by advances in communication technologies, NGOs want to scale up advocacy activities. NGOs have increasingly globalised, and are working through transnational networks, forging new enduring alliances and networks to maximise their impact, and engage in global policy processes. There is an acknowledgement that various actors at the national and international level need to cooperate to achieve commonly accepted goals and to address global challenges and that the global governance system should be flexible enough to adapt to changing environments and challenges especially in the context of the current global financial crisis.

NGO campaigning (sometimes also through media) has led to NGOs being represented at important UN conferences and its contact with the World Bank has grown rapidly over the 1990s especially in the context of policy debates, for example, environment, debt relief, gender, participation strategies and service delivery. NGOs are now also invited for formal consultations as part of civil society, on various sector policies and on the themes of the annual World Development Reports. NGOs have also been very critical of World Bank financed infrastructural projects (e.g. involuntary resettlement, energy, forestry and dams) and have contributed to policy changes.

As economic inequalities have grown throughout the world, the political conditions required to overcome democratic deficits in decision making have becomes very crucial especially in the context of discrimination on the grounds of race, colour, sex, language, religion, political or other opinions. Universal international human rights provide hope for many of the oppressed. It is also about generating accountability of states and security to citizens and upholds national and international standards in the context of the right to development.

10.1 Foreign aid in a changing world

Stephen Brown

Foreign aid, like the world in which it operates, is in a state of flux. Since the end of the cold war, the sources, goals and modalities of aid have all shifted to a certain extent, but it is not yet clear what the result will be.

Foreign aid during the cold war

During the cold war, there were two main donor groups, centred on the United States and the Soviet Union. The main Western donors formed the Development Assistance Committee (DAC) within the Organisation for Economic Co-operation and Development (OECD), the main coordination body of the Western industrialized countries. The Soviet Bloc established its own version, the Council for Mutual Economic Assistance (CMEA, also known as COMECON). Each provided assistance to its allied states in the developing world. Many DAC members also provided assistance to officially non-aligned socialist countries. Much of Western foreign aid during this period served as a tool of donor country foreign policy, especially in the case of the United States (Morgenthau 1962). Still, many donors also provided significant amounts for more altruistic purposes (Lumsdaine 1993).

Most foreign aid was – and still is – provided directly from government to government, known as bilateral aid, but a significant amount was channeled through multilateral institutions such as the United Nations and the World Bank. The focus of the assistance varied according to trends in development thinking. For instance, an initial emphasis on large infrastructural projects, such as hydroelectric dams, in the 1960s and 1970s failed to produce the expected 'trickle-down' benefits for the poor. As a result, donors placed greater emphasis in the 1970s on meeting basic needs. In the 1980s, facing persistent poverty, donors focused on macroeconomic stability and liberalization, but that approach has also proved disappointing.

Early reconfigurations: Aid in the 1990s

The collapse of the Soviet Bloc and the end of the cold war presented the West with a tremendous opportunity to reallocate massive defence spending into more productive areas, including international development, and usher in the much heralded New World Order, based on democracy, good governance and free markets. However, the 'peace dividend' never materialized. Faced with severe fiscal deficits, Western governments, rather than reallocate military budgets funds to aid, actually cut their budgets in both areas. Knowing that their former rivals could not take advantage of the situation, they reduced aid spending without fear. As Figure 10.1.1 illustrates, total DAC aid disbursements fell after 1992.

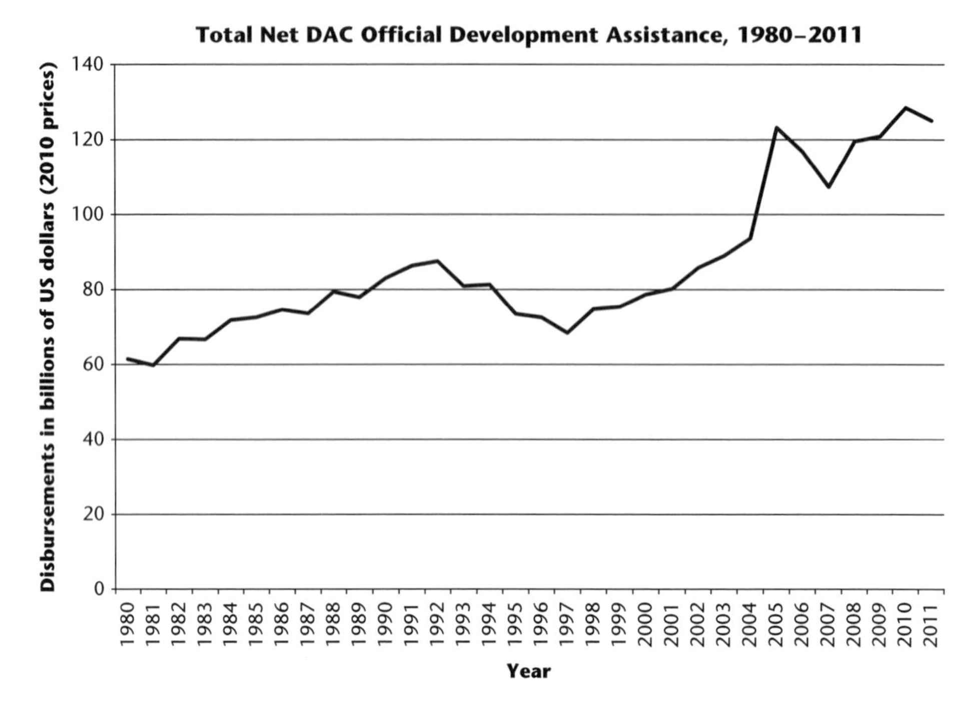

Figure 10.1.1 Twenty-first century development cooperation

Source: OECD Query Wizard for International Development Statistics, http://stats.oecd.org/qwids (accessed 6 May 2012)

The US's 'unipolar moment' (Krauthammer 1990/91) also changed the nature of foreign aid. For instance, donors placed greater emphasis on democratization and peacebuilding. However, those efforts soon proved disappointing as well. Though donors helped liberalize political systems to a certain degree in recipient countries, the results often fell short of liberal democracy. In many cases, donors' economic and security priorities meant that they did not translate their pro-democracy rhetoric into action (Brown 2005). Though several civil wars, which had been fed by superpower rivalry, did end in the 1990s, such as Mozambique's, a number of new conflicts also emerged, including the Rwandan genocide and two deadly civil wars in Zaire/Democratic Republic of Congo. Indeed, during this time some jokingly referred to the New World Disorder.

Changing context

Early in the new millennium, two events profoundly reshaped the context of foreign aid and justified increased expenditure. In 2000, the United Nations' member states unanimously adopted eight Millennium Development Goals (MDGs), which committed them to achieving ambitious poverty-reducing targets by 2015. This helped refocus foreign aid away from macroeconomic reforms towards measures with more concrete short- to medium-term effects on the quality of life of billions of people living in poverty, including by reducing child mortality and achieving universal primary education.

The following year, however, the MDGs were overshadowed by a new threat that replaced communism in the Western mind set: terrorism. The attacks of 11 September 2001 led many donors, especially the United States, to reorient their development assistance to sectors and countries that played a central part in the 'war on terror'. After US-led invasions, Iraq and Afghanistan became top aid recipients of foreign aid. Donors used aid – with limited success – to try to 'win hearts and minds' in conflict areas, rather than fight poverty per se, often administered by military actors instead of civilians. Increasingly, donors recast underdevelopment as a source of terrorism and a threat to Western countries, rather than a black mark on human conscience that needed to be reduced or eliminated for altruistic, ethical reasons.

Later in the decade, in 2007–08, a third event sharply affected the foreign aid context: the global economic crisis. The latter hit OECD countries especially hard, leading many DAC members to cut their aid levels, although some protected their aid budgets. (Figure 10.1.1 shows a drop in total expenditure in 2011.) The slow and uneven recovery not only threatened global aid flows but also discredited Western development models, especially when rapid-growing non-Western countries like China remained relatively unaffected (Birdsall 2012).

New actors and modalities

The stagnation of OECD aid after 2011 accelerated the rise in importance of non-DAC donors, already an important trend (Woods 2008). Often referred to as 'emerging donors', the growth in aid from countries such as Brazil, China, India, Saudi Arabia, Turkey and Venezuela not only multiplied the number of donors but also provided assistance on different terms from 'traditional' ones.

Following the end of the cold war, especially after the early 2000s, DAC donors increasingly recognized their own part in the often disappointing results of development assistance. They recognized that not only was *more* aid needed, but also – perhaps more importantly – *better* aid. For instance, they have progressively implemented the principle that foreign aid should not be 'tied' to the purchase of goods from the donor country, as this practice adds to the costs of aid without any commensurate benefit from the recipient's perspective (Jepma 1991). Donors have also increasingly worked together to provide joint assistance to government programs, for instance in the education sector, rather than working in isolation and supporting uncoordinated projects, such as the construction of schools.

In 2005, DAC members agreed to five basic principles, laid out in the Paris Declaration on Aid Effectiveness: a developing country's lead role in designing and implementing its own strategies (known as ownership), donors' alignment with the latter, coordination among donors (harmonization), mutual accountability and a focus on results. Despite donors' enthusiastic embrace of these principles, to date they have largely failed to put them into practice (OECD 2011).

However, emerging donors, even if they endorsed the Paris principles as aid recipients, do not want to be bound by them when providing assistance of their own. In fact, they do not all share the basic concept of altruism that has to various degrees characterized the theory, if not the practice, of six decades of Western foreign aid. Whereas DAC members generally separate 'official development assistance' (defined as government assistance whose main purpose is improving economic or social well-being in developing countries) from other financial flows, Southern donors frame development cooperation (as distinct from aid) as being of mutual benefit. They emphasize the value of bundling of aid and non-aid instruments, including investment, loans and trade, often relying on tied aid.

Coordination problems and the future of the international aid architecture

An additional challenge to coordination in the field of development assistance is not only the growing number of actors but also of *types* of actors involved, many of which are reluctant to coordinate with traditional donors. Non-state development actors have become prominent, especially in the health sector, and often have a different approach to development. Like emerging donors, private foundations, such as the Bill and Melinda Gates Foundation, and 'vertical funds', including the Global Fund to Fight AIDS, Tuberculosis and Malaria, often operate outside traditional donor coordination mechanisms, as does the private sector. The lack of cooperation can lead to duplication of efforts, as well as contradictory approaches and activities.

In the past, DAC members and, to a lesser extent, the subset that comprise the G8 industrialized countries set foreign aid norms. To date, they have driven discussions of aid effectiveness, but this dominance is being increasingly challenged. Though the DAC and G8's role in coordinating foreign aid is waning, it is not yet clear who or what will play a similar role for development cooperation in the future. The G20 and the UN Development Cooperation Forum are possible successors, but the international aid 'architecture' is still under construction – without a designated architect.

With the current focus on state-led development strategies, non-governmental organizations from both the global North and the South tend to be marginalized in development policy discussions. They are increasingly vocal and organized transnationally, which may help them claim a larger role in the future.

The implications of all these changes for foreign aid remain unknown. A convergence might develop among the practices and principles of the various actors in the field. Some of these new trends could help reduce poverty and inequality, but the effects of others may not be so positive. In addition, it is unclear what the main focus of development efforts will be after 2015, the target year for achieving the MDGs.

References

Birdsall, Nancy. 2012. 'The Global Financial Crisis: The Beginning of the End of the "Development" Agenda?'. CGD Policy Paper 003. Washington, DC: Center for Global Development. Available online at www.cgdev.org/content/publications/detail/1426133 (accessed on 6 May 2012).

Brown, Stephen. 2005. 'Foreign Aid and Democracy Promotion: Lessons from Africa'. *European Journal of Development Research*, vol. 17, no. 2 (June), pp. 179–98.

Jepma, Catrinus J. 1991. *The Tying of Aid.* Paris: OECD.

Krauthammer, Charles. 1990/91. 'The Unipolar Moment'. *Foreign Affairs*, vol. 70, no. 1, pp. 23–33.

Lumsdaine, David Halloran. 1993. *Moral Vision in International Politics: The Foreign Aid Regime, 1949–89.* Princeton, NJ: Princeton University Press.

Morgenthau, Hans. 1962. 'A Political Theory of Foreign Aid'. *American Political Science Review*, vol. 56, no. 2 (June), pp. 301–9.

Organisation for Economic Co-operation and Development (OECD). 2011. *Aid Effectiveness 2005–10: Progress in Implementing the Paris Declaration.* Paris: OECD.

Woods, Ngaire. 2008. 'Whose aid? Whose influence? China, emerging donors and the silent revolution in development assistance'. *International Affairs*, vol. 84, no. 6, pp. 1205–21.

10.2

The rising powers as development donors and partners

Emma Mawdsley

Perhaps the single most important trend within the development sector in the last five to ten years has been the growth in the number, visibility and impacts of Southern, Gulf and Central/East European countries as donors and development partners. They are now estimated to contribute around 12–15 per cent of global overseas development assistance (Park 2011), although these figures are open to all sorts of questions about definitions and measurements. Nonetheless, it is clear that the rising powers are becoming larger and more important as development donors and partners. These actors have profound implications for poorer people and poorer countries; and for global development norms, governance and the international aid and cooperation architecture (Mawdsley 2012).

The (re)emerging development actors include global powers like Brazil, China, India and Russia; regional hegemons like Saudi Arabia and South Africa; Central and East European countries like Poland, Hungary and Slovenia; and a multitude of 'second tier' rising powers, such as Indonesia, Thailand and Turkey (these categories are problematic, of course, and also overlap). Some of these states have been engaged in foreign aid and/or development cooperation (a distinction we come to below) for decades. China started giving assistance to North Korea in the 1950s, expanding to other Asian and African countries soon after (Bräutigam 2009); while states like Vietnam and Czechoslovakia were embedded in what Bayly (2009) calls a 'socialist ecumene' of assistance and exchange. Many of the oil-rich Gulf states have been donors since the late 1960s, and historically have been exceptionally generous in terms of the share of GNI directed towards aid. Other countries are more recent arrivals. Thailand, for example, announced in 2003 that it was ready to transition from recipient to donor. A large number of smaller and poorer states also provide aid and assistance on a sporadic basis, usually in response to humanitarian crises. Sri Lanka, for example, sent aid to the city of New Orleans after Hurricane Katrina in 2005. Like their Western counterparts, these states are motivated by altruism, the pursuit of diplomatic alliances, encouraging trade and investment, projecting soft power, and in support of security and geopolitical agendas. But, as we shall see, there are also important differences in the construction, practices and discourses of aid and development cooperation.

Given these diverse but long-standing histories, why did the vast majority of Western development textbooks, academic analyses, media reports and policy discussions almost entirely overlook these other actors until very recently? Postcolonial scholars argue that Western development imaginations are profoundly anchored in essentialised dichotomies of North/South, developed/underdeveloped, donors/recipients. Non-Western donors transgress these cognitive boundaries (Six 2009), leading to a (sub-)conscious politics of suppression of the Other as donor rather than recipient. But whereas in earlier decades they could be largely overlooked, the deep shifts now taking place within the global geographies of wealth and power associated with the 'rising powers' mean that this neglect is no longer possible. That said, most non-Western donors remain underreported, while the new awareness of the (re) emerging donors has often been accompanied by considerable ignorance and prejudice. Deborah Bräutigam's blog (see below) demonstrates the nature and extent of weak reporting and analysis of China in Africa, for example.

This brings us to terminological challenges. Many of these states avoid the label of 'donor' given its associations with North–South power relations, and prefer the term 'development partners'. Some, such as India, reject the term 'foreign aid' for the same reason. This is not just a matter of different terms – there are very real conceptual differences in their diverse constructions of these relationships and flows. This can make comparisons hard, and in some cases, invalid, as commentators do not always compare like for like. Most of these states engage in what the (so-called) 'mainstream' donors would recognise as foreign aid or (with slightly more latitude), aid-like activities. This includes grants, loans which meet certain concessional criteria and which are intended to promote economic development and welfare, food aid, debt relief, humanitarian assistance, technical assistance and so on. Where this aid sometimes differs (in theory at least, often less in practice) from 'mainstream' aid is that it is often tied to the purchase of donor goods and services. But many of these actors – and notably the Southern and Gulf states – also mobilise a wider reaching and more complex notion of 'development cooperation'. This can encompass diplomatic events, cultural exchanges, and critically, a variety of official development financing instruments that the OECD–DAC currently seeks to exclude from its definition of 'foreign aid', such as export credits (although this is currently under debate). Development cooperation clearly blurs with commercial interests, and indeed, geostrategic goals such as energy security and resource access. Aid flows and activities are often packaged up with these other official development cooperation flows and agreements. Critics argue that they distort market competition and also undermine genuine development effectiveness as profit imperatives take priority. However, supporters of the development cooperation model argue that its more holistic agenda and integrated approach has more effective results, something we now turn to in more detail.

The (re)emerging donors are involved in a huge array of projects, sectors and activities. Brazil, for example, is starting to export its very successful innovations in social protection programmes through various forms of conditional cash transfers (CCTs) to combat poverty and social exclusion; and the provision of medical personnel and technical assistance is a long-standing and well-respected form of cooperation. But compared to the 'traditional' bilateral donors, a striking characteristic of most of the (re)emerging partners is their focus on enhancing economic growth and productivity within their partner countries. This includes a strong focus on technical support, funding and/or building transport and energy infrastructure (roads, railways, airports; dams, pipelines, refineries, substations), and on supporting large-scale agricultural and manufacturing growth. Aid and aid-like activities play a part, as do wider development cooperation flows and agreements. In contrast, over the last ten to fifteen years, the mainstream development community has tended to concentrate increasingly on good governance and social inclusion. While some of these trends have been progressive in terms of widening out understandings of both poverty and development, a growing criticism has been that this has neglected the critical engine of growth (Chang 2010). In contrast, the (re)emerging partners appear to be re-animating the modernisation theories of the 1950s and 1960s, in which economic growth is the primary and prior requirement of 'development'. In theory, poverty reduction will follow.

Recipient states and peoples certainly have reservations about aspects of the (re)emerging actors, and the ability or desire to engage strategically and with longer-term developmental benefits in mind varies enormously between the different partners in Africa, Asia and elsewhere. But by and large, recipient governments have welcomed the (re)emerging partners with open arms (Woods 2008). They do not just provide more financing, but just as importantly, alternative sources of financing, breaking the monopoly of the 'traditional' partners. Some, notably China, also seem to demonstrate different routes to economic growth to those imposed by the market fundamentalists of the North. Many poorer countries are currently experiencing accelerating growth rates, which are being driven in large part by the resource demands and economic dynamism of the rising powers.

The key issue is then whether these growth rates and economic successes will translate into sustainable and inclusive *development*. Governments and social and political elites are likely to prosper from the rising tide, but will ordinary people, and will the poorest and/or the least politically powerful – pastoralists, small farmers, forest dwellers and those living in informal urban settlements? What of non-capitalist social structures and cultures – subsistence farmers, communal land owners, local non-monetary collectives? Many of the (re) emerging donors and development partners, especially those in the global South, assert win–win outcomes and mutual benefits of growing aid, trade and investment relations. This may well often be the case – building roads, developing raw materials and 'modernising' agriculture will bring benefits to many. But it will also bring costs, particularly to indigenous peoples, small farmers, forest-reliant people and the poorest. Win–win assertions are often founded on a simplistic construction of the 'national interest' of both partners, obscuring the contested and dislocating nature of 'development'. The uneven social and economic consequences of such modernisation and economic growth are glossed over beneath a symbolic regime of striving nations seeking to contest inequalities and injustices within the international hierarchy of states (Prashad 2007). The question is whether today's social movements, labour organisations, civil societies and media in poorer countries will be able to hold their own business people, policymakers and parliamentarians to account, in an effort to formulate and manage strategies that make the best of the growing relationships with the rising powers, and indeed, their relationships with the 'traditional' powers.

This discussion has taken us beyond aid and even development cooperation to the broader sweep of trade, investment and geopolitical impacts of the rising powers. Let us return then specifically to aid and aid-like activities amongst the diverse and numerous field of (re)emerging donors, and in this final section look briefly at changes within the international development architecture. Some of the (re)emerging powers are essentially converging with the mainstream development regime, although by no means entirely sacrificing their policy space. South Korea, for example, joined the OECD–DAC in 2010, making it only the second non-Western member (Japan was the first). The ten 'new member states' of the EU are obliged to adopt EU standards and rules regarding donor activities, although they are also maintaining room for manoeuvre in pursuit of their national interests (Lightfoot 2010). At the other end of the spectrum are a few who actively reject the Western-dominated mainstream aid infrastructure. In 2006, for example, the Venezuelan government announced that it would switch its multilateral funding from the World Bank's International Development Association (IDA) and the Inter-American Development Bank (IADB) to Southern-led institutions like the Banco del Sur and the Organization of the Petroleum Exporting Countries (OPEC) Fund. Most of the (re)emerging partners fall somewhere in between, cooperating or coordinating in some contexts but not in others. Some simply do not have the administrative infrastructure or resources to manage the transaction costs of coordination – whether this is recording and reporting aid channels and activities, or contributing to national, regional and global aid forums. Many do not wish to have their policy space constrained by signing up to international or regional agreements, especially as these are often Western-dominated. Many Southern partners especially also have long-standing ideological differences with the Western-dominated mainstream – as we have seen they articulate a different political and discursive construction of aid and development cooperation (Mawdsley 2012). While they may be willing to cooperate and coordinate in some contexts, they refuse to be co-opted.

The current aid architecture is being redesigned with the intention of creating a new global development partnership that reflects a radically changed development landscape. In particular, it aims to include the rising powers as development donors and partners. It is likely to be a far more inclusive and representative body than the existing dominant aid institutions, but it remains to be seen whether it will manage to work effectively, or with sufficient voice and protection for weaker countries and sections. Whatever the case, it is clear that we are currently witnessing a revolutionary shift in the development landscape, with profound consequences for the norms, practices and theories of aid and development, in which the complex and highly varied (re)emerging development actors are key actors and agents.

References

Bayly, S. (2009) 'Vietnamese narratives of tradition, exchange and friendship in the worlds of the global socialist ecumene', in H. West and P. Raman (eds) *Enduring Socialism: Explorations of Revolution, Transformation and Restoration*. Oxford: Berghahn Books, pp. 125–147.

Bräutigam, D. (2009) *The Dragon's Gift: The Real Story of China in Africa*. Oxford: Oxford University Press.

Chang, H.-J. (2010) 'Hamlet without the Prince of Denmark: How development has disappeared from today's "development" discourse', in S. Khan and J. Christiansen (eds) *Towards New Developmentalism: Market as Means rather than Master*. Abingdon, UK: Routledge.

Lightfoot, S. (2010) 'The Europeanisation of international development policies: The case of Central and Eastern European States'. *Europe-Asia Studies*, 62, 329–350.

Mawdsley, E. (2012) *From Recipients to Donors: The Emerging Powers and the Changing Development Landscape*. London: Zed.

Park, K. (2011) 'New development partners and a global development partnership', in H. Kharas, K. Makino and W. Jung (eds) *Catalysing Development: A New Vision for Aid*. Washington, DC: Brookings Institute, pp. 38–60.

Prashad, V. (2007) *The Darker Nations: A People's History of the Third World*. New York: The New Press.

Six, C. (2009) 'The rise of postcolonial states as donors: A challenge to the development paradigm?' *Third World Quarterly*, 30(6): 1103–1121.

Woods, N. (2008) 'Whose aid? Whose influence? China, emerging donors and the silent revolution in development assistance'. *International Affairs*, 84(6): 1205–1221.

Further reading

Chaturvedi, S., Fues, T. and Sidiropoulos, E. (2012) *Development Cooperation and Emerging Powers: New Partners or Old Patterns?* New York: Zed.

Mawdsley, E. (2012) *From Recipients to Donors: The Emerging Powers and the Changing Development Landscape*. London: Zed.

Smith, K. (2011) Non-DAC Donors and Humanitarian Aid: Shifting Structures, Changing Trends. Global Humanitarian Assistance Briefing Paper.

Zimmermann, F. and Smith, K. (2011) 'More Actors, More Money, More Ideas for International Development Co-operation'. *Journal of International Development*, 23(5), 722–738.

Websites

China in Africa – The real story: www.chinaafricarealstory.com/
The Rising Powers – China as a new 'shaper' of global development: http://risingpowers.open.ac.uk/
The South–South Opportunity: www.southsouth.info/

10.3 Aid conditionality

Jonathan R. W. Temple

Aid donors sometimes impose conditions on grants, loans or technical assistance. The donors demand that, in return for aid, certain actions should be carried out by the recipient government. This practice has been sufficiently important to acquire a name, conditionality. As a term, it sounds dry, technical and unexceptionable, but it has long been one of the most controversial aspects of international development policy, and its implementation has sometimes led to fury directed at the World Bank and IMF.

There are some who think conditionality represents aid at its worst, allowing powerful donors to impose their will on poorer countries, often for reasons of national self-interest. There are others who think aid is undermined because conditionality rarely goes far enough, or is not properly enforced. The introduction of structural adjustment lending in the 1980s intensified the debate. The World Bank and IMF sought to make loan disbursement conditional on macroeconomic stabilization, and often on wider reforms. That approach was not a conspicuous success, and conditionality has evolved significantly over time.

The case for conditionality

Attaching conditions to aid starts from a reasonable assumption, namely that recipient governments typically have multiple objectives, not all of which are aligned with those of donors. This should not be controversial. Even in long-standing democracies, politicians often have mixed motives, and do not simply seek the national interest. The objectives of political leaders are likely to be similarly compromised in aid recipients, many of which are autocracies of various kinds, or countries where democracy is fragile and not long established.

For now, assume (a little naively) that the donor is altruistic, while the recipient government has differing objectives. How should the donor proceed? The conflict of interest corresponds to the "principal–agent" problem in the economic theory of incentives. The donor is the principal, and sets the terms for transfers to the recipient government, the agent. Usually, the optimal policy for the principal is to make transfers that are conditional on either the actions of the agent (if these are observed) or final outcomes that are linked to those actions. In the case of aid, conditionality seeks to make aid more effective: the donor induces or "buys" certain policy actions or reforms which, if the donor is altruistic and well informed, should benefit the poor.

The arguments against conditionality

At this point, the case against conditionality can proceed in several different directions. One set of arguments is that, whatever the technical case for imposing conditions, it is morally illegitimate for donors to use their economic strength to influence the choices of sovereign nations. This argument has more force than economists typically allow, but the importance of sovereignty needs to be carefully weighed. Since relatively few recipient countries are stable, effective democracies, a distinction can be made between the sovereignty and interests of recipient governments, and the interests of their populations. In principle, if conditions on aid help to improve the effectiveness of a recipient government, or its responsiveness to the domestic population, it might be a mistake to give the loss of sovereignty too much weight.

Some would reject the assumption that the donor is altruistic, however. In this view, donors seek major reforms, such as trade liberalization or privatization, primarily to serve their own economic interests. Conditionality is characterized as a new form of economic imperialism. A full assessment of this view is beyond the scope of this chapter, but many economists would be sceptical that reforms in recipient countries, at least of the type usually promoted, will have major economic benefits for donors. The technical case that some economists make for specific policies, such as trade liberalization, is usually based on projected benefits for the reforming country.

The use of "projected" raises a further set of objections, however. It would be wrong to assume that the donor is necessarily well informed, or that there is a technical consensus over growth strategies. Since there is typically room for controversy about major policy reforms, the logic of policy conditionality risks hubris. The appropriate mix of policies will vary across countries and over time, and recipient governments will have their own view of development priorities.

These arguments are important, but the extent to which they apply varies. There is general agreement that the use of policy conditionality has sometimes gone too far, imposing conditions that were questionable or intrusive. But some conditions are designed to forestall the damaging long-run consequences of sustained budget deficits and high inflation in circumstances where recipient governments are failing to address these problems. Other conditions may be relatively innocuous, such as those relating to the internal organization of the recipient state, in areas such as budgetary administration, financial planning and other aspects of public management.

Another set of objections to conditionality is less obvious. There is a widespread perception that traditional conditionality fails to influence the policies and actions of recipient governments. One explanation is that donors find it hard to commit to their conditions. In theory, if the conditions are not met, the aid should not be disbursed. In practice, that decision may be hard to sustain. A donor may want to disburse the aid in any case, perhaps because this will lead to the best remaining outcome for the country's poor. A recipient government that understands this has no incentive to reform, and conditionality becomes ineffective. This problem is compounded by "budget pressure" effects, in which the officials of aid agencies face significant institutional pressures to disburse aid, regardless of whether or not conditions have been met. In some cases, so-called defensive lending is likely, where new loans have to be disbursed to help a country avoid default on past loans.

The modern view is that much depends on the political economy of the recipient country. Conditionality may sometimes strengthen the hand of a reforming government, but in other cases, a donor's threat to terminate aid flows may not have much leverage, relative to the other pressures and constraints on political leaders. Recipient governments may pay lip service to the need for reform, and even pursue it temporarily, but lasting change requires a domestic constituency that is willing to champion a given set of policy measures and priorities.

Country ownership

The political economy perspective leads naturally to the idea of "country ownership". This is sometimes characterized as an action to be taken by recipient governments. From the point of view of donors, it is important for recipient governments to take ownership of a set of policy measures, and commit to seeing them through to a successful conclusion. As Khan and Sharma (2003) put it, the country carrying out reforms has to be committed to their spirit as well as their letter.

But, from an alternative perspective, that commitment could look more like a consequence of ownership than the thing itself. Ownership is not something that recipient governments can simply will into existence. Instead, it must emerge through a deliberative process. Recipient governments are more likely to be committed to a policy agenda if they have played a role in shaping it. Hence, recipient governments must retain some degree of autonomy, and the concepts of ownership and autonomy are linked. But a donor might want to add that even autonomous individuals and governments should be open to advice, persuasion and the exchange of information.

The major donors have sought a role for country ownership partly through the Poverty Reduction Strategy Papers (PRSPs). Initiated in 1999, these have been led by the IMF and especially the World Bank, with the support of other donors. Each aid recipient draws up an explicit long-term strategy for addressing poverty, and links this to an overall development strategy. But it has been argued that, in practice, many PRSPs are remarkably alike. This has prompted the charge of "ventriloquism": aid recipients are led to understand which policies will be needed to gain donor support. If countries lack the capacity to formulate development strategies, or donors remain too prescriptive, the stated commitments to autonomy are no more than window dressing. The approach could even be harmful: the contributors to Gould (2005) argue that it works against the consolidation of democratic forces and civil society, partly by side-lining parliaments and the role of political parties.

The evolution of conditionality

Conditionality has evolved over time, not least in the approach taken by the IMF. For obvious reasons, the IMF attaches macroeconomic policy conditions to its adjustment loans. More controversially, the conditions have sometimes included detailed structural reforms. This could be counter-productive. The attempt to impose unpopular structural reforms can undermine the perceived legitimacy of an overall programme, threatening its chances of success. Multiple conditions also increase commitment problems: should the donor disburse when the conditions have been only partially fulfilled? In some cases, the conditions became overly demanding and intrusive. In the wake of criticism, the IMF has sought a more streamlined approach, with fewer conditions. In principle, this should emphasize only those conditions seen as critical to the IMF's core responsibilities and to the repayment of its adjustment loans. But successful implementation remains difficult, not least when decisions have to be taken over the timing and extent of fiscal austerity.

The appropriate route is even less clear for long-term development lending, such as that undertaken by the World Bank, sometimes the IMF, and bilateral donors. As we have seen, traditional policy conditionality is often thought to have failed. The World Bank's landmark 1998 report *Assessing Aid* advocated selective aid allocation: instead of imposing ineffective conditions, aid should be targeted at countries with relatively good institutions and policies. The Millennium Challenge Account of the US is one example of this approach.

Some economists would argue that abandoning policy conditionality is premature, and that sustained pressure has led to long-term improvements in the policies of recipient governments. New instruments, such as "floating tranche" conditionality, introduce greater flexibility in the timing of reforms and disbursement. And the recognition of the importance of domestic politics has encouraged a shift towards "process" or governance conditionality. Instead of emphasizing specific economic reforms, donors attach broader conditions to governance, public management and policymaking. Governance conditionality could be used to promote simple, concrete changes that are likely to lead to better outcomes, such as presidential term limits. This approach seems more legitimate than attempts to micro-manage economic policy, especially when it strengthens the accountability of leaders to their domestic populations.

A more radical idea is that conditionality should shift from input and process variables, to an emphasis on measured performance. Aid flows could be based on evidence of improvements in final outcomes such as immunization rates, infant mortality or primary school enrolment. In January 2012 the World Bank announced its Program for Results initiative, which will adopt this approach.

Conclusion

Arguably, the debate on conditionality too often assumes that the same strategy should be adopted across the board. Instead, Bourguignon and Sundberg (2007) argue that current policies increasingly amount to a three-track model. Countries with good governance and policies receive largely unconditional budget support, intermediate countries face something more like traditional conditionality but with greater emphasis on governance and performance, and fragile states are aided through a combination of humanitarian assistance and aid that bypasses the state, for example, by allocation to NGOs.

The three-track model recognizes that there are reasons to be wary of unconditional aid. Arguably, we already have a natural experiment, namely the resource rents that some developing countries receive. The conventional wisdom is that these unconditional transfers have led to diminished accountability and poor outcomes, at least in countries with weaknesses in governance. Given this observation and the technical case for imposing conditions, it seems likely that donors will continue to use conditionality, but in ways that acknowledge some of the weaknesses of the past.

References

Bourguignon, F. and Sundberg, M. (2007). Aid effectiveness – Opening the black box. *American Economic Review*, 97(2), pp. 316–321.

Gould, J. (ed.) (2005). *The new conditionality: The politics of poverty reduction strategies*. Zed Books, London.

Khan, M. S. and Sharma, S. (2003). IMF conditionality and country ownership of adjustment programs. *World Bank Research Observer*, 18(2), pp. 227–248.

World Bank (1998). *Assessing Aid*. Oxford University Press, New York.

Further reading

Grant, R. W. (2012). *Strings attached.* Princeton University Press, Princeton. Chapter 6 of this book includes an especially useful discussion of the circumstances in which IMF loan conditions are legitimate.

Koeberle, S., Bedoya, H., Silarszky, P. and Verheyen, G. (eds) (2005). *Conditionality revisited: Concepts, experiences and lessons.* World Bank, Washington, DC. The essays in this book cover a wide range of relevant issues.

Temple, J. R. W. (2010). Aid and conditionality. In Dani Rodrik and Mark R. Rosenzweig (eds) *Handbook of Development Economics*, Vol. 5. North-Holland, Amsterdam, pp. 4415–4523. The above chapter draws on the longer discussion in this literature review.

10.4
Aid effectiveness

Jonathan Glennie

Aid donors are concerned with the effectiveness of the money they spend in foreign countries, that is, how well it is achieving its objectives. While this has always been the case, a special focus on the effectiveness of aid has emerged since the turn of the century, as donors, recipients and civil society alike became more concerned that the funds invested in overseas aid were not leading to the desired results. It is uncommon nowadays to hear the demand for more aid without its sister demand that it also be 'better'.

The Paris agenda for aid effectiveness

In 2003 governments and civil society from around the world met in Rome, under the auspices of the OECD, to begin work on some agreed guidelines to improve aid mechanisms. Two years later, in March 2005, the Paris Declaration on Aid Effectiveness was published, encapsulating several 'aid effectiveness' concerns and setting out targets to improve aid in a number of areas.

Paris emerged at a time when the main concern in the development aid industry was the undermining of 'ownership' in recipient countries brought about by two decades of 'policy lending', that is, lending to poor countries on condition that they comply with certain economic and political conditions. While it was recognised that sharing control and responsibility for development interventions with local (notably state) actors might lead to a range of new risks and challenges, when donors force their own priorities on countries, they run the risk of working counter to, rather than along with, the grain of local actors, thus making progress less likely. It was therefore thought that integration of donor and local development objectives would lead on the whole to a higher probability of development objectives being fulfilled.

Box 1 The Paris Declaration on Aid Effectiveness, five principles

1. *Ownership*: Developing countries should take the lead in deciding their own policies.
2. *Alignment*: Donors should support national development strategies, institutions and procedures.
3. *Harmonisation*: Donors should reduce transaction costs for recipient governments by reforming reporting requirements and working better together.
4. *Managing for results*: Both donors and developing country governments should improve monitoring, decision making and resource management.
5. *Mutual accountability*: Donors can hold developing countries to account for their performance but developing country governments should also be able to hold donors to account for whether they have delivered on their commitments.

Apart from achieving better results, two other motivating factors can be identified for this shift in emphasis, sometimes inter-connected, and sometimes even apparently contradictory: sustainability of progress (through institutional strengthening) and reduced costs.

The logic behind Paris can be understood by looking at the 15 indicators against which donors and recipients are judged. First the recipient country has a strong development strategy, often developed in consultation with civil society (this 'ownership' criterion is judged, ironically, by the World Bank) (Indicator 1). It then demonstrates that it has reliable public financial and procurement systems (2a/2b). With those in place, donors align their aid to the priorities set out by the government in its development strategy (3), coordinate technical assistance with other donors (4), use the recipient country's public financial management (PFM) and procurement systems (5a/5b), close down parallel project implementation units (PIUs, which donors often set up when they want to manage development spending themselves rather than rely on governments they may not trust enough) (6), spend aid as planned (7), untie 100 per cent of aid (8), and use common arrangements and joint missions and country analytic work (9/10a/10b). Finally the recipient has a good results frameworks (this is judged, again, by the World Bank) (11) and develops a mutual accountability framework, whereby recipients are able to hold donors to account for their commitments, as well as vice versa (12).

As is clear, a particular focus began to be placed on the importance of a well-functioning state, an area considered to have been marginalised in the theories and practices of some development actors in previous years. The failure of the state was seen by many as the primary cause of slow development, and the theory of 'strengthening country institutions by using them' became commonplace in aid effectiveness discussions. One significant change in direction in the new era of aid, used by donors to demonstrate progressive reform in the aid system, has been the increase of general budget support which now accounts for about 5 per cent of global aid, with some governments receiving up to 20–30 per cent of their aid via GBS. The idea is that by providing aid directly to the finance ministry donors can get more of it on budget (thus supporting national procedures) and support the recipient government's broad political direction (rather than picking certain pet projects).

The Paris Declaration focuses much of its attention on some of the technical aspects of aid giving: how money could be transferred more efficiently; how its (direct) impacts should be monitored; how best to get value for money; how known bottlenecks can be eliminated. Examples of inefficiency in aid giving were abundant at the time:

- The Tanzanian government famously received 541 donor monitoring missions in 2006, of which less than a fifth were joint with one or more other donors.
- There were 35,000 aid transactions a year worldwide, 85 per cent of them worth less than US$1 million; most African countries were submitting around 10,000 quarterly donor reports every year.
- Tying aid to the purchase of goods and services from the donor country, another key area of interest in the Paris framework, was estimated to make it between 15 and 40 per cent less efficient.
- According to the World Bank there were more donors per country than ever before, with the average number tripling from 12 in the 1960s to 33 in the 2000s.
- Donor activities had tripled from around 20,000 in 1997 to nearer 60,000 in 2004, while their average size had been cut from over $2.5 million to a little over $1.5 million.

Critiques of Paris

There were from the outset a number of critiques of the Paris theory of aid effectiveness. One of them centred around its failure to respond to the most important issues in aid. While limited efforts to increase ownership are likely to have some positive effects, attempts to make aid processes 'mutually accountable' are unlikely to succeed given the intrinsic and profound power imbalances between donor and recipient. According to one Niger government official, 'These negotiations are by their nature unequal as we need the money.' One of the virtues of Paris is its ability to focus on technical issues, rather than get caught up in the politics of aid. But ultimately the politics matter, and insofar as Paris has focused on efficiency and cutting costs rather than the core issues of conditionality and dependency, which continue to be significant problems, it may have caused some damage.

Harmonisation is a cornerstone of Paris, but while it may seem obvious that donors should coordinate their aid giving, such coordination might be entrenching and continuing the system of consolidated donor pressure that has done damage in the past. According to the Reality of Aid Network, 'Many of the reforms suggested by the Paris Declaration on their own may in fact further undermine [the ability of poor people to claim their] rights and the promotion of democratic processes, the rule of law, and parliamentary processes'. While less bureaucracy means more time spent doing more important things, variety among donors can sometimes be a good thing if it allows innovation and even competition, whereby recipients might refuse some offers of aid and accept others depending on the modalities.

Busan – responding to 7 years of Paris

At the fourth High Level Forum on Aid Effectiveness at Busan in November 2011, these criticisms and others were discussed, along with concerns that even in its own terms little progress was being made on Paris principles. The orthodox analyses of the impacts of the Paris process to date are found in the official OECD analysis based on a monitoring survey involving about 80 recipient countries, and an independent evaluation based on more in-depth analyses carried out in 30 countries. Both reports substantially agree that when it has been implemented the Paris Agenda has worked (to varying degrees), but there has been very little implementation. Of the 15 targets, significant progress has been made on just four since 2007, significant worsening on one, and no significant change on the remaining nine. The number of parallel initiatives, for example, is growing, not falling, mostly in the health sector, and still well under 50 per cent of donor assistance is considered to be aligned to country priorities, a statistic that is largely unchanged in the last decade.

The outcome of Busan can be split into three categories:

1. *Business as usual* – The Busan monitoring framework is predominantly focused on the unfinished business from Paris, with six of the twelve Paris monitoring themes remaining.
2. *Newly emerging agendas* – Some significant new issues emerged in the Busan monitoring framework that have not been given prominence before, including transparency, the role of the private sector (both of which were strongly promoted by donors) and enabling environment for civil society (the top priority for civil society organisations at Busan).
3. *Yesterday's issues* – The issues dropped in the transition from Paris to Busan are dominated by harmonisation (PIU's and joint missions and analysis) and specific aid modalities (programme-based approaches and technical assistance). There seems to be a consensus that significant efforts to harmonise have achieved little and future efforts are unlikely to be transformative. Programme-based approaches and technical cooperation may well have been victims of the significant definitional challenges around monitoring these issues.

Future of aid effectiveness

In Busan adherents reaffirmed their commitment to continuing this broad shift to strengthening and supporting local actors, systems and processes, but they did so in a context that is significantly different today from when the commitment was first made in Paris in 2005. First, in a time of fiscal austerity in OECD countries, donors cannot allow aid to be perceived by taxpayers as ineffective or misdirected. The risk analysis carried out by donors may therefore have shifted somewhat. A strong focus on (short-term) results and 'value for money' has emerged. Second, the rise of the emerging powers (particularly China) as aid givers, along with a re-emerging theory of 'South–South Cooperation', has challenged traditional donors to reassess their aid methodologies. With millions of people still living in poverty in their own countries, the aid they give is logically linked to strategic interests in other countries and the development of their exports whether by their private sector or state-owned enterprises. There is no reason why untying aid should be a priority for them – on the contrary, they see aid as mutually beneficial. Decent levels of growth in many developing countries, combined with alternative financing options, are the third and final ingredient in a changed context. Africa in particular is experiencing its longest income boom for 30 years, with gross domestic product growth rates averaging about 5 per cent annually over the past decade. Africa will have the world's fastest-growing economy during the next five years of any continent. Trade between Africa and the rest of the globe increased by 200 per cent between 2000 and 2011. In this context, most countries are increasingly less reliant on aid which is seen as both humiliating and unpredictable. As a consequence, donor influence in these countries may wane further.

As these geopolitical shifts settle in, one central misunderstanding at the heart of Paris is likely to become clearer in the years to come: Paris is not a blueprint. Sometimes the Paris process is misconceived as a journey towards an ideal aid relationship. But no such ideal relationship exists. In fact, the Paris Declaration and the bureaucratic process accompanying it, has been, in essence, a response to a particular set of problems that have dogged particular aid relationships for the past couple of decades, namely the aid relationship between Western OECD countries and aid dependent recipient countries, mostly in Africa. The issues that form the Paris Agenda were chosen because they were important in particular aid relationships, and because they were considered possible to improve through an international bureaucratic process. But this group of sound corrective measures has become in the minds of some development practitioners a kind of platonic ideal of an aid relationship, a blueprint. The actual indicators used to judge progress on these ideals have come in for serious criticism. But that is not the point here – it is accepted that indicators will always have weaknesses and be at best proxies for progress. The point is that the ideal itself is unhelpful. There is a recognised tendency in international relations for context specific declarations or decisions to gradually metamorphose into policy blueprints. When this happens the corrective qualities of such declarations, which serve an important temporal purpose, themselves become the problem, as they are applied to situations for which they were not devised, and to which they do not pertain.

Bibliography

ActionAid (2011) *Real Aid 3: Ending aid dependency*, London: ActionAid International.
Booth, D. (2011) *Aid effectiveness: Bringing country ownership (and politics) back in*, London: Overseas Development Institute.
Eurodad (2008) *Turning the tables: Aid and accountability under the Paris Framework*, Brussels: Eurodad.
Glennie, J. (2008) *The trouble with aid: Why less could mean more for Africa*, London: Zed Books.
OECD (2005) *The Paris Declaration on Aid Effectiveness*, Paris: OECD.
OECD (2011) Aid effectiveness 2005–10: Progress in implementing the Paris Declaration, Paris: OECD.
Raffer, K. and Singer, H. (1997) *The foreign aid business: Economic assistance and development cooperation*, Cheltenham, UK; Edward Elgar.
Riddell, R. (2007) *Does Foreign Aid Really Work?* Oxford: Oxford University Press.
Wood, B. and Dansk Institut for Internationale Studier (2011) *The Evaluation of the Paris Declaration, Final Report*, Copenhagen: Danish Institute for International Studies.

10.5

Global governance issues and the current crisis

Isabella Massa and José Brambila-Macias

Background

Global governance is a process through which actors at the national and international level cooperate to achieve commonly accepted goals and to address global challenges. An ideal system of global governance should be: (a) effective; (b) efficient; (c) legitimate; (d) transparent and accountable; and (e) adaptive. In other words, it should be able to achieve its intended objectives by avoiding unnecessary costs and improving collective outcomes. Second, it should be inclusive and represent all stakeholders. Rules and procedures should also be clear and transparent, and decision-making bodies should be held accountable for their actions. Finally, a global governance system should be flexible enough to adapt to changing environments and challenges.

Global governance has recently become the subject of intense debate, in particular during the 2008–9 global financial turmoil and the ongoing eurozone crisis. However, the concept is not new and difficulties in creating an effective and inclusive system of global governance date back to the beginning of the twentieth century. After the First World War, in 1919 the League of Nations was created as an intergovernmental organization with the mission to maintain world peace. The league counted 42 founding members, but four countries (i.e. France, Italy, the United Kingdom, and the United States) outweighed the others. Although successful at the beginning, the league failed to prevent the Second World War and once the conflict ended, new global governance institutions were set up to replace it and deal with future issues. These institutions include the United Nations (UN), the so-called Bretton Woods institutions (the World Bank and International Monetary Fund (IMF)), and the General Agreement on Tariffs and Trade (GATT) – subsequently replaced by the World Trade Organization (WTO), which dominated the global governance scene for most of the second half of the twentieth century.

Over the last two decades, the international context has changed dramatically with the advent of globalisation and the consequent intensification of interdependence among countries, as well as with the rise of new global powers such as emerging economies. The global governance system has failed to keep up with these changes. Its institutions, indeed, have remained largely unchanged over time being unable to fully adapt their structures to the new global environment, to enhance their legitimacy by providing more voice to developing countries, and to act effectively to respond to new global challenges. For example, the five permanent members of the UN Security Council have remained unchanged since their designation more than six decades ago, and emerging and developing countries are still under-represented on the IMF's Executive Board.

This chapter analyses in depth global governance issues in the context of the current crisis. It also highlights that the economic and financial turmoil represents not only a challenge but also a unique opportunity for policymakers to reform the global governance architecture. The crisis has proved that decisions made by developed countries may have significant impacts on developing economies (Massa *et al.* 2012). So, it is the right time to increase the voice and influence of poor countries in global governance structures and to promote collective and coordinated solutions through expanded global fora which recognize the role and participation of developing and emerging economies. This can only be achieved by making the existing global governance system more effective and inclusive.

Key issues

In the context of the current crisis, the global governance architecture is facing four main challenges: (a) lack of representativeness; (b) increased fragmentation; (c) inability to match the scope, scale and nature of new global challenges; and (d) waning relevance.

Legitimacy

Emerging and developing economies have recently become key players in the global scenario. At the end of 2010 they jointly accounted for 48 per cent of global output (measured in purchasing power parity terms), while emerging markets alone accounted for about two-thirds of the total global output growth in 2009–10 compared with only one-third in the 1960s (Zhu 2011). Moreover, emerging economies now account for over half of the global consumption of most commodities, world exports, and foreign direct inflows (*The Economist* 2011). Emerging markets have also been the engine of the global recovery after the 2008–9 global financial crisis.

Notwithstanding this, the global governance architecture has been slow to adapt to the shift in economic power from developed to developing economies. This has left emerging and developing countries under-represented thus calling into question the legitimacy of existing international organizations. For example, the United Nations Security Council's permanent members which have veto power (China, Russia, France, the United States, and the United Kingdom) do not include any African or Latin American country. This is particularly worrying for Africa since in 2011 alone, 38 of the 60 resolutions passed by the Security Council were directed at African countries (Kimenyi and Moyo 2012). Moving to the IMF, its board of governors decided in 2008 to engage in a series of governance reforms to increase the voting shares of emerging economies. As a result, Brazil, India, China, Mexico, Turkey, Korea, and Singapore increased their voting power. However, from a developing countries' perspective this is still not enough to provide adequate representation to the whole developing world. In the case of the WTO, no sub-Saharan African country (with the exception of South Africa) has been involved in the Dispute Settlement Mechanism, that is, the process through which trade disputes in the WTO, which may originate when WTO agreements are violated or obligations are not fulfilled by one country, are settled by the Dispute Settlement Body composed of representatives of all WTO members (Kimenyi and Moyo 2012; WTO 2011). In a context of rising protectionism after the global financial crisis, this may pose important challenges to African countries many of which are heavily dependent on exports.

In the wake of the 2008–9 global financial crisis, developed economies attempted to overcome the legitimacy issue by coordinating a global response to the crisis through the G-20 forum – a forum for international cooperation on international economic and financial issues which brings together the world's major advanced and emerging economies (19 member countries and the European Union), which has maintained a predominant role also during the eurozone crisis. This has been seen as a step forward in achieving a more inclusive and participatory system of global governance. Indeed, even though more than 170 countries remained unrepresented, the G-20 still accounts for two-thirds of the world population and more than half of world's poor people. This is certainly an improvement but it still leaves aside many of the poorest countries that are also the most vulnerable to global downturns. Among African countries, for example, only South Africa has membership in the G-20, while countries such as Mozambique, Kenya, Niger, Cape Verde, and Cameroon which have been found to be among the most vulnerable to the eurozone crisis (Massa *et al.* 2012) do not have a place in the G-20 forum. Therefore, in the global governance system, there is room for further progress into achieving more adequate representation.

Fragmentation

Another important issue affecting global governance is the excessive fragmentation of decision-making bodies. Indeed, international organizations have so far been characterized by a high degree of specialization in one or few areas, and tended to focus on their own objectives and priorities. For example, the IMF monitors the international monetary and financial system. The World Bank oversees development by providing adequate funds, while the World Health Organisation (WHO) focuses on health issues. On the other hand, the WTO regulates global trade, and the UN Security Council is in charge of maintaining global peace. In addition to this, several new actors have emerged in the global governance scene exacerbating further the issue of fragmentation. This is particularly evident in the international aid system where there are more than 260 multilateral aid organizations, in addition to government donors, international NGOs, private sector actors and even emerging economies which are promoting South–South cooperation (Barder *et al.* 2010).

Such a high degree of fragmentation poses huge coordination challenges for the existing global governance architecture, and it is at odds with the increased economic, financial, and social interconnectivity among countries as well as with the global nature of the current crisis.

Effectiveness

The existing global governance structures prevent international organizations from acting effectively in the event of global challenges such as the 2008–9 global financial crisis or the ongoing eurozone crisis. For example, it often occurs that international organizations delay their actions since they need first to secure the backing of key countries which usually have the majority of votes and provide the biggest share of funds. Moreover, some international organizations face difficulties in delivering effective solutions since they are financially constrained. For example, in the context of the global financial crisis, the IMF was in need of finding new financial sources to be able to fulfil its mission, and only after the G-20 took action and agreed to provide to the IMF the necessary funds, the institution was able to intervene.

Therefore, international organizations need to be endowed with the adequate financial means and authority to provide fair, quick, and effective solutions. This is essential in a globalized world where crises spread at an astonishing speed affecting developed and developing countries alike.

Relevance

Prior to the 2008–9 global financial crisis, global governance was progressively loosing importance: the IMF was in a profound financial crisis of its own, the World Bank was experiencing huge legitimacy problems, while international trade and climate change negotiations were on a dead track (Woods 2010).

It was only when the crisis hit the global economy that it became clear that countries on their own would not have been able to weather the crisis shock waves, and that global governance institutions were essential to provide a collective and coordinated response. As a result, the IMF seemed to regain a central role in the global economic scenario.

However, this renewed relevance of multilateralism might be more apparent than real. Indeed, Woods (2010) highlights that the recent increase of financial resources made available to the IMF came in the form of credit lines rather than actual fund transfers (capital increase). This shows that the IMF might not be regaining protagonism but rather maintaining a limited role at the dependence of its main creditors. Moreover, the eurozone crisis is leading some countries to seek unilateral or regional solutions thus weakening further the role of global governance institutions.

Concluding remarks

In the context of the current crisis, global governance is at the crossroads. On one hand, a number of issues related to the legitimacy, fragmentation, effectiveness and relevance of the global governance architecture clearly points out to the urgent need of reforming the system. The increasingly important role of emerging and developing countries should be recognized by providing them with adequate representation within the global governance decision-making bodies. Excessive fragmentation should be avoided in favour of a more coordinated system which prevents duplication of efforts and ensures an effective and collective response to global challenges. The effectiveness and relevance of global governance actors should also be safeguarded by enhancing their financial and political independence.

On the other hand, however, when designing new governance mechanisms it is inevitable to encounter some trade-offs. In particular, creating a more participatory system of global governance may affect the effectiveness of decision-making processes.

Therefore, the main issue for global governance in the wake of the current crisis remains to strike an adequate balance between inclusiveness (legitimacy) and effectiveness (ability to deliver). This is essential to respond effectively to the current and future global challenges from the perspective of both rich and poor economies.

Note

The views presented in this paper are those of the authors and do not necessarily represent the views of ODI and OECD.

References

Barder, O., Gavas, M., Maxwell, S. and Johnson, D. (2010) Governance of the aid system and the role of the EU, paper prepared for the Conference on Development Cooperation in Times of Crisis and on Achieving the MDGs, 9–10 June 2010, Madrid.

The Economist (2011) Emerging vs developed economies. Power shift, 4 August 2011.

Kimenyi, M. S. and Moyo, N. (2012) Enhancing Africa's voice in global governance, in: *Foresight Africa. Top Priorities for the Continent in 2012*, Brookings Africa Growth Initiative Report, January 2012.

Massa, I., Keane, J. and Kennan, J. (2012) The eurozone crisis and developing countries, ODI Working Papers 345, May 2012.

Woods, N. (2010) Global governance after the financial crisis: A new multilateralism or the last gasp of the great powers? *Global Policy*, Vol. 1, No. 1, January.

WTO (2011) Understanding the WTO, July, Geneva: World Trade Organization.

Zhu, M. (2011) Straight Talk. Emerging Challenges, in: *Finance & Development*, Vol. 48, No. 2, June.

10.6

Change agents

A history of hope in NGOs, civil society, and the 99%

Alison Van Rooy

This chapter talks about the changing fashions for imbuing hope in development actors – and what happens in the backlash that follows. Starting from "NGOs", then "civil society," we now focus on variations on the social media-empowered "99%." How do those changes help or hinder our understanding of change?

It is worth remembering that in that same timeframe, the "backward countries" turned into the "third world," then the "developing world," then "the South" (or "majority world") and then "emerging markets" with the BRIC steamroller. Each of those terms says something about the worldview of the speaker: what matters most and who matters in making it happen. As we shall see, the scene is changing rapidly. The worldview of decision-makers in one era do not always carry well into another.

1970s and the NGOs

As the modernization school of thinking waned, development pundits in the West ceased to see the state as the only – or even the most important – driver of developmental change. The hero in the 1970s was the NGO: typically, an organization formed in a rich country to address poverty by somehow making up for the absence of a functioning state in a poor country (see Figure 10.6.1). Of course, host country communities were already working to cope on their own, and many domestic organizations already existed to provide services or advocate for political change. But there was a special celebrity status for NGOs in the English-speaking West, seen to be closer to the people, more honest, more effective.

Backlash against the favourite status arose first in the mid-1980s when real-life examples showed that power inequalities, grandstanding, over-claiming and other ordinary aspects of human behaviour existed in the NGO community as well as in the state (Bond 2000; Bob 2002). The wholesale criticism seriously undermined the efforts of those whose work was genuinely transformative, but it did spur a new – and corrective – focus on increasing accountability to those whose interests are supposedly served (Jordan and Van Tuijl 2006).

1990s, the fall of the Iron Curtain, and the rediscovery of civil society

The 1990s saw the adoption of a new hero: civil society. This one was trickier to define: the story of the re-application of a Roman concept to the travails of Eastern Europe is its own amazing story (Van Rooy 1998; Hall 1995; Spurk 2010). The short version is that Eastern European pundits, trained in Marxism and having read Hegel, took to explaining the rapid dismantling of the Iron Curtain in terms of the rise of civil society. Westerners adopted the term (without its intended intellectual history) and defined it in multiple (sometimes conflicting)

Purpose	Alleviation of poverty/promotion of community development in other countries
Origin	Solidarity or charity organizations in wealthy countries
Mode	Projects or programs funded by foreign donations or ODA, often channelled to local counterpart organizations
Claim to fame	Closer to the poor than official aid could reach
Critique	Not always accountable to "beneficiaries"
Future if . . .	. . . if seen to be Southern-led with high accountability standards and respected results

Figure 10.6.1 Describing the NGO

ways, without much consideration of its applicability in other parts of the world. Perhaps the two most prominent definitions were:

1 *Civil society as a collective noun.* Civil society is most commonly defined as a collective noun: synonymous with the voluntary sector (or the Third Sector), and with advocacy groups, non-governmental organizations (NGOs), social movement agents, human rights organizations and others actors explicitly involved in change work. The definition most often excludes those groups belonging to the marketplace and the state, and further specifies that civil society organizations do not include those groups interested in acquiring political power, hence the usual exclusion of political parties.
2 *Civil society as a space for action.* Civil society has also been used to describe the sphere or arena in which civil organizations prosper – not the organizations themselves. This idea of a space also implied a *function* – civil society was seen to have a crucial check-and-balance role in relation to The State and The Marketplace. The new international CSO umbrella group took on that ethos in creating its name, *CIVICUS*, and in designing its index of civil society health (CIVICUS 2011).

One implication that followed was a resurrection of attention to the state as a (usually detrimental) actor in the making of social justice. The almost apolitical language around the NGO was now a highly politicized language that pitted civil society against The Market and against The State (Edwards 2004). The NGOs who continued to deliver services and projects were joined by those with a democratization agenda, all under a highly desirable civil society banner (Van Rooy 2004) (see Figure 10.6.2).

Helpfully, the fashion suited proponents on both left and right. Civil society could be read as a "check-and-balance" phenomenon, a populist surge, a democratic trend, or an aspiration of global citizenship. Other ideas became associated with the concept, most importantly "good governance and democratic development" (a new wave of pro-democracy aid programs; Carothers 1999) and "social capital" (re-popularized by Harvard professor Robert Putnam, which looked at measurable economic and social outcomes of a strong civil society; Putnam 2000). Civil society was a deeply satisfying idea.

Yet the normative glow around "civil society" came under attack. The 2001 and subsequent al-Qaeda attacks changed the whole tone of the civil society conversation: were some

Purpose	Alleviation of poverty/promotion of community development and possibly democratic development and good governance in own country. Term expands to encompass all of foreign NGO groups as well
Origin	Solidarity or charity organizations *as well as* groups with a democratization agenda
Mode	Projects, programs *and* campaigns funded by foreign donations or ODA, paid directly to local organizations or funnelled through Northern ones
Claim to fame	Even closer to the poor than NGOs could reach; active in generating democratic change as well
Critique	Not always accountable to "beneficiaries," fronts for political actors (or at worst, for "terrorist" organizations)
Future if . . .	. . . if seen to be Southern led with high accountability standards and respected results

Figure 10.6.2 Describing the CSO

civil society organizations possible fronts for terrorist activities? The backlash was swift and wholesale, often an excuse for delegitimizing organizations that repressive governments simply wished to silence. Worldwide, a range of retractions of rights to association were put into place and new surveillance triggered (ICNL 2012; CIVICUS 2012). While the service delivery role continued to grow (an increasing portion of a larger global aid budget continued via NGOs), the political role was under challenge.

2010s, the Arab Spring and the 99%

Then, something remarkable happened. The Arab Spring, begun in Tunisia in the winter of 2010, became a trigger to a new – much messier – understanding of change agency. There were no NGOs, few projects, little ODA, fewer Northerners, but rather a sudden social movement that tipped the balance across the Middle East (see Figure 10.6.3).

The Arab Spring threw three core ideas back into the global spotlight, shifting the discourse away from the old heroes:

- The first important idea was that change agents were ordinary people acting in their own countries. Connected up, ordinary people could do remarkable things. Time magazine, for instance, even named "The Protester" its Person of the Year "for capturing and highlighting a global sense of restless promise, for upending governments and conventional wisdom, for combining the oldest of techniques with the newest of technologies to shine a light on human dignity and, finally, for steering the planet on a more democratic though sometimes more dangerous path for the 21st century" (*Time* 2011).
- The second idea was that ordinary people were similar worldwide. Even though the occupy movements of the summer of 2011 were triggered by other factors (widespread discontent with wealthy financiers who caused the 2008 financial crisis), they were part of the same movement (*Time* 2012). The Protester was a universal force – extending to occupations in 1,500 cities (Van Gelder *et al.* 2011) – a force whose currency was social media-enabled organizational skills more than development funding.
- The third idea was that weak democracy (including government independence from elites) was *the* development problem – both North and South. The language of change was universally framed in terms of democracy (along with class most notably, but also the impacts of government action on the environment, education, and other outcomes).

Purpose	The Arab Spring protests sought regime changes; its success contributed to movements elsewhere with critiques of government management or complicity with the wealthy in the 2008 financial collapse
Origin	Begun in Tunisia, spread to middle east, contributed to existing concerns in the US, UK, and other countries
Mode	Protest
Claim to fame	Mass participation, leaderless, non-violent, social media-enabled, worldwide
Critique	For those in "Northern" countries: unclear message, unorganized for developing alternatives
Future if . . .	?

Figure 10.6.3 Defining the 99%

The backlash arose before the revolutions even settled into the ordinary work of organizing better governments. In some parts of the world – Syria most brutally – the uprisings triggered civil war rather than bloodless regime change; in others, elections followed that returned leaders from the old regimes; and in North America, the occupy movements eventually abandoned their camps having lost initial public approval and without leaving an organizational apparatus to promote the changes flagged.

The narrative of hope

Of course, all through these forty or so years there have been local people working for change, sometimes with others in neighbouring communities or faraway countries, and sometimes with remarkable changes to life and livelihood and political regime (Gills and Gray 2012). There have always been efforts by governments to serve their peoples, and there has always been abuse of government power – although the number of democracies has satisfyingly increased. For those same forty years, foreign aid has been a factor: growing in volume, and in the proportion spent via NGOs (OECD 2012). And, in those same years, there has always been protest.

But what does change is our *story* about that jumble of players and forces. Narratives are constructed that identify good guys and bad guys, that frame particular kinds of solutions, that seek the magic bullet. Narratives about the roles of NGOs, CSOs, and The Protester/99% matter: they focus our attention, rally our efforts and dollars, and later serve as lightning rods for backlash. The study of development must also be a study of the *narratives* about what development is. In simplifying the story, in imbuing too much hope in one place, we may miss a deeper understanding of the real world jumble.

Bibliography

Bob, Clifford (2002) "Merchants of Morality," *Foreign Policy* 129 (March/April): 36–45.

Bond, Michael (2000) "The Backlash Against NGOs," *Prospect Magazine*, April.

Carothers, Thomas (1999) *Aiding Democracy Abroad: The Learning Curve*, Washington DC: Carnegie Endowment for International Peace.

CIVICUS (2011) *State of Civil Society Report*, Johannesburg.

CIVICUS (2012) *Civil Society Watch*, www.civicus.org/csw.

Cohen, Jean L. and Arato, Andrew (1992) *Civil Society and Political Theory*, Cambridge, MA and London: MIT Press.

Edwards, Michael (2004, 2nd edition 2009) *Civil Society*, Cambridge: Polity.

Gills, Barry K. and Gray, Kevin (2012) "People Power in the Era of Global Crisis: Rebellion, resistance, and liberation," *Special Issue: People Power in the Era of Global Crisis, Third World Quarterly*, 33(2), 205–224.

Hall, John A. (ed.) (1995) *Civil Society: Theory, History, Comparison*, Cambridge, MA: Polity Press.

ICNL (International Center for Not-for-Profit Law) (2012) *Global Trends in NGO Law*, www.icnl.org/research/trends/index.html.

Jordan, Lisa and Van Tuijl, Peter (2006) *NGO Accountability: Politics, Principles and Innovations*, London: Earthscan.

OECD (2012) *OECD Stat Extracts*, creditor reporting system, http://stats.oecd.org/Index.aspx#.

Putnam, Robert D. (2000) *Bowling Alone: The Collapse and Revival of American Community*, New York: Simon and Schuster.

Spurk, Christoph (2010) "Understanding Civil Society", *Civil Society and Peacebuilding: A Critical Assessment* (ed. Thania Paffenholz), Boulder, CO: Lynne Reinner.

Time (2011), Person of the Year edition, 14 December, www.time.com/time/specials/packages/article/0,28804,2101745_2102139,00.html.

Time (2012) What is Occupy? Inside the Global Movement, Time Home Entertainment Inc.

Van Gelder, Sarah, Staff of Yes! Magazine (eds) (2011), *This Changes Everything: Occupy Wall Street and the 99% Movement*, San Francisco: Barrett-Koehler.
Van Rooy, Alison (ed.) (1998) *Civil Society and the Aid Industry: The Politics and Promise: London and the North–South Institute*, Ottawa: Earthscan.
Van Rooy, Alison (2004) *The Global Legitimacy Game: Civil Society, Globalisation, and Protest*, London: Macmillan Palgrave.

10.7

Corruption and development

Susan Rose-Ackerman

Economic development depends upon both good policies and effective institutions to carry out those policies. A country that tries to follow the prescriptions of macroeconomists will not succeed in promoting growth if its public and private institutions are very corrupt and dysfunctional. True, some countries are able to grow in spite of pervasive corruption, but extremely corrupt countries are not in that group and even corrupt countries that manage to grow would do better with more effective institutions that need not be circumvented with bribes.

Tracking the connection between corruption and growth, however, is a difficult empirical exercise because the perpetrators of corrupt acts understandably wish to keep them secret. Nevertheless, a start has been made. Economists have generated a range of productive research focusing mostly on bribery and illegal kickbacks.

Conceptual underpinnings

Corruption occurs where private wealth and public power overlap. It represents the illicit use of willingness-to-pay as a decision-making criterion. Frequently, bribes induce officials to take actions that are against the interest of their principals, who may be bureaucratic superiors, politically appointed ministers, or multiple principals such as the general public. Pathologies in the agency/principal relation are at the heart of the corrupt transaction. I differentiate between low-level opportunistic payoffs, on the one hand, and systemic corruption, on the other, that implicates an entire bureaucratic hierarchy, electoral system, or governmental structure from top to bottom (Rose-Ackerman, 1999).

Low-level corruption occurs within a framework where basic laws and regulations are in place, and officials and private individuals seize upon opportunities to benefit personally. There are several generic situations.

First, a public benefit may be scarce, and officials may have discretion to assign it to applicants. Then the qualified applicants with the highest willingness to pay and the fewest scruples will get the benefit in a corrupt system.

Second, suppose that low-level officials are required to select only qualified applicants and that their exercise of discretion cannot be perfectly monitored. The overall supply may be scarce (for example, government-subsidized apartments), or open-ended (for example, drivers' licenses). In either case, the officials' discretion permits them to collect bribes from both the qualified and the unqualified. Incentives for payoffs will also depend upon the ability of superiors to monitor allocations. It will also depend upon the options for the qualified. For example, can they approach another, potentially honest, official?

Third, the bureaucratic process itself may be a source of delay. Incentives for corruption arise as applicants try to get to the head of the queue. To further exploit their corrupt opportunities, officials may create or threaten to create more delay as a means of extracting bribes.

Some government programs impose costs – for example, tax collection or the possibility of arrest by the police. Officials can then extract payoffs in return for overlooking the illegal underpayment of taxes or for tolerating illegal activities. They can demand payoffs in exchange for refraining from arresting people on trumped-up charges.

Low-level corruption can lead to the inefficient and unfair distribution of scarce benefits, undermine the purposes of public programs, encourage officials to create red tape, increase the cost of doing business, and lower state legitimacy.

"Grand" corruption shares some features with low-level payoffs, but it can be more deeply destructive of state functioning – bringing the state to the edge of outright failure and undermining the economy. I distinguish three varieties.

First, a branch of the public sector may be organized as a bribe generating machine. For example, top police officials may organize large scale corrupt systems in collaboration with organized crime groups. Tax collection agencies and regulatory inspectorates can also degenerate into corrupt systems managed by high-level officials.

Second, a nominal democracy may have a corrupt electoral system. Corruption can undermine limits on spending, get around limits on the types of spending permitted, and subvert controls on the sources of funds.

Third, governments regularly contract for major construction projects, allocate natural resource concessions, and privatize state-owned firms. High-level politicians can use their influence to collect kickbacks from private firms.

Consequences of corruption

Corruption is generally defined as the abuse of public power for private gain. This is an umbrella definition, covering behavior as varied as a head of state embezzling public funds or a police officer extorting bribes in the street. Most cross-country data do not distinguish between varieties of corruption, limiting the relevance of these measures as guides to policy. Nevertheless, research suggests that the many types of corruption are highly correlated so that countries can be characterized as more or less corrupt (Treisman, 2007). Citizens perceive that corruption, however defined, is on the rise. According to data from Transparency International's 2010 Global Corruption Barometer, 56 per cent of citizens worldwide believe corruption has increased in the past three years.

Countries with higher levels of corruption have lower levels of human development. Highly corrupt countries tend to under-invest in human capital by spending less on education, over-investing in public infrastructure relative to private investment, and degrading environmental quality. However, some countries have managed to have high levels of human development despite high levels of corruption, showing that the relationship is far from deterministic. In general, richer countries and those with high growth rates have less reported corruption and better functioning governments (Kaufmann, 2003). Aidt (2011) constructs a broader index of sustainable development and shows that corruption has a detrimental effect.

Corrupt countries tend to suffer from more bureaucratic red tape, which may be intentionally created by rent-seeking bureaucrats. According to Wei (2000), an increase in the corruption level from relatively clean Singapore to relatively corrupt Mexico is the equivalent of an increase in the tax rate of over 20 percentage points. Lambsdorff (2003) finds that if Colombia could attain the level of integrity of the United Kingdom, net yearly capital inflows would increase by 3 percent of GDP.

Much of this work leaves unclear whether low levels of income and growth are a consequence or a cause of corruption. Most likely, the causal arrow runs both ways, creating vicious or virtuous spirals (Treisman, 2007).

Reform proposals

Much has been made of the importance of moral leadership from the top, but this is not sufficient. Too much moralizing risks degenerating into empty rhetoric – or worse, witch hunts against political opponents. Policy must address the underlying conditions that create corrupt incentives, or it will have no long-lasting effects.

Some argue that the main cure for corruption is economic growth and that economic growth is furthered by good policies, especially the promotion of education (Glaeser and Saks 2006). However, that claim reflects an overly simple view of the roots both of economic growth and of corruption. Corruption is a symptom that state/society relations undermine the legitimacy of the state and lead to wasteful public policies. The options for institutional reform fall into several broad categories: program redesign, polices that increase transparency and accountability, and, in severe cases, constitutional change.

The first line of policy response is the redesign of programs to limit the underlying incentives for payoffs. This might mean eliminating highly corrupt programs, but, of course, the state cannot abandon its responsibilities in many areas where corruption is pervasive. One response is to limit official discretion by, for example, streamlining and simplifying regulations, expanding the supply of benefits, making eligibility criteria clear, introducing legal payments for services, giving officials overlapping jurisdictions to give citizens choices, or redesigning systems to limit delays. Reformers should consider if cleanups in one area will just shift corruption to another part of the government. Programs may need to be comprehensive to have any impact. In addition, service delivery can be improved by civil service reforms that provide better salaries, improved monitoring, and the use of incentives.

The second collection of reform strategies focuses on the accountability and transparency of government actions. For example, a freedom-of-information law can give people access to government information, and many government decision-making processes should be open to public scrutiny and participation. Other options to improve accountability are the creation of independent oversight agencies and the use of external and internal benchmarks. Ongoing experiments with grassroots democracy need more study to determine their impact and their transferability to other contexts. Open government also depends upon a vigorous and free media that can perform a watchdog function.

Third, some countries may need to consider more radical reforms in government structure. Democracy is valuable for many reasons, but, taken by itself, is hardly a cure for corruption. Elections are not sufficient. The state must protect civil liberties and establish the rule of law. Rules must be clear and fair and be administered competently and fairly. This implies an honest, professional and independent judiciary, and police and prosecutors who have similar level of integrity and competence.

Cross-country research suggests that the gains from reducing corruption and improving governance are large. The main problem is tracing specific links from particular, concrete policies to desirable outcomes. The few studies that do exist suggest a substantial net benefit for targeted anti-corruption interventions (Olken and Pande 2012).

Some options look promising because benefits seem clear and the costs are minimal. Hence, even if the benefits cannot be precisely measured, the rates of return appear large. The release of information to citizens, for example, may require little more than a website or a well-placed newspaper story. Civil service reforms, with the exception of wage increases, may require only a thoughtful reshuffling of personnel and recruitment practices. Audits and heightened monitoring do require resources but have proven effective in many contexts.

Improving top-down monitoring and punishment, fostering transparency and citizen involvement, adjusting bureaucratic incentives through civil service reforms, improving the competiveness of government asset sales and large purchases, and privatizing certain government services may provide the shock needed to push a country or sector towards a self-fulfilling cycle of good governance. It is likely that the initiatives would prove most effective when bundled together, signaling a firm commitment to anti-corruption for all would-be corrupt officials.

Obviously, individuals and firms, many with political power, benefit from the status quo and will oppose change. A major challenge for governance reform is to overcome or co-opt entrenched interests. Political calculations can derail even the most well-conceived initiatives. Some reforms may be blocked directly, but equally pernicious are corrupt leaders who pose as reformers – expressing a superficial commitment to good governance as they continue to gain at public expense. A crackdown on low-level corruption may just push the illicit rents up the hierarchy where they can be captured by the top officials.

Clever technical solutions, based on economic incentives, are not enough (Johnston 2005). If corruption is one of the pillars supporting a political system, it cannot be substantially reduced unless an alternative source of revenue replaces it. Powerful groups that lose one source of patronage will search for another vulnerable sector. Tough political and policy choices need to be faced squarely. It is little wonder that effective and long-lasting corruption control is a rare and precious achievement. But it is not beyond the power of determined and intelligent political reformers.

References

Aidt, T., 2011: Corruption and Sustainability, in S. Rose-Ackerman and T. Søreide (eds), *The International Handbook on the Economics of Corruption*, Vol. 2, Cheltenham UK: Edward Elgar Publishing, pp. 3–51.

Glaeser, Edward L., and Raven Saks, 2006: Corruption in America, *Journal of Public Economics*, 90 (6–7), 1053–72.

Johnston, M., 2005: *Syndromes of Corruption: Wealth, Power, and Democracy*, Cambridge: Cambridge University Press.

Kaufmann, D., 2003: Rethinking Governance: Empirical Lessons Challenge Orthodoxy, Discussion draft, World Bank, Washington DC, www.worldbank.org/wbi/governance/pdf/ rethink_gov_standford.pdf.

Lambsdorff, J. Graf, 2003: How Corruption Affects Persistent Capital Flows, *Economics of Governance*, 4, 229–43.

Olken, B. A., and R. Pande, 2012: Corruption in Developing Countries, *Annual Review of Economics*. 4, 479–509.

Rose-Ackerman, S., 1999: *Corruption and Government*, Cambridge: Cambridge University Press.

Treisman, D., 2007: What Have We Learned About the Causes of Corruption from Ten Years of Cross-National Empirical Research? *Annual Review of Political Science*, 10, 211–244.
Wei, S.-J., 2000: How Taxing is Corruption on International Investors? *Review of Economics and Statistics*, 82, 1–11.

Further reading

Glaeser, E. L., and C. Goldin, 2006: *Corruption and Reform: Lessons from America's Economic History*, Chicago: Chicago University Press for the National Bureau of Economic Research.
Rose-Ackerman, S., and Tina Søreide, (eds), 2011: *International Handbook on the Economics of Corruption*, Vol. II, Cheltenham, UK: Edward Elgar.
Rose-Ackerman, S. and R. Truex, 2013: Corruption and Policy Reform, in Bjørn Lomborg (ed.), *Global Crisis., Global Solutions*, Cambridge: Cambridge University Press, pp. 632–73.
Svensson, J., 2005: Eight questions about corruption, *Journal of Economic Perspectives* 19 (3), 19–42.
Transparency International, http://www.transparency.org

10.8

The role of non-governmental organizations (NGOs)

Vandana Desai

The growth of the NGO sector

The term NGO is applied to many kinds of organization, ranging from large Northern-based charities such as Oxfam to local self-help organizations in the South with an aim to improve the quality of life of disadvantaged people. They are mainly private initiatives, involved in development issues on a non-profit basis. The term 'NGO' is understood to refer to those autonomous, non-membership, relatively permanent or institutionalized (but not always voluntary) intermediary organizations, staffed by professionals or the educated elite, which work with grassroots organizations in a supportive capacity. Grassroots organizations (GROs) on the other hand are issue-based, often ephemeral, membership organizations; they may coalesce around particular goals and interests, and dissipate once their immediate concerns have been addressed. Non-governmental organizations (NGOs) have become an important and vocal platform for the involvement of civil society in public affairs.

NGO partnerships

Since the 1950s, NGOs have come to play an increasingly important part in the formulation and implementation of development policy, becoming key actors in the political economy of development. There has been increased collaboration both with governments and aid agencies based on a growing belief over the period that the promotion of NGOs could offer an alternative model of development and play a key role in processes of democratization (see Mercer 2002) and decentralization policies.

NGOs are popular because they are perceived to be administratively flexible, able to access the poor through work at the grassroots level, innovative in problem solving, adaptable to local context, more cost effective than corresponding state partners and their grassroots representation brings legitimacy and community mobilization to programmes and projects.

Donor pressure towards structural reform and privatization in the 1990s underlies the increased interest in NGOs as 'service deliverers' – part of a wider and explicit objective to facilitate productive NGO–state partnership. There is a realization among donor countries that aid is becoming ever more complex with new instruments and players with increasing criticisms over the effectiveness of aid. Private financial flows increased rapidly in the wake of increased liberalization, driven by market reforms, rise in global trade, lowering investment barriers in developing countries, and the fall in communications and transport costs. This also led to the expansion in partnership between Northern and Southern NGOs in changing attitudes and educating constituencies in the North towards development interventions and the underlying causes of poverty, drawing people into active lobbying and campaigning for change in the South. A view took hold that merely transferring resources in the form of tools or funds was not an adequate response to alleviating poverty which was rooted in deeper structural problems. Indeed such transfers could just preserve the situation by creating financial dependency.

Northern NGOs also have an humanitarian function and respond to emergencies, short-term relief and long-term rehabilitation, such as for victims of war and of natural disasters or man-made disasters. They raise money in the North, from the general public, private sector and governments to pay for their work, and share as much as possible with their Southern counterparts in building their capacity.

Southern NGOs have the basic responsibility among NGOs for leading the development process in developing countries and the expertise to do so (as outlined above). It is expected that relationship between Northern and Southern NGOs is based on an equal partnership incorporating transparency, mutual accountability and risk-sharing. Concerns have been raised recently on the impact of the unequal relationship between donors, Northern NGOs (see Bebbington 2005) and Southern NGOs on issues of accountability (see section on accountability).

Roles of NGOs

NGOs play three main roles: service delivery, advocacy on behalf of the poor, and empowerment, that is, enabling the poor to become advocates for themselves. As service delivery agents, NGOs provide welfare, technical, legal and financial services to the poor or work with community organizations in basic service and infrastructure provision. This is frequently a matter of filling gaps left by the partial service delivery of governments withdrawing from involvement in provision. In the past, governments of developing countries were seen as spearheading the development process. However, such paternalism reached its limits when it became clear that government did not have the financial resources to pay for the essential services of the poor and lacked the organizational expertise to be effective. In such an environment, the important role for NGOs in the last few decades has been in mitigating the adverse costs of structural adjustment and promoting donor reform packages in offering insurance against a political backlash against harsh adjustment regimes. Such a role raises important questions. Patterns of service delivery through the voluntary sector may lack compatibility and coordination. In so far as such efforts rely on government funding, their ultimate sustainability is questioned. At a deeper level, there are worries about the long-term impact of NGO service provision on the sustainability of national health and education systems (rather than programmes) and access to quality services for all. NGOs also constantly face the challenge of the need to adopt 'best fit' approaches within the context of the particular country and existing institutions rather than import 'best practices' which may not fit or apply to the local context.

It is also believed that NGOs provide support, in an ad hoc manner but sometimes through more structured programmes. For example, recently NGOs have been very active in promoting local formal savings or microfinance groups in communities supported by or linked to external organizations such as credit unions and banks especially in promoting groups for women as women are least likely to have access to more formal means of saving and borrowing. Women are also considered to be more reliable at repayment than men and to spend the income they control on things that are more likely to benefit family welfare. In addition to their financial objectives, these NGOs may also promote increased collective action by their members, for example, investing in community development projects such as construction of wells, school or lobbying other actors such as larger Northern NGOs to undertake larger projects beyond the means of the community (see Dodman and Mitlin 2013).

Networking and building social movements

The other frequent role for NGOs is policy advocacy, seeking social change by influencing attitudes, policy and practice, seeking to reform state services on the basis of NGO experiences and to lobby directly for the policy changes and reform. This is involvement in participatory, public-interest politics, and NGOs engaging in such activity realize the increasing importance of information (which they gather through their own experience of working at the grassroots) as they begin to utilize the power of ideas and information to promote positive change in the wider structures of government and the official aid community. These NGOs often play a catalytic or seeding role – demonstrating the efficacy of a new idea, publicizing it, perhaps persuading those with access to greater power and budgets to take notice, and then encouraging the widespread adoption by others of the idea.

Neither of these roles need exclude the other. Some NGOs naturally progress from filling a gap in service delivery to recognizing the need to look outward to the wider context in which the need arises and find themselves drawn, possibly through involvement in NGO networks (Bebbington 2003), into national or global policy advocacy.

NGOs work with grassroots community organizations often involves poor and marginalized groups and has been important in mobilizing these large numbers of people against either entrenched elites or state interests, campaigning on their behalf and seeking to influence public policy (e.g. cancellation of debt of heavily indebted nations, placing it on the agenda of G8 summit in Gleneagles in July 2005) through citizen participation. This type of 'bottom-up democracy' has been so successful in many instances that it might eventually lead to 'top-down political change'. NGOs have become key actors in a process of transformatory development. They can affect norm changes that lead to regime change or restructuring of world politics. They do this (especially Northern NGOs) through the communicative power of information, lobbying, research, campaigning or media work, acting on the basis of their moral authority (for example, in the area of human rights, free trade, debt relief, child labour, etc.). NGOs create alliances and networks to place pressure on the state which are growing in their complexities.

In considering the role of NGOs and their links to social movements, there is a growing set of work which conceptualizes two meanings of the term 'development' (Hart 2001; Mitlin *et al.* 2007; Banks and Hulme 2012). One is a historical process of social change, while the other refers to specific interventions, particularly those that fit broadly within the post-World War II project of aid and development. Hart (2001) labels these as 'little d development' and 'Big D Development' respectively. These classifications are not independent: NGOs are all acting in interventionist Development, while also being part of 'little d development' (Mitlin *et al.* 2007).

For Banks and Hulme (2012), service provision activities and advocacy undertaken by NGOs on behalf of the poor remain within the domain of 'Big D Development' and depoliticized approaches. In contrast, approaches where the poor act as advocates for themselves should be seen as part of 'little d development' and a way for NGOs to engage with promoting alternatives to 'Big D Development'. It proposes that NGOs should shift further towards radical, system-changing alternatives and be more aligned with social movements rather than merely seeking reforms within existing systems.

NGOs and the role of civil society

There is an increasing interest in the role of NGOs in promoting democratic development by virtue of their existence as autonomous actors. NGOs are said to pluralize and therefore to strengthen and expand the institutional arena and bring more democratic actors into the political sphere. More civic actors mean more opportunities for a wider range of interest groups to have a 'voice', more autonomous organizations to act in a 'watchdog' role (especially by international donors) vis-à-vis the state, and more opportunities for networking and creating alliances of civic actors to place pressure on the state (e.g. Jubilee 2000 debt initiative) to channel public opinion into policy making and bring about reform. It is believed that NGOs strengthen the state through their participation in improving the efficiency in government services, acting as strategic partners for reform-oriented ministries, filling in gaps in service provision, and helping the government forge ties with the grassroots. It is believed that NGOs have a key role to play during and after democratic transitions. For example, in Chile, NGOs played a vital role in opposing the Pinochet regime throughout the late 1970s and 1980s.

NGOs have become inextricably implicated in civil society, promoting democracy, good governance and building of social capital. Clarke (1998a) examines the role of NGOs in the politics (see Devine 2006) of development across the developing world, opines that the failure to theorize the political impact of NGOs has led to an overly 'inadequate, explicitly normative interpretation of NGO ideology' (Clarke 1998a: 40). This failure has encouraged a tendency to take NGOs' political role as natural/self-evident. Hence most development aid aimed at promoting such linkages between civil society and the state has actually been directed to strengthening central governments. NGOs are a key part of civil society, sometimes to the extent that NGOs are conflated with civil society (Mercer 2002), a move that appeals to donors since it permits them to fund NGOs and then claim to be promoting civil society.

It is also important to understand that NGOs are inherently constrained in the extent to which they can act politically or have power to engage with processes of institutional change, then it makes sense to focus on specific problems where there may be some possibility of small positive steps, even if these are likely to be incremental rather than more extensive reforms. The impact of states upon NGOs is absolutely central in defining the role NGOs can play in national development, for it is governments which give NGOs the space and the autonomy to organize, network and campaign (Clarke, 1998a, 1998b). Of course, it is difficult to generalize about state–NGO relations, as local political networks are always diverse.

NGOs and accountability

NGOs are increasingly being funded by states and official aid agencies, raising questions about what impact this trend has on NGO accountability. Concerns are raised by many regarding internal and external accountability of NGOs. NGOs have downward accountability to members and upward accountability to donors/governments (patrons). NGOs are internally accountable to beneficiaries, donors, boards of directors, trustees and advisory committees. They are externally accountable to organizations and actors with which they affiliate, international government organizations, states, and people throughout the world. Accountability is crucial for NGOs as NGOs have only their reputation for credibility on which to base their action. Accountability in NGOs needs to be appropriate for their work, for the needs of the beneficiaries (clients) and the values of the organization itself. There is *functional accountability* in relation to accounting for resources and their impacts and *strategic accountability*, which relates to the wider implications of an NGO's work (especially in the context of impacts on other organizations or the wider environment in which NGOs operate). These are basically mechanisms for assessing effectiveness, monitoring and evaluation of NGOs. Accountability in NGOs is becoming quite complex (see Edwards and Hulme 1995).

Conclusion

The key issue for the future is whether and how NGOs will adapt to the global changes which are currently under way. NGOs are constantly having to link both local and global agendas if they are to be effective, and they will increasingly be forced to learn from, and adapt to, changing demands and opportunities.

The increased availability of large-scale funding has been one of the primary factors driving NGO growth in the 1980s, encouraging the proliferation of social welfare organizations which often had little or no political agenda. The 'inherent' advantages of the NGOs themselves are gradually worn away by increased funding, professionalization, bureaucracy and the shifting of objectives away from 'social mobilization' (which might be less attractive to donors) towards service delivery. This process may lead to a widening rift between well-resourced service providers and poorly funded social mobilization organizations. This highlights that NGOs exhibit potentially illuminating contrasts in emphasis and packaging of activities, in client groups and organizational style. Considerable diversity exists in relation to how autonomous NGOs are from the influence of funding agencies or management influences of donors. Increasingly questions have been asked. Can NGOs deliver all that is expected from them? Is the glowing image realistic? How effective are NGOs? Attention must also be paid to understanding the role of NGOs in influencing institutional change. This is a reminder of the importance of politics in public services. What are the opportunities and constraints related to influencing policies?

There seems to be more concentration on success stories, and there seems to be a gap emerging between rhetoric and practice, which raises issues of objective monitoring and evaluation of NGOs' projects, effectiveness, legitimacy, performance and accountability. Despite growing interest in evaluation, there is still a lack of reliable evidence on the impact of NGO development projects and programmes. If NGOs want to continue to 'sustain their claim to moral authority' NGOs need to maintain attributes such as impartiality and independence, veracity and reliability, representativeness, and accountability and transparency.

References and Guide to Further Reading

Banks, N. and Hulme, D. (2012) *The role of NGOs and civil society in development and poverty reduction*, Manchester, UK: Brooks World Poverty Institute, University of Manchester.
Bebbington, A. (2003) 'Global networks and local developments: Agendas for development geography', *Tijdschrift voor Economische en Sociale Geografie* 94(3): 297–309.
Bebbington, A. (2005) 'Donor–NGO relations and representations of livelihood in nongovernmental aid chains', *World Development* 33(6): 937–50.
Bebbington, A., Hickey, Samuel and Mitlin, D. (eds) (2008) *Can NGOs make a difference? The challenge of development alternatives*. London: Zed Books Ltd.
Clarke, G. (1998a) *The politics of NGOs in South-East Asia: Participation and protest in the Philippines*, London: Routledge.
Clarke, G. (1998b) 'Non-governmental organisations (NGOs) and politics in the developing world', *Political Studies* XLVI: 36–52.
Devine, J. (2006) 'NGOs, politics and grassroots mobilisation: Evidence from Bangladesh', *Journal of South Asian Development* 1(1): 77–99.
Dodman, D. and Mitlin, D (2013) 'Challenges for community-based adaptation: Discovering the potential for transformation', *Journal of International Development* 25(5): 640–659.
Edwards, M. and Hulme, D. (eds) (1995) *Beyond the magic bullet: NGO performance and accountability in the post-cold war world*, London: Macmillan.
Hart, G. (2001) 'Development critiques in the 1990s: *Culs de sac* and promising paths', *Progress in Human Geography* 25: 649–658
Hulme, D. and Edwards, M. (eds) (1997) *NGOs, States and donors: Too close for comfort?* London: Macmillan.
Mercer, C. (2002) 'NGOs, civil society and democratization in the developing world: A critical review of the literature', *Progress in Development Studies* 2(1): 5–22.
Mitlin, D., Hickey, S. and Bebbington, A. (2007) 'Reclaiming development? NGOs and the challenge of alternatives', *World Development* 35(10): 1699–1720.

10.9

Non-government public action networks and global policy processes

Barbara Rugendyke

With their primary mandate to contribute to improved quality of life for people in disadvantaged communities in Southern nations, non-government development assistance organisations (NGOs) based in Northern donor nations historically delivered assistance, largely at the local scale. Through the 'development project', they provided services like water supplies, sanitation systems, or encouraged income generation and alternative livelihood strategies. In recent decades though, NGOs have devoted their energies to advocacy and campaigning. Drawing on new opportunities afforded by growth in their numbers, size and available resources, and by the revolution in communications technologies, NGOs have globalised, working through transnational networks to maximise their impact on global policy processes.

Development practitioners and theorists have persistently debated the most appropriate forms of development intervention. In the 1980s, radical development theorists contributed to understandings that 'underdevelopment' was a condition maintained by inequitable global structures, such as entrenched unfair terms of trade, poor commodity prices, oppressive debt obligations and inequitable resource distribution between different social groups and nations. Simultaneously, NGOs grappled with the uncertainties of development praxis in seeking the most effective means to improve quality of life for those suffering the ravages of poverty and disadvantage. One common criticism was that NGOs' focus on local level development and empowerment meant they failed to consider wider structural factors which perpetuated poverty. Decades of local level development efforts also demonstrated that project work in communities is inadequate to promote broader changes to address ongoing causes of poverty. Responding to this and to the new theoretical understandings, NGOs sought to tackle these causal factors through increased commitment to advocacy, thus aiming to maximise the impacts and cost-effectiveness of their work.

This trend also reflected Northern NGOs' reaction to calls from their Southern partners to increase their engagement in advocacy. Questioned about how to best help the poor in Tanzania, Julius Nyerere's response in the early 1980s was: 'change public opinion in your own country' (Burnell 1991: 240). Northern NGOs' greater access to power and resources enabled them to engage with global institutions and influence their policies, so Southern NGOs consistently urged Northern NGOs to lobby and campaign about the global crises of poverty and environmental degradation. Concurrently, an evolving emphasis on the importance of empowering local communities to participate in all stages of the development process resulted in recognition that growing numbers of Southern NGOs were best placed to initiate and manage development work in local communities. Northern NGOs thus responded by looking for new ways to contribute to poverty alleviation.

Since the 1990s, therefore, advocacy as a poverty alleviation strategy has been given higher priority by Northern NGOs. Advocacy broadly includes campaigning (aiming to change public opinion) and lobbying (attempting to change policy). The intent of both is to influence policy, particularly those with potential to impact positively on people's lives. Facilitated by the revolution in communications technology, rapid expansion in opportunities for sharing information and accessing and contacting the public broadened the spheres of influence of activist organisations. Harnessing public support gives NGOs the potential to impact on the policies of global and national institutions, which determine access to resources and power.

In recent decades, Northern NGOs' financial commitment to advocacy, reflected in budgets, increased staff time and establishment of departments dedicated to research and lobbying, has grown. Anderson tracked the financial commitment of 14 international NGOs to policy research, lobbying and public campaigning and found that this increased dramatically. In 1996, just 4.1 per cent of total expenditure by international NGOs was dedicated to advocacy. By 2005, only three had increased resource allocation to advocacy by less than 100 per cent; over half by three or six times (Anderson 2007). Thus, in recent decades, NGOs have given greater priority to advocacy, prompted by the success of earlier campaigns.

Reflecting broader trends in economic globalisation, in seeking to optimise the reach of their advocacy, NGOs forged new alliances and increased their engagement in global networks. Arguably, NGO operations were always 'global', with Northern donors traditionally operating through expatriates in Southern nations and, later, through 'Southern partners'. NGOs therefore worked through networks 'through which people, ideas and resources circulate and in which material interventions in particular locations are conceptualised and executed' (Bebbington 2003: 300). However, in attempting to influence national and global affairs, NGOs formed national and international networks. By 2003, for example, 90 per cent of Australian NGOs had participated in campaigns with other Australian NGOs, 75 per cent had either participated in global alliances or cooperated with international NGOs in advocacy and 37 per cent had joined international advocacy networks. Common broad themes of advocacy campaigns conducted within alliances included debt, aid, banning land mines, the environment, and the extension of education (Rugendyke and Ollif 2007). Bond, the UK's membership body for non-governmental organisations (NGOs) working in international development, established in 1993, consisted of 258 members by March 2011. Listed first amongst its activities, Bond 'facilitates collective action by its members and wider UK civil society to influence the policies and practice of governments and institutions at UK, European and international levels' (Bond 2012). Recent Bond campaigning focussed on G8 Summits, aid effectiveness (in conjunction with the Fourth High Level Forum on Aid Effectiveness (HLF4) in Busan, South Korea, in November 2011), achievement of the Millennium Development Goals, and on recommendations related to food security, green growth, transparency and accountability, and inclusive growth and the reduction of inequality, all presented to the G20 Summit in Mexico, in June 2012 (Bond 2012).

NGOs not only joined national networks, but also globalised their activities by formalising and strengthening existing alliances. For example, influenced by the globalisation of economic and political systems, and to facilitate global coordination and expansion of previously independent Oxfam affiliates, nine previously independent national Oxfam affiliates merged to form Oxfam International (OI) in January 1996 (Anderson 2007). Seventeen affiliates now belong to Oxfam International. In 1995, Oxfam International established its Washington Advocacy Office (WAO) to coordinate the development of joint policy and strategies for the Oxfams. Its location in Washington DC facilitated direct access to Washington based Multilateral Agencies (MLAs), the World Bank, the International Monetary Fund (IMF) and the United Nations (UN), headquartered in New York. By 2012, Oxfam International supported advocacy and campaigning staff in Washington DC, Geneva, New York, Brussels, Brasilia and Addis Ababa (Oxfam International 2012).

Working collaboratively to broaden their impact, NGOs were drawn into networks of growing complexity. The primary impetus for this was growing awareness of the power of public action through experience of its success. Although arguably 'little progress at the level of ideology and global systems' (Nyamagasira 2002: 8) results from NGO action, through uniting in alliances NGOs 'materially contributed to policy changes by Northern and Southern governments' (Anderson 2002: 84). These included campaigning successes related to baby milk marketing codes, drafting of an international essential drugs lists, trade liberalisation relating to clothing manufacture in Southern nations, action against global warming and rainforest destruction, debt relief for African nations, and the imposition of sanctions to combat apartheid (Anderson 2002).

The first global campaign through which development NGOs united over a single issue, debt relief for the world's most heavily indebted nations, the Jubilee 2000 Campaign mobilised over 24 million people from over 60 nations to sign a petition (Mayo 2005: 174), drawing supporters of all ages and occupations from faith-based and secular organisations, trade unions and businesses. With 110 member organisations, within four years of its launch in 1997, 'Jubilee 2000's campaign for the remission of unpayable Third World debt made a remarkable impact, raising the issue's profile internationally as well as locally' (Mayo 2005: 172). After a concerted lobbying campaign, in 1999 at the Cologne G8 Summit, world leaders agreed to cancel $100 billion of debts owed by the poorest nations (Bedell 2005: 19). The campaign is credited with having contributed to poverty alleviation, enabling diversion of resources from debt repayment to increased spending on health care, housing provision and education.

NGOs' attempts to mobilise the public into widespread social movements clearly had an impact and realised the calls of commentators to 'scale up' advocacy activity (Edwards and Hulme 2000). The ongoing Global Call to Action against Poverty (GCAP), a global coalition of national anti-poverty campaigns claiming to be 'the world's largest civil society movement calling for an end to poverty and inequality' (GCAP, see www.whiteband.org), epitomises this trend (Mati 2009). Formed in 2004, GCAP consists of national coalitions from over 100 countries, representing over nineteen hundred organisations. These collaborate in global campaigns in the fight against poverty and in lobbying for further debt cancellation for heavily indebted nations, for trade justice and for increased quantity and quality of aid.

At the launch of the affiliated Make Poverty History campaign in London in 2005, Nelson Mandela called for a 'public movement' to eliminate poverty, likening this to international solidarity movements to abolish slavery and apartheid, and urging that 'poverty is man-made and it can be overcome and eradicated by the actions of human beings' (in Bedell 2005: 10). With over 540 organisations in the United Kingdom alone committed within six months of its launch, 87 per cent of the UK's population were aware of the campaign and eight million people wore a white wristband signifying support (www.makepovertyhistory.org). Concerted lobbying in the UK culminated in an estimated 250,000 people marching through Edinburgh on 6 July 2005, demanding that world leaders act in the interests of poverty reduction. That debt relief was so firmly on the agenda at the Gleneagles G8 meeting that month is widely attributed to the campaign.

GCAP members voted to extend the campaigning alliance until at least 2015. While acknowledging its members' great diversity, GCAP (GCAP, 2012) argues: 'we know we will be more effective when we work together. We do not endeavour to reach absolute agreement on detailed policy, but we want to pressure governments to eradicate poverty, dramatically lessen inequality, and achieve the Millennium Development Goals'. GCAPs demands include public accountability, just governance and the fulfilment of human rights, women's rights and gender justice, trade justice, aid and financing for development, debt cancellation and peace and security.

The expanding influence of NGOs is similarly traceable through their history of engagement with the United Nations (UN). In 1997, in response to their growing influence, the United Nations Security Council held its first meeting with NGOs. Represented at important UN conferences, including the 1992 Earth Summit, NGOs became '*incorporated* into the UN system' (Martens 2006: 692). In 1968, 377 NGOs had consultative status with the UN; by 2001 this extended to over 1,550 organisations (Opoku-Mensah 2001). By November 2011, 3,535 NGOs had attained consultative status (UN 2012), which invites participation in functional commissions of the Economic and Social Council of the UN, including Commissions for Social Development, the Status of Women, Population and Development, Sustainable Development, the UN Forum on Forests and the Permanent Forum on Indigenous Issues (UN 2012). Thus, increasing integration of NGOs into global decision-making processes is evident.

Development NGOs continue to contribute directly to material change at the local level, while also providing through their advocacy 'a conduit for voluntary participation in global issues outside the formal political realm of the states' (Taylor 2004: 273). Simultaneously, NGOs' participation in global advocacy alliances provides 'a legitimizing platform for dissident and diverse voices from regions where economic and political power is lacking' (Taylor 2004: 273). Recent commentators suggest the need for 'multi-level, multi-country, multi-sector alliances . . . to respond to the challenges that assail and often divide the global community . . . and growing questions about humanity's capacity and will to managed its expanding global interdependence' (Brown *et al.* 2012). Following rapid growth in size and reach, non-governmental global action networks' impacts on global policy have been far reaching; while their *raison d'être* remains undiminished, such networks will continue to be an important global political force.

References

Anderson, I. (2002) 'Northern NGO Advocacy: Perceptions, Reality, and the Challenge' in D. Eade (ed.) *Development and Advocacy*, Oxford: Oxfam Great Britain.

Anderson, I. (2007) 'Global Actions: International NGOs and Advocacy' in B. Rugendyke (ed.) *NGOs as Advocates for Development in a Globalising World*, Abingdon and New York: Routledge.

Bebbington, A. (2003) 'Global Networks and Local Developments: Agendas for Development Geography', *Tijdschrift voor economische en sociale geografie*, 94 (3): 297–309.

Bedell, G. (2005) *Makepovertyhistory: How You Can Help Defeat World Poverty in Seven Easy Steps*, London: Penguin Books.

Bond (2012) Bond for International Development, About us, available online at www.bond.org.uk/pages/about-us.html (accessed 10 June 2012).

Brown, L. D., Ebrahim, A. and Batliwala, S. (2012) 'Governing International Advocacy NGOs' in *World Development*, 40 (6): 1098–1108.

Burnell, P. (1991) *Charity, Politics and the Third World*, London: Harvester Wheatsheaf.

Edwards, M. and Hulme, D. (2000) 'Scaling up NGO Impact on Development: Learning from Experience', in D. Eade (ed.) *Development, NGOs and Civil Society*, Oxford: Oxfam Great Britain.

Martens, D. (2006) 'NGOs in the United Nations System: Evaluating Theoretical Approaches', *Journal of International Development*, 18: 691–700.

Mati, J. (2009) 'A Cartography of a Global Civil Society Advocacy Alliance: The Case of the Global Call to Action Against Poverty', *Journal of Civil Society*, 5 (1): 83–105.

Mayo, M. (2005) 'Learning from Jubilee 2000: Mobilizing for Debt Relief' in *Global Citizens: Social Movements and the Challenge of Globalization*, London and New York: Zed Books.

Nyamagasira, W (2002) 'NGOs and Advocacy: How Well Are the Poor Represented?' in D. Eade (ed.) *Development and Advocacy*, Oxford: Oxfam Great Britain.

Opoku-Mensah, P. (2001) 'The Rise and Rise of NGOs: Implications for Research', *Tidsskrift ved Institutt vor sosiologi og statsvitenskap*, 1: 1–3.

Oxfam International (2012) 'About us – Offices', available online at www.oxfam.org/en/about/offices (accessed 24 June 2012).

Rugendyke, B. and Ollif, C. (2007) 'Charity to Advocacy: Changing Agendas of Australian NGOs' in *NGOs as Advocates for Development in a Globalising World*, Abingdon and New York: Routledge.

Taylor, P. (2004) 'NGOs in the World City Network', *Globalizations*, 1: 265–277.

United Nations (UN) (2012) 'Basic Facts about ECOSOC Status', available online at www.csonet.org/index.php?menu=17 (accessed 20 June 2012) and 'Participating in a UN Event', available online at www.csonet.org/index.php?menu=113 (accessed 20 June 2012).

Further reading

Brown, L. D., Ebrahim, A. and Batliwala, S. (2012) 'Governing International Advocacy NGOs' in *World Development*, 40 (6): 1098–1108. This article examines governance, organisational structures and the effectiveness of international advocacy NGOs.

Mitlin, D., Hickey, S. and Bebbington, A. (2007) 'Reclaiming Development? NGOs and the Challenge of Alternatives', *World Development*, 35 (10): 1699–1720. This article explores the notion that NGOs would benefit from rethinking development alternatives, focussing on the political economy of social change.

Rugendyke, B. (ed.) (2007) *NGOs as Advocates for Development in a Globalising World*, Abingdon and New York: Routledge. Devoting increasing resources to advocacy campaigns directed at global actors, NGOs have globalised. These changes, and the strengths, limitations and impacts of NGO advocacy at a range of scales, are discussed.

Websites

Global Call to Action against Poverty: www.whiteband.org/
Jubilee Debt Campaign: www.jubileedebtcampaign.org.uk/
Make Poverty History: www.makepovertyhistory.org/

10.10

Multilateral institutions

'Developing countries' and 'emerging markets' – stability or change?

Morten Bøås

Introduction

In November 2010, the Executive Board of the International Monetary Fund (IMF) approved far-reaching reforms signalling the end of the power structure that had prevailed ever since this multilateral institution was conceived in 1944. Speaking at a press conference immediately after the board's decision, IMF's managing director at that time, Dominique Strauss-Kahn hailed this as 'a historical reform, increasing the voice and representation of emerging markets and developing countries in the IMF' (IMF 2010: 1). In practice, what this means is that the four BRICs (Brazil, Russia, India and China) now will be among the IMF's top shareholders due to a shift of 6 per cent of the total quota shares. As 80 per cent of this shift comes from advanced economies (United States and Europe mainly), this means that there will be two fewer seats for European countries in the IMF Executive Board and two more for emerging market countries. As this will have significant consequences also for the power structure of other important multilateral institutions, the World Bank, included, this is a decision of systemic magnitude that reflects how much the world has changed since the IMF and the World Bank was established as the institutional anchors of the post-Second World War order established by the United States and its allies at the Bretton Woods conference in 1944.

The World Bank has yet to see equally significant changes, but also here emerging economies play a larger role today than only some years ago. Currently there are eight member countries in the World Bank selecting their executive director directly. These are the five largest shareholders: United States, Japan, Germany, France and the United Kingdom; and China, Russia and Saudi Arabia.[1]

Multilateral institutions as a term is used to describe many different global institutions, however, in this chapter it includes the major multilateral development banks (MDBs) – the World Bank, the IMF, and the three largest regional development banks, the African Development Bank (AfDB), the Asian Development Bank (ADB) and the Inter-American Development Bank (IDB). As these institutions not only are very complex organisations, but also perform a wide range of different tasks they cannot be approached as unitary actors. What is needed is an approach that highlights its internal processes and politics. The first part of this chapter looks at power and organisation in such institutions, whereas the second part addresses the changes and challenges they have encountered.

Multilateral institutions: Hierarchy, organisation and financing

Multilateral institutions represent a form of international cooperation the world had not experienced prior to 1945, an evolution from the League of Nations and United Nations models, in which all member countries formally have an equal voice and vote. Their structure is inspired by the joint-stock model of private capitalist corporations, in which member countries are shareholders whose voting powers vary with their relative economic importance. In other words, each member country's share of the votes is weighted in accordance with the combined amount of capital it has paid in and is guaranteeing for. All member countries are organised in a country constituency headed by an executive director who controls the combined votes of his or her constituency and sits on the institution's board of directors. The size and composition of the country constituencies vary across the universe of multilateral institutions, but the principle of organisation is similar (Bøås and McNeill 2003, 2004).

The capital construction of all multilateral institutions was established with the formation of the World Bank. It was established as an institution, which was to be owned, and whose capital would be provided, by government and not by private sources. Its initial authorised capitalisation of US$10 billion consisted of 20 per cent in the form of paid-in capital and 80 per cent in the form of guaranteed capital. This distinction is crucial for our understanding of these institutions. Each country subscription is divided into two parts. The larger one is the so-called *guaranteed capital*. This amount is not actually paid by the member countries to the multilateral institution, but each member guarantees for a certain sum of money. The credit rating of the World Bank and the other multilateral institutions is based on the amount guaranteed for by rich industrialised member countries. This gives these institutions the best possible international credit rating (e.g. triple A) and makes it possible for them to lend money on international capital markets and re-lend it to poorer countries with a lower credit rating. This would not be possible without rich member countries, and therefore also constitutes the backbone of their power. They could not function without them, and this knowledge that is shared by all actors in such institutions also implies that votes rarely are used. They are simply not needed very often as the power relations on the Board of Directors are so transparent, and the consequences of constantly working against them so obvious to everybody concerned.

Multilateral institutions therefore reflect the power relations prevailing at their point of origin and tend, at least initially, to facilitate worldviews and beliefs (for instance in the merits of neoliberal economics) in accordance with these power relations (cf. Cox 1992; Wade 2002). Outcomes, however, are determined not simply by the distribution of power among the members that constitute the institution in question, but also by the multilateral institution itself, which can affect how choices are framed and outcomes reached. All multilateral institutions should therefore be approached first and foremost as social constructions (see Kratochwil and Ruggie 1986). The actors involved are political, economic and social actors, operating not just through the state's foreign policy apparatus, but also transnationally.

Changes and challenges

All multilateral institutions are originally established in order to solve problems. After the completion of the reconstruction of Europe, the 'problem' was development (or the lack of it). President Truman's inaugural speech on 20 January 1949 is commonly held to mark the beginning of the modern development practice (Nustad 2004). In this speech, the transfer and transfusion of scientific and expert knowledge was presented as the solution to poverty and misery. This was the means, and the objective – increased prosperity and closer resemblance to Western societies – was the original goal of multilateral institutions, and despite all the new policies and approaches that have emerged this has remained at the heart of their activities.

What have changed are means and not the ends. And the changes that have taken place have been incremental; most often without any attempt to place new objectives in a logical, prioritised order. The process of change that has taken place in multilateral institutions thus resembles what Ernst Haas (1990) has called 'change by adaptation'.

In comparison to other social units, multilateral institutions confront rather special challenges when faced with demands to incorporate new issue areas. Their mission is never simple and straightforward because both member states and other actors in their external environment may disagree on the interpretation of the mission (the ends) as well as on the tasks (the means) that need to be done if the mission is to be completed. In social units that function under such circumstances, organisational routines and standard operating procedures will be preferred to substantive change. Multilateral institutions will therefore favour one particular way of arranging and routinising their activities. Since they have to satisfy different constituencies (that is, borrowing countries, donor countries and increasingly also NGOs and civil society), multilateral institutions will try to avoid articulating competing views. Consensus therefore becomes an objective in itself, but the kind of consensus established in multilateral institutions is constructed on the power relationships prevailing in the institution in question. This means that consensus in multilateral institutions is usually artificial.

This way of reasoning also help us understand why the favoured approach of multilateral institutions in promoting development was that of the 'engineer'. Development (or the lack of it) was seen as a technical issue, and not as a political question. If the challenges of development, and the new ideas supposed to resolve them, could be defined in technical terms, this increased the possibility of getting a proposal for action approved by staff and borrowing-country governments. Over time, a limited re-examination of the means utilised to reach their ends was made possible when new issue-areas were presented to the multilateral institutions in the same language as the old and familiar knowledge. By applying such a strategy of depoliticisation, new and potentially challenging discussions were kept within the framework of already existing standard operating procedures. It was therefore possible to treat potentially highly political issues, such as governance, as technical issues, and thereby the underlying political conflicts could, at least partly, be controlled. For example, it was the ability to define governance in strictly economic and technical terms that facilitated this issue area's incorporation into multilateral institutions.

Projects have been modified by the inclusion of new social and environmental components and regulatory safeguards in order to ensure that environmental and social damage is avoided as much as possible. The approach of multilateral institutions, however, is still of an engineering problem-solving type, with policies and project papers written in the technical language that staff, management and the boards of multilateral institutions are used to.

In the 1950s and 1960s this strategy worked remarkably well. But in the 1970s and 1980s it was gradually called into question, and by the mid-1990s it was fully apparent that new development challenges could no longer be tackled by narrow technical approaches, and multilateral institutions started to experience more severe difficulties. The issue was no longer just a matter of finding the right technical solution to a functional problem. Today, the challenge is to construct some sort of consensus around an increasingly politicised agenda constituted around a whole range of new cross-cutting themes such as governance, involuntary resettlement and indigenous people. The technocratic consensus has reached its limits. It is no longer in any credible way possible to define development solely in a technical and functional manner. As a consequence, the internal artificial consensus is disappearing, not only between donor and borrowing member countries of multilateral institutions, but also internally in these institutions. An increasingly political agenda will make the process of political manoeuvring between donor and recipient countries and other stakeholders (civil societies and the private sectors) increasingly difficult for multilateral institutions. This trend is further impacted on by the changes in the global economy.

A new global economy?

The IMF and the World Bank are among the world's most powerful international organisations. As the very institutional 'anchors' of the post-Second World War order, the IMF and the World Bank reflected and still reflects the power relations prevailing at their time of origin. They have therefore also tended to facilitate the worldviews in accordance with these power relations. This may be about to change in the IMF and the World Bank as they primarily are economic institutions and thereby bound to reflect global economic realities. These are clearly currently changing the world in a multipolar direction economically speaking. However, it is also important to recognise that even if power relations are important for our understanding of the projects, policies and approaches of the World Bank and the IMF, power is most often exercised carefully and through more subtle means that the direct power of money that capital subscriptions and quotas may allow for. Votes are very rarely used, rather it has been the totality of US hegemony that for the better part of the post-1945 era that has made it possible for the United States to exercise the considerable influence that it does.

Votes are important as they constitute the basis for the artificial consensus, but just as important for our understanding of the diplomacy and statecraft that revolves around the multilateral institutions are the more informal pathways of American hegemony such as their distinctively American Charters, the US hold on the position of the presidency in the World Bank and the first deputy managing director in the IMF, and the US as the centre of gravity for their conceptual debates – the great majority of staff tend to have a degree from a

North American University. As both the IMF and the World Bank are located in downtown Washington DC, they are not only just a few blocks away from the US Treasury, but this also entails that staff is constantly exposed to and influenced by American worldviews. In sum, this may suggest that the United States will continue to be the most important member country in the World Bank and the IMF even if global economic realities may be pointing towards China and Asia, thereby also illustrating that power relations prevailing at time of origin in this type of international institutions only will change gradually and over a considerable time span. The argument is not that these power relations are cemented for ever, but rather to point out that the exercise of influence in institutions such as the IMF and the World Bank is nested in sources of power of a material and an ideational nature.

Concluding remarks

The questions concerning the future of multilateral institutions and how they respond to changes and new challenges are many, and the answers too few. However, what is more important than finding answers is to ask the right questions, and they can only be coined if we grasp the realities of these institutions. They do reflect the power relations prevailing when they were established, but they have also been subjected to change, and the new issue areas that they have had to tackle have challenged the old technocratic mode of these institutions. They have also been forced to deal with criticism from a wide range of civil society actors. However, these changes apart, they still operate in a way that give power to rich member countries, and this is not likely to change as the entire *modus operandi* of multilateral institutions is based on their ability to secure loans from international capital markets on the basis of the credit rating they achieve from the guaranteed capital of the rich member countries.

Note

1 The United States controls 16.36 per cent of the total number of votes, and Japan 7.85 per cent, Germany 4.48 per cent, France 4.30 per cent and the United Kingdom 4.30 per cent, whereas Russia, China and Saudi Arabia each controls 2.78 per cent of the votes.

References

Bøås, Morten and Desmond McNeill (2003) *Multilateral Institutions: A Critical Introduction*, London: Pluto Press.

Bøås, Morten and Desmond McNeill (2004) 'Introduction: Power and ideas in multilateral institutions: towards an interpretative framework', in Morten Bøås and Desmond McNeill (eds) *Global Institutions and Development: Framing the World?* London: Routledge, pp. 1–12.

Cox, Robert (1992) 'Multilateralism and world order', *Review of International Studies*, vol. 18, no. 2, pp. 161–80.

Haas, Ernst (1990) *When Knowledge is Power – Three Models of Change in International Organizations*, Berkeley: University of California Press.

IMF (2010) *IMF Board Approves Far-Reaching Governance Reforms*, Washington, DC: The IMF.

Kratochwil, Friedrich and John G. Ruggie (1986) 'International organization: A state-of-the-art or an art of the state', *International Organization*, vol. 40, no. 3, pp. 753–75.

Nustad, Knut (2004) 'The development discourse in the multilateral system', in Morten Bøås and Desmond McNeill (eds) *Global Institutions and Development: Framing the World?* London: Routledge, pp. 13–23.

Wade, Robert (2002) 'US hegemony and the World Bank: The fight over people and ideas', *Review of International Political Economy*, vol. 9, no. 2, pp. 201–29.

Websites

Bank Information Centre (BIC) www.bicusa.org
Bretton Woods Project (BWP) www.brettonwoodsproject.org
World Bank (WB) www.worldbank.org
Structural Adjustment Participatory Review International Network (SAPRIN) www.saprin.org

10.11

Is there a legal right to development?

Radha D'Souza

Introduction

The question forms part of a wider question about the status of socioeconomic rights in law. Rights in law entail two aspects: normative standards for society and the institutional mechanisms for enforcement. Before the World Wars the right to property was the only economic right that was recognised as part of basic freedoms. Enlightenment thinkers put property rights on par with conventional human rights that existed in pre-modern societies such as right to life, liberty and conscience. Enlightenment thinkers also argued that state interference in the economy must be minimal. Adam Smith, the father of the discipline of economics, for example, argued in the eighteenth century that the role of the state must be that of a 'night watchman'. The primacy of property rights and its equal status alongside life, liberty and conscience created economic polarisation, financial crises and social unrest. By late nineteenth century the economic and social inequalities that property rights introduced in societies made socialist thinkers like Karl Marx argue that the state was 'the executive committee of the bourgeoisie'. The first half of the twentieth century was engulfed in political and social upheaval including the revolutions in Russia and Eastern Europe, the World Wars and the anti-colonial struggles. When the World Wars ended the victorious Allies proposed the formation of the United Nations. The UN Charter for the first time recognised economic rights beyond property rights.

The UN has two goals: promoting peace (Article 1.1) and cooperation for economic, social and cultural problems (Article 1.3). The UN Charter includes harmonisation as a related goal (Art. 1(4), 13) and allows it to codify international law and develop global standards. Chapters IX and X provide the legal basis for the expanded economic role of the UN that we see today. Chapter IX expands on economic goals to include higher standards of living, health and social progress including education, as well as human rights including non-discrimination. It empowers the UN to sign agreements known as the Specialised

Agency Agreements (SAA) with the International Economic Organisations (IEO) like the World Bank, the IMF and standard setting organisations like the WHO, ILO and others. The SAAs define the relationship of each Specialised Agency (SA) to the UN. Most importantly the UN Charter created a new economic organ, the Economic and Social Council (ECOSOC) to enable the UN to perform its economic role. The ECOSOC's main role is coordinating the work of the SAs, the states (through its regional commissions), the international standard setting organisations and the NGOs. International law until the UN Charter was almost entirely about war and peace between states. The vastly expanded economic role for international law and its institutionalisation under the UN Charter were radical departures by any measure from the League of Nations or international law before the League.

Socioeconomic rights

On 6 December 1986 the UN General Assembly (UNGA) adopted GA Resolution 41/128 on the UN Declaration on the Right to Development (DRtD). The DRtD forms a continuum in a series of developments in socioeconomic rights following the expanded economic role of the UN in the Charter. The first of these was the Universal Declaration on Human Rights (UDHR) adopted in 1948. The UDHR recognises conventional civil liberties and individual freedoms (Art. 1–21) together with property rights (Art. 17) alongside international cooperation for the realisation of social security (Art. 22), right to work without discrimination (Art. 23), fair conditions of work (Art. 24), standards of living (Art. 25) and education (Art. 26). There were differences between the Big Four (US, Britain, USSR and China) on the scope of human rights when the UN Charter was drafted. These differences were by no means resolved in the UDHR.

Briefly summarising, the approach of Euro-American states was grounded in classical liberal philosophy where human rights were primarily personal freedoms, non-discrimination by the state, fair process and guarantees of property rights. The Soviet bloc states argued human freedoms were contingent on the capacities of states to meet basic needs of its populations, that is, food, clothing, health, shelter and education. Recognition of socioeconomic rights was necessary for human rights to be realised. The Third World states, emerging from colonial rule, inherited institutions that were authoritarian and economies that continued to be dominated by colonial/imperial economic actors. Restricting the role of the state in the Third World would, they argued, make de-colonisation impossible. The priority for Third World states were rights to self-determination and affirmative economic actions. The UDHR was at best a compromise document that straddled diverse approaches to civil and political rights on the one hand and socioeconomic rights on the other. The UDHR remained a non-binding declaration. Unlike treaties and covenants, declarations in international law are not binding on the signatories. They are at best statements of intent that guide international cooperation between states. The UDHR became the tool for engagement between the states throughout the cold war.

In 1966 the UN adopted two binding covenants relevant to DRtD: the International Covenant on Civil and Political Rights (ICCPR) and the International Covenant on Economic, Social and Cultural Rights (ICESCR). The two covenants share similar preambles and recognise some common rights such as the right to self-determination and non-discrimination. If the UDHR was a compromise document the ICCPR and ICESCR was a breakdown of that compromise. The ICCPR expanded on the Western nations' approaches to human rights as civil liberties, individual freedoms and property rights in Articles 1–21 of the UDHR and the ICESCR expanded on the socioeconomic rights in Articles 22–29. Against the backdrop of the cold war, not surprisingly, the ICCPR was canvassed by Western states and the ICESCR by the Soviet Bloc. Unlike the UDHR both covenants created binding obligations on the signatories.

The Third World states mounted a concerted effort to rewrite international law. They argued that international law, including the UN Charter, was unjust because it was written by colonial and imperial powers (Anghie, 2004). Instead they called for a New International Economic Order. In 1974 the UNGA adopted Resolution 3281 which became a charter of economic rights. The Western states were not supportive of the resolution. The DRtD and its emergence within the Commission on Human Rights must be seen against these attempts by Third World states for economic rights. The DRtD came at the tail end of the three UN Development Decades from 1960 to 1990. By that stage it was apparent that the UN's development agenda was not delivering and there was widespread disillusionment.

The DRtD comprises ten articles that guide state actions. It states RtD is an inalienable human right (Art. 1) and the human person is the focus of RtD (Art. 2). States must create conditions for the realisation of RtD (Art. 3) cooperate to achieve RtD (Arts 4–6), ensure peace security and disarmament to achieve RtD (Art. 7) and take steps to realise equal opportunities, basic needs and democracy (Art. 8). Lastly it pronounces that RtD was indivisible and interdependent (Art. 9) bridging the rift in the ICCPR and ICESCR. There is no enforcement or reporting mechanisms in the DRtD.

Globalisation and DRtD

Three years after the DRtD was adopted the architecture of the post-World War world changed dramatically. Two significant moments in the change was the Washington Consensus and the fall of the Berlin Wall both in 1989. Both these events set in motion the most comprehensive reform of the international legal order since the end of the World Wars. The main thrust of the change was neoliberal transformation of the international order – that is, rolling back states and rolling in private actors in the global arena. The newly formed WTO took the lead in initiating neoliberal transformation of International Organisations and interstate relations through its trade regime (D'Souza, 2010). With the fall of the Berlin Wall the walls that separated civil and political rights from economic and social rights and both from RtD also fell. The Vienna Declaration and Programme of Action (VDPA) was adopted in 1993 at the World Conference on Human Rights and endorsed by the UNGA Resolution 48/121. The VDPA effectively merged the UDHR, the ICCPR and ICESCR and DRtD into a single programme. The VDPA in the new international legal regime became part of the Good Governance requirements proposed under the Washington Consensus and underpinned by IEOs (Woods, 1999).

Until the neoliberal reforms the IEOs interpreted development primarily as lending for infrastructure, industrial and agricultural production and building capital markets. Managing social, political and human consequences of development projects, they argued, was the legal responsibility of the states and beyond the remit of their Articles of Association. The neoliberal reforms of the international economic regime changed all that. An important part of the VDPA's plan of action was 'mainstreaming' human rights (Part II). The scope of the renamed Human Rights Council (HRC) included 'institution-building package', Universal Periodic Review Mechanism, a new Advisory Committee, a revised Complaints Procedure, and participation of private experts. Within the overall framework of Good Governance 'mainstreaming' aligned RtD to the goals of the IEOs. Good Governance programmes require Third World states to remove restrictions on private economic actors, domestic and international. Under the overall remit of the IEOs compliance with human rights standards is seen as a risk-mitigation strategy by private industries, states and financial institutions (Likosky, 2006). Good Governance programmes envisage wide ranging law reforms within Third World states. These requirements form part of contractual agreements between the IEOs and states leading some scholars to argue that IEOs increasingly function as law makers (Alvarez, 2005).

Does that make DRtD enforceable through international contracts with IEOs? The VDPA like the DRtD is also normative. The constitutions of many Third World states recognise socioeconomic rights as Directive Principles that ought to guide state actions and policies. They are not justiciable in that they cannot be enforced by courts. Generally courts have taken the view that economic entitlements under RtD are contingent on the states' economic capacity to provide for its population and cannot be enforced through courts. Courts have been more open to enforcing social aspects of DRtD like non-discrimination. In the US the Alien Tort Claims Act has been used by activists to hold US corporations to account with mixed results. In the UK courts have held tort claims must have a direct nexus between the action and wrong doer. Responding to criticisms that lending policies breached human rights the WB set up the Inspection Panel as a forum to address grievances arising from WB funded projects. The credibility of the mechanism is limited by the in-house procedures and management.

Justiciability of economic rights eludes modern law. 'Where there is a right, there must be a remedy' is a basic common law principle. If RtD is not justiciable, if a person without food or clothing cannot claim them as of 'right', RtD challenges the commonplace understandings of 'rights'. RtD sits on a conceptual fault line: the meaning of development. If development is what states and IEOs say it is, namely that development is about building dams, railways, industries, banks, then, invariably it is followed by displacement, dispossession and disenfranchisement, and winners and losers. If development is understood as human well-being then it challenges the concept of modernisation. The fault line that runs through RtD sits on a deeper rift on which modern law stands. That rift is the idea that property rights are inalienable human rights to be valued on par with life, liberty and conscience. RtD challenges us to think about which of these two ought to be privileged when the two do not work in tandem.

References

Alvarez, J E. (2005) *International Organizations as Law-Makers*, Oxford: Oxford University Press.

Anghie, A. (2004) *Imperialism, Sovereignty and the Making of International Law*, Cambridge: Cambridge University Press.

D'Souza, R. (2010) Law and Development Discourse About Water: Understanding Agency in Regime Changes. In Cullet, P. Gowlland-Gualtieri, A., Madhav, R. and Ramanathan, U. (eds) *Water Governance in Motion: Towards Socially and Environmentally Sustainable Water Laws*, New Delhi: Foundation Books, p. 491.

Likosky, M. B. (2006) *Law, Infrastructure, and Human Rights*, Cambridge: Cambridge University Press.

Woods, N. (1999) Good Governance in International Organizations. *Global Governance* 5: 1, 39.

Further readings

International Commission of Jurists (2008) *Courts and the Legal Enforcement of Economic, Social and Cultural Rights: Comparative Experiences of Justiciability*, Geneva: ICJ. Provides information on socioeconomic law.

California Western International Law Journal (1985). A special symposium on Right to Development is published in this journal, see p. 429.

Index

Page numbers in *italics* denote an illustration